W9-CPF-886

# Your steps to success.

## STEP 1: Register

All you need to get started is a valid email address and the access code below. To register, simply:

1. Go to www.aw-bc.com/chaisson
2. Click the appropriate book cover.
   *Cover must match the textbook edition being used for your class.*
3. Click "**Register**" under "**First-Time User?**"
4. Leave "**No, I Am a New User**" selected.
5. Using a coin, scratch off the silver coating below to reveal your access code.
   *Do not use a knife or other sharp object, which can damage the code.*
6. Enter your access code in lowercase or uppercase, without the dashes.
7. Follow the on-screen instructions to complete registration.
   *During registration, you will establish a personal login name and password to use for logging into the website. You will also be sent a registration confirmation email that contains your login name and password.*

### Your Access Code is:

*Note:* If there is no silver foil covering the access code, it may already have been redeemed, and therefore may no longer be valid. In that case, you can purchase access online using a major credit card. To do so, go to www.aw-bc.com/chaisson, click the cover of your textbook, click "**Buy Now**", and follow the on-screen instructions.

## STEP 2: Log in

1. Go to www.aw-bc.com/chaisson and click the appropriate book cover.
2. Under "**Established User?**" enter the login name and password that you created during registration. *If unsure of this information, refer to your registration confirmation email.*
3. Click "**Log In**".

## STEP 3: (Optional) Join a class

Instructors have the option of creating an online class for you to use with this website. If your instructor decides to do this, you'll need to complete the following steps using the Class ID your instructor provides you. By "joining a class," you enable your instructor to view the scored results of your work on the website in his or her online gradebook.

To join a class:

1. Log into the website. For instructions, see "STEP 2: Log in."
2. Click "**Join a Class**" near the top right.
3. Enter your instructor's "**Class ID**" and then click "**Next**".
4. At the Confirm Class page you will see your instructor's name and class information. If this information is correct, click "**Next**".
5. Click "**Enter Class Now**" from the Class Confirmation page.

• *To confirm your enrollment in the class, check for your instructor and class name at the top right of the page. You will be sent a class enrollment confirmation email.*

• *As you complete activities on the website from now through the class end date, your results will post to your instructor's gradebook, in addition to appearing in your personal view of the Results Reporter.*

To log into the class later, follow the instructions under "STEP 2: Log in."

## Got technical questions?

Customer Technical Support: To obtain support, please visit us online anytime at http://247.aw.com where you can search our knowledgebase for common solutions, view product alerts, and review all options for additional assistance.

### SITE REQUIREMENTS

For the latest updates on Site Requirements, go to www.aw-bc.com/chaisson, choose your text cover, and click Site Reqs.

WINDOWS
OS: Windows 2000, XP
Resolution: 1024 x 768
Plugins: Latest Version of Flash/QuickTime/Shockwave (as needed)
Browsers: Internet Explorer 6.0; Firefox 1.0

MACINTOSH
OS: 10.2.4, 10.3.2
Resolution: 1024 x 768
Plugins: Latest Version of Flash/QuickTime/Shockwave (as needed)
Browsers: Firefox 1.0; Safari 1.3

Internet Connection: 56k modem minimum

Important: Please read the Subscription and End-User License agreement, accessible from the book website's login page, before using the Companion WebSite with GradeTracker. By using the website, you indicate that you have read, understood, and accepted the terms of this agreement.

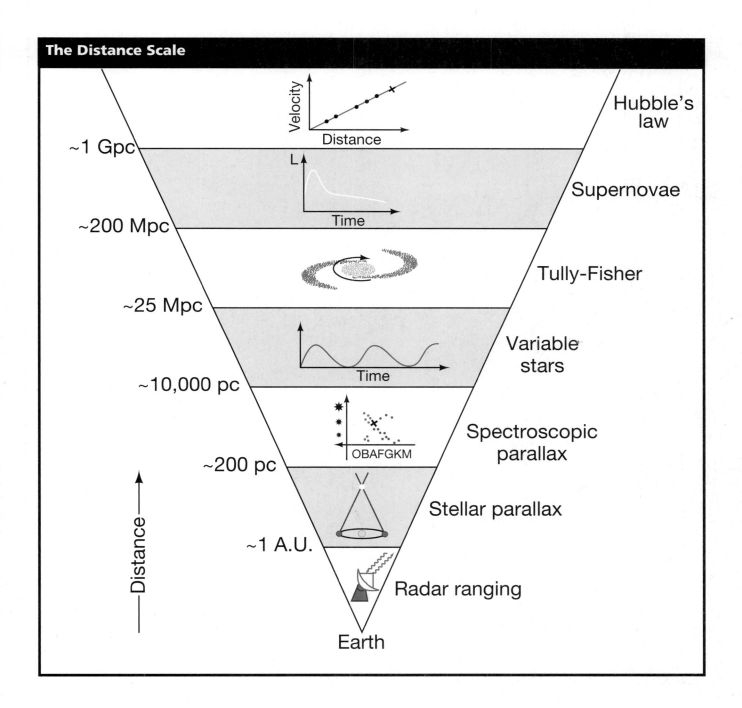

**The Distance Scale**

Hubble's law

~1 Gpc

Supernovae

~200 Mpc

Tully-Fisher

~25 Mpc

Variable stars

~10,000 pc

Spectroscopic parallax

~200 pc

Stellar parallax

~1 A.U.

Radar ranging

Earth

Distance

# ASTRONOMY TODAY

SIXTH EDITION

## The Solar System

Volume I

## Eric Chaisson
*Tufts University*

## Steve McMillan
*Drexel University*

PEARSON

Addison
Wesley

San Francisco   Boston   New York
Cape Town   Hong Kong   London   Madrid   Mexico City
Montreal   Munich   Paris   Singapore   Sydney   Tokyo   Toronto

Sponsoring Editor: Christian Botting
Publisher: Jim Smith
Executive Editor: Nancy Whilton
Development Manager: Michael Gillespie
Editorial Manager: Laura Kenney
Development Editors: John Haber and Elaine Page
Director of Marketing: Christy Lawrence
Executive Marketing Manager: Scott Dustan
Managing Editor: Corinne Benson
Production Supervisor: Shannon Tozier
Compositor: Progressive Information Technologies
Production Service: Progressive Publishing Alternatives
Illustrations: ArtWorks
Text Design: Maureen Eide and John Christiana
Cover Design: Devenish Design
Manufacturing Manager: Pam Augspurger
Director, Image Resource Center: Melinda Patelli
Manager, Rights and Permissions: Zina Arabia
Manager, Visual Research: Elaine Soares
Manager, Cover Visual Research & Permissions: Karen Sanatar
Image Permission Coordinator: Debbie Latronica
Photo Research: Truitt and Marshall
Cover Printer: Phoenix Color
Printer and Binder: R.R. Donnelley, Willard

Cover Image: NASA/JPL/Cornell

**Library of Congress Cataloging-in-Publication Data**

Chaisson, Eric.
  Astronomy today / Eric Chaisson, Steve McMillan.—6th ed.
    p.   cm.
  Includes index.
  ISBN 0-13-240085-5 (hardcover)—ISBN 0-13-615549-9 (pbk.: v.1)—ISBN 0-13-615550-2 (pbk.: v.2)
  1. Astronomy—Textbooks.  I. ᕫ  McMillan, S. (Stephen)  II. Title.
  QB43.3.C48 2007
  520—dc22                                                    2007017107

ISBN 10-digit 0-13-240085-5; 13-digit 978-0-13-240085-5 (Student edition)
ISBN 10-digit 0-321-46115-0; 13-digit 978-0-321-46115-5 (Professional copy)
ISBN 10-digit 0-13-615549-9; 13-digit 978-0-13-615549-2 (Volume 1)
ISBN 10-digit 0-13-615550-2; 13-digit 978-0-13-615550-8 (Volume 2)
ISBN 10-digit 0-13-241817-7; 13-digit 978-0-13-241817-1 (NASTA)

# BRIEF CONTENTS

# CONTENTS

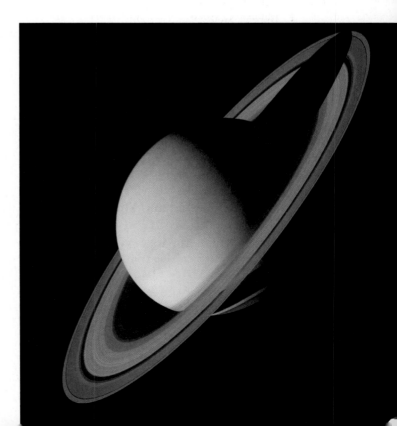

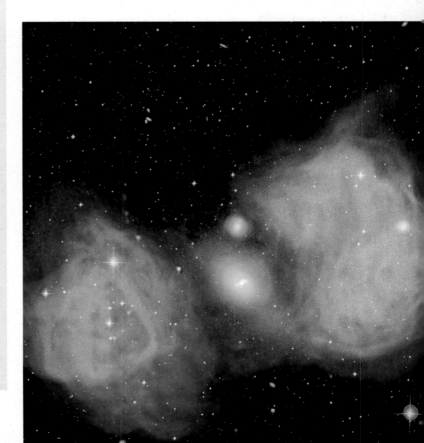

# ABOUT THE AUTHORS

## Eric Chaisson

Eric holds a doctorate in astrophysics from Harvard University, where he spent ten years on the faculty of Arts and Sciences. For several years thereafter, he was a senior scientist and director of educational programs at the Space Telescope Science Institute and adjunct professor of physics at Johns Hopkins University. For the past decade, Eric has been at Tufts University, where he is a research professor in the Department of Physics and in the School of Education, and director of the Wright Center for Innovative Science Education. He has written 12 books on astronomy, which have received such literary awards as the Phi Beta Kappa Prize, two American Institute of Physics Awards, the Kistler Book Award, and Harvard's Smith-Weld Prize for Literary Merit. He has published more than 100 scientific papers in professional journals and has also received Harvard's Bok Prize for original contributions to astrophysics.

## Steve McMillan

Steve holds a bachelor's and master's degree in mathematics from Cambridge University and a doctorate in astronomy from Harvard University. He held post-doctoral positions at the University of Illinois and Northwestern University, where he continued his research in theoretical astrophysics, star clusters, and numerical modeling. Steve is currently Distinguished Professor of Physics at Drexel University and a frequent visiting researcher at Princeton's Institute for Advanced Study and the University of Amsterdam. He has published more than 50 scientific papers in professional journals.

# PREFACE

Astronomy is a science that thrives on new discoveries. Fueled by new technologies and novel theoretical insights, the study of the cosmos continues to change our understanding of the universe. We are pleased to have the opportunity to present in this book a representative sample of the known facts, evolving ideas, and frontier discoveries in astronomy today.

*Astronomy Today* has been written for students who have taken no previous college science courses and who will likely not major in physics or astronomy. It is intended for use in a one- or two-semester, nontechnical astronomy course. We present a broad view of astronomy, straightforwardly descriptive and without complex mathematics. The absence of sophisticated mathematics, however, in no way prevents discussion of important concepts. Rather, we rely on qualitative reasoning as well as analogies with objects and phenomena familiar to the student to explain the complexities of the subject without oversimplification. We have tried to communicate the excitement we feel about astronomy and to awaken students to the marvelous universe around us.

We are very gratified that the first five editions of this text have been so well received by many in the astronomy education community. In using those earlier texts, many teachers and students have given us helpful feedback and constructive criticisms. From these, we have learned to communicate better both the fundamentals and the excitement of astronomy. Many improvements inspired by these comments have been incorporated into this new edition.

## Focus of the Sixth Edition

From the first edition, we have tried to meet the challenge of writing a book that is both accurate and approachable. To the student, astronomy sometimes seems like a long list of unfamiliar terms to be memorized and repeated. Many new terms and concepts will be introduced in this course, but we hope students will also learn and remember how science is done, how the universe works, and how things are connected. In the sixth edition, we have taken particular care to show how astronomers know what they know, and to highlight both the scientific principles underlying their work and the process used in discovery.

## New and Revised Material

Astronomy is a rapidly evolving field and, in the three years since the publication of the fifth edition of *Astronomy Today*, has seen many new discoveries covering the entire spectrum of astronomical research. Almost every chapter in the sixth edition has been substantially updated with new information. Several chapters have also seen significant internal reorganization in order to streamline the overall presentation, strengthen our focus on the process of science, and reflect new understanding and emphases in contemporary astronomy. Among the many changes are:

- Streamlining of the art program to provide more direct and accurate representations of astronomical objects.
- Restructuring of the text to further emphasize the process of science and explain how astronomers "know what they know."
- Reworking of the end-of-chapter summaries to tie directly to the learning goals, and to link them to key figures in the chapter.
- Newly expanded and more explicit presentation of Newtonian mechanics in Chapter 2.
- New *More Precisely* box in Chapter 3 on applications of the Doppler Effect.
- More streamlined presentation of spectroscopic analysis in Chapter 4.
- New coverage of adaptive optics, the CHARA array, SST, and the future of HST in Chapter 5.
- Expanded treatment of comparative planetology in Chapter 6.
- Introduction of Eris and discussion of Pluto's demotion in Chapter 6.
- Updated discussion of planetary exploration in Chapter 6, with coverage of new missions now at Venus, Mars, and Saturn.
- New results from *Venus Express* in Chapter 9.
- Significant updates on Martian water from the *Mars Exploration Rovers*, *Mars Express*, and *Mars Reconnaissance Orbiter* in Chapter 10.
- Expanded treatment of climate evolution on Mars and Venus in Section 10.6.
- Expanded discussion of weather and the "Red Spot Junior" in Chapter 11.
- Updates on the *Cassini-Huygens* missions to Saturn and Titan.
- Updates in Chapter 14 on the *Stardust* and *Deep Impact* missions.
- Updates on observations and properties of extrasolar planets in Chapter 15.
- Update on gamma-ray burst observations and models in Chapter 22.
- Updates on the Galactic center in Chapter 23.
- New presentation of the uniform model for all active galaxies, following the discussion of their properties in Chapter 24.

- Restructured discussion of galaxy evolution in Chapter 25 to better reflect current thinking.

- Restructured discussion of black holes in Chapter 25 to emphasize the connection between normal and active galaxies, including quasar feedback as a mechanism coupling black holes and galaxy evolution.

- New results on mapping dark matter by gravitational lensing in Chapter 25.

- Updated *Discovery 25-1* on the Sloan Digital Sky Survey.

- New coverage in Chapter 26 of current ideas on the connection between density, geometry, and dynamics of the universe.

- Updated *Discovery 26-1* with the Hubble Ultra-Deep Field.

- Expanded discussion of the cosmological constant and introduced quintessence as another dark energy candidate in Chapter 26 and *Discovery 26-2*.

- Integration of dark energy into the discussion in Sections 27.1 and 27.2.

- Cosmic time line throughout revised to reflect new data from WMAP and elsewhere.

- Revised discussion of inflation in Chapter 27 reflecting the current view of the physics involved, including the introduction of scalar fields as part of the process.

- Expanded discussion of extremophilic life and stellar and galactic habitable zones in Chapter 28.

## The Illustration Program

Visualization plays an important role in both the teaching and the practice of astronomy, and we continue to place strong emphasis on this aspect of our book. We have tried to combine aesthetic beauty with scientific accuracy in the artist's conceptions that adorn the text, and we have sought to present the best and latest imagery of a wide range of cosmic objects. Each illustration has been carefully crafted to enhance student learning; each is pedagogically sound and tied tightly to the nearby discussion of important scientific facts and ideas. For this edition, the illustration program has been extensively revised and updated, resulting in more than **300** revised figures that show the latest imagery and the results learned from them.

## Full Spectrum Coverage and Spectrum Icons

Increasingly, astronomers exploit the full range of the electromagnetic spectrum to gather information about the cosmos. Throughout this book, images taken at radio, infrared, ultraviolet, X-ray, or gamma-ray wavelengths are used to supplement visible-light images. As it is sometimes difficult (even for a professional) to tell at a glance which images are visible-light photographs and which are false-color images created with other wavelengths, each photo in the text is accompanied by an icon that identifies the wavelength of electromagnetic radiation used to capture the image.

### ▶ Compound Art

It is rare that a single image, be it a photograph or an artist's conception, can capture all aspects of a complex subject. Wherever possible, multiple-part figures are used in an attempt to convey the greatest amount of information in the most vivid way:

- Visible images are often presented along with their counterparts captured at other wavelengths.

- Interpretive line drawings are often superimposed on or juxtaposed with real astronomical photographs, helping students to really "see" what the photographs reveal.

- Breakouts—often multiple ones—are used to zoom in from wide-field shots to close-ups so that detailed images can be understood in their larger context.

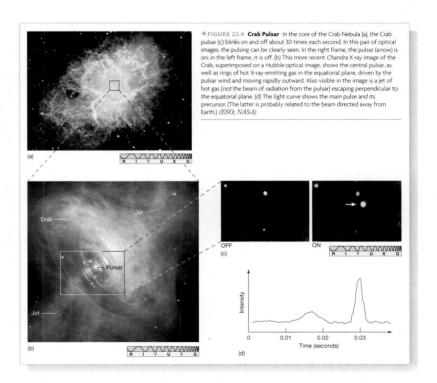

◀ FIGURE 22.4  **Crab Pulsar** In the core of the Crab Nebula (a), the Crab pulsar (c) blinks on and off about 30 times each second. In this pair of optical images, the pulsing can be clearly seen. In the right frame, the pulsar (arrow) is on; in the left frame, it is off. (b) This more recent *Chandra* X-ray image of the Crab, superimposed on a *Hubble* optical image, shows the central pulsar, as well as rings of hot X-ray-emitting gas in the equatorial plane, driven by the pulsar wind and moving rapidly outward. Also visible in the image is a jet of hot gas (*not* the beam of radiation from the pulsar) escaping perpendicular to the equatorial plane. (d) The light curve shows the main pulse and its precursor. (The latter is probably related to the beam directed away from Earth.) *(ESO; NASA)*

▲ **Explanatory Captions.** Students often review a chapter by "looking at the pictures." For this reason, the captions in this book are often a bit longer and more detailed than those in other texts.

## Other Pedagogical Features

As with many other parts of our text, instructors have helped guide us toward what is most helpful for effective student learning. With their assistance, we have revised both our in-chapter and end-of-chapter pedagogical apparatus to increase its utility to students.

**Learning Goals.** Studies indicate that beginning students have trouble prioritizing textual material. For this reason, a few (typically five or six) well-defined Learning Goals are provided at the start of each chapter. These help students structure their reading of the chapter and then test their mastery of key concepts. The Goals are numbered and keyed to the items in the Chapter Summary, which in turn refer back to passages in the text. This highlighting of the most important aspects of the chapter helps students prioritize information and also aids in their review. The Goals are organized and phrased in such a way as to make them objectively testable, affording students a means of gauging their own progress.

**∞ Concept-Links.** In astronomy, as in many scientific disciplines, almost every topic seems to have some bearing on almost every other. In particular, the connection between the astronomical material and the physical principles set forth early in the text is crucial. Practically everything in Chapters 6–28 of this text rests on the foundation laid in the first five chapters. For example, it is important that students, when they encounter the discussion of high-redshift objects in Chapter 25, recall not only what they just learned about Hubble's law in Chapter 24 but also refresh their memories, if necessary, about the inverse-square law (Chapter 17), stellar spectra (Chapter 4), and the Doppler shift (Chapter 3). Similarly, the discussions of the mass of binary-star components (Chapter 17) and of galactic rotation (Chapter 23) both depend on the discussion of Kepler's and Newton's laws in Chapter 2. Throughout, discussions of new astronomical objects and concepts rely heavily on comparison with topics introduced earlier in the text.

It is important to remind students of these links so that they recall the principles on which later discussions rest and, if necessary, review them. To this end, we have inserted "concept-links" throughout the text—symbols that mark key intellectual bridges between material in different chapters. The links, denoted by the symbol ∞ together with a section reference, signal that the topic under discussion is related in some significant way to ideas developed earlier, and provide direction to material to review before proceeding.

**Key Terms.** Like all subjects, astronomy has its own specialized vocabulary. To aid student learning, the most important astronomical terms are boldfaced at their first appearance in the text. Boldfaced Key Terms in the Chapter Summary are linked with the page number where the term was defined. In addition, an expanded alphabetical glossary, defining each Key Term and locating its first use in the text, appears at the end of the book.

**▼ Concept Checks.** We incorporate into each chapter a number of "Concept Checks"—key questions that require the reader to reconsider some of the material just presented or attempt to place it into a broader context. Answers to these in-chapter questions are provided at the back of the book.

---

CONCEPT CHECK

✔ In what ways might observations of extrasolar planets help us understand our own solar system?

---

**▼ H–R Diagrams and Acetate Overlays.** All of the book's H–R diagrams are drawn in a uniform format, using real data. In addition, a unique set of transparent acetate overlays dramatically demonstrates to students how the H–R diagram helps us to organize our information about the stars and track their evolutionary histories.

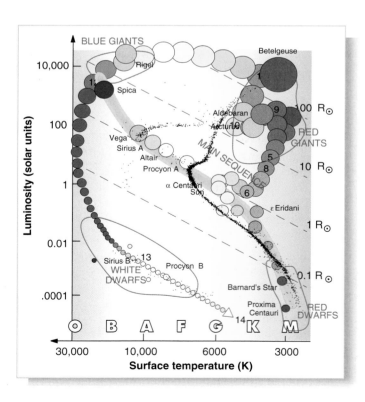

## MORE PRECISELY 17-3

### Measuring Stellar Masses in Binary Stars

As discussed in the text, most stars are members of binary systems—where two stars orbit one another, bound together by gravity. Here we describe—in an idealized case where the relevant orbital parameters are known—how we can use the observed orbital data, together with our knowledge of basic physics, to determine the masses of the component stars.

Consider the nearby visual binary system made up of the bright star Sirius A and its faint companion Sirius B, sketched in the accompanying figure. The binary's orbital period can be measured simply by watching the stars orbit one another, or alternatively by following the back-and-forth velocity wobbles of Sirius A due to its faint companion. It is almost exactly 50 years. The orbital semimajor axis can also be obtained by direct observation of the orbit, although in this case we must use some additional knowledge of Kepler's laws to correct for the binary's 46° inclination to the line of sight. ∞ (Sec. 2.5) It is 20 AU—a measured angular size of 7.5″ at a distance of 2.7 pc. ∞ (More Precisely 1-3) Once we know these two key orbital parameters, we can use the modified version of Kepler's third law to calculate the sum of the masses of the two stars. The result is $20^3/50^2 = 3.2$ times the mass of the Sun. ∞ (Sec. 2.8)

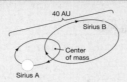

Further study of the orbit allows us to determine the individual stellar masses. Doppler observations show that Sirius A moves at approximately half the speed of its companion relative to their center of mass. ∞ (Secs. 2.8, 3.5) This implies that Sirius A must have twice the mass of Sirius B. It then follows that the masses of Sirius A and Sirius B are 2.1 and 1.1 solar masses, respectively.

Often the calculation of the masses of binary components is complicated by the fact that only partial information is available—we might only be able to see one star, or perhaps only spectroscopic velocity information is available (see Section 17.7). Nevertheless, this technique of combining elementary physical principles with detailed observations is how virtually every stellar mass quoted in this text has been determined.

▲ **More Precisely Boxes.** These boxes provide more quantitative treatments of subjects discussed qualitatively in the text. Removing these more challenging topics from the main flow of the narrative and placing them within a separate modular element of the chapter design (so that they can be covered in class, assigned as supplementary material, or simply left as optional reading for those students who find them of interest) will allow instructors greater flexibility in setting the level of their coverage.

## DISCOVERY 16-1

### *SOHO:* Eavesdropping on the Sun

Throughout the few decades of the Space Age, various nations, led by the United States, have sent spacecraft to all but one of the major bodies in the solar system. That unexplored body is the Sun. Currently, the next best thing to a dedicated reconnoitering spacecraft is the *Solar and Heliospheric Observatory* (*SOHO*), which has radioed back to Earth volumes of new data—and more than a few new puzzles—about our parent star since the spacecraft's launch in 1995.

*SOHO* is a billion-dollar mission operated primarily by the European Space Agency. The 2-ton robot is now on station about 1.5 million km sunward of Earth—about 1 percent of the distance from Earth to the Sun. This is the so-called $L_1$ Lagrangian point, where the gravitational pull of the Sun and Earth are precisely equal—a good place to park a monitoring platform. ∞ (Sec. 14.1) There, *SOHO* looks unblinkingly at our star 24 hours a day, eavesdropping on the Sun's surface, atmosphere, and interior. The automated vehicle carries a dozen instruments capable of measuring almost everything from the Sun's corona and magnetic field to its solar wind and internal vibrations. The accompanying figure shows a false-color image of the Sun's lower corona, obtained by combining *SOHO* data captured at three different ultraviolet wavelengths.

*SOHO* is positioned just beyond Earth's magnetosphere, so its instruments can cleanly study the charged particles of the solar wind—high-speed matter escaping from the Sun, flowing outward from the corona. Coordinating these on-site measurements with *SOHO* images of the Sun itself, astronomers now think they can follow solar magnetic field loops expanding and breaking as the Sun prepares itself for mass ejections several days before they actually occur (see Section 16.5). Given that such coronal storms can wreak havoc on communications, power grids, satellite electronics, and other human activities, the prospect of having accurate forecasts of disruptive solar events is a welcome development.

Section 16.2 discusses how astronomers can "take the pulse" of the Sun by measuring its complex rhythmic motions. *SOHO* also has the ability to study the weak sound waves that echo and resonate inside the Sun and can map these vibrations with much higher resolution than was previously possible. It does so not by sensing sound itself, but by watching the Sun's surface move up and down ever so slightly. The Sun's "loudest" vibrations are extremely low pitched (0.003 Hz, or one oscillation every 5 minutes), more like rolling rumbles, just as we might expect from such a huge and massive object.

As of late 2006, *SOHO* is still operating, 8 years beyond its planned lifetime, having survived assaults from both the Sun and Earth. Twice so far, skillful engineers have brought the spacecraft back from apparent death after mission controllers mistakenly sent incorrect commands from the ground. It is a good thing that they did, for this remarkable spacecraft has radioed back to Earth a wealth of new scientific insight into our parent star.

(NASA/ESA)    R I V U X G

▲ **Discovery Boxes.** Exploring a wide variety of interesting supplementary topics, *Discovery* boxes provide the reader with insight into how scientific knowledge evolves and emphasizes the process of science.

▶ **Chapter Summaries.** The Chapter Summaries, a primary review tool, have been expanded and improved for the sixth edition. All Key Terms introduced in each chapter are listed again, in context and in boldface, along with key figures and page references to the text discussion.

## CHAPTER REVIEW

### Summary

❶ The interior orbit of Venus with respect to Earth's means that Venus never strays far from the Sun in the sky. Because of its highly reflective cloud cover, Venus is brighter than any star in the sky, as seen from Earth. The planet's mass and radius are similar to those of Earth. The planet's rotation is slow and retrograde, most likely because of a collision between Venus and some other solar system body during the late stages of the planet's formation.

❷ The extremely thick atmosphere of Venus is nearly opaque to visible radiation, making the planet's surface invisible at optical wavelengths from the outside. Venus's atmosphere is nearly 100 times denser than Earth's and consists mainly of carbon dioxide. The temperature of the upper atmosphere is much like that of Earth's upper atmosphere, but the surface temperature of Venus is a searing 730 K. The planet's high-level winds circulate rapidly around the planet, and there are peculiarly shaped **polar vortices (p. 243)** at both poles.

❸ Venus's surface has been thoroughly mapped by radar from Earth-based radio telescopes and orbiting satellites. The planet's surface is mostly smooth, resembling rolling plains with modest highlands and lowlands. Two elevated continent-sized regions are called Ishtar Terra and Aphrodite Terra. There is no evidence for plate tectonic activity as on Earth. Features called **coronae (p. 240)** are thought to have been caused by an upwelling of mantle material. For unknown reasons, the upwelling never developed into full convective motion. The surface of the planet appears to be relatively young, resurfaced by volcanism within the past few hundred million years. Many **lava domes (p. 240)** and **shield volcanoes (p. 239)** were found by the *Magellan* orbiter on Venus's surface, but none of the volcanoes has yet proved to be currently active.

❹ Some craters on Venus are due to meteoritic impact, but the majority appear to be volcanic in origin. The evidence for currently active volcanoes on Venus includes surface features resembling those produced in Earthly volcanism, fluctuating levels of sulfur dioxide in Venus's atmosphere, and bursts of radio energy similar to those produced by lightning discharges that often occur in the plumes of erupting volcanoes on Earth. However, no actual eruptions have been seen. Soviet spacecraft that landed on Venus photographed surface rocks with sharp edges and a slablike character. Some rocks appear predominantly basaltic in nature, implying a volcanic past, others resemble terrestrial granite and are probably part of the planet's ancient crust.

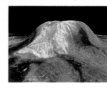

**End-of-Chapter Questions and Problems.** Many elements of the end-of-chapter material have seen substantial reorganization:

- Each Chapter incorporates 20 **Review and Discussion Questions,** which may be used for in-class review or for assignment. As with the Self-Test Questions, the material needed to answer Review Questions may be found within the chapter. The Discussion Questions explore particular topics more deeply, often asking for opinions, not just facts. As with all discussions, these questions usually have no single "correct" answer.

- Each Chapter also contains 20 **Conceptual Self-Test Questions,** equally divided between "true/false" and multiple choice formats, including select questions that are tied directly to a specific figure or diagram in the text, allowing students to assess their understanding of the chapter material. Answers to all these questions appear at the end of the book.

- The End-of-Chapter material includes 15 **Problems,** based on the chapter contents and requiring some numerical calculation. In many cases the problems are tied directly to quantitative statements made (but not worked out in detail) in the text. The solutions to the problems are not contained verbatim within the chapter, but the information necessary to solve them has been presented in the text. Answers to odd-numbered Problems appear at the end of the book.

### Instructor Resources

*Instructor Resource Manual.* Revised by James Heath (Austin Community College), this manual provides: sample syllabi and course schedules; an overview of each chapter; pedagogical tips; useful analogies; suggestions for classroom demonstrations; writing questions, selected readings, and answers/solutions to the end-of-chapter Review and Discussion Questions and Problems; and additional references and resources. ISBN 0-13-240095-2

*Test Bank.* An extensive file of approximately 2800 test questions, newly compiled and revised for the sixth edition by Lauren Jones (Ohio Department of Education, Office of Curriculum and Assessment). The questions are organized and referenced by chapter section and by question type. The sixth edition *Test Bank* has been thoroughly revised and includes many new Multiple Choice and Essay questions for added conceptual emphasis. This manual is available in both printed and electronic formats (see description of *Media Manager*). ISBN 0-13-240096-0

*Media Manager.* This DVD-ROM provides virtually every electronic asset professors will need in and out of the classroom. The disc contain all text figures in jpeg and PowerPoint formats, as well as the animations and videos from the *Companion Website with GradeTracker*. The Media Manager also contains TestGenerator, an easy to use, fully networkable program for creating tests ranging from short quizzes to long exams. Questions from the *Test Bank* are supplied, and professors can use the Question Editor to modify existing questions or create new questions. This disc set also contains chapter-by-chapter lecture outlines in PowerPoint, conceptual "clicker" questions in PowerPoint, Word documents of the *Test Bank*, and Word and PDF versions of the *Instructor Resource Manual*. ISBN 0-13-240087-1

***Transparency Acetates.*** Approximately 260 images from the text are available as a package of color acetates and are available free to qualified adopters. ISBN 0-13-240093-6

### *Learner-Centered Astronomy Teaching: Strategies for ASTRO 101*

Timothy F. Slater, *University of Arizona*
Jeffrey P. Adams, *Montana State University*
*Strategies for ASTRO 101* is a guide for instructors of the introductory astronomy course for nonscience majors. Written by two leaders in astronomy education research, this book details various techniques instructors can use to increase students' understanding and retention of astronomy topics, with an emphasis on making the lecture a forum for active student participation. Drawing from the large body of recent research to discover how students learn, this guide describes the application of multiple classroom-tested techniques to the task of teaching astronomy to predominantly nonscience students. ISBN 0-13-046630-1

### *Peer Instruction for Astronomy*

Paul Green, *Harvard Smithsonian Center for Astrophysics*
Peer Instruction is a simple yet effective method for teaching science. Techniques of peer instruction for introductory physics were developed primarily at Harvard and have aroused interest and excitement in the Physics Education community. This approach involves students in the teaching process, making science more accessible to them. This book is an important vehicle for providing a large number of thought-provoking, conceptual short-answer questions aimed at a variety of class levels. While significant numbers of such questions have been published for use in physics, *Peer Instruction for Astronomy* provides the first such compilation for astronomy. ISBN 0-13-026310-9

## Course and Homework Management Tools
### Blackboard

**Blackboard**™ is a comprehensive and flexible e-learning software platform that delivers a course management system, customizable institution-wide portals, online communities, and an advanced architecture that allows for Web-based integration of multiple administrative systems.

**Features:**
- Course management system, including progress tracking, class and student management, gradebook, communication, assignments, and reporting tools.
- Testing programs that allow instructors to create online quizzes and tests and automatically grade and track results. All tests can be entered into the gradebook for easy course management.
- Communications tools such as the Virtual Classroom™ (chat rooms, WhiteBoard, slides), document sharing, and bulletin boards.

### CourseCompass
### *Powered by* Blackboard

With the highest level of service, support, and training available today, CourseCompass combines tested, quality, online course resources with easy-to-use online course management tools. CourseCompass is designed to address the individual needs of instructors, who will be able to create an online course without any special technical skills or training.

**Features:**
- Flexibility—Instructors can adapt *Astronomy Today, Sixth Edition* content to match their own teaching goals.
- Assessment, customization, class administration, and communication tools.
- Point-and-click access—The publisher's vast educational resources are available to instructors at a click of the mouse.
- A nationally hosted and fully supported system that relieves individuals and institutions of the burdens of troubleshooting and maintenance.

### WebCT

WebCT offers a powerful set of tools that enables professors to create practical Web-based educational programs—the ideal resources to enhance a campus course or to construct one entirely online. The WebCT shell and tools, integrated with the *Astronomy Today, Sixth Edition* content, results in a versatile, course-enhancing teaching and learning system.

**Features:**
- Page tracking, progress tracking, class and student management, gradebook, communication, calendar, reporting tools, and more.
- Communication tools including chat rooms, bulletin boards, private e-mail, and WhiteBoard.
- Testing tools that help create and administer timed online quizzes and tests and automatically grade and track all results.

Along with all the material from the Companion Website (www.aw-bc.com/chaisson), our course management cartridges include additional materials available only through Blackboard, CourseCompass, or WebCT, and include

- Test bank questions, converted from our TestGen *Test Bank*
- Instructor Resources from the *Media Manager*

## WebAssign (www.webassign.com)

WebAssign's homework delivery service harnesses the power of the Internet and puts it to work for you. Collecting and grading homework becomes the province of the WebAssign service, which gives you the freedom to get back to teaching. Create assignments from a database of end-of-chapter problems and *Test Bank* questions from *Astronomy Today, Sixth Edition* or write and customize your own. You have complete control over the homework your students receive, including due date, content, feedback, and question formats.

### Student Resources
### Companion Website With GradeTracker (www.aw-bc.com/chaisson)

The text-specific Companion Website with GradeTracker for *Astronomy Today, Sixth Edition* organizes material from a variety of sources on a chapter-by-chapter basis, providing students with a variety of inquiry, visual, and interactive tools of each chapter's topics, and provides gradebook and syllabus manager functionality for instructors who wish to use it. The Companion Website for the sixth edition includes:

- The newly assembled **Animations and Videos** gallery. This collection of the most up-to-date, illustrative, and dynamic pieces on key concepts and recent events helps to create a more scientifically literate astronomy student. See the associated icons within the text at point of use.
- A collection of the most current and astonishing **Images** from the best sources on the World Wide Web.
- **Multiple Choice** and **True/False** quizzes with hints and instant feedback.
- Algorithmically generated versions of select end-of-chapter **Practice Problems** from the text.
- **Labeling** exercises, where students identify various components of a fundamental figure from the textbook.
- Now online for the sixth edition, the **eBook** is a full electronic version of the text, with key term hyperlinks and imbedded media elements.
- An interactive Glossary.

▼ The revamped and amplified **Tutorials** gallery. These animated, interactive Flash™ files bring concepts from the text to life, allowing students to explore ideas in depth. Students are engaged in the thought process as they answer questions and change parameters in these exploratory activities. Also included are gradable follow-up multiple choice **Tutorials quiz questions,** which will check the student's understanding of the tutorial. See the associated icons within the text at point of use.

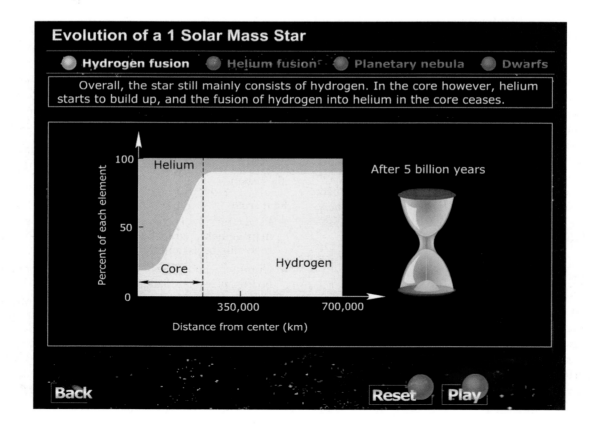

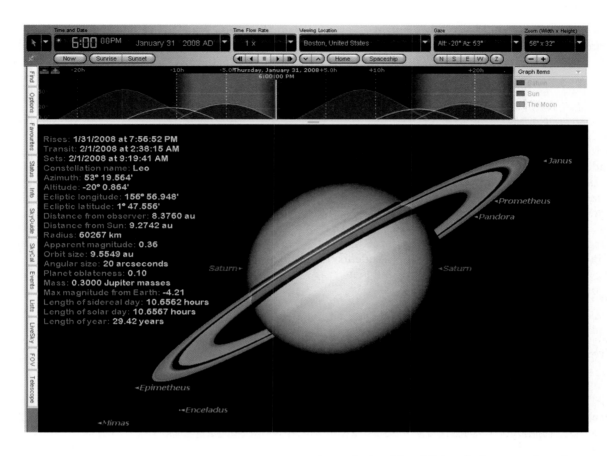

▲ **Starry Night Pro™** *6 Student DVD*

This best selling planetarium software lets you escape the Milky Way and travel within 700 million light years of space. View more than 16 million stars in stunningly realistic star fields. Zoom in on thousands of galaxies, nebulae, and star clusters. Move through 200,000 years of time to see key celestial events in a dynamic and ever-changing universe. Blast off from Earth and see the motions of the planets from a new perspective. Hailed for its breathtaking realism, powerful suite of features, and intuitive ease-of-use, **Starry Night Pro™** lives up to its reputation as astronomy software's brightest . . . night after night. ISBN 0-321-53973-7

**Starry Night Pro™** *Activities, and Research & Observation Projects*

This supplement contains activities for the **Starry Night Pro™** planetarium software by Erin O'Connor (Santa Barbara City College), as well as observation and research projects authored by Steve McMillan. ISBN 0-13-240097-9

*Edmund Scientific Star and Planet Locator*

The famous rotating roadmap of the heavens shows the location of the stars, constellations, and planets relative to the horizon for the exact hour and date you determine. This 8 square star chart was plotted by the late astronomer and cartographer George Lovi. The reverse side of the locator is packed with additional data on the planets, meteor showers, and bright stars. Included with each star chart is a 16-page, fully-illustrated, pocket-size instruction booklet.

*Norton's Star Atlas and Reference Handbook, 20th Edition*

Ian Ridpath

Now in a superbly redesigned, two-color landmark 20th edition, the first of a new century, this combination star atlas and reference work has no match in the field. First published in 1910, coinciding with the first of two appearances by Halley's Comet during the book's life, *Norton's* owes much of its legendary success to its unique maps, arranged in slices known as gores, each covering approximately one-fifth of the sky. Every star visible to the naked eye under the clearest skies—down to magnitude 6.5—is charted along with star clusters, nebulae and galaxies. Extensive tables of data on interesting objects for observation accompany each of the precision drawn maps. Preceding the maps is the unique and authoritative reference handbook covering timekeeping and positional measurements on the celestial sphere; the Sun, Moon, and other bodies of the Solar System; telescopes and other equipment for observing and imaging the sky; and stars, nebulae, and galaxies. Throughout, succinct fundamental principles and practical tips guide the reader into the night sky. The appendices Units and Notation, Astronomical Constants, Symbols and Abbreviations, and Useful Addresses complete what has long been the only essential reference for the stargazer. ISBN 0-13-145164-2

### Lecture Tutorials for Introductory Astronomy, 2nd Edition

Edward E. Prather, *University of Arizona*
Timothy F. Slater, *University of Arizona*
Jeffrey P. Adams, *Montana State University*
Gina Brissenden, *NASA Center for Astronomy Education (CAE) at the University of Arizona*
CAPER, *Conceptual Astronomy and Physics Education Research Team*

Funded by the National Science Foundation, *Lecture-Tutorials for Introductory Astronomy* are designed to help make large lecture-format courses more interactive. The second edition features new Tutorials on the Doppler Shift, Extrasolar Planets, Kepler's 2nd Law, and Seasons. Each of the 38 Lecture-Tutorials is presented in a classroom-ready format, asks students to work in groups of 2–3, takes between 10–15 minutes, challenges students with a series of carefully designed questions that spark classroom discussion, engages students in critical reasoning, and requires no equipment. ISBN 0-13-239226-7

## Acknowledgments

Throughout the many drafts that have led to this book, we have relied on the critical analysis of many colleagues. Their suggestions ranged from the macroscopic issue of the book's overall organization to the minutiae of the technical accuracy of each and every sentence. We have also benefited from much good advice and feedback from users of the first five editions of the text. To these many helpful colleagues, we offer our sincerest thanks.

## Reviewers of the Sixth Edition

William Alexander
*James Madison University*

Malcolm Cleaveland
*University of Arkansas*

Richard Gelderman
*Western Kentucky University*

James Heath
*Austin Community College*

Matthew Lister
*Purdue University*

Mark Moldwin
*UCLA*

Lawrence Pinsky
*University of Houston*

Dwight Russell
*Baylor University*

Larry Sessions
*Metropolitan State College of Denver*

George Stanley, Jr.
*San Antonio College*

Donald Terndrup
*Ohio State University*

Robert Zimmerman
*University of Oregon*

## Reviewers of Previous Editions

Stephen G. Alexander
*Miami University of Ohio*

Robert H. Allen
*University of Wisconsin, La Crosse*

Cecilia Barnbaum
*Valdosta State University*

Peter A. Becker
*George Mason University*

Timothy C. Beers
*University of Evansville*

William J. Boardman
*Birmingham Southern College*

Donald J. Bord
*University of Michigan, Dearborn*

Elizabeth P. Bozyan
*University of Rhode Island*

Anne Cowley
*Arizona State University*

Bruce Cragin
*Richland College*

Ed Coppola
*Community College of Southern Nevada*

David Curott
*University of North Alabama*

Norman Derby
*Bennington College*

John Dykla
*Loyola University, Chicago*

Kimberly Engle
*Drexel University*

Michael N. Fanelli
*University of North Texas*

Harold A. Geller
*George Mason University*

David Goldberg
*Drexel University*

Martin Goodson
*Delta College*

David G. Griffiths
*Oregon State University*

Donald Gudehus
*Georgia State University*

Marilynn Harper
*Delaware County Community College*

Clint D. Harper
*Moorpark College*

Susan Hartley
*University of Minnesota, Duluth*

Joseph Heafner
*Catawaba Valley Community College*

Fred Hickok
*Catonsville Community College*

Darren L. Hitt
*Loyola College, Maryland*

F. Duane Ingram
*Rock Valley College*

Steven D. Kawaler
*Iowa State University*

William Keel
*University of Alabama*

Marvin Kemple
*Indiana University-Purdue University, Indianapolis*

Mario Klairc
*Midlands Technical College*

Kristine Larsen
*Central Connecticut State University*

Andrew R. Lazarewicz
*Boston College*

Robert J. Leacock
*University of Florida*

Larry A. Lebofsky
*University of Arizona*

M. A. Lohdi
*Texas Tech University*

Michael C. LoPresto
*Henry Ford Community College*

Phillip Lu
*Western Connecticut State University*

Fred Marschak
*Santa Barbara College*

Matthew Malkan
*University of California, Los Angeles*

Chris Mihos
*Case Western Reserve University*

Milan Mijic
*California State University, Los Angeles*

Scott Miller
*Pennsylvania State University*

Richard Nolthenius
*Cabrillo College*

Edward Oberhofer
*University of North Carolina, Charlotte*

Andrew P. Odell
*Northern Arizona University*

Gregory W. Ojakangas
*University of Minnesota, Duluth*

Ronald Olowin
*Saint Mary's College of California*

Robert S. Patterson
*Southwest Missouri State University*

Cynthia W. Peterson
*University of Connecticut*

Andreas Quirrenback
*University of California, San Diego*

Richard Rand
*University of New Mexico*

James A. Roberts
*University of North Texas*

Gerald Royce
*Mary Washington College*

Vicki Sarajedini
*University of Florida*

Malcolm P. Savedoff
*University of Rochester*

John C. Schneider
*Catonsville Community College*

Harry L. Shipman
*University of Delaware*

C. G. Pete Shugart
*Memphis State University*

Stephen J. Shulik
*Clarion University*

Tim Slater
*University of Arizona*

Don Sparks
*Los Angeles Pierce College*

Maurice Stewart
*Williamette University*

Jack W. Sulentic
*University of Alabama*

Andrew Sustich
*Arkansas State University*

Craig Tyler
*Fort Lewis College*

Stephen R. Walton
*California State University, Northridge*

Peter A. Wehinger
*University of Arizona*

Louis Winkler
*Pennsylvania State University*

We would also like to express our gratitude to Ray Villard, Scott Miller, Carl Adler, and David Zeigler for updating and maintaining the media resources on the Companion Website.

The publishing team at Addison-Wesley/Benjamin Cummings has assisted us at every step along the way in creating this text. Many thanks to Christian Botting, who managed the many variables that go into a multifaceted publication such as this. John Haber, our Development Editor, has read every sentence in the book from the student perspective. Production editors Crystal Clifton and Linda Kern have done an excellent job of tying together the threads of this very complex project, made all the more complex by the necessity of combining text, art, and electronic media into a coherent whole. Special thanks are also in order to interior designers John Christiana and Maureen Eide, and cover designers Mark Ong and Lisa Devenish for making the sixth edition look spectacular.

Finally, we would like to express our gratitude to renowned space artist Dana Berry for allowing us to use many of his beautiful renditions of astronomical scenes, and to Lola Judith Chaisson for assembling and drawing all the H–R diagrams (including the acetate overlays) for this edition.

Eric Chaisson
Steve McMillan

# PART 1
# ASTRONOMY and the UNIVERSE

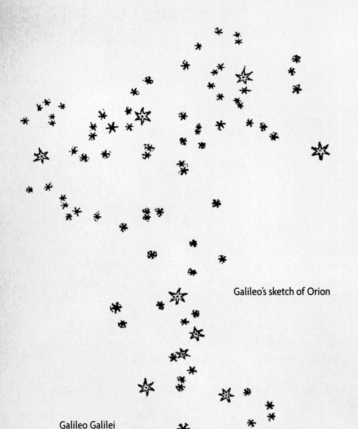

Galileo's sketch of Saturn

Galileo's sketch of Orion

Galileo Galilei

It is often said that we live in a golden age of astronomy. Yet the dawn of the 21st century is actually the second such period of rich discovery and rapid exploration. The late Renaissance began the first era of stunning scientific advancement, when modern astronomy was born. Most notable among those spearheading the rebirth of astronomy at the time was the Italian scientist Galileo Galilei (1564–1642). By turning his telescope to the heavens, he changed radically and forever humankind's view of the universe in which we live.

Although he did not invent the telescope, in 1610 Galileo was the first to record what he saw when he aimed a small (5-centimeter-diameter) lens at the sky. His findings created nothing less than a revolution in astronomy. Viewing for the first time dark blemishes on the Sun, rugged mountains on the Moon, and whole new worlds orbiting Jupiter, he demolished the Aristotelian view of cosmic immutability—the notion that the heavens were perfect and unchanging. To be sure, it was with the philosophers of the day, as much as with the theologians, that Galileo had trouble. In championing the scientific method, he used a tool to test his ideas, and what he found disagreed greatly with the leading thoughts and beliefs of the time.

Galileo's advance was simple yet profound: He used a telescope to focus, magnify, and study radiation reaching Earth from the heavens—in particular, light from the Sun, the Moon, and the planets. Light is the most familiar kind of radiation to humans on Earth, since it enables us to get around on the surface of our planet. But light also enables telescopes to see objects deep in space, allowing us to probe farther than the eye can alone. With his simple optical telescope, Galileo changed completely the way that the oldest science—astronomy—is pursued.

Among other "wondrous things" he found were crowded star clusters along the Milky Way, moons and rings around the big planets, and colorful nebulae aglow unlike anything anyone had seen before. Some of Galileo's sketches are reproduced here at the left and compared with modern views at the right.

Galileo's sketch of the Pleiades

Saturn in the ultraviolet *(STScI)*

Today, we are again in the midst of another period of unsurpassed scientific achievement—a revolution in which modern astronomers are revealing the invisible universe as Galileo once spied the visible universe. We have learned how to detect, measure, and analyze invisible radiation streaming to us from dark objects in space. And once again our perceptions are changing.

Astronomy no longer evokes visions of plodding intellectuals peering through long telescope tubes. Nor does the cosmos any longer refer to that seemingly inactive, immutable regime seen visually when we gaze at the nighttime sky. Modern astronomers now decipher a more vibrant, changing universe—one in which stars emerge and perish much like living things, galaxies spew forth vast quantities of energy, and life itself is thought to be a natural consequence of the evolution of matter.

New discoveries are rapidly advancing our understanding of the universe, but they also raise new questions. Astronomers will encounter many problems in the decades ahead, but this should neither dismay nor frustrate us, for it is precisely how science operates. Each discovery adds to our storehouse of information, generating a host of questions that lead in turn to more discoveries, and so on, causing an acceleration of basic knowledge.

Most notably, we are beginning to perceive the universe in all its multivaried ways. A single generation—not the generation of our parents and not that of our children, but our generation—has opened up the whole electromagnetic spectrum beyond visible light. And what we, too, have found are "wondrous things."

Emerging largely from studies of the invisible universe, our view of the cosmos in its full splendor is one of many new scientific insights that we have recently been privileged to attain. Historians of the future may well regard our generation as the one that took a great leap forward, providing a whole new glimpse of our richly endowed universe. In all of history, there have been only two periods in which our perception of the universe has been so revolutionized within a single human lifetime. The first occurred four centuries ago at the time of Galileo; the second is now underway.

Orion in the infrared *(Caltech)*

Pleiades in the optical *(AURA)*

CHAPTER

1

*Stars are perhaps the most fundamental component of the universe that we inhabit. Roughly as many stars reside in the observable universe as grains of sand in all the beaches of the world. Here, we see displayed one of the most easily recognizable star fields in the winter sky—the familiar constellation Orion. Near center are three blue-white diagonally aligned stars hotter than our Sun, below which is a red-white nebulosity where stars are now forming. At top left is an old giant red star probably about to explode. The entire field of view spans about 15 degrees on the sky.* (J. SANFORD/ASTROSTOCK-SANFORD) ▶

# CHARTING the HEAVENS

The Foundations of Astronomy

N ature offers no greater splendor than the starry sky on a clear, dark night. Silent and jeweled with the constellations of ancient myth and legend, the night sky has inspired wonder throughout the ages—a wonder that leads our imaginations far from the confines of Earth and the pace of the present day and out into the distant reaches of space and cosmic time itself.

Astronomy, born in response to that wonder, is built on two of the most basic traits of human nature: the *need to explore* and the *need to understand*. Through the interplay of curiosity, discovery, and analysis—the keys to exploration and understanding—people have sought answers to questions about the universe since the earliest times. Astronomy is the oldest of all the sciences, yet never has it been more exciting than it is today.

## LEARNING GOALS

*Studying this chapter will enable you to*

1  Describe how scientists combine observation, theory, and testing in their study of the universe.

2  Explain the concept of the celestial sphere and how we use angular measurement to locate objects in the sky.

3  Describe how and why the Sun and the stars appear to change their positions from night to night and from month to month.

4  Explain why Earth's rotation axis shifts slowly with time, and say how this affects Earth's seasons.

5  Tell how our clocks and calendars are linked to Earth's rotation and orbit around the Sun.

6  Show how the relative motions of Earth, the Sun, and the Moon lead to eclipses.

7  Explain the simple geometric reasoning that allows astronomers to measure the distances and sizes of otherwise inaccessible objects.

 Visit www.aw-bc.com/chaisson for additional images, animations, quizzes, and eBook for this chapter.

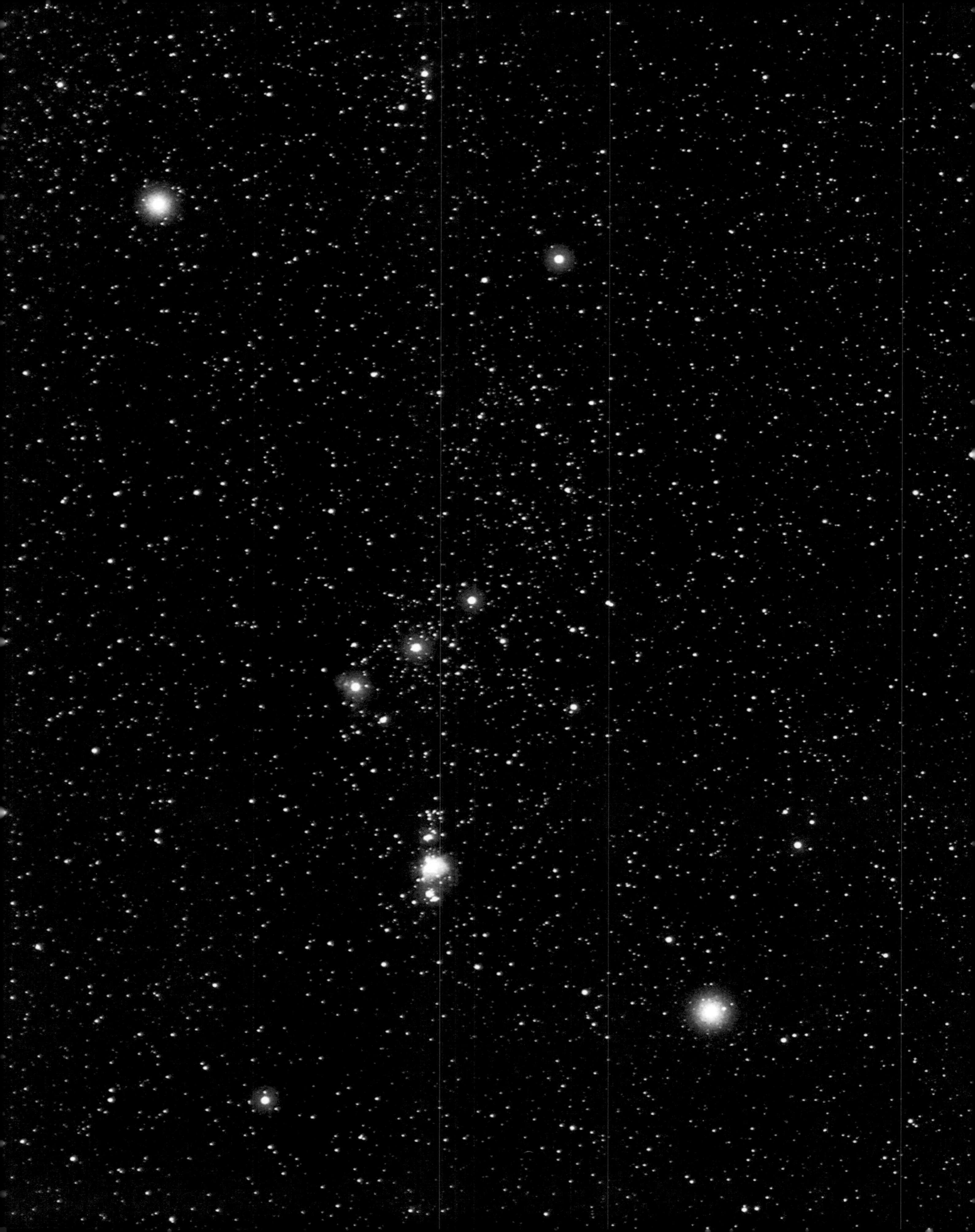

## 1.1 Our Place in Space

Of all the scientific insights attained to date, one stands out boldly: Earth is neither central nor special. We inhabit no unique place in the universe. Astronomical research, especially within the past few decades, strongly suggests that we live on what seems to be an ordinary rocky *planet* called Earth, one of eight known planets orbiting an average *star* called the Sun, a star near the edge of a huge collection of stars called the Milky Way galaxy, which is one *galaxy* among billions of others spread throughout the observable universe. To begin to get a feel for the relationships among these very different objects, consult Figures 1.1 through 1.4; put them in perspective by studying Figure 1.5.

We are connected to the most distant realms of space and time not only by our imaginations, but also through a common cosmic heritage. Most of the chemical elements that make up our bodies (hydrogen, oxygen, carbon, and many more) were created billions of years ago in the hot centers of long-vanished stars. Their fuel supply spent, these giant stars died in huge explosions, scattering the elements created deep within their cores far and wide. Eventually, this matter collected into clouds of gas that slowly collapsed to give birth to new generations of stars. In this way, the Sun and its family of planets formed nearly five billion years ago. Everything on Earth embodies atoms from other parts of the universe and from a past far more remote than the beginning of human evolution. Elsewhere, other beings—perhaps with intelligence much greater than our own—may at this very moment be gazing in wonder at their own night sky. Our own Sun may be nothing more than an insignificant point of light to them—if it is visible at all. Yet if such beings exist, they must share our cosmic origin.

Simply put, the **universe** is the totality of all space, time, matter, and energy. **Astronomy** is the study of the universe. It is a subject unlike any other, for it requires us to profoundly change our view of the cosmos and to consider matter on scales totally unfamiliar from everyday experience. Look again at the galaxy in Figure 1.3. It is a swarm of about a hundred billion stars—more stars than the number of people who have ever lived on Earth. The entire assemblage is spread across a vast expanse of space 100,000 **light-years** in diameter. Although it sounds like a unit of time, a light-year is in fact the *distance* traveled by light in a year, at a speed of about 300,000 kilometers per second. Multiplying out, it follows that a light-year is equal to 300,000 kilometers/second × 86,400 seconds/day × 365 days or about 10 trillion kilometers, or roughly 6 trillion miles. Typical galactic systems are truly "astronomical" in size. For

15,000 kilometers

▲ FIGURE 1.1 **Earth** Earth is a planet, a mostly solid object, although it has some liquid in its oceans and core, and gas in its atmosphere. In this view, the North and South American continents are clearly visible. (*NASA*)

1,500,000 kilometers

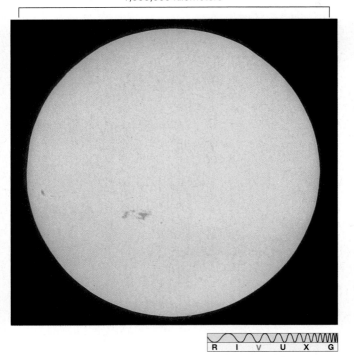

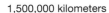

▲ FIGURE 1.2 **The Sun** The Sun is a star, a very hot ball of gas composed mainly of hydrogen and helium. Much bigger than Earth—more than 100 times larger in diameter—the Sun is held together by its own gravity. (*AURA*)

comparison, Earth's roughly 13,000-km diameter is less than one-twentieth of a light-*second*.

The light-year is a unit introduced by astronomers to help them describe immense distances. We will encounter many such custom units in our studies. As discussed in more detail in Appendix 2, astronomers frequently augment the standard SI (Système Internationale) metric system with additional units tailored to the particular problem at hand.

A thousand (1000), a million (1,000,000), a billion (1,000,000,000), and even a trillion (1,000,000,000,000)—these words occur regularly in everyday speech. But let's take a moment to understand the magnitude of the numbers and appreciate the differences among them. One thousand is easy enough to understand: At the rate of one number per second, you could count to a thousand in 1000 seconds—about 16 minutes. However, if you wanted to count to a million, you would need more than two weeks of counting at the rate of one number per second, 16 hours per day (allowing 8 hours per day for sleep). To count from one to a billion at the same rate of one number per second and 16 hours per day would take nearly 50 years—the better part of an entire human lifetime.

In this book, we consider *distances* in space spanning not just billions of kilometers, but billions of light-years; *objects* containing not just trillions of atoms, but trillions of stars; and *time intervals* of not just billions of seconds or hours, but billions of years. You will need to become familiar—and comfortable—with such enormous numbers. A good way to begin is learning to recognize just how much larger than a thousand is a million, and how much larger still is a billion. Appendix 1 explains the convenient method used by scientists for writing and manipulating very large and very small numbers. If you are unfamiliar with this method, please read that appendix carefully—the *scientific notation* described there will be used consistently throughout our text, beginning in Chapter 2.

Lacking any understanding of the astronomical objects they observed, early skywatchers made up stories to explain them: The Sun was pulled across the heavens by a chariot drawn by winged horses, and patterns of stars traced heroes and animals placed in the sky by the gods. Today, of course, we have a radically different conception of the universe. The stars we see are distant, glowing orbs hundreds of times larger than our entire planet, and the patterns they form span hundreds of light-years. In this first chapter we present some basic methods used by astronomers to chart the space around us. We describe the slow progress of scientific knowledge, from chariots and gods to today's well-tested theories and physical laws, and explain why we now rely on science rather than on myth to help us explain the universe.

About 1000 quadrillion kilometers, or 100,000 light-years

R I V U X G

▲ FIGURE 1.3 **Galaxy** A typical galaxy is a collection of a hundred billion stars, each separated by vast regions of nearly empty space. Our Sun is a rather undistinguished star near the edge of another such galaxy, called the Milky Way. (*NASA*)

About 1,000,000 light-years

R I V U X G

▲ FIGURE 1.4 **Galaxy Cluster** This photograph shows a typical cluster of galaxies, spread across roughly a million light-years of space. Each galaxy contains hundreds of billions of stars, probably planets, and, possibly, living creatures. (*NASA*)

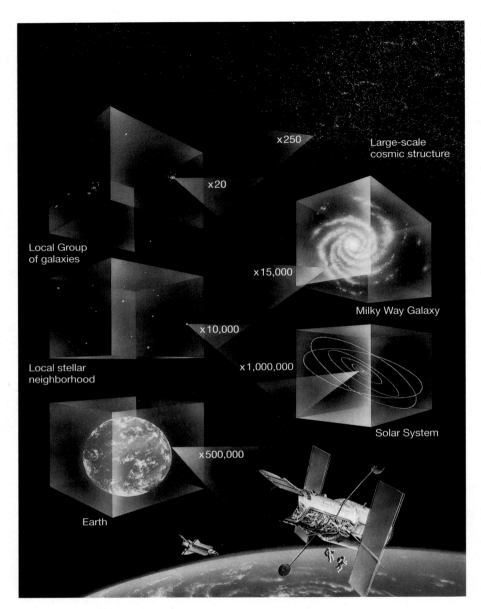

◀ FIGURE 1.5 **Size and Scale** This artist's conception puts each of the previous four figures in perspective. The bottom of this figure shows spacecraft (and astronauts) in Earth orbit, a view that widens progressively in each of the next five cubes drawn from bottom to top—Earth, the planetary system, the local neighborhood of stars, the Milky Way Galaxy, and the closest cluster of galaxies. The top image depicts the distribution of galaxies in the universe on extremely large scales—the field of view in this final frame is hundreds of millions of light-years across. The numbers indicate approximately the increase in scale between successive images: Earth is 500,000 times larger than the spacecraft in the foreground, the solar system in turn is some 1,000,000 times larger than Earth, and so on. *(D. Berry)*

## 1.2 Scientific Theory and the Scientific Method

How have we come to know the universe around us? How do we know the proper perspective sketched in Figure 1.5? The earliest known descriptions of the universe were based largely on imagination and mythology and made little attempt to explain the workings of the heavens in terms of known earthly experience. However, history shows that some early scientists did come to realize the importance of careful observation and testing to the formulation of their ideas. The success of their approach changed, slowly but surely, the way science was done and opened the door to a fuller understanding of nature. As knowledge from all sources was sought and embraced for its own sake, the influence of logic and reasoned argument grew and the

power of myth diminished. People began to inquire more critically about themselves and the universe. They realized that *thinking* about nature was no longer sufficient—that *looking* at it was also necessary. Experiments and observations became a central part of the process of inquiry.

To be effective, a **theory**—the framework of ideas and assumptions used to explain some set of observations and make predictions about the real world—must be continually tested. Scientists accomplish this by using a theory to construct a **theoretical model** of a physical object (such as a planet or a star) or phenomenon (such as gravity or light), that accounts for its known properties. The model then makes further predictions about the object's properties, or perhaps how it might behave or change under new circumstances. If experiments and observations favor those predictions, the theory can be further developed and refined. If they do not, the theory must be reformulated or rejected, no matter how appealing it originally seemed. This approach to investigation, combining thinking and doing—that is, theory and experiment—is known as the **scientific method.** The process, combining theoretical reasoning with experimental testing, is illustrated schematically in Figure 1.6. It lies at the heart of modern science, separating science from pseudoscience, fact from fiction.

The notion that theories must be tested and may be proven wrong sometimes leads people to dismiss their importance. We have all heard the expression, "Of course, it's only a theory," used to deride or dismiss an idea that someone finds unacceptable. Don't be fooled by this abuse of the concept! Gravity (see Section 2.7) is "only" a theory, but calculations based on it have guided human spacecraft throughout the solar system. Electromagnetism (Chapter 3) and quantum mechanics (Chapter 4) are theories, too, yet they form the foundation for most of 20th- (and 21st) century technology. Facts about the universe are a dime a dozen. Theories are the intellectual "glue" that combine seemingly unrelated facts into a coherent and interconnected whole.

Notice that there is no end point to the process depicted in Figure 1.6. A theory can be invalidated by a single wrong prediction, but no amount of observation or experimentation can ever prove it "correct." Theories simply become more and more widely accepted as their predictions are repeatedly confirmed. As a class, modern scientific theories share several important defining characteristics:

- They must be *testable*—that is, they must admit the possibility that their underlying assumptions and their predictions can, in principle, be exposed to experimental verification. This feature separates science from, for example, religion, since, ultimately, divine revelations or scriptures cannot be challenged within a religious framework—we can't design an experiment to "verify the mind of God." Testability also distinguishes science from a pseudoscience such as astrology, whose underlying assumptions and predictions have been repeatedly tested and never verified, with no apparent impact on the views of those who continue to believe in it!

- They must continually be *tested*, and their consequences tested, too. This is the basic circle of scientific progress depicted in Figure 1.6.

- They should be *simple*. Simplicity is less a requirement than a practical outcome of centuries of scientific experience—the most successful theories tend to be the simplest ones that fit the facts. This viewpoint is often encapsulated in a principle known as *Occam's razor:* If two competing theories both explain the facts and make the same predictions, then the simpler one is better. Put another way—"Keep it simple!" A good theory should contain no more complexity than is absolutely necessary.

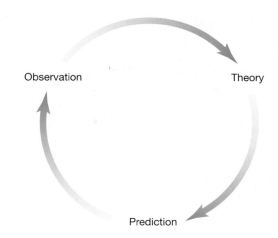

▲ FIGURE 1.6 **Scientific Method** Scientific theories evolve through a combination of observation, theoretical reasoning, and prediction, which in turn suggests new observations. The process can begin at any point in the cycle (although it usually starts with observations), and it continues forever—or until the theory fails to explain an observation or makes a demonstrably false prediction.

- Finally, most scientists have the additional bias that a theory should in some sense be *elegant*. When a clearly stated simple principle naturally ties together and explains several phenomena previously thought to be completely distinct, this is widely regarded as a strong point in favor of the new theory.

You may find it instructive to apply these criteria to the many physical theories—some old and well established, others much more recent and still developing—we will encounter throughout the text.

The birth of modern science is usually associated with the Renaissance, the historical period from the late 14th to the mid-17th century that saw a rebirth (*renaissance* in French) of artistic, literary, and scientific inquiry in European culture following the chaos of the Dark Ages. However, one of the first documented uses of the scientific method in an astronomical context was made by Aristotle (384–322 B.C.) some 17 centuries earlier. Aristotle is not normally remembered as a strong proponent of this approach—many of his best known ideas were based on pure thought, with no attempt at experimental test or verification. Nevertheless, his brilliance extended into many areas now thought of as modern science. He noted that, during a lunar eclipse (Section 1.6), Earth casts a curved shadow onto the surface of the Moon. Figure 1.7 shows a series of photographs taken during a recent lunar eclipse. Earth's shadow, projected onto the Moon's surface, is indeed slightly curved. This is what Aristotle must have seen and recorded so long ago.

Because the observed shadow seemed always to be an arc of the same circle, Aristotle theorized that Earth, the cause of the shadow, must be round. Don't underestimate the scope of this apparently simple statement. Aristotle also had to reason that the dark region was indeed a shadow and

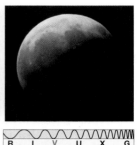

▲ FIGURE 1.7 **A Lunar Eclipse** This series of photographs show Earth's shadow sweeping across the Moon during an eclipse. By observing this behavior, Aristotle reasoned that Earth was the cause of the shadow and concluded that Earth must be round. His theory has yet to be disproved. *(G. Schneider)*

that Earth was its cause—facts we regard as obvious today, but far from clear 25 centuries ago. On the basis of this *hypothesis*—one possible explanation of the observed facts— he then predicted that any and all future lunar eclipses would show Earth's shadow to be curved, regardless of our planet's orientation. That prediction has been tested every time a lunar eclipse has occurred. It has yet to be proved wrong. Aristotle was not the first person to argue that Earth is round, but he was apparently the first to offer observational proof using the lunar eclipse method.

This basic reasoning forms the basis of all modern scientific inquiry. Armed only with naked-eye observations of the sky (the telescope would not be invented for almost another 2000 years), Aristotle first made an observation. Next, he formulated a hypothesis to explain that observation. Then he tested the validity of his hypothesis by making predictions that could be confirmed or refuted by further observations. *Observation, theory, and testing*—these are the cornerstones of the scientific method, a technique whose power will be demonstrated again and again throughout our text.

Today, scientists throughout the world use an approach that relies heavily on testing ideas. They gather data, form a working hypothesis that explains the data, and then proceed to test the implications of the hypothesis using experiment and observation. Eventually, one or more "well-tested" hypotheses may be elevated to the stature of a physical law and come to form the basis of a theory of even broader applicability. The new predictions of the theory will in turn be tested, as scientific knowledge continues to grow. Experiment and observation are integral parts of the process of scientific inquiry. Untestable theories, or theories unsupported by experimental evidence, rarely gain any measure of acceptance in scientific circles. Used properly over a period of time, this rational, methodical approach enables us to arrive at conclusions that are mostly free of the personal bias and human values of any one scientist. The scientific method is designed to yield an objective view of the universe we inhabit.

CONCEPT CHECK

✔ Can a theory ever become a "fact," scientifically speaking?

## 1.3 The "Obvious" View

To see how astronomers have applied the scientific method to understand the universe around us, let's start with some very basic observations. Our study of the cosmos, the modern science of astronomy, begins simply, with looking at the night sky. The overall appearance of the night sky is not so different now from what our ancestors would have seen hundreds or even thousands of years ago, but our *interpretation* of what we see has changed immeasurably as the science of astronomy has evolved and grown.

### Constellations in the Sky

Between sunset and sunrise on a clear night, we can see about 3000 points of light. If we include the view from the opposite side of Earth, nearly 6000 stars are visible to the unaided eye. A natural human tendency is to see patterns and relationships among objects even when no true connection exists, and people long ago connected the brightest stars into configurations called **constellations,** which ancient astronomers named after mythological beings, heroes, and animals—whatever was important to them. Figure 1.8 shows a constellation especially prominent in the nighttime sky from October through March: the hunter named Orion. Orion was a mythical Greek hero famed, among other things, for his amorous pursuit of the Pleiades, the seven daughters of the giant Atlas. According to Greek mythology, to protect the Pleiades from Orion, the gods placed them among the stars, where Orion nightly stalks them across the sky. Many constellations have similarly fabulous connections with ancient lore.

Perhaps not surprisingly, the patterns have a strong cultural bias—the astronomers of ancient China saw mythical figures different from those seen by the ancient Greeks, the Babylonians, and the people of other cultures, even though they were all looking at the same stars in the night sky. Interestingly, different cultures often made the same basic *groupings* of stars, despite widely varying interpretations of what they saw. For example, the group of seven stars usually known in North America as "the Dipper" is known as "the Wagon" or "the Plough" in western Europe. The ancient Greeks regarded these same stars as the tail of "the Great Bear," the Egyptians saw them as the

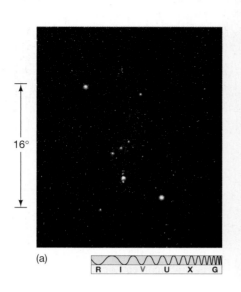

(a)

(b)

R I V U X G

leg of an ox, the Siberians as a stag, and some Native Americans as a funeral procession.

Early astronomers had very practical reasons for studying the sky. Some constellations served as navigational guides. The star Polaris (part of the Little Dipper) indicates north, and the near constancy of its location in the sky, from hour to hour and night to night, has aided travelers for centuries. Other constellations served as primitive calendars to predict planting and harvesting seasons. For example, many cultures knew that the appearance of certain stars on the horizon just before daybreak signaled the beginning of spring and the end of winter.

In many societies, people came to believe that there were other benefits in being able to trace the regularly changing positions of heavenly bodies. The relative positions of stars and planets at a person's birth were carefully studied by *astrologers*, who used the data to make predictions about that person's destiny. Thus, in a sense, astronomy and astrology arose from the same basic impulse—the desire to "see" into the future—and, indeed, for a long time they were indistinguishable from one another. Today, most people recognize that astrology is nothing more than an amusing diversion (although millions still study their horoscope in the newspaper every morning!). Nevertheless, the ancient astrological terminology—the names of the constellations and many terms used to describe the locations and motions of the planets—is still used throughout the astronomical world.

Generally speaking, as illustrated in Figure 1.9 for the case of Orion, the stars that make up any particular constellation are not actually close to one another in space, even by astronomical standards. They merely are bright enough to observe with the naked eye and happen to lie in roughly the same direction in the sky as seen from Earth. Still, the constellations provide a convenient means for astronomers to specify large areas of the sky, much as geologists use continents or politicians use voting precincts to identify certain localities on planet Earth. In all, there are

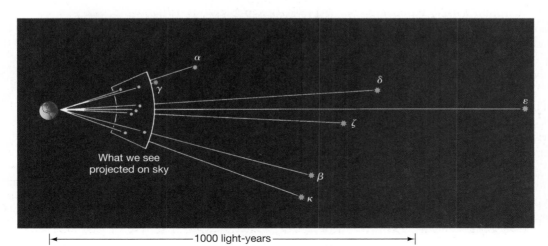

What we see projected on sky

|◀—————————— 1000 light-years ——————————|

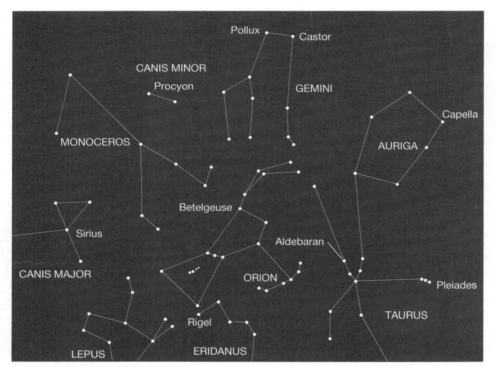

◀ FIGURE 1.10 **Constellations Near Orion** The region of the sky conventionally associated with the constellation Orion, together with some neighboring constellations (labeled in all capital letters). Some prominent stars are also labeled in lowercase letters. The 88 constellations span the entire sky, so that every astronomical object lies in precisely one of them.

88 constellations, most of them visible from North America at some time during the year. Figure 1.10 shows how the conventionally defined constellations cover a portion of the sky in the vicinity of Orion.

## The Celestial Sphere

Over the course of a night, the constellations seem to move smoothly across the sky from east to west, but ancient skywatchers were well aware that the *relative* locations of stars remained unchanged as this nightly march took place.* It was natural for those observers to conclude that the stars must be firmly attached to a **celestial sphere** surrounding Earth—a canopy of stars resembling an astronomical painting on a heavenly ceiling. Figure 1.11 shows how early astronomers pictured the stars as moving with this celestial sphere as it turned around a fixed, unmoving Earth. Figure 1.12 shows how all stars appear to move in circles around a point very close to the star Polaris (better known as the Pole Star or North Star). To the ancients, this point represented the axis around which the entire celestial sphere turned.

Today we recognize that the apparent motion of the stars is the result of the spin, or **rotation,** not of the celestial sphere, but of Earth. Polaris indicates the direction—due north—in which Earth's rotation axis points. Even though we

now know that the celestial sphere is an incorrect description of the heavens, we still use the idea as a convenient fiction that helps us visualize the positions of stars in the sky. The points where Earth's axis intersects the celestial sphere are called the **celestial poles.** In the Northern Hemisphere, the north celestial pole lies directly above Earth's North Pole. The extension of Earth's axis in the opposite direction defines the south celestial pole, directly above Earth's South Pole. Midway between the

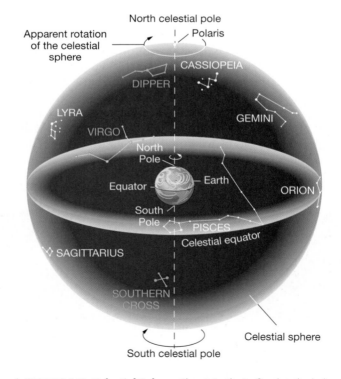

▲ FIGURE 1.11 **Celestial Sphere** Planet Earth sits fixed at the hub of the celestial sphere, which contains all the stars. This is one of the simplest possible models of the universe, but it doesn't agree with all the facts that astronomers now know about the universe.

*We now know that stars do in fact move relative to one another, but this* proper motion *across the sky is too slow to be discerned with the naked eye (see Section 17.1).*

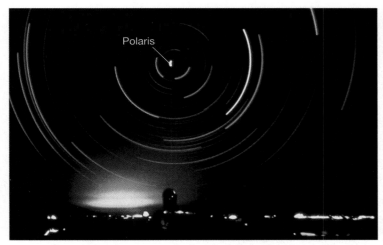

◀ FIGURE 1.12 **Northern Sky** A time-lapse photograph of the northern sky. Each trail is the path of a single star across the night sky. The duration of the exposure is about 5 hours, since each star traces out approximately 20 percent of a circle. The concentric circles are centered near the North Star, Polaris, whose short, bright arc is labeled. *(AURA)*

R I U V X G

north and south celestial poles lies the **celestial equator,** representing the intersection of Earth's equatorial plane with the celestial sphere. These parts of the celestial sphere are marked on Figure 1.11.

When discussing the locations of stars "on the sky," astronomers naturally talk in terms of *angular* positions and separations. *More Precisely 1-1* presents some basic information on angular measure. *More Precisely 1-2* discusses in more detail some systems of coordinates used to specify stellar positions.

## MORE PRECISELY 1-1

### Angular Measure

Size and scale are often specified by measuring lengths and angles. The concept of length measurement is fairly intuitive to most of us. The concept of *angular measurement* may be less familiar, but it, too, can become second nature if you remember a few simple facts:

- A full circle contains 360 *degrees* (360°). Thus, the half-circle that stretches from horizon to horizon, passing directly overhead and spanning the portion of the sky visible to one person at any one time, contains 180°.

- Each 1° increment can be further subdivided into fractions of a degree, called *arc minutes*. There are 60 arc minutes (written 60′) in one degree. (The term "arc" is used to distinguish this angular unit from the unit of time.) Both the Sun and the Moon project an angular size of 30 arc minutes (half a degree) on the sky. Your little finger, held at arm's length, has a similar angular size, covering about a 40′ slice of the 180° horizon-to-horizon arc.

- An arc minute can be divided into 60 *arc seconds* (60″). Put another way, an arc minute is $\frac{1}{60}$ of a degree, and an arc second is $\frac{1}{60} \times \frac{1}{60} = \frac{1}{3600}$ of a degree. An arc second is an extremely small unit of angular measure—the angular size of a centimeter-sized object (a dime, say) at a distance of about 2 kilometers (a little over a mile).

The accompanying figure illustrates this subdivision of the circle into progressively smaller units.

Don't be confused by the units used to measure angles. Arc minutes and arc seconds have nothing to do with the measurement of time, and degrees have nothing to do with temperature. Degrees, arc minutes, and arc seconds are simply ways to measure the size and position of objects in the universe.

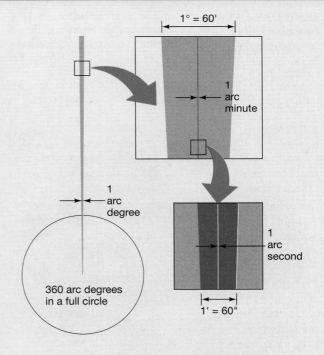

The angular size of an object depends both on its actual size and on its distance from us. For example, the Moon at its present distance from Earth has an angular diameter of 0.5°, or 30′. If the Moon were twice as far away, it would appear half as big—15′ across—even though its actual size would be the same. Thus, *angular size by itself is not enough to determine the actual diameter of an object—the distance to the object must also be known.* We return to this topic in more detail in *More Precisely 1-3*.

# MORE PRECISELY 1-2

## Celestial Coordinates

The simplest method of locating stars in the sky is to specify their constellation and then rank the stars in it in order of brightness (see *Discovery 17.1* for more on how stars are identified and named). The brightest star is denoted by the Greek letter α (alpha), the second brightest by β (beta), and so on (see Figure 1.8). Thus, the two brightest stars in the constellation Orion—Betelgeuse and Rigel—are also known as α Orionis and β Orionis, respectively. (More recently, precise observations show that Rigel is actually brighter than Betelgeuse, but the names are now permanent.) Similarly, Sirius, the brightest star in the sky (see Appendix 3), which lies in the constellation Canis Major (the Great Dog), is denoted α Canis Majoris (or α CMa for short); the (present) Pole Star, in Ursa Minor (the Little Bear), is also known as α Ursae Minoris (or α UMi), and so on. Because there are many more stars in any given constellation than there are letters in the Greek alphabet, this method is of limited utility. However, for naked-eye astronomy, where only bright stars are involved, it is quite satisfactory.

For more precise measurements, astronomers find it helpful to lay down a system of **celestial coordinates** on the sky. If we think of the stars as being attached to the celestial sphere centered on Earth, then the familiar system of latitude and longitude on Earth's surface extends naturally to cover the sky. The celestial analogs of latitude and longitude on Earth's surface are called *declination* and *right ascension*, respectively. The accompanying figure illustrates the meaning of right ascension and declination on the celestial sphere and compares them with longitude and latitude on Earth. Note the following points:

- Declination (dec) is measured in degrees (°) north or south of the celestial equator, just as latitude is measured in degrees north or south of Earth's equator. (See *More Precisely 1-1* for a discussion of angular measure.) Thus, the celestial equator is at a declination of 0°, the north celestial pole is at +90°, and the south celestial pole is at −90° (the minus sign here just means "south of the celestial equator").

- Right ascension (RA) is measured in units called *hours*, *minutes*, and *seconds*, and it increases in the eastward direction. The angular units used to measure right ascension are constructed to parallel the units of time in order to assist astronomical observation. The two sets of units are connected by the rotation of Earth (or of the celestial sphere). In 24 hours, Earth rotates once on its axis, or through 360°. Thus, in a period of 1 hour, Earth rotates through $360°/24 = 15°$, or $1^h$. In 1 minute of time, Earth rotates through an angle of $1^m = 15°/60 = 0.25°$, or 15 arc minutes (15′). In 1 second of time, Earth rotates through an angle of $1^s = 15′/60 = 15$ arc seconds (15″). As with longitude, the choice of zero right ascension (the celestial equivalent of the Greenwich Meridian) is quite arbitrary—it is conventionally taken to be the position of the Sun in the sky *at the instant of the vernal equinox.*

Right ascension and declination specify locations on the sky in much the same way as longitude and latitude allow us to locate a point on Earth's surface. For example, to find Washington on Earth, look 77° west of the Greenwich Meridian (the line on Earth's surface with a longitude of zero) and 39° north of the equator. Similarly, to locate the star Betelgeuse on the celestial sphere, look $5^h52^m0^s$ east of the vernal equinox (the line on the sky with a right ascension of zero) and 7°24′ north of the celestial equator. The star Rigel, also mentioned earlier, lies at $5^h13^m36^s$ (RA), −8°13′ (dec). Thus, we have a quantitative alternative to the use of constellations in specifying the positions of stars in the sky. Just as latitude and longitude are tied to Earth, right ascension and declination are fixed on the celestial sphere. Although the stars appear to move across the sky because of Earth's rotation, their celestial coordinates remain *constant* over the course of a night.

Actually, because right ascension is tied to the position of the vernal equinox, the celestial coordinates of any given star slowly drift due to Earth's *precession* (see Section 1.4). Since one cycle of precession takes 26,000 years to complete, the shift on a night-to-night basis is a little over 0.1″—a small angle, but one that must be taken into account in high-precision astronomical measurements. Rather than deal with slowly changing coordinates for every object in the sky, astronomers conventionally correct their observations to the location of the vernal equinox at some standard epoch (such as January 1, 1950, or January 1, 2000).

## 1.4 Earth's Orbital Motion

### Day-to-Day Changes

We measure time by the Sun. Because the rhythm of day and night is central to our lives, it is not surprising that the period from one noon to the next, the 24-hour **solar day,** is our basic social time unit. The daily progress of the Sun and the other stars across the sky is known as *diurnal motion.* As we have just seen, it is a consequence of Earth's rotation. But the stars' positions in the sky do *not* repeat themselves exactly from one night to the next. Each night, the whole celestial sphere appears to be shifted a little relative to the horizon compared with the night before. The easiest way to confirm this difference is by noticing the stars that are visible just after sunset or just before dawn. You will find that they are in slightly different locations from those of the previous night. Because of this shift, a day measured by the stars—called a **sidereal day** after the Latin word *sidus,* meaning "star"—differs in length from a solar day. Evidently, there is more to the apparent motion of the heavens than simple rotation.

The reason for the difference between a solar day and a sidereal day is sketched in Figure 1.13. It is a result of the fact that Earth moves in two ways simultaneously: It rotates on its central axis while at the same time **revolving** around the Sun. Each time Earth rotates once on its axis, it also moves a small distance along its orbit about the Sun. Earth therefore has to rotate through slightly more than 360° (360 degrees—see *More Precisely 1-1*) for the Sun to return to the same apparent location in the sky. Thus, the interval of time between noon one day and noon the next (a solar day) is slightly greater than one true rotation period (one sidereal day). Our planet takes 365 days to orbit the Sun, so the additional angle is 360°/365 = 0.986°. Because Earth, rotating at a rate of 15° per hour, takes about 3.9 minutes to rotate through this angle, the solar day is 3.9 minutes longer than the sidereal day (i.e., 1 sidereal day is roughly 23$^h$56$^m$ long).

### Seasonal Changes

Figure 1.14(a) illustrates the major stars visible from most locations in the United States on clear summer evenings. The brightest stars—Vega, Deneb, and Altair—form a conspicuous triangle high above the constellations Sagittarius and Capricornus, which are low on the southern

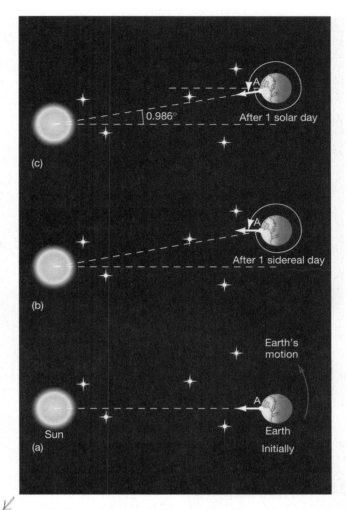

▲ **FIGURE 1.13 Solar and Sidereal Days** A sidereal day is Earth's true rotation period—the time taken for our planet to return to the same orientation in space relative to the distant stars. A solar day is the time from one noon to the next. The difference in length between the two is easily explained once we understand that Earth revolves around the Sun at the same time as it rotates on its axis. Frames (a) and (b) are one sidereal day apart. During that time, Earth rotates exactly once on its axis and also moves a little in its solar orbit— approximately 1°. Consequently, between noon at point A on one day and noon at the same point the next day, Earth actually rotates through about 361° (frame c), and the solar day exceeds the sidereal day by about four minutes. Note that the diagrams are not drawn to scale; the true 1° angle is in reality much smaller than shown here.

horizon. In the winter sky, however, these stars are replaced as shown in Figure 1.14(b) by several other, well-known constellations, including Orion, Leo, and Gemini. In the constellation Canis Major lies Sirius (the Dog Star), the brightest star in the sky. Year after year, the same stars and constellations return, each in its proper season. Every winter evening, Orion is high overhead; every summer, it is gone. (For more detailed maps of the

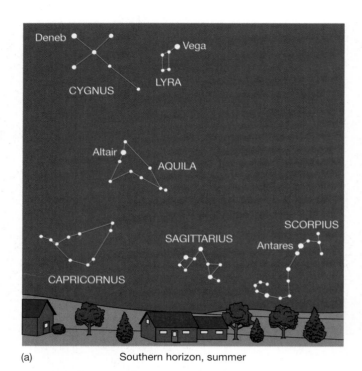

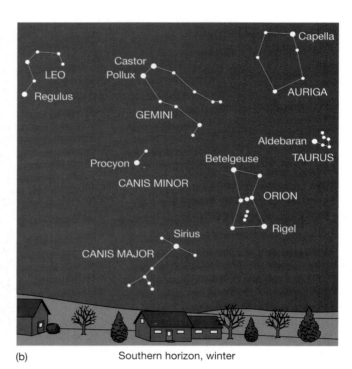

(a) Southern horizon, summer

(b) Southern horizon, winter

▲ FIGURE 1.14 **Typical Night Sky** (a) A typical summer sky above the United States. Some prominent stars (labeled in lowercase letters) and constellations (labeled in all capital letters) are shown. (b) A typical winter sky above the United States.

sky at different seasons, consult the star charts at the end of the book.)

These regular seasonal changes occur because of Earth's revolution around the Sun: Earth's darkened hemisphere faces in a slightly different direction in space each evening. The change in direction is only about 1° per night (Figure 1.13)—too small to be easily noticed with the naked eye from one evening to the next, but clearly noticeable over the course of weeks and months, as illustrated in Figure 1.15. After 6 months, Earth has reached

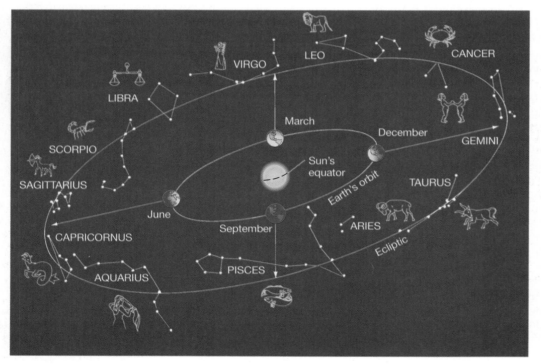

◄ FIGURE 1.15 **The Zodiac** The view of the night sky changes as Earth moves in its orbit about the Sun. As drawn here, the night side of Earth faces a different set of constellations at different times of the year. The 12 constellations named here make up the astrological zodiac. The arrows indicate the most prominent zodiacal constellations in the night sky at various times of year. For example, in June, when the Sun is "in" Gemini, Sagittarius and Capricornus are visible at night.

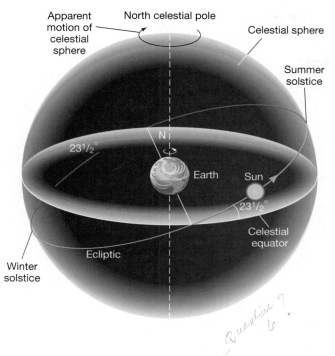

Apparent motion of celestial sphere

North celestial pole

Celestial sphere

Summer solstice

23½°

N

Earth

Sun

23½°

Celestial equator

Ecliptic

Winter solstice

◀ FIGURE 1.16 **Ecliptic** The apparent path of the Sun on the celestial sphere over the course of a year is called the ecliptic. As indicated on the diagram, the ecliptic is inclined to the celestial equator at an angle of 23.5°. In this picture of the heavens, the seasons result from the changing height of the Sun above the celestial equator. At the summer solstice, the Sun is at its northernmost point on its path around the ecliptic; it is therefore highest in the sky, as seen from Earth's Northern Hemisphere, and the days are longest. The reverse is true at the winter solstice. At the vernal and autumnal equinoxes, when the Sun crosses the celestial equator, day and night are of equal length.

the opposite side of its orbit, and we face an entirely different group of stars and constellations at night. Because of this motion, the Sun appears (to an observer on Earth) to move relative to the background stars over the course of a year. This apparent motion of the Sun on the sky traces out a path on the celestial sphere known as the **ecliptic.**

The 12 constellations through which the Sun passes as it moves along the ecliptic—that is, the constellations we would see looking in the direction of the Sun if they weren't overwhelmed by the Sun's light—had special significance for astrologers of old. These constellations are collectively known as the **zodiac.**

As illustrated in Figure 1.16, the ecliptic forms a great circle on the celestial sphere, inclined at an angle of 23.5° to the celestial equator. In reality, as illustrated in Figure 1.17, the plane of the ecliptic is *the plane of Earth's orbit around the Sun.* Its tilt is a consequence of the *inclination* of our planet's rotation axis to the plane of its orbit.

The point on the ecliptic where the Sun is at its northernmost point above the celestial equator is known as the **summer solstice** (from the Latin words *sol,* meaning "sun," and *stare,* "to stand"). As indicated in Figure 1.17, it represents the location in Earth's orbit where our planet's North Pole comes closest to pointing in the direction of the Sun. This occurs on or near June 21—the exact date varies slightly from year to year because the actual length of a year is not a whole number of days. As Earth rotates, points north of the equator spend the greatest

▶ FIGURE 1.17 **Seasons** In reality, the Sun's apparent motion along the ecliptic is a consequence of Earth's orbital motion around the Sun. The seasons result from the inclination of our planet's rotation axis with respect to its orbit plane. The summer solstice corresponds to the point on Earth's orbit where our planet's North Pole points most nearly toward the Sun. The opposite is true of the winter solstice. The vernal and autumnal equinoxes correspond to the points in Earth's orbit where our planet's axis is perpendicular to the line joining Earth and the Sun. The insets show how rays of sunlight striking the ground at an angle (e.g., during northern winter) are spread over a larger area than rays coming nearly straight down (e.g., during northern summer). As a result, the amount of solar heat delivered to a given area of Earth's surface is greatest when the Sun is high in the sky.

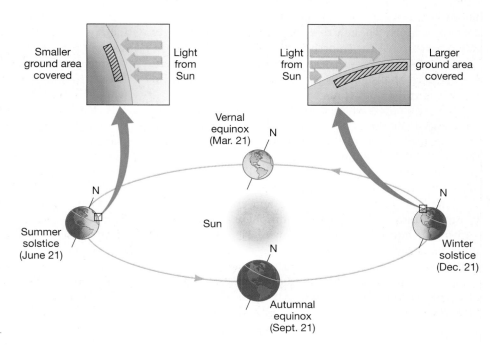

Smaller ground area covered

Light from Sun

Light from Sun

Larger ground area covered

Vernal equinox (Mar. 21)

N

N

Summer solstice (June 21)

Sun

N

Winter solstice (Dec. 21)

N

Autumnal equinox (Sept. 21)

fraction of their time in sunlight on that date, so the summer solstice corresponds to the longest day of the year in the Northern Hemisphere and the shortest day in the Southern Hemisphere.

Six months later, the Sun is at its southernmost point below the celestial equator (Figure 1.16)—or, equivalently, the North Pole points farthest from the Sun (Figure 1.17). We have reached the **winter solstice** (December 21), the shortest day in Earth's Northern Hemisphere and the longest in the Southern Hemisphere.

The tilt of Earth's rotation axis relative to the ecliptic is responsible for the **seasons** we experience—the marked difference in temperature between the hot summer and cold winter months. As illustrated in Figure 1.17, two factors combine to cause this variation. First, there are more hours of daylight during the summer than in winter. To see why this is, look at the yellow lines on the surfaces of the drawings of Earth in the figure. (For definiteness, they correspond to a latitude of 45°—roughly that of the Great Lakes or the south of France.) A much larger fraction of the line is sunlit in the summertime, and more daylight means more solar heating. Second, as illustrated in the insets in Figure 1.17, when the Sun is high in the sky in summer, rays of sunlight striking Earth's surface are more concentrated—spread out over a smaller area—than in winter. As a result, the Sun feels hotter. Therefore summer, when the Sun is highest above the horizon and the days are longest, is generally much warmer than winter, when the Sun is low and the days are short.

A popular misconception is that the seasons have something to do with Earth's distance from the Sun. Figure 1.18 illustrates why this is *not* the case. It shows Earth's orbit "face on," instead of almost edge-on, as in Figure 1.17. Notice that the orbit is almost perfectly circular, so the distance from Earth to the Sun varies very little (in fact, by only about 3 percent) over the course of a year—not nearly enough to explain the seasonal changes in temperature. What's more, Earth is actually *closest* to the Sun in early January, the dead of winter in the Northern Hemisphere, so distance from the Sun cannot be the main factor controlling our climate.

The two points where the ecliptic intersects the celestial equator (Figure 1.16)—that is, where Earth's rotation axis is perpendicular to the line joining Earth to the Sun (Figure 1.17)—are known as **equinoxes.** On those dates, day and night are of equal duration. (The word *equinox* derives from the Latin for "equal night.") In the fall (in the Northern Hemisphere), as the Sun crosses from the Northern into the Southern Hemisphere, we have the **autumnal equinox** (on September 21). The **vernal equinox** occurs in northern spring, on or near March 21, as the Sun crosses the celestial equator moving north. Because of its association with the end of winter and the start of a new growing season, the vernal equinox was particularly important to early astronomers and astrologers. It

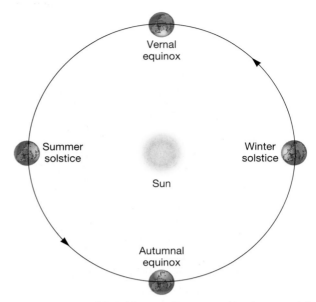

▲ **FIGURE 1.18 Earth's Orbit** Seen face on, Earth's orbit around the Sun is almost indistinguishable from a perfect circle. The distance from Earth to the Sun varies only slightly over the course of a year and is *not* the cause of the seasonal temperature variations we experience.

also plays an important role in human timekeeping: The interval of time from one vernal equinox to the next— 365.2422 mean solar days (see Section 1.5)—is known as one **tropical year.**

## Long-Term Changes

Earth has many motions—it spins on its axis, it travels around the Sun, and it moves with the Sun through our Galaxy. We have just seen how some of these motions can account for the changing nighttime sky and the changing seasons. In fact, the situation is even more complicated. Like a spinning top that rotates rapidly on its own axis while that axis slowly revolves about the vertical, Earth's axis changes its *direction* over the course of time (although the angle between the axis and a line perpendicular to the plane of the ecliptic always remains close to 23.5°). Illustrated in Figure 1.19, this change is called **precession.** It is caused by torques (twisting forces) on Earth due to the gravitational pulls of the Moon and the Sun, which affect our planet in much the same way as the torque due to Earth's own gravity affects a top. During a complete cycle of precession—about 26,000 years—Earth's axis traces out a cone.

The time required for Earth to complete exactly one orbit around the Sun, relative to the stars, is called a **sidereal year.** One sidereal year is 365.256 mean solar days long—about 20 minutes longer than a tropical year.

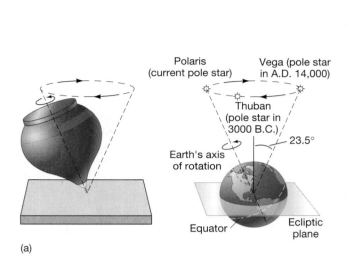

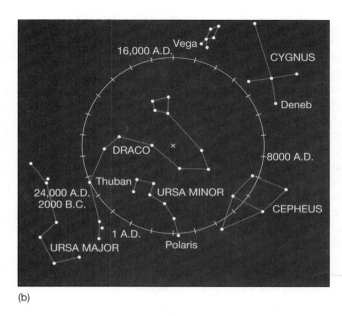

▲ FIGURE 1.19 **Precession** (a) Earth's axis currently points nearly toward the star Polaris. About 12,000 years from now—almost halfway through one cycle of precession—Earth's axis will point toward a star called Vega, which will then be the "North Star." Five thousand years ago, the North Star was a star named Thuban in the constellation Draco. (b) The circle shows the precessional path of the north celestial pole among some prominent northern stars. Tick marks indicate intervals of a thousand years.

The reason for this slight difference is Earth's precession. Recall that the vernal equinox occurs when Earth's rotation axis is perpendicular to the line joining Earth and the Sun, and the Sun is crossing the celestial equator moving from south to north. In the absence of precession, this combination of events would occur exactly once per sidereal orbit, and the tropical and sidereal years would be identical. However, because of the slow precessional shift in the orientation of Earth's rotation axis, the instant when the axis is next perpendicular to the line from Earth to the Sun occurs slightly *sooner* than we would otherwise expect. Consequently, the vernal equinox drifts slowly around the zodiac over the course of the precession cycle.

The tropical year is the year that our calendars measure. If our timekeeping were tied to the sidereal year, the seasons would slowly march around the calendar as Earth precessed—13,000 years from now, summer in the Northern Hemisphere would be at its height in late February! By using the tropical year, we ensure that July and August will always be (northern) summer months. However, in 13,000 years' time, Orion will be a summer constellation.

CONCEPT CHECK

✔ In astronomical terms, what are *summer* and *winter,* and why do we see different constellations during those seasons?

## 1.5 Astronomical Timekeeping

Well-defined standards of measurement are essential to modern science and technology. Earth's rotation and revolution around the Sun define the basic time units—the day and the year—by which we measure our lives. Today the SI unit of time, the second (see Appendix 2), is defined by ultra-high-precision atomic clocks maintained in the National Institute of Standards and Technology, in Gaithersburg, Maryland, and in other sites around the world. However, the time those clocks measure is based squarely on astronomical events, and astronomers (or astrologers) have traditionally borne the responsibility for keeping track of the days and the seasons.

As we have just seen, at any location on Earth, a solar day is defined as the time between one noon and the next. Here, *noon* means the instant when the Sun crosses the **meridian**—an imaginary line on the celestial sphere through the north and south celestial poles, passing directly overhead at the given location. This is the time that a sundial would measure. Unfortunately, this most direct measure of time has two serious drawbacks: The length of the solar day varies throughout the year, and the time at which noon occurs varies from place to place.

Recall that the solar day is the result of a "competition" between Earth's *rotation* and its *revolution* around the Sun. Earth's revolution means that our planet must rotate through a little more than 360° between one noon and the next (see Figure 1.13). However, while Earth's rotation rate is virtually constant, the rate of revolution—or, more specifically, the rate at which the Sun traverses the celestial sphere as it moves along the ecliptic—is not constant, for two reasons, illustrated and exaggerated in Figure 1.20.

First, as we will see in Chapter 2, Earth's orbit is not exactly circular, and our orbital speed is not constant— Earth moves more rapidly than average when closer to the Sun, more slowly when farther away—so the speed at which the Sun appears to move along the ecliptic varies with time (Figure 1.20a). Second, because the ecliptic is inclined to the celestial equator, the *eastward* component of the Sun's motion on the celestial sphere depends on the time of year (and note that this would be the case even if Earth's orbit were circular and the first point did not apply). As illustrated in Figure 1.20b, at the equinoxes the Sun's path is inclined to the equator, and only the Sun's eastward progress across the sky contributes to the motion. At the solstices, however, the motion is entirely eastward.

The combination of these effects means that the solar day varies by roughly half a minute over the course of the year—not a large variation, but unacceptable for astronomical and many other purposes. The solution is to define a **mean solar day,** which, in effect, is just the average (or *mean*) of all the solar days over an entire year. This is the day our clocks (atomic or otherwise) measure, and it is constant by definition. One second is $\frac{1}{24} \times \frac{1}{60} \times \frac{1}{60} = \frac{1}{86,400}$ mean solar days.

Now all of our clocks tick at a constant rate, but we are not quite out of the woods in our search for a standard of time. The preceding definition is still *local*—observers at different longitudes see noon at different times, so even though their clocks keep pace, they all tell different times. In 1883, driven by the need for uniform and consistent times in long-distance travel and communications (the railroads and the telegraph), the continental United States was divided into four standard **time zones.** Within each zone, everyone adopted the mean solar time corresponding to a specific meridian inside the zone. Since 1884, **standard time** has been used around the world. The global standard time zones are shown in Figure 1.21. By convention, the reference meridian is taken to be 0° longitude (the Greenwich meridian). In the United States, Eastern, Central, Mountain, and Pacific time zones keep the mean solar time of longitudes 75° W, 90° W, 105° W, and 120° W, respectively. Hawaii and Alaska keep the time of longitude 150° W. In some circumstances, it is even more convenient to adopt a single time zone: **Universal time** (formerly known as *Greenwich mean time*) is simply the mean solar time at the Greenwich meridian.

Having defined the day, let's now turn our attention to how those days fit into the year—in other words, to

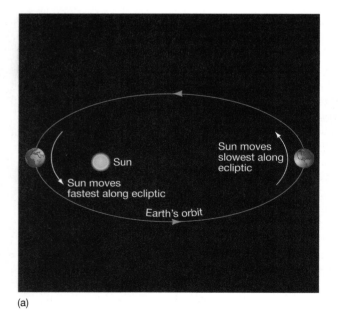

(a)

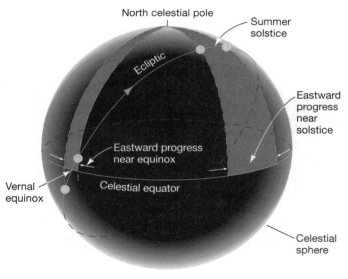

(b)

▲ FIGURE 1.20 **Variations in the Solar Day** (a) Because Earth does not orbit the Sun at a constant speed, the rate at which the Sun appears to move across the sky, and hence the length of the solar day, also varies. (b) In addition, the length of the solar day changes with the season, because the eastward progress of the Sun in, say, 1 hour (indicated here), is greater at the solstices than at the equinoxes. The eccentricity of Earth's orbit in part (a) and the angle between the ecliptic and the celestial equator in part (b) are greatly exaggerated here to illustrate the effect more clearly.

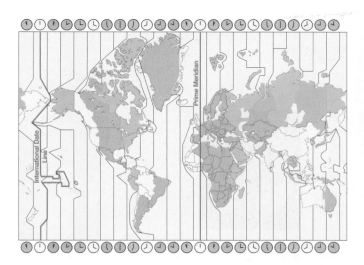

▲ FIGURE 1.21  **Time Zones** The standard time zones were adopted to provide consistent times around the world. Within each time zone, all clocks show the same time, the local solar time of one specific longitude inside the zone.

the construction of a calendar. The year in question is the *tropical year*, tied to the changing seasons. It is conventionally divided into *months*, which were originally defined by the phases of the Moon. (See Section 1.6; the word month derives from *Moon*.) The basic problem with all this is that a lunar month does not contain a whole number of mean solar days and the tropical year does not contain a whole number of either months or days. Many ancient calendars were lunar, and the variable number of days in modern months may be traced back to attempts to approximate the 29.5-day lunar month by alternating 29- and 30-day periods. The Chinese and Islamic calendars in use today are still based squarely on the lunar cycle, each month beginning with the new Moon. The modern Western calendar retains months as convenient subdivisions, but they have no particular lunar significance.

One tropical year is 365.2422 mean solar days long and hence cannot be represented by a whole number of calendar days. The solution is to vary the number of days in the year to ensure the correct *average* length of the year. In 46 B.C., the Roman emperor Julius Caesar decreed that the calendar would include an extra day every fourth year—a **leap year**—ensuring that the average year would be $(3 \times 365 + 366)/4 = 365.25$ days long. (Nowadays, the extra day is inserted at the end of February.) This *Julian calendar* was a great improvement over earlier lunar calendars, which had $354 (= 12 \times 29.5)$ days in the year, necessitating the inclusion of an extra month every 3 years!

Still, the Julian year was not exactly equal to the tropical year, and over time the calendar drifted relative to the

seasons. By A.D. 1582, the difference amounted to 10 days, and Pope Gregory XIII instituted another reform, first skipping the extra 10 days (resetting the vernal equinox back to March 21) and then changing the rule for leap years to omit the extra day from years that were multiples of 100, except for those that are multiples of 400, which remain leap years. The effect of Pope Gregory's reform was that the average year became 365.2425 mean solar days long—good to 1 day in 3300 years. The idea of "losing" 10 days at the behest of the Pope was unacceptable to many countries, with the result that the *Gregorian calendar* was not fully accepted for several centuries. Britain and the American colonies finally adopted it in 1752. Russia abandoned the Julian calendar only in 1917, after the Bolshevik Revolution, at which time the country had to skip 13 days to come into agreement with the rest of the world.

The most recent modern correction to the Gregorian calendar has been to declare that the years 4000, 8000, etc., will *not* be leap years, improving the accuracy of the calendar to 1 day in 20,000 years—good enough for most of us to make it to work on time!

## CONCEPT CHECK

✔ Is the Sun always directly overhead at noon, according to your watch (assuming it tells standard time at your location)?

# 1.6  The Motion of the Moon

The Moon is our nearest neighbor in space. Apart from the Sun, it is the brightest object in the sky. Like the Sun, the Moon appears to move relative to the background stars. Unlike the Sun, however, the Moon really does revolve around Earth. It crosses the sky at a rate of about 12° per day, which means that it moves an angular distance equal to its own diameter—30 arc minutes—in about an hour.

## Lunar Phases

The Moon's appearance undergoes a regular cycle of changes, or **phases,** taking roughly 29.5 days to complete. Figure 1.22 illustrates the appearance of the Moon at different times in this monthly cycle. Starting from the *new Moon*, which is all but invisible in the sky, the Moon appears to *wax* (or grow) a little each night and is visible as a growing *crescent* (photo 1 of Figure 1.22). One week after new Moon, half of the lunar disk can be seen (photo 2). This phase is known as a *quarter Moon*. During the next week, the Moon continues to wax, passing through the *gibbous* phase (photo 3) until, 2 weeks after new Moon, the

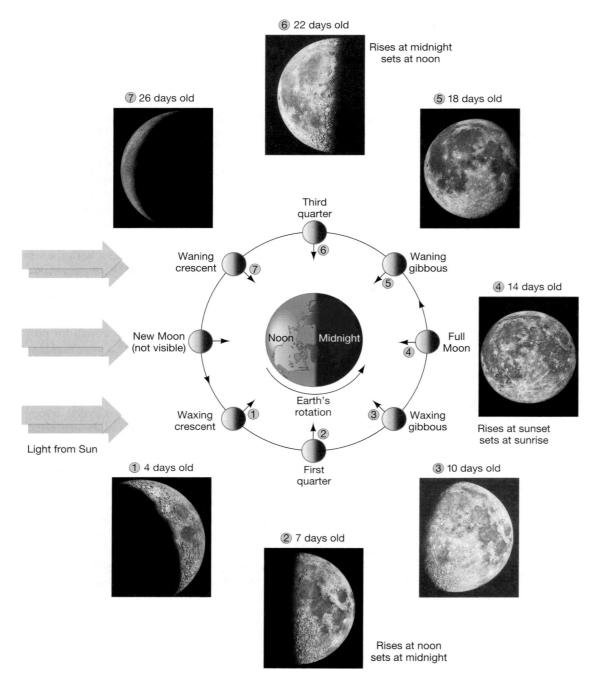

**6** 22 days old

Rises at midnight
sets at noon

**7** 26 days old

**5** 18 days old

Third
quarter

Waning
crescent

Waning
gibbous

**4** 14 days old

New Moon
(not visible)

Noon    Midnight

Full
Moon

Light from Sun

Earth's
rotation

Waxing
crescent

Waxing
gibbous

Rises at sunset
sets at sunrise

**1** 4 days old

First
quarter

**3** 10 days old

**2** 7 days old

Rises at noon
sets at midnight

▲ **FIGURE 1.22  Lunar Phases**  Because the Moon orbits Earth, the visible fraction of the lunar sunlit face varies from night to night, although the Moon always keeps the same face toward our planet. (Note the location of the small, straight arrows, which mark the same point on the lunar surface at each phase shown.) The complete cycle of lunar phases, shown here starting at the waxing crescent phase and following the Moon's orbit counterclockwise, takes 29.5 days to complete. Rising and setting times for some phases are also indicated. *(UC/Lick Observatory)*

Phases of the Moon

TUTORIAL

*full Moon* (photo 4) is visible. During the next 2 weeks, the Moon *wanes* (or shrinks), passing in turn through the gibbous, quarter, crescent phases (photos 5–7) and eventually becoming new again.

The position of the Moon in the sky relative to the Sun, as seen from Earth, varies with lunar phase. For example, the full Moon rises in the east as the Sun sets in the west, while the first quarter Moon actually rises at noon, but sometimes becomes visible only late in the day as the Sun's light fades. By this time the Moon is already high in the sky. Some connections between the lunar phase and the rising and setting times of the Moon are indicated in Figure 1.22.

The Moon doesn't actually change its size and shape from night to night, of course. Its full circular disk is present at all times. Why, then, don't we always see a full Moon? The answer to this question lies in the fact that, unlike the Sun and the other stars, the Moon emits no light of its own. Instead, it shines by reflected sunlight. As illustrated in Figure 1.22, half of the Moon's surface is illuminated by the Sun at any instant. However, not all of the Moon's sunlit face can be seen because of the Moon's position with respect to Earth and the Sun. When the Moon is full, we see the entire "daylit" face because the Sun and the Moon are in opposite directions from Earth in the sky. In the case of a new Moon, the Moon and the Sun are in almost the same part of the sky, and the sunlit side of the Moon is oriented away from us. At new Moon, the Sun must be almost behind the Moon, from our perspective.

As the Moon revolves around Earth, our satellite's position in the sky changes with respect to the stars. In 1 **sidereal month** (27.3 days), the Moon completes one revolution and returns to its starting point on the celestial sphere, having traced out a great circle in the sky. The time required for the Moon to complete a full cycle of phases, one **synodic month,** is a little longer—about 29.5 days. The synodic month is a little longer than the sidereal month for the same reason that a solar day is slightly longer than a sidereal day: Because of Earth's motion around the Sun, the Moon must complete slightly more than one full revolution to return to the same phase in its orbit (Figure 1.23).

## Eclipses

From time to time—but only at new or full Moon—the Sun and the Moon line up precisely as seen from Earth, and we observe the spectacular phenomenon known as an **eclipse.** When the Sun and the Moon are in exactly *opposite* directions, as seen from Earth, Earth's shadow sweeps across the Moon, temporarily blocking the Sun's light and darkening the Moon in a **lunar eclipse,** as illustrated in Figure 1.24. From Earth, we see the curved edge

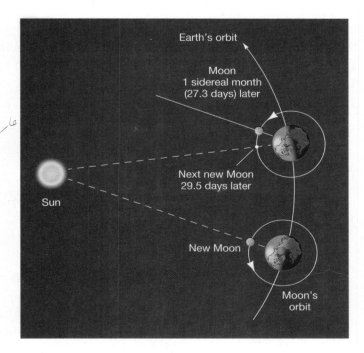

▲ FIGURE 1.23 **Sidereal Month** The difference between a *synodic* and a *sidereal* month stems from the motion of Earth relative to the Sun. Because Earth orbits the Sun in 365 days, in the 29.5 days from one new Moon to the next (1 synodic month), Earth moves through an angle of approximately 29°. Thus, the Moon must revolve more than 360° between new Moons. The sidereal month, which is the time taken for the Moon to revolve through exactly 360°, relative to the stars, is about 2 days shorter.

of Earth's shadow begin to cut across the face of the full Moon and slowly eat its way into the lunar disk. Usually, the alignment of the Sun, Earth, and Moon is imperfect, so the shadow never completely covers the Moon. Such an occurrence is known as a **partial lunar eclipse.** Occasionally, however, the entire lunar surface is obscured in a **total lunar eclipse,** such as that shown in the inset of Figure 1.24. Total lunar eclipses last only as long as is needed for the Moon to pass through Earth's shadow—no more than about 100 minutes. During that time, the Moon often acquires an eerie, deep red coloration—the result of a small amount of sunlight reddened by Earth's atmosphere (for the same reason that sunsets appear red—see *More Precisely 7-1*) and refracted (bent) onto the lunar surface, preventing the shadow from being completely black.

When the Moon and the Sun are in exactly the *same* direction, as seen from Earth, an even more awe-inspiring event occurs. The Moon passes directly in front of the Sun, briefly turning day into night in a **solar eclipse.** In a *total solar eclipse,* when the alignment is perfect, planets and some stars become visible in the daytime as the Sun's light is reduced to nearly nothing. We

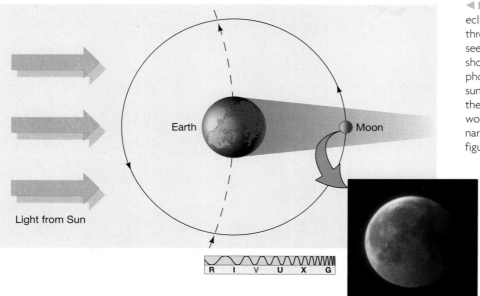

ANIMATION/VIDEO    Solar Eclipse in Indiana

◀ FIGURE 1.24 **Lunar Eclipse** A lunar eclipse occurs when the Moon passes through Earth's shadow. At these times, we see a darkened, copper-colored Moon, as shown by the partial eclipse in the inset photograph. The red coloration is caused by sunlight deflected by Earth's atmosphere onto the Moon's surface. An observer on the Moon would see Earth surrounded by a bright, but narrow, ring of orange sunlight. Note that this figure is not drawn to scale, and only Earth's umbra (see text and Figure 1.26) is shown. *(Inset: G. Schneider)*

can also see the Sun's ghostly outer atmosphere, or *corona* (Figure 1.25).* In a *partial solar eclipse*, the Moon's path is slightly "off center," and only a portion of the Sun's face is covered. In either case, the sight of the Sun apparently being swallowed up by the black disk of the Moon is disconcerting even today. It must surely have inspired fear in early observers. Small wonder that the ability to predict such events was a highly prized skill.

Unlike a lunar eclipse, which is simultaneously visible from all locations on Earth's night side, a total solar eclipse can be seen from only a small portion of Earth's daytime side. The Moon's shadow on Earth's surface is about 7000 kilometers wide—roughly twice the diameter of the Moon. Outside of that shadow, no eclipse is seen. However, within the central region of the shadow, called the **umbra,** the eclipse is total. Within the shadow, but outside the umbra, in the **penumbra,** the eclipse is partial, with less and less of the Sun obscured the farther one travels from the shadow's center.

The connections among the umbra, the penumbra, and the relative locations of Earth, Sun, and Moon are illustrated in Figure 1.26. The umbra is always very small. Even under the most favorable circumstances, its diameter never exceeds 270 kilometers. Because the shadow sweeps across Earth's surface at over 1700 kilometers per hour, the duration of a total eclipse at any given point on our planet can never exceed 7.5 minutes.

The Moon's orbit around Earth is not exactly circular. Thus, the Moon may be far enough from Earth at the moment of an eclipse that its disk fails to cover the disk of the Sun completely, even though their centers coincide. In that case, there is no region of totality—the umbra never

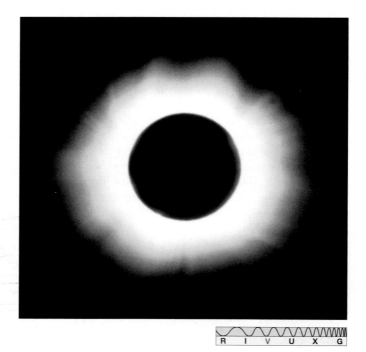

▲ FIGURE 1.25 **Total Solar Eclipse** During a total solar eclipse, the Sun's corona becomes visible as an irregularly shaped halo surrounding the blotted-out disk of the Sun. This was the August 1999 eclipse, as seen from the banks of the Danube River near Sofia, Bulgaria. *(M. Tsavkova/B. Angelov)*

*Actually, although a total solar eclipse is undeniably a spectacular occurrence, the visibility of the corona is probably the most important astronomical aspect of such an event today. It enables us to study this otherwise hard-to-see part of our Sun (see Chapter 16).*

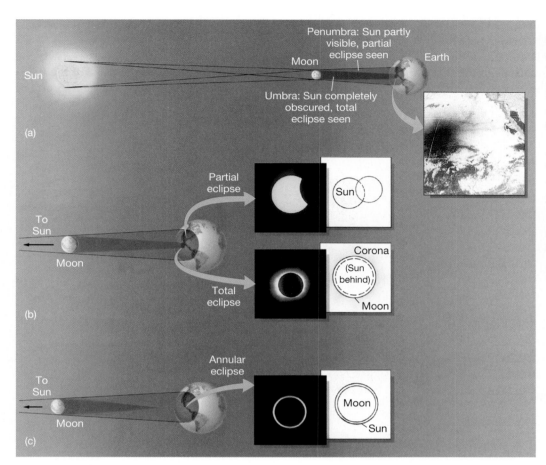

(a)

(b)

(c)

Penumbra: Sun partly visible, partial eclipse seen

Sun

Moon

Earth

Umbra: Sun completely obscured, total eclipse seen

To Sun

Moon

Partial eclipse

Sun

Total eclipse

Corona (Sun behind)

Moon

To Sun

Moon

Annular eclipse

Moon

Sun

◄ FIGURE 1.26 **Types of Solar Eclipse** (a) The Moon's shadow consists of two parts: the umbra, where no sunlight is seen, and the penumbra, where a portion of the Sun is visible. (b) If we are in the umbra, we see a total eclipse; in the penumbra, we see a partial eclipse. (c) If the Moon is too far from Earth at the moment of the eclipse, the umbra does not reach Earth and there is no region of totality; instead, an annular eclipse is seen. (Note that these figures are not drawn to scale.) *(Insets: NOAA; G. Schneider)*

reaches Earth at all, and a thin ring of sunlight can still be seen surrounding the Moon. Such an occurrence, called an **annular eclipse,** is illustrated in Figure 1.26(c) and shown more clearly in Figure 1.27. Roughly half of all solar eclipses are annular.

## Eclipse Seasons

Why isn't there a solar eclipse at every new Moon and a lunar eclipse at every full Moon? That is, why doesn't the Moon pass directly between Earth and the Sun once per orbit and directly through Earth's shadow two weeks later?

The answer is that the Moon's orbit is slightly inclined to the ecliptic (at an angle of 5.2°), so the chance

R I V U X G

◄ FIGURE 1.27 **Annular Solar Eclipse** During an annular solar eclipse, the Moon fails to completely hide the Sun, so a thin ring of light remains. No corona is seen in this case because even the small amount of the Sun still visible completely overwhelms the corona's faint glow. This was the December 1973 eclipse, as seen from Algiers. (The gray fuzzy areas at the top left and right are clouds in Earth's atmosphere.) *(G. Schneider)*

that a new (or full) Moon will occur just as the Moon happens to cross the plane of the ecliptic (with Earth, Moon, and Sun perfectly aligned) is quite low. Figure 1.28 illustrates some possible configurations of the three bodies. If the Moon happens to lie above or below the plane of the ecliptic when new (or full), a solar (or lunar) eclipse cannot occur. Such a configuration is termed *unfavorable* for producing an eclipse. In a *favorable* configuration, the Moon is new or full just as it crosses the plane of the ecliptic, and eclipses are seen. Unfavorable configurations are

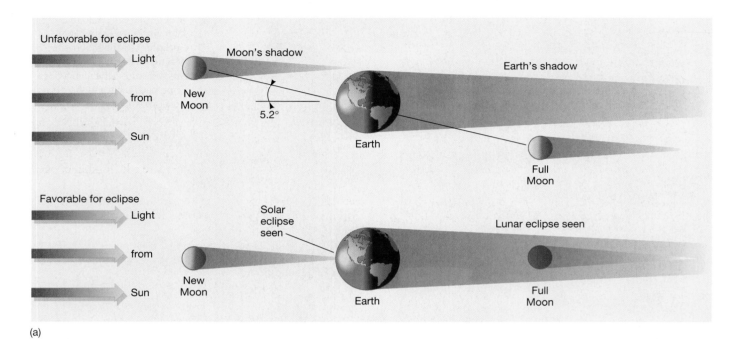

(a)

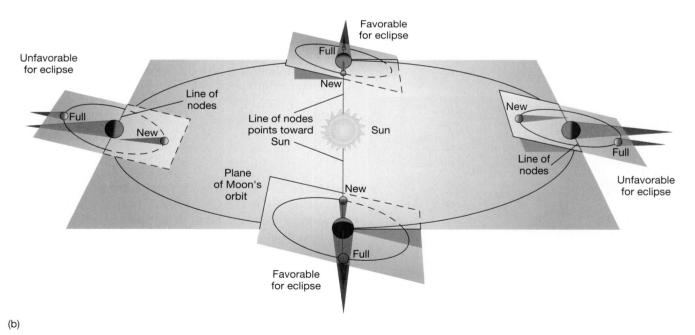

(b)

▲ **FIGURE 1.28 Eclipse Geometry** (a) An eclipse occurs when Earth, Moon, and Sun are precisely aligned. If the Moon's orbital plane lay in exactly the plane of the ecliptic, this alignment would occur once a month. However, the Moon's orbit is inclined at about 5° to the ecliptic, so not all configurations are favorable for producing an eclipse. (b) For an eclipse to occur, the line of intersection of the two planes must lie along the Earth–Sun line. Thus, eclipses can occur just at specific times of the year. Only the umbra of each shadow is shown, for clarity (see Figure 1.26).

much more common than favorable ones, so eclipses are relatively rare events.

As indicated on Figure 1.28(b), the two points on the Moon's orbit where it crosses the plane of the ecliptic are known as the *nodes* of the orbit. The line joining the nodes, which is also the line of intersection of Earth's and the Moon's orbital planes, is known as the *line of nodes.* When the line of nodes is not directed toward the Sun, conditions are unfavorable for eclipses. However, when the line of nodes briefly lies along Earth–Sun line, eclipses are possible. These two periods, known as **eclipse seasons,** are the only times at which an eclipse can occur. Notice that there is no guarantee that an eclipse *will* occur. For a solar eclipse, we must have a new Moon during an eclipse season. Similarly, a lunar eclipse can occur only at full Moon during an eclipse season.

Because we know the orbits of Earth and the Moon to great accuracy, we can predict eclipses far into the future. Figure 1.29 shows the location and duration of all total and annular eclipses of the Sun between 2000 and 2020. It is interesting to note that the eclipse tracks run from west to east—just the opposite of more familiar phenomena such as sunrise and sunset, which are seen earlier by observers located farther east. The reason is that the Moon's shadow sweeps across Earth's surface faster than our planet rotates, so the eclipse actually *overtakes* observers on the ground.

The solar eclipses that we do see highlight a remarkable cosmic coincidence. Although the Sun is many times farther away from Earth than is the Moon, it is also much larger. In fact, the ratio of distances is almost exactly the same as the ratio of sizes, so the Sun and the Moon both have roughly the *same* angular diameter—about half a degree, seen from Earth. Thus, the Moon covers the face of the Sun almost exactly. If the Moon were larger, we would never see annular eclipses, and total eclipses would be much more common. If the Moon were a little smaller, we would see only annular eclipses.

The gravitational tug of the Sun causes the Moon's orbital orientation, and hence the direction of the line of

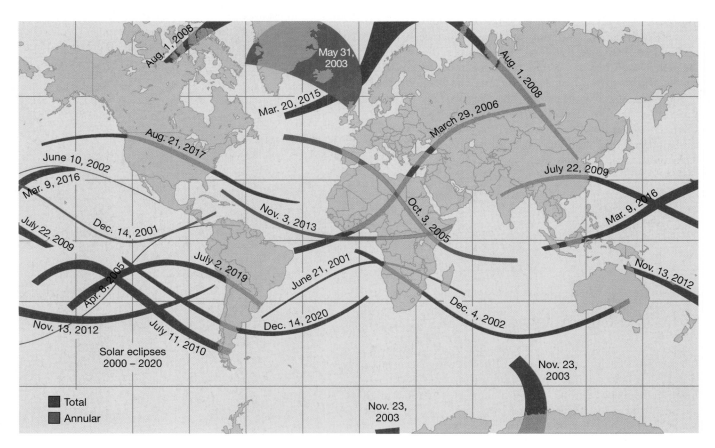

▲ **FIGURE 1.29 Eclipse Tracks** Regions of Earth that saw or will see total (red) or annular (blue) solar eclipses between the years 2000 and 2020. Each track represents the path of the Moon's umbra across Earth's surface during an eclipse. The width of the track depends upon the latitude on Earth and the distance from Earth to the Moon during the eclipse. High-latitude tracks are broader because sunlight strikes Earth's surface at an oblique angle near the poles (and also because of the projection of the map). The closer the Moon is to Earth during a total eclipse, the wider is the umbra (see Figure 1.26).

nodes, to change slowly with time. As a result, the time between one orbital configuration with the line of nodes pointing at the Sun and the next (with the Moon crossing the ecliptic in the same sense in each case) is not exactly 1 year, but instead is 346.6 days—sometimes called one *eclipse year*. Thus, the eclipse seasons gradually progress backward through the calendar, occurring about 19 days earlier each year. For example, in 1999 the eclipse seasons were in February and August, and on August 11 much of Europe and southern Asia was treated to the last total eclipse of the millennium (Figure 1.25). By 2002, those seasons had drifted into December and June, and eclipses actually occurred on June 10 and December 4 of that year. By studying Figure 1.29, you can follow the progression of the eclipse seasons through the calendar. (Note that two partial eclipses in 2004 and two in 2007 are not shown in the figure.)

The combination of the eclipse year and the Moon's synodic period leads to an interesting long-term cycle in solar (and lunar) eclipses. A simple calculation shows that 19 eclipse years is almost exactly 223 lunar months. Thus, every 6585 solar days (actually 18 years, 11.3 days) the "same" eclipse recurs, with Earth, the Moon, and the Sun in the same relative configuration. Several such repetitions are evident in Figure 1.29—see, for example, the similarly shaped December 4, 2002, and December 14, 2020, tracks. (Note that we must take leap years properly into account to get the dates right!) The roughly 120° offset in longitude corresponds to Earth's rotation in 0.3 day. This recurrence is called the *Saros cycle*. Well known to ancient astronomers, it undoubtedly was the key to their "mystical" ability to predict eclipses!

---

CONCEPT CHECK

✔ What types of solar eclipses would you expect to see if Earth's distance from the Sun were to double? What if the distance became half its present value?

---

## 1.7 The Measurement of Distance

We have seen a little of how astronomers track and record the positions of the stars in the sky. But knowing the direction to an object is only part of the information needed to locate it in space. Before we can make a systematic study of the heavens, we must find a way of measuring *distances*, too. One distance-measurement method, called **triangulation,** is based on the principles of Euclidean geometry and finds widespread application today in both terrestrial and astronomical settings. Surveyors use these age-old geometric ideas to measure the distance to faraway objects indirectly. Triangulation forms the foundation of the family of distance-measurement techniques making up the **cosmic distance scale.**

### Triangulation and Parallax

Imagine trying to measure the distance to a tree on the other side of a river. The most direct method is to lay a tape across the river, but that's not the simplest way (nor, because of the current, may it even be possible). A smart surveyor would make the measurement by visualizing an *imaginary* triangle (hence *triangulation*), sighting the tree on the far side of the river from two positions on the near side, as illustrated in Figure 1.30. The simplest possible triangle is a right triangle, in which one of the angles is exactly 90°, so it is usually convenient to set up one observation position directly opposite the object, as at point A. The surveyor then moves to another observation position at point B, noting the distance covered between points A and B. This distance is called the **baseline** of the imaginary triangle. Finally, the surveyor, standing at point B, sights toward the tree and notes the angle at point B between this line of sight and the baseline. Knowing the value of one side (AB) and two angles (the right angle at point A and the angle at point B) of the right triangle, the surveyor geometrically constructs the remaining sides and angles and establishes the distance from A to the tree.

To use triangulation to measure distances, a surveyor must be familiar with *trigonometry*, the mathematics of geometrical angles and distances. However, even if we knew no

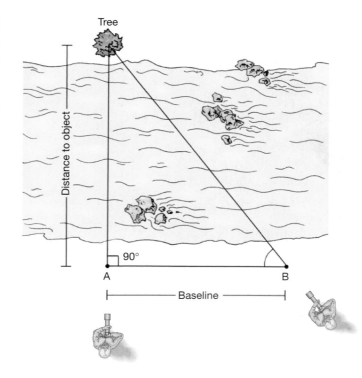

▲ FIGURE 1.30 **Triangulation** Surveyors often use simple geometry and trigonometry to estimate the distance to a faraway object by triangulation. By measuring the angles at A and B and the length of the baseline, the distance can be calculated without the need for direct measurement.

trigonometry at all, we could still solve the problem by graphical means, as shown in Figure 1.31. Suppose that we pace off the baseline AB, measuring it to be 450 meters, and measure the angle between the baseline and the line from B to the tree to be 52°, as illustrated in the figure. We can transfer the problem to paper by letting one box on our graph represent 25 meters on the ground. Drawing the line AB on paper and completing the other two sides of the triangle, at angles of 90° (at A) and 52° (at B), we measure the distance on paper from A to the tree to be 23 boxes—that is, 575 meters. We have solved the real problem by *modeling* it on paper. The point to remember here is this: Nothing more complex than basic geometry is needed to infer the distance, the size, and even the shape of an object that is too far away or inaccessible for direct measurement.

Obviously, for a fixed baseline the triangle becomes longer and narrower as the tree's distance from A increases. Narrow triangles cause problems, because it becomes hard to measure the angles at A and B with sufficient accuracy. The measurements can be made easier by "fattening" the triangle—that is, by lengthening the baseline—but there are limits on how long a baseline we can choose in astronomy. For example, consider an imaginary triangle extending from Earth to a nearby object in space, perhaps a neighboring planet. The triangle is now extremely long and narrow, even for a relatively nearby object (by cosmic standards). Figure 1.32(a) illustrates a case in which the longest baseline possible on Earth—Earth's diameter, measured from point A to point B—is used.

In principle, two observers could sight the planet from opposite sides of Earth, measuring the triangle's angles at

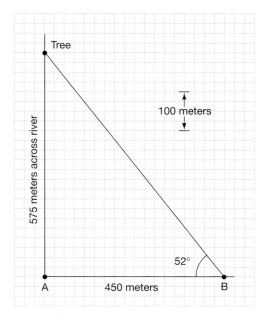

▲ FIGURE 1.31 **Geometric Scaling** Not even trigonometry is needed to estimate distances indirectly. Scaled estimates, like this one on a piece of graph paper, often suffice.

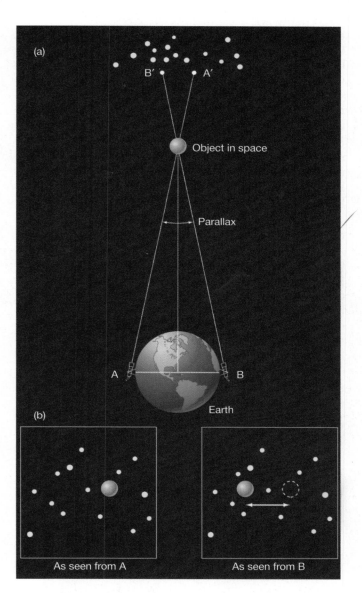

▲ FIGURE 1.32 **Parallax** (a) This imaginary triangle extends from Earth to a nearby object in space (such as a planet). The group of stars at the top represents a background field of very distant stars. (b) Hypothetical photographs of the same star field showing the nearby object's apparent displacement, or shift, relative to the distant undisplaced stars.

A and B. However, in practice it is easier to measure the third angle of the imaginary triangle. Here's how. The observers sight toward the planet, taking note of its position *relative to some distant stars* seen on the plane of the sky. The observer at point A sees the planet at apparent location A′ relative to those stars, as indicated in Figure 1.32(a). The observer at B sees the planet at point B′. If each observer takes a photograph of the appropriate region of the sky, the planet will appear at slightly different places in the two images. The planet's photographic image is slightly displaced, or shifted, relative to the field of distant background stars, as shown in Figure 1.32(b). The background stars themselves appear undisplaced because of their much greater distance from the observer.

This apparent displacement of a foreground object relative to the background as the observer's location changes is known as **parallax.** The size of the shift in Figure 1.32(b), measured as an angle on the celestial sphere, is the third, small angle in Figure 1.32(a). In astronomical contexts, the parallax is usually very small. For example, the parallax of a point on the Moon, viewed using a baseline equal to Earth's diameter, is about 2°; the parallax of the planet Venus at closest approach (45 million km), is just 1′ (see *More Precisely 1-3*).

The closer an object is to the observer, the larger is the parallax. Figure 1.33 illustrates how you can see this for yourself. Hold a pencil vertically in front of your nose and concentrate on some far-off object—a distant wall, perhaps. Close one eye, and then open it while closing the other. You should see a large shift in the apparent position of the pencil projected onto the distant wall—a large parallax. In this example, one eye corresponds to point A, the other eye to point B, the distance between your eyeballs to the baseline, the pencil to the planet, and the distant wall to a remote field of stars. Now hold the pencil at arm's

length, corresponding to a more distant object (but still not as far away as the even more distant stars). The apparent shift of the pencil will be less. You might even be able to verify that the apparent shift is inversely proportional to the distance to the pencil. By moving the pencil farther away, we are narrowing the triangle and decreasing the parallax (and also making accurate measurement more difficult). If you were to paste the pencil to the wall, corresponding to the case where the object of interest is as far away as the background star field, blinking would produce no apparent shift of the pencil at all.

The amount of parallax is thus inversely proportional to an object's distance. Small parallax implies large distance, and large parallax implies small distance. Knowing the amount of parallax (as an angle) and the length of the baseline, we can easily derive the distance through triangulation. *More Precisely 1-3* explores the connection between angular measure and distance in more detail, showing how we can use elementary geometry to determine both the distances and the dimensions of faraway objects.

Surveyors of the land routinely use such simple geometric techniques to map out planet Earth. As surveyors of the sky, astronomers use the same basic principles to chart the universe.

## Sizing Up Planet Earth

Now that we have studied some of the tools available to astronomers, let's end the chapter with a classic example of how the scientific method, combined with the basic geometric techniques just described, enabled an early scientist to perform a calculation of truly "global" proportions.

In about 200 B.C., a Greek philosopher named Eratosthenes (276–194 B.C.) used simple geometric reasoning to calculate the size of our planet. He knew that at noon on the first day of summer observers in the city of Syene (now called Aswan) in Egypt saw the Sun pass directly overhead. This was evident from the fact that vertical objects cast no shadows and sunlight reached to the very bottoms of deep wells, as shown in the insets in Figure 1.34. However, at noon of the same day in Alexandria, a city 5000 *stadia* to the north, the Sun was seen to be displaced slightly from the vertical. (The *stadium* was a Greek unit of length, roughly equal to 0.16 km—the modern town of Aswan lies about 780 km, or 490 miles, south of Alexandria.) By measuring the length of the shadow of a vertical stick and applying elementary trigonometry, Eratosthenes determined the angular displacement of the Sun from the vertical at Alexandria to be 7.2°.

What could have caused this discrepancy between the two measurements? It was not the result of measurement

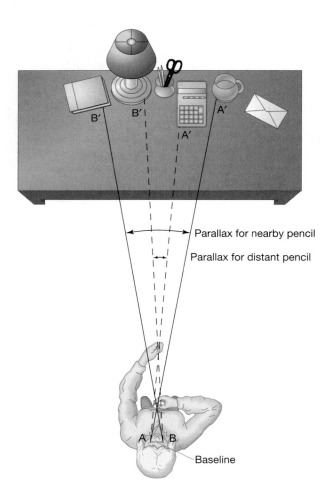

Parallax for nearby pencil

Parallax for distant pencil

A    B

Baseline

◄ FIGURE 1.33 **Parallax Geometry** Parallax is inversely proportional to an object's distance. An object near your nose has a much larger parallax than an object held at arm's length.

# MORE PRECISELY 1-3

## Measuring Distances with Geometry

Simple geometrical reasoning forms the basis for almost every statement made in this book about size and scale in the universe. In a very real sense, our modern knowledge of the cosmos depends on the elementary mathematics of ancient Greece. Let's take a moment to look in a little more detail at how astronomers use geometry to measure the distances to, and sizes of, objects near and far.

We can convert baselines and parallaxes into distances, and vice versa, by using arguments made by the Greek geometer Euclid. The first figure represents Figure 1.32(a), but we have changed the scale and added the circle centered on the target planet and passing through our baseline on Earth:

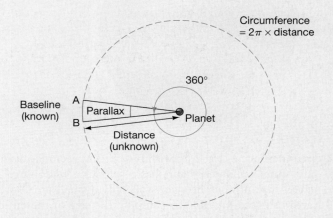

To see how the planet's parallax relates to its distance, we note that the ratio of the baseline AB to the circumference of the large circle shown in the figure must be equal to the ratio of the parallax to one full revolution, 360°. Recall that the circumference of a circle is always $2\pi$ times its radius (where $\pi$—the Greek letter "pi"—is approximately equal to 3.142). Applying this relation to the large circle in the figure, we find that

$$\frac{\text{baseline}}{2\pi \times \text{distance}} = \frac{\text{parallax}}{360°},$$

from which it follows that

$$\text{parallax} = (360°/2\pi) \times \frac{\text{baseline}}{\text{distance}}.$$

The angle $360°/2\pi \approx 57.3°$ in the preceding equation is usually called 1 *radian*.

**EXAMPLE 1** The planet Venus lies roughly 45,000,000 km from Earth at closest approach. Two observers 13,000 km apart (i.e., at opposite ends of Earth's diameter) looking at the planet would measure a parallax of $57.3° \times (13{,}000 \text{ km}/45{,}000{,}000 \text{ km}) = 0.017° = 1.0$ arc minutes, as stated in the text.

Alternatively, if we know the parallax (from direct measurement, such as the photographic technique described in Section 1.7), we can rearrange the above equation to tell us the distance to the planet:

$$\text{distance} = \text{baseline} \times \frac{57.3°}{\text{parallax}}.$$

**EXAMPLE 2** Two observers 1000 km apart looking at the Moon might measure a parallax of 9.0 arc minutes—that is, 0.15°. It then follows that the distance to the Moon is 1000 km $\times$ (57.3/0.15) $\approx$ 380,000 km. (More accurate measurements, based on laser ranging using equipment left on the lunar surface by *Apollo* astronauts, yield a mean distance of 384,000 km.)

Knowing the distance to an object, we can determine many other properties. For example, by measuring the object's *angular diameter*—the angle from one side of the object to the other as we view it in the sky—we can compute its size. The second figure illustrates the geometry involved:

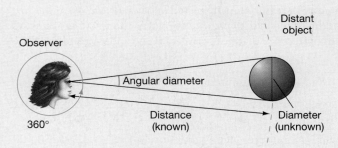

Notice that this is basically the same diagram as the previous one, except that now the angle (the angular diameter) and distance are known, instead of the angle (the parallax) and baseline. Exactly the same reasoning as before then allows us to calculate the diameter. We have

$$\frac{\text{diameter}}{2\pi \times \text{distance}} = \frac{\text{angular diameter}}{360},$$

so

$$\text{diameter} = \text{distance} \times \frac{\text{angular diameter}}{57.3°}.$$

**EXAMPLE 3** The Moon's angular diameter is measured to be about 31 arc minutes—a little over half a degree. From the preceding discussion, it follows that the Moon's actual diameter is 380,000 km $\times$ (0.52°/57.3°) $\approx$ 3450 km. A more precise measurement gives 3476 km.

Study the foregoing reasoning carefully. We will use these simple arguments, in various forms, many times throughout this text.

▶ FIGURE 1.34 **Measuring Earth's Radius** The Sun's rays strike different parts of Earth's surface at different angles. The Greek philosopher Eratosthenes realized that the difference was due to Earth's curvature, enabling him to determine Earth's radius by using simple geometry.

error—the same results were obtained every time the observations were repeated. Instead, as illustrated in Figure 1.34, the explanation is simply that Earth's surface is not flat, but *curved.* Our planet is a sphere. Eratosthenes was not the first person to realize that Earth is spherical—the philosopher Aristotle had done that over 100 years earlier (see Section 1.2), but he was apparently the first to build on this knowledge, combining geometry with direct measurement to infer the size of our planet. Here's how he did it.

Rays of light reaching Earth from a very distant object, such as the Sun, travel almost parallel to one another. Consequently, as shown in the figure, the angle measured at Alexandria between the Sun's rays and the vertical (i.e., the line joining Alexandria to the center of Earth) is equal to the angle between Syene and Alexandria, as seen from Earth's center. (For the sake of clarity, the angle has been exaggerated in the figure.) As discussed in *More Precisely 1-3,* the size of this angle in turn is proportional to the fraction of Earth's circumference that lies between Syene and Alexandria:

$$\frac{7.2° \text{ (angle between Syene and Alexandria)}}{360° \text{ (circumference of a circle)}}$$
$$= \frac{5000 \text{ stadia}}{\text{Earth's circumference}}.$$

Earth's circumference is therefore $50 \times 5000$, or 250,000 stadia, or about 40,000 km, so Earth's radius is $250,000/2\pi$ stadia, or 6366 km. The correct values for Earth's circumference and radius, now measured accurately by orbiting spacecraft, are 40,070 km and 6378 km, respectively.

Eratosthenes' reasoning was a remarkable accomplishment. More than 20 centuries ago, he estimated the

circumference of Earth to within 1 percent accuracy, using only simple geometry and basic scientific reasoning. A person making measurements on only a small portion of Earth's surface was able to compute the size of the entire planet on the basis of observation and pure logic—an early triumph of the scientific method.

CONCEPT CHECK

✔ Why is elementary geometry essential for measuring distances in astronomy?

# CHAPTER REVIEW

## Summary

❶ The **scientific method (p. 7)** is a methodical approach employed by scientists to explore the universe around us in an objective manner. A **theory (p. 6)** is a framework of ideas and assumptions used to explain some set of observations and construct **theoretical models (p. 6)** that make predictions about the real world. These predictions in turn are amenable to further observational testing. In this way, the theory expands and science advances.

Observation  Theory  Prediction

❷ Early observers grouped the thousands of stars visible to the naked eye into patterns called **constellations (p. 8),** which they imagined were attached to a vast **celestial sphere (p. 10)** centered on Earth. Constellations have no physical significance, but are still used to label regions of the sky. The points where Earth's axis of rotation intersects the celestial

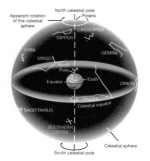

sphere are called the north and south **celestial poles (p. 10)**. The line where Earth's equatorial plane cuts the celestial sphere is the **celestial equator (p. 11)**. **Celestial coordinates (p. 12)** are a more precise way of specifying a star's location on the celestial sphere.

❸ The nightly motion of the stars across the sky is the result of Earth's **rotation (p. 10)** on its axis. The time from one noon to the next is called a **solar day (p. 13)**. The time between successive risings of any given star is one **sidereal**

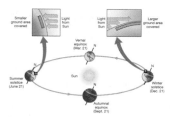

**day (p. 13)**. Because of Earth's **revolution (p. 13)** around the Sun, we see different stars at night at different times of the year, and the Sun appears to move relative to the stars. The Sun's apparent yearly path around the celestial sphere (or the plane of Earth's orbit around the Sun) is called the **ecliptic (p. 15)**. We experience **seasons (p. 16)** because Earth's rotation axis is inclined to the ecliptic plane. At the **summer solstice (p. 15)**, the Sun is highest in the sky and the length of the day is greatest. At the **winter solstice (p. 16)**, the Sun is lowest and the day is shortest. At the **vernal (p. 16)** and **autumnal equinoxes (p. 16)**, Earth's axis of rotation is perpendicular to the line joining Earth to the Sun, so day and night are of equal length. The interval of time from one vernal equinox to the next is one **tropical year (p. 16)**.

❹ In addition to its rotation about its axis and its revolution around the Sun, Earth has many other motions. One of the most important of these is **precession (p. 16)**, the slow "wobble" of Earth's axis due to the influence of the Moon. As a result, the **sidereal year (p. 16)** is slightly longer than

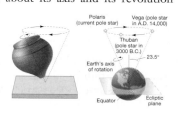

the tropical year, and the particular constellations that happen to be visible during any given season change over the course of thousands of years.

❺ The speed at which the Sun appears to traverse the ecliptic varies around the year, so the noon-to-noon solar day is not constant. The 24-hour **mean solar day (p. 18)** is the average solar day over the entire year. Since noon occurs when the Sun crosses the local **meridian (p. 17)**, observers at different locations will dis-

agree on the time at which that happens. To standardize timekeeping, Earth is divided into 24 standard **time zones (p. 18)**. Within each zone, all clocks keep the same **standard time (p. 18)**. Standard time in the time zone of 0° longitude is **universal time (p. 18)**. One tropical year is not a whole number of mean solar days. In order to keep the average length of a calendar year equal to 1 tropical year, an extra day is inserted into the calendar on **leap years (p. 19)**.

❻ The Moon emits no light of its own, but instead shines by reflected sunlight. As the Moon orbits Earth, we see **lunar phases (p. 19)** as

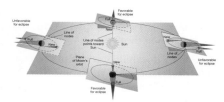

the amount of the Moon's sunlit face visible to us varies. A **lunar eclipse (p. 21)** occurs when the Moon enters Earth's shadow. A **solar eclipse (p. 21)** occurs when the Moon passes between Earth and the Sun. An eclipse may be **total (p. 21)** if the body in question (Moon or Sun) is completely obscured, or **partial (p. 21)** if only a portion of the surface is affected. If the Moon happens to be too far from Earth for its disk to completely hide the Sun, an **annular eclipse (p. 23)** occurs. Because the Moon's orbit around Earth is slightly inclined with respect to the ecliptic, solar and lunar eclipses do not occur every month, but only during **eclipse seasons (p. 25)** (twice per year).

❼ Surveyors on Earth use **triangulation (p. 26)** to determine the distances to faraway objects. Astronomers use the same technique to measure the distances to planets and stars. The **cosmic distance scale (p. 26)** is the family of distance-measurement techniques by which astronomers chart the universe. **Parallax (p. 28)** is the apparent motion of a foreground object relative to a distant background as the observer's position changes. The larger the **baseline (p. 26)**—the distance between the two observation points— the greater is the parallax. The same basic

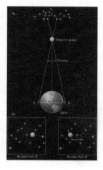

geometric reasoning is used to determine the sizes of objects whose distances are known. The Greek philosopher Eratosthenes used elementary geometry to determine Earth's radius.

## Review and Discussion

1. Compare the size of Earth with that of the Sun, the Milky Way galaxy, and the entire universe.

2. What does an astronomer mean by "the universe"?

3. How big is a light-year?

4. What is the scientific method, and how does science differ from religion?

5. What is a constellation? Why are constellations useful for mapping the sky?

6. Why does the Sun rise in the east and set in the west each day? Does the Moon also rise in the east and set in the west? Why? Do stars do the same? Why?

7. How and why does a day measured with respect to the Sun differ from a day measured with respect to the stars?

8. How many times in your life have you orbited the Sun?

9. Why do we see different stars at different times of the year?

10. Why are there seasons on Earth?

11. What is precession, and what causes it?

12. Why do we need standard time zones and leap years?

13. If one complete hemisphere of the Moon is always lit by the sun, why do we see different phases of the Moon?

14. What causes a lunar eclipse? A solar eclipse?

15. Why aren't there lunar and solar eclipses every month?

16. Do you think an observer on another planet might see eclipses? Why or why not?

17. What is parallax? Give an everyday example.

18. Why is it necessary to have a long baseline when using triangulation to measure the distances to objects in space?

19. What two pieces of information are needed to determine the diameter of a faraway object?

20. If you traveled to the outermost planet in our solar system, do you think the constellations would appear to change their shapes? What would happen if you traveled to the next-nearest star? If you traveled to the center of our Galaxy, could you still see the familiar constellations found in Earth's night sky?

## Conceptual Self-Test: True or False/Multiple Choice

1. The light-year is a measure of distance.

2. The stars in a constellation are physically close to one another.

3. The constellations lying along the ecliptic are collectively referred to as the zodiac.

4. The seasons are caused by the precession of Earth's axis.

5. At the winter solstice, the Sun is at its southernmost point on the celestial sphere.

6. When the Sun, Earth, and Moon are positioned to form a right angle at Earth, the Moon is seen in the new phase.

7. A lunar eclipse can occur only during the full phase.

8. An annular eclipse is a type of eclipse that occurs every year.

9. The parallax of an object is inversely proportional to the object's distance from us.

10. If we know the distance of an object from Earth, we can determine the size of the object by measuring its parallax.

11. If Earth rotated twice as fast as it currently does, but its motion around the Sun stayed the same, then (a) the night would be twice as long; (b) the night would be half as long; (c) the year would be half as long; (d) the length of the day would be unchanged.

12. A long, thin cloud that stretched from directly overhead to the western horizon would have an angular size of (a) 45°; (b) 90°; (c) 180°; (d) 360°.

13. According to Figure 1.15 ("The Zodiac"), in January the Sun is in the constellation (a) Cancer; (b) Gemini; (c) Leo; (d) Aquarius.

14. If Earth orbited the Sun in 9 months instead of 12, then, compared with a sidereal day, a solar day would be (a) longer; (b) shorter; (c) unchanged.

15. When a thin crescent of the Moon is visible just before sunrise, the Moon is in its (a) waxing phase; (b) new phase; (c) waning phase; (d) quarter phase.

16. If the Moon's orbit were a little larger, solar eclipses would be (a) more likely to be annular; (b) more likely to be total; (c) more frequent; (d) unchanged in appearance.

17. If the Moon orbited Earth twice as fast, the frequency of solar eclipses would (a) double; (b) be cut in half; (c) stay the same.

18. In Figure 1.30 ("Triangulation"), using a longer baseline would result in (a) a less accurate distance to the tree; (b) a more accurate distance to the tree; (c) a smaller angle at point B; (d) a greater distance across the river.

19. In Figure 1.32 ("Parallax"), a smaller Earth would result in (a) a smaller parallax angle; (b) a shorter distance measured to the object; (c) a larger apparent displacement; (d) stars appearing closer together.

20. Today, distances to stars are measured by (a) bouncing radar signals; (b) reflected laser beams; (c) travel time by spacecraft; (d) geometry.

## Problems

 *Algorithmic versions of these Problems are available in the Practice Problems module of the Companion Website.* *The number of dots preceding each Problem indicates its approximate level of difficulty.*

1. • In 1 second, light leaving Los Angeles reaches approximately as far as (a) San Francisco, about 500 km; (b) London, roughly 10,000 km; (c) the Moon, 384,000 km; (d) Venus, 45,000,000 km from Earth at closest approach; or (e) the nearest star, about 4 light-years from Earth. Which is correct?

2. • (a) Write the following numbers in scientific notation (see Appendix 1 if you are unfamiliar with this notation): 1000; 0.000001; 1001; 1,000,000,000,000,000; 123,000; 0.000456. (b) Write the following numbers in "normal" numerical form: $3.16 \times 10^7$; $2.998 \times 10^5$; $6.67 \times 10^{-11}$; $2 \times 10^0$. (c) Calculate: $(2 \times 10^3) + 10^{-2}$; $(1.99 \times 10^{30})/(5.98 \times 10^{24})$; $(3.16 \times 10^7) \times (2.998 \times 10^5)$.

3. •• How, and by roughly how much, would the length of the solar day change if Earth's rotation were suddenly to reverse direction?

4. • The vernal equinox is now just entering the constellation Aquarius. In what constellation will it lie in the year A.D. 10,000?

5. •• What would be the length of the synodic month if the Moon's sidereal orbital period were (a) 1 week (7 solar days); (b) 1 (sidereal) year?

6. • Through how many degrees, arc minutes, or arc seconds does the Moon move in (a) 1 hour of time; (b) 1 minute; (c) 1 second? How long does it take for the Moon to move a distance equal to its own diameter?

7. •• Given the data presented in the text, estimate the speed (in kilometers per second) at which the Moon moves in its orbit around Earth.

8. • A surveyor wishes to measure the distance between two points on either side of a river, as illustrated in Figure 1.30. She measures the distance AB to be 250 m and the angle at B to be 30°. What is the distance between the two points?

9. • At what distance is an object if its parallax, as measured from either end of a 1000-km baseline, is (a) 1°; (b) 1′; (c) 1″?

10. • Given that the angular size of Venus is 55″ when the planet is 45,000,000 km from Earth, calculate Venus's diameter (in kilometers).

11. • Calculate the parallax, using Earth's diameter as a baseline, of the Sun's nearest neighbor, Proxima Centauri, which lies 4.3 light-years from Earth.

12. • Estimate the angular diameter of your thumb, held at arm's length.

13. • The Moon lies roughly 384,000 km from Earth and the Sun lies 150,000,000 km away. If both have the same angular size as seen from Earth, how many times larger than the Moon is the Sun?

14. • Given that the distance from Earth to the Sun is 150,000,000 km, through what distance does Earth move in (a) a second, (b) an hour, (c) a day?

15. • What angle would Eratosthenes have measured (see Section 1.7) had Earth been flat?

*The Companion Website at www.aw-bc.com/chaisson provides algorithmically generated versions of each chapter's Problems, along with additional quizzes, an Animations & Videos gallery, an Images gallery, an interactive Glossary, and a full eBook.*

CHAPTER

2

# THE COPERNICAN REVOLUTION

## The Birth of Modern Science

Living in the Space Age, we have become accustomed to the modern view of our place in the universe. Images of our planet taken from space leave little doubt that Earth is round, and no one seriously questions the idea that we orbit the Sun. Yet there was a time, not so long ago, when some of our ancestors maintained that Earth was flat and lay at the center of all things.

Our view of the universe—and of ourselves—has undergone a radical transformation since those early days. Earth has become a planet like many others, and humankind has been torn from its throne at the center of the cosmos and relegated to a rather unremarkable position on the periphery of the Milky Way galaxy. But we have been amply compensated for our loss of prominence: We have gained a wealth of scientific knowledge in the process. The story of how all this came about is the story of the rise of the scientific method and the genesis of modern astronomy.

## LEARNING GOALS

*Studying this chapter will enable you to*

1 Describe how some ancient civilizations attempted to explain the heavens in terms of Earth-centered models of the universe.

2 Explain how the observed motions of the planets led to our modern view of a Sun-centered solar system.

3 Describe the major contributions of Galileo and Kepler to our understanding of the solar system.

4 State Kepler's laws of planetary motion.

5 Explain how astronomers have measured the true size of the solar system.

6 State Newton's laws of motion and universal gravitation and explain how they account for Kepler's laws.

7 Explain how the law of gravitation enables us to measure the masses of astronomical bodies.

 Visit www.aw-bc.com/chaisson for additional images, animations, quizzes, and eBook for this chapter.

## 2.1    Ancient Astronomy

Many ancient cultures took a keen interest in the changing nighttime sky. The records and artifacts that have survived until the present make that abundantly clear. But unlike today, the major driving force behind the development of astronomy in those early societies was probably neither scientific nor religious. Instead, it was decidedly practical and down to earth. Seafarers needed to navigate their vessels, and farmers had to know when to plant their crops. In a real sense, then, human survival depended on knowledge of the heavens. The ability to predict accurately the arrival of the seasons, as well as other astronomical events, was undoubtedly a highly prized, perhaps jealously guarded, skill.

In Chapter 1, we saw that the human mind's ability to perceive patterns in the stars led to the "invention" of constellations as a convenient means of labeling regions of the celestial sphere. ∞ (Sec. 1.3) The realization that these patterns returned to the night sky at the same time each year met the need for a practical means of tracking the seasons. Widely separated cultures all over the world built elaborate structures to serve, at least in part, as primitive calendars. Often the keepers of the secrets of the sky enshrined their knowledge in myth and ritual, and these astronomical sites were also used for religious ceremonies.

Perhaps the best-known such site is *Stonehenge*, located on Salisbury Plain in England, and shown in Figure 2.1. This ancient stone circle, which today is one of the most popular tourist attractions in Britain, dates from the Stone Age. Researchers think it was an early astronomical observatory of sorts—not in the modern sense of the term (a place for making new observations and discoveries pertaining to the heavens), but rather a kind of three-dimensional calendar or almanac, enabling its builders and their descendants to identify important dates by means of specific celestial events. Its construction apparently spanned a period of about 17 centuries, beginning around 2800 B.C. Additions and modifications continued to about 1100 B.C., indicating its ongoing importance to the Stone Age and, later, Bronze Age people who built, maintained, and used Stonehenge. The largest stones shown in Figure 2.1 weigh up to 50 tons and were transported from quarries many miles away.

Many of the stones are aligned so that they point toward important astronomical events. For example, the line joining the center of the inner circle to the so-called heel stone, set off some distance from the rest of the structure, points in the direction of the rising Sun on the summer solstice. Other alignments are related to the rising and setting of the Sun and the Moon at other times of the year. The accurate alignments (within a degree or so) of the stones of Stonehenge were first noted in the 18th century, but it was only relatively recently—in the second half of the 20th century, in fact—that the scientific community began to credit Stone Age technology with the ability to carry out such a precise feat of engineering. Although some of Stonehenge's purposes remain uncertain and controversial, the site's function as an astronomical almanac seems well established. Although Stonehenge is

◀ FIGURE 2.1 **Stonehenge** This remarkable site in the south of England was probably constructed as a primitive calendar or almanac. The inset shows sunrise at Stonehenge at the summer solstice. As seen from the center of the stone circle, the Sun rose directly over the "heel stone" on the longest day of the year. (*English Heritage*)

the most impressive and the best preserved, other stone circles, found all over Europe, are thought to have performed similar functions.

Many North American cultures were interested in the heavens. The Big Horn Medicine Wheel in Wyoming (Figure 2.2a) is similar to Stonehenge in design—and, perhaps, intent—although it is somewhat simpler in execution. Some researchers have identified alignments between the Medicine Wheel's spokes and the rising and setting Sun at solstices and equinoxes, and with some bright stars, suggesting that its builders—the Plains Indians—had much more than a passing familiarity with the changing nighttime sky. Other experts disagree, however, arguing that the alignments are quite inaccurate and consistent with pure chance and that the Medicine Wheel's purpose was more likely symbolic, rather than practical. A similar controversy swirls around the Caracol temple (Figure 2.2b) in the famous Mayan city of Chitzen Itza, built around A.D. 1000 on Mexico's Yucatán peninsula. Was it an observatory, as

some suggest, perhaps tied to human sacrifices when Venus appeared in the morning or evening sky? Or are the claimed alignments of its windows just wishful thinking and the temple's purpose simply religious, rather than astronomical?

Experts do seem to agree—for now, at least—that the Sun Dagger (Figure 2.2c), in Chaco Canyon, New Mexico, is a genuine astronomical calendar. It is constructed so that the sliver of light passes precisely through the center of the carved stone spiral at noon on the summer solstice. Numerous similar sites have been found throughout the American southwest.

The ancient Chinese also observed the heavens. Their astrology attached particular importance to "omens" such as comets and "guest stars"—stars that appeared suddenly in the sky and then slowly faded away—and they kept careful and extensive records of such events. Twentieth-century astronomers still turn to the Chinese records to obtain observational data recorded during the Dark Ages (roughly from the 5th to the 10th century A.D.), when turmoil in

(a)

(c)

(b)

▲ FIGURE 2.2 **Observatories in the Americas** (a) The Big Horn Medicine Wheel in Wyoming, built by the Plains Indians, has spokes and other features that roughly align with risings and settings of the Sun and other stars. (b) The Caracol temple in Mexico, built by the Mayan civilization, has some windows that seem to align with astronomical events, suggesting that at least part of Caracol's function may have kept track of the seasons and the heavens. (c) This thin streak of light and shadow, created by the Sun's rays playing off the cliffs in Chaco Canyon of America's southwest, aligns exactly with a carved rock pattern at noon on the summer solstice—almost certainly an intentional alignment for astronomical or agricultural purposes. (*G. Gerster/Comstock*)

▲ FIGURE 2.3 **Turkish Astronomers at Work** During the Dark Ages, much scientific information was preserved and new discoveries were made by astronomers in the Islamic world, as depicted in this illustration from a 16th-century manuscript. (*The Granger Collection*)

Europe largely halted the progress of Western science. Perhaps the best-known guest star was one that appeared in A.D. 1054 and was visible in the daytime sky for many months. We now know that the event was actually a *supernova:* the explosion of a giant star, which scattered most of its mass into space (see Chapter 21). It left behind a remnant that is still detectable today, nine centuries later. The Chinese data are a prime source of historical information for supernova research.

A vital link between the astronomy of ancient Greece and that of medieval Europe was provided by astronomers in the Muslim world (see Figure 2.3). For six centuries, from the depths of the Dark Ages to the beginning of the Renaissance, Islamic astronomy flourished and grew, preserving and augmenting the knowledge of the Greeks. Its influence on modern astronomy is subtle, but quite pervasive. Many of the mathematical techniques involved in trigonometry were developed by Islamic astronomers in response to practical problems, such as determining the precise dates of holy days or the direction of Mecca from any given location on Earth. Astronomical terms such as *zenith* and *azimuth* and the names of many stars—for example, Rigel, Betelgeuse, and Vega—all bear witness to this extended period of Muslim scholarship.

Astronomy is not the property of any one culture, civilization, or era. The same ideas, the same tools, and even the same misconceptions have been invented and reinvented by human societies all over the world in response to the same basic driving forces. Astronomy came into being because people knew that there was a practical benefit in being able to predict the positions of the stars, but its roots go much deeper than that. The need to understand where we came from and how we fit into the cosmos is an integral part of human nature.

## 2.2 The Geocentric Universe

The Greeks of antiquity, and undoubtedly civilizations before them, built models of the universe. The study of the workings of the universe on the largest scales is called *cosmology.* Today, cosmology entails looking at the universe on scales so large that even entire galaxies can be regarded as mere points of light scattered throughout space. To the Greeks, however, the universe was basically the *solar system*—the Sun, Earth, and Moon, and the planets known at that time. The stars beyond were surely part of the universe, but they were considered to be fixed, unchanging beacons on the celestial sphere. The Greeks did not consider the Sun, the Moon, and the planets to be part of this mammoth celestial dome, however. Those objects had patterns of behavior that set them apart.

### Observations of the Planets

Greek astronomers observed that over the course of a night, the stars slid smoothly across the sky. Over the course of a month, the Moon moved smoothly and steadily along its path on the sky relative to the stars, passing through its familiar cycle of phases. Over the course of a year, the Sun progressed along the ecliptic at an almost constant rate, varying little in brightness from day to day. In short, the behavior of both Sun and Moon seemed fairly simple and orderly. But ancient astronomers were also aware of five other bodies in the sky—the planets Mercury, Venus, Mars, Jupiter, and Saturn—whose behavior was not so easy to grasp. Their motions ultimately led to the downfall of an entire theory of the solar system and to a fundamental change in humankind's view of the universe.

To the naked eye (or even through a telescope), planets do not behave in as regular and predictable a fashion as the Sun, Moon, and stars. They vary in brightness, and they don't maintain a fixed position in the sky. Unlike the Sun and Moon, the planets seem to wander around the celestial sphere—indeed, the word *planet* derives from the Greek word *planetes*, meaning "wanderer." Planets never stray far from the ecliptic and generally traverse the celestial sphere from west to east, like the Sun. However, they seem to speed up and slow down during their journeys, and at times they even appear to loop back and forth relative to the stars, as shown in Figure 2.4. In other words, there are periods when a planet's eastward motion (relative to the stars) stops, and the planet appears to move westward in

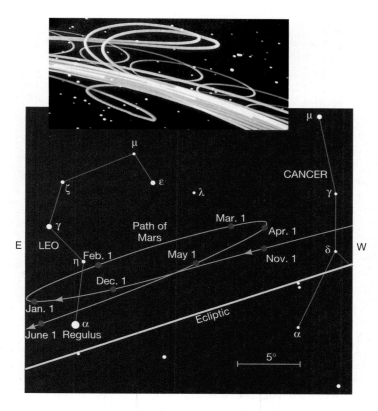

◀ **FIGURE 2.4 Planetary Motion** Most of the time, planets move from west to east relative to the background stars. Occasionally—roughly once per year—however, they change direction and temporarily undergo retrograde motion (east to west) before looping back. The main illustration shows an actual retrograde loop in the motion of the planet Mars. The inset depicts the movements of several planets over the course of several years, as reproduced on the inside dome of a planetarium. The motion of the planets relative to the stars (represented as unmoving points) produces continuous streaks on the planetarium "sky." *(Boston Museum of Science)*

the sky for a month or two before reversing direction again and continuing on its eastward journey. Motion in the eastward sense is usually referred to as *direct*, or *prograde*, motion; the backward (westward) loops are known as **retrograde motion.**

Ancient astronomers knew well that the periods of retrograde motion were closely correlated with other planetary properties, such as apparent brightness and position in the sky. Figure 2.5 (a modern view of the solar system, note!) shows three schematic planetary orbits and defines some time-honored astronomical terminology describing a planet's location relative to Earth and the Sun. Mercury and Venus are referred to as *inferior* ("lower") *planets* because their orbits lie between Earth and the Sun. Mars, Jupiter, and Saturn, whose orbits lie outside Earth's, are known as *superior* ("higher") *planets*. For early astronomers, the key observations of planetary orbits were the following:

- An inferior planet never strays too far from the Sun, as seen from Earth. As illustrated in the inset to Figure 2.5, because its path on the celestial sphere is close to the ecliptic, an inferior planet makes two *conjunctions* (or close approaches) with the Sun during each orbit. (It doesn't actually come close to the Sun, of course. Conjunction is simply the occasion when the planet and the Sun are in the same direction in the sky.) At *inferior conjunction*, the planet is closest to Earth and moves past the Sun from east to west—that is, in the retrograde sense. At *superior conjunction*, the planet is farthest from Earth and passes the Sun in the opposite (prograde) direction.

- Seen from Earth, the superior planets are not "tied" to the Sun as the inferior planets are. The superior planets make one prograde conjunction with the Sun during each trip around the celestial sphere. However, they exhibit retrograde motion (Figure 2.4) when they are at *opposition*, diametrically opposite the Sun on the celestial sphere.

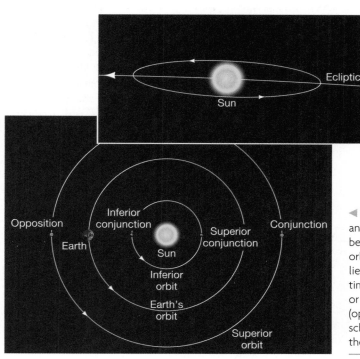

◀ **FIGURE 2.5 Inferior and Superior Orbits** Diagram of Earth's orbit and two other possible planetary orbits. An "inferior" orbit lies between Earth's orbit and the Sun. Mercury and Venus move in such orbits. A "superior" orbit (such as the orbit of Mars, Jupiter, or Saturn) lies outside that of Earth. The points noted on the orbits indicate times when a planet appears to come close to the Sun (conjunction) or is diametrically opposite the Sun on the celestial sphere (opposition); they are discussed further in the text. The inset is a schematic representation of the orbit of an inferior planet relative to the Sun, as seen from Earth.

- The superior planets are brightest at opposition, during retrograde motion. By contrast, the inferior planets are brightest a few weeks before and after inferior conjunction.

The challenge facing astronomers—then as now—was to find a solar system model that could explain all the existing observations and that could also make testable and reliable predictions of future planetary motions. ∞ (Sec. 1.2)

Ancient astronomers correctly reasoned that the changing brightness of a planet in the night sky is related to the planet's distance from Earth. Like the Moon, the planets produce no light of their own. Instead, they shine by reflected sunlight and, generally speaking, appear brightest when closest to us. Looking at Figure 2.5, you may already be able to discern the basic reasons for some of the planetary properties just listed; we'll return to the "modern" explanation in the next section. However, as we now discuss, the ancients took a very different path in their attempts to explain planetary motion.

## A Theoretical Model

The earliest models of the solar system followed the teachings of the Greek philosopher Aristotle (384–322 B.C.) and were **geocentric,** meaning that Earth lay at the center of the universe and all other bodies moved around it. ∞ (Sec. 1.3) The celestial sphere, shown in Figures 1.11 and 1.16, illustrates the basic geocentric view. These models employed what Aristotle, and Plato before him, had taught was the perfect form: the circle. The simplest possible description—uniform motion around a circle with Earth at its center—provided a fairly good approximation to the orbits of the Sun and the Moon, but it could not account for the observed variations in planetary brightness or the retrograde motion of the planets. A more complex model was needed to describe these heavenly "wanderers."

In the first step toward this new model, each planet was taken to move uniformly around a small circle, called an **epicycle,** whose *center* moved uniformly around Earth on a second and larger circle, known as a **deferent** (Figure 2.6). The motion was now composed of two separate circular orbits, creating the possibility that, at some times, the planet's apparent motion could be retrograde. Also, the distance from the planet to Earth would vary, accounting for changes in brightness. By tinkering with the relative sizes of the epicycle and deferent, with the planet's speed on the epicycle, and with the epicycle's speed along the deferent, early astronomers were able to bring this "epicyclic" motion into fairly good agreement with the observed paths of the planets in the sky. Moreover, the model had good predictive power, at least to the accuracy of observations at the time.

However, as the number and the quality of observations increased, it became clear that the simple epicyclic model was not perfect. Small corrections had to be introduced to bring it into line with new observations. The center of the deferents had to be shifted slightly from

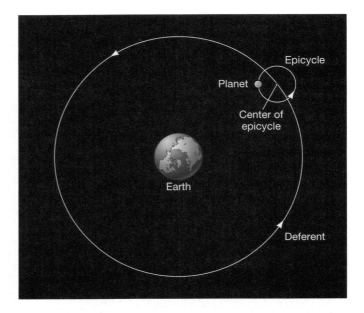

▲ FIGURE 2.6 **Geocentric Model** In the geocentric model of the solar system, the observed motions of the planets made it impossible to assume that they moved on simple circular paths around Earth. Instead, each planet was thought to follow a small circular orbit (the epicycle) about an imaginary point that itself traveled in a large, circular orbit (the deferent) about Earth.

Earth's center, and the motion of the epicycles had to be imagined uniform with respect to yet another point in space, not Earth. Furthermore, in order to explain the motions of the inferior planets, the model simply had to assume that the deferents of Mercury and Venus were, for some (unknown) reason, tied to that of the Sun. Similar assumptions also applied to the superior planets, to ensure that their retrograde motion occurred at opposition.

Around A.D. 140, a Greek astronomer named Ptolemy constructed perhaps the most complete geocentric model of all time. Illustrated in simplified form in Figure 2.7, it explained remarkably well the observed paths of the five planets then known, as well as the paths of the Sun and the Moon. However, to achieve its explanatory and predictive power, the full **Ptolemaic model** required a series of no fewer than 80 distinct circles. To account for the paths of the Sun, Moon, and all eight planets (and their moons) that we know today would require a vastly more complicated set. Nevertheless, Ptolemy's comprehensive text on the topic, *Syntaxis* (better known today by its Arabic name, *Almagest,* "the greatest"), provided the intellectual framework for all discussion of the universe for well over a thousand years.

## Evaluating the Geocentric Model

Today, our scientific training leads us to seek simplicity, because, in the physical sciences, simplicity has so often proved to be an indicator of truth. We would regard the intricacy of a model as complicated as the Ptolemaic system as a clear sign of a fundamentally flawed theory.

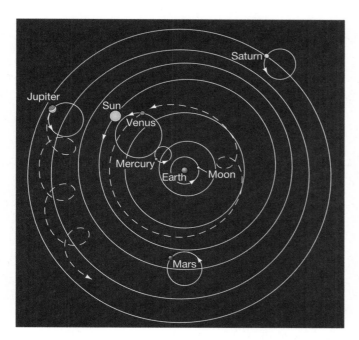

▲ FIGURE 2.7 **Ptolemaic Model** The basic features, drawn roughly to scale, of Ptolemy's geocentric model of the inner solar system, a model that enjoyed widespread popularity prior to the Renaissance. Only the five planets visible to the naked eye and hence known to the ancients—Mercury, Venus, Mars, Jupiter, and Saturn—are shown. The planets' deferents were considered to move on spheres lying within the celestial sphere that held the stars. The celestial sphere carried all interior spheres around with it, but the planetary (and solar) spheres had additional motions of their own, causing the Sun and planets to move relative to the stars. To avoid confusion, partial paths (dashed) of only two planets—Venus and Jupiter—are drawn here.

∞ (Sec. 1.2) Why was the Ptolemaic model so complex? With the benefit of hindsight, we now recognize that its major error lay in its assumption of a geocentric universe. This misconception was compounded by the insistence on uniform circular motion, whose basis was largely philosophical, rather than scientific, in nature.

Actually, history records that some ancient Greek astronomers reasoned differently about the motions of heavenly bodies. Foremost among them was Aristarchus of Samos (310–230 B.C.), who proposed that all the planets, including Earth, revolve around the Sun and, furthermore, that Earth rotates on its axis once each day. This combined revolution and rotation, he argued, would create an *apparent* motion of the sky—a simple idea that is familiar to anyone who has ridden on a merry-go-round and watched the landscape appear to move past in the opposite direction. However, Aristarchus's description of the heavens, though essentially correct, did not gain widespread acceptance during his lifetime. Aristotle's influence was too strong, his followers too numerous, and his writings too comprehensive. The geocentric model went largely unchallenged until the 16th century A.D.

The Aristotelian school did present some simple and (at the time) compelling arguments in favor of their views.

First, of course, Earth doesn't *feel* as if it's moving—and if it were moving, wouldn't there be a strong wind as the planet revolves at high speed around the Sun? Also, considering that the vantage point from which we view the stars changes over the course of a year, why don't we see stellar parallax? ∞ (Sec. 1.4)

Nowadays we might dismiss the first points as merely naive, but the last is a valid argument and the reasoning essentially sound. Indeed, we now know that there *is* stellar parallax as Earth orbits the Sun. However, because the stars are so distant, it amounts to less than 1 arc second (1′), even for the closest stars. Early astronomers simply would not have noticed it. (In fact, stellar parallax was conclusively measured only in the middle of the 19th century.)

We will encounter many other instances in astronomy wherein correct reasoning led to the wrong conclusions because it relied on inadequate data. Even when the scientific method is properly applied and theoretical predictions are tested against reality, a theory can be only as good as the observations on which it is based. ∞ (Sec. 1.2)

## 2.3 The Heliocentric Model of the Solar System

The Ptolemaic picture of the universe survived, more or less intact, for almost 14 centuries, until a 16th-century Polish cleric, Nicolaus Copernicus (Figure 2.8), rediscovered Aristarchus's **heliocentric** (Sun-centered) model and showed how, in its harmony and organization, it provided a more natural explanation of the observed facts than did

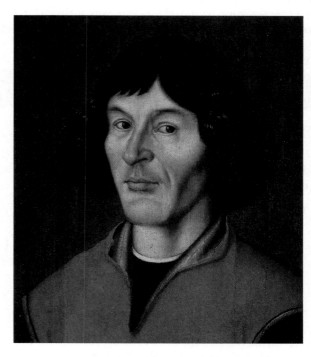

▲ FIGURE 2.8 **Nicolaus Copernicus (1473–1543).** *(E. Lessing/Art Resource, NY)*

the tangled geocentric cosmology. Copernicus asserted that Earth spins on its axis and, like the other planets, orbits the Sun. Only the Moon, he said, orbits Earth. As we will see, not only does this model explain the observed daily and seasonal changes in the heavens, but it also naturally accounts for retrograde motion and variations in brightness of the planets. ∞ (Sec. 1.4)

The critical realization that Earth is not at the center of the universe is now known as the **Copernican revolution.** The seven crucial statements that form its foundation are summarized in *Discovery 2-1.*

Figure 2.9 shows how the Copernican view explains the varying brightness of a planet (in this case, Mars), its observed looping motions, and the fact that the retrograde motion of a superior planet occurs at opposition. If we suppose that Earth moves faster than Mars, then every so often Earth "overtakes" that planet. Mars will then appear to move backward in the sky, in much the same way as a car we overtake on the highway seems to slip backward relative to us. Replace Mars by Earth and Earth by Venus, and you should also be able to extend the explanation to the inferior planets. (To complete the story with a full explanation of their apparent brightnesses, however, you'll have to wait until Section 9.1!) Notice that, in the Copernican picture, the planet's looping motions are only apparent. In the Ptolemaic view, they are real.

Copernicus's major motivation for introducing the heliocentric model was simplicity. Even so, he was still influenced by Greek thinking and clung to the idea of circles to model the planets' motions. To bring his theory into agreement with observations of the night sky, he was forced to retain the idea of epicyclic motion, although with the deferent centered on the Sun rather than on Earth and with smaller epicycles than in the Ptolemaic

picture. Thus, he retained unnecessary complexity and actually gained little in predictive power over the geocentric model. The heliocentric model did rectify some small discrepancies and inconsistencies in the Ptolemaic system, but for Copernicus, the primary attraction of heliocentricity was its simplicity—its being "more pleasing to the mind." His theory was more something he *felt* than he could *prove.* To the present day, scientists still are guided by simplicity, symmetry, and beauty in modeling all aspects of the universe.

Despite the support of some observational data, neither his fellow scholars nor the general public easily accepted Copernicus's model. For the learned, heliocentricity went against the grain of much previous thinking and violated many of the religious teachings of the time, largely because it relegated Earth to a noncentral and undistinguished place within the solar system and the universe. And Copernicus's work had little impact on the general populace of his time, at least in part because it was published in Latin (the standard language of academic discourse at the time), which most people could not read. Only long after Copernicus's death, when others—notably Galileo Galilei—popularized his ideas, did the Roman Catholic Church take them seriously enough to bother banning them. Copernicus's writings on the heliocentric universe were placed on the Church's *Index of Prohibited Books* in 1616, 73 years after they were first published. They remained there until the end of the 18th century.

CONCEPT CHECK

✔ How do the geocentric and heliocentric models of the solar system differ in their explanations of planetary retrograde motion?

# DISCOVERY 2-1

## The Foundations of the Copernican Revolution

The following seven points are essentially Copernicus's own words, with the italicized material providing additional explanation:

1. The celestial spheres do not have just one common center. *Specifically, Earth is not at the center of everything.*

2. The center of Earth is not the center of the universe, but is instead only the center of gravity and of the lunar orbit.

3. All the spheres revolve around the Sun. *By spheres, Copernicus meant the planets.*

4. The ratio of Earth's distance from the Sun to the height of the firmament is so much smaller than the ratio of Earth's radius to the distance to the Sun that the distance to the Sun is imperceptible compared with the height of

the firmament. *By firmament, Copernicus meant the distant stars. The point he was making is that the stars are very much farther away than the Sun.*

5. The motions appearing in the firmament are not its motions, but those of Earth. Earth performs a daily rotation around its fixed poles, while the firmament remains immobile as the highest heaven. *Because the stars are so far away, any apparent motion we see in them is the result of Earth's rotation.*

6. The motions of the Sun are not its motions, but the motion of Earth. *Similarly, the Sun's apparent daily and yearly motion are actually due to the various motions of Earth.*

7. What appears to us as retrograde and forward motion of the planets is not their own, but that of Earth. *The heliocentric picture provides a natural explanation for retrograde planetary motion, again as a consequence of Earth's motion.*

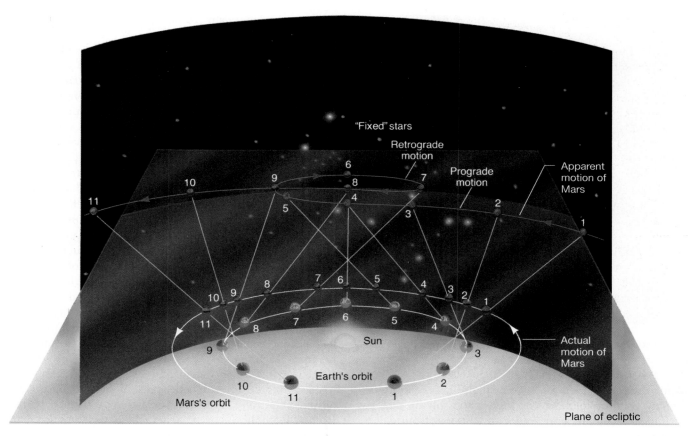

▲ FIGURE 2.9 **Retrograde Motion** The Copernican model of the solar system explains both the varying brightnesses of the planets and the phenomenon of retrograde motion. Here, for example, when Earth and Mars are relatively close to one another in their respective orbits (as at position 6), Mars seems brighter. When they are farther apart (as at position 1), Mars seems dimmer. Also, because the (light blue) line of sight from Earth to Mars changes as the two planets orbit the Sun, Mars appears to loop back and forth in retrograde motion. Follow the lines in numerical order, and note how the line of sight moves backward relative to the stars between locations 5 and 7. The line of sight changes because Earth, on the inside track, moves faster in its orbit than does Mars. The actual planetary orbits are shown as white curves. The apparent motion of Mars, as seen from Earth, is indicated by the red curve.

## 2.4 The Birth of Modern Astronomy

In the century following the death of Copernicus and the publication of his theory of the solar system, two scientists—Galileo Galilei and Johannes Kepler—made indelible imprints on the study of astronomy. Contemporaries, they were aware of each other's work and corresponded from time to time about their theories. Each achieved fame for his discoveries and made great strides in popularizing the Copernican viewpoint, yet in their approaches to astronomy they were as different as night and day.

### Galileo's Historic Observations

Galileo Galilei (Figure 2.10) was an Italian mathematician and philosopher. By his willingness to perform experiments to test his ideas—a rather radical approach in those days—and by embracing the brand-new technology of the telescope, he revolutionized the way in which science was done, so much so that he is now widely regarded as the father of experimental science.

The telescope was invented in Holland in the early 17th century. Hearing of the invention (but without having seen one), Galileo built a telescope for himself in 1609 and aimed it at the sky. What he saw conflicted greatly with the philosophy of Aristotle and provided much new data to support the ideas of Copernicus.*

Using his telescope, Galileo discovered that the Moon had mountains, valleys, and craters—terrain in many ways reminiscent of that on Earth. Looking at the Sun (something that should *never* be done directly and that may have eventually blinded Galileo), he found imperfections—dark

*In fact, Galileo had already abandoned Aristotle in favor of Copernicus, although he had not published his opinions at the time he began his telescopic observations.

▲ FIGURE 2.10 **Galileo Galilei (1564–1642).** *(Art Resource, NY)*

blemishes now known as *sunspots*. These observations ran directly counter to the orthodox wisdom of the day. By noting the changing appearance of sunspots from day to day, Galileo inferred that the Sun *rotates*, approximately once per month, around an axis roughly perpendicular to the ecliptic plane.

Galileo also saw four small points of light, invisible to the naked eye, orbiting the planet Jupiter and realized that they were moons. Figure 2.11 shows some sketches

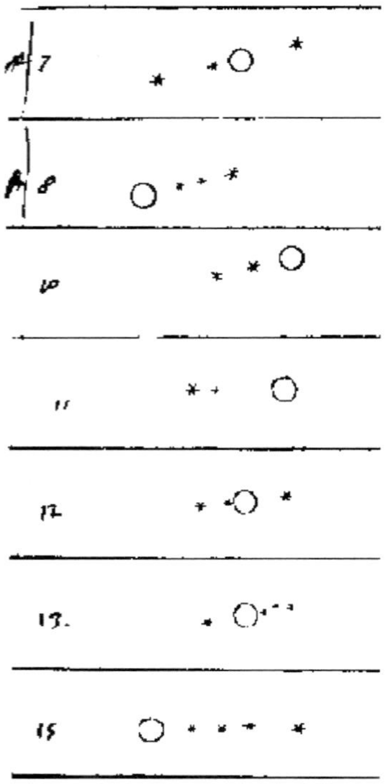

◀ FIGURE 2.11 **Galilean Moons** The four Galilean moons of Jupiter, as sketched by Galileo in his notebook. The sketches show what Galileo saw on seven nights between January 7 and 15, 1610. The orbits of the moons (sketched here as asterisks, and now called Io, Europa, Ganymede, and Callisto) around the planet (open circle) can clearly be seen. More of Galileo's remarkable sketches of Saturn, star clusters, and the Orion constellation can be seen on the first page of Part 1. (From *Sidereus Nuncius*)

of these moons, taken from Galileo's notes. To Galileo, the fact that another planet had moons provided the strongest support for the Copernican model. Clearly, Earth was not the center of all things. He also found that Venus varied in apparent size and showed a complete cycle of phases, like those of our Moon (Figure 2.12), findings that could be explained only by the planet's motion around the Sun. These observations were more strong evidence that Earth is not the center of all things and that at least one planet orbited the Sun. For more of Galileo's sketches, with comparisons to modern photographs, see the Part 1 Opener on p. 1.

In 1610, Galileo published a book called *Sidereus Nuncius* (*The Starry Messenger*), detailing his observational findings and his controversial conclusions supporting the Copernican theory. In reporting and interpreting the wondrous observations made with his new telescope, Galileo was directly challenging both the scientific orthodoxy and the religious dogma of his day. He was (literally) playing with fire—he must certainly have been aware that only a few years earlier, in 1600, the astronomer Giordano Bruno had been burned at the stake in Rome, in part for his heretical teaching that Earth orbited the Sun. However, by all accounts, Galileo delighted in publicly ridiculing and irritating his Aristotelian colleagues. In 1616 his ideas were judged heretical, Copernicus's works were banned by the Roman Catholic Church, and Galileo was instructed to abandon his astronomical pursuits.

But Galileo would not desist. In 1632 he raised the stakes by publishing *Dialogue Concerning the Two Chief World Systems*, which compared the Ptolemaic and Copernican models. The book presented a discussion among three people, one of them a dull-witted Aristotelian whose views (which were in fact the stated opinions of the then Pope, Urban VIII) time and again were roundly defeated by the arguments of one of his two companions, an articulate proponent of the heliocentric system. To make the book accessible to a wide popular audience, Galileo wrote it in Italian rather than Latin. These actions brought Galileo into direct conflict with the authority of the Church. Eventually, the Inquisition forced him, under threat of torture, to retract his claim that Earth orbits the Sun, and he was placed under house arrest in 1633. He remained imprisoned for the rest of his life. Not until 1992 did the Church publicly forgive Galileo's "crimes." But the damage to the orthodox view of the universe was done, and the Copernican genie was out of the bottle once and for all.

## The Ascendancy of the Copernican System

Although Renaissance scholars were correct, they could not *prove* that our planetary system is centered on the Sun or even that Earth moves through space. The observational consequences of Earth's orbital motion were just too small for the technology of the day to detect. Direct evidence of Earth's motion was obtained only in 1728, when English astronomer James Bradley discovered the **aberration of starlight**—a

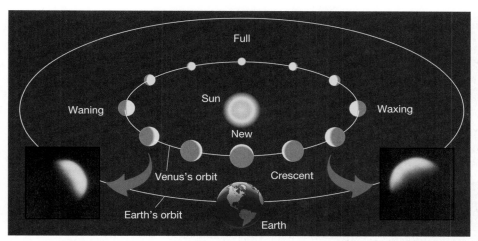

(a) Sun-centered model

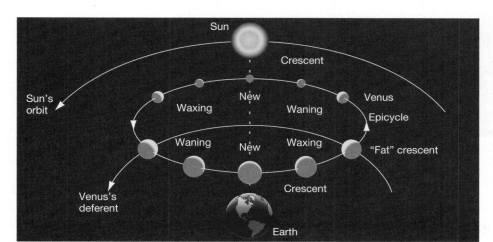

(b) Ptolemy's model

◀ **FIGURE 2.12  Venus Phases**  Both the Ptolemaic and the Copernican models of the solar system predict that Venus should show phases as it moves in its orbit. (a) In the Copernican picture, when Venus is directly between Earth and the Sun, its unlit side faces us and the planet is invisible to us. As Venus moves in its orbit (at a faster speed than Earth moves in its orbit), progressively more of its illuminated face is visible from Earth. Note the connection between the orbital phase and the apparent size of the planet: Venus seems much larger in its crescent phase than when it is full because it is much closer to us during its crescent phase. This is the behavior actually observed. (The insets at bottom left and right are actual photographs of Venus taken at two of its crescent phases.) (b) The Ptolemaic model (see also Figure 2.7) is unable to account for these observations. In particular, the full phase of the planet cannot be explained. Seen from Earth, Venus reaches only a "fat crescent" phase, then begins to wane as it nears the Sun. (Both these views are from a sideways perspective; from overhead, both orbits are very nearly circular, as shown in Figure 2.19.) *(Images from R. Beebe/ Courtesy of New Mexico State University)*

slight (roughly 20″) shift in the observed direction to a star, caused by Earth's motion perpendicular to the line of sight. Figure 2.13 illustrates the phenomenon and compares it with a much more familiar example. Bradley's observation was the first proof that Earth revolves around the Sun; subsequent observations of many stars in many different directions have repeatedly confirmed the effect. Additional proof of Earth's orbital motion came in 1838, with the first unambiguous determination of stellar parallax (see Figure 1.32) by German astronomer Friedrich Bessel. ∞ (Sec. 1.7)

Following those early measurements, support for the heliocentric solar system has grown steadily, as astronomers have subjected the theory to more and more sophisticated observational tests, culminating in the interplanetary expeditions of our unmanned space probes of the 1960s, 1970s, and 1980s. Today, the evidence is overwhelming. The development and eventual acceptance of the heliocentric model were milestones in human thinking. This removal of Earth from any position of great cosmic significance is generally known, even today, as the *Copernican principle*. It has become a cornerstone of modern astrophysics.

The Copernican revolution is a prime example of how the scientific method, though affected at any given time by the subjective whims, human biases, and even sheer luck of researchers, does ultimately lead to a definite degree of objectivity. ∞ (Sec. 1.2) Over time, many groups of scientists checking, confirming, and refining experimental tests can neutralize the subjective attitudes of individuals. Usually, one generation of scientists can bring sufficient objectivity to bear on a problem, although some especially revolutionary concepts are so swamped by tradition, religion, and politics that more time is necessary. In the case of heliocentricity, objective confirmation was not obtained until about three centuries after Copernicus published his work and more than 2000 years after Aristarchus had proposed the concept. Nonetheless, objectivity *did in fact* eventually prevail, and our knowledge of the universe has expanded immeasurably as a result.

CONCEPT CHECK

✔ In what ways did Galileo's observations of Venus and Jupiter conflict with the prevailing view of the time?

(a)

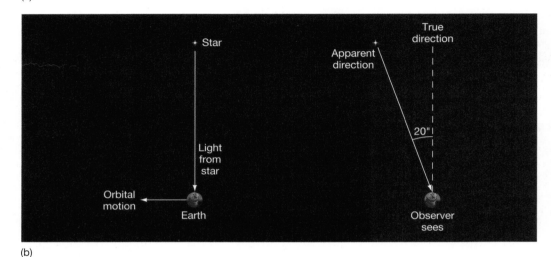

(b)

**◀ FIGURE 2.13 Aberration of Starlight** (a) Rain falling vertically onto the windows of a moving train appears to a passenger on the train to be falling at an angle. The train's motion relative to the raindrops changes the passenger's perception of the direction in which the rain seems to fall. (b) Similarly, because of Earth's motion through space, light from a distant star appears to come from a slightly different direction in the sky than if Earth were at rest. Because the speed of light is large compared with Earth's orbital velocity, the displacement is small—never more than about 20″—but it is easily measurable using a telescope. The maximum displacement occurs when Earth's orbital velocity is exactly perpendicular to the line of sight to the star.

## 2.5  The Laws of Planetary Motion

At about the same time as Galileo was becoming famous—or notorious—for his pioneering telescopic observations and outspoken promotion of the Copernican system, Johannes Kepler (Figure 2.14), a German mathematician and astronomer, was developing the laws of planetary motion that now bear his name. Galileo was in many ways the first "modern" observer. He used emerging technology, in the form of the telescope, to achieve new insights into the universe. In contrast, Kepler was a pure theorist. His groundbreaking work that so clarified our knowledge of planetary motion was based almost entirely on the observations of others, principally an extensive collection of data compiled by Tycho Brahe (1546–1601), Kepler's employer and arguably one of the greatest observational astronomers that has ever lived.

### Brahe's Complex Data

Tycho, as he is often called, was both an eccentric aristocrat and a skillful observer. Born in Denmark, he was educated at some of the best universities in Europe, where he studied astrology, alchemy, and medicine. Most of his observations, which predated the invention of the telescope

**▲ FIGURE 2.14  Johannes Kepler (1571–1630).** *(E. Lessing/Art Resource, NY)*

by several decades, were made at his own observatory, named *Uraniborg*, in Denmark (Figure 2.15). There, using instruments of his own design, Tycho maintained meticulous and accurate records of the stars, planets, and other noteworthy celestial events, including a comet and a supernova (see Chapter 21), the appearance of which helped convince him that the Aristotelean view of the universe could not be correct.

In 1597, having fallen out of favor with the Danish court, Tycho moved to Prague as Imperial Mathematician of the Holy Roman Empire. Prague happens to be fairly close to Graz, in Austria, where Kepler lived and worked. Kepler joined Brahe in Prague in 1600 and was put to work trying to find a theory that could explain Brahe's planetary data. When Tycho died a year later, Kepler inherited not only his position, but also his most priceless possession: the accumulated observations of the planets, spanning several decades. Tycho's observations, though made with the naked eye, were nevertheless of very high quality. In most cases, his measured positions of stars and planets were accurate to within about 1'. Kepler set to work seeking a unifying principle to explain in detail the motions of the planets, without the need for epicycles. The effort was to occupy much of the remaining 29 years of his life.

Kepler had already accepted the heliocentric picture of the solar system. His goal was to find a simple and elegant description of planetary motion within the Copernican framework that fit Brahe's complex mass of detailed observations. In the end, he found it necessary to abandon Copernicus's simple idea of circular planetary orbits. However, even greater simplicity emerged as a result. After long years of studying Brahe's planetary data, and after many false starts and blind alleys, Kepler developed the laws that now bear his name.

Kepler determined the shape of each planet's orbit by triangulation—not from different points on Earth, but from different points on Earth's orbit, using observations made at many different times of the year. ∞ (Sec. 1.7) By using a portion of Earth's orbit as a baseline for his triangle, Kepler was able to measure the relative sizes of the other planetary orbits. Noting where the planets were on successive nights, he found the speeds at which the planets move. We do not know how many geometric shapes Kepler tried for the orbits before he hit upon the correct one. His difficult task was made even more complicated because he had to determine Earth's own orbit, too. Nevertheless, he eventually succeeded in summarizing the motions of all the known planets, including Earth, in just three laws: the **laws of planetary motion.**

## Kepler's Simple Laws

Kepler's first law of planetary motion has to do with the *shapes* of the planetary orbits:

> The orbital paths of the planets are elliptical (*not* circular), with the Sun at one focus.

An **ellipse** is simply a flattened circle. Figure 2.16 illustrates a means of constructing an ellipse with a piece of string and two thumbtacks. Each point at which the string is pinned is called a **focus** (plural: *foci*) of the ellipse. The long axis of the ellipse, containing the two foci, is known as the *major axis*. Half the length of this long axis is referred to as the **semimajor axis,** a measure of the ellipse's size. A circle is a special case in which the two foci happen to coincide; its semimajor axis is simply its radius.

The **eccentricity** of an ellipse is simply a measure of how flattened it is. Technically, eccentricity is defined as the ratio of the distance between the foci to the length of the major axis, but the most important thing to remember here is that an eccentricity of zero corresponds to no flattening—a perfect circle—whereas an eccentricity of one means that the circle has been squashed all the way down

▲ FIGURE 2.15 **Tycho Brahe** The astronomer in his observatory Uraniborg, on the island of Hveen in Denmark. Brahe's observations of the positions of stars and planets on the sky were the most accurate and complete set of naked-eye measurements ever made. *(Royal Ontario Museum)*

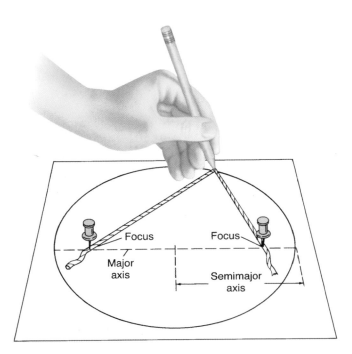

▲ FIGURE 2.16 **Ellipse** An ellipse can be drawn with the aid of a string, a pencil, and two thumbtacks. The wider the separation of the foci, the more elongated, or eccentric, is the ellipse. In the special case where the two foci are at the same place, the curve drawn is a circle.

to a straight line. Note that, while the Sun resides at one focus of the elliptical orbit, the other focus is empty and has no particular physical significance. (However, we can still figure out where it is, because the two foci are symmetrically placed about the center, along the major axis.)

The length of the semimajor axis and the eccentricity are all we need to describe the size and shape of a planet's orbital path (see *More Precisely 2-1*). In fact, no planet's elliptical orbit is nearly as elongated as the one shown in Figure 2.16. With one exception (the orbit of Mercury), planetary orbits in our solar system have such small eccentricities that our eyes would have trouble distinguishing them from true circles. Only because the orbits are so nearly circular were the Ptolemaic and Copernican models able to come as close as they did to describing reality.

Kepler's substitution of elliptical for circular orbits was no small advance. It amounted to abandoning an aesthetic bias—the Aristotelian belief in the perfection of the circle—that had governed astronomy since Greek antiquity. Even Galileo Galilei, not known for his conservatism in scholarly matters, clung to the idea of circular motion and never accepted the notion that the planets move in elliptical paths.

The second law, illustrated in Figure 2.17, addresses the *speed* at which a planet traverses different parts of its orbit:

An imaginary line connecting the Sun to any planet sweeps out equal areas of the ellipse in equal intervals of time.

While orbiting the Sun, a planet traces the arcs labeled A, B, and C in the figure in equal times. Notice, however, that the distance traveled by the planet along arc C is greater than the distance traveled along arc A or arc B. Because the time is the same and the distance is different, the speed must vary. When a planet is close to the Sun, as in sector C, it moves much faster than when farther away, as in sector A.

By taking into account the relative speeds and positions of the planets in their elliptical orbits about the Sun, Kepler's first two laws explained the variations in planetary brightness and some observed peculiar nonuniform motions that could not be accommodated within the assumption of circular motion, even with the inclusion of epicycles. Gone at last were the circles within circles that rolled across the sky. Kepler's modification of the Copernican theory to allow the possibility of elliptical orbits both greatly simplified the model of the solar system and at the same time provided much greater predictive accuracy than had previously been possible. Note, too, that these laws are not restricted to planets. They apply to *any* orbiting object. Spy satellites, for example, move very rapidly as they swoop close to Earth's surface, not because they are propelled with powerful onboard rockets, but because their highly eccentric orbits are governed by Kepler's laws.

Kepler published his first two laws in 1609, stating that he had proved them only for the orbit of Mars. Ten years later, he extended them to all the then-known planets (Mercury, Venus, Earth, Mars, Jupiter, and Saturn) and added a third law relating the size of a planet's orbit to its sidereal orbital **period**—the time needed for the planet to complete one circuit around the Sun:

The square of a planet's orbital period is proportional to the cube of its semimajor axis.

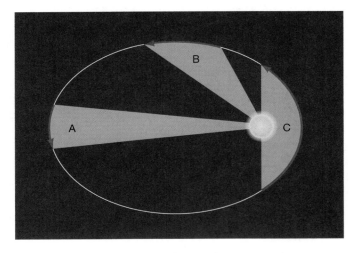

▲ FIGURE 2.17 **Kepler's Second Law** A line joining a planet to the Sun sweeps out equal areas in equal intervals of time. The three shaded areas A, B, and C are equal. Any object traveling along the elliptical path would take the same amount of time to cover the distance indicated by the three red arrows. Therefore, planets move faster when closer to the Sun.

# MORE PRECISELY 2-1

## Some Properties of Planetary Orbits

Two numbers—*semimajor axis* and *eccentricity*—are all that are needed to describe the size and shape of a planet's orbital path. From them, we can derive many other useful quantities. Two of the most important are the planet's *perihelion* (its point of closest approach to the Sun) and its *aphelion* (point of greatest distance from the Sun). From the definitions presented in the text, it follows that if the planet's orbit has semimajor axis $a$ and eccentricity $e$, the planet's perihelion is at a distance $a(1 - e)$ from the Sun, while its aphelion is at $a(1 + e)$. These points and distances are illustrated in the following figure:

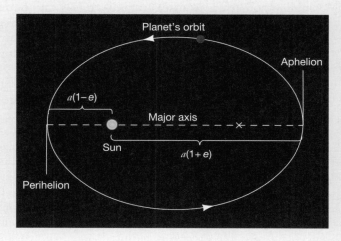

**EXAMPLE 1** We can locate the other focus of the ellipse in the diagram and hence determine the eccentricity from the definition in the text, quite simply. The second focus is placed symmetrically along the major axis, at the point marked with an "X." With a ruler, measure (1) the length of the major axis and (2) the distance between the two foci. Dividing the second distance by the first, you should find an eccentricity of 3.4 cm/6.8 cm = 0.5. Alternatively, we could use the formula given for the perihelion. Measure the perihelion distance to be $a (1 - e) = 1.7$ cm. Dividing this by $a = 3.4$ cm, we obtain $1 - e = 0.5$, so once again, $e = 0.5$.

**EXAMPLE 2** A (hypothetical) planet with a semimajor axis of 400 million km and an eccentricity of 0.5 (i.e., with an orbit as shown in the diagram) would range between 400 × (1 − 0.5) = 200 million km and 400 × (1 + 0.5) = 600 million km from the Sun over the course of one complete orbit. With $e = 0.9$, the range in distances would be 40 to 760 million km, and so on.

No planet has an orbital eccentricity as large as 0.5—the planet with the most eccentric orbit is Mercury, with $e = 0.206$ (see Table 2.1). However, many meteoroids and all comets (see Chapter 14) have eccentricities considerably greater than that. In fact, most comets visible from Earth have eccentricities very close to $e = 1$. Their highly elongated orbits approach within a few astronomical units of the Sun at perihelion, yet these tiny frozen worlds spend most of their time far beyond the orbit of Pluto.

This law becomes particularly simple when we choose the (Earth sidereal) year as our unit of time and the *astronomical unit* as our unit of length. One **astronomical unit** (AU) is the semimajor axis of Earth's orbit around the Sun—essentially the average distance between Earth and the Sun. Like the light-year, the astronomical unit is custom made for the vast distances encountered in astronomy. Using these units for time and distance, we can rewrite Kepler's third law for any planet as

$$P^2 \text{ (in Earth years)} = a^3 \text{ (in astronomical units)},$$

where $P$ is the planet's sidereal orbital period and $a$ is the length of its semimajor axis. The law implies that a planet's "year" $P$ increases more rapidly than does the size of its orbit, $a$. For example, Earth, with an orbital semimajor axis of 1 AU, has an orbital period of 1 Earth year. The planet Venus, orbiting at a distance of roughly 0.7 AU, takes only 0.6 Earth year—about 225 days—to complete one circuit. By contrast, Saturn, almost 10 AU from the Sun, takes considerably more than 10 Earth years—in fact, nearly 30 years—to orbit the Sun just once.

Table 2.1 presents basic data describing the orbits of the eight planets now known. Renaissance astronomers knew these properties for the innermost six planets and used them to construct the currently accepted heliocentric model of the solar system. The second column presents each planet's orbital semimajor axis, measured in astronomical units; the third column gives the orbital period, in Earth years. The fourth column lists the planets' orbital eccentricities. For purposes of verifying

| TABLE 2.1 | Some Solar System Dimensions | | |
|---|---|---|---|
| **Planet** | **Orbital Semimajor Axis, $a$ (AU)** | **Orbital Period, $P$ (years)** | **Orbital Eccentricity, $e$** | **$P^2/a^3$** |
| Mercury | 0.387 | 0.241 | 0.206 | 1.002 |
| Venus | 0.723 | 0.615 | 0.007 | 1.001 |
| Earth | 1.000 | 1.000 | 0.017 | 1.000 |
| Mars | 1.524 | 1.881 | 0.093 | 1.000 |
| Jupiter | 5.203 | 11.86 | 0.048 | 0.999 |
| Saturn | 9.537 | 29.42 | 0.054 | 0.998 |
| Uranus | 19.19 | 83.75 | 0.047 | 0.993 |
| Neptune | 30.07 | 163.7 | 0.009 | 0.986 |

Kepler's third law, the fifth column lists the ratio $P^2/a^3$. As we have just seen, the third law implies that this number should always be unity in the units used in the table.

The main points to be grasped from Table 2.1 are these: (1) With the exception of Mercury, the planets' orbits are nearly circular (i.e., their eccentricities are close to zero); and (2) the farther a planet is from the Sun, the greater is its orbital period, in agreement with Kepler's third law to within the accuracy of the numbers in the table. (The small, but significant, deviations of $P^2/a^3$ from unity in the cases of Uranus and Neptune are caused by the gravitational attraction between those two planets; see Chapter 13.) Most important, note that Kepler's laws are obeyed by *all* the known planets, *not just by the six on which he based his conclusions.*

The laws developed by Kepler were far more than mere fits to existing data. They also made definite, testable predictions about the future locations of the planets. Those predictions have been borne out to high accuracy every time they have been tested by observation—the hallmark of any credible scientific theory. ∞ (Sec. 1.2)

CONCEPT CHECK

✔ Why is it significant that Kepler's laws also apply to Uranus and Neptune?

## 2.6 The Dimensions of the Solar System

Kepler's laws allow us to construct a scale model of the solar system, with the correct shapes and *relative* sizes of all the planetary orbits, but they do not tell us the *actual* size of any orbit. We can express the distance to each planet only in terms of the distance from Earth to the Sun. Why is this? Because Kepler's triangulation measurements all used a portion of Earth's orbit as a baseline, distances could be expressed only relative to the size of that orbit, which was itself not determined by the method. Thus, our model of the solar system would be like a road map of the United States showing the *relative* positions of cities and towns, but lacking the all-important scale marker indicating distances in kilometers or miles. For example, we would know that Kansas City is about three times more distant from New York than it is from Chicago, but we would not know the actual mileage between any two points on the map.

If we could somehow determine the value of the astronomical unit—in kilometers, say—we would be able to add the vital scale marker to our map of the solar system and compute the precise distances between the Sun and each of the planets. We might propose using triangulation to measure the distance from Earth to the Sun directly. However, we would find it impossible to measure the Sun's parallax using Earth's diameter as a baseline. The Sun is too bright, too big, and too fuzzy for us to distinguish any apparent displacement relative to a field of distant stars. To measure the Sun's distance from Earth, we must resort to some other method.

Before the middle of the 20th century, the most accurate measurements of the astronomical unit were made by using triangulation on the planets Mercury and Venus during their rare *transits* of the Sun—that is, during the brief periods when those planets passed directly between the Sun and Earth (as shown for the case of Mercury in Figure 2.18). Because the time at which a transit occurs can be measured with great precision, astronomers can use this information to make accurate measurements of a planet's position in the sky. They can then employ simple geometry to compute the distance to the planet by combining observations made from different locations on Earth, as discussed earlier in Chapter 1. ∞ (Sec. 1.7) For example, the parallax of Venus at closest approach to Earth, as seen from two diametrically opposite points on Earth (separated by about 13,000 km), is about $1'$ (1/60°)—at the limit of naked-eye capabilities, but easily measurable telescopically. Using the second formula presented in *More Precisely 1-3*, we find that this parallax represents a distance of 13,000 km $\times$ 57.3°/(1/60°), or approximately 45,000,000 km.

Knowing the distance to Venus, we can compute the magnitude of the astronomical unit. Figure 2.19 is an idealized diagram of the Sun–Earth–Venus orbital geometry. The planetary orbits are drawn as circles here, but in reality they are slight ellipses. This is a subtle difference, and we can correct for it using detailed knowledge of orbital motions. Assuming for the sake of simplicity

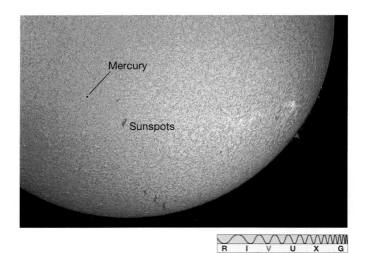

▲ FIGURE 2.18 **Solar Transit** The transit of Mercury across the face of the Sun. Such transits happen only about once per decade because Mercury's orbit does not quite coincide with the plane of the ecliptic. Transits of Venus are even rarer, occurring only about twice per century. The most recent took place in 2006. (*P. Jones*)

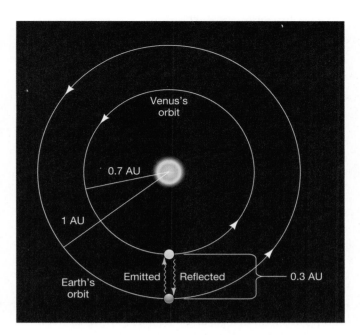

**▲ FIGURE 2.19 Astronomical Unit** Simplified geometry of the orbits of Earth and Venus as they move around the Sun. The wavy blue lines represent the paths along which radar signals are transmitted toward Venus and received back at Earth at the particular moment (chosen for simplicity) when Venus is at its minimum distance from Earth. Because the radius of Earth's orbit is 1 AU and that of Venus is about 0.7 AU, we know that this distance is 0.3 AU. Thus, radar measurements allow us to determine the astronomical unit in kilometers.

The round-trip travel time (for example, at closest approach, as indicated by the wavy lines in Figure 2.19) can be measured with high precision—in fact, well enough to determine the planet's distance to an accuracy of about 1 km. In this way, the astronomical unit is now known to be 149,597,870 km. We will use the rounded-off value of $1.5 \times 10^8$ km in this text.

Having determined the value of the astronomical unit, we can reexpress the sizes of the other planetary orbits in terms of more familiar units, such as miles or kilometers. The entire scale of the solar system can then be calibrated to high precision.

## CONCEPT CHECK

✔ Why don't Kepler's laws tell us the value of the astronomical unit?

## 2.7 Newton's Laws

Kepler's three laws, which so simplified the solar system, were discovered *empirically*. In other words, they resulted solely from the analysis of observational data and were not derived from any theory or mathematical model. Indeed, Kepler did not have any appreciation of the physics underlying his laws. Nor did Copernicus understand *why* his heliocentric model of the solar system worked. Even Galileo, often called the father of modern physics, failed to understand why the planets orbit the Sun (although Galileo's work laid vital groundwork for Newton's theories).

What prevents the planets from flying off into space or from falling into the Sun? What causes them to revolve about the Sun, apparently endlessly? To be sure, the motions of the planets obey Kepler's three laws, but only by considering something more fundamental than those laws can we really understand planetary motion. The heliocentric system was secured when, in the 17th century, the British mathematician Isaac Newton (Figure 2.20) developed a deeper understanding of the way *all* objects move and interact with one another.

### The Laws of Motion

Isaac Newton was born in Lincolnshire, England, on Christmas Day in 1642, the year Galileo died. Newton studied at Trinity College of Cambridge University, but when the bubonic plague reached Cambridge in 1665, he returned to the relative safety of his home for 2 years. During that time he made probably the most famous of his discoveries, the law of gravity (although it is but one of the many major scientific advances for which Newton was responsible). However, either because he regarded the theory as incomplete or possibly because he was afraid that he would be attacked or plagiarized by his colleagues, he

that the orbits are perfect circles, we see from the figure that the distance from Earth to Venus at closest approach is approximately 0.3 AU. Knowing that 0.3 AU is 45,000,000 km makes determining 1 AU straightforward—the answer is 45,000,000/0.3, or 150,000,000 km.

The modern method for deriving the absolute scale (that is, the scale expressed in kilometers, rather than just relative to Earth's orbit) of the solar system uses radar rather than triangulation. The word **radar** is an acronym for **ra**dio **d**etection **a**nd **r**anging. In this technique, radio waves are transmitted toward an astronomical body, such as a planet. (We cannot use radar ranging to measure the distance to the Sun directly, because radio signals are absorbed at the solar surface and are not reflected to Earth.) The returning echo indicates the body's direction and range, or distance, in absolute terms—that is, in kilometers rather than in astronomical units. Multiplying the 300-second round-trip travel time of the radar signal (the time elapsed between transmission of the signal and reception of the echo) by the speed of light (300,000 km/s, which is also the speed of radio waves), we obtain twice the distance to the target planet.

Venus, whose orbit periodically brings it closest to Earth, is the most common target for radar ranging.

▲ FIGURE 2.20 **Isaac Newton (1642–1727).** *(The Granger Collection)*

The first law simply states that a moving object will move forever in a straight line, unless some external **force**—a push or a pull—changes its speed or direction of motion. For example, the object might glance off a brick wall or be hit with a baseball bat; in either case, a force changes the original motion of the object. Another example of a force, well known to most of us, is **weight**—the force (commonly measured in pounds in the United States) with which gravity pulls you toward Earth's center.

The tendency of an object to keep moving at the same speed and in the same direction unless acted upon by a force is known as **inertia.** Newton's first law implies that it requires no force to maintain motion in a straight line with constant speed. This contrasts sharply with the view of Aristotle, who maintained (incorrectly) that the natural state of an object was to be *at rest*—most probably an opinion based on Aristotle's observations of the effect of friction. In our discussion, we will neglect friction—the force that slows balls rolling along the ground, blocks sliding across tabletops, and baseballs moving through the air. In any case, it is not an issue for the planets because there is no appreciable friction in outer space—there is no air or any other matter to impede a planet's motion. The fallacy in Aristotle's argument was first realized and exposed by Galileo, who conceived of the notion of inertia long before Newton formalized it into a law.

A familiar measure of an object's inertia is its **mass**—loosely speaking, the total amount of matter the object contains. The greater an object's mass, the more inertia it has, and the greater is the force needed to change its state of motion.

Newton's first law describes motion in a straight line with constant speed—that is, motion with constant **velocity.** An object's velocity includes both its speed (in miles per hour or meters per second, say) *and* its direction

did not tell anyone of his monumental achievement for almost 20 years. It was not until 1684, when Newton was discussing the leading astronomical problem of the day—*Why* do the planets move according to Kepler's laws?—with Edmund Halley (of Halley's comet fame) that he astounded his companion by revealing that he had solved the problem in its entirety nearly two decades before!

Prompted by Halley, Newton published his theories in perhaps the most influential physics book ever written: *Philosophiae Naturalis Principia Mathematica* (*The Mathematical Principles of Natural Philosophy*—what we would today call "science"), usually known simply as Newton's *Principia.* The ideas expressed in that work form the basis for what is now known as **Newtonian mechanics.** Three basic laws of motion, the law of gravity, and a little calculus (which Newton also developed) are sufficient to explain and quantify virtually all the complex dynamic behavior we see on Earth and throughout the universe.

Figure 2.21 illustrates Newton's first law of motion:

Every body continues in a state of rest or in a state of uniform motion in a straight line, unless it is compelled to change that state by a force acting on it.

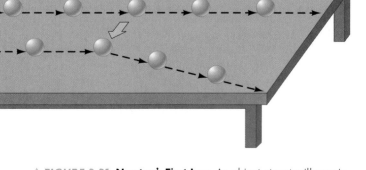

▲ FIGURE 2.21 **Newton's First Law** An object at rest will remain at rest (a) until some force acts on it. When a force (represented by the red arrow) does act (b), the object will remain in that state of uniform motion until another force acts on it. When a second force (green arrow) acts in a direction different from the first force (c), the object changes its direction of motion.

in space (up, down, northwest, and so on). In everyday speech, we tend to use the terms "speed" and "velocity" more or less interchangeably, but we must realize that they are actually different quantities and that Newton's laws are always stated in terms of the latter. As a specific illustration of the difference, consider a rock tied to a string, moving at a constant rate in a circle as you whirl it around your head. The rock's *speed* is constant, but its *direction* of motion, and hence its velocity, is continually changing. Thus, according to Newton's first law, a force must be acting. That force is the tension you feel in the string. In a moment, we'll see reasoning similar to this applied to the problem of planetary motion.

The rate of change of the velocity of an object—speeding up, slowing down, or simply changing direction—is called the object's **acceleration** and is the subject of Newton's second law, which states that the acceleration of an object is directly proportional to the applied force and inversely proportional to its mass:

> When a force $F$ acts on a body of mass $m$, it produces in it an acceleration $a$ equal to the force divided by the mass. Thus, $a = F/m$, or $F = ma$.

Hence, the greater the force acting on the object or the smaller the mass of the object, the greater is the acceleration of the object. If two objects are pulled with the same force, the more massive one will accelerate less; if two identical objects are pulled with different forces, the one acted on by the greater force will accelerate more.

Acceleration is the rate of change of velocity, so its units are velocity units per unit of time, such as meters per second *per second* (usually written as m/s$^2$). In honor of Newton, the SI unit of force is named after him. By definition, 1 newton (N) is the force required to cause a mass of 1 kilogram to accelerate at a rate of 1 meter per second every second (1 m/s$^2$). One newton is approximately 0.22 pound.

At Earth's surface, the force of gravity produces a downward acceleration of approximately 9.8 m/s$^2$ on *all* bodies, regardless of mass. According to Newton's second law, this means that your weight (in newtons) is directly proportional to your mass (in kilograms). We will return to this very important point later.

Finally, Newton's third law simply tells us that forces cannot occur in isolation:

> To every action, there is an equal and opposite reaction.

In other words, if body $A$ exerts a force on body $B$, then body $B$ necessarily exerts a force on body $A$ that is equal in magnitude, but oppositely directed. For example, when a baseball player hits a home run, Newton's third law says that the bat and the ball exert equal and opposite forces on one another during the instant they are in contact. According to the second law, the ball subsequently moves away much faster than the bat because the *mass* of the ball is much less than the combined mass of the bat plus batter (whose body absorbs much of the reaction force), so the ball's *acceleration* is much greater.

Only in extreme circumstances—when speeds approach the speed of light—do Newton's laws break down, and this fact was not realized until the 20th century, when Albert Einstein's theories of relativity once again revolutionized our view of the universe (see Chapter 22). Most of the time, however, Newtonian mechanics provides an excellent description of the motions of planets, stars, and galaxies through the cosmos.

## Gravity

Forces may act *instantaneously* or *continuously*. The force from the baseball bat that hits the home run can reasonably be thought of as being instantaneous. A good example of a continuous force is the one that prevents the baseball from zooming off into space—**gravity,** the phenomenon that started Newton on the path to the discovery of his laws. Newton hypothesized that any object having mass always exerts an attractive **gravitational force** on all other massive objects. The more massive an object, the stronger is its gravitational pull.

Consider a baseball thrown upward from Earth's surface, as illustrated in Figure 2.22. In accordance with Newton's first law, the downward force of Earth's gravity

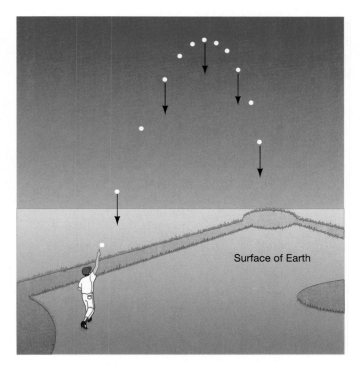

Surface of Earth

▲ FIGURE 2.22 **Gravity** A ball thrown up from the surface of a massive object, such as a planet, is pulled continuously downward (arrows) by the gravity of that planet (and, conversely, the gravity of the ball continuously pulls the planet).

continuously modifies the baseball's velocity, slowing the initial upward motion and eventually causing the ball to fall back to the ground. Of course, the baseball, having some mass of its own, also exerts a gravitational pull on Earth. By Newton's third law, this force is equal and opposite to the weight of the ball (the force with which Earth attracts it). But, by Newton's second law, Earth has a much greater effect on the light baseball than the baseball has on the much more massive Earth. The ball and Earth act upon each other with the same gravitational force, but Earth's acceleration is much smaller.

Now consider the trajectory of the same baseball batted from the surface of the Moon. The pull of gravity is about one-sixth as great on the Moon as on Earth, so the

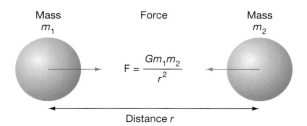

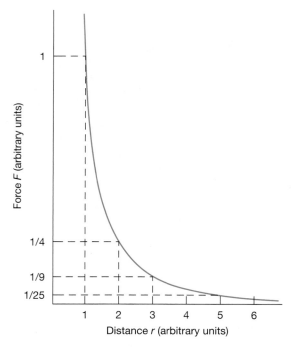

▲ **FIGURE 2.23 Gravitational Force** (a) The gravitational force between two bodies is proportional to the mass of each and is inversely proportional to the square of the distance between them (b). Inverse-square forces rapidly weaken with distance from their source. The strength of the Sun's gravitational force decreases with the square of the distance from the Sun, but never quite reaches zero, no matter how far away from the Sun.

baseball's velocity changes more slowly—a typical home run in a ballpark on Earth would travel nearly half a mile on the Moon. Less massive than Earth, the Moon has less gravitational influence on the baseball. The magnitude of the gravitational force, then, depends on the *masses* of the attracting bodies. In fact, the force is *directly proportional* to the product of the two masses.

Studying the motions of the planets uncovers a second aspect of the gravitational force. At locations equidistant from the Sun's center, the gravitational force has the same strength and is always directed toward the Sun. Furthermore, detailed calculation of the planets' accelerations as they orbit the Sun reveals that the strength of the Sun's gravitational pull decreases in proportion to the *square* of the distance from the Sun. The force of gravity is said to obey an **inverse-square law.** (See *More Precisely 2-2* for more on the line of reasoning that originally led Newton to this conclusion.) As shown in Figure 2.23, inverse-square forces decrease rapidly with distance from their source. For example, tripling the distance makes the force $3^2 = 9$ times weaker, whereas multiplying the distance by five results in a force that is $5^2 = 25$ times weaker. Despite this rapid decrease, the force never quite reaches zero. The gravitational pull of an object having some mass can never be completely extinguished.

We can combine the preceding statements about mass and distance to form a law of gravity that dictates the way in which *all* massive objects (i.e., objects having some mass) attract one another:

> Every particle of matter in the universe attracts every other particle with a force that is directly proportional to the product of the masses of the particles and inversely proportional to the square of the distance between their centers.

As a proportionality, the law of gravity may be written as

$$\text{gravitational force} \propto \frac{\text{mass of object 1} \times \text{mass of object 2}}{\text{distance}^2}$$

(The symbol $\propto$ here means "is proportional to.") The rule for computing the force $F$ between two bodies of masses $m_1$ and $m_2$, separated by a distance $r$, is usually written more compactly as

$$F = \frac{Gm_1m_2}{r^2}.$$

The quantity $G$ is known as the *gravitational constant*, or, often, simply as Newton's constant. It is one of the fundamental constants of the universe. The value of $G$ has been measured in extremely delicate laboratory experiments as $6.67 \times 10^{-11}$ newton meter$^2$/kilogram$^2$ (N·m$^2$/kg$^2$).

# MORE PRECISELY 2-2

## The Moon Is Falling!

The story of Isaac Newton seeing an apple fall to the ground and "discovering" gravity is well known, in one form or another, to most high school students. However, the real importance of Newton's observation was his realization that, by observing falling bodies on Earth and elsewhere, he could *quantify* the properties of the gravitational force and deduce the mathematical form of his law of gravitation.

Galileo Galilei had demonstrated some years earlier, by the simple experiment of dropping different objects from a great height (the top of the Tower of Pisa, according to lore) and noting that they hit the ground at the same time (at least, to the extent that air resistance was unimportant), that gravity causes the *same* acceleration in all bodies, regardless of mass. Since acceleration is proportional to force divided by mass (Newton's second law), this meant that the gravitational force on one body due to another had to be directly proportional to the first body's mass. Applying the same reasoning to the other body and using Newton's third law, we find that the force must also be proportional to the mass of the second body. This is the origin of the two "mass" terms in the law of gravity on p. 54. (The experimental finding that the gravitational force is precisely proportional to mass is now known as the *equivalence principle*. This principle forms an essential part of the modern theory of gravity; see Section 22.6).

What about the inverse-square part of Newton's law? At Earth's surface, the acceleration due to gravity, denoted by the letter *g*, is approximately 9.80 m/s². Where else other than on Earth could a falling body be seen? As illustrated in the accompanying figures, Newton realized that he could tell how gravity varies with distance by studying another object influenced by our planet's gravity: the Moon. Here's how he did it:

Let's assume for the sake of simplicity that the Moon's orbit around Earth is circular. As shown in the second figure, even though the Moon's orbital *speed* is constant, its *velocity* (red arrows) is not—the direction of the Moon's motion is steadily changing. In other words, the Moon is *accelerating*, constantly falling toward Earth. In fact, the acceleration of any body moving with speed *v* in a circular orbit of radius *r* may be shown to be

$$a = \frac{v^2}{r},$$

always directed toward the center of the circle (toward Earth in the case of the Moon). This acceleration is sometimes called *centripetal* ("center-seeking") acceleration. You probably already have an intuitive experience of this kind of acceleration—just think of the acceleration you feel as you take a tight corner (small *r*) at high speed (large *v*) in your car.

**EXAMPLE** Knowing the Moon's distance $r = 384,000$ km (measured by triangulation) and the Moon's sidereal orbit period $P = 27.3$ days, Newton computed the Moon's orbital speed to be (distance $2\pi r$ in time *P*) $v = 2\pi r/P = 1.02$ km/s and hence determined its acceleration to be $a = 0.00272$ m/s², or 0.000278 g. ∞ (*More Precisely 1-3*)

Thus Newton found the Moon, lying 60 times farther from Earth's center than the apple falling from the tree in his garden (taking Earth's radius to be 6400 km), experiences an acceleration 3600, or $60^2$, times smaller. In other words, *the acceleration due to gravity is inversely proportional to the square of the distance*. Isaac Newton's application of simple geometric reasoning and some very basic laws of motion resulted in a breakthrough that would revolutionize astronomers' view of the solar system and ultimately, through its influence on spaceflight and interplanetary navigation, pave the way for humanity's exploration of the universe.

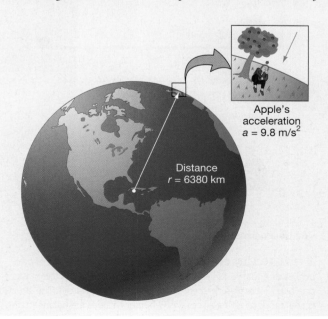

Apple's acceleration
a = 9.8 m/s²

Distance
r = 6380 km

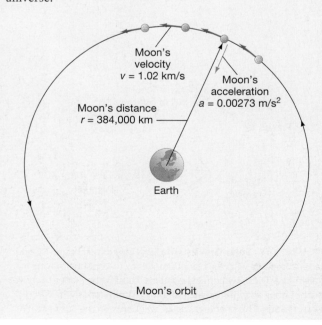

Moon's velocity
v = 1.02 km/s

Moon's distance
r = 384,000 km

Moon's acceleration
a = 0.00273 m/s²

Earth

Moon's orbit

## 2.8 Newtonian Mechanics

Newton's three laws of motion and the law of gravitation provide a solid theoretical foundation upon which we can base a deeper understanding of planetary orbits, the laws of planetary motion, and many other important aspects of orbital motion. With the development of Newtonian mechanics, the transition from geocentric lore to heliocentric fact was complete.

### Planetary Motion

The mutual gravitational attraction between the Sun and the planets, as expressed by Newton's law of gravity, is responsible for the observed planetary orbits. As depicted in Figure 2.24, this gravitational force continuously pulls each planet toward the Sun, deflecting its forward motion into a curved orbital path. Because the Sun is much more massive than any of the planets, it dominates the interaction. We might say that the Sun "controls" the planets, not the other way around.

The Sun–planet interaction sketched here is analogous to our earlier example of the rock whirling on a string. The Sun's gravitational pull is your hand and the string, and the planet is the rock at the end of that string. The tension in the string provides the force necessary for the rock to move in a circular path. If you were suddenly to release the string—which would be like eliminating the Sun's gravity—the rock would fly away along a tangent to the circle, in accordance with Newton's first law.

In the solar system, at this very moment, Earth is moving under the combined influence of gravity and inertia. The net result is a stable orbit, despite our continuous rapid

motion through space. (In fact, Earth orbits the Sun at a speed of about 30 km/s, or approximately 70,000 mph. You can verify this for yourself by calculating how fast Earth must move to complete a circle of radius 1 AU—and hence of circumference $2\pi$ AU, or 940 million km—in 1 year, or $3.2 \times 10^7$ seconds. The answer is $9.4 \times 10^8$ km/$3.2 \times 10^7$ s, or 29.4 km/s.) *More Precisely 2-3* describes how astronomers can use Newtonian mechanics and the law of gravity to quantify planetary motion and measure the masses of Earth, the Sun, and many other astronomical objects by studying the orbits of objects near them.

### Kepler's Laws Reconsidered

Newton's laws of motion and law of universal gravitation provide a theoretical explanation for Kepler's empirical laws of planetary motion. Kepler's three laws follow directly from Newtonian mechanics, as solutions of the equations describing the motion of a body moving in response to an inverse-square force. However, just as Kepler modified the Copernican model by introducing ellipses rather than circles, so, too, did Newton make corrections to Kepler's first and third laws. It turns out that a planet does not orbit the exact center of the Sun. Instead, both the planet and the Sun orbit their common **center of mass**—the "average" position of all the matter making up the two bodies (Figure 2.25). Because the Sun and the planet are acted upon by equal and opposite gravitational forces (by Newton's third law), the Sun must also move (by Newton's first law), driven by the gravitational influence of the planet. The Sun, however, is so much more massive than any planet that the center of mass of the planet–Sun system is very

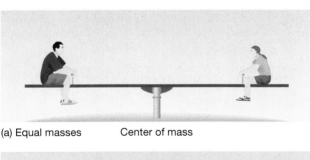

(a) Equal masses      Center of mass

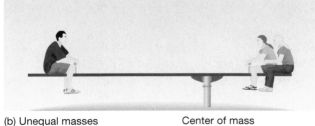

(b) Unequal masses      Center of mass

▲ **FIGURE 2.24 Solar Gravity** The Sun's inward pull of gravity on a planet competes with the planet's tendency to continue moving in a straight line. These two effects combine, causing the planet to move smoothly along an intermediate path, which continuously "falls around" the Sun. This unending tug-of-war between the Sun's gravity and the planet's inertia results in a stable orbit.

▲ **FIGURE 2.25 Center of Mass** (a) The center of mass of two bodies of equal mass lies midway between them. (b) As the mass of one body increases, the center of mass moves toward it. The situation is analogous to riding a seesaw—when both sides are balanced, the center of mass is at the pivot point.

## MORE PRECISELY 2-3

### Weighing the Sun

We can use Newtonian mechanics to calculate some useful formulae relating the properties of planetary orbits to the mass of the Sun. Again for simplicity, let's assume that the orbits are circular (not a bad approximation in most cases, and Newton's laws easily extend to cover the more general case of eccentric orbits). Consider a planet of mass $m$ moving at speed $v$ in an orbit of radius $r$ around the Sun, of mass $M$. The planet's acceleration (see *More Precisely 2-2*) is

$$a = \frac{v^2}{r},$$

so, by Newton's second law, the force required to keep the planet in orbit is

$$F = ma = \frac{mv^2}{r}.$$

Setting this equation equal to the gravitational force due to the Sun, we obtain

$$\frac{mv^2}{r} = \frac{GmM}{r^2},$$

so the speed of the planet in the circular orbit is

$$v = \sqrt{\frac{GM}{r}}.$$

Now let's turn the problem around. Because we have measured $G$ in the laboratory on Earth, and because we know the length of a year and the size of the astronomical unit, we can use Newtonian mechanics to *weigh* the Sun—that is, find its mass by measuring its gravitational influence on another body (in this case, Earth). Rearranging the last equation to read

$$M = \frac{rv^2}{G}$$

and substituting the known values of $v = 30$ km/s, $r = 1$ AU $= 1.5 \times 10^{11}$ m, and $G = 6.7 \times 10^{-11}$ Nm$^2$/kg$^2$, we calculate the mass of the Sun to be $2.0 \times 10^{30}$ kg—an enormous mass by terrestrial standards.

---

**EXAMPLE** Similarly, knowing the distance from Earth to the Moon ($r = 384{,}000$ km) and the length of the (sidereal) month ($P = 27.3$ days), we can calculate the Moon's orbital speed to be $v = 2\pi r/P = 1.02$ km/s, and hence, using the preceding formula, measure Earth's mass to be $6.0 \times 10^{24}$ kg.

---

In fact, this is basically how *all* masses are measured in astronomy. Because we can't just go out and attach a scale to an astronomical object when we need to know its mass, we must look for its gravitational influence on something else. This principle applies to planets, stars, galaxies, and even clusters of galaxies—very different objects, but all subject to the same physical laws.

close to the center of the Sun, which is why Kepler's laws are so accurate. Thus, Kepler's first law becomes

> The orbit of a planet around the Sun is an ellipse, with the *center of mass of the planet–Sun system* at one focus.

As shown in Figure 2.25, the center of mass of two objects of comparable mass does not lie within either object. For identical masses orbiting one another (Figure 2.26a), the orbits are identical ellipses, with a common focus located midway between the two objects. For unequal masses (as in Figure 2.26b), the elliptical orbits still share a focus, and both have the same eccentricity, but the more massive

object moves more slowly and on a tighter orbit. (Note that Kepler's second law, as stated earlier, continues to apply without modification to each orbit separately, but the *rates* at which the two orbits sweep out areas are different.)

The change to Kepler's third law is also small in the case of a planet orbiting the Sun, but very important in other circumstances, such as the orbital motion of two stars that are gravitationally bound to each other. Following through the mathematics of Newton's theory, we find that

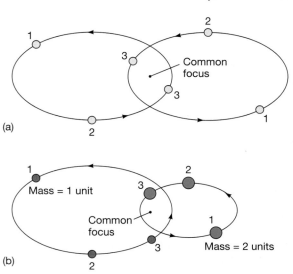

▶ FIGURE 2.26 **Orbits** (a) The orbits of two bodies (stars, for example) with equal masses, under the influence of their mutual gravity, are identical ellipses with a common focus. That focus is not at the center of either star, but instead at their center of mass, located midway between them. The pairs of numbers indicate the positions of the two bodies at three different times. (Note that a line joining the bodies at any give time always passes through the common focus.) (b) The orbits of two bodies, one of which is twice as massive as the other. Again, the elliptical orbits have a common focus (at the center of mass of the two-body system), and the two ellipses have the same eccentricity. However, in accordance with Newton's laws of motion, the more massive body moves more slowly and in a smaller orbit. In this particular case, the larger ellipse is twice the size of the smaller one.

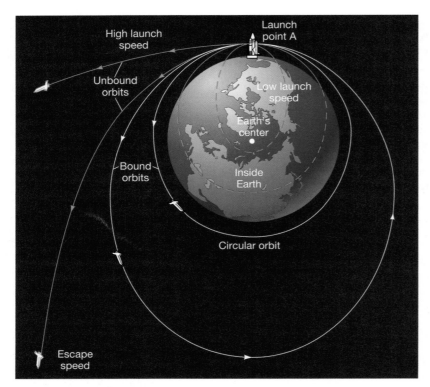

**▲ FIGURE 2.27 Escape Speed** The effect of launch speed on the trajectory of a satellite. With too low a speed at point A, the satellite will simply fall back to Earth. Given enough speed, however, the satellite will go into orbit—it "falls around Earth." As the initial speed at point A is increased, the orbit will become more and more elongated. When the initial speed exceeds the escape speed, the satellite will become unbound from Earth and will escape along a hyperbolic trajectory.

the true relationship between the semimajor axis $a$ (measured in astronomical units) of the planet's orbit relative to the Sun and its orbital period $P$ (in Earth years) is

$$P^2 \text{ (in Earth years)} = \frac{a^3 \text{ (in astronomical units)}}{M_{\text{total}} \text{ (in solar units)}},$$

where $M_{\text{total}}$ is the *combined* mass of the two objects. Notice that Newton's restatement of Kepler's third law preserves the proportionality between $P^2$ and $a^3$, but now the proportionality includes $M_{\text{total}}$, so it is *not* quite the same for all the planets. The Sun's mass is so great, however, that the differences in $M_{\text{total}}$ among the various combinations of the Sun and the other planets are almost unnoticeable, so Kepler's third law, as originally stated, is a very good approximation. This modified form of Kepler's third law is true in all circumstances, inside or outside the solar system.

## Escaping Forever

The law of gravity that describes the orbits of planets around the Sun applies equally well to natural moons and artificial satellites orbiting any planet. All our Earth-orbiting, human-made satellites move along paths governed by a combination of the inward pull of Earth's gravity and the forward motion gained during the rocket launch. If the rocket initially imparts enough speed to the satellite, it can go into orbit. Satellites not given enough speed at launch, by accident or design (e.g., intercontinental ballistic missiles) fail to achieve orbit and fall back to Earth (see Figure 2.27).

Some space vehicles, such as the robot probes that visit other planets, attain enough speed to escape our planet's gravity and move away from Earth forever. This speed, known as the **escape speed,** is about 41% greater (actually, $\sqrt{2} = 1.414...$ times greater) than the speed of an object traveling in a circular orbit at any given radius.*At less than the escape speed, the old adage "What goes up must come down" (or at least stay in orbit) still applies. At more than the escape speed, however, a spacecraft will leave Earth for good. Planets, stars, galaxies—all gravitating bodies—have escape speeds. No matter how massive the body, gravity decreases with distance. As a result, the escape speed diminishes with increasing separation. The farther we go from Earth (or any gravitating body), the easier it becomes to escape.

The speed of a satellite in a circular orbit just above Earth's atmosphere is 7.9 km/s (roughly 18,000 mph). The satellite would have to travel at 11.2 km/s (about 25,000 mph) to escape from Earth altogether. If an object exceeds the escape speed, its motion is said to be **unbound,** and the orbit is no longer an ellipse. In fact, the path of the spacecraft relative to Earth is a related geometric figure called a *hyperbola*. If we simply change the word *ellipse* to *hyperbola*, the modified version of Kepler's first law still applies, as does Kepler's second law. (Kepler's third law does not extend to unbound orbits, because it doesn't make sense to talk about a period in those cases.)

## The Circle of Scientific Progress

The progression from the complex Ptolemaic model of the universe to the elegant simplicity of Newton's laws is a case study in the scientific method. ∞ (Sec. 1.2) Copernicus made a radical conceptual leap away from the Ptolemaic view, gaining much in insight but little in predictive power. Kepler made critical changes to the Copernican picture and gained both accuracy and predictive power but still fell short of a true physical explanation of planetary motion within the solar system, or of orbital motion in general.

*In terms of the formula presented in *More Precisely 2-3*, the escape speed is given by $v_{\text{escape}} = \sqrt{2GM/r}$.

Eventually, Newton showed how all known planetary motion could be explained in detail by the application of four, simple, fundamental laws—the three laws of motion and the law of gravity. The process was slow, with many starts and stops and a few wrong turns, but it worked!

In a sense, then, the development of Newton's laws and their application to planetary motion represented the end of the first "loop" around the schematic diagram presented in Figure 1.6. The long-standing practical and conceptual questions raised by ancient observations of retrograde motion were finally resolved, and new predictions, themselves amenable to observational testing, became possible. And the laws are still being tested today. Every time a comet appears in the night sky right on schedule, or a spacecraft reaches the end of a billion-kilometer journey within meters

of its target and seconds of the predicted arrival time, our confidence in Newtonian mechanics is further strengthened. But unlike the essentially descriptive models of Ptolemy, Copernicus, and Kepler, Newtonian mechanics is not limited to the motions of planets, or even to events occurring within our own solar system. They apply to moons, comets, spacecraft, stars, and even the most distant galaxies, extending the range of our scientific inquiries across the observable universe—as well as apples falling to the ground.

## CONCEPT CHECK

✔ Explain, in terms of Newton's laws of motion and gravity, why planets orbit the Sun.

# CHAPTER REVIEW

## Summary

**❶ Geocentric (p. 40)** models of the universe were based on the view that the Sun, the Moon, and the planets all orbit Earth. The most successful and long lived of these was the **Ptolemaic model (p. 40)**. Unlike the Sun and the Moon, planets sometimes appear to temporarily reverse their direction of motion (from night to night) relative to the stars and then resume their normal "forward" course. This phenomenon is called **retrograde motion (p. 39)**. To account for retrograde motion within the geocentric picture, it was necessary to suppose that planets moved on small circles called **epicycles (p. 40)**, whose centers orbited Earth on larger circles called **deferents (p. 40)**.

**❷ The heliocentric (p. 41)** view of the solar system holds that Earth, like all the planets, orbits the Sun. This model accounts for retrograde motion and the observed sizes and variations in brightness of the planets in a much more straightforward way than the geocentric model. The widespread realization during the Renaissance that the solar system is Sun centered, and not Earth centered, is known as the **Copernican revolution (p. 42)**, in honor of Nicolaus Copernicus, who laid the foundations of the modern heliocentric model.

**❸** Galileo Galilei is often regarded as the father of experimental science. His telescopic observations of the Moon, the Sun, Venus, and Jupiter played a crucial role in supporting and strengthening the Copernican picture of the solar system. Johannes Kepler improved on Copernicus's model by condensing the observational data of Tycho Brahe into three **laws of planetary motion (p. 47)**.

**❹** Kepler's Laws state: (1) Planetary orbits are **ellipses (p. 47)** with the Sun at one **focus (p. 47)**; (2) a planet moves faster as its orbit takes it closer to the Sun; (3) the **semimajor axis (p. 47)** of the orbit is

related in a simple way to the planet's orbital **period (p. 48)**. Most planets move on orbits whose **eccentricities (p. 47)** are quite small, so their paths differ only slightly from perfect circles.

**❺** Kepler's laws tell us only the relative sizes of the planetary orbits. To determine the true scale of the solar system, astronomers must find an independent measure of the distance between two bodies. The distance from Earth to the Sun is called the **astronomical unit (p. 49)**. Nowadays, the astronomical unit is determined by bouncing **radar (p. 51)** signals off the planet Venus and measuring the time the signal takes to return.

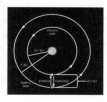

**❻** Isaac Newton succeeded in explaining planetary motion in terms of a few general physical principles, now known as **Newtonian mechanics (p. 52)**. Newton's laws of motion are: (1) To change the body's velocity, a **force (p. 52)** must be applied; (2) the rate of change of velocity, called **acceleration (p. 53)**, is equal to the applied force divided by the body's **mass (p. 52)**; (3) when bodies interact, the forces between them are always equal and opposite to one another. To explain Kepler's laws, Newton postulated that **gravity (p. 53)** attracts the planets to the Sun. Every object with any mass exerts a **gravitational force (p. 53)** on all other objects in the universe. The strength of this force decreases with increasing distance according to an **inverse-square law (p. 54)**.

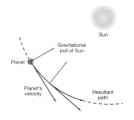

**❼** For one object to escape from the gravitational pull of another, its speed must exceed the **escape speed (p. 58)** of that other object. By determining the gravitational force needed to keep one body orbiting another, Newton's laws allow astronomers to measure the masses of distant objects.

## Review and Discussion

1. What contributions to modern astronomy were made by Chinese and Islamic astronomers during the Dark Ages of medieval Europe?

2. Briefly describe the geocentric model of the universe.

3. The benefit of our current knowledge lets us see flaws in the Ptolemaic model of the universe. What is its basic flaw?

4. What was the great contribution of Copernicus to our knowledge of the solar system? What was still a flaw in the Copernican model?

5. What is a theory? Can a theory ever be proved to be true?

6. When were Copernicus's ideas finally accepted?

7. What is the Copernican principle?

8. What discoveries of Galileo helped confirm the views of Copernicus, and how did they do so?

9. Briefly describe Kepler's three laws of planetary motion.

10. How did Tycho Brahe contribute to Kepler's laws?

11. If radio waves cannot be reflected from the Sun, how can radar be used to find the distance from Earth to the Sun?

12. How did astronomers determine the scale of the solar system prior to the invention of radar?

13. What does it mean to say that Kepler's laws are empirical?

14. What are Newton's laws of motion and gravity?

15. List the two modifications made by Newton to Kepler's laws.

16. Why do we say that a baseball falls toward Earth, and not Earth toward the baseball?

17. Why would a baseball go higher if it were thrown up from the surface of the Moon than if it were thrown with the same velocity from the surface of Earth?

18. In what sense is the Moon falling toward Earth?

19. What is the meaning of the term *escape speed*?

20. What would happen to Earth if the Sun's gravity were suddenly "turned off"?

## Conceptual Self-Test: True or False/Multiple Choice

1. Aristotle was the first to propose that all planets revolve around the Sun.

2. Retrograde motion is the apparent "backward" (westward) motion of a planet relative to the stars.

3. The heliocentric model of the universe holds that Earth is at the center and everything else moves around it.

4. Galileo's observations of the sky were made with the naked eye.

5. Kepler discovered that the shape of an orbit is a circle, not an ellipse, as had previously been thought.

6. Kepler's discoveries regarding the orbital motion of the planets were based mainly on observations made by Copernicus.

7. The Sun lies at the center of a planet's orbit.

8. Kepler never knew the true distances between the planets and the Sun, only their relative distances.

9. According to Newton's laws, a moving object will come to rest unless acted upon by a force.

10. You throw a baseball to someone; before the ball is caught, it is temporarily in orbit around Earth's center.

11. Planets near opposition (a) rise in the east; (b) rise in the west; (c) do not rise or set; (d) have larger deferents.

12. A major flaw in Copernicus's model was that it still had (a) the Sun at the center; (b) Earth at the center; (c) retrograde loops; (d) circular orbits.

13. As shown in Figure 2.12 ("Venus Phases"), Galileo's observations of Venus demonstrated that Venus must be (a) orbiting Earth; (b) orbiting the Sun; (c) about the same diameter as Earth; (d) similar to the Moon.

14. An accurate sketch of Jupiter's orbit around the Sun would show (a) the Sun far off center; (b) an oval twice as long as it is wide; (c) a nearly perfect circle; (d) phases.

15. A calculation of how long it takes a planet to orbit the Sun would be most closely related to Kepler's (a) first law of orbital shapes; (b) second law of orbital speeds; (c) third law of planetary distances; (d) first law of inertia.

16. An asteroid with an orbit lying entirely inside Earth's (a) has an orbital semimajor axis of less than 1 AU; (b) has a longer orbital period than Earth's; (c) moves more slowly than Earth; (d) must have a highly eccentric orbit.

17. If Earth's orbit around the Sun were twice as large as it is now, the orbit would take (a) less than twice as long; (b) two times longer; (c) more than two times longer to traverse.

18. Figure 2.22 ("Gravity"), showing the motion of a ball near Earth's surface, depicts how gravity (a) increases with altitude; (b) causes the ball to accelerate downward; (c) causes the ball to accelerate upward; (d) has no effect on the ball.

19. If the Sun *and its mass* were suddenly to disappear, Earth would (a) continue in its current orbit; (b) suddenly change its orbital speed; (c) fly off into space; (d) stop spinning.

20. Figure 2.26(b) ("Orbits") shows the orbits of two stars of unequal masses. If one star has twice the mass of the other, then the more massive star (a) moves more slowly than; (b) moves more rapidly than; (c) has half the gravity of; (d) has twice the eccentricity of the less massive star.

## Problems

 *Algorithmic versions of these Problems are available in the Practice Problems module of the Companion Website. The number of dots preceding each Problem indicates its approximate level of difficulty.*

1. • Tycho Brahe's observations of the stars and planets were accurate to about 1 arc minute (1′). To what distance does this angle correspond at the distance of (a) the Moon; (b) the Sun; and (c) Saturn (at closest approach)?

2. •• To an observer on Earth, through what angle will Mars appear to move relative to the stars over the course of 24 hours when the two planets are at closest approach? Assume for simplicity that Earth and Mars move on circular orbits of radii 1.0 AU and 1.5 AU, respectively, in exactly the same plane. Will the apparent motion be prograde or retrograde?

3. • Using the data in Table 2.1, estimate the minimum possible distance between Mars and Earth.

4. •• An asteroid has a perihelion distance of 2.0 AU and an aphelion distance of 4.0 AU. Calculate its orbital semimajor axis, eccentricity, and period.

5. •• A spacecraft has an orbit that just grazes Earth's orbit at aphelion and just grazes Venus's orbit at perihelion. Assuming that Earth and Venus are in the right places at the right times, how long will the spacecraft take to travel from Earth to Venus?

6. •• Halley's comet has a perihelion distance of 0.6 AU and an orbital period of 76 years. What is the aphelion distance of Halley's comet from the Sun?

7. •• What is the maximum possible parallax of Mercury during a solar transit, as seen from either end of a 3000-km baseline on Earth?

8. • How long would a radar signal take to complete a round-trip between Earth and Mars when the two planets are 0.7 AU apart?

9. • Jupiter's moon Callisto orbits the planet at a distance of 1.88 million kilometers. Callisto's orbital period about Jupiter is 16.7 days. What is the mass of Jupiter? [Assume that Callisto's mass is negligible compared with that of Jupiter, and use the modified version of Kepler's third law (Section 2.7).]

10. •• The Sun moves in a roughly circular orbit around the center of the Milky Way galaxy at a distance of 26,000 light-years. The orbit speed is approximately 220 km/s. Calculate the Sun's orbital period and centripetal acceleration, and use these numbers to estimate the mass of our galaxy.

11. •• At what distance from the Sun would a planet's orbital period be 1 million years? What would be the orbital period at a distance of 1 light-year?

12. • The acceleration due to gravity at Earth's surface is $9.80 \text{ m/s}^2$. What is the gravitational acceleration at altitudes of (a) 100 km; (b) 1000 km; (c) 10,000 km? Take Earth's radius to be 6400 km.

13. •• What would be the speed of a spacecraft moving in a circular orbit at each of the three altitudes listed in the previous problem? In each case, how does the centripetal acceleration (*More Precisely 2-2*) compare with the gravitational acceleration?

14. • Use Newton's law of gravity to calculate the force of gravity between you and Earth. Convert your answer, which will be in newtons, to pounds, using the conversion 4.45 N equals 1 (1 lb). What do you normally call this force?

15. • The Moon's mass is $7.4 \times 10^{22}$ kg and its radius is 1700 km. What is the speed of a spacecraft moving in a circular orbit just above the lunar surface? What is the escape speed from the Moon?

*The Companion Website at www.aw-bc.com/chaisson provides algorithmically generated versions of each chapter's Problems, along with additional quizzes, an Animations & Videos gallery, an Images gallery, an interactive Glossary, and a full eBook.*

CHAPTER

# RADIATION
## Information from the Cosmos

Astronomical objects are more than just things of beauty in the night sky. Planets, stars, and galaxies are of vital significance if we are to fully understand our place in the big picture—the "grand design" of the universe. Each object is a source of information about the material aspects of our universe—its state of motion, its temperature, its chemical composition, and even its past history.

This information comes to us in the form of light. When we look at the stars, the light we see actually began its journey to Earth decades, centuries—even millennia—ago. The faint rays from the most distant galaxies have taken billions of years to reach us. The stars and galaxies in the night sky show us the far away and the long ago. In this chapter, we begin our study of how astronomers extract information from the light emitted by astronomical objects. These basic concepts of radiation are central to modern astronomy.

## LEARNING GOALS

*Studying this chapter will enable you to*

1  Describe the basic properties of wave motion.
2  Tell how electromagnetic radiation transfers energy and information through interstellar space.
3  Describe the major regions of the electromagnetic spectrum and explain how Earth's atmosphere affects our ability to make astronomical observations at different wavelengths.
4  Explain what is meant by the term "blackbody radiation" and describe the basic properties of such radiation.
5  Tell how we can determine the temperature of an object by observing the radiation it emits.
6  Show how the relative motion between a source of radiation and its observer can change the perceived wavelength of the radiation, and explain the importance of this phenomenon to astronomy.

 Visit www.aw-bc.com/chaisson for additional images, animations, quizzes, and eBook for this chapter.

## 3.1    Information from the Skies

Figure 3.1 shows a galaxy in the constellation Andromeda. On a dark, clear night, far from cities or other sources of light, the Andromeda galaxy, as it is generally called, can be seen with the naked eye as a faint, fuzzy patch on the sky, comparable in diameter to the full Moon. Yet the fact that it is visible from Earth belies this galaxy's enormous distance from us: It lies roughly 2.5 million *light-years* away.

An object at such a distance is truly inaccessible in any realistic human sense. Even if a space probe could miraculously travel at the speed of light, it would need 2.5 million years to reach this galaxy and 2.5 million more to return with its findings. Considering that civilization has existed on Earth for less than 10,000 years, and its prospects for the next 10,000 are far from certain, even this unattainable technological feat would not provide us with a practical means of exploring other galaxies. Even the farthest reaches of our own galaxy, "only" a few tens of thousands of light-years distant, are effectively off limits to visitors from Earth, at least for the foreseeable future.

Given the practical impossibility of traveling to such remote parts of the universe, how do astronomers know anything about objects far from Earth? How do we obtain detailed information about any planet, star, or galaxy too distant for a personal visit or any kind of controlled experiment? The answer is that we use the laws of physics, as we know them here on Earth, to interpret the **electromagnetic radiation** emitted by those objects.

### Light and Radiation

*Radiation* is any way in which energy is transmitted through space from one point to another without the need for any physical connection between the two locations. The term *electromagnetic* just means that the energy is carried in the form of rapidly fluctuating *electric* and *magnetic* fields (to be discussed in more detail later in Section 3.2). Virtually all we know about the universe beyond Earth's atmosphere has been gleaned from painstaking analysis of electromagnetic radiation received from afar. Our understanding depends completely on our ability to decipher this steady stream of data reaching us from space.

How bright are the stars (or galaxies, or planets), and how hot? What are their masses? How rapidly do they spin, and what is their motion through space? What are they made of, and in what proportion? The list of questions is long, but one fact is clear: Electromagnetic theory is vital to providing the answers—without it, we would have no way of testing our models of the cosmos, and the modern science of astronomy simply would not exist. ∞ (Sec. 1.2)

**Visible light** is the particular type of electromagnetic radiation to which our human eyes happen to be sensitive. As light enters our eye, small chemical reactions triggered by the incoming energy send electrical impulses to the brain, producing the sensation of sight. But modern instruments (see Chapter 5) can also detect many forms of *invisible* electromagnetic radiation, which goes completely unnoticed by our eyes. **Radio, infrared,** and **ultraviolet** waves, as well as **X rays** and **gamma rays,** all fall into this category.

Note that, despite the different names, the words *light*, *rays*, *radiation*, and *waves* all really refer to the same thing. The names are just historical accidents, reflecting the fact that it took many years for scientists to realize that these apparently very different types of radiation are in reality one and the same physical phenomenon. Throughout this text, we will use the general terms *light* and *electromagnetic radiation* more or less interchangeably.

### Wave Motion

Despite the early confusion still reflected in current terminology, scientists now know that all types of electromagnetic radiation travel through space in the form of **waves.** To understand the behavior of light, then, we must know a little about wave motion.

Simply stated, a wave is a way in which energy is transferred from place to place without the physical movement of

R  I  V  U  X  G

▲ **FIGURE 3.1 Andromeda Galaxy**  The pancake-shaped Andromeda Galaxy lies about 2.5 million light-years away and contains a few hundred billion stars. *(R. Gendler)*

material from one location to another. In wave motion, the energy is carried by a *disturbance* of some sort. This disturbance, whatever its nature, occurs in a distinctive repeating pattern. Ripples on the surface of a pond, sound waves in air, and electromagnetic waves in space, despite their many obvious differences, all share this basic defining property.

Imagine a twig floating in a pond (Figure 3.2). A pebble thrown into the pond at some distance from the twig disturbs the surface of the water, setting it into up-and-down motion. This disturbance will move outward from the point of impact in the form of waves. When the waves reach the twig, some of the pebble's energy will be imparted to it, causing the twig to bob up and down. In this way, both energy and *information*— the fact that the pebble entered the water—are transferred from the place where the pebble landed to the location of the twig. We could tell that a pebble (or, at least, some object) had entered the water just by observing the twig. With a little additional physics, we could even estimate the pebble's energy.

A wave is not a physical object. No water traveled from the point of impact of the pebble to the twig—at any location on the surface, the water surface simply moved up and down as the wave passed. What, then, *did* move across the surface of the pond? As illustrated in the figure, the answer is that the wave was the *pattern* of up-and-down motion. This pattern was transmitted from one point to the next as the disturbance moved across the water.

Figure 3.3 shows how wave properties are quantified and illustrates some standard terminology. The wave's **period** is the number of seconds needed for the wave to repeat itself at any given point in space. The **wavelength** is the number of meters needed for the wave to repeat itself at a given moment in time. It can be measured as the distance between two adjacent wave *crests*, two adjacent wave *troughs*, or any other two similar points on adjacent wave cycles (e.g., the points marked $\times$ in the figure). A

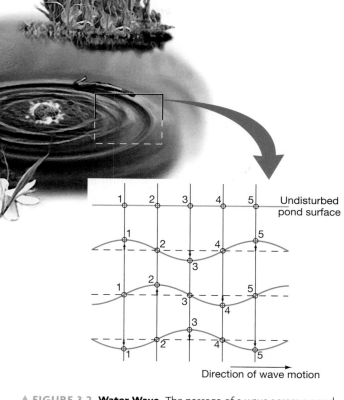

▲ FIGURE 3.2 **Water Wave** The passage of a wave across a pond causes the surface of the water to bob up and down, but there is no movement of water from one part of the pond to another. Here waves ripple out from the point where a pebble has hit the water to the point where a twig is floating. The inset shows a simplified series of "snapshots" of the pond surface as the wave passes by. The points numbered 1 through 5 represent surface locations that move up and down with the passage of the wave.

wave moves a distance equal to one wavelength in one period. The maximum departure of the wave from the undisturbed state—still air, say, or a flat pond surface—is called the wave's **amplitude.**

The number of wave crests passing any given point per unit time is called the wave's **frequency.** If a wave of a given wavelength moves at high speed, then many crests pass per second and the frequency is high. Conversely, if

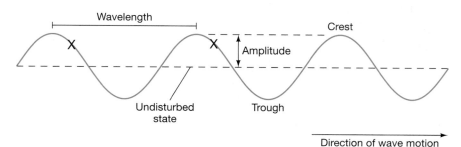

◄ FIGURE 3.3 **Wave Properties** Representation of a typical wave, showing its direction of motion, wavelength, and amplitude. In one wave period, the entire pattern shown here moves one wavelength to the right.

the same wave moves slowly, then its frequency will be low. The frequency of a wave is just the reciprocal of the wave's period; that is,

$$\text{frequency} = \frac{1}{\text{period}}.$$

Frequency is expressed in units of inverse time (that is, 1/second, or cycles per second), called hertz (Hz) in honor of the 19th-century German scientist Heinrich Hertz, who studied the properties of radio waves. Thus, a wave with a period of 5 seconds (5 s) has a frequency of (1/5) cycles/s = 0.2 Hz, meaning that one wave crest passes a given point in space every 5 seconds.

Because a wave travels one wavelength in one period, it follows that the *wave velocity* is simply equal to the wavelength divided by the period:

$$\text{velocity} = \frac{\text{wavelength}}{\text{period}}.$$

Since the period is the reciprocal of the frequency, we can equivalently (and more commonly) write this relationship as

$$\text{velocity} = \text{wavelength} \times \text{frequency}.$$

Thus, if the wave in our earlier example had a wavelength of 0.5 m, its velocity would be (0.5 m)/(5 s), or (0.5 m) × (0.2 Hz) = 0.1 m/s. In the case of electromagnetic radiation, the velocity is the speed of light. Notice that wavelength and wave frequency are *inversely* related—doubling one halves the other.

## The Components of Visible Light

White light is a mixture of colors, which we conventionally divide into six major hues: red, orange, yellow, green, blue, and violet. As shown in Figure 3.4, we can separate a beam of white light into a rainbow of these basic colors—called a *spectrum* (plural, *spectra*)—by passing it through a prism. This experiment was first reported by Isaac Newton over 300 years ago. In principle, the original beam of white light could be recovered by passing the spectrum through a second prism to recombine the colored beams.

What determines the color of a beam of light? The answer is its frequency (or alternatively, its wavelength). We see different colors because our eyes react differently to electromagnetic waves of different frequencies. A prism splits a beam of light up into separate colors because light rays of different frequencies are bent, or *refracted*, slightly differently as they pass through the prism—red light the least, violet light the most. Red light has a frequency of roughly $4.3 \times 10^{14}$ Hz, corresponding to a wavelength of about $7.0 \times 10^{-7}$ m. Violet light, at the other end of the visible range, has nearly double the frequency—$7.5 \times 10^{14}$ Hz—and (since the speed of light is always the same) just over half the wavelength—$4.0 \times 10^{-7}$ m. The other colors we see have frequencies and wavelengths intermediate between these two extremes, spanning the entire *visible spectrum* shown in Figure 3.4. Radiation outside this range is invisible to human eyes.

Scientists often use a unit called the *nanometer* (nm) in describing the wavelength of light (see Appendix 2). There are $10^9$ nanometers in 1 meter. An older unit called the

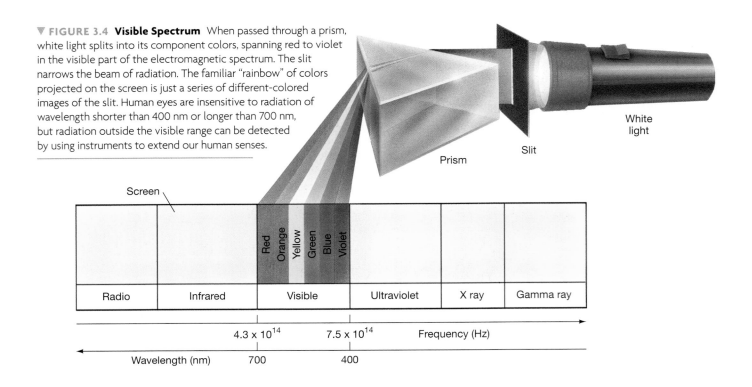

▼ FIGURE 3.4 **Visible Spectrum** When passed through a prism, white light splits into its component colors, spanning red to violet in the visible part of the electromagnetic spectrum. The slit narrows the beam of radiation. The familiar "rainbow" of colors projected on the screen is just a series of different-colored images of the slit. Human eyes are insensitive to radiation of wavelength shorter than 400 nm or longer than 700 nm, but radiation outside the visible range can be detected by using instruments to extend our human senses.

*angstrom* (1 Å = $10^{-10}$ m = 0.1 nm) is also widely used. (The unit is named after the 19th-century Swedish physicist Anders Ångstrom—pronounced "ong·strem,") However, in SI units, the nanometer is preferred. Thus, the visible spectrum covers the range of wavelengths from 400 nm to 700 nm (4000 Å to 7000 Å). The radiation to which our eyes are most sensitive has a wavelength near the middle of this range, at about 550 nm (5500 Å), in the yellow-green region of the spectrum. It is no coincidence that this wavelength falls within the range of wavelengths at which the Sun emits most of its electromagnetic energy—our eyes have evolved to take greatest advantage of the available light.

## 3.2 Waves in What?

Waves of radiation differ fundamentally from water waves, sound waves, or any other waves that travel through a material medium. Radiation needs *no* such medium. When light travels from a distant galaxy, or from any other cosmic object, it moves through the virtual vacuum of space. Sound waves, by contrast, cannot do this, despite what you have probably heard in almost every sci-fi movie ever made! If we were to remove all the air from a room, conversation would be impossible (even with suitable breathing apparatus to keep our test subjects alive!) because sound waves cannot exist without air or some other physical medium to support them. Communication by flashlight or radio, however, would be entirely feasible.

The ability of light to travel through empty space was once a great mystery. The idea that light, or any other kind of radiation, could move as a wave through nothing at all seemed to violate common sense, yet it is now a cornerstone of modern physics.

### Interactions Between Charged Particles

To understand more about the nature of light, consider for a moment an *electrically charged* particle, such as an **electron** or a **proton.** Like mass, electrical charge is a fundamental property of matter. Electrons and protons are elementary particles—"building blocks" of atoms and all matter—that carry the basic unit of charge. Electrons are said to carry a *negative* charge, whereas protons carry an equal and opposite *positive* charge.

Just as a massive object exerts a gravitational force on every other massive body, an electrically charged particle exerts an *electrical* force on every other charged particle in the universe. ∞ (Sec. 2.7) The buildup of electrical charge (a net excess of positive over negative, or vice versa) is what causes "static cling" on your clothes when you take them out of a hot clothes dryer; it also causes the shock you sometimes feel when you touch a metal door frame on a particularly dry day.

Unlike the gravitational force, which is always attractive, electrical forces can be either attractive or repulsive. As illustrated in Figure 3.5(a), particles with *like* charges (i.e., both negative or both positive—for example, two electrons or two protons) repel one another. Particles with *unlike* charges (i.e., having opposite signs—an electron and a proton, say) attract.

How is the electrical force transmitted through space? Extending outward in all directions from any charged particle is an **electric field,** which determines the electrical force exerted by the particle on all other charged particles in the universe (Figure 3.5b). The strength of the electric field, like the strength of the gravitational field, decreases with increasing distance from the charge according to an inverse-square law. By means of the electric field, the particle's presence is "felt" by all other charged particles, near and far.

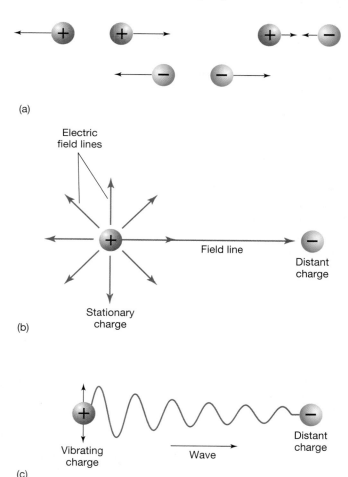

▲ **FIGURE 3.5 Charged Particles** (a) Particles carrying like electrical charges repel one another, whereas particles with unlike charges attract. (b) A charged particle is surrounded by an electric field, which determines the particle's influence on other charged particles. We represent the field by a series of field lines. (c) If a charged particle begins to vibrate, its electric field changes. The resulting disturbance travels through space as a wave, eventually interacting with distant charged particles.

Now, suppose our particle begins to vibrate, perhaps because it becomes heated or collides with some other particle. Its changing position causes its associated electric field to change, and this changing field in turn causes the electrical force exerted on other charges to vary (Figure 3.5c). If we measure the change in the force on these other charges, we learn about our original particle. Thus, *information about the particle's state of motion is transmitted through space via a changing electric field.* This *disturbance* in the particle's electric field travels through space as a wave.

## Electromagnetic Waves

The laws of physics tell us that a **magnetic field** must accompany every changing electric field. Magnetic fields govern the influence of *magnetized* objects on one another, much as electric fields govern interactions among charged particles. The fact that a compass needle always points to magnetic north is the result of the interaction between the magnetized needle and Earth's magnetic field (Figure 3.6). Magnetic fields also exert forces on *moving* electric charges (i.e., electric currents)—electric meters and motors rely on this basic fact. Conversely, moving charges *create* magnetic fields (electromagnets are a familiar example). In short, electric and magnetic fields are inextricably linked to one another: A change in either one necessarily creates the other.

Thus, as illustrated in Figure 3.7, the disturbance produced by the moving charge in Figure 3.5(c) actually consists of vibrating electric *and* magnetic fields, moving together through space. Furthermore, as shown in the diagram, these fields are always oriented *perpendicular* to one another and to the direction in which the wave is traveling. The fields do not exist as independent entities; rather, they are different aspects of a single physical phenomenon: **electromagnetism.** Together, they constitute an *electromagnetic wave* that carries energy and information from one part of the universe to another.

Now consider a real cosmic object—a star, say. When some of its charged contents move around, their electric fields change, and we can detect that change. The resulting electromagnetic ripples propagate (travel) outward as waves through space, requiring no material medium in which to move. Small charged particles, either in our eyes or in our experimental equipment, eventually respond to the electromagnetic field changes by vibrating in tune with the radiation that is received. This response is how we detect the radiation—and how we see. Figure 3.8 shows a more familiar example of information being transferred by electromagnetic radiation. A cellular transmitter causes electric charges to oscillate up and down in a metal rod mounted at the top of a tower,

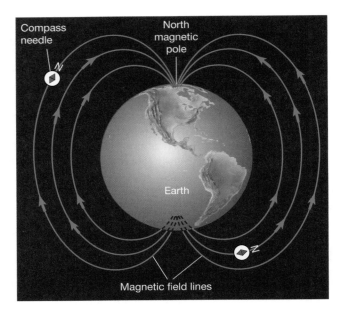

▲ FIGURE 3.6 **Magnetism** Earth's magnetic field interacts with a magnetic compass needle, causing the needle to become aligned with the field—that is, to point toward Earth's north (magnetic) pole. The north magnetic pole actually lies at latitude 80° N, longitude 107° W, some 1140 km from the geographic North Pole.

thereby generating electromagnetic radiation. This radiation is detected by the antenna in your cell phone. Within the metal core of the receiving antenna, electric charges respond to the incoming radiation by vibrating in time with the transmitted wave frequency. The information

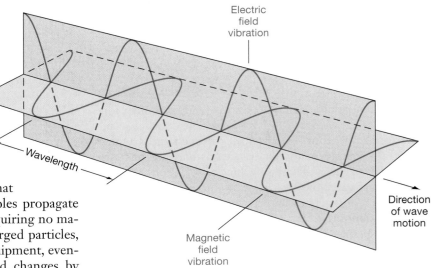

▲ FIGURE 3.7 **Electromagnetic Wave** Electric and magnetic fields vibrate perpendicularly to each other. Together they form an electromagnetic wave that moves through space at the speed of light in the direction perpendicular to both the electric and the magnetic fields comprising it.

Cellular tower

Incoming wave

◀ FIGURE 3.8 **Cellular Signal** Charged particles in a cell phone antenna vibrate in response to electromagnetic radiation broadcast by a distant transmitter. The radiation is produced when electric charges are made to oscillate in the transmitter's emitting antenna. The vibrations in the receiving antenna "echo" the oscillations in the transmitter, allowing the original information to be retrieved. This rude student probably missed the day's lesson.

carried by the pattern of vibrations is then reconverted into sound and images by your phone.

How *quickly* is one charge influenced by the change in the electromagnetic field when another charge begins to move? This is an important question because it is equivalent to asking how fast an electromagnetic wave travels. Does it propagate at some measurable speed, or is it instantaneous? Both theory and experiment tell us that all electromagnetic waves move at a very specific speed—the **speed of light** (always denoted by the letter $c$). Its exact value is 299,792.458 km/s in a vacuum (and somewhat less in material substances such as air or water). We will round this value off to $c = 3.00 \times 10^5$ km/s, an extremely high speed. In the time needed to snap your fingers (about a tenth of a second), light can travel three-quarters of the way around our planet! If the currently known laws of physics are correct, then the speed of light is the fastest speed possible (see *More Precisely 22-1*).

The speed of light is very large, but it is still *finite*. That is, light does not travel instantaneously from place to place. This fact has some interesting consequences for our study of distant objects. It takes time—often lots of time—for light to travel through space. The light we see from the nearest large galaxy—the Andromeda Galaxy, shown in Figure 3.1—left that object about 2.5 million years ago, around the time our first human ancestors appeared on planet Earth. We can

know nothing about this galaxy as it exists today. For all we know, it may no longer even exist! Only our descendants, 2.5 million years into the future, will know whether it exists now. So as we study objects in the cosmos, remember that the light we see left those objects long ago. We can never observe the universe as it is—only as it was.

## The Wave Theory of Radiation

The description presented in this chapter of light and other forms of radiation as electromagnetic waves traveling through space is known as the **wave theory of radiation**. It is a spectacularly successful scientific theory, full of explanatory and predictive power and deep insight into the complex interplay between light and matter—a cornerstone of modern physics.

Two centuries ago, however, the wave theory stood on much less solid scientific ground. Before about 1800, scientists were divided in their opinions about the nature of light. Some thought that light was a wave phenomenon (although at the time, electromagnetism was unknown), whereas others maintained that light was in reality a stream of particles that moved in straight lines. Given the experimental apparatus available at the time, neither camp could find conclusive evidence to disprove the other's theory. *Discovery 3-1* discusses some wave properties that are of particular importance to modern astronomers and describes how their detection in experiments using visible light early in the 19th century tilted the balance of scientific opinion in favor of the wave theory.

But that's not the end of the story. The wave theory, like all good scientific theories, can and must continually be tested by experiment and observation. ∞ (Sec. 1.2) Around the turn of the 20th century, physicists made a series of discoveries about the behavior of radiation and matter on very small (atomic) scales that simply could not be explained by the "classical" wave theory just described. Changes had to be made. As we will see in Chapter 4, the modern theory of radiation is actually a hybrid of the once-rival wave and particle views, combining key elements of each in a unified and—for now—undisputed whole.

CONCEPT CHECK

✔ What is light? List some similarities and differences between light waves and waves on water or in air.

## 3.3 The Electromagnetic Spectrum

Figure 3.9 plots the entire range of electromagnetic radiation, illustrating the relationships among the different types of electromagnetic radiation listed earlier. Notice that the only characteristic distinguishing one from another is wavelength, or frequency. To the low-frequency, long-wavelength side of visible light lie *radio* and *infrared* radiation. Radio frequencies include radar, microwave radiation, and the familiar AM, FM, and TV bands. We perceive infrared radiation as heat. At higher frequencies (shorter wavelengths) are the domains of *ultraviolet, X-ray*, and *gamma-ray* radiation. Ultraviolet radiation, lying just beyond the violet end of the visible spectrum, is responsible for suntans and sunburns. The shorter-wavelength X-rays are perhaps best known for their ability to penetrate human tissue and reveal the state of our insides without resorting to surgery. Gamma rays are the shortest-wavelength radiation. They are often associated with radioactivity and are invariably damaging to living cells they encounter.

### The Spectrum of Radiation

All these spectral regions, including the visible spectrum, collectively make up the **electromagnetic spectrum.** Remember that, despite their greatly differing wavelengths and the different roles they play in everyday life on Earth, all are basically the same phenomenon, and all move at the same speed—the speed of light, *c.*

Figure 3.9 is worth studying carefully, as it contains a great deal of information. Note that wave frequency (in hertz) increases from left to right, and wavelength (in meters) increases from right to left. Scientists often disagree on the "correct" way to display wavelengths and frequencies in diagrams of this type. In picturing

wavelengths and frequencies, this book consistently adheres to the convention that frequency increases toward the *right.*

Notice also that the wavelength and frequency scales in Figure 3.9 do not increase by equal increments of 10. Instead, successive values marked on the horizontal axis differ by *factors of 10*—each is 10 times greater than its neighbor. This type of scale, called a *logarithmic* scale, is often used in science to condense a large range of some quantity into a manageable size. Had we used a linear scale for the wavelength range shown in the figure, it would have been many light-years long! Throughout the text, we will often find it convenient to use a logarithmic scale to compress a wide range of some quantity onto a single, easy-to-view plot.

Figure 3.9 shows that wavelengths extend from the size of mountains (radio radiation) to the size of an atomic nucleus (gamma-ray radiation). The box at the upper right emphasizes how small the visible portion of the electromagnetic spectrum is. Most objects in the universe emit large amounts of invisible radiation. Indeed, many of them emit only a tiny fraction of their total energy in the visible range. A wealth of extra knowledge can be gained by studying the invisible regions of the electromagnetic spectrum.

To remind you of this important fact and to identify the region of the electromagnetic spectrum in which a particular observation was made, we have attached the following spectrum icon—an idealized version of the wavelength scale in Figure 3.9—to every astronomical image presented in this text ⌇⌇⌇⌇. Hence, we can tell at a glance from the highlighted "V" that, for example, Figure 3.1 (p. 64) is an image made with the use of visible light, whereas the first image in Figure 3.12 (p. 76) was captured in the radio ("R") part of the spectrum. Chapter 5 discusses in more detail how astronomers actually make such observations, using telescopes and sensitive detectors tailored to different electromagnetic waves.

### Atmospheric Opacity

Only a small fraction of the radiation produced by astronomical objects actually reaches Earth's surface, because of the **opacity** of our planet's atmosphere. Opacity is the extent to which radiation is blocked by the material through which it is passing—in this case, air. The more opaque an object is, the less radiation gets through it: Opacity is just the opposite of transparency. Earth's atmospheric opacity is plotted along the wavelength and frequency scales at the bottom of Figure 3.9. The extent of shading is proportional to the opacity. Where the shading is greatest (such as at the X-ray or "far" infrared regions of the spectrum), no radiation can get in or out. Where there is no shading at all (in the optical and part of the radio domain), the atmosphere is almost completely transparent. In some parts of the spectrum (e.g., the microwave band and much of the infrared portion), Earth's atmosphere is partly

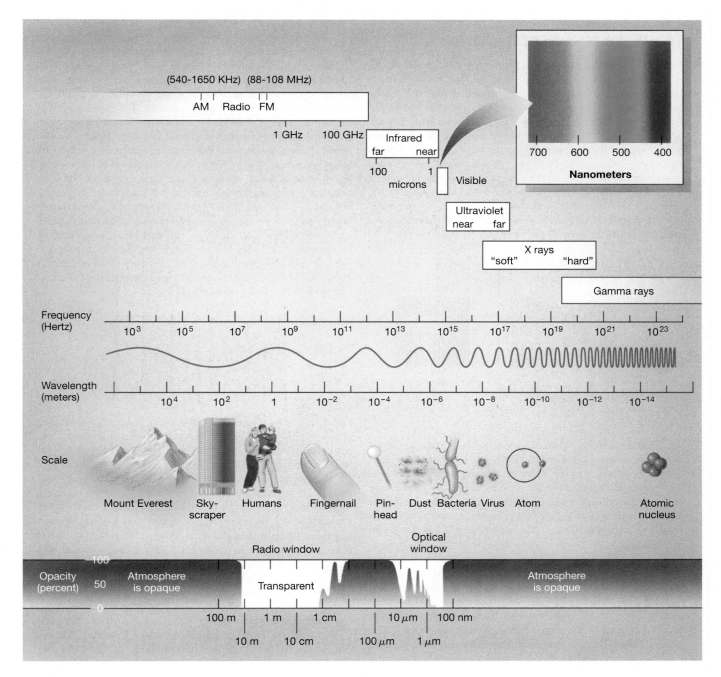

▲ FIGURE 3.9 **Electromagnetic Spectrum** The entire electromagnetic spectrum, running from long-wavelength, low-frequency radio waves to short-wavelength, high-frequency gamma rays.

transparent, meaning that some, but not all, incoming radiation makes it to the surface.

The effect of atmospheric opacity is that there are only a few *spectral windows* at well-defined locations in the electromagnetic spectrum where Earth's atmosphere is transparent. In much of the radio domain and in the visible portions of the spectrum, the opacity is low and we can study the universe at those wavelengths from ground level. In parts of the infrared range, the atmosphere is partially transparent, so we can make certain infrared observations

from the ground. Moving to the tops of mountains, above as much of the atmosphere as possible, improves observations. In the rest of the spectrum, however, the atmosphere is opaque: Ultraviolet, X-ray, and gamma-ray observations can be made only from above the atmosphere, from orbiting satellites.

What causes opacity to vary along the spectrum? Certain atmospheric gases absorb radiation very efficiently at some wavelengths. For example, water vapor ($H_2O$) and oxygen ($O_2$) absorb radio waves having wavelengths

## The Wave Nature of Radiation

Until the early 19th century, debate raged in scientific circles regarding the true nature of light. On the one hand, the particle, or *corpuscular*, theory, first expounded in detail by Isaac Newton, held that light consisted of tiny particles moving in straight lines at the speed of light. Different colors were presumed to correspond to different particles. On the other hand, the *wave* theory, championed by the 17th-century Dutch astronomer Christian Huygens, viewed light as a wave phenomenon, in which color was determined by frequency, or wavelength. During the first few decades of the 19th century, growing experimental evidence that light displayed three key wave properties—*diffraction*, *interference*, and *polarization*—argued strongly in favor of the wave theory.

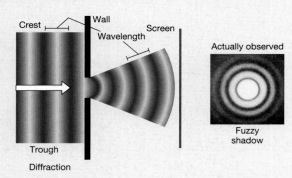

**Diffraction** is the deflection, or "bending," of a wave as it passes a corner or moves through a narrow gap. We might expect that light passing through a sharp-edged hole in a barrier would produce a sharp shadow, especially if radiation were composed of rays or particles moving in perfectly straight lines. As depicted in the first figure, however, closer inspection reveals that the shadow actually has a "fuzzy" edge, as shown in the photograph at the right—the diffraction pattern produced by a small circular opening. We are not normally aware of such effects in everyday life, because diffraction is generally very small for visible light. For any wave, the amount of diffraction is proportional to the ratio of the wavelength to the width of the gap. The longer the wavelength or the smaller the gap, the greater is the angle through which the wave is diffracted. Thus, visible light, with its extremely short wavelengths, shows perceptible diffraction only when passing through very narrow openings. (The effect is much more noticeable for sound waves, however: No one thinks twice about our ability to hear people even when they are around a corner and out of our line of sight.)

**Interference** is the ability of two or more waves to reinforce or diminish each other. The second figure shows two waves moving through the same region of space. Both waves are positioned so that their crests and troughs exactly coincide; the waves have the same wavelength, but the green one has twice the amplitude in the opposite direction to the orange one. The net effect is that the two wave motions interfere with each other, resulting in the wave at the right. This phenomenon is known as *destructive interference*. When, by contrast, the waves reinforce each other, the effect is known as *constructive interference*. As with diffraction, interference between waves of visible light is not noticeable in everyday experience; however, today it is easily measured in the laboratory. The final photograph shows the characteristic *interference pattern* that results when two identical light sources are placed side by side. The light and dark bands are formed by constructive and destructive interference of the beams from the two sources. This classic experiment, first performed by English physicist Thomas Young around 1805, was instrumental in establishing the wave nature of radiation.

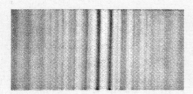

Finally, the phenomenon known as **polarization** of light is also readily understood in terms of the description of electromagnetic waves presented in the text. Normally, light waves are randomly oriented—the electric field in Figure 3.7 may vibrate in any direction perpendicular to the direction of wave motion—and we say that the radiation is unpolarized. Most natural objects emit unpolarized radiation. Under some circumstances, however, the electric fields can become aligned, all vibrating in the same plane as the radiation moves through space, and the radiation is said to be polarized. On Earth, we can produce polarized light by passing unpolarized light through a Polaroid™ filter, which has specially aligned elongated molecules that allow the passage of only those waves having electric fields oriented in some specific direction. Reflected light is often polarized, which is why sunglasses constructed with suitably oriented Polaroid™ filters can be effective in blocking reflected glare.

Diffraction and interference play critical roles in many areas of observational astronomy, including telescope design (Chapter 5). The polarization of starlight provides astronomers with an important technique for probing the properties of interstellar gas (Chapter 18). All three phenomena are predicted by the wave theory of light. The particle theory did not predict them; in fact, it predicted that they should *not* occur.

Until the early 1800s, the technology was inadequate to resolve the issue. However, by 1830, experimenters had reported clear measurements of each, convincing most scientists that the wave theory was the proper description of electromagnetic radiation. It would be almost a century before the particle description of radiation would resurface, but in a radically different form, as we will see in Chapter 4.

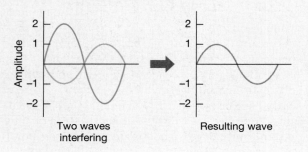

less than about a centimeter, whereas water vapor and carbon dioxide ($CO_2$) are strong absorbers of infrared radiation. Ultraviolet, X-ray, and gamma-ray radiation are completely blocked by the *ozone* ($O_3$) *layer* high in Earth's atmosphere (see Section 7.3). A passing, but unpredictable, source of atmospheric opacity in the visible part of the spectrum is the blockage of light by atmospheric clouds.

In addition, the interaction between the Sun's ultraviolet radiation and the upper atmosphere produces a thin, electrically conducting layer at an altitude of about 100 km. The *ionosphere*, as this layer is known, reflects long-wavelength radio waves (wavelengths greater than about 10 m) as well as a mirror reflects visible light. In this way, extraterrestrial waves are kept out, and terrestrial waves—such as those produced by AM radio stations—are kept in. (That's why it is possible to transmit some radio frequencies beyond the horizon—the broadcast waves bounce off the ionosphere.)

## CONCEPT CHECK

✔ In what sense are radio waves, visible light, and X rays one and the same phenomenon?

## 3.4 Thermal Radiation

*All* macroscopic objects—fires, ice cubes, people, stars—emit radiation at all times, regardless of their size, shape, or chemical composition. They radiate mainly because the microscopic charged particles they are made up of are in constantly varying random motion, and whenever charges interact ("collide") and change their state of motion, electromagnetic radiation is emitted. The **temperature** of an object is a direct measure of the amount of microscopic motion within it (see *More Precisely 3-1*). The hotter the object—that is, the higher its temperature—the faster its component particles move, the more violent are their collisions, and the more energy they radiate.

## MORE PRECISELY 3-1

### The Kelvin Temperature Scale

The atoms and molecules that make up any piece of matter are in constant random motion. This motion represents a form of energy known as *thermal energy*, or, more commonly, *heat*. The quantity we call *temperature* is a direct measure of an object's internal motion: The higher the object's temperature, the faster, on average, is the random motion of its constituent particles. Note that the two concepts, though obviously related, are different. The temperature of a piece of matter specifies the *average* thermal energy of the particles it contains.

Our familiar Fahrenheit temperature scale, like the archaic English system in which length is measured in feet and weight in pounds, is of somewhat dubious value. In fact, the "degree Fahrenheit" is now a peculiarity of American society. Most of the world uses the Celsius scale of temperature measurement (also called the centigrade scale). In the Celsius system, water freezes at 0 degrees (0°C) and boils at 100 degrees (100°C), as illustrated in the accompanying figure.

There are, of course, temperatures below the freezing point of water. In principle, temperatures can reach as low as −273.15°C (although we know of no matter anywhere in the universe that is actually that cold). Known as *absolute zero*, this is the temperature at which, theoretically, all thermal atomic and molecular motion ceases. Since no object can have a temperature below that value, scientists find it convenient to use a temperature scale that takes absolute zero as its starting point. This scale is called the *Kelvin scale*, in honor of the 19th-century British physicist Lord Kelvin. Since it starts at absolute zero, the Kelvin scale differs from the Celsius scale by 273.15°. In this book, we round off the decimal places and simply use

$$\text{kelvins} = \text{degrees Celsius} + 273.$$

Thus,

- all thermal motion ceases at 0 kelvins (0 K).
- water freezes at 273 kelvins (273 K).
- water boils at 373 kelvins (373 K).

Note that the unit is *kelvins*, or *K*, but not *degrees kelvin* or *°K*. (Occasionally, the term *degrees absolute* is used instead.)

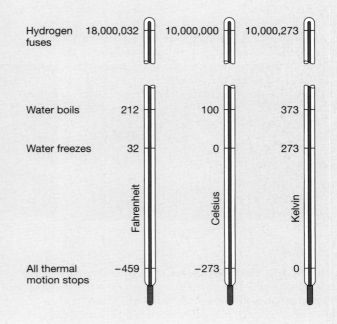

| | Fahrenheit | Celsius | Kelvin |
|---|---|---|---|
| Hydrogen fuses | 18,000,032 | 10,000,000 | 10,000,273 |
| Water boils | 212 | 100 | 373 |
| Water freezes | 32 | 0 | 273 |
| All thermal motion stops | −459 | −273 | 0 |

## The Blackbody Spectrum

*Intensity* is a term often used to specify the amount or strength of radiation at any point in space. Like frequency and wavelength, intensity is a basic property of radiation. No natural object emits all its radiation at just one frequency. Instead, because particles collide at many different speeds—some gently, others more violently—the energy is generally spread out over a range of frequencies. By studying how the intensity of this radiation is distributed across the electromagnetic spectrum, we can learn much about the object's properties.

Figure 3.10 sketches the distribution of radiation emitted by an object. The curve peaks at a single, well-defined frequency and falls off to lesser values above and below that frequency. Note that the curve is not shaped like a symmetrical bell that declines evenly on either side of the peak. Instead, the intensity falls off more slowly from the peak to lower frequencies than it does on the high-frequency side. This overall shape is characteristic of the thermal radiation emitted by *any* object, regardless of its size, shape, composition, or temperature.

The curve drawn in Figure 3.10(a) is the radiation-distribution curve for a mathematical idealization known as a *blackbody*—an object that absorbs all radiation falling on it. In a steady state, a blackbody must reemit the same amount of energy it absorbs. The **blackbody curve** shown in the figure describes the distribution of that reemitted radiation. (The curve is also known as the *Planck curve*, after Max Planck, the German physicist whose mathematical analysis of such thermal emission in 1900 played a key role in the development of modern physics.)

No real object absorbs and radiates as a perfect blackbody. For example, the Sun's actual curve of emission is shown in Figure 3.10(b). However, in many cases, the blackbody curve is a good approximation to reality, and the properties of blackbodies provide important insights into the behavior of real objects.

## The Radiation Laws

The blackbody curve shifts toward higher frequencies (shorter wavelengths) and greater intensities as an object's temperature increases. Even so, the *shape* of the curve remains the same. This shifting of radiation's peak frequency with temperature is familiar to us all: Very hot glowing objects, such as toaster filaments or stars, emit visible light. Cooler objects, such as warm rocks, household radiators, or people, produce invisible radiation—warm to the touch, but not glowing hot to the eye. These latter objects emit most of their radiation in the lower frequency infrared part of the electromagnetic spectrum (Figure 3.9).

Imagine a piece of metal placed in a hot furnace. At first, the metal becomes warm, although its visual appearance doesn't change. As it heats up, the metal begins to glow dull red, then orange, brilliant yellow, and finally white. How do we explain this phenomenon? As illustrated in Figure 3.11, when the metal is at room temperature (300 K—see *More Precisely 3-1* for a discussion of the Kelvin temperature scale), it emits only invisible infrared radiation. As the metal becomes hotter, the peak of its blackbody curve shifts toward higher frequencies. At 1000 K, for instance, most of the emitted radiation is still infrared, but now there is also a small amount of visible (dull red) radiation being emitted. (Note that the high-frequency portion of the 1000 K curve just overlaps the visible region of the graph.)

As the temperature continues to rise, the peak of the metal's blackbody curve moves through the visible spec-

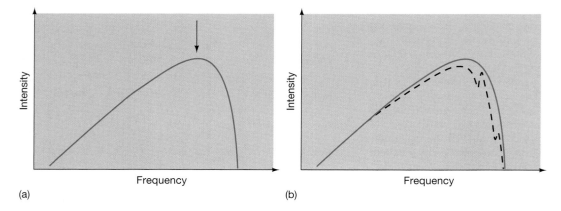

(a)    (b)

▲ **FIGURE 3.10 Blackbody Curves, Ideal vs. Reality** The blackbody, or Planck, curve represents the spread of the intensity of radiation emitted by any object over all possible frequencies. The arrow indicates the frequency of peak emission. Note the contrast between the clean, "textbook" case (a) with a real graph (dashed) of the Sun's emission (b). Absorption in the atmospheres of the Sun and Earth causes the difference.

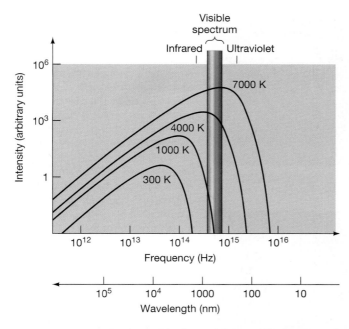

▲ FIGURE 3.11 **Multiple Blackbody Curves** As an object is heated, the radiation it emits peaks at higher and higher frequencies. Shown here are curves corresponding to temperatures of 300 K (room temperature), 1000 K (beginning to glow dull red), 4000 K (red hot), and 7000 K (white hot).

trum, from red (the 4000 K curve) through yellow. Eventually, the metal becomes white hot because, when its blackbody curve peaks in the blue or violet part of the spectrum (the 7000 K curve), the low-frequency tail of the curve extends through the entire visible spectrum (to the left in the figure), meaning that substantial amounts of green, yellow, orange, and red light are also emitted. Together, all these colors combine to produce white.

From studies of the precise form of the blackbody curve, we obtain a very simple connection between the wavelength at which most radiation is emitted and the absolute temperature (i.e., the temperature measured in kelvins) of the emitting object:

$$\text{wavelength of peak emission} \propto \frac{1}{\text{temperature}}.$$

(Recall that the symbol "∝" here just means "is proportional to.") This relationship is called **Wien's law,** after Wilhelm Wien, the German scientist who formulated it in 1897.

Simply put, Wien's law tells us that the hotter the object, the bluer is its radiation. For example, an object with a temperature of 6000 K emits most of its energy in the visible part of the spectrum, with a peak wavelength of 480 nm. At 600 K, the object's emission would peak at a wavelength of 4800 nm, well into the infrared portion of the spectrum. At a temperature of 60,000 K, the peak would move all the way through the visible spectrum to a wavelength of 48 nm, in the ultraviolet range (see Figure 3.12).

It is also a matter of everyday experience that, as the temperature of an object increases, the *total* amount of energy it radiates (summed over all frequencies) increases rapidly. For example, the heat given off by an electric heater increases very sharply as it warms up and begins to emit visible light. Careful experimentation leads to the conclusion that the total amount of energy radiated per unit time is actually proportional to the fourth power of the object's temperature:

$$\text{total energy emission} \propto \text{temperature}^4.$$

This relation is called **Stefan's law,** after the 19th-century Austrian physicist Josef Stefan. From the form of Stefan's law, we can see that the energy emitted by a body rises dramatically as its temperature increases. Doubling the temperature causes the total energy radiated to increase by a factor of $2^4 = 16$; tripling the temperature increases the emission by $3^4 = 81$, and so on.

The radiation laws are presented in more detail in *More Precisely 3-2.*

## Astronomical Applications

No known natural terrestrial objects reach temperatures high enough to emit very high frequency radiation. Only human-made thermonuclear explosions are hot enough for their spectra to peak in the X-ray or gamma-ray range. (Most human inventions that produce short-wavelength, high-frequency radiation, such as X-ray machines, are designed to emit only a specific range of wavelengths and do not operate at high temperatures. They are said to produce a *nonthermal* spectrum of radiation.) Many extraterrestrial objects, however, do emit large amounts of ultraviolet, X-ray, and even gamma-ray radiation. Figure 3.13 shows a familiar object—our Sun—as it appears when viewed with the use of radiation from different parts of the electromagnetic spectrum.

Astronomers often use blackbody curves as thermometers to determine the temperatures of distant objects. For example, an examination of the solar spectrum indicates the temperature of the Sun's surface. Observations of the radiation from the Sun at many frequencies yield a curve shaped somewhat like that shown in Figure 3.10. The Sun's curve peaks in the visible part of the electromagnetic spectrum; the Sun also emits a lot of infrared and a little ultraviolet radiation. Using Wien's law, we find that the temperature of the Sun's surface is approximately 6000 K. (A more precise measurement, applying Wien's law to the blackbody curve that best fits the solar spectrum, yields a temperature of 5800 K.)

Other cosmic objects have surfaces very much cooler or hotter than the Sun's, emitting most of their radiation in invisible parts of the spectrum. For example, the relatively cool surface of a very young star may measure 600 K and emit mostly infrared radiation. Cooler still is the interstellar gas cloud from which the star formed; at a temperature

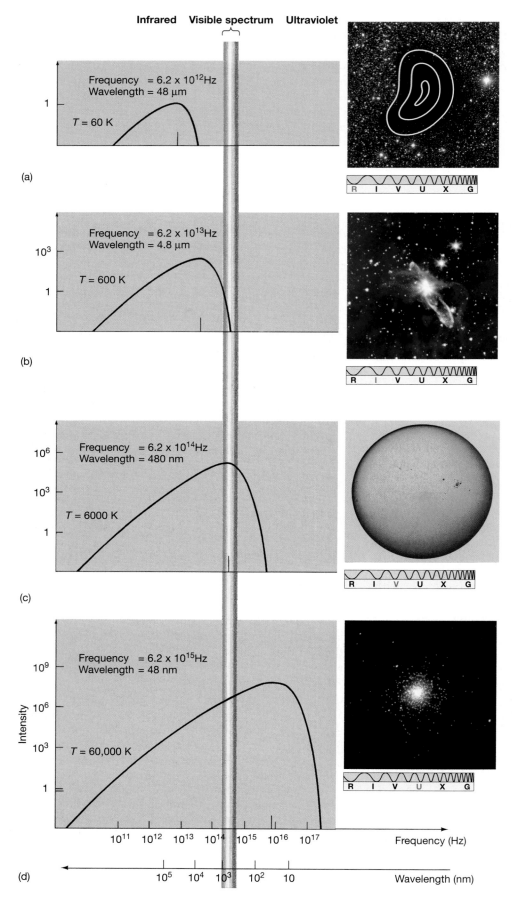

**Infrared    Visible spectrum    Ultraviolet**

(a)

Frequency  = 6.2 x 10$^{12}$Hz
Wavelength = 48 μm

$T = 60$ K

(b)

Frequency  = 6.2 x 10$^{13}$Hz
Wavelength = 4.8 μm

$T = 600$ K

(c)

Frequency  = 6.2 x 10$^{14}$Hz
Wavelength = 480 nm

$T = 6000$ K

(d)

Frequency  = 6.2 x 10$^{15}$Hz
Wavelength = 48 nm

$T = 60,000$ K

Intensity

10$^{11}$  10$^{12}$  10$^{13}$  10$^{14}$  10$^{15}$  10$^{16}$  10$^{17}$    Frequency (Hz)

10$^5$   10$^4$   10$^3$   10$^2$   10    Wavelength (nm)

◀ FIGURE 3.12 **Astronomical Thermometer** Comparison of blackbody curves of four cosmic objects. The frequencies and wavelengths corresponding to peak emission are marked. (a) A cool, dark galactic gas cloud called Barnard 68. At a temperature of 60 K, it emits mostly radio radiation, shown here as overlaid contours. (b) A dim, young star (shown white in the inset photograph) called Herbig-Haro 46. The star's atmosphere, at 600 K, radiates mainly in the infrared. (c) The Sun's surface, at approximately 6000 K, is brightest in the visible region of the electromagnetic spectrum. (d) Some very hot, bright stars in a cluster called Messier 2, as observed by an orbiting space telescope above Earth's atmosphere. At a temperature of 60,000 K, these stars radiate strongly in the ultraviolet. *(ESO; AURA; SST; GALEX)*

of 60 K, such a cloud emits mainly long-wavelength radiation in the radio and infrared parts of the spectrum. The brightest stars, by contrast, have surface temperatures as high as 60,000 K and hence emit mostly ultraviolet radiation (see Figure 3.12).

Multispectral View of Orion Nebula   ANIMATION/VIDEO   Earth Aurora in X-rays   ANIMATION/VIDEO

## CONCEPT CHECK

✔ Describe, in terms of the radiation laws, how and why the appearance of an incandescent lightbulb changes as you turn a dimmer switch to increase its brightness from "off" to "maximum."

# MORE PRECISELY 3-2

## More About the Radiation Laws

As mentioned in Section 3.4, *Wien's law* relates the temperature $T$ of an object to the wavelength $\lambda_{max}$ at which the object emits the most radiation. (The Greek letter $\lambda$—lambda—is conventionally used to denote wavelength.) Mathematically, if we measure $T$ in kelvins and $\lambda_{max}$ in millimeters, we can determine the constant of proportionality in the relation presented in the text, to find that

$$\lambda_{max} = \frac{2.9 \text{ mm}}{T}.$$

We could also convert Wien's law into an equivalent statement about frequency $f$, using the relation $f = c/\lambda$ (see Section 3.1), where $c$ is the speed of light, but the law is most commonly stated in terms of wavelength and is probably easier to remember that way.

**EXAMPLE 1** For a blackbody with the same temperature $T$ as the surface of the Sun ($\approx 6000$ K), the wavelength of maximum intensity is $\lambda_{max} = (2.9/6000)$ mm, or 480 nm, corresponding to the yellow-green part of the visible spectrum. A cooler star with a temperature of $T = 3000$ K has a peak wavelength of $\lambda_{max} = (2.9/3000)$ mm $\approx 970$ nm, just beyond the red end of the visible spectrum, in the near infrared. The blackbody curve of a hotter star with a temperature of 12,000 K peaks at 242 nm, in the near ultraviolet, and so on.

In fact, this application—simply looking at the spectrum and determining where it peaks—is an important way of estimating the temperature of planets, stars, and other objects throughout the universe and will be used extensively throughout the text.

We can also give *Stefan's law* a more precise mathematical formulation. With $T$ measured in kelvins, the total amount of energy emitted per square meter of the body's surface per second (a quantity known as the *energy flux F*) is given by

$$F = \sigma\, T^4.$$

Energy per unit area | Temperature to the fourth power

Constant

This equation is usually referred to as the *Stefan-Boltzmann* equation. Stefan's student, Ludwig Boltzmann, was an Austrian physicist who played a central role in the development of the laws of thermodynamics during the late 19th and early 20th centuries. The constant $\sigma$ (the Greek letter sigma) is known as the Stefan-Boltzmann constant.

The SI unit of energy is the *joule* (J). Probably more familiar is the closely related unit called the *watt* (W), which measures power—the *rate* at which energy is emitted or expended by an object. One watt is the emission of 1 J per second. For example, a 100-W lightbulb emits energy (mostly in the form of infrared and visible light) at a rate of 100 J/s. In SI units, the Stefan-Boltzmann constant has the value $\sigma = 5.67 \times 10^{-8}$ W/m$^2 \cdot$ K$^4$.

**EXAMPLE 2** Notice just how rapidly the energy flux increases with increasing temperature. A piece of metal in a furnace, when at a temperature of $T = 3000$ K, radiates energy at a rate of $\sigma T^4 \times (1 \text{ cm})^2 = 5.67 \times 10^{-8}$ W/m$^2 \cdot$ K$^4 \times (3000 \text{ K})^4 \times (0.01 \text{ m})^2 = 460$ W for every square centimeter of its surface area. Doubling this temperature to 6000 K, (so that the metal becomes yellow hot, by Wien's law), the surface temperature of the Sun, increases the energy emitted by a factor of 16 (four "doublings"), to 7.3 *kilo*watts (7300 W) per square centimeter.

Finally, note that Stefan's law relates to energy emitted *per unit area*. The flame of a blowtorch is considerably hotter than a bonfire, but the bonfire emits far more energy *in total* because it is much larger. Thus, in computing the total energy emitted from a hot object, both the object's temperature *and* its surface area must be taken into account. This fact is of great importance in determining the "energy budget" of planets and stars, as we will see in later chapters. The next example illustrates the point in the case of the Sun.

**EXAMPLE 3** The Sun's temperature is approximately $T = 5800$ K. (The earlier example used a rounded-off version of this number.) Thus, by Stefan's law, each square meter of the Sun's surface radiates energy at a rate of $\sigma T^4 = 6.4 \times 10^7$ W (64 megawatts). By measuring the Sun's angular size and knowing its distance, we can employ simple geometry to determine the solar radius. ∞ (Secs. 1.7, 2.6) The answer is $R = 700,000$ km, or $7 \times 10^8$ m, allowing us to calculate the Sun's total surface area as $4\pi R^2 = 6.2 \times 10^{18}$ m$^2$. Multiplying by the energy emitted per unit area, we find that the Sun's total energy emission (or *luminosity*) is $4 \times 10^{26}$ W—a remarkable number that we obtained without ever leaving Earth!

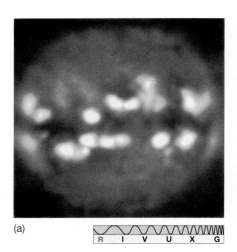 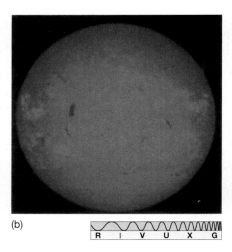 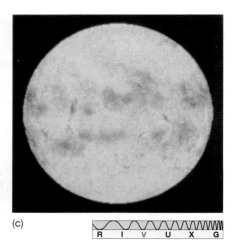

(a)     R I V U X G     (b)     R I V U X G     (c)     R I V U X G

▲ FIGURE 3.13 **The Sun at Many Wavelengths** Three images of the Sun, obtained on the same day using telescopes sensitive to (a) radio waves (note the highlighted "R" in the spectrum icon), (b) infrared radiation ("I"), and (c) visible light ("V"). These images are shown here in "false color," a technique commonly used for displaying intensity, especially with nonvisible radiation. By studying the similarities and differences among various views of the same object acquired on the same day, astronomers can find important clues to the object's structure, composition, and surface activity. Although most sunlight is emitted in the form of infrared and visible radiation, a wealth of information about our parent star can be obtained by studying it in other regions of the electromagnetic spectrum. (*NRAO; AURA*)

## 3.5 The Doppler Effect

Imagine a rocket ship launched from Earth with enough fuel to allow it to accelerate to speeds approaching that of light. As the ship's speed increased, a remarkable thing would happen (Figure 3.14). Passengers would notice that the light from the star system toward which they were traveling seemed to be getting *bluer*. In fact, *all* stars in front of the ship would appear bluer than normal, and the greater the ship's speed, the greater the color change would be. Furthermore, stars behind the vessel would seem *redder* than normal, but stars to either side would be unchanged in appearance. As the spacecraft slowed down and came to rest relative to Earth, all stars would resume their usual appearance.

The travelers would have to conclude that the stars had changed their colors, not because of any real change in their physical properties, but because of the spacecraft's own *motion*. This motion-induced change in the observed frequency of a wave is known as the **Doppler effect,** in honor of Christian Doppler, the 19th-century Austrian physicist who first explained it in 1842. This phenomenon is not restricted to electromagnetic radiation and fast-moving spacecraft. Waiting at a railroad crossing for an express train to pass, most of us have had the experience of hearing the pitch of a train whistle change from high shrill (high frequency, short wavelength) to low blare (low frequency, long wavelength) as the train approaches and then recedes. The explanation is basically the same. Applied to cosmic sources of electromagnetic radiation, the Doppler effect has become one of the most important measurement techniques in all of modern astronomy. Here's how it works.

Imagine a wave moving from the place where it is created toward an observer who is not moving with respect to the source of the wave, as shown in Figure 3.15(a). By noting the distances between successive crests, the observer can determine the wavelength of the emitted wave. Now suppose that not just the wave, but the *source* of the wave, also is moving. As illustrated in Figure 3.15(b), because the source moves between the times of emission of one crest and the next, successive crests in the direction of motion of the source will be seen to be *closer together* than normal, whereas crests behind the source will be more widely spaced. An observer in front of the source will therefore measure a *shorter* wavelength than normal, whereas one behind will see a *longer* wavelength. The numbers indicate (a) successive crests emitted by the source and (b) the location of the source at the instant each crest was emitted.

The greater the relative speed between source and observer, the greater is the observed shift. If the other velocities involved are not too large compared with the wave speed—less than a few percent, say—we can write down a particularly simple formula for what the observer sees. In terms of the net velocity of *recession* between source and observer, the apparent wavelength and frequency (measured by the observer) are related to the true quantities (emitted by the source) as follows:

$$\frac{\text{apparent wavelength}}{\text{true wavelength}} = \frac{\text{true frequency}}{\text{apparent frequency}}$$

$$= 1 + \frac{\text{recession velocity}}{\text{wave speed}}.$$

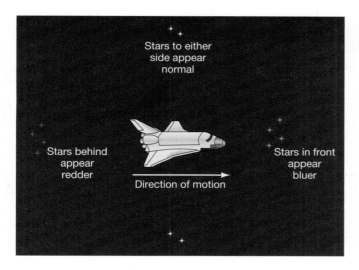

▲ FIGURE 3.14 **High-Speed Observers** Observers in a fast-moving spacecraft see the stars ahead of them bluer than normal, while those behind are reddened. The stars have not actually changed their properties—the color changes result from the observers' motion relative to the stars.

The recession velocity measures the rate at which the distance between the source and the observer is changing. A positive recession velocity means that the two are moving apart; a negative velocity means that they are approaching.

The wave speed is the speed of light, $c$, in the case of electromagnetic radiation. For most of this text, the assumption that the recession velocity is small compared to the speed of light will be a good one. Only when we discuss the properties of black holes (Chapter 22) and the structure of the universe on the largest scales (Chapters 25 and 26) will we have to reconsider this formula.

Note that in Figure 3.15 the *source* is shown in motion (as in our train analogy), whereas in our earlier spaceship example (Figure 3.14) the *observers* were in motion. For electromagnetic radiation, the result is the same in either case—only the *relative* motion between source and observer matters. Note also that only motion along the line joining source and observer—known as *radial* motion—appears in the foregoing equation. Motion that is *transverse* (perpendic-

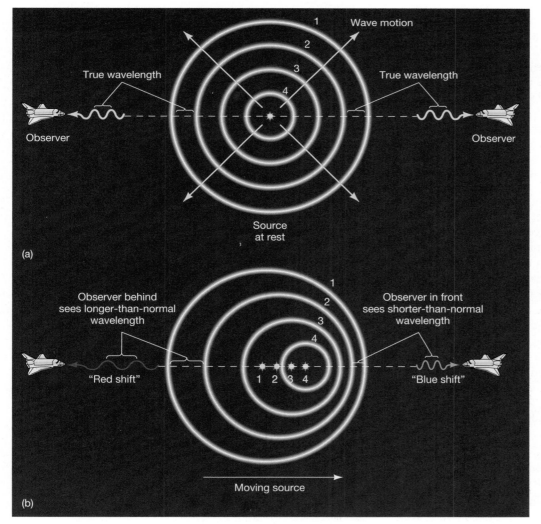

(a)

(b)

◀ FIGURE 3.15 **Doppler Effect** (a) Wave motion from a source toward an observer at rest with respect to the source. The four numbered circles represent successive wave crests emitted by the source. At the instant shown, the fifth crest is just about to be emitted. As seen by the observer, the source is not moving, so the wave crests are just concentric spheres (shown here as circles). (b) Waves from a moving source tend to "pile up" in the direction of motion and be "stretched out" on the other side. (The numbered points indicate the location of the source at the instant each wave crest was emitted.) As a result, an observer situated in front of the source measures a shorter-than-normal wavelength—a *blueshift*—while an observer behind the source sees a *redshift*. In this diagram, the source is shown in motion. However, the same general statements hold whenever there is any relative motion between source and observer.

# MORE PRECISELY 3-3

## Measuring Velocities with the Doppler Effect

Because the speed of light, $c$, is so large—300,000 km/s—the Doppler effect is extremely small for everyday terrestrial velocities. For example, consider a source receding from the observer at Earth's orbital speed of 30 km/s, a velocity much greater than any encountered in day-to-day life. Using the formula in the text, we find that the shift in wavelength of a beam of blue light would be just

$$\frac{\text{change in wavelength}}{\text{true wavelength}} = \frac{\text{recession velocity}}{\text{wave speed}}$$

$$= \frac{30 \text{ km/s}}{300,000 \text{ km/s}} = 0.01 \text{ percent.}$$

That is, the wavelength would lengthen from 400 nm to 400.04 nm—a very small change indeed, and one that the human eye cannot distinguish. However, it *is* easily detectable with modern instruments.

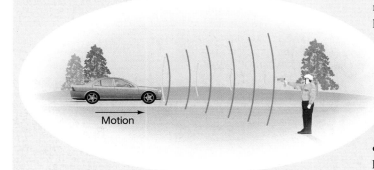

Motion

Astronomers can use the Doppler effect to find the line-of-sight speed of any cosmic object simply by measuring the extent to which its light is redshifted or blueshifted. To see how, let's use a simple example.

---

**EXAMPLE**   Suppose that the beam of blue light just mentioned is observed to have a wavelength of 401 nm, instead of the 400 nm with which it was emitted. (Let's defer until the next chapter the question of *how* an observer might know the wavelength of the emitted light.) Using the earlier equation, rewritten as

$$\frac{\text{recession velocity}}{\text{wave speed, } c} = \frac{\text{change in wavelength}}{\text{true wavelength}}$$

$$= \frac{1 \text{ nm}}{400 \text{ nm}} = 0.0025,$$

the observer could calculate the source's recession velocity to be 0.0025 times 300,000 km/s, or 750 km/s.

---

The basic reasoning is simple, but very powerful. The motions of nearby stars and distant galaxies—even the expansion of the universe itself—have all been measured in this way.

Motorists stopped for speeding on the highway have experienced another, much more down-to-earth, application: As illustrated in the accompanying figure, police radar measures speed by means of the Doppler effect, as do the radar guns used to clock the velocity of a pitcher's fastball or a tennis player's serve. As shown in the illustration, the reflected radiation (blue crests) from the oncoming car is shifted to shorter wavelengths by an amount proportional to the car's speed. Gotcha!

---

ular) to the line of sight has no significant effect.* Notice, incidentally, that the Doppler effect depends only on the *relative motion* between source and observer; it does not depend on the *distance* between them in any way.

A wave measured by an observer situated in front of a moving source is said to be *blueshifted*, because blue light has a shorter wavelength than red light. Similarly, an observer situated behind the source will measure a longer-than-normal wavelength—the radiation is said to be *redshifted*. This terminology is used even for invisible radiation, for which "red" and "blue" have no meaning. Any shift toward shorter wavelengths is called a blueshift, and any shift toward longer wavelengths is called a redshift. For example, ultraviolet radiation might be blueshifted into the X-ray part of the spectrum or redshifted into the visible; infrared

radiation could be redshifted into the microwave range, and so on. *More Precisely 3-3* describes how the Doppler effect is used in practice to measure velocities in astronomy.

In practice, it is hard to measure the Doppler shift of an entire blackbody curve, simply because it is spread over many wavelengths, making small shifts hard to determine with any accuracy. However, if the radiation is more narrowly defined and takes up just a narrow "sliver" of the spectrum, then precise measurements of Doppler effect *can* be made. We will see in the next chapter that in many circumstances this is precisely what does happen, making the Doppler effect one of the observational astronomer's most powerful tools.

CONCEPT CHECK
✔ Astronomers observe two stars orbiting one another. How might the Doppler effect be useful in determining the masses of the stars?

*In fact, Einstein's theory of relativity (see Chapter 22) implies that when the transverse velocity is comparable to the speed of light, a change in wavelength, called the transverse Doppler shift, does occur. For most terrestrial and astronomical applications, however, this shift is negligibly small, and we will ignore it here.*

# CHAPTER REVIEW

## Summary

**1** **Visible light (p. 64)** is a particular type of **electromagnetic radiation (p. 64)** and travels through space in the form of a **wave (p. 64)**. A

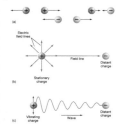

wave is characterized by its **period (p. 65)**, the length of time taken for one complete cycle; its **wavelength (p. 65)**, the distance between successive wave crests; and its **amplitude (p. 65)**, which measures the size of the disturbance associated with the wave. A wave's **frequency (p. 65)** is the reciprocal of the period—it counts the number of wave crests that pass a given point in one second.

**2** **Electrons (p. 67)** and **protons (p. 67)** are elementary particles that carry equal and opposite electrical charges. Any electrically charged object is surrounded by an **electric field (p. 67)** that determines the force the object exerts on other charged objects. When a charged particle moves, information about its motion is transmitted throughout the universe by the particle's changing electric and magnetic fields. According to the **wave theory of radiation (p. 69)**, the information travels through space in the form of a wave at the **speed of light (p. 69)**. Because both electric and **magnetic fields (p. 68)** are involved, the phenomenon is known as **electromagnetism (p. 68)**. **Diffraction (p. 72)**, **interference (p. 72)**, and **polarization (p. 72)** are properties of radiation that mark it as a wave phenomenon.

**3** A beam of white light is bent, or refracted, as it passes through a prism. Different frequencies of light within the beam are refracted by different amounts, so the beam is split up into its component colors—the visible spectrum. The color of visible light is simply a measure of its wavelength—red light has a longer wavelength than blue light. The entire **electromagnetic spectrum (p. 70)** consists of (in order of

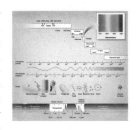

increasing frequency) **radio waves, infrared radiation, visible light, ultraviolet radiation, X rays,** and **gamma rays (p. 64)**. Only radio waves, some infrared wavelengths, and visible light can penetrate the atmosphere and reach the ground from space.

**4** The **temperature (p. 73)** of an object is a measure of the speed with which its constituent particles move. The intensity of radiation of different frequencies emitted by a hot object has a characteristic distribution, called a **blackbody curve (p. 74)**, that depends only on the object's temperature.

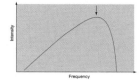

**5** **Wien's law (p. 75)** tells us that the wavelength at which the object radiates most of its energy is inversely proportional to the temperature of the object. Measuring that peak wavelength tells us the object's temperature. **Stefan's law (p. 75)** states that the total amount of energy radiated is proportional to the fourth power of the temperature.

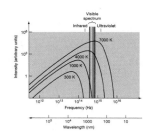

**6** Our perception of the wavelength of a beam of light can be altered by our velocity relative to the source. This motion-induced change in the observed frequency of a wave is called the **Doppler effect (p. 78)**. Any net motion away from the source causes a redshift—a shift to lower frequencies—in the received beam. Motion toward the source causes a blueshift. The extent of the shift is directly proportional to the observer's recession velocity relative to the source, providing a way to measure the velocity of a distant object by observing the radiation it emits.

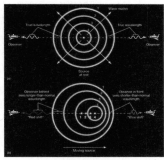

## Review and Discussion

1. What is a wave?
2. Define the following wave properties: period, wavelength, amplitude, and frequency.
3. What is the relationship between wavelength, wave frequency, and wave velocity?
4. What is diffraction, and how does it relate to the behavior of light as a wave?
5. What's so special about $c$?
6. Name the colors that combine to make white light. What is it about the various colors that causes us to perceive them differently?

7. What effect does a positive charge have on a nearby negatively charged particle?
8. Compare and contrast the gravitational force with the electric force.
9. Describe the way in which light leaves a star, travels through the vacuum of space, and finally is seen by someone on Earth.
10. Why is light referred to as an electromagnetic wave?
11. What do radio waves, infrared radiation, visible light, ultraviolet radiation, X-rays, and gamma rays have in common? How do they differ?

12. In what regions of the electromagnetic spectrum is the atmosphere transparent enough to allow observations from the ground?

13. What is a blackbody? What are the main characteristics of the radiation it emits?

14. What does Wien's law reveal about stars in the sky?

15. What does Stefan's law tell us about the radiation emitted by a blackbody?

16. In terms of its blackbody curve, describe what happens as a red-hot glowing coal cools.

17. What is the Doppler effect, and how does it alter the way in which we perceive radiation?

18. How do astronomers use the Doppler effect to determine the velocities of astronomical objects?

19. A source of radiation and an observer are traveling through space at precisely the same velocity, as seen by a second observer. Would you expect the first observer to measure a Doppler shift in the light received from the source?

20. If Earth were completely blanketed with clouds and we couldn't see the sky, could we learn about the realm beyond the clouds? What forms of radiation might be received?

## Conceptual Self-Test: True or False/Multiple Choice

1. Electromagnetic waves can travel through a perfect vacuum.

2. Sound is a familiar type of electromagnetic wave.

3. Interference occurs when one wave is brighter than another and the fainter wave cannot be observed.

4. A blackbody emits all its radiation at a single frequency.

5. Earth's atmosphere is transparent to all forms of electromagnetic radiation.

6. The peak of an object's emitted radiation occurs at a frequency determined by the object's temperature.

7. The lowest possible temperature is 0 K.

8. Two otherwise identical objects have temperatures of 1000 K and 1200 K, respectively. The object at 1200 K emits roughly twice as much radiation as the object at 1000 K.

9. As you drive away from a radio transmitter, the radio signal you receive from the station is shifted to longer wavelengths.

10. The Doppler effect occurs for all types of wave motion.

11. Compared with ultraviolet radiation, infrared radiation has a greater (a) wavelength; (b) amplitude; (c) frequency; (d) energy.

12. Compared with red light, blue wavelengths of visible light travel (a) faster; (b) slower; (c) at the same speed.

13. An electron that collides with an atom will (a) cease to have an electric field; (b) produce an electromagnetic wave; (c) change its electric charge; (d) become magnetized.

14. A wavelength of green light is about the size of (a) an atom; (b) a bacterium; (c) a fingernail; (d) a skyscraper.

15. An X-ray telescope located in Antarctica would not work well because of (a) the extreme cold; (b) the ozone hole; (c) continuous daylight; (d) Earth's atmosphere.

16. In Figure 3.11 ("Multiple Blackbody Curves"), an object at 1000 K emits mostly (a) infrared light; (b) red light; (c) multiple green light; (d) blue light.

17. According to Wien's law, the hottest stars also have (a) the longest peak wavelength; (b) the shortest peak wavelength; (c) maximum emission in the infrared region of the spectrum; (d) the largest diameters.

18. Stefan's law says that if the Sun's temperature were to double, its energy emission would (a) become half its present value; (b) double; (c) increase four times; (d) increase 16 times.

19. A star much cooler than the Sun would appear (a) red; (b) blue; (c) smaller; (d) larger.

20. The blackbody curve of a star moving toward Earth would have its peak shifted (a) to a higher intensity; (b) toward higher energies; (c) toward longer wavelengths; (d) to a lower intensity.

# Problems

*Algorithmic versions of these Problems are available in the Practice Problems module of the Companion Website. The number of dots preceding each Problem indicates its approximate level of difficulty.*

1. • A sound wave moving through water has a frequency of 256 Hz and a wavelength of 5.77 m. What is the speed of sound in water?

2. • What is the wavelength of a 100-MHz ("FM 100") radio signal?

3. • What would be the frequency of an electromagnetic wave having a wavelength equal to Earth's diameter? In what part of the electromagnetic spectrum would such a wave lie?

4. • Estimate the frequency of an electromagnetic wave having a wavelength equal to the size of the period at the end of this sentence. In what part of the electromagnetic spectrum would such a wave lie?

5. • What would be the wavelength of an electromagnetic wave having a frequency equal to the clock speed of a 3.2-GHz personal computer? In what part of the electromagnetic spectrum would such a wave lie?

6. • The blackbody emission spectrum of object *A* peaks in the ultraviolet region of the electromagnetic spectrum at a wavelength of 200 nm. That of object *B* peaks in the red region, at 650 nm. Which object is hotter, and, according to Wien's law, how many times hotter is it? According to Stefan's law, how many times more energy per unit area does the hotter body radiate per second?

7. • Normal human body temperature is about 37°C. What is this temperature in kelvins? What is the peak wavelength emitted by a person with this temperature? In what part of the spectrum does this lie?

8. •• Estimate the total amount of energy you radiate to your surroundings.

9. • The Sun has a temperature of 5800 K, and its blackbody emission peaks at a wavelength of approximately 500 nm. At what wavelength does a protostar with a temperature of 1000 K radiate most strongly?

10. • Two otherwise identical bodies have temperatures of 300 K and 1500 K, respectively. Which one radiates more energy, and by what factor does its emission exceed the emission of the other body?

11. •• According to the Stefan-Boltzmann law, how much energy is radiated into space per unit time by each square meter of the Sun's surface? (See *More Precisely 3-2*.) If the Sun's radius is 696,000 km, what is the total power output of the Sun?

12. • Radiation from the nearby star Alpha Centauri is observed to be reduced in wavelength (after correction for Earth's orbital motion) by a factor of 0.999933. What is the recession velocity of Alpha Centauri relative to the Sun?

13. • At what velocity and in what direction would a spacecraft have to be moving for a radio station on Earth transmitting at 100 MHz to be picked up by a radio tuned to 99.9 MHz?

14. • A space traveler is approaching the Sun at a speed of 100 km/s and is observing a 700-nm red laser beam coming from Earth. If the traveler's trajectory lies in the same plane as Earth's orbit, what will be the minimum and maximum wavelengths he observes as Earth orbits the Sun?

15. ••• Imagine that you are observing a spacecraft moving in a circular orbit of radius 100,000 km around a distant planet. You happen to be located in the plane of the spacecraft's orbit. You find that the spacecraft's radio signal varies periodically in wavelength between 2.99964 m and 3.00036 m. Assuming that the radio is broadcasting normally, at a constant wavelength, what is the mass of the planet?

*The Companion Website at www.aw-bc.com/chaisson provides algorithmically generated versions of each chapter's Problems, along with additional quizzes, an Animations & Videos gallery, an Images gallery, an interactive Glossary, and a full eBook.*

CHAPTER 4

Spectroscopy is a powerful observational technique enabling scientists to infer the nature of matter by the way it emits or absorbs radiation. Not only can spectroscopy reveal the chemical composition of distant stars and yield knowledge of how they shine, it can also provide a wealth of information about the origin, evolution, and destiny of stars in the universe. Here, part of the incredibly rich spectrum of the star Procyon, wrapped around row after row from left to right and top to bottom across the visible spectrum from red to blue, shows myriad dark lines caused by the absorption of light in the hot star's cooler atmosphere. (AURA) ▶

# SPECTROSCOPY

## The Inner Workings of Atoms

The wave description of radiation allowed 19th-century astronomers to begin to decipher the information reaching Earth from the cosmos in the form of visible and invisible light. However, early in the 20th century, it became clear that the wave theory of electromagnetic phenomena was incomplete—some aspects of light simply could not be explained in purely wave terms. When radiation interacts with matter on atomic scales, it does so not as a continuous wave, but in a jerky, discontinuous way—in fact, as a particle. With this discovery, scientists quickly realized that atoms, too, must behave in a discontinuous way, and the stage was set for a scientific revolution—*quantum mechanics*—that has affected virtually every area of modern life.

The collection of observational and theoretical techniques that enable researchers to determine the nature of distant atoms by the way they emit and absorb radiation is called *spectroscopy*. It is the indispensable foundation of modern astrophysics.

## LEARNING GOALS

*Studying this chapter will enable you to*

1 Describe the characteristics of continuous, emission, and absorption spectra and the conditions under which each is produced.

2 Explain the relation between emission and absorption lines and what we can learn from those lines.

3 Specify the basic components of the atom and describe our modern conception of its structure.

4 Discuss the observations that led scientists to conclude that light has particle as well as wave properties.

5 Explain how electron transitions within atoms produce unique emission and absorption features in the spectra of those atoms.

6 Describe the general features of spectra produced by molecules.

7 List and explain the kinds of information that can be obtained by analyzing the spectra of astronomical objects.

 Visit www.aw-bc.com/chaisson for additional images, animations, quizzes, and eBook for this chapter.

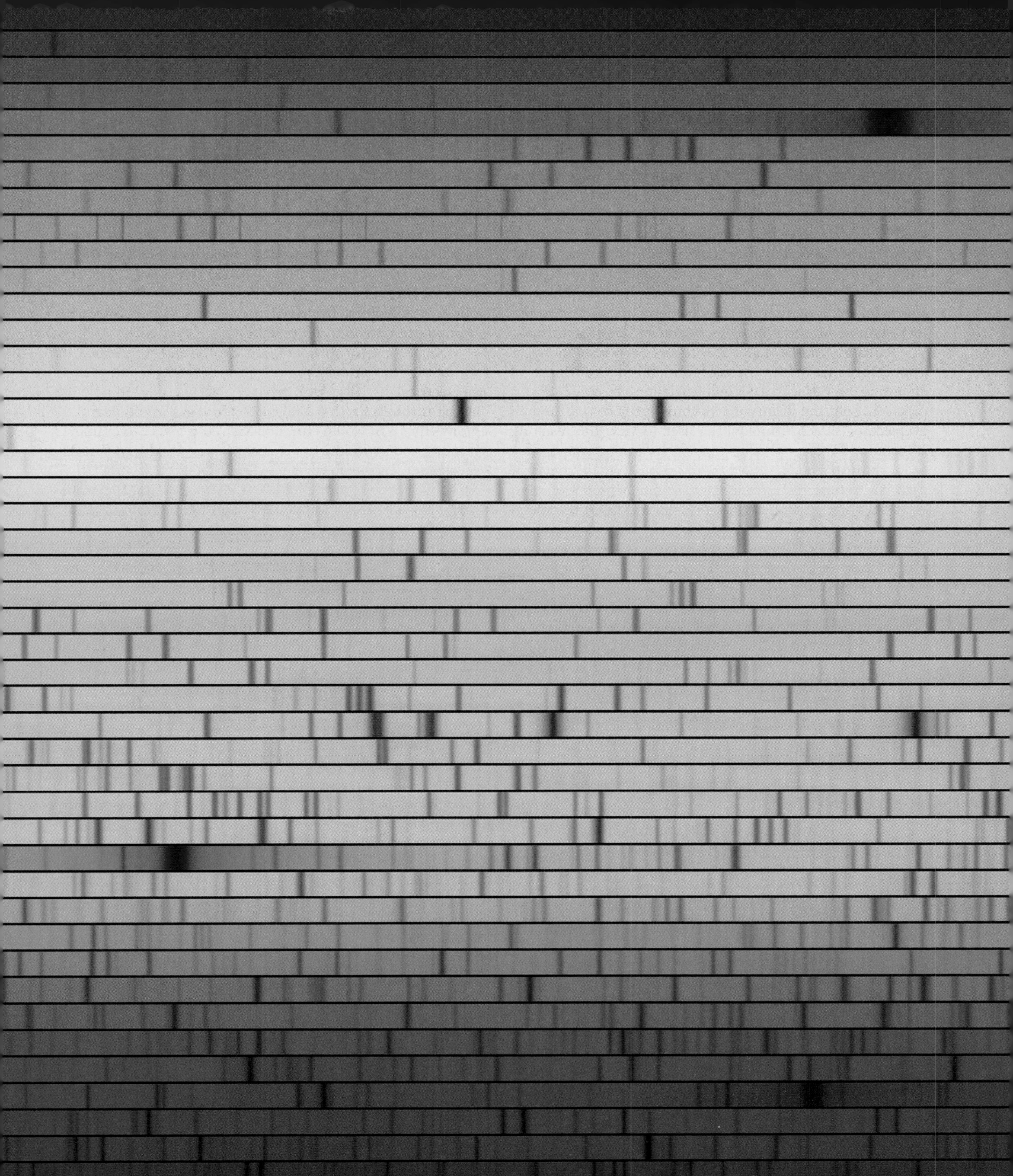

# 4.1 Spectral Lines

In Chapter 3, we saw something of how astronomers can analyze electromagnetic radiation received from space to obtain information about distant objects. A vital step in this process is the formation of a *spectrum*—a splitting of the incoming radiation into its component wavelengths. But in reality, *no* cosmic object emits a perfect blackbody spectrum like those discussed earlier. ∞ (Sec. 3.4) All spectra deviate from this idealized form—some by only a little, others by a lot. Far from invalidating our earlier studies, however, these deviations contain a wealth of detailed information about physical conditions within the source of the radiation. Because spectra are so important, let's examine how astronomers obtain and interpret them.

Radiation can be analyzed with an instrument known as a **spectroscope.** In its most basic form, this device consists of an opaque barrier with a slit in it (to define a beam of light), a prism (to split the beam into its component colors), and an eyepiece or screen (to allow the user to view the resulting spectrum). Figure 4.1 shows such an arrangement. The research instruments called *spectrographs*, or *spectrometers*, used by professional astronomers are rather more complex, consisting of a telescope (to capture the radiation), a dispersing device (to spread the radiation out into a spectrum), and a detector (to record the result). Despite their greater sophistication, their basic operation is conceptually similar to the simple spectroscope shown in the figure.

In many large instruments, the prism is replaced by a device called a *diffraction grating*, consisting of a sheet of transparent material with numerous closely spaced parallel lines ruled on it. The spacing between the lines is typically a few microns ($10^{-6}$ m), comparable to the wavelength of visible light. The spaces act as many tiny openings, and light is diffracted as it passes through the grating (or is reflected from it, depending on the design of the device). ∞ (*Discovery 3-1*) Because different wavelengths of electromagnetic radiation are diffracted by different amounts on encountering the grating, the effect is to split a beam of light into its component colors. You are probably more familiar with diffraction gratings than you think—the "rainbow" of colors seen in light reflected from a compact disk is the result of precisely this process.

## Emission Lines

The spectra we encountered in Chapter 3 are examples of **continuous spectra.** A lightbulb, for example, emits radiation of all wavelengths (mostly in the visible range), with an intensity distribution that is well described by the blackbody curve corresponding to the bulb's temperature. ∞ (Sec. 3.4) Viewed through a spectroscope, the spectrum of the light from the bulb would show the familiar rainbow of colors, from red to violet, without interruption, as presented in Figure 4.2(a).

Not all spectra are continuous, however. For instance, if we took a glass jar containing pure hydrogen gas and passed an electrical discharge through it (a little like a lightning bolt arcing through Earth's atmosphere), the gas would begin to glow—that is, it would emit radiation. If we were to examine that radiation with our spectroscope, we would find that its spectrum consists of only a few bright lines on an otherwise dark background, quite unlike the continuous spectrum described for the incandescent lightbulb. Figure 4.2(b) shows the experimental arrangement and its result schematically. (A more detailed rendering of the spectrum of hydrogen appears in the top panel of Figure 4.3.) Note that the light produced by the hydrogen in this experiment does *not* consist of all possible colors, but instead includes only a few narrow, well-defined **emission lines**—thin "slices" of the continuous spectrum. The black background represents all the wavelengths *not* emitted by hydrogen.

After some experimentation, we would also find that, although we could alter the *intensity* of the lines—for example, by changing the amount of hydrogen in the jar or the strength of the electrical discharge—we could not alter their *color* (in other words, their frequency or wavelength). The pattern of spectral emission lines shown is a property of the element hydrogen. Whenever we perform this experiment, the same characteristic colors result.

By the early 19th century, scientists had carried out similar experiments on many different gases. By vaporizing solids and liquids in a flame, they extended their inquiries to include materials that are not normally found in the gaseous state. Sometimes the pattern of lines was fairly simple,

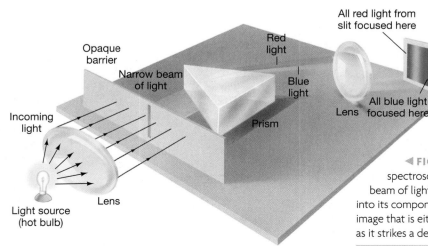

◀ **FIGURE 4.1  Spectroscope** Diagram of a simple spectroscope. A thin slit in the barrier at the left allows a narrow beam of light to pass. The light continues through a prism and is split into its component colors. A lens then focuses the light into a sharp image that is either projected onto a screen, as shown here, or analyzed as it strikes a detector.

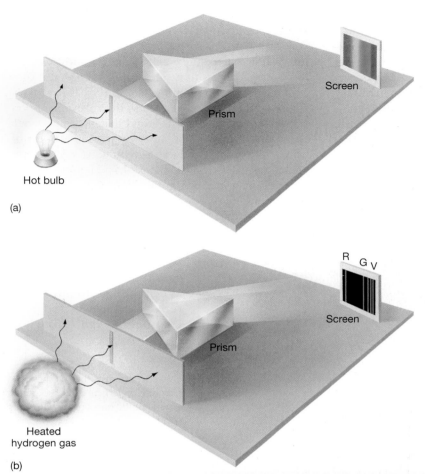

(a)

(b)

◀ FIGURE 4.2 **Continuous and Emission Spectra** When passed through a slit and split up by a prism, light from a source of continuous radiation (a) gives rise to the familiar rainbow of colors. By contrast, the light from excited hydrogen gas (b) consists of a series of distinct bright spectral lines called emission lines. (The focusing lenses have been omitted for clarity—see Section 5.1.)

and sometimes it was complex, but it was always *unique* to that element. Even though the origin of the lines was not understood, researchers quickly realized that the lines provided a one-of-a-kind "fingerprint" of the substance under investigation. They could detect the presence of a particular atom or molecule (a group of atoms held together by chemical bonds—see Section 4.4) solely through the study of the light it emitted. Scientists have accumulated extensive catalogs of the specific wavelengths at which many different hot gases emit radiation. The particular pattern of light emitted by a gas of a given chemical composition is known as the **emission spectrum** of the gas. The emission spectra of some common substances are shown in Figure 4.3.

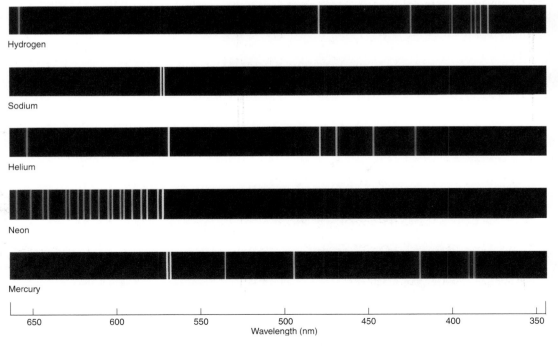

▲ FIGURE 4.3 **Elemental Emission** The emission spectra of some well-known elements. In accordance with the convention adopted throughout this text, frequency increases to the right. Note that wavelengths shorter than approximately 400 nm, shown here in shades of purple, are actually in the ultraviolet part of the spectrum and are not visible to the human eye. *(Wabash Instrument Corp.)*

## Absorption Lines

When sunlight is split by a prism, at first glance it appears to produce a continuous spectrum. However, closer scrutiny with a spectroscope shows that the solar spectrum is interrupted vertically by a large number of narrow dark lines, as shown in Figure 4.4. We now know that many of these lines represent wavelengths of light that have been removed (absorbed) by gases present either in the outer layers of the Sun or in Earth's atmosphere. These gaps in the spectrum are called **absorption lines.**

The English astronomer William Wollaston first noticed the solar absorption lines in 1802. They were studied in greater detail about 10 years later by the German physicist Joseph von Fraunhofer, who measured and cataloged over 600 of them. They are now referred to collectively as *Fraunhofer lines*. Although the Sun is by far the easiest star to study, and so has the most extensive set of observed absorption lines, similar lines are known to exist in the spectra of all stars.

At around the same time as the solar absorption lines were discovered, scientists found that such lines could also be produced in the laboratory by passing a beam of light from a source that produces a continuous spectrum through a cool gas, as shown in Figure 4.5. The scientists quickly observed an intriguing connection between emission and absorption lines: The absorption lines associated with a given gas occur at precisely the *same* wavelengths as the emission lines produced when the gas is heated.

As an example, consider the element sodium, whose emission spectrum appears in Figure 4.6. When heated to high temperatures, a sample of sodium vapor emits visible light strongly at just

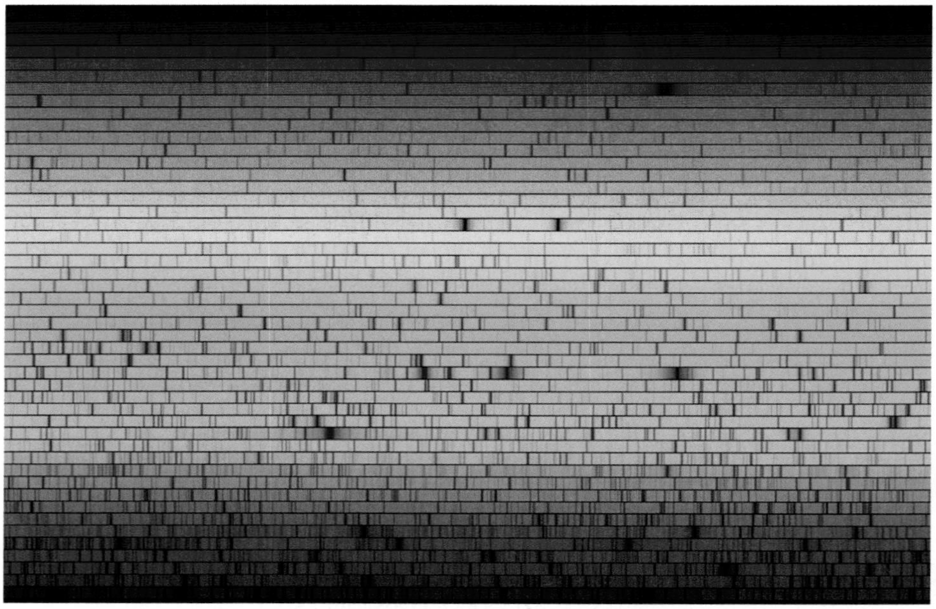

◀ FIGURE 4.4 **Solar Spectrum** This visible spectrum of the Sun shows hundreds of vertical dark absorption lines superimposed on a bright continuous spectrum. The high-resolution spectrum is displayed in a series of 48 horizontal strips stacked vertically; each strip covers a small portion of the entire spectrum from left to right. If the strips were placed side by side, the full spectrum would be some 6 meters (20 feet) across! The scale extends from long wavelengths (red) at the upper left to short wavelengths (blue) at the lower right. (*AURA*)

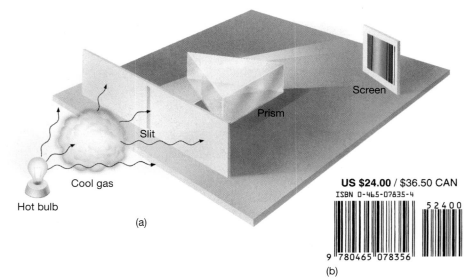

(a)

(b)

◀ FIGURE 4.5 **Absorption Spectrum** (a) When cool gas is placed between a source of continuous radiation (such as a hot lightbulb) and a detector/screen, the resulting color spectrum is crossed by a series of dark absorption lines. These lines are formed when the intervening cool gas absorbs certain wavelengths (colors) from the original beam of light. The absorption lines appear at precisely the same wavelengths as the emission lines that would be produced if the gas were heated to high temperatures (see Figure 4.2). (b) An everyday analogy for any of these line spectra is a supermarket bar code that uniquely determines the cost of some product. (*(b) E. Chaisson*)

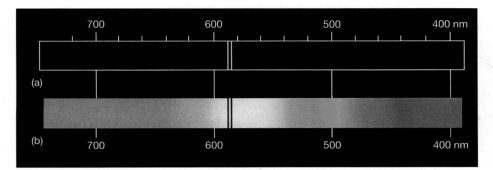

(a) The characteristic emission lines of sodium. The two bright lines in the center appear in the yellow part of the spectrum. (b) The absorption spectrum of sodium. The two dark lines appear at exactly the same wavelengths as the bright lines in the sodium emission spectrum.

two wavelengths—589.9 nm and 589.6 nm—lying in the yellow part of the spectrum. When a continuous spectrum is passed through some relatively cool sodium vapor, two sharp, dark absorption lines appear at precisely the same wavelengths. The emission and absorption spectra of sodium are compared in Figure 4.6, clearly showing the relation between emission and absorption features.

## Kirchhoff's Laws

The analysis of the ways in which matter emits and absorbs radiation is called **spectroscopy.** One early spectroscopist, the German physicist Gustav Kirchhoff, summarized the observed relationships among the three types of spectra—continuous, emission line, and absorption line—in 1859. He formulated three spectroscopic rules, now known as **Kirchhoff's laws,** governing the formation of spectra:

1.  A luminous solid or liquid, or a sufficiently dense gas, emits light of all wavelengths and so produces a *continuous spectrum* of radiation.

2.  A low-density, hot gas emits light whose spectrum consists of a series of bright *emission lines* that are characteristic of the chemical composition of the gas.

3.  A cool, thin gas absorbs certain wavelengths from a continuous spectrum, leaving dark *absorption lines* in their place, superimposed on the continuous spectrum. Once again, these lines are characteristic of the composition of the intervening gas—they occur at precisely the same wavelengths as the emission lines produced by that gas at higher temperatures.

Figure 4.7 illustrates Kirchhoff's laws and the relationship between absorption and emission lines. Viewed directly, the light source, a hot solid (the filament of the bulb), has a continuous (blackbody) spectrum. When the light source is viewed through a cloud of cool hydrogen gas, a series of dark absorption lines appear, superimposed on the spectrum at wavelengths characteristic of hydrogen. The lines appear because the light at those wavelengths is absorbed by the hydrogen. As we will see later in this chapter, the absorbed energy is subsequently reradiated

into space—but in all directions, not just the original direction of the beam. Consequently, when the cloud is viewed from the side against an otherwise dark background, a series of faint emission lines is seen. These lines contain the energy lost by the forward beam. If the gas was heated to incandescence, it would produce stronger emission lines at precisely the same wavelengths.

## Identifying Starlight

By the late 19th century, spectroscopists had developed a formidable arsenal of techniques for interpreting the radiation received from space. Once astronomers knew that spectral lines were indicators of chemical composition, they set about identifying the observed lines in the solar spectrum. Almost all the lines in light from extraterrestrial sources could be attributed to known elements. For example, many of the Fraunhofer lines in sunlight are associated with the element iron, a fact first recognized by Kirchhoff and coworker Robert Bunsen (of Bunsen burner fame) in 1859. However, some unfamiliar lines also appeared in the solar spectrum. In 1868, astronomers realized that those lines must correspond to a previously unknown element. It was given the name helium, after the Greek word *helios*, meaning "Sun." Not until 1895, almost three decades after its detection in sunlight, was helium discovered on Earth! (A laboratory spectrum of helium is included in Figure 4.3.)

Yet, for all the information that 19th-century astronomers could extract from observations of stellar spectra, they still lacked a theory explaining how the spectra themselves arose. Despite their sophisticated spectroscopic equipment, they knew scarcely any more about the physics of stars than did Galileo or Newton. To understand how spectroscopy can be used to extract detailed information about astronomical objects from the light they emit, we must delve more deeply into the processes that produce line spectra.

## CONCEPT CHECK

✔ What are absorption and emission lines, and what do they tell us about the composition of the gas producing them?

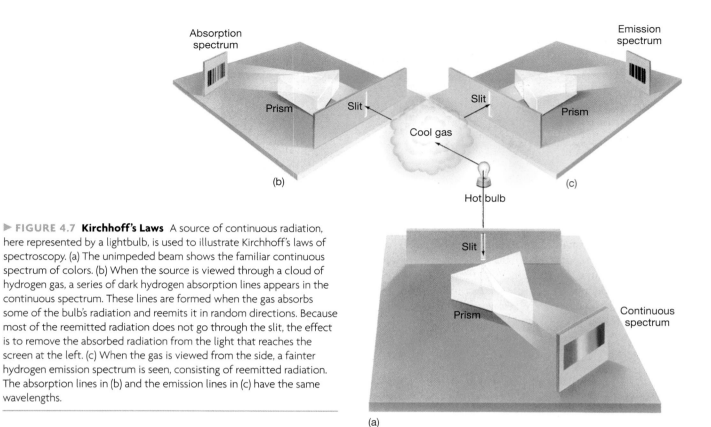

▶ FIGURE 4.7 **Kirchhoff's Laws** A source of continuous radiation, here represented by a lightbulb, is used to illustrate Kirchhoff's laws of spectroscopy. (a) The unimpeded beam shows the familiar continuous spectrum of colors. (b) When the source is viewed through a cloud of hydrogen gas, a series of dark hydrogen absorption lines appears in the continuous spectrum. These lines are formed when the gas absorbs some of the bulb's radiation and reemits it in random directions. Because most of the reemitted radiation does not go through the slit, the effect is to remove the absorbed radiation from the light that reaches the screen at the left. (c) When the gas is viewed from the side, a fainter hydrogen emission spectrum is seen, consisting of reemitted radiation. The absorption lines in (b) and the emission lines in (c) have the same wavelengths.

## 4.2 Atoms and Radiation

By the start of the 20th century, physicists had accumulated substantial evidence that light sometimes behaves in a manner that cannot be explained by the wave theory. As we have just seen, the production of absorption and emission lines involves only certain very specific frequencies or wavelengths of light. This would not be expected if light behaved like a continuous wave and matter always obeyed the laws of Newtonian mechanics. Other experiments conducted around the same time strengthened the conclusion that the notion of radiation as a wave was incomplete. It became clear that when light interacts with matter on very small scales, it does so not in a continuous way, but in a discontinuous, "stepwise" manner. The challenge was to find an explanation for this unexpected behavior. The eventual solution revolutionized our view of nature and now forms the foundation for all of physics and astronomy—indeed, for virtually all modern science.

### Atomic Structure

To explain the formation of emission and absorption lines, we must understand not just the nature of light, but also the structure of **atoms**—the microscopic building blocks from which all matter is constructed. Let's start with the simplest atom of all: hydrogen. A hydrogen atom consists of an electron with a negative electrical charge orbiting a proton carrying a positive charge. The proton forms the central

nucleus (plural: nuclei) of the atom. The hydrogen atom as a whole is electrically neutral. The equal and opposite charges of the proton and the orbiting electron produce an electrical attraction that binds them together within the atom.

How does this picture of the hydrogen atom relate to the characteristic emission and absorption lines associated with hydrogen gas? If an atom absorbs some energy in the form of radiation, that energy must cause some internal change. Similarly, if the atom emits energy, that energy must come from somewhere within the atom. It is reasonable (and correct) to suppose that the energy absorbed or emitted by the atom is associated with changes in the motion of the orbiting electron.

The first theory of the atom to provide an explanation of hydrogen's observed spectral lines was set forth by the Danish physicist Niels Bohr in 1912. Now known simply as the *Bohr model* of the atom, its essential features are as follows:

1. There is a state of lowest energy—the **ground state**—which represents the "normal" condition of the electron as it orbits the nucleus.
2. There is a maximum energy that the electron can have and still be part of the atom. Once the electron acquires more than that maximum energy, it is no longer bound to the nucleus, and the atom is said to be **ionized;** an atom missing one or more of its electrons is called an *ion*.

**3.** Most important (and also least intuitive), between those two energy levels, the electron can exist only in certain sharply defined energy states, often referred to as **orbitals.**

This description of the atom contrasts sharply with the predictions of Newtonian mechanics, which would permit orbits with *any* energy, not just at certain specific values. ∞ (Sec. 2.8) In the atomic realm, such discontinuous behavior is the norm. In the jargon of the field, the orbital energies are said to be **quantized.** The rules of **quantum mechanics,** the branch of physics governing the behavior of atoms and subatomic particles, are far removed from everyday experience.

In Bohr's original model, each electron orbital was pictured as having a specific radius, much like a planetary orbit in the solar system, as shown in Figure 4.8. However, the modern view is not so simple. Although each orbital *does* have a precise energy, the orbits are not sharply defined, as indicated in the figure. Rather, the electron is now envisioned as being smeared out in an "electron cloud" surrounding the nucleus, as illustrated in Figure 4.9. We cannot tell "where" the electron is—we can only speak of the *probability* of finding it in a certain location within the cloud. It is common to speak of the average distance from the cloud to the nucleus as the "radius" of the electron's orbit. When a hydrogen atom is in its ground state, the radius of the orbit is about 0.05 nm (0.5 Å). As the orbital energy increases, the radius increases, too.

For the sake of clarity in the diagrams that follow, we will represent electron orbitals in this chapter as solid lines. (See *More Precisely 4-1* on p. 92 for a more detailed rendition of hydrogen's energy levels.) However, you should always bear in mind that Figure 4.9 is a more accurate depiction of reality.

Atoms do not always remain in their ground state. An atom is said to be in an **excited state** when an electron occupies an orbital at a greater-than-normal distance from its parent nucleus. An atom in such an excited state has a greater-than-normal amount of energy. The

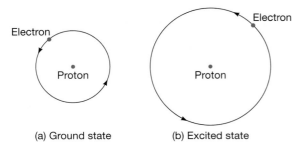

**▲ FIGURE 4.8 Classical Atom** An early-20th-century conception of the hydrogen atom—the Bohr model—pictured its electron orbiting the central proton in a well-defined orbit, rather like a planet orbiting the Sun. Two electron orbitals of different energies are shown: (a) the ground state and (b) an excited state.

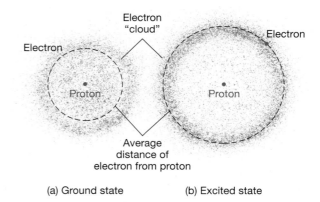

**▲ FIGURE 4.9 Modern Atom** The modern view of the hydrogen atom sees the electron as a "cloud" surrounding the nucleus. The same two energy states are shown as in Figure 4.8.

excited state with the lowest energy (that is, the state closest in energy to the ground state) is called the *first excited state*, that with the second-lowest energy is the *second excited state*, and so on. An atom can become excited in one of two ways: by absorbing some energy from a source of electromagnetic radiation or by colliding with some other particle—another atom, for example. However, the electron cannot stay in a higher orbital forever; the ground state is the only level where it can remain indefinitely. After about $10^{-8}$ s, an excited atom returns to its ground state.

## Radiation as Particles

Because electrons can exist only in orbitals having specific energies, atoms can absorb only specific amounts of energy as their electrons are boosted into excited states. Likewise, atoms can emit only specific amounts of energy as their electrons fall back to lower energy states. Thus, the amount of light energy absorbed or emitted in these processes *must correspond precisely to the energy difference between two orbitals.* The atom's quantized energy levels require that light be absorbed and emitted in the form of distinct "packets" of electromagnetic radiation, each carrying a specific amount of energy. We call these packets **photons.** A photon is, in effect, a "particle" of electromagnetic radiation.

The idea that light sometimes behaves not as a continuous wave, but as a stream of particles, was proposed by Albert Einstein in 1905 to explain a number of experimental results (especially the *photoelectric effect*—see *Discovery 4-1*) then puzzling physicists. Furthermore, Einstein was able to quantify the relationship between the two aspects of light's double nature. He found that the energy carried by a photon had to be proportional to the *frequency* of the radiation:

$$\text{photon energy} \propto \text{radiation frequency.}$$

For example, a "deep red" photon having a frequency of $4 \times 10^{14}$ Hz (or a wavelength of approximately 750 nm) has half the energy of a violet photon of frequency of

# MORE PRECISELY 4-1

## The Hydrogen Atom

By observing the emission spectrum of hydrogen and using the connection between photon energy and color first suggested by Einstein (Section 4.2), Niels Bohr determined early in the 20th century what the energy differences between the various energy levels must be. Using that information, he was then able to infer the actual energies of the excited states of hydrogen.

A unit of energy often used in atomic physics is the *electron volt* (eV). (The name actually has a rather technical definition: the amount of energy gained by an electron when it accelerates through an electric potential of 1 volt. For our purposes, however, it is just a convenient quantity of energy.) One electron volt (1 eV) is equal to $1.60 \times 10^{-19}$ J (joule)—roughly half the energy carried by a single photon of red light. The minimum amount of energy needed to ionize hydrogen from its ground state is 13.6 eV. Bohr numbered the energy levels of hydrogen, with level 1 the ground state, level 2 the first excited state, and so on. He found that, by assigning zero energy to the ground state, the energy of any state (the *n*-th, say) could then be written as follows:

$$E_n = 13.6 \left( 1 - \frac{1}{n^2} \right) \text{eV.}$$

Thus, the ground state ($n = 1$) has energy $E_1 = 0$ eV (by our definition), the first excited state ($n = 2$) has energy $E_2 = 13.6 \times (1 - 1/4)$ eV = 10.2 eV, the second excited state has energy $E_3 = 13.6 \times (1 - 1/9)$ eV = 12.1 eV, and so on. Notice that there are infinitely many excited states between the ground state and the energy at which the atom is ionized, crowding closer and closer together as *n* becomes large and $E_n$ approaches 13.6 eV.

---

**EXAMPLE**  Using Bohr's formula for the energy of each electron orbital, we can reverse his reasoning and calculate the energy associated with a transition between any two given states. To boost an electron from the first excited state to the second, an atom must be supplied with $E_3 - E_2 =$ 12.1 eV − 10.2 eV = 1.9 eV of energy, or $3.0 \times 10^{-19}$ J. Now, from the formula $E = hf$ presented in the text, we find that this energy corresponds to a photon with a frequency of $4.6 \times 10^{14}$ Hz, having a wavelength of 656 nm, and lying in the red portion of the spectrum. (A more precise calculation gives the value 656.3 nm reported in the text.) Similarly, the jump from level $n = 3$ to level $n = 4$ requires $E_4 - E_3 =$ $13.6 \times (1/3^2 - 1/4^2)$ eV = $13.6 \times (1/9 - 1/16)$ eV = 0.66 eV of

energy, corresponding to an infrared photon with a wavelength of 1880 nm, and so on. A handy conversion between photon energies $E$ in electron volts and wavelengths $\lambda$ in nanometers is

$$E \, (\text{eV}) = \frac{1240}{\lambda} \, (\text{nm}).$$

---

The accompanying diagram summarizes the structure of the hydrogen atom. The various energy levels are depicted as a series of circles of increasing radius, representing increasing energy. The electronic transitions between these levels (indicated by arrows) are conventionally grouped into families, named after their discoverers, that define the terminology used to identify specific spectral lines. (Note that the spacings of the energy levels are not drawn to scale here to provide room for all labels on the diagram. In reality, the circles should become more and more closely spaced as we move outward.)

Transitions starting from or ending at the ground state (level 1) form the *Lyman series*, named after American spectroscopist Theodore Lyman, who discovered these lines in 1914. The first is *Lyman alpha* (Lyα), corresponding to the transition between the first excited state (level 2) and the ground state. As we have seen, the energy difference is 10.2 eV, and the Lyα photon has a wavelength of 121.6 nm (1216 Å). The Lyβ (beta) transition, between level 3 (the second excited state) and the ground state, corresponds to an energy change of 12.10 eV and a photon of wavelength 102.6 nm (1026 Å). Ly γ (gamma) corresponds to a jump from level 4 to level 1, and so on. The accompanying table shows how we can calculate the energies, frequencies, and wavelengths of the photons in the Lyman series, using the formulae given previously. All Lyman-series energies lie in the ultraviolet region of the spectrum.

The next series of lines, the *Balmer series*, involves transitions down to (or up from) level 2, the first excited state. The series is named after the Swiss mathematician Johann Balmer, who didn't discover these lines (they were well known to spectroscopists early in the 19th century), but who published a mathematical formula for their wavelengths in 1885. This formula laid the foundation for a series of experimental and theoretical breakthroughs that culminated in 1913 with Bohr's more general (and more famous) formula, presented earlier. All the Balmer series lines lie in or close to the visible portion of the electromagnetic spectrum.

Because they form the most easily observable part of the hydrogen spectrum and were the first to be discovered, the

$8 \times 10^{14}$ Hz (wavelength = 375 nm) and 500 times the energy of an $8 \times 10^{11}$ Hz (wavelength = 375 $\mu$m) microwave photon.

The constant of proportionality in the preceding relation is now known as *Planck's constant*, in honor of the German physicist Max Planck, who determined its numerical value. It is always denoted by the symbol *h*, and the

equation relating the photon energy $E$ to the radiation frequency *f* is usually written

$$E = hf.$$

Like the gravitational constant $G$ and the speed of light, $c$, Planck's constant is one of the fundamental physical constants of the universe.

Balmer lines are often referred to simply as the "*Hydrogen series*," denoted by the letter *H*. As with the Lyman series, the individual transitions are labeled with Greek letters. An Hα photon (level 3 to level 2) has a wavelength of 656.3 nm, in the red part of the visible spectrum, Hβ (level 4 to level 2) has a wavelength of 486.1 nm (green), Hγ (level 5 to level 2) has a wavelength of 434.1 nm (blue), and so on. We will use these designations (especially Hα and Hβ) frequently in later chapters. The most energetic Balmer series photons have energies that place them just beyond the blue end of the visible spectrum, in the near ultraviolet.

The classification continues with the *Paschen series* (transitions down to or up from the second excited state, discovered in 1908), the *Brackett series* (third excited state; 1922), and the *Pfund series* (fourth excited state; 1924). All lie in the infrared. Beyond that point, infinitely many other families exist, moving farther and farther into the infrared and radio regions of the spectrum, but they are not referred to by any special names. A few of the transitions making up the Lyman and Balmer (Hydrogen) series are marked on the figure. Astronomically, these are the most important sequences.

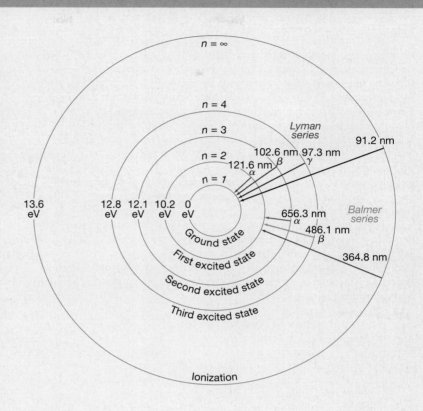

| Transition | Energy (eV) | Frequency (10^15 Hz) | Wavelength (nm) |
|---|---|---|---|
| Ly α (2 ↔ 1) | $13.6 \times \left(1 - \frac{1}{4}\right) = 10.2$ | 2.46 | 122 |
| Ly β (3 ↔ 1) | $13.6 \times \left(1 - \frac{1}{9}\right) = 12.1$ | 2.92 | 103 |
| Ly γ (4 ↔ 1) | $13.6 \times \left(1 - \frac{1}{16}\right) = 12.75$ | 3.08 | 97.3 |
| Ly δ (5 ↔ 1) | $13.6 \times \left(1 - \frac{1}{25}\right) = 13.1$ | 3.15 | 95.0 |
| Ionization | $13.6 \times (1 - 0) = 13.6$ | 3.28 | 91.2 |

In SI units, the value of Planck's constant is a very small number: $h = 6.63 \times 10^{-34}$ joule seconds (J · s). Consequently, the energy of a single photon is tiny. Even a very high-frequency gamma ray (the most energetic type of electromagnetic radiation) with a frequency of $10^{22}$ Hz has an energy of just $(6.63 \times 10^{-34}) \times 10^{22} \approx 7 \times 10^{-12}$ J—about

the same energy carried by a flying gnat. Nevertheless, this energy is more than enough to damage a living cell. The basic reason that gamma rays are so much more dangerous to life than visible light is that each gamma-ray photon typically carries millions, if not billions, of times more energy than a photon of visible radiation.

ANIMATION/VIDEO  Classical Hydrogen Atom I, Classical Hydrogen Atom II

# DISCOVERY 4-1

## The Photoelectric Effect

Einstein developed his breakthrough insight into the nature of radiation partly as a means of explaining a puzzling experimental result known as the **photoelectric effect.** This effect can be demonstrated by shining a beam of light on a metal surface (as shown in the accompanying figure). When high-frequency ultraviolet light is used, bursts of electrons are dislodged from the surface by the beam, much as when one billiard ball hits another, knocking it off the table. However, the speed with which the particles are ejected from the metal is found to depend only on the *color* of the light, and not on its intensity. For lower-frequency light—blue, say—an electron detector still records bursts of electrons, but now their speeds, and hence their *energies*, are less. For even lower frequencies—red or infrared light—*no* electrons are kicked out of the metal surface at all.

These results are difficult to reconcile with a wave model of light, which would predict that the energies of the ejected electrons should increase steadily with increasing intensity at any frequency. Instead, the detector shows an abrupt cutoff in ejected electrons as the frequency of the incoming radiation drops below a certain level. Einstein realized that the only way to explain the cutoff, and the increase in electron speed with frequency above the cutoff, was to envision radiation as traveling as "bullets," or particles, or *photons.* Furthermore, to account for the experimental findings, the energy of any photon had to be proportional to the *frequency* of the radiation. Low-frequency, long-wavelength photons carry less energy than high-frequency, short-wavelength ones.

If we also suppose that some minimum amount of energy is needed just to "unglue" the electrons from the metal, then we can see why no electrons are emitted below some critical frequency: The photons associated with red light in the diagram just don't carry enough energy. Above the critical frequency, photons do have enough energy to dislodge the electrons. Moreover, any energy they possess above the necessary minimum is imparted to the electrons as *kinetic energy,* the energy of motion. Thus, as the frequency of the radiation increases, so, too, does the photon's energy and hence the speed of the electrons that they liberate from the metal.

The realization and acceptance of the fact that light can behave both as a wave and as a particle is another example of the scientific method at work. Despite the enormous success of the wave theory of radiation in the 19th century, the experimental evidence led 20th-century scientists to the inevitable conclusion that the theory was incomplete—it had to be modified to allow for the fact that light sometimes acts like a particle. Although Einstein is perhaps best known today for his theories of relativity, in fact his 1919 Nobel prize was for his work on the photoelectric effect. In addition to bringing about the birth of a whole new branch of physics—the field of quantum mechanics—Einstein's explanation of the photoelectric effect radically changed the way physicists view light and all other forms of radiation.

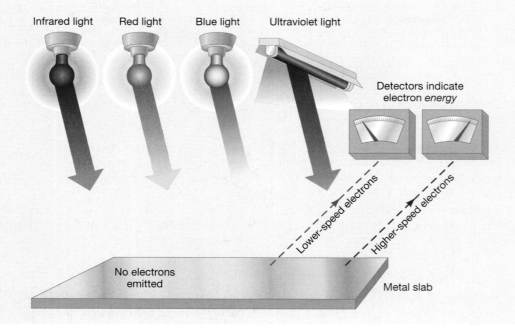

Infrared light　Red light　Blue light　Ultraviolet light

Detectors indicate electron *energy*

Lower-speed electrons

Higher-speed electrons

No electrons emitted

Metal slab

The equivalence between the energy and frequency (or inverse wavelength) of a photon completes the connection between atomic structure and atomic spectra. Atoms absorb and emit radiation at characteristic wavelengths determined by their own particular internal structure. Because this structure is *unique* to each element, the colors of the absorbed and emitted photons—that is, the spectral lines we observe—are characteristic of that element *and only that element*. The spectrum we see is thus a unique identifier of the atom involved.

Many people find it confusing that light can behave in two such different ways. To be truthful, modern physicists don't yet fully understand *why* nature displays this wave–particle duality. Nevertheless, there is irrefutable experimental evidence for both of these aspects of radiation. Environmental conditions ultimately determine which description—wave or stream of particles—better fits the behavior of electromagnetic radiation in a particular instance. As a general rule of thumb, in the macroscopic realm of everyday experience, radiation is more usefully described as a wave, whereas in the microscopic domain of atoms, it is best characterized as a series of particles.

CONCEPT CHECK

✔ In what ways do electron orbits in an atom differ from planetary orbits around the Sun?

## 4.3 The Formation of Spectral Lines

With quantum mechanics as our guide to the internal structure of atoms, we can now explain quantitatively the spectral lines we see. Let's start with hydrogen, the simplest element, then move on to more complex systems.

### The Spectrum of Hydrogen

The full spectrum of hydrogen consists of many lines, spread across much of the electromagnetic spectrum, from ultraviolet to radio; we focus here on just a few of those lines. The energy levels and spectrum of hydrogen are discussed in more detail in *More Precisely 4-1.*

Figure 4.10 illustrates schematically the absorption and emission of photons by a hydrogen atom. Figure 4.10(a) shows the atom absorbing a photon and making a transition from the ground state to the first excited state. It then emits a photon of precisely the same energy and drops back to the ground state. The energy difference between the two states corresponds to an ultraviolet photon of wavelength 121.6 nm (1216 Å).

Absorption may also boost an electron into an excited state higher than the first excited state. Figure 4.10(b) depicts the absorption of a more energetic (higher frequency, shorter wavelength) ultraviolet photon, one with a wavelength of 102.6 nm (1026 Å). The absorption of this photon causes the atom to jump to the *second* excited state. As before, the atom returns rapidly to the ground state, but this time, because there are two states lying below the excited state, the atom can do so in one of two possible ways:

1. It can proceed directly back to the ground state, in the process emitting an ultraviolet photon identical to the one that excited the atom in the first place.

2. Alternatively, the electron can *cascade* down, one orbital at a time. If this occurs, the atom will emit *two* photons: one with an energy equal to the difference between the second and first excited states and the other with an energy equal to the difference between the first excited state and the ground state.

Either possibility can occur, with roughly equal probability. The second step of the cascade process produces a 121.6-nm ultraviolet photon, just as in Figure 4.10(a). However, the first transition—the one from the second to the first excited state—produces a photon with a wavelength of 656.3 nm (6563 Å), which is in the visible part of the spectrum. This photon is seen as red light. An individual atom—if one could be isolated—would emit a momentary red flash. This is the origin of the red line in the hydrogen spectrum shown in Figure 4.3.

The absorption of additional energy can boost the electron to even higher orbitals within the atom. As the excited electron cascades back down to the ground state, the atom may emit many photons, each with a different energy and hence a different wavelength, and the resulting spectrum shows many spectral lines. In a sample of heated hydrogen gas, at any instant atomic collisions ensure that atoms are found in many different excited states. The complete emission spectrum therefore consists of wavelengths corresponding to all possible transitions between those states and states of lower energy.

In the case of hydrogen, all transitions ending at the ground state produce ultraviolet photons. However, downward transitions ending at the *first* excited state give rise to spectral lines in or near the visible portion of the electromagnetic spectrum (Figure 4.3; *More Precisely 4-1*). Other transitions ending in higher states generally give rise to infrared and radio spectral lines.

The inset in Figure 4.10 shows an astronomical object whose red coloration is the result of precisely the process mentioned in step 2 above. As ultraviolet photons from a young, hot star pass through the surrounding cool hydrogen gas out of which the star recently formed, some photons are absorbed by the gas, boosting its atoms into excited states or ionizing them completely. The 656.3-nm red glow characteristic of excited hydrogen gas results as the atoms cascade back to their ground states. The phenomenon is called *fluorescence.*

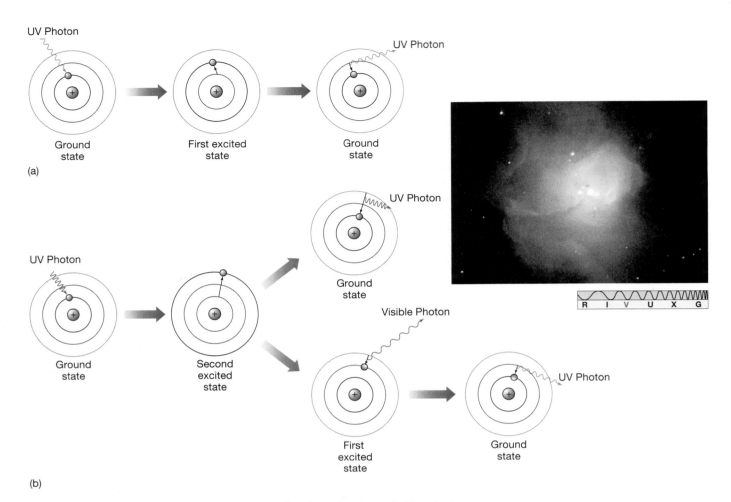

(a)

(b)

▲ **FIGURE 4.10** **Atomic Excitation** (a) Absorption of an ultraviolet photon (left) by a hydrogen atom causes the momentary excitation of the atom into its first excited state (center). After about $10^{-8}$ s, the atom returns to its ground state (right), in the process emitting a photon having exactly the same energy as the original photon. (b) Absorption of a higher-energy ultraviolet (UV) photon may boost the atom into a higher excited state, from which there are several possible paths back to the ground state. (Remember, the sharp lines used for the orbitals here and in similar figures that follow are intended merely as a schematic representation of the electron energy levels and are not meant to be taken literally. In actuality, electron orbitals are "clouds," as shown in Figure 4.9.) At the top, the electron falls immediately back to the ground state, emitting a photon identical to the one it absorbed. At the bottom, the electron initially falls into the first excited state, producing visible radiation of wavelength 656.3 nm—the characteristic (Hα) red glow of excited hydrogen. Subsequently, the atom emits another photon (having the same energy as in part (a) as it falls back to the ground state. The object shown in the inset, designated N81, is an emission nebula—an interstellar cloud made mostly of hydrogen gas excited by absorbing radiation emitted by some extremely hot stars (the white areas near the center). (*Inset: NASA*)

## Kirchhoff's Laws Explained

Let's consider again our earlier discussion of emission and absorption lines in terms of the model just presented. In Figure 4.7, a beam of continuous radiation shines through a cloud of hydrogen gas. The beam contains photons of all energies, but most of them cannot interact with the gas—the gas can absorb only those photons having just the right energy to cause a change in an electron's orbit from one state to another. All other photons in the beam—with energies that cannot produce a transition—do not interact with the gas at all, but pass through it unhindered. Photons having the right energies are absorbed, excite the gas, and are removed from the beam. This sequence is the cause of the dark absorption lines in the spectrum of Figure 4.7(b). The lines are direct indicators of the energy differences between orbitals in the atoms making up the gas.

The excited gas atoms return rapidly to their original states, each emitting one or more photons in the process. We might think, then, that, although some photons from the beam are absorbed by the gas, they are quickly replaced

◂ FIGURE 4.11 **Helium and Carbon** (a) A helium atom in its ground state. Two electrons occupy the lowest-energy orbital around a nucleus containing two protons and two neutrons. (b) A carbon atom in its ground state. Six electrons orbit a six-proton, six-neutron nucleus, two of the electrons in an inner orbital, the other four at a greater distance from the center.

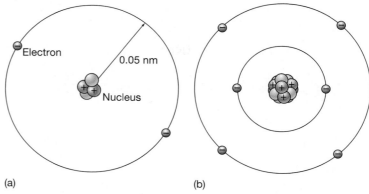

by reemitted photons, with the result that we could never observe the effects of absorption. In fact, this is not the case, for two reasons. First, although the photons not absorbed by the gas continue on directly to the detector, the reemitted photons can leave in *any* direction. In effect, between absorption and reemission, the atom "forgets" the direction from which the original incoming photon came. Consequently, most of the reemitted photons leave at angles that do not take them through the slit and on to the detector, so they are effectively lost from the original beam. Second, as we have just seen, electrons may cascade back to the ground state, emitting several photons of lower energy instead of a single photon equal in energy to that of the one originally absorbed.

The net result of these processes is that some of the original energy is channeled into photons with energies associated with many different colors and moving in many different directions. A second detector looking at the cloud from the side would record the reemitted energy as an emission spectrum, as in Figure 4.7(c). (A spectrum of the object shown in the inset of Figure 4.10, called an *emission nebula*, would show the same thing.) Like the absorption spectrum, the emission spectrum is characteristic of the gas, not of the original beam.

Absorption and emission spectra are created by the same atomic processes. They correspond to the same atomic transitions. They contain the same information about the composition of the gas cloud. In the laboratory, we can move our detector and measure both. In astronomy, we cannot easily change our vantage point (on or near Earth), so the type of spectrum we see depends on our chance location with respect to both the source and the intervening gas cloud.

## More Complex Spectra

All hydrogen atoms have basically the same structure—a single electron orbiting a single proton—but, of course, there are many other kinds of atoms, each kind having a unique internal structure. The number of protons in the nucleus of an atom determines the **element** that it represents. Just as all hydrogen atoms have a single proton, all oxygen atoms have 8 protons, all iron atoms have 26 protons, and so on.

The next simplest element after hydrogen is helium. The central nucleus of the most common form of helium is made up of two protons and two **neutrons** (another kind of elementary particle having a mass slightly larger than that of a proton, but having no electrical charge). Two electrons

orbit this nucleus. As with hydrogen and all other atoms, the "normal" condition for helium is to be electrically neutral, with the negative charge of the orbiting electrons exactly canceling the positive charge of the nucleus (Figure 4.11a).

More complex atoms contain more protons (and neutrons) in the nucleus and have correspondingly more orbiting electrons. For example, an atom of carbon, shown in Figure 4.11(b), consists of six electrons orbiting a nucleus containing six protons and six neutrons. As we progress to heavier and heavier elements, the number of orbiting electrons increases, and the number of possible electron transitions rises rapidly. The result is that very complicated spectra can be produced. The complexity of atomic spectra generally reflects the complexity of the atoms themselves. A good example is the element iron, which contributes nearly 800 of the Fraunhofer absorption lines seen in the solar spectrum (Figure 4.4).

Atoms of a single element such as iron can yield many lines for two main reasons. First, the 26 electrons of a normal iron atom can make an enormous number of different transitions among available energy levels. Second, many iron atoms are *ionized*, with some of their 26 electrons stripped away. The removal of electrons alters an atom's electromagnetic structure, and the energy levels of ionized iron are quite different from those of neutral iron. Each new level of ionization introduces a whole new set of spectral lines. Besides iron, many other elements, also in different stages of excitation and ionization, absorb photons at visible wavelengths. When we observe the entire Sun, all these atoms and ions absorb simultaneously, yielding the rich spectrum we see.

The power of spectroscopy is most apparent when a cloud contains many different gases mixed together, because it enables us to study one kind of atom or ion to the exclusion of all others simply by focusing on specific wavelengths of radiation. By identifying the superimposed absorption and emission spectra of many different atoms, we can determine the cloud's composition (and much more—see Section 4.4). Figure 4.12 shows an actual spectrum observed from a real cosmic object. As in Figure 4.10, the characteristic red glow of this emission nebula comes from the Hα transition in hydrogen, the nebula's main constituent.

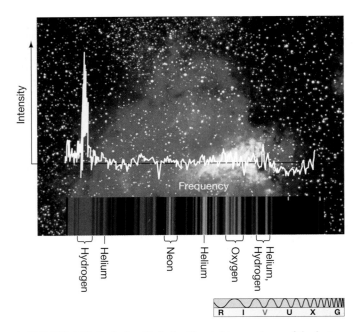

▲ FIGURE 4.12 **Emission Nebula** The visible spectrum of the hot gases in a nearby gas cloud known as the Omega Nebula (M17). (The word *nebula* means "gas cloud"—one of many sites in our Galaxy where new stars are forming today.) Shining by the light of several very hot stars, the gas in the nebula produces a complex spectrum of bright and dark lines (bottom). That same spectrum can also be displayed, as shown here, as a white graph of intensity versus frequency, spanning the spectrum from red to blue. (*Adapted from ESO*)

Spectral lines occur throughout the entire electromagnetic spectrum. Usually, electron transitions among the lowest orbitals of the lightest elements, such as hydrogen and helium, produce visible and ultraviolet spectral lines. Transitions among very highly excited states of hydrogen and other elements can produce spectral lines in the infrared and radio parts of the electromagnetic spectrum. Conditions on Earth make it all but impossible to detect these radio and infrared features in the laboratory, but they are routinely observed by radio and infrared telescopes (see Chapter 5) in radiation coming from space. Electron transitions among lower energy levels in heavier, more complex elements produce X-ray spectral lines, which have been observed in the laboratory. Some have also been observed in stars and other cosmic objects.

▶ FIGURE 4.13 **Molecular Emission** Molecules can change in three ways while emitting or absorbing electromagnetic radiation. The colors and wavelengths of the emitted photons represent the relative energies involved. Sketched here is the molecule carbon monoxide (CO) undergoing (a) a change in which an electron in the outermost orbital of the oxygen atom drops to a lower energy state (emitting a photon of shortest wavelength, in the visible or ultraviolet range), (b) a change in vibrational state (of intermediate wavelength, in the infrared), and (c) a change in rotational state (of longest wavelength, in the radio range).

CONCEPT CHECK
✔ How does the structure of an atom determine the atom's emission and absorption spectra?

## 4.4 Molecules

A **molecule** is a tightly bound group of atoms held together by interactions among their orbiting electrons—interactions that we call *chemical bonds*. Much like atoms, molecules can exist only in certain well-defined energy states, and again like atoms, molecules produce characteristic emission or absorption spectral lines when they make a transition from one state to another. Because molecules are more complex than individual atoms, the rules of molecular physics are also more complex. Nevertheless, as with atomic spectral lines, painstaking experimental work over many decades has determined the precise frequencies (or wavelengths) at which millions of molecules emit and absorb radiation.

In addition to the lines resulting from electron transitions, molecular lines result from two other kinds of change not possible in atoms: Molecules can *rotate*, and they can *vibrate*. Figure 4.13 illustrates these basic molecular

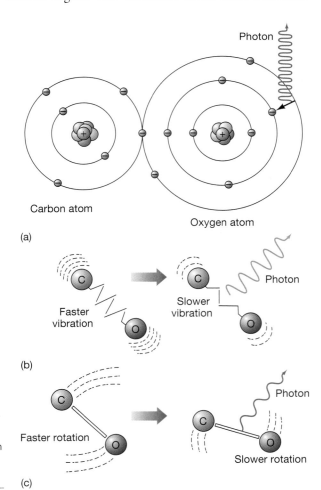

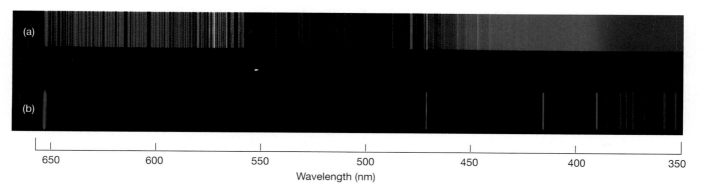

▲ FIGURE 4.14 **Hydrogen Spectra** (a) The emission spectrum of molecular hydrogen. Notice how it differs from the spectrum of the simpler atomic hydrogen (b). (© *Bausch & Lomb, Inc.*)

motions. Molecules rotate and vibrate in specific ways. Just as with atomic states, only certain spins and vibrations are allowed by the rules of molecular physics. When a molecule *changes* its rotational or vibrational state, a photon is emitted or absorbed. Spectral lines characteristic of the specific kind of molecule result. Like their atomic counterparts, these lines are unique molecular fingerprints, enabling researchers to identify and study one kind of molecule to the exclusion of all others. As a rule of thumb,

- *electron transitions* within molecules produce visible and ultraviolet spectral lines (the largest energy changes).

- changes in molecular *vibration* produce infrared spectral lines.

- changes in molecular *rotation* produce spectral lines in the radio part of the electromagnetic spectrum (the smallest energy changes).

Molecular lines usually bear little resemblance to the spectral lines associated with their component atoms. For example, Figure 4.14(a) shows the emission spectrum of the simplest molecule known: molecular hydrogen. Notice how different it is from the spectrum of atomic hydrogen shown in part (b) of the figure.

CONCEPT CHECK

✔ What kinds of internal changes within a molecule can cause radiation to be emitted or absorbed?

## 4.5 Spectral-Line Analysis

Astronomers apply the laws of spectroscopy in analyzing radiation from beyond Earth. A nearby star or a distant galaxy takes the place of the lightbulb in our previous examples. An interstellar cloud or a stellar (or even planetary) atmosphere plays the role of the intervening cool gas, and a spectrograph attached to a telescope replaces our simple prism and detector. We began our study of electro-

magnetic radiation by stating that virtually all we know about planets, stars, and galaxies is gleaned from studies of the light we receive from them, and we have presented some of the ways in which that knowledge is obtained. Here, we describe a few of the ways in which the properties of emitters and absorbers can be determined by careful analysis of radiation received on (or near) Earth (see Table 4.1 for a summary). We will encounter other important examples as our study of the cosmos unfolds.

### A Spectroscopic Thermometer

Stars are very hot, especially deep down in their cores, where the temperature is measured in millions of kelvins. Because of the intense heat in the core, atoms are fully ionized. Electrons travel freely through the gas, unbound to any nucleus, and the spectrum of radiation is continuous. However, at the relatively cool stellar surface, some atoms retain a few, or even most, of their orbital electrons. As discussed previously, by matching the spectral lines we see with the laboratory spectra of known atoms, ions, and molecules, we can determine a star's chemical composition.

| TABLE 4.1 Spectral Information Derived from Starlight | |
|---|---|
| **Observed Spectral Characteristic** | **Information Provided** |
| Peak frequency or wavelength (continuous spectra only) | Temperature (Wien's law) |
| Lines present | Composition, temperature |
| Line intensities | Composition, temperature |
| Line width | Temperature, turbulence, rotation speed, density, magnetic field |
| Doppler shift | Line-of-sight velocity |

The strength of a spectral line (brightness or darkness, depending on whether the line is seen in emission or absorption) depends in part on the *number* of atoms giving rise to the line: The more atoms there are to emit or absorb photons of the appropriate frequency, the stronger is the line. But the strength of line also depends on the *temperature* of the gas containing the atoms, because temperature determines how many atoms at any instant are in the right orbital to undergo any particular transition.

Consider the absorption of radiation by hydrogen atoms in an interstellar gas cloud or in the outer atmosphere of a star. If all the hydrogen were in its ground state—as it would be if the temperature were relatively low—then the only transitions that could occur would be the Lyman series (see *More Precisely 4-1*), resulting in absorption lines in the ultraviolet portion of the spectrum. Thus, astronomers would observe *no* visible hydrogen absorption lines (for example, the Balmer series) in the spectrum of the object, not because there was no hydrogen, but because there would be no hydrogen atoms in the first excited state (as is required to produce visible absorption features).

The spectrum of our own Sun is a case in point. Because the temperature of the Sun's atmosphere is a relatively cool 5800 K (as we saw in Chapter 3), few hydrogen atoms have electrons in any excited state. ∞ (Sec. 3.4) Hence, in the Sun, visible hydrogen lines are quite weak— that is, of low intensity compared with the same lines in many other stars—even though hydrogen is by far the most abundant element there.

As the temperature rises, atoms move faster and faster. More and more energy becomes available through atomic collisions, and more and more electrons are boosted into an excited state. At any instant, then, some atoms are temporarily in an excited state and so are capable of absorbing at visible or even longer wavelengths. As the number of atoms in the first excited state increases, lines in the Balmer series become more and more evident in the spectrum. Eventually, a temperature is reached at which *most* of the atoms are in the first excited state, simply because of their frequent energetic collisions with other atoms in the gas. At this point, the Balmer lines are at their strongest (and the Lyman lines are much weaker).

At even higher temperatures, most atoms are kicked beyond the first excited state into higher energy orbitals, and new series of absorption lines are seen, while the strength of the Balmer series declines again. Eventually, the temperature becomes so high that most hydrogen is ionized, and no spectral lines are seen at all.

Spectroscopists have developed mathematical formulas that relate the number of emitted or absorbed photons to the energy levels of the atoms involved and the temperature of the gas. Once an object's spectrum is measured, astronomers can interpret it by matching the observed intensities of the spectral lines with those predicted by the formulas. In this way, astronomers can refine their measurements of both the composition *and* the temperature of the gas producing the lines. These temperature measurements are generally much more accurate than crude estimates based on the radiation laws and the assumption of blackbody emission. ∞ (Sec. 3.4)

## Measurement of Radial Velocity

The Doppler effect—the apparent shift in the frequency of a wave due to the motion of the source relative to the observer—is a classical phenomenon common to all waves. ∞ (Sec. 3.5) However, by far its most important astronomical application comes when it is combined with observations of atomic and molecular spectral lines.

As we have seen, the spectra of many atoms, ions, and molecules are well known from laboratory measurements. Often, however, a familiar pattern of lines appears, but the lines are displaced from their usual locations. In other words, as illustrated in Figure 4.15, a set of spectral lines may be recognized as belonging to a particular element, except that the lines are all offset—blueshifted or redshifted—by the same fractional amount from their normal wavelengths. These shifts are due to the Doppler effect, and they allow astronomers to measure how fast the source of the radiation is moving along the line of sight from the observer (the *radial velocity* of the source).

For example, in Figure 4.15, the 486.1-nm Hβ line of hydrogen in the spectrum of a distant galaxy is received on Earth at a wavelength of 485.1 nm—blueshifted to a slightly shorter wavelength. (Remember, we know that it is the Hβ line because *all* the hydrogen lines are observed to have the same fractional shift—the characteristic line pattern identifies the spectrum as that of hydrogen.) We can compute the galaxy's line-of-sight velocity relative to Earth by using the Doppler equation presented in Section 3.5. The calculation is essentially the same as that presented in *More Precisely 3-3:* The change in wavelength (apparent minus true) is 485.1 nm – 486.1 nm = –1.0 nm (the negative sign simply indicating that the wavelength has decreased). It then follows that the recession velocity is

$$\frac{-1.0 \text{ nm}}{486.1 \text{ nm}} \times c = -620 \text{ km/s}.$$

In other words, the galaxy is *approaching* us (this is the meaning of the negative sign) at a speed of 620 km/s.

This book will have a lot to say about the motions of planets, stars, and galaxies throughout the universe. Just bear in mind that almost all of that information is derived from telescopic observations of Doppler-shifted spectral lines in many different parts of the electromagnetic spectrum.

## Line Broadening

The structure of the lines themselves reveals still more information. At first glance the emission lines shown earlier may seem uniformly bright, but more careful study

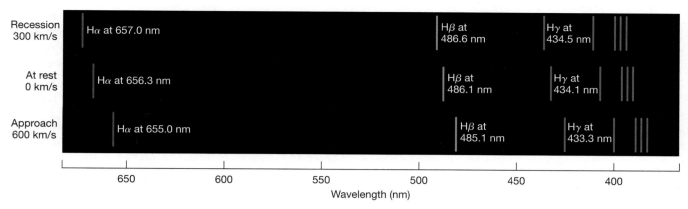

**▲ FIGURE 4.15 Doppler Shift** Because of the Doppler effect, the entire spectrum of a moving object is shifted to higher or lower frequencies. The spectrum at the center is the unshifted emission spectrum of pure hydrogen, corresponding to the object at rest. The top spectrum shows the slight redshift of the hydrogen lines from an object moving at a speed of 300 km/s away from the observer. Once we recognize the spectrum as that of hydrogen, we can identify specific lines and measure their shifts. The amount of the shift (0.1 percent here) tells us the object's recession velocity—0.001c. The spectrum at the bottom shows the blueshift of the same set of lines from an object approaching us at 600 km/s. The shift is twice as large (0.2 percent), because the speed has doubled, and in the opposite sense because the direction has reversed.

shows that this is in fact not the case. As illustrated in Figure 4.16, the brightness of a line is greatest at the center and falls off toward either side. Earlier, we stressed that photons are emitted and absorbed at very precise energies, or frequencies. Why, then, aren't spectral lines extremely narrow, occurring only at specific wavelengths? This *line broadening* is not the result of some inadequacy of our experimental apparatus; rather, it is caused by the *environment* in which the emission or absorption occurs— the physical state of the gas or star in which the line is formed. For definiteness, we have drawn Figure 4.16 and subsequent figures to refer to emission lines, but realize that the ideas apply equally well to absorption features.

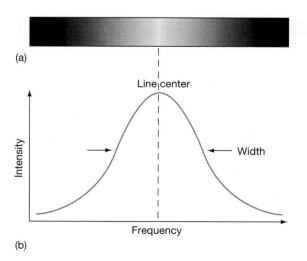

**▲ FIGURE 4.16 Line Profile** By tracing the changing brightness across a typical emission line (a) and expanding the scale, we obtain a graph of the line's intensity versus its frequency (b).

Several processes can broaden spectral lines. The most important again involve the Doppler effect. Imagine a hot cloud of gas containing individual atoms in random thermal motion in every possible direction, as illustrated in Figure 4.17(a). If an atom happens to be moving away from us as it emits a photon, that photon is redshifted by the Doppler effect—we do not record it at the precise wavelength predicted by atomic physics, but rather at a slightly longer wavelength. The extent of this redshift is proportional to the atom's instantaneous velocity away from the detector. Similarly, if the atom is moving toward us at the instant of emission, its light is blueshifted. In short, because of thermal motion within the gas, emission and absorption lines are observed at frequencies slightly different from those we would expect if all atoms in the cloud were motionless.

Most atoms in a typical cloud have small thermal velocities, so in most cases the line is Doppler shifted just a little. Only a few atoms have large shifts. As a result, the center of a spectral line is much more pronounced than its "wings," producing a bell-shaped spectral feature like that shown in Figure 4.17(b). Thus, even if all atoms emitted and absorbed photons at only one precise wavelength, the effect of their thermal motion would be to smear the line out over a range of wavelengths. The hotter the gas, the larger is the spread of Doppler motions and the greater is the width of the line. ∞ (*More Precisely 3-1*) By measuring a line's width, astronomers can estimate the average speed of the particles, and hence the temperature of the gas, producing the line.

*Rotation* causes a similar effect. Consider an astronomical object (a planet, a star, or even an entire galaxy) oriented so that we see it spinning, as sketched in Figure 4.18. Photons emitted from the side spinning toward us are

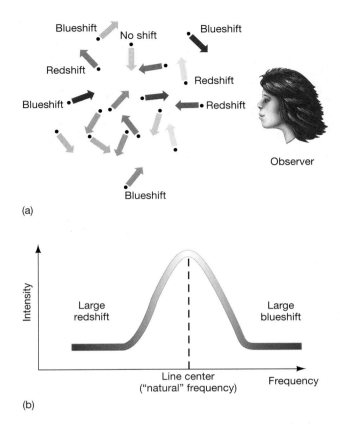

(a)

(b)

▲ FIGURE 4.17 **Thermal Broadening** Atoms moving randomly (a) produce broadened spectral lines (b) as their individual redshifted and blueshifted emission lines merge in our detector. The hotter the gas, the greater is the degree of thermal broadening.

blueshifted by the Doppler effect. Photons emitted from the side spinning away from us are redshifted. Very often, the object under study is so small or far away that our equipment cannot distinguish, or *resolve*, different parts from one another—all the emitted light is blended together in our detector, producing a net broadening of the observed spectral lines. (If we could resolve the object, of course, we would be able to distinguish the redshifted and blueshifted sides and measure the rotation rate directly.) Note that this broadening has *nothing* to do with the temperature of the gas producing the lines and is generally superimposed on the thermal broadening just discussed. The faster the object spins, the more rotational broadening we see.

Yet another mechanism that can cause line broadening is gas *turbulence*, which exists when the gas in an interstellar cloud is

not at rest or flowing smoothly, but instead is seething and churning in eddies and vortices of many sizes. Turbulence may occur, for example, when a cloud starts to collapse under its own gravitational pull or perhaps when it collides with another cloud. Motion of this type also causes Doppler shifts of spectral lines, but lines from different parts of the cloud are shifted more or less randomly. Just as in the case of rotation, if our equipment is unable to resolve the cloud, a net broadening of its observed spectral lines results. The faster the internal motion, the greater is the broadening observed. Remember again that the internal motion has nothing to do with the temperature of the gas.

Other broadening mechanisms do not depend on the Doppler effect at all. For example, if electrons are moving between orbitals while their parent atom is colliding with another atom, the energy of the emitted or absorbed photons changes slightly, blurring the spectral lines. This mechanism, which occurs most often in dense gases, where collisions are most frequent, is usually referred to as *collisional broadening*. The amount of broadening increases as the density of the emitting or absorbing gas rises.

Finally, *magnetic fields* can also broaden spectral lines, by a process called the *Zeeman effect*. The electrons and nuclei within atoms behave as tiny spinning magnets, and the basic emission and absorption rules of atomic physics change slightly whenever atoms are immersed in a magnetic field, as is the case in many stars to greater or lesser extents. The result is a slight splitting of a spectral line, which then blurs into an overall line broadening. Generally, the stronger the magnetic field, the more pronounced is the broadening.

## Information from Spectral Lines

As described in the text, the Doppler effect can broaden spectral lines in several ways. Here we look at how

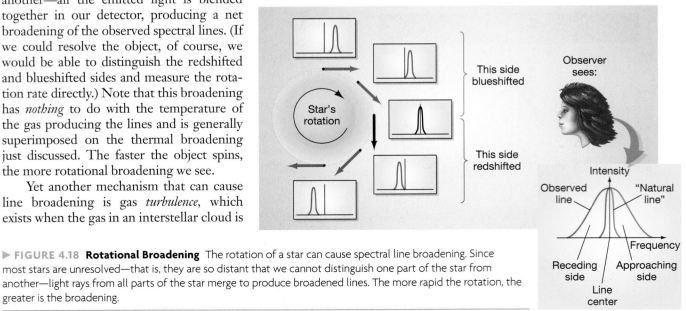

▶ FIGURE 4.18 **Rotational Broadening** The rotation of a star can cause spectral line broadening. Since most stars are unresolved—that is, they are so distant that we cannot distinguish one part of the star from another—light rays from all parts of the star merge to produce broadened lines. The more rapid the rotation, the greater is the broadening.

measurements of that broadening can be used to estimate the internal velocities responsible for the broadening.

Imagine that the 656.3-nm Hα line of hydrogen in some distant star is observed to have a width of 0.06 nm (a tiny amount, but still far greater than the "natural" width of the line). We can estimate the average (thermal or rotational) speed in the star's atmosphere as follows: The Doppler equation in *More Precisely 3-3* may be rewritten in this context to read

$$\frac{\text{spread in wavelength (0.06 nm)}}{\text{true wavelength (656.3 nm)}} \approx \frac{2 \times \text{average speed}}{\text{speed of light, } c},$$

where the "2" on the right-hand side takes into account the fact that the spread in wavelength is caused by a combination of motion toward us and motion away from us. It then follows that the typical line-of-sight velocity, averaged over the star's surface, is $\frac{1}{2} \times (0.06/656.3) \times c$, or 14 km/s. If this broadening is thermal, it corresponds to a gas temperature of about 7600 K (see *More Precisely 8-1*). If the broadening is rotational, we can estimate the rotation rate if we know the star's radius. For a star like the Sun, the velocity of 14 km/s would correspond to a rotation rate of a little under one revolution per day.

Note that even once this velocity information is available it may still not be obvious which process—thermal motion, rotation, or turbulence—actually caused the observed broadening. It is the astronomer's task to use whatever other knowledge she has about the object under study to decipher the details of its internal motion.

## The Message of Starlight

Given sufficiently sensitive equipment, there is almost no end to the wealth of data that can be obtained from starlight. Table 4.1 lists some basic measurable properties of an incoming beam of radiation and indicates what sort of information can be obtained from them.

It is important to realize, however, that deciphering the extent to which each of the factors just described influences a spectrum can be a very difficult task. Typically, the spectra of many elements are superimposed on one another, and often several competing physical effects are occurring simultaneously, each modifying the spectrum in its own way. Further analysis is generally required to disentangle them. For example, if we know the temperature of the emitting gas (perhaps by comparing intensities of different spectral lines, as discussed earlier), then we can calculate how much of the broadening is due to thermal motion and therefore how much is due to the other mechanisms just described. In addition, it is often possible to distinguish between the various broadening mechanisms by studying the detailed *shapes* of the lines.

The challenge facing astronomers is to decode spectral-line profiles to obtain meaningful information about the sources of the lines. In the next chapter, we will discuss some of the means by which astronomers obtain the raw data they need in their quest to understand the cosmos.

### CONCEPT CHECK

✔ Why is it so important for astronomers to analyze spectral lines in detail?

# CHAPTER REVIEW

## Summary

① A **spectroscope (p. 86)** is a device for splitting a beam of radiation into its component frequencies and delivering them to a screen or detector for detailed study.  Many hot objects emit a **continuous spectrum (p. 86)** of radiation, containing light of all wavelengths. A hot gas may instead produce an **emission spectrum (p. 87)**, consisting of only a few well-defined **emission lines (p. 86)** of specific frequencies, or colors. Passing a continuous beam of radiation through cool gas will produce **absorption lines (p. 88)** at precisely the same frequencies as are present in the gas's emission spectrum.

② **Kirchhoff's laws (p. 89)** describe the relationships among these different types of spectra. The emission and absorption lines produced by each element are unique—they provide a "fingerprint" of that element. The study of the spectral  lines produced by different substances is called **spectroscopy (p. 89)**. Spectroscopic studies of the Fraunhofer lines in the solar spectrum yield detailed information about the Sun's composition.

③ **Atoms (p. 90)** are made up of negatively charged electrons orbiting a positively charged heavy **nucleus (p. 90)** consisting of positively charged protons and  electrically neutral **neutrons (p. 97)**. Usually, the number of orbiting electrons equals the number of protons in the nucleus, and the atom as a whole is electrically neutral. The number of protons in the nucleus determines the particular **element (p. 97)** of which the atom is a constituent. An atom has a minimum-energy **ground state (p. 90)**, representing its "normal" condition. If an orbiting electron is given enough energy, it can escape from the atom, which is then **ionized (p. 90)**. Between these two states, the electron can exist only in certain well-defined **excited states (p. 91)**, each with a specific energy—the electron's energy is **quantized (p. 91)**.

In the modern view, the electron is envisaged as being spread out in a "cloud" around the nucleus, but still with a sharply defined energy.

4 Electromagnetic radiation exhibits both wave and particle properties. Particles of radiation are called **photons (p. 91)**. In order to explain the **photoelectric effect (p. 94)**, Einstein found that the energy of a photon must be directly proportional to the photon's frequency.

5 As electrons move between energy levels within an atom, the difference in energy between the states is emitted or absorbed in the form of photons. Because the energy levels have definite energies, the photons also have definite energies, and hence colors, that are characteristic of the type of atom involved. More complex atoms generally produce more complex spectra.

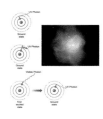

6 **Molecules (p. 98)** are groups of two or more atoms bound together by electromagnetic forces. Like atoms, molecules exist in energy states that obey rules similar to those governing the internal structure of atoms. Again like atoms, when molecules make transitions between energy states, they emit or absorb a characteristic spectrum of radiation that identifies them uniquely.

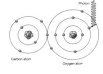

7 Astronomers apply the laws of spectroscopy in analyzing radiation from beyond Earth. Several physical mechanisms can broaden spectral lines. The most important is the Doppler effect, which occurs because stars are hot and their atoms are in motion or because the object being studied is rotating or in turbulent motion.

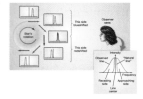

## Review and Discussion

1. What is spectroscopy? Explain how astronomers might use spectroscopy to determine the composition and temperature of a star.

2. Describe the basic components of a simple spectroscope.

3. What is a continuous spectrum? An absorption spectrum?

4. Why are gamma rays generally harmful to life-forms, but radio waves generally harmless?

5. What is a photon?

6. In the particle description of light, what is color?

7. In what ways does the Bohr model of atomic structure differ from the modern view?

8. Give a brief description of a hydrogen atom.

9. What does it mean to say that a physical quantity is quantized?

10. What is the normal condition for atoms? What is an excited atom? What are orbitals?

11. Why do excited atoms absorb and reemit radiation at characteristic frequencies?

12. How are absorption and emission lines produced in a stellar spectrum? What information might absorption lines in the spectrum of a star reveal about a cloud of cool gas lying between us and the star?

13. According to Kirchhoff's laws, what are the necessary conditions for a continuous spectrum to be produced?

14. Explain how a beam of light passing through a diffuse cloud may give rise to both absorption and emission spectra.

15. Why is the Hα absorption line of hydrogen in the Sun relatively weak, even though the Sun has abundant hydrogen?

16. How do molecules produce spectral lines unrelated to the movement of electrons between energy levels?

17. How does the intensity of a spectral line yield information about the source of the line?

18. How can the Doppler effect cause broadening of a spectral line?

19. Describe what happens to a spectral line from a star as the star's rotation rate increases.

20. List three properties of a star that can be determined from observations of its spectrum.

## Conceptual Self-Test: True or False/Multiple Choice

1. Imagine an emission spectrum produced by a container of hydrogen gas. Changing the amount of hydrogen in the container will change the colors of the lines in the spectrum.

2. In the previous question, changing the gas in the container from hydrogen to helium will change the colors of the lines occurring in the spectrum.

3. The wavelengths of the emission lines produced by an element are different from the wavelengths of the absorption lines produced by the same element.

4. The energy of a photon is inversely proportional to the wavelength of the radiation.

5. An electron can have any energy within an atom, so long as it is above the ground-state energy.

6. Spectral lines of hydrogen are relatively weak in the Sun because the Sun contains relatively little hydrogen.

7. Fraunhofer lines are emission lines in the solar spectrum.

8. Light behaves both as a wave and as a particle.

9. An electron moves to a higher energy level in an atom after absorbing a photon of a specific energy.

10. High temperatures, rotation, and magnetic fields all tend to broaden spectral lines.

11. Compared with a spectrum from a ground-based observation, the spectrum of a star observed from above Earth's atmosphere would show (a) no absorption lines; (b) fewer emission lines; (c) slightly fewer absorption lines; (d) many more absorption lines.

12. The visible spectrum of sunlight reflected from Saturn's cold moon Titan would be expected to be (a) continuous; (b) an emission spectrum; (c) an absorption spectrum.

13. Figure 4.3 ("Elemental Emission") shows the emission spectrum of neon gas. If the temperature of the gas were increased, we would observe (a) fewer red lines and more blue lines; (b) even more red lines; (c) some faint absorption features; (d) no significant change.

14. Compared with a star having many blue absorption lines, a star with many red and blue absorption lines must be (a) moving away from the observer; (b) cooler than the other star; (c) of different composition than that of the other star; (d) moving away from the other star.

15. An atom that has been ionized (a) has equal numbers of protons and electrons; (b) has more protons than electrons; (c) is radioactive; (d) is electrically neutral.

16. Compared with an electron transition from the first excited state to the ground state, a transition from the third excited state to the second excited state emits a photon of (a) greater energy; (b) lower energy; (c) identical energy.

17. Compared with a complex atom like neon, a simple atom such as hydrogen has (a) more excited states; (b) fewer excited states; (c) the same number of excited states.

18. Compared with cooler stars, the hottest stars have absorption lines that are (a) thin and distinct; (b) broad and fuzzy; (c) identical to the lines in the cooler stars.

19. Compared with slowly rotating stars, the fastest spinning stars have absorption lines that are (a) thin and distinct; (b) broad and fuzzy; (c) identical to the lines in the slowly rotating stars.

20. Astronomers analyze starlight to determine a star's (a) temperature; (b) composition; (c) motion; (d) all of the above.

## Problems

 *Algorithmic versions of these Problems are available in the Practice Problems module of the Companion Website. The number of dots preceding each Problem indicates its approximate level of difficulty.*

1. • What is the energy (in electron volts—see *More Precisely 4-1*) of a 450-nm blue photon? A 200-nm ultraviolet photon?

2. • What is the energy (in electron volts) of a 100-GHz (1 gigahertz = $10^9$ Hz) microwave photon?

3. • What is the wavelength of a 2-eV red photon? Repeat your calculation for an 0.1-eV infrared photon and a 5000-eV (5-keV) X-ray.

4. • How many times more energy has a 1-nm gamma ray than a 10-MHz radio photon?

5. • How many times longer in wavelength is a 100-MHz radio photon than a 100-eV X-ray?

6. • Calculate the energy change in the transition responsible for the left-hand yellow sodium line shown in Figure 4.6.

7. •• Calculate the wavelength and frequency of the radiation emitted by the electronic transition from the 10th to the 9th excited state of hydrogen. In what part of the electromagnetic spectrum does this radiation lie? Repeat the question for transitions from the 100th to the 99th excited state and from the 1000th to the 999th excited state.

8. •• How many different photons (i.e., photons of different frequencies) can be emitted as a hydrogen atom in the third excited state falls back, directly or indirectly, to the ground state? What are the wavelengths of those photons?

9. •• List all the spectral lines of hydrogen that lie in the visible range (taken to run from 400 to 700 nm in wavelength).

10. • A distant galaxy is receding from Earth with a radial velocity of 3000 km/s. At what wavelength would its Lyα line be received by a detector above Earth's atmosphere?

11. • The Hα line of a certain star is received on Earth at a wavelength of 656 nm. What is the star's radial velocity with respect to Earth?

12. •• In a demonstration of the photoelectric effect, suppose that a minimum energy of $5 \times 10^{-19}$ J (3.1 eV) is required to dislodge an electron from a metal surface. What is the minimum frequency (and longest wavelength) of radiation for which the detector registers a response?

13. •• At a temperature of 5800 K, hydrogen atoms in the solar atmosphere have typical random speeds of about 12 km/s. Assuming that spectral-line broadening is simply the result of atoms moving toward us or away from us at this random speed, estimate the thermal width (in nanometers) of the 656.3-nm solar Hα line.

14. •• Turbulent motion in a radio-emitting gas cloud broadens a 1.2-GHz (1 gigahertz = $10^9$ Hz) radio line to a width of 0.5 MHz. Estimate the average speed (see Problem 13) at which the gas within the cloud is moving.

15. •• Estimate how fast, in revolutions per day, the Sun would have to rotate in order for rotational broadening to be comparable to thermal broadening in determining line widths. (See Problem 13; the radius of the Sun is roughly 700,000 km.)

*The Companion Website at www.aw-bc.com/chaisson provides algorithmically generated versions of each chapter's Problems, along with additional quizzes, an Animations & Videos gallery, an Images gallery, an interactive Glossary, and a full eBook.*

This composite photograph shows two of the premier optical telescopes available to astronomers today. At the top. overflying Earth is HST—the Hubble Space Telescope—in orbit some 600 kilometers above the surface of our planet; operating well above Earth's atmosphere. it has superb angular resolution. At the bottom, in the Atacama Desert high in the Chilean Andes is the new VLT—the Very Large Telescope—which is actually an array of four large telescopes working together to view the universe with equally superb resolution. (NASA; ESO) ▶

# TELESCOPES
## The Tools of Astronomy

At its heart, astronomy is an observational science. More often than not, observations of cosmic phenomena precede any clear theoretical understanding of their nature. As a result, our detecting instruments—our telescopes—have evolved to observe as broad a range of wavelengths as possible.

Until the middle of the 20th century, telescopes were limited to collecting visible light. Since then, technological advances have expanded our view of the universe to all regions of the electromagnetic spectrum. Some telescopes are sited on Earth, whereas others must be placed in space, and design considerations vary widely from one part of the spectrum to another. Whatever the details of their construction, however, telescopes are devices whose basic purpose is to collect electromagnetic radiation and deliver it to a detector for detailed study.

## LEARNING GOALS

*Studying this chapter will enable you to*

1 Sketch and describe the basic designs of the major types of optical telescopes used by astronomers.

2 Explain the particular advantages of reflecting telescopes for astronomical use, and specify why very large telescopes are needed for most astronomical studies.

3 Explain the purpose of some of the detectors used in astronomical telescopes.

4 Describe how Earth's atmosphere affects astronomical observations, and discuss some of the current efforts to improve ground-based astronomy.

5 Discuss the advantages and disadvantages of radio astronomy compared with optical observations.

6 Explain how interferometry can enhance the usefulness of astronomical observations.

7 Explain why some astronomical observations are best done from space, and discuss the advantages and limitations of space-based astronomy.

8 Say why it is important to make astronomical observations in different regions of the electromagnetic spectrum.

 Visit www.aw-bc.com/chaisson for additional images, animations, quizzes, and eBook for this chapter.

# 5.1 Optical Telescopes

In essence, a **telescope** is a "light bucket" whose primary function is to capture as many photons as possible from a given region of the sky and concentrate them into a focused beam for analysis. Much like a water bucket that collects only the rain falling into it, a telescope intercepts only that radiation falling onto it.

*Optical* telescopes are designed specifically to collect the wavelengths that are visible to the human eye. These telescopes have a long history, stretching back to the days of Galileo in the early 17th century, and for most of the past four centuries astronomers have built their instruments primarily for use in the narrow, visible, portion of the electromagnetic spectrum. ∞ (Sec. 3.3) Optical telescopes are probably also the best-known type of astronomical hardware, so it is fitting that we begin our study with them.

Although the various telescope designs presented in this section all come to us from optical astronomy, the discussion applies equally well to many instruments designed to capture *invisible* radiation, particularly in the infrared and ultraviolet regimes. Many large ground-based optical facilities are also used extensively for infrared work.* Indeed, many ground-based observatories have recently been constructed with infrared observing as their principal function.

## Refracting and Reflecting Telescopes

Optical telescopes fall into two basic categories: *refractors* and *reflectors*. **Refraction** is the bending of a beam of light as it passes from one transparent medium (e.g., air) into

---

*\*Recall from Chapter 3 that, while Earth's atmosphere effectively blocks all ultraviolet, and most infrared, radiation, there remain several fairly broad spectral windows through which ground-based infrared observations can be made. ∞ (Sec. 3.3)*

another (e.g., glass). Consider for example how a straw that is half immersed in a glass of water looks bent (Figure 5.1). The straw is straight, of course, but the light by which we see it is bent—refracted—as that light leaves the water and enters the air. When the light then enters our eyes, we perceive the straw as being bent.

A **refracting telescope** uses a *lens* to gather and concentrate a beam of light. Figure 5.2(a) shows how refraction at two faces of a prism can be used to change the direction of a beam of light. As illustrated in Figure 5.2(b), we can think of a lens as a series of prisms combined in such a way that all light rays arriving parallel to its axis (the imaginary line through the center of the lens), regardless of their distance from that axis, are refracted to pass through a single point, called the *focus*. The distance between the primary mirror and the focus is the *focal length*.

Figure 5.3 shows how a **reflecting telescope** uses a curved *mirror* instead of a lens to focus the incoming light. As shown in Figure 5.3(a), light striking a polished surface is reflected back, leaving the mirror at the same angle at which it arrived. The mirror in a reflecting telescope is constructed so that all light rays arriving parallel to its axis are reflected to pass through the focus (Figure 5.3b). In astronomical contexts, the mirror that collects the incoming light is usually called the *primary mirror*, because telescopes often contain more than one mirror. The focus of the primary mirror is referred to as the **prime focus.**

Astronomical telescopes are often used to make **images** of their field of view (simply, the portion of the sky that the telescope "sees"). Figure 5.4 illustrates how that is accomplished, in this case by the mirror in a reflecting telescope. Light from a distant object (here, a comet) reaches us as parallel, or very nearly parallel, rays. Any ray of light entering the instrument parallel to the telescope's axis strikes the mirror and is reflected through the prime focus. Light coming

(a)

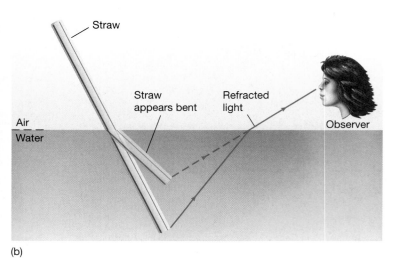

(b)

▲ **FIGURE 5.1 Refraction** A straw placed in a bowl of water appears bent (a) because the light from the part of the straw under the surface is refracted as it leaves the water and enters the air. (b) Consequently, the image formed in our eyes is displaced relative to the true position of the straw. *(R. Megna/Fundamental Photographs, NYC)*

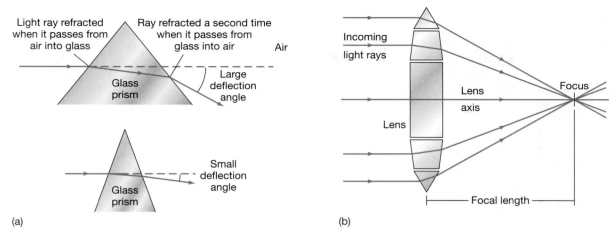

**▲ FIGURE 5.2 Refracting Lens** (a) Refraction by a prism changes the direction of a light ray by an amount that depends on the angle between the prism's faces. When the angle between the faces is large, the deflection is large; when the angle is small, so is the deflection. (b) A lens can be thought of as a series of prisms. A light ray traveling along the axis of a lens is undeflected as it passes through the lens. Parallel rays arriving at progressively greater distances from the axis are refracted by increasing amounts in such a way that they all pass through a single point—the focus.

from a slightly different direction—inclined slightly to the axis—is focused to a slightly different point. In this way, an image is formed near the prime focus. Each point on the image corresponds to a different point in the field of view.

The prime-focus images produced by large telescopes are actually quite small—the image of the entire field of view may be as little as 1 cm across. Often, the image is magnified with a lens known as an *eyepiece* before being observed by eye or, more likely, recorded as a photograph or digital image. The angular diameter of the magnified image is much greater than the telescope's field of view, allowing much more detail to be discerned. Figure 5.5(a) shows the basic design of a simple refracting telescope, illustrating how a small eyepiece is used to view the image focused by the lens. Figure 5.5(b) shows how a reflecting telescope accomplishes the same function.

## Comparing Refractors and Reflectors

The two telescope designs shown in Figure 5.5 achieve the same result: Light from a distant object is captured and focused to form an image. On the face of it, then, it might appear that there is little to choose between the two in deciding which type to buy or build. However, as the sizes of telescopes have increased steadily over the years (for reasons to be discussed in Section 5.3), a number of important factors have tended to favor reflecting instruments over refractors:

1. The fact that light must pass through the lens is a major disadvantage of refracting telescopes. Just as a prism disperses white light into its component colors, the lens in a refracting telescope tends to focus red and blue light differently. This deficiency is

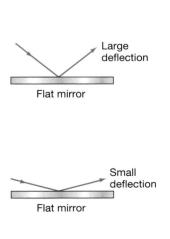

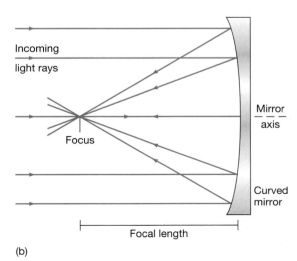

**◄ FIGURE 5.3 Reflecting Mirror** (a) Reflection of light from a flat mirror occurs when light is deflected, depending on its angle of incidence. (b) A curved mirror can be used to focus to a single point all rays of light arriving parallel to the mirror axis. Light rays traveling along the axis are reflected back along the axis, as indicated by the arrowheads pointing in both directions. Off-axis rays are reflected through greater and greater angles the farther they are from the axis, so that they all pass through the focus.

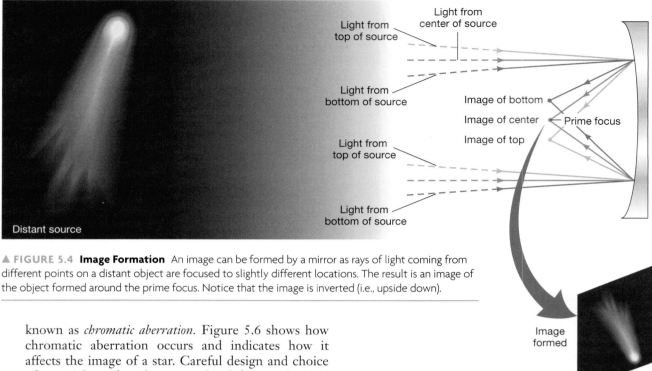

▲ FIGURE 5.4 **Image Formation** An image can be formed by a mirror as rays of light coming from different points on a distant object are focused to slightly different locations. The result is an image of the object formed around the prime focus. Notice that the image is inverted (i.e., upside down).

known as *chromatic aberration.* Figure 5.6 shows how chromatic aberration occurs and indicates how it affects the image of a star. Careful design and choice of materials can largely correct this deficiency, but it is very difficult to eliminate entirely. Obviously, the problem does not occur with mirrors.

2. As light passes through the lens, some of it is absorbed by the glass. This absorption is a relatively minor problem for visible radiation, but it can be severe for infrared and ultraviolet observations because glass blocks most of the radiation in those regions of the electromagnetic spectrum. Again, the problem does not affect mirrors.

3. A large lens can be quite heavy. Because it can be supported only around its edge (so as not to block the incoming radiation), the lens tends to deform under its own weight. A mirror does not have this drawback because it can be supported over its entire back surface.

4. A lens has two surfaces that must be accurately machined and polished—a task that can be very difficult indeed—but a mirror has only one.

For these reasons, *all* large modern telescopes use mirrors as their primary light gatherers. The largest refractor ever built, installed in 1897 at the Yerkes Observatory in Wisconsin and still in use today, has a lens diameter of just over 1 m (40 inches). By contrast, many

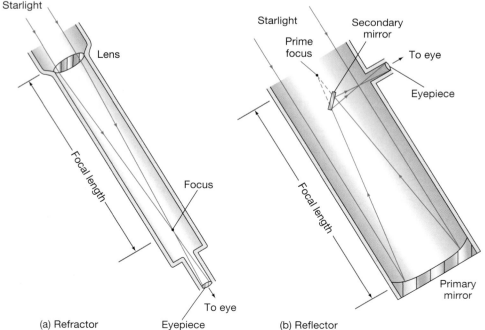

(a) Refractor

(b) Reflector

◄ FIGURE 5.5 **Refractors and Reflectors** Comparison of (a) refracting and (b) reflecting telescopes. Both types are used to gather and focus electromagnetic radiation—to be observed by human eyes or recorded on photographs or in computers. In both cases, the image formed at the focus is viewed with a small magnifying lens called an eyepiece.

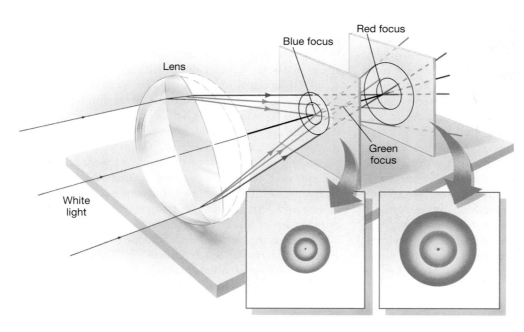

Lens

Blue focus

Red focus

Green focus

White light

◀ FIGURE 5.6 **Chromatic Aberration** A prism bends blue light more than it bends red light, so the blue component of light passing through a lens is focused slightly closer to the lens than the red component is. As a result, the image of an object acquires a colored "halo," no matter where we place our detector.

at the prime focus; however, it can be inconvenient, or even impossible, to suspend bulky pieces of equipment there. More often, the light is intercepted on its path to the focus by a *secondary mirror* and redirected to a more convenient location, as in Figure 5.7(b) through (d).

recently constructed reflecting telescopes have mirror diameters in the 10-m range, and still larger instruments are on the way.

## Types of Reflecting Telescope

Figure 5.7 shows some basic reflecting telescope designs. Radiation from a star enters the instrument, passes down the main tube, strikes the primary mirror, and is reflected back toward the prime focus, near the top of the tube. Sometimes astronomers place their recording instruments

In a **Newtonian telescope** (named after Sir Isaac Newton, who invented this particular design), the light is intercepted before it reaches the prime focus and then is deflected by 90°, usually to an eyepiece at the side of the instrument. This is a popular design for smaller reflecting telescopes, such as those used by amateur astronomers, but it is relatively uncommon in large instruments. On a large telescope, the Newtonian focus may be many meters above the ground, making it an inconvenient place to attach equipment (or place an observer).

Alternatively, astronomers may choose to work on a rear platform where they can use equipment, such as a spectroscope, that is too heavy to hoist to the prime focus. In this case, light reflected by the primary mirror toward the prime focus is intercepted by a smaller secondary mirror, which reflects it

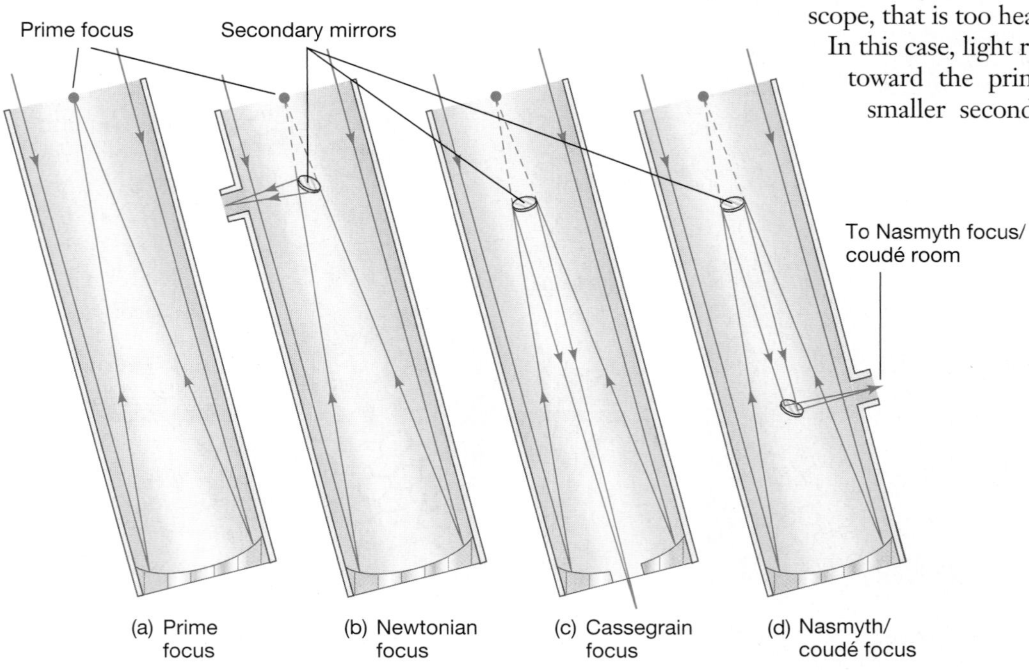

Prime focus

Secondary mirrors

To Nasmyth focus/ coudé room

(a) Prime focus

(b) Newtonian focus

(c) Cassegrain focus

(d) Nasmyth/ coudé focus

◀ FIGURE 5.7 **Reflecting Telescopes** Four reflecting telescope designs: (a) prime focus, (b) Newtonian focus, (c) Cassegrain focus, and (d) Nasmyth/coudé focus. Each design uses a primary mirror at the bottom of the telescope to capture radiation, which is then directed along different paths for analysis. Notice that the secondary mirrors shown in (c) and (d) are actually slightly diverging, so that they move the focus outside the telescope.

back down through a small hole at the center of the primary mirror. This arrangement is known as a **Cassegrain telescope** (after Guillaume Cassegrain, a French lensmaker). The point behind the primary mirror where the light from the star finally converges is called the *Cassegrain focus*.

A more complex observational configuration requires starlight to be reflected by several mirrors. As in the Cassegrain design, light is first reflected by the primary mirror toward the prime focus and is then reflected back down the tube by a secondary mirror. Next, a third, much smaller, mirror reflects the light out of the telescope, where (depending on the details of the telescope's construction) the beam may be analyzed by a detector mounted alongside, at the *Nasmyth focus*, or it may be directed via a further series of mirrors into an environmentally controlled laboratory known as the *coudé* room (from the French word for "bent"). This laboratory is separate from the telescope itself, enabling astronomers to use very heavy and finely tuned equipment that cannot be placed at any of the other foci (all of which necessarily move with the telescope). The arrangement of mirrors is such that the light path to the coudé room does not change as the telescope tracks objects across the sky.

To illustrate some of these points, Figure 5.8(a) shows the twin 10-m-diameter optical/infrared telescopes of the Keck Observatory on Mauna Kea in Hawaii, operated jointly by the California Institute of Technology and the University of California. The diagram in part (b) illustrates the light paths and some of the foci. Observations may be made at the Cassegrain, Nasmyth, or coudé focus, depending on the needs of the user. As the size of the person in part (c) indicates, this is indeed a very large telescope—in fact, the two mirrors are currently the largest on Earth. We will see numerous examples throughout this text of Keck's many important discoveries.

Perhaps the best-known telescope on (or, rather, near) Earth is the *Hubble Space Telescope (HST)*, named for one of America's most notable astronomers, Edwin Hubble. The device was placed in Earth orbit by NASA's space shuttle *Discovery* in 1990 and is still in operation (as of 2007; see *Discovery 5-1*). As illustrated in Figure 5.9, *HST* is a Cassegrain telescope in which all the instruments are located directly behind the primary mirror. The telescope's detectors are capable of making measurements in the optical, infrared, and ultraviolet parts of the spectrum. The inset in the figure shows a distant galaxy in which a star has exploded, an example of the important work

◄ **FIGURE 5.8 Keck Telescope** (a) The two 10-m telescopes of the Keck Observatory. (b) Artist's illustration of the telescope, the path taken by an incoming beam of starlight, and some of the locations where instruments may be placed. (c) One of the 10-m mirrors. (The odd shape is explained in Section 5.3.) Note the technician in orange coveralls at center. *(W. M. Keck Observatory)*

(c)

(a)

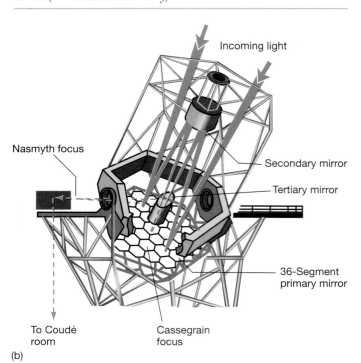

(b)

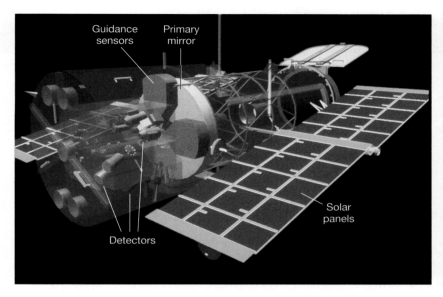

▲ FIGURE 5.9 *HST* **Detectors** The principal features inside the *Hubble Space Telescope* are displayed in this "see-through" illustration. The telescope's Cassegrain design focuses light through a small opening in the center of the primary mirror (large bluish disk at center), from where it can be directed to any of several instruments arrayed behind the mirror (shown here in various colors at left). The large red objects are sensors that guide the pointing of the telescope, and the huge blue panels collect light from the Sun to power everything onboard. The inset shows an example of one of *HST*'s most important findings: exploding stars (lower left) in distant galaxies, this one called NGC 4526, some 100 million light-years away. See also the chapter opening photo on page 107. (*D. Berry; NASA*)

R I V U X G

(as we shall see later in this book) that *HST* has done to map the size and scale of the universe.

CONCEPT CHECK

✔ Why do all modern telescopes use mirrors to gather and focus light?

## 5.2 Telescope Size

Modern astronomical telescopes have come a long way from Galileo's simple apparatus (see the Part 1 opening text). Their development over the years has seen a steady increase in *size*, for two main reasons. The first has to do with the amount of light a telescope can collect—its *light-gathering power*. The second is related to the amount of detail that can be seen—the telescope's *resolving power*. Simply put, large telescopes can gather and focus more radiation than can their smaller counterparts, allowing astronomers to study fainter objects and to obtain more detailed information about bright ones. This fact has played a central role in determining the design of contemporary instruments.

### Light-Gathering Power

One important reason for using a larger telescope is simply that it has a greater **collecting area,** which is the total area capable of gathering radiation. The larger the telescope's

reflecting mirror (or refracting lens), the more light it collects, and the easier it is to measure and study an object's radiative properties. Astronomers spend much of their time observing very distant—and hence very *faint*—cosmic sources. In order to make detailed observations of such objects, very large telescopes are essential. Figure 5.10 illustrates the effect of increasing the size of a telescope by comparing images of the Andromeda Galaxy taken with two different instruments. A large collecting area is particularly important for spectroscopic work, as the radiation received in that case must be split into its component wavelengths for further analysis.

The observed brightness of an astronomical object is directly proportional to the area of our telescope's mirror and therefore to the *square* of the mirror diameter. Thus, a 5-m telescope will produce an image 25 times as bright as a 1-m instrument, because a 5-m mirror has $5^2 = 25$ times the collecting area of a 1-m mirror. We can also think of this relationship in terms of the length of *time* required for a telescope to collect enough energy to create a recognizable image on a photographic plate. Our 5-m telescope will produce an image 25 times faster than the 1-m device because it gathers energy at a rate 25 times greater. Put another way, a 1-hour exposure with a 1-m telescope is roughly equivalent to a 2.4-minute exposure with a 5-m instrument.

Until the 1980s, the conventional wisdom was that telescopes with mirrors larger than 5 or 6 m in diameter were simply too expensive and impractical to build. The problems involved in casting, cooling, and polishing a huge

# DISCOVERY 5-1

## The Hubble Space Telescope

The *Hubble Space Telescope (HST)* is the largest, most complex, most sensitive observatory ever deployed in space. At over $9 billion (including the cost of several missions to service and refurbish the system), it is also one of the most expensive scientific instrument ever built and operated. A joint project of NASA and the European Space Agency, *HST* was designed to allow astronomers to probe the universe with at least 10 times finer resolution and some 30 times greater sensitivity to light than existing Earth-based devices. *HST* is operated remotely from the ground; there are no astronauts aboard the telescope, which orbits Earth about once every 95 minutes at an altitude of some 600 km (380 miles).

The telescope's overall dimensions approximate those of a city bus or railroad tank car: 13 m (43 feet) long, 12 m (39 feet) across with solar arrays extended, and 11,000 kg (12.5 tons when weighed on Earth). At the heart of *HST* is a 2.4-m-diameter mirror designed to capture optical, ultraviolet, and infrared radiation before it reaches Earth's murky atmosphere. The accompanying figure shows the telescope being lifted out of the cargo bay of the space shuttle *Discovery* in the spring of 1990.

*(NASA)*

The telescope reflects light from its large mirror back to a smaller, 0.3-m secondary mirror, which sends the light through a small hole in the main mirror and into the rear portion of the spacecraft. There, any of five major scientific instruments wait to analyze the incoming radiation. Most of these instruments (see Figure 5.9) are about the size of a telephone booth. They are designed to be maintained by NASA astronauts, and indeed, most of the telescope's instruments have been upgraded or replaced since *HST* was launched. The current detectors on the

telescope span the visible, near-infrared, and near-ultraviolet regions of the electromagnetic spectrum, from about 100 nm (UV) to 2200 nm (IR).

Soon after the launch of *HST*, astronomers discovered that the telescope's primary mirror had been polished to the wrong shape. The mirror was too flat by 2 $\mu$m, about 1/50 the width of a human hair, making it impossible to focus light as well as expected. The optical flaw (known as *spherical aberration*) produced by this design error meant that *HST* was not as sensitive as designed, although it could still see many objects in the universe with unprecedented resolution. In 1993, astronauts aboard the space shuttle *Endeavour* visited *HST* and succeeded in repairing some of its ailing equipment. They replaced *Hubble*'s gyroscopes to help the telescope point more accurately, installed sturdier versions of the solar panels that power the telescope's electronics, and—most importantly—inserted an intricate set of small mirrors (each about the size of a coin) to compensate for the faulty primary mirror. *Hubble*'s resolution is now close to the original design specifications, and the telescope has regained much of its lost sensitivity. Additional service missions were performed in 1997, 1999, and 2002 to replace instruments and repair faulty systems.

A good example of *Hubble*'s scientific capabilities today can be seen by comparing the two images of the spiral galaxy M100, shown in the accompanying figure. On the left is perhaps the best ground-based photograph of this beautiful galaxy, showing rich detail and color in its spiral arms. On the right, to the same scale and orientation, is an *HST* image showing improvement in both resolution and sensitivity. (The chevron-shaped field of view is caused by the corrective optics inserted into the telescope; an additional trade-off is that *Hubble*'s field of view is smaller than those of ground-based telescopes.) The inset shows *Hubble*'s exquisite resolution of small fields of view.

During its decade and a half of operation, *Hubble* has revolutionized our view of the sky, helping to rewrite some theories of the universe along the way. The space-based instrument has studied newborn galaxies almost at the limit of the observable universe with unprecedented clarity, allowing astronomers to see the interactions and collisions that may have shaped the evolution of our Milky Way. Turning its gaze to the hearts of galaxies closer to home, *Hubble* has provided strong evidence for supermassive black holes in their cores. Within our own Galaxy, it has given astronomers stunning new insights into the physics of star formation and the evolution of stellar systems and stars of all sizes, from superluminous giants to

block of quartz or glass to very high precision (typically less than the width of a human hair) were just too great. However, new, high-tech manufacturing techniques, coupled with radically new mirror designs, make the construction of telescopes in the 8- to 12-m range almost a routine matter. Experts can now make large mirrors much lighter for their size than had previously been thought feasible and can combine many smaller mirrors into the equivalent of a much larger single-mirror telescope. Several large-diameter instruments now exist, and many more are planned.

The Keck telescopes, shown in detail in Figure 5.8 and in a larger view in Figure 5.11, are a case in point. Each telescope combines 36 hexagonal 1.8-m mirrors into the equivalent collecting area of a single 10-m reflector. The first Keck telescope became fully operational in 1992; the second was completed in 1996. The large size of these devices and the high altitude at which they operate make them particularly well suited to detailed spectroscopic studies of very faint objects, in both the optical and infrared parts of the spectrum. Mauna Kea's 4.2-km

(NASA)

(D. Malin/Angio-Australian Telescope)

objects barely more massive than planets. Finally, in our solar system, *Hubble* has afforded scientists new views of the planets, their moons, and the tiny fragments from which they formed long ago. Many spectacular examples of the telescope's remarkable capabilities appear throughout this book.

With *Hubble* now well into its second decade of a highly successful career, NASA scientists are planning the telescope's successor. The Next Generation Space Telescope (now known as the *James Webb Space Telescope*, or *JWST*, after the administrator who led NASA's *Apollo* program during the 1960s and 1970s) will dwarf *Hubble* in both scale and capability. Sporting a 6-m segmented mirror with almost six times the collecting area of *Hubble*'s mirror, and containing a formidable array of detectors optimized for use at visible and infrared wavelengths, *JWST* will orbit at a distance of 1.5 million km from Earth, far beyond the Moon. NASA hopes to launch the instrument by 2013. The telescope's primary mission aims to study the formation of the first stars and galaxies, measuring the large-scale structure of the universe, and investigating the evolution of planets, stars, and galaxies.

In early 2004 NASA announced that, for budgetary and safety reasons, they had canceled the planned 2006 servicing mission that would have extended the telescope's lifetime. Not surprisingly, the decision was actively, and vocally, opposed by many astronomers who feared that, without servicing, *Hubble* could become inoperative as early as 2007. However, with *JWST*'s launch window slipping and still uncertain, and in the face of stiff budgetary competition from other NASA missions, manned and unmanned, NASA has now agreed to reconsider the possibility of a fourth service mission by 2009. If current safety concerns with the shuttle fleet can be addressed, and if the construction schedule of the International Space Station permits, *Hubble* may yet receive a new lease on life and continue operating past 2010. Stay tuned!

(13,800 feet) altitude minimizes atmospheric absorption of infrared radiation, making this site one of the finest locations on Earth for infrared astronomy.

Numerous other large telescopes can be seen in the figure. Some are designed exclusively for infrared work; others, like Keck, operate in both the optical and the infrared. To the right of the Keck domes is the 8.3-m Subaru (the Japanese name for the Pleiades) telescope, part of the National Astronomical Observatory of Japan. Its mirror, shown in Figure 5.11(b), is the largest single mirror (as opposed to the segmented design used in Keck) yet built. Subaru saw "first light" in 1999. In the distance is another large single-mirror instrument: the 8.1-m Gemini North telescope, completed in 1999 by a consortium of seven nations, including the United States. Its twin, Gemini South, in the Chilean Andes, went into service in 2002.

In terms of total available collecting area, the largest telescope currently available is the European Southern Observatory's optical-infrared Very Large Telescope (VLT), located at Cerro Paranal, in Chile (Figure 5.12).

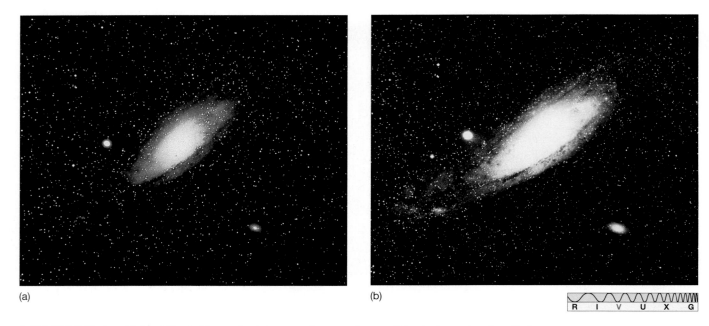

(a)

(b)

R I V U X G

▲ FIGURE 5.10 **Sensitivity** Effect of increasing telescope size on an image of the Andromeda galaxy. Both photographs had the same exposure time, but image (b) was taken with a telescope twice the size of that used to make image (a). Fainter detail can be seen as the diameter of the telescope mirror increases because larger telescopes are able to collect more photons per unit time. (*Adapted from AURA*)

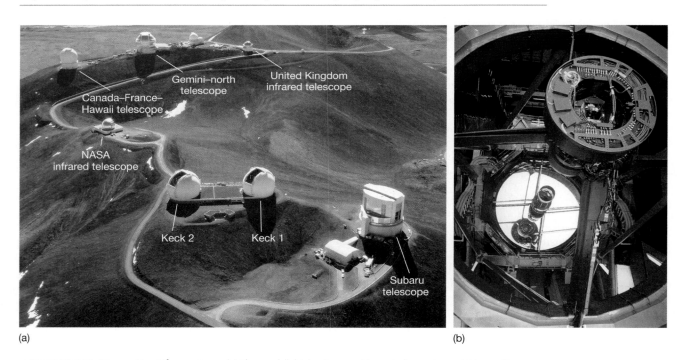

(a)

(b)

▲ FIGURE 5.11 **Mauna Kea Observatory** (a) The world's highest ground-based observatory, at Mauna Kea, Hawaii, is perched atop a dormant volcano more than 4 km (nearly 14,000 feet) above sea level. Among the domes visible in the picture are those housing the Canada–France–Hawaii 3.6-m telescope, the 8.1-m Gemini North instrument, the 2.2-m telescope of the University of Hawaii, Britain's 3.8-m infrared facility, and the twin 10-m Keck telescopes. To the right of the twin Kecks is the Japanese 8.3-m Subaru telescope. The thin air at this high-altitude site guarantees less atmospheric absorption of incoming radiation and hence a clearer view than at sea level, but the air is so thin that astronomers must occasionally wear oxygen masks while working. (b) The mirror in the Subaru telescope, currently the largest one-piece mirror in any astronomical instrument. (*R. Wainscoat; R. Underwood/Keck Observatory; NAOJ*)

▲ FIGURE 5.12 **VLT Observatory** Located at the Paranal Observatory in Atacama, Chile, the European Southern Observatory's Very Large Telescope (VLT) is the world's largest optical telescope. It comprises four 8.2-m reflecting telescopes, which can be used in tandem to create the effective area of a single 16-m mirror. See also chapter opening photo on page 107. *(ESO)*

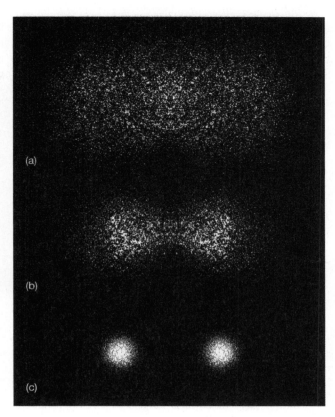

▲ FIGURE 5.13 **Resolving Power** Two comparably bright light sources become progressively clearer when viewed at finer and finer angular resolution. When the angular resolution is much poorer than the separation of the objects, as in (a), the objects appear as a single fuzzy "blob." As the resolution improves, through (b) and (c), the two sources become discernible as separate objects.

The VLT consists of four separate 8.2-m mirrors that can function as a single instrument. The last of its four mirrors was completed in 2001.

## Resolving Power

A second advantage of large telescopes is their finer **angular resolution.** In general, *resolution* refers to the ability of any device, such as a camera or telescope, to form distinct, separate images of objects lying close together in the field of view. The finer the resolution, the better we can distinguish the objects and the more detail we can see. In astronomy, where we are always concerned with angular measurement, "close together" means "separated by a small angle in the sky," so angular resolution is the factor that determines our ability to see fine structure. Figure 5.13 illustrates how the appearance of two objects—stars, say—might change as the angular resolution of our telescope varies. Figure 5.14 shows the result of increasing resolving power with views of the Andromeda galaxy at several different resolutions.

What limits a telescope's resolution? One important factor is *diffraction*, the tendency of light—and all other waves, for that matter—to bend around corners. ∞ *(Discovery 3-1)* Because of diffraction, when a parallel beam of light enters a telescope, the rays spread out slightly, making it impossible to focus the beam to a sharp point, even with a perfectly constructed mirror. Diffraction introduces a certain "fuzziness," or loss of resolution, into any optical system. The degree of fuzziness—the minimum angular separation that can be distinguished—determines the angular resolution of the telescope.

As discussed further in *More Precisely 5-1*, the amount of diffraction is proportional to the wavelength of the radiation and inversely proportional to the diameter of the telescope mirror. Thus, for light of any given wavelength, large telescopes produce less diffraction than small ones. A 5-m telescope observing in blue light would have a diffraction-limited resolution five times finer than the 1-m telescope just discussed—about 0.02″. A 0.1-m (10-cm) telescope would have a diffraction limit of 1″, and so on. For comparison, the angular resolution of the human eye in the middle of the visual range is about 0.5′.

## CONCEPT CHECK

✔ Give two reasons why astronomers need to build very large telescopes.

(a)

(b)

(c)

(d)

R  I  V  U  X  G

◀ **FIGURE 5.14** **Resolution** Detail becomes clearer in the Andromeda galaxy as the angular resolution is improved some 600 times, from (a) 10′, to (b) 1′, (c) 5″, and (d) 1″. *(Adapted from AURA)*

## 5.3 Images and Detectors

In the previous section, we saw how telescopes gather and focus light to form an image of their field of view. In fact, most large observatories use many different instruments to analyze the radiation received from space—including detectors sensitive to many different wavelengths of light, spectroscopes to study emission and absorption lines, and other custom-made equipment designed for specialized studies. These devices may be placed at various points along the light path outside the telescope—see, for example, the multiple foci and light paths in Figure 5.8(b), or the more compact arrangement of detectors within the *Hubble Space Telescope* in Figure 5.9. In this section, we look in a little more detail at how telescopic images are actually produced and at some other types of detectors that are in widespread use.

### Image Acquisition

Computers play a vital role in observational astronomy. Most large telescopes today are controlled either by computers or by operators who rely heavily on computer assistance, and images and data are recorded in a form that can be easily read and manipulated by computer programs.

It is becoming rare for photographic equipment to be used as the primary means of data acquisition at large observatories. Instead, electronic detectors known as **charge-coupled devices,** or **CCDs,** are in widespread use. Their output goes directly to a computer. A CCD (Figure 5.15a and b) consists of a wafer of silicon divided into a two-dimensional array of many tiny picture elements, known as **pixels.** When light strikes a pixel, an electric charge builds up on it. The amount of charge is directly proportional to the number of photons striking each pixel—in other words, to the intensity of the light at that point. The buildup of charge is monitored electronically, and a two-dimensional image is obtained (Figure 5.15c and d).

A CCD is typically a few square centimeters in area and may contain several million pixels, generally arranged on a square grid. As the technology improves, both the areas of CCDs and the number of pixels they contain continue to increase. Incidentally, the technology is not limited to astronomy: Many home video cameras contain CCD chips similar in basic design to those in use at the great astronomical observatories of the world.

CCDs have two important advantages over photographic plates, which were the staple of astronomers for over a century. First, CCDs are much more *efficient* than photographic plates, recording as many as 90 percent of the photons striking them, compared with less than 5 percent for photographic methods. This difference means that a CCD image can show objects 10 to 20 times fainter than a photo-

# MORE PRECISELY 5-1

## Diffraction and Telescope Resolution

As described in the text, a major factor limiting the resolution of telescopes is *diffraction*, the process whereby light tends to spread out slightly as it passes a corner or through an opening. ∞ (*Discovery 3-1*) Theory and experiment agree that the amount of diffraction—that is, the degree of "fuzziness" introduced into an instrument—is *directly proportional* to the wavelength of the radiation and *inversely proportional* to the size of the opening (in our case, the diameter of the primary mirror). The accompanying diagrams illustrate these dependencies.

As a result of diffraction, the resolution of a telescope depends both on the wavelength of the radiation being studied and on the diameter of the mirror used. Detailed calculations indicate that, for a circular mirror and otherwise perfect optics, the angular resolution of a telescope may be written (in convenient units) as

$$\text{angular resolution (arcsec)} = 0.25 \, \frac{\text{wavelength } (\mu m)}{\text{mirror diameter (m)}},$$

where 1 $\mu m$ (1 micron) = $10^{-6}$ m (see Appendix 2). Thus, the amount of diffraction increases in proportion to the wavelength used and inversely with the mirror diameter.

**EXAMPLE** Observations in the infrared or radio range are often limited by the effects of diffraction. For example, according to the preceding formula, in an otherwise perfect observing environment the best possible angular resolution of blue light (with a wavelength of 400 nm) that can be obtained using a 1-m telescope is about $0.25'' \times (0.4/1) = 0.1''$ But if we were to use our 1-m telescope to make observations in the near infrared, at a wavelength of 10 $\mu m$ (10,000 nm), the best resolution we could obtain would be only 2.5″. A 1-m radio telescope operating at a wavelength of 1 cm would have an angular resolution of just under 1°.

The resolution implied by the above equation is known as the **diffraction-limited resolution** of the telescope. We will see later that other factors may degrade the resolution of a telescope, resulting in considerably poorer performance than we would expect based on this formula. With modern technology, many of these factors can be controlled, and some eliminated almost entirely, but realize that diffraction limit still places a fundamental limit on the *best possible resolution* a telescope can achieve.

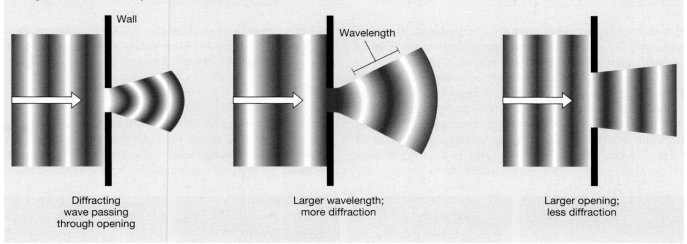

Diffracting wave passing through opening

Larger wavelength; more diffraction

Larger opening; less diffraction

Wall

Wavelength

graph made with the same telescope and the same exposure time. Alternatively, a CCD can record the same level of detail in less than a tenth of the time required by photographic techniques, or it can record that detail with a much smaller telescope. Second, CCDs produce a faithful representation of an image in a digital format that can be placed directly on magnetic tape or disk or, more commonly, sent across a computer network to an observer's home institution.

## Image Processing

Computers are also widely used to reduce *background noise* in astronomical images. Noise is anything that corrupts the integrity of a message, such as static on an AM radio or

"snow" on a television screen. The noise corrupting telescopic images has many causes. In part, it results from faint, unresolved sources in the telescope's field of view and from light scattered into the line of sight by Earth's atmosphere. It can also be caused by imperfections within the detector itself, which may result in an electronic "hiss" similar to the faint background hiss you might hear when you listen to a particularly quiet piece of music on your stereo.

Even though astronomers often cannot determine the origin of the noise in their observations, they can at least measure its characteristics. For example, if we observe a part of the sky where there are no known sources of radiation, then whatever signal we do receive is (almost by definition) noise. Once the properties of the signal have been measured,

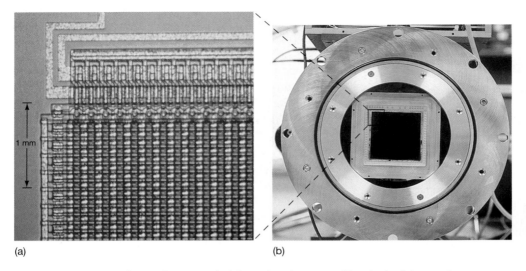

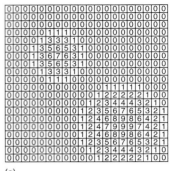

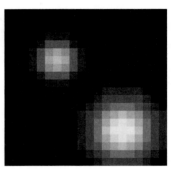

▲ FIGURE 5.15 **CCD Chip** A charge-coupled device (CCD) consists of hundreds of thousands, or even millions, of tiny light-sensitive cells called pixels, usually arranged in a square array. Light striking a pixel causes an electrical charge to build up on it. By electronically reading out the charge on each pixel, a computer can reconstruct the pattern of light—the image—falling on the chip. (a) Detail of a CCD array. (b) A CCD chip mounted for use at the focus of a telescope. (c) Typical data from the chip consist of an array of numbers, running from 0 to 9 in this simplified example. Each number represents the intensity of the radiation striking that particular pixel. (d) When interpreted as intensity levels on a computer screen, an image of the field of view results. (*MIT Lincoln Lab*)

the effects of noise can be partially removed with the aid of high-speed computers, allowing astronomers to see features in their data that would otherwise remain hidden.

Using computer processing, astronomers can also compensate for known instrumental defects. In addition, the computer can often carry out many of the relatively simple, but tedious and time-consuming, chores that must be performed before an image (or spectrum) reaches its final "clean" form. Figure 5.16 illustrates how computerized image-processing techniques were used to correct for

known instrumental problems in *HST*, allowing much of the planned resolution of the telescope to be recovered even before its repair in 1993 (see *Discovery 5-1*).

## Wide-Angle Views

Large reflectors are very good at forming images of narrow fields of view, wherein all the light that strikes the mirror surface moves almost parallel to the axis of the instrument. However, as the angle at which the light enters increases,

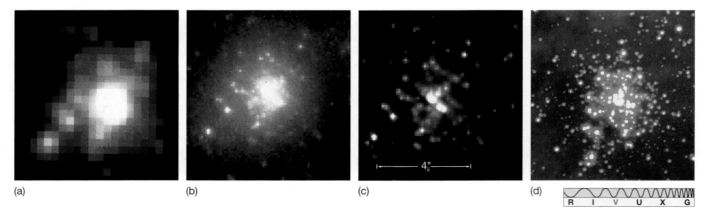

▲ FIGURE 5.16 **Image Processing** (a) Ground-based view of the star cluster R136, a group of stars in the Large Magellanic Cloud (a nearby galaxy). (b) The "raw" image of this same region as seen by the *Hubble Space Telescope* in 1990, before its first repair mission. (c) The same image after computer processing that partly compensated for imperfections in the mirror. (d) The same region as seen by the repaired *HST* in 1994, here observed at a somewhat bluer wavelength. (*AURA/NASA*)

the accuracy of the focus decreases, degrading the overall quality of the image. The effect (called *coma*) worsens as we move farther from the center of the field of view. Eventually, the quality of the image is reduced to the point where the image is no longer usable. The distance from the center to where the image becomes unacceptable defines the useful field of view of the telescope—typically, only a few arc minutes for large instruments.

A design that overcomes this problem is the *Schmidt telescope*, named after its inventor, Bernhard Schmidt, who built the first such instrument in the 1930s. The telescope uses a correcting lens that sharpens the final image of the entire field of view. Consequently, a Schmidt telescope is well suited to producing wide-angle photographs, covering several degrees of the sky. The design of the Schmidt telescope results in a curved image that is not suitable for viewing with an eyepiece, so the image is recorded on a specially shaped piece of photographic film. For this reason, the instrument is often called a *Schmidt camera* (Figure 5.17).

## Photometry

When a CCD is placed at the focus of a telescope to record an image of the instrument's field of view, the telescope is acting, in effect, as a high-powered camera. However, astronomers often want to carry out more specific measurements of the radiation received from space.

One very fundamental property of a star (or any other astronomical object) is its *brightness*—the amount of light energy from the star striking our detector every second. The measurement of brightness is called **photometry** (literally, "light measurement"). In principle, determining a star's brightness is just a matter of adding up the values in all the CCD pixels corresponding to the star (see Figure 5.15c). However, in practice, the process is more complicated, as stellar images may overlap, and computer assistance is generally need to disentangle them.

Astronomers often combine photometric measurements with the use of colored *filters* in order to limit the wavelengths they measure. (A filter simply blocks out all incoming radiation, except in some specific range of wavelengths; see Section 17.3 for a more detailed discussion.) Many standard filters exist, covering various "slices" of the spectrum, from near-infrared through visible to near-ultraviolet wavelengths. By confining their attention to these relatively narrow ranges, astronomers can often estimate the shape of an object's blackbody curve and hence determine, at least approximately, the object's temperature. ∞ (Sec. 3.4) Filters are also used with CCD images in order to simulate natural color. For example, most of the visible-light *HST* images in this text are actually composites of *three* raw images, taken through red, green, and blue filters, respectively, and combined afterwards to reconstruct a single color frame.

Astronomical objects are generally faint, and most astronomical images entail long exposures—minutes to hours—in order to see fine detail (Section 5.2). Thus, the brightness we measure from an image is really an average over the entire exposure. Short-term fluctuations (if any) cannot be seen. When highly accurate and rapid measurements of light intensity are required, a specialized device known as a **photometer** is used. The photometer measures the total amount of light received in all or part of the field of view. When only a portion of the field is of interest, that region is selected simply by masking (blocking) out the rest of the field of view. Using a photometer often means "throwing away" spatial detail—usually no image is produced—but in return, more information is obtained about the intensity and time variability of a source, such as a pulsating star or a supernova explosion.

## Spectroscopy

Often, astronomers want to study the *spectrum* of the incoming light. Large **spectrometers** work in tandem with optical telescopes. Light collected by the primary mirror may be redirected to the coudé room, defined by a narrow slit, dispersed (split into its component colors) by means of a prism or a diffraction grating, and then sent on to a detector—a process not so different in concept from the operation of the simple spectroscope described in Chapter 4. ∞ (Sec. 4.1) The spectrum can be studied in real time (i.e., as it is being received at the telescope) or recorded using a CCD (or, less commonly nowadays, a photographic plate) for

◄ **FIGURE 5.17  Schmidt Telescope**  The Schmidt camera of the European Southern Observatory in Chile. The astronomer in the photograph is not looking through the optics of the main telescope, but instead is using a smaller "finding telescope" to verify the direction in which the instrument is pointed. *(ESO)*

later analysis. Astronomers can then apply the analysis techniques discussed in Chapter 4 to extract detailed information from the spectral lines they record. ∞ (Sec. 4.5)

---

CONCEPT CHECK

✔ Why aren't astronomers satisfied with just taking photographs of the sky?

---

## 5.4 High-Resolution Astronomy

Even large telescopes have limitations. For example, according to the discussion in the preceding section, the 10-m Keck telescope should have an angular resolution of around 0.01″ in blue light. In practice, however, without further technological advances, it could not do better than about 1″. In fact, apart from instruments using special techniques developed to examine some particularly bright stars, *no* ground-based optical telescope built before 1990 can resolve astronomical objects to much better than 1″. The reason is *turbulence* in Earth's atmosphere—small-scale eddies of swirling air all along the line of sight, which blur the image of a star even before the light reaches our instruments.

### Atmospheric Blurring

As we observe a star, atmospheric turbulence produces continual small changes in the optical properties of the air between the star and our telescope (or eye). As a result, the light from the star is refracted slightly as it travels toward us, so the stellar image dances around on our detector (or on our retina). This continual deflection is the cause of the well-known "twinkling" of stars. It occurs for the same reason that objects appear to shimmer when viewed across a hot roadway on a summer day: The constantly shifting rays of light reaching our eyes produce the illusion of motion.

On a good night at the best observing sites, the maximum amount of deflection produced by the atmosphere is slightly less than 1″. Consider taking a photograph of a star.

After a few minutes of exposure (long enough for the intervening atmosphere to have undergone many small random changes), the image of the star has been smeared out over a roughly circular region an arc second or so in diameter. Astronomers use the term **seeing** to describe the effects of atmospheric turbulence. The circle over which a star's light (or the light from any other astronomical source) is spread is called the **seeing disk.** Figure 5.18 illustrates the formation of the seeing disk for a small telescope.*

Atmospheric turbulence has less effect on light of longer wavelengths—ground-based astronomers generally "see" better in the infrared. However, offsetting this improvement in image quality is the fact that the atmosphere is wholly or partially opaque over much of the infrared range. ∞ (Sec. 3.3) For these reasons, to achieve the best possible observing conditions, telescopes are sited on mountaintops (to get above as much of the atmosphere as possible) in regions of the world where the atmosphere is known to be fairly stable and relatively free of dust and moisture. Another reason for the choice of remote locations is the growing problem of **light pollution** in populated areas—unwanted upward-directed light from streets, parking lots, homes, and businesses, that scatters back from dust in the air into our telescopes, literally drowning out the faint signals from distant stars and galaxies that astronomers want to observe.

In the continental United States, these sites tend to be in the desert Southwest. The U.S. National Observatory for optical astronomy in the Northern Hemisphere,

---

*In fact, for a large instrument—more than about 1 m in diameter—the situation is more complicated, because rays striking different parts of the mirror have actually passed through different turbulent regions of the atmosphere. The end result is still a seeing disk, however.

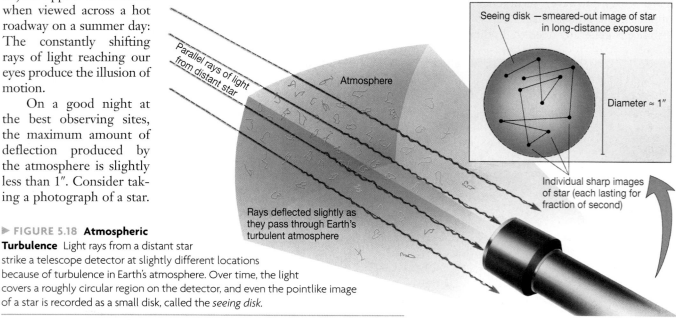

▶ FIGURE 5.18 **Atmospheric Turbulence** Light rays from a distant star strike a telescope detector at slightly different locations because of turbulence in Earth's atmosphere. Over time, the light covers a roughly circular region on the detector, and even the pointlike image of a star is recorded as a small disk, called the *seeing disk.*

▲ FIGURE 5.19 **European Southern Observatory** Located in the Andes Mountains of Chile, the European Southern Observatory at La Silla is run by a consortium of European nations. Numerous domes house optical telescopes of different sizes, each with varied support equipment, making this one of the most versatile observatories south of the equator. The largest telescope at La Silla—the square building to the right of center—is the New Technology Telescope, a 3.5-m state-of-the-art active-optics device. *(ESO)*

completed in 1973, is located high on Kitt Peak near Tucson, Arizona. The site was chosen because of its many dry, clear nights. Seeing less than 1″ from such a location is regarded as good, and seeing a few arc seconds is tolerable for many purposes. Even better conditions are found on Mauna Kea, Hawaii (Figure 5.11), and at numerous sites in the Andes Mountains of Chile (Figures 5.12 and 5.19), which is why many large telescopes have recently been constructed at these exceptionally clear locations.

An optical telescope placed in orbit about Earth or on the Moon could obviously overcome the limitations imposed by the atmosphere on ground-based instruments. Without atmospheric blurring, extremely fine resolution—close to the diffraction limit—can be achieved, subject only to the engineering restrictions associated with building or placing large structures in space. The 2.4-m mirror in the *Hubble Space Telescope* has a (blue-light) diffraction limit of only 0.05″, giving astronomers a view of the universe as much as 20 times sharper than that normally available from even much larger ground-based instruments.

## Active Optics

The latest techniques for producing ultrasharp images take the ideas of computer control and image processing (see Section 5.2) several stages further. By analyzing the image formed by a telescope *while the light is still being collected*, it is now possible to adjust the telescope from moment to moment to avoid or compensate for the effects of mirror distortion, temperature changes in the dome, and even atmospheric turbulence. By these means, some recently constructed telescopes have achieved resolutions very close to their theoretical (diffraction) limits.

Even under conditions of perfect seeing, most telescopes would not achieve diffraction-limited resolution.

The temperature of the mirror or in the dome may fluctuate slightly during the many minutes or even hours required for the image to be exposed, and the precise shape of the mirror may change slightly as the telescope tracks a source across the sky. The effect of these changes is that the mirror's focus may shift from minute to minute, blurring the eventual image in much the same way as atmospheric turbulence creates a seeing disk (Figure 5.18). At the best observing sites, the seeing is often so good that these tiny effects may be the main cause of image blurring. The collection of techniques aimed at controlling such environmental and mechanical fluctuations is known as **active optics.**

The first telescope designed to incorporate active optics was the New Technology Telescope (NTT), constructed in 1989 at the European Southern Observatory in Chile and upgraded in 1997. (NTT is the most prominent instrument visible in Figure 5.19.) This 3.5-m instrument, employing the latest in real-time telescope controls, can achieve a resolution as sharp as 0.2″ by making minute modifications to the tilt of its mirror as its temperature and orientation change, thus maintaining the best possible focus at all times. Figure 5.20 shows how active optics can dramatically improve the resolution of an image. Active-optics techniques now include improved dome design to control airflow, precise control of the mirror temperature, and the use of pistons behind the mirror to maintain its precise shape. All of the large telescopes described earlier include active-optics systems, improving their resolution to a few tenths of an arc second.

## Real-Time Control

With active-optics systems in place, Earth's atmosphere once again becomes the main agent limiting a telescope's resolution. Remarkably, even this problem can now be

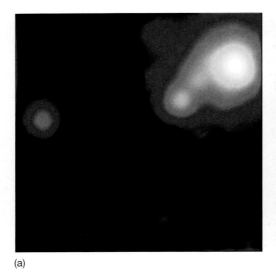

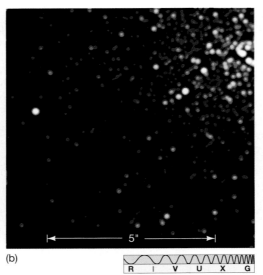

(a)

(b)

◄ FIGURE 5.20 **Active Optics** These false-color infrared photographs of part of the star cluster R136—the same object shown in Figure 5.16—contrast the resolution obtained (a) without and (b) with an active-optics system. Both images were taken with the New Technology Telescope shown in Figure 5.19. *(ESO)*

addressed, using an approach known as **adaptive optics,** a technique that actually deforms the shape of a mirror's surface under computer control, while the image is being exposed, in order to undo the effects of atmospheric turbulence. The mirror in question is generally not the large primary mirror of the telescope. Rather, for both economic and technical reasons, a much smaller (typically 20- to 50-cm-diameter) mirror is inserted into the light path and manipulated to achieve the desired effect. Adaptive optics presents formidable theoretical and practical problems, but the rewards are so great that it has been the subject of intense research since the 1970s.

The effort received an enormous boost in the 1990s from declassified military technology from the Strategic Defensive Initiative, a Reagan-era missile defense program (dubbed "Star Wars" by its detractors) intended to target and shoot down incoming ballistic missiles. In the system shown in Figure 5.21, a laser probes the atmosphere above

the telescope, creating an "artificial star" that allows astronomers to gauge atmospheric conditions and pass that information to a computer that modifies the telescope mirror thousands of times per second to compensate for poor seeing.

Adaptive corrections are somewhat easier to apply in the infrared than in the optical, because atmospheric distortions are smaller (stars "twinkle" less in the infrared) and because the longer infrared wavelengths impose less stringent requirements on the precise shape of the mirror. Infrared adaptive-optics systems already exist in many large telescopes. For example, Gemini and Subaru have reported adaptive-optics resolutions of around 0.06″ in the near infrared—not quite at the diffraction limit (0.03″ for an 8-m telescope at 1 μm, according to the equation in *More Precisely 5-1*), but already better than the resolution of *HST* at the same wavelengths (see Figure 5.22a). Both Keck and the VLT incorporate

◄ FIGURE 5.21 **Adaptive-Optics System** In this daytime photo, a test is being conducted at the Lick Observatory 3-m Shane telescope in California. A laser is used to create an "artificial star" (light reflected from the atmosphere back into the telescope) to improve guiding. The laser beam probes the atmosphere above the telescope, allowing tiny computer-controlled changes to be made in the shape of the mirror surface thousands of times each second. *(Lick Observatory)*

ANIMATION/VIDEO  Adaptive Optics

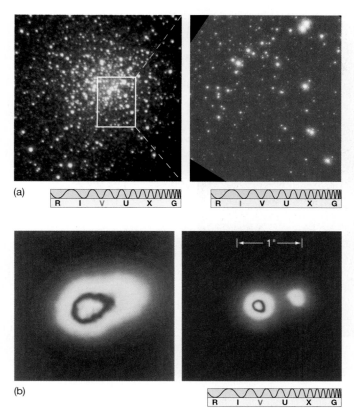

(a)

(b)

◀ FIGURE 5.22 **Adaptive Optics in Action** (a) The improvement in image quality produced by such systems can be seen in these images acquired by the 8-m Gemini telescope atop Mauna Kea in Hawaii. The uncorrected visible-light image (left) of the star cluster NGC 6934 is resolved to a little less than 1". With adaptive optics applied (right), the resolution in the infrared is improved by nearly a factor of 10, allowing more stars to be seen more clearly. (b) These visible-light images were acquired at a military observatory atop Mount Haleakala in Maui, Hawaii. The uncorrected image (left) of the double star Castor is a blur spread over several arc seconds, giving only a hint of its binary nature. With adaptive compensation applied (right), the resolution is improved to a mere 0.1", and the two stars are clearly separated. *(NOAO; MIT Lincoln Laboratory)*

adaptive-optics instrumentation that will ultimately be capable of producing diffraction-limited images at near-infrared wavelengths.

Visible-light adaptive optics has been demonstrated experimentally, and some astronomical telescopes may incorporate the technology within the next few years. Figure 5.22(b) compares a pair of visible-light observations of a nearby double star called Castor. The observations were made with a relatively modest 1.5-m telescope. The adaptive-optics system clearly distinguishes the two stars. Remarkably, it may soon be possible to have the "best of both worlds," achieving with large ground-based optical telescopes the kind of resolution presently attainable only from space.

## CONCEPT CHECK

✔ What steps do optical astronomers take to overcome the obscuring and blurring effects of Earth's atmosphere?

## 5.5 Radio Astronomy

In addition to the visible radiation that penetrates Earth's atmosphere on a clear day, radio radiation also reaches the ground. Indeed, as indicated in Figure 3.9, the radio window in the electromagnetic spectrum is much wider than the optical window. ∞ (Sec. 3.3) Because the atmosphere is no hindrance to long-wavelength radiation, radio astronomers have built many ground-based **radio telescopes** capable of detecting radio waves reaching us from space. These devices have all been constructed since the 1950s—radio astronomy is a much younger subject than optical astronomy.

### Early Observations

The field of radio astronomy originated with the work of Karl Jansky at Bell Labs in 1931. Jansky was engaged in a study of shortwave-radio interference when he discovered a faint static "hiss" that had no apparent terrestrial (Earthly) source. He noticed that the strength of the hiss varied in time and that its peak occurred about four minutes earlier each day. He soon realized that the peaks were coming exactly one *sidereal day* apart, and he concluded that the hiss was indeed not of terrestrial origin, but came from a definite direction in space. ∞ (Sec. 1.4) That direction is now known to correspond to the center of our Galaxy.

Some astronomers were intrigued by Jansky's discovery, but with the limited technology of the day—and even more limited Depression-era budgets—progress was slow. Jansky himself was moved to another project at Bell Labs and never returned to astronomical studies. However, by 1940, the first systematic surveys of the radio sky were being carried out. After a series of technological breakthroughs made during World War II, these studies rapidly grew into a distinct branch of astronomy.

During the 1930s, astronomers became aware that the space between the stars in our Galaxy is not empty, but instead is filled with extremely diffuse (low-density) gas (see Chapter 18). The growing realization in the 1940s that this otherwise completely invisible part of the Galaxy could be observed and mapped in detail at radio wavelengths established the true importance of Jansky's pioneering work. Today he is regarded as the father of radio astronomy.

### Essentials of Radio Telescopes

Figure 5.23(a) shows the world's largest steerable radio telescope: the large 105-m-diameter (340-foot-diameter) telescope located at the National Radio Astronomy Observatory in West Virginia. Although much larger than reflecting optical telescopes, most radio telescopes are built

(a)                                                                                              (b)

▲ FIGURE 5.23 **Radio Telescope** (a) The 105-m-diameter device at the National Radio Astronomy Observatory in Green Bank, West Virginia, is 150 m tall—taller than the Statue of Liberty and nearly as tall as the Washington Monument. (b) Schematic diagram of the telescope, showing the path taken by an incoming beam of radio radiation. (*NRAO*)

in basically the same way. They have a large, horseshoe-shaped mount supporting a huge curved metal dish that serves as the collecting area. As illustrated in Figure 5.23(b), the dish captures incoming radio waves and reflects them to the focus, where a receiver detects the signals and channels them to a computer.

Conceptually, the operation of a radio telescope is similar to the operation of an optical reflector with the detecting instruments placed at the prime focus (Figure 5.7a). However, unlike optical instruments, which can detect all visible wavelengths simultaneously, radio detectors normally register only a narrow band of wavelengths at any one time. To observe radiation at another radio frequency, we must retune the equipment, much as we tune a television set to a different channel.

Radio telescopes must be built large partly because cosmic radio sources are extremely faint. In fact, the total amount of radio energy received by Earth's entire surface is less than a trillionth of a watt. Compare this with the roughly 10 *million* watts our planet's surface receives in the form of infrared and visible light from any of the bright stars seen in the night sky. To capture enough radio energy to allow detailed measurements to be made, a large collecting area is essential.

Because of diffraction, the angular resolution of radio telescopes is generally quite poor compared with that of their optical counterparts. Typical wavelengths of radio waves are about a million times longer than those of

visible light, and these longer wavelengths impose a corresponding crudeness in angular resolution. (Recall from Section 5.3 that the longer the wavelength, the greater the amount of diffraction.) Even the enormous sizes of radio dishes only partly offset this effect. The radio telescope shown in Figure 5.23 can achieve a resolution of about 1′ when receiving radio waves having wavelengths of around 3 cm. However, it was designed to operate most efficiently (i.e., it is most sensitive to radio signals) at wavelengths closer to 1 cm, where the resolution is approximately 20″. The best angular resolution obtainable with a single radio telescope is about 10″ (for the largest instruments operating at millimeter wavelengths)—at least 100 times coarser than the capabilities of some large optical systems.

Radio telescopes can be built so much larger than their optical counterparts because their reflecting surfaces need not be as smooth as is necessary for light waves of shorter wavelength. Provided that surface irregularities (dents, bumps, and the like) are much smaller than the wavelength of the waves to be detected, the surface will reflect them without distortion. Because the wavelength of visible radiation is short (less than $10^{-6}$ m), extremely smooth mirrors are needed to reflect the waves properly, and it is difficult to construct very large mirrors to such exacting tolerances. However, even rough metal surfaces can focus 1-cm waves accurately, and radio waves of wavelength a meter or more can be reflected and focused perfectly well by surfaces having irregularities even as large as your fist.

▲ FIGURE 5.24 **Arecibo Observatory** An aerial photograph of the 300-m-diameter dish at the National Astronomy and Ionospheric Center near Arecibo, Puerto Rico. The receivers that detect the focused radiation are suspended nearly 150 m (about 45 stories) above the center of the dish. The left inset shows a close-up of the radio receivers hanging high above the dish. The right inset shows technicians adjusting the dish surface to make it smoother. (*D. Parker/T. Acevedo/NAIC; Cornell*)

Figure 5.24 shows an even larger, but unmovable radio telescope, strung among the hills of Arecibo, Puerto Rico. Approximately 300 m (1000 feet) in diameter, the surface of the Arecibo telescope spans nearly 20 acres. Constructed in 1963 in a natural depression in the hillside, the dish was originally surfaced with chicken wire, which was lightweight and cheap. Although fairly rough, the chicken wire was adequate for proper reflection because the openings between adjacent strands of wire were much smaller than the long-wavelength radio waves that were to be detected.

The entire Arecibo dish was resurfaced with thin metal panels in 1974 and was upgraded in 1997 so that it can now be used to study radio radiation of shorter wavelength. Since the 1997 upgrade, the panels can be adjusted to maintain a precise spherical shape to an accuracy of about 3 mm over the entire surface. At a frequency of 5 GHz (corresponding to a wavelength of 6 cm—the shortest wavelength that can be studied, given the properties of the dish surface), the telescope's angular resolution is about 1′. The huge size of the dish creates one distinct disadvantage, however: The

Arecibo telescope cannot be pointed very well to follow cosmic objects across the sky. The detectors can move roughly 10° on either side of the focus, restricting the telescope's observations to those objects that happen to pass within about 20° of overhead as Earth rotates.

Arecibo is an example of a rough-surfaced telescope capable of detecting long-wavelength radio radiation. At the other extreme, Figure 5.25 shows the 36-m-diameter Haystack dish in northeastern Massachusetts. Constructed of polished aluminum, this telescope maintains a parabolic curve to an accuracy of about a millimeter all the way across its solid surface. It can reflect and accurately focus radio radiation with wavelengths as short as a few millimeters. The telescope is contained within a protective shell, or radome, that protects the surface from the harsh New England weather. The radome acts much like the protective dome of an optical telescope, except that there is no slit through which the telescope "sees." Incoming cosmic radio signals pass virtually unimpeded through the radome's fiberglass construction.

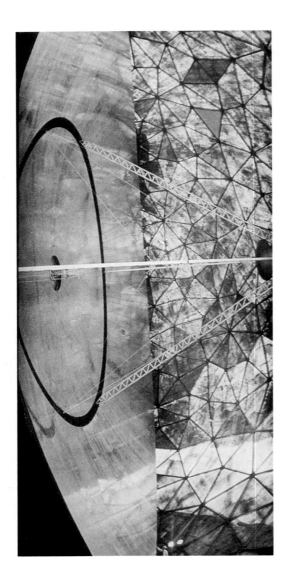

◀ **FIGURE 5.25 Haystack Observatory** Photograph of the Haystack dish, inside its protective radome. For scale, note the engineer standing at the bottom. Note also the dull shine on the telescope surface, indicating its smooth construction. Haystack is a poor optical mirror, but a superb radio telescope. Accordingly, it can be used to reflect and accurately focus radiation having short radio wavelengths, even as small as a fraction of a centimeter. *(MIT)*

nonvisible regions of the electromagnetic spectrum) is that it opens up a whole new window on the universe. There are three main reasons for this. First, just as objects that are bright in the visible part of the spectrum (e.g., the Sun) are not necessarily strong radio emitters, many of the strongest radio sources in the universe emit little or no visible light. Second, visible light may be strongly absorbed by interstellar dust along the line of sight to a source. Radio waves, by contrast, are generally unaffected by intervening matter. Third, as mentioned earlier, many parts of the universe cannot be seen at all by optical means, but are easily detectable at longer wavelengths. The center of the Milky Way Galaxy is a prime example of such a totally invisible region—our knowledge of the Galactic center is based almost entirely on radio and infrared observations. Thus, these observations not only afford us the opportunity to study the same objects at different wavelengths, but also allow us to see whole new classes of objects that would otherwise be completely unknown.

Figure 5.26 shows an optical photograph of the Orion Nebula (a huge cloud of interstellar gas) taken with the 4-m telescope on Kitt Peak. Superimposed on the optical image is a radio map of the same region, obtained by scanning the Haystack radio telescope (Figure 5.25) back and forth across the nebula and taking many measurements of radio intensity. The map is drawn as a series of contour lines connecting locations of equal radio brightness, similar to pressure contours drawn by meteorologists on weather maps or height contours drawn by cartographers on topographic maps. The inner contours represent stronger radio signals, the outside contours weaker signals. Note, however, how the radio map is much less detailed than its optical counterpart; that's because the acquired radio radiation has such a long wavelength compared to light.

The radio map in Figure 5.26 has many similarities to the visible-light image of the nebula. For instance, the radio emission is strongest near the center of the optical image and declines toward the nebular edge. But there are also subtle differences between the radio and optical images. The two differ chiefly toward the upper left of the main cloud, where visible light seems to be absent, despite the existence of radio waves. How can radio waves be detected from locations not showing any emission of light? The answer is that this particular nebular region is known to be especially dusty in its top left quadrant. The dust obscures the short-wavelength visible light, but not the long-wavelength radio radiation. Thus, we have a trade-off typical of many in astronomy:

## The Value of Radio Astronomy

Despite the inherent disadvantage of relatively poor angular resolution, radio astronomy enjoys many advantages. Radio telescopes can observe 24 hours a day. Darkness is not needed for receiving radio signals, because the Sun is a relatively weak source of radio energy, so its emission does not swamp radio signals arriving at Earth from elsewhere in the sky. In addition, radio observations can often be made through cloudy skies, and radio telescopes can detect the longest-wavelength radio waves even during rain or snowstorms. Poor weather causes few problems, because the wavelength of most radio waves is much larger than the typical size of atmospheric raindrops or snowflakes. Optical astronomy cannot be performed under these conditions, because the wavelength of visible light is smaller than a raindrop, a snowflake, or even a minute water droplet in a cloud.

However, perhaps the greatest value of radio astronomy (and, in fact, of all astronomies concerned with

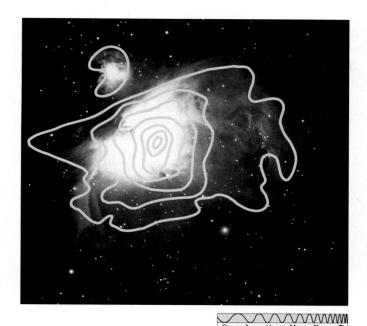

▲ **FIGURE 5.26 Orion Nebula in Radio and Visible** The Orion Nebula is a star-forming region about 1500 light-years from Earth. (The nebula is located in the constellation Orion and can be seen in Figure 1.8.) The bright regions in this photograph are stars and clouds of glowing gas. The dark regions are not empty, but their visible emission is obscured by interstellar matter. Superimposed on the optical image is a radio contour map (blue lines) of the same region. Each curve of the contour map represents a different intensity of radio emission. The resolution of the optical image is about 1″; that of the radio map is 1′. *(Background photo: AURA)*

Although the long-wave radio signals provide a less resolved map of the region, those same radio signals can pass relatively unhindered through dusty regions, in this case allowing us to see the true extent of the Orion Nebula.

CONCEPT CHECK

✔ In what ways does radio astronomy complement optical observations?

## 5.6 Interferometry

The main disadvantage of radio astronomy compared with optical work is its relatively poor angular resolution. However, in some circumstances, radio astronomers can overcome this limitation with a technique known as **interferometry.** This technique makes it possible to produce radio images of angular resolution higher than can be achieved with even the best optical telescopes, on Earth or in space.

In interferometry, two or more radio telescopes are used in tandem to observe the *same* object at the *same* wave-length and at the *same* time. The combined instruments together make up an **interferometer.** Figure 5.27 shows a large interferometer—many separate radio telescopes working together as a team. By means of electronic cables or radio links, the signals received by each antenna in the array making up the interferometer are sent to a central computer that combines and stores the data.

### A Radio Interferometer

Interferometry works by analyzing how the signals interfere with each other when added together. ∞ *(Discovery 3-1)* Consider an incoming wave striking two detectors (Figure 5.28). Because the detectors lie at different distances from the source, the signals they record will, in general, be out of step with one another. In that case, when the signals are combined, they will interfere destructively, partly canceling each other out. Only if the detected radio waves happen to be exactly in step will the signals combine constructively to produce a strong signal. Notice that the amount of interference depends on the *direction* in which the wave is traveling relative to the line joining the detectors. Thus—in principle, at least—careful analysis of the strength of the combined signal can provide an accurate measurement of the source's position in the sky.

As Earth rotates and the antennae track their target, the interferometer's orientation relative to the source changes, and a pattern of peaks and troughs emerges. In practice, extracting positional information from the data is a complex task, as multiple antennae and several sources are usually involved. Suffice it to say that, after extensive computer processing, the interference pattern translates into a high-resolution image of the target object.

An interferometer is, in essence, a substitute for a single huge antenna. As far as resolving power is concerned, the effective diameter of an interferometer is the distance between its outermost dishes. In other words, two small dishes can act as opposite ends of an imaginary, but huge, single radio telescope, dramatically improving angular resolution. For example, a resolution of a few arc seconds can be achieved at typical radio wavelengths (such as 10 cm), either by using a single radio telescope 5 km in diameter (which is impossible to build) or by using two or more much smaller dishes separated by the same 5 km, but connected electronically. The larger the distance separating the telescopes—that is, the longer the *baseline* of the interferometer—the better is the resolution attainable.

Large interferometers like the instrument shown in Figure 5.27 now routinely attain radio resolution comparable to that of optical images. Figure 5.29 shows an interferometric radio map of a nearby galaxy and a photograph of that same galaxy, made with a large optical telescope. The radio clarity is much better than in the contour map of Figure 5.26—in fact, the radio resolution in part (a) is comparable to that of the optical image in part (b).

(a)

◀ FIGURE 5.27 **VLA Interferometer** (a) This large interferometer, located on the Plain of San Augustin in New Mexico, comprises 27 separate dishes spread along a Y-shaped pattern about 30 km across. The most sensitive radio device in the world, it is called the Very Large Array, or VLA for short. (b) A close-up view from ground level shows how some of the VLA dishes are mounted on railroad tracks so that they can be repositioned easily. *(NRAO)*

(b)

antenna in orbit, together with several antennae on the ground, to construct an even longer baseline and achieve still better resolution. Proposals exist to place interferometers entirely in Earth orbit and even on the Moon.

## Interferometry at Other Wavelengths

Although the technique was originally developed by radio astronomers, interferometry is no longer restricted to the radio domain. Radio interferometry became feasible when electronic equipment and computers achieved speeds great enough to combine and analyze radio signals from separate radio detectors without loss of data.

As the technology has improved, it has become possible to apply the same methods to radiation of higher frequency. Millimeter-wavelength

Astronomers have created radio interferometers spanning great distances, first across North America and later between continents. A typical very-long-baseline interferometry experiment (usually known by its abbreviation, VLBI) might use radio telescopes in North America, Europe, Australia, and Russia to achieve an angular resolution on the order of 0.001″. It seems that even Earth's diameter is no limit: Radio astronomers have successfully used an

▶ FIGURE 5.28 **Interferometry** Two detectors, A and B, record different signals from the same incoming wave because of the time it takes the radiation to traverse the distance between them. When the signals are combined, the amount of interference depends on the wave's direction of motion, providing a means of measuring the position of the source in the sky. Here, the dark-blue waves come from a source high in the sky and interfere destructively when captured by antennas A and B. But when the same source has moved because of Earth's rotation (light-blue waves), the interference can be constructive.

A — Waves arrive out of step
B — = destructive interference

A — Waves arrive in step
B — = constructive interference

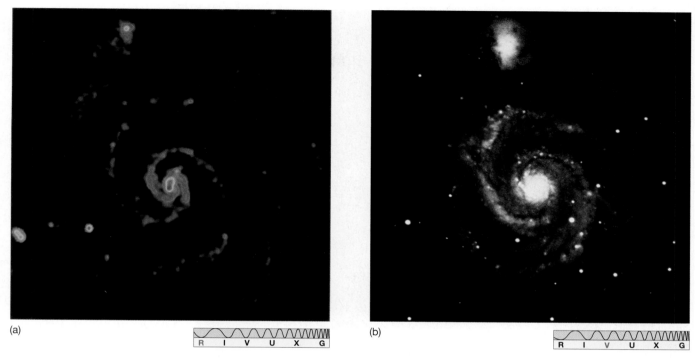

▲ FIGURE 5.29  **Radio–Optical Comparison** (a) VLA radio "image" (or radiograph) of the spiral galaxy M51, observed at radio frequencies with an angular resolution of a few arc seconds. (b) Visible-light photograph of that same galaxy, made with the 4-m Kitt Peak optical telescope and displayed on the same scale as (a). (*NRAO; AURA*)

interferometry has already become an established and important observational technique, and both the Keck telescopes and the VLT are expected to be used for near-infrared interferometry within the next few years.

Optical interferometry is currently the subject of intensive research. In 1997, a group of astronomers in Cambridge, England, succeeded in combining the light from three small optical telescopes to produce a single, remarkably clear, image. Each telescope of the Cambridge Optical Aperture Synthesis Telescope (COAST) was only 0.4 m in diameter, but with the mirrors positioned 6 m apart, the resulting resolution was a stunning 0.01″—better than the resolution of the best adaptive-optics systems on the ground or of *HST* operating above Earth's atmosphere. The three telescopes working together enabled astronomers to "split" the binary star Capella, whose two member stars are separated by only 0.05″. Normally Capella is seen from the ground only as a slightly oblong blur. With the COAST array, the two stars that make up the system are cleanly and individually separated.

Perhaps the highest resolution interferometric instrument currently in operation is the six-telescope optical array operated by the Center for High Angular Resolution Astronomy (CHARA) on Mount Wilson in California (Figure 5.30). Although each telescope is only 1 m in diameter, the placement of the array over the mountain results in a combined light beam having resolution equivalent to a single telescope 300 m across. CHARA is not

designed to produce images of the stars it studies; however, it can resolve details as small as 0.0002″ across, allowing the positions, orbits, and even radii of some stars to be measured with exquisite accuracy.

CONCEPT CHECK

✔ What is the main reason for the poor angular resolution of radio telescopes? How do radio astronomers overcome this problem?

## 5.7  Space-Based Astronomy

Optical and radio astronomy are the oldest branches of astronomy, but since the 1970s there has been a virtual explosion of observational techniques spanning the rest of the electromagnetic spectrum. Today, all portions of the spectrum are studied, from radio waves to gamma rays, to maximize the amount of information available about astronomical objects.

As noted earlier, the types of astronomical objects that can be observed differ quite markedly from one wavelength range to another. Full-spectrum coverage is essential not only to see things more clearly, but even to see some things at all. Because of the transmission characteristics of Earth's atmosphere, astronomers must study practically all regions of the electromagnetic spectrum, from gamma rays through X-rays to visible light, and on down to infrared

◀ FIGURE 5.30 **Optical Interferometry** This aerial photo shows the facilities of the CHARA array intermingled with the existing equipment of the historic Mount Wilson Observatory, in California. The small 1-m telescopes in the array are numbered. *(E. Simison/Sea West Enterprises)*

## Infrared Astronomy

Infrared studies are a vital component of modern observational astronomy. Infrared astronomy spans a broad range of phenomena in the universe, from planets and their parent stars to the vast regions of interstellar space where new stars are forming, to explosive events occurring in faraway galaxies. Generally, **infrared telescopes** resemble optical telescopes, but their detectors are designed to be sensitive to radiation of longer wavelengths. Indeed, as we have seen, many ground-based "optical" telescopes are also used for infrared work, and some of the most useful infrared observing is done from the ground (e.g., from Mauna Kea—see Figure 5.11), even though the radiation is somewhat diminished in intensity by our atmosphere.

As with radio observations, the longer wavelength of infrared radiation often enables us to perceive objects that are partially hidden from optical view. As a terrestrial example of the penetrating properties of infrared radiation, Figure 5.31(a) shows a dusty and hazy region in California, hardly viewable optically, but easily seen in the infrared (Figure 5.31b). Figures 5.31(c) and (d) show a similar comparison for an astronomical object—the dusty regions of the Orion Nebula, where visible light is blocked by interstellar clouds, but which is clearly distinguishable in the infrared.

Astronomers can make better infrared observations if they can place their instruments above most or all of Earth's atmosphere, using balloon-, aircraft-, rocket-, and satellite-based telescopes. However, as might be expected, the

and radio waves, from space. The rise of these "other astronomies" has therefore been closely tied to the development of the space program.

infrared telescopes that can be carried above the atmosphere are considerably smaller than the massive instruments found in ground-based observatories. Figure 5.32 shows an infrared image of the Orion region as seen by the 0.6-m British–Dutch–U.S. *Infrared Astronomy Satellite (IRAS)* in 1983 and compares it with the same region made in visible light. Figure 5.32(a) is a composite of IRAS images taken at three different infrared wavelengths (12 $\mu$m, 60 $\mu$m, and 100 $\mu$m). It is represented in *false color*, a technique commonly used for displaying images taken in nonvisible light. The colors do not represent the actual wavelength of the radiation emitted, but instead some other property of the source, in this case temperature, descending from white to orange to black. The whiter regions thus denote higher temperatures and hence greater strength of infrared radiation. At about 1′ angular resolution, the fine details of the Orion nebula evident in the visible portion of Figure 5.26 cannot be perceived. Nonetheless, the infrared image shows clouds of warm dust and gas thought to play a critical role in the formation of stars and also shows groups of bright young stars that are completely obscured at visible wavelengths.

Much of the material between the stars has a temperature between a few tens and a few hundreds of kelvins. Accordingly, Wien's law tells us that, the infrared domain is the natural portion of the electromagnetic spectrum in which to study it. ∞ (Sec. 3.4) Unfortunately, also by Wien's law, telescopes themselves also radiate strongly in the infrared, unless they are cooled to nearly absolute zero. The end of *IRAS*'s mission came not because of any equipment malfunction or unexpected mishap, but simply because its supply of liquid helium coolant ran out. *IRAS*'s own thermal emission then overwhelmed the radiation it was built to detect.

In August 2003, NASA launched the 0.85-m *Space Infrared Telescope Facility (SIRTF,* shown in Figure 5.32b). In December of that year, when the first images from the instrument were made public, NASA renamed the spacecraft the *Spitzer Space Telescope (SST)*, in honor of Lyman Spitzer, Jr., renowned astrophysicist and the first

▲ FIGURE 5.31 **Smog Revealed** An optical photograph (a) taken near San Jose, California, and an infrared photo (b) of the same area taken at the same time. Infrared radiation of long wavelength can penetrate smog much better than short-wavelength visible light. The same advantage pertains to astronomical observations: An optical view (c) of an especially dusty part of the central region of the Orion Nebula is more clearly revealed in this infrared image (d) showing a cluster of stars behind the obscuring dust. *(Lick Observatory; NASA)*

person to propose (in 1946) that a large telescope be located in space. The facility's detectors are designed to operate at wavelengths in the 3-$\mu$m to 200-$\mu$m range. Unlike previous space-based observatories, the *SST* does not orbit Earth, but instead follows our planet in its orbit around the Sun, trailing millions of kilometers behind in order to minimize Earth's heating effect on the detectors. The spacecraft is currently drifting away from Earth at the rate of 0.1 AU per year. The detectors are cooled to near absolute zero in order to observe infrared signals from space without interference from the telescope's own heat. Figure 5.33 shows some examples of the spectacular imagery obtained from NASA's latest eye on the universe.

The present instrument package aboard the *Hubble Space Telescope (see Discovery 5-1)* also includes a high-resolution (0.1″) near-infrared camera and spectroscope.

## Ultraviolet Astronomy

On the short-wavelength side of the visible spectrum lies the ultraviolet domain. This region of the spectrum, extending in wavelength from 400 nm (blue light) down to a few nanometers ("soft" X-rays), has only recently begun to be explored. Because Earth's atmosphere is partially opaque to radiation below 400 nm and is totally opaque below about 300 nm (due in part to the ozone layer), astronomers cannot conduct any useful ultraviolet observations from the ground, not even from the highest mountaintop. Rockets, balloons, or satellites are therefore essential to any **ultraviolet telescope**—a device designed to capture and analyze that high-frequency radiation.

One of the most successful ultraviolet space missions was the *International Ultraviolet Explorer (IUE)*, placed in Earth orbit in 1978 and shut down for budgetary reasons

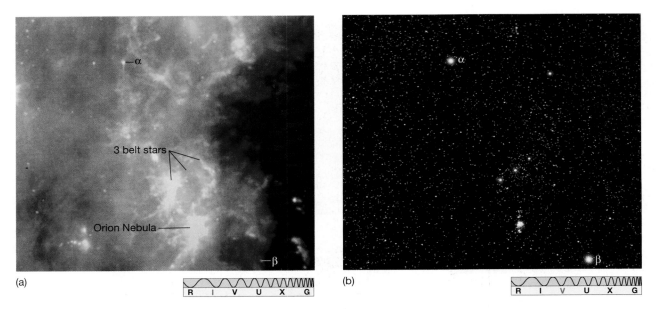

▲ FIGURE 5.32 **Infrared Image** (a) This infrared image of the Orion region was made by the *Infrared Astronomy Satellite*. In this false-color image, colors denote different temperatures, descending from white to orange to black. The resolution is about 1'. (b) The same region photographed in visible light with 1" resolution. The labels $\alpha$ and $\beta$ refer, respectively, to Betelgeuse and Rigel, the two brightest stars in the constellation. Note how the red star Betelgeuse is easily seen in the infrared (part a), whereas the blue star Rigel is very faint. (*NASA; J. Sanford*)

in late 1996 (see *Discovery 18-1*). Like all ultraviolet telescopes, its basic appearance and construction were quite similar to those of optical and infrared devices. Several hundred astronomers from all over the world used *IUE*'s near-ultraviolet spectroscopes to explore a variety of phenomena in planets, stars, and galaxies. In subsequent chapters, we will learn what this relatively new window on the universe has shown us about the activity and even the violence that seems to pervade the cosmos.

Figure 5.34 shows images captured by two more recent ultraviolet satellites. Figure 5.34(a) shows an image of a supernova remnant—the remains of a violent stellar explosion that occurred some 12,000 years ago—obtained by the *Extreme Ultraviolet Explorer (EUVE)* satellite, launched in 1992. Since its launch, *EUVE* has mapped out our local cosmic neighborhood as it presents itself in the far ultraviolet and has radically changed astronomers' conception of interstellar space in the vicinity of the Sun. Figure 5.34(b) shows an image of two relatively nearby galaxies, called M81 and M82, captured by the *Galaxy Evolution Explorer (GALEX)* satellite, launched in 2003. *HST*, best known as an optical telescope, is also a superb imaging and spectroscopic ultraviolet instrument.

## High-Energy Astronomy

*High-energy* astronomy studies the universe as it presents itself to us in X-rays and gamma rays—the types of radiation whose photons have the highest frequencies and hence the

greatest energies. How do we detect radiation of such short wavelengths? First, it must be captured high above Earth's atmosphere, because none of it reaches the ground. Second, its detection requires the use of equipment fundamentally different in design from that used to capture the relatively low-energy radiation discussed up to this point.

The difference in the design of **high-energy telescopes** comes about because X-rays and gamma rays cannot be reflected easily by any kind of surface. Rather, these rays tend to pass straight through, or be absorbed by, any material they strike. When X-rays barely graze a surface, however, they can be reflected from it in a way that yields an image, although the mirror design is fairly complex. As illustrated in Figure 5.35, to ensure that all incoming rays are reflected at grazing angles, the telescope is constructed as a series of nested cylindrical mirrors, carefully shaped to bring the X-rays to a sharp focus. For gamma rays, no such method of producing an image has yet been devised; present-day gamma-ray telescopes simply point in a specified direction and count the photons they collect.

In addition, detection methods using photographic plates or CCD devices do not work well for hard (high-frequency) X-rays and gamma rays. Instead, individual photons are counted by electronic detectors on board an orbiting device, and the results are then transmitted to the ground for further processing and analysis. Furthermore, the number of photons in the universe seems to be inversely related to their frequency. Trillions of visible (starlight) photons reach the detector of an optical telescope on Earth each second, but hours or even days are sometimes needed for a single gamma-ray photon to be recorded. Not only are these photons hard to focus and measure, but they are also very scarce.

The *Einstein Observatory*, launched by NASA in 1978, was the first X-ray telescope capable of forming an image of its field of view. During its two-year lifetime, this spacecraft drove major advances in our understanding of high-energy phenomena throughout the universe. Its observational database is still heavily used. More recently, the German *ROSAT* (short for "Röntgen Satellite," after Wilhelm Röntgen, the discoverer of X-rays) was launched in 1991. During its 7-year lifetime, this instrument gener-

(a)

(b)

▲ **FIGURE 5.33** *Spitzer* **Images** Sample images from the *Spitzer Space Telescope*, now in orbit around the Sun, clearly show its camera's capabilities. (a) The magnificent spiral galaxy, M81, about 12 million light-years away. (b) Its companion, M82, is not so serene, rather resembling a "smoking hot cigar." *(JPL)*

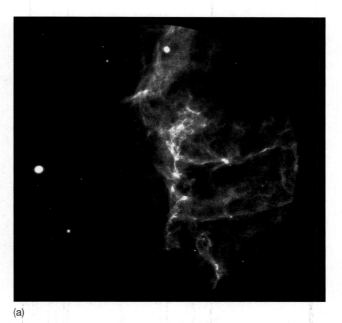

(a)

(b)

▶ **FIGURE 5.34** **Ultraviolet Images** (a) A camera on board the *Extreme Ultraviolet Explorer* satellite captured this image of the Cygnus loop supernova remnant, the result of a massive star that blew itself virtually to smithereens. The release of energy was enormous and the afterglow has lingered for centuries. The glowing field of debris shown here within the telescope's circular field of view lies some 1500 light-years from Earth. Based on the velocity of the outflowing debris, astronomers estimate that the explosion itself must have occurred about 12,000 years ago. (b) This false-color image of the spiral galaxy M81 and its companion M82, (see also Figure 5.33) made by the *Galaxy Evolution Explorer* satellite, reveals stars forming in the blue arms well away from the galaxy's center. *(NASA; GALEX)*

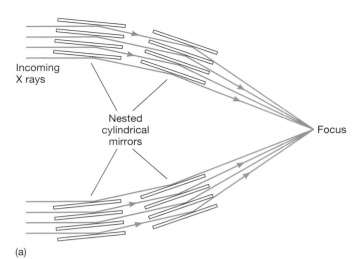

(a)

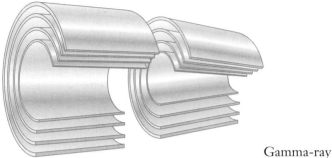

(b)

▲ FIGURE 5.35 **X-Ray Telescope** (a) The arrangement of nested mirrors in an X-ray telescope allows the rays to be reflected at grazing angles and focused to form an image. (b) A cutaway 3-D rendition of the mirrors, showing their shape more clearly.

ated a wealth of high-quality observational data. It was turned off in 1999, a few months after its electronics were irreversibly damaged when the telescope was accidentally pointed too close to the Sun.

In July 1999, NASA launched the *Chandra X-Ray Observatory* (named in honor of the Indian astrophysicist Subramanyan Chandrasekhar and shown in Figure 5.36). With greater sensitivity, a wider field of view, and better resolution than either *Einstein* or *ROSAT*, *Chandra* is providing high-energy astronomers with new levels of observational detail. Figure 5.37 shows the first image returned by *Chandra*: a supernova remnant in the constellation Cassiopeia. Known as Cas A, the ejected gas is all that now remains of a star that was observed to explode about 320 years ago. The false-color image shows 50-million-K gas in wisps of ejected stellar material; the bright white point at the very center of the debris may be a black hole. The *XMM-Newton* satellite is more sensitive to X-rays than is *Chandra* (that is, it can detect fainter X-ray sources), but it has significantly poorer angular resolution (5″, compared to 0.5″ for *Chandra*), making the two missions complementary to one another.

▶ FIGURE 5.36 *Chandra* **Observatory** The *Chandra* X-ray telescope is shown here during the final stages of its construction in 1998. The left end of the mirror arrangement, depicted in Figure 5.36, is at the bottom of the satellite as oriented here. *Chandra's* effective angular resolution is 1″, allowing this spacecraft to produce images of quality comparable to that of optical photographs. *Chandra* now occupies an elliptical orbit high above Earth; its farthest point from our planet, 140,000 km out, reaches almost one-third of the way to the Moon. Such an orbit takes *Chandra* above most of the interfering radiation belts girdling Earth (see Section 7.5) and allows more efficient observations of the cosmos. *(NASA)*

Gamma-ray astronomy is the youngest entrant into the observational arena. As just mentioned, imaging gamma-ray telescopes do not exist, so only fairly coarse (1° resolution) observations can be made. Nevertheless, even at this resolution, there is much to be learned. Cosmic gamma rays were originally detected in the 1960s by the U.S. *Vela* series of satellites, whose primary mission was to monitor illegal nuclear detonations on Earth. Since then, several X-ray telescopes have also been equipped with gamma-ray detectors.

By far the most advanced instrument was the *Compton Gamma-Ray Observatory (CGRO)*, launched by the space shuttle in 1991 (Figure 5.38a). At 17 tons, it was at the time the largest astronomical instrument ever launched by NASA. *CGRO* scanned the sky and studied individual

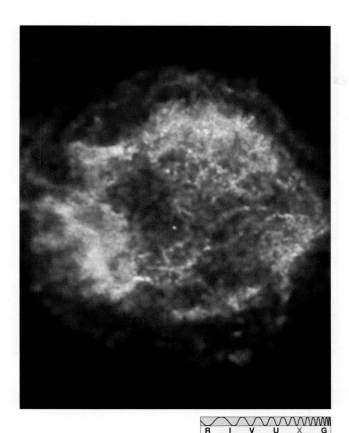

◀ **FIGURE 5.37** **X-Ray Image** A false-color *Chandra* X-ray image of the supernova remnant Cassiopeia A, a debris field of scattered, hot gases that were once part of a massive star. Here, color represents the intensity of the X-rays observed, from white (brightest) through red (faintest). Roughly 10,000 light-years from Earth and barely visible in the optical part of the spectrum, Cas A is now awash in brilliantly glowing X-rays spread across some 10 light-years. *(NASA)*

following a failure of one of the satellite's three gyroscopes, NASA opted for a controlled reentry and dropped *CGRO* into the Pacific Ocean.

CONCEPT CHECK

✔ List some scientific benefits of placing telescopes in space. What are the drawbacks of space-based astronomy?

## 5.8 Full-Spectrum Coverage

Table 5.1 lists the basic regions of the electromagnetic spectrum and describes objects typically studied in each frequency range. Bear in mind that the list is far from exhaustive and that many astronomical objects are now routinely observed at many different electromagnetic wavelengths. As we proceed through the text, we will discuss more fully the wealth of information that high-precision astronomical instruments can provide us.

objects in much greater detail than had previously been attempted. Figure 5.38(b) shows a false-color gamma-ray image of a highly energetic outburst in the center of a distant galaxy. The mission ended on June 4, 2000, when,

It is reasonable to suppose that the future holds many further improvements in both the quality and the availability of astronomical data and that many new dis-

(a)

(b)

▲ **FIGURE 5.38** **Gamma-Ray Astronomy** (a) This photograph of the 17-ton *Compton Gamma-Ray Observatory (CGRO;* named after an American pioneer in gamma-ray telescopy) was taken by an astronaut during the satellite's deployment from the space shuttle *Atlantis* over the Pacific Coast of the United States. (b) A typical false-color gamma-ray image—this one showing a violent event in the distant galaxy 3C279, also known as a "gamma-ray blazar." *(NASA)*

coveries will be made. The current and proposed pace of technological progress presents us with the following exciting prospect: Within the next decade, if all goes according to plan, it will be possible, for the first time ever, to make *simultaneous* high-quality measurements of any astronomical object at *all* wavelengths, from radio ray to gamma ray. The consequences of this development for our understanding of the workings of the universe may be little short of revolutionary.

As a preview of the comparison that full-spectrum coverage allows, Figure 5.39 shows a series of images of our own Milky Way Galaxy. The images were made by several different instruments, at wavelengths ranging from radio to gamma rays, over a period of about 5 years. By comparing the features visible in each, we immediately see how multiwavelength observations can complement one another, greatly extending our perception of the dynamic universe around us.

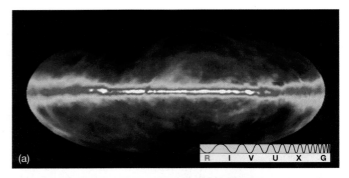

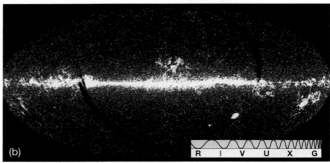

| | General Considerations | Common Applications (Chapter Reference) |
|---|---|---|
| **TABLE 5.1 Astronomy at Many Wavelengths** | | |
| *Radio* | Can penetrate dusty regions of interstellar space | Radar studies of planets (2,9) |
| | Earth's atmosphere largely transparent to these wavelengths | Planetary magnetic fields (11) |
| | Can be detected in the daytime as well as at night | Interstellar gas clouds and molecules (18) |
| | High resolution at long wavelengths requires very large telescopes or interferometers | Galactic structure (23, 24) |
| | | Galactic nuclei and active galaxies (23, 24) |
| | | Cosmic background radiation (27) |
| *Infrared* | Can penetrate dusty regions of interstellar space | Star formation (19,20); cool stars (20) |
| | Earth's atmosphere only partially transparent to infrared radiation, so some observations must be made from space | Center of the Milky Way Galaxy (23) |
| | | Active galaxies (24) |
| | | Large-scale structure of the universe (25, 27) |
| *Visible* | Earth's atmosphere transparent to visible light | Planets (7–14) |
| | | Stars and stellar evolution (17, 20, 21) |
| | | Normal and active galaxies (23, 24) |
| | | Large-scale structure of the universe (25) |
| *Ultraviolet* | Earth's atmosphere is opaque to ultraviolet radiation, so observations must be made from space | Interstellar medium (19) |
| | | Hot stars (21) |
| *X-ray* | Earth's atmosphere is opaque to X-rays, so observations must be made from space. | Stellar atmospheres (16) |
| | Special mirror configurations are needed to form images | Neutron stars and black holes (22) |
| | | Active galactic nuclei (24) |
| | | Hot gas in galaxy clusters (25) |
| *Gamma ray* | Earth's atmosphere is opaque to gamma rays, so observations must be made from space. | Neutron stars (22) |
| | Cannot form images | Active galactic nuclei (24) |

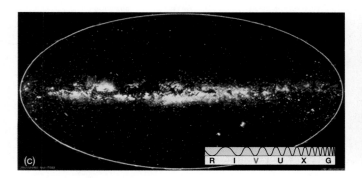

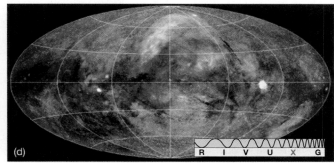

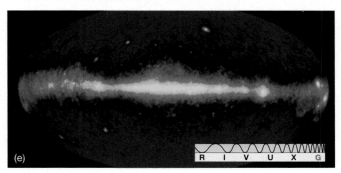

◀ **FIGURE 5.39 Multiple Wavelengths** The Milky Way Galaxy as it appears at (a) radio, (b) infrared, (c) visible, (d) X-ray, and (e) gamma-ray wavelengths. Each frame is a panoramic view covering the entire sky. The center of our Galaxy, which lies in the direction of the constellation Sagittarius, is at the center of each map. *(NRAO; NASA; Lund Observatory; MPI; NASA)*

# CHAPTER REVIEW

## Summary

❶ A **telescope (p. 108)** is a device designed to collect as much light as possible from some distant source and deliver it to a detector for detailed study. A **refracting telescope (p. 108)** uses a lens to concentrate and focus the light; **reflecting telescopes (p. 108)** use mirrors. The **Newtonian (p. 111)** and **Cassegrain (p. 112)** reflecting telescope designs employ secondary mirrors to avoid placing detectors at the prime focus. More complex light paths are also used to allow the use of large or heavy equipment that cannot be placed near the telescope.

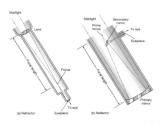

❷ All large astronomical telescopes are reflectors, because large mirrors are lighter and much easier to construct than large lenses, and they also suffer from fewer optical defects. The light-gathering power of a telescope depends on its **collecting area (p. 113)**, which is proportional to the square of the mirror diameter. To study the faintest sources of radiation, astronomers must use large telescopes. Large telescopes also suffer least from the effects of diffraction and hence can achieve better **angular resolution (p. 117)** once the blurring effects of Earth's atmosphere are overcome. The amount of diffraction is proportional to the wavelength of the radiation under study and inversely proportional to the size of the mirror.

❸ Most modern telescopes use **charge-coupled devices, or CCDs (p. 118),** instead of photographic plates, to collect their data. The field of view is divided into an array of millions of **pixels (p. 118)** that accumulate an electric charge when light strikes them. CCDs are many times more sensitive than photographic plates, and the resultant data are easily saved directly in digital form image processing. The light collected by a telescope may be processed in a number of ways. It can be made to form an **image (p. 108)**. **Photometry (p. 121)** may be performed either on a stored image or during the observation itself, using a specialized detector, or a **spectrometer (p. 121)** may be used to analyze the spectrum of the radiation received.

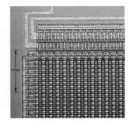

❹ The resolution of most ground-based optical telescopes is limited by **seeing (p. 122)**—the blurring effect of Earth's turbulent atmosphere, which smears the pointlike images of stars out into **seeing disks (p. 122)** a few arc seconds in diameter. Radio and space-based telescopes do not suffer from this effect, so their resolution is determined by the effects of diffraction. Astronomers can greatly improve a telescope's resolution by using **active optics (p. 123)**, in which a telescope's environment and focus are carefully monitored and controlled, and **adaptive optics (p. 124)**, in which the blurring effects of atmospheric turbulence are corrected for in real time.

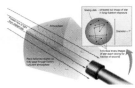

**⑤ Radio telescopes (p. 125)** are conceptually similar in construction to optical reflectors. However, they are generally much larger than optical instruments, in part because so little radio radiation reaches Earth from space, so a large collecting area is essential. Their main disadvantage is that diffraction of long-wavelength radio waves limits their resolution, even for very large radio telescopes. Their principal advantage is that they allow astronomers to explore a whole new part of the electromagnetic spectrum and of the universe—many radio emitters are completely undetectable in visible light. In addition, radio observations are largely unaffected by Earth's atmosphere and weather or by the position of the Sun.

**⑥** In order to increase the effective area of a telescope, and hence improve its resolution, several separate instruments may be combined into a device called an **interferometer (p. 129)**, in which the interference pattern of radiation received by two or more detectors is used to reconstruct a high-precision map of the source. Using **interferometry (p. 129)**, radio telescopes can produce images sharper than those from the best optical telescopes. Infrared and optical interferometric systems are now in use at many observatories.

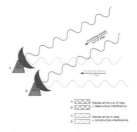

**⑦ Infrared telescopes (p. 132)** and **ultraviolet telescopes (p. 133)** are generally similar in design to optical systems. Studies undertaken in some parts of the infrared range can be carried out using large ground-based systems. Ultraviolet astronomy *must* be carried out from space. **High-energy telescopes (p. 134)** study the X-ray and gamma-ray regions of the electromagnetic spectrum. X-ray telescopes can form images of their field of view, although the mirror design is more complex than that of optical instruments. Gamma-ray telescopes simply point in a certain direction and count the photons they collect. Because the atmosphere is opaque at these short wavelengths, both types of telescopes must be placed in space.

**⑧** Different physical processes can produce very different types of electromagnetic radiation, and the appearance of an object at one wavelength may bear little resemblance to its appearance at another. Observations at wavelengths spanning the spectrum are essential to a complete understanding of astronomical events.

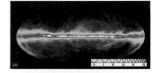

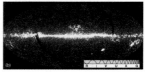

## Review and Discussion

1. Cite two reasons that astronomers are continually building larger and larger telescopes.

2. List three advantages of reflecting telescopes over reflectors.

3. What and where are the largest optical telescopes in use today?

4. How does Earth's atmosphere affect what is seen through an optical telescope?

5. What advantages does the *Hubble Space Telescope (HST)* have over ground-based telescopes? List some disadvantages.

6. What are the advantages of a CCD over a photograph?

7. What is image processing?

8. What determines the resolution of a ground-based telescope?

9. How do astronomers use active optics to improve the resolution of telescopes?

10. How do astronomers use adaptive optics to improve the resolution of telescopes?

11. Why do radio telescopes have to be very large?

12. Which astronomical objects are best studied with radio techniques?

13. What is interferometry, and what problem in radio astronomy does it address?

14. Is interferometry limited to radio astronomy?

15. Compare the highest resolution attainable with optical telescopes with the highest resolution attainable with radio telescopes (including interferometers).

16. Why do infrared satellites have to be cooled?

17. Are there any ground-based ultraviolet observatories?

18. In what ways do the mirrors in X-ray telescopes differ from those found in optical instruments?

19. What are the main advantages of studying objects at many different wavelengths of radiation?

20. Our eyes can see light with an angular resolution of 1′. Suppose our eyes detected only infrared radiation, with 1° angular resolution. Would we be able to make our way around on Earth's surface? To read? To sculpt? To create technology?

## Conceptual Self-Test: True or False/Multiple Choice

1. The primary purpose of any telescope is to produce an enormously magnified image of the field of view.

2. The main advantage to using the *HST* is the increased amount of "night time" viewing it affords.

3. The Keck telescopes contain the largest single mirrors ever constructed.

4. The term *seeing* is used to describe how faint an object can be detected by a telescope.

5. Radio telescopes are large in part to improve their angular resolution, which is poor because of the long wavelengths at which they are used to observe the skies.

6. As a rule, larger telescopes can detect fainter objects.

7. The angular resolution of a 2-m ground-based optical telescope is limited more by seeing than by diffraction.

8. An object having a temperature of 300 K would be best observed with an infrared telescope.

9. X-ray telescopes cannot form images of their fields of view.

10. Gamma-ray telescopes employ the same basic design that optical instruments use.

11. The thickest lenses deflect and bend light (a) the fastest; (b) the slowest; (c) the most; (d) the least.

12. The main reason that most professional research telescopes are reflectors is that (a) mirrors produce sharper images than lenses do; (b) their images are inverted; (c) they do not suffer from the effects of seeing; (d) large mirrors are easier to build than large lenses.

13. If telescope mirrors could be made of odd sizes, the one with the *most* light-gathering power would be (a) a triangle with 1-m sides; (b) a square with 1-m sides; (c) a circle 1 m in diameter; (d) a rectangle with two 1-m sides and two 2-m sides.

14. The image shown in Figure 5.10 ("Resolution") is sharpest when the ratio of wavelength to telescope size is (a) large; (b) small; (c) close to unity; (d) none of these.

15. The primary reason professional observatories are built on the highest mountaintops is to (a) get away from city lights; (b) be above the rain clouds; (c) reduce atmospheric blurring; (d) improve chromatic aberration.

16. Compared with radio telescopes, optical telescopes can (a) see through clouds; (b) be used during the daytime; (c) resolve finer detail; (d) penetrate interstellar dust.

17. When multiple radio telescopes are used for interferometry, resolving power is most improved by increasing (a) the distance between telescopes; (b) the number of telescopes in a given area; (c) the diameter of each telescope; (d) the electrical power supplied to each telescope.

18. The *Spitzer Space Telescope (SST)* is stationed far from Earth because (a) this increases the telescope's field of view; (b) the telescope is sensitive to electromagnetic interference from terrestrial radio stations; (c) doing so avoids the obscuring effects of Earth's atmosphere; (d) Earth is a heat source and the telescope must be kept very cool.

19. The best way to study young stars hidden behind interstellar dust clouds would be to use (a) X-rays; (b) infrared light; (c) ultraviolet light; (d) blue light.

20. Table 5.1 ("Astronomy at Many Wavelengths") suggests that the best frequency range in which to study the hot (million-kelvin) gas found among the galaxies in the Virgo cluster would be (a) the radio frequencies; (b) the infrared region of the electromagnetic spectrum; (c) the X-ray region of the spectrum; (d) the gamma-ray region of the spectrum.

## Problems

*Algorithmic versions of these Problems are available in the Practice Problems module of the Companion Website.*

*The number of dots preceding each Problem indicates its approximate level of difficulty.*

1. • A certain telescope has a $10' \times 10'$ field of view that is recorded using a CCD chip having $2048 \times 2048$ pixels. What angle on the sky corresponds to 1 pixel? What would be the diameter, in pixels, of a typical seeing disk ($1''$ radius)?

2. • The *SST*'s planned operating temperature is 5.5 K. At what wavelength (in micrometers) does the telescope's own blackbody emission peak? How does this wavelength compare with the wavelength range in which the telescope is designed to operate? ∞ *(More Precisely 3-2)*

3. • A 2-m telescope can collect a given amount of light in 1 hour. Under the same observing conditions, how much time would be required for a 6-m telescope to perform the same task? A 12-m telescope?

4. • A space-based telescope can achieve a diffraction-limited angular resolution of $0.05''$ for red light (wavelength 700 nm). What would the resolution of the instrument be (a) in the infrared, at 3.5 $\mu$m, and (b) in the ultraviolet, at 140 nm?

5. • Based on the numbers given in the text, estimate the angular resolution of the Gemini North telescope (a) in red light (700 nm) and (b) in the near infrared (2 $\mu$m).

6. •• Two identical stars are moving in a circular orbit around one another, with an orbital separation of 2 AU. ∞ *(Sec. 2.6)* The system lies 200 light-years from Earth. If we happen to view the orbit head-on, how large a telescope would we need to resolve the stars, assuming diffraction-limited optics at a wavelength of 2 $\mu$m?

7. •• What is the greatest distance at which *HST*, in blue light (400 nm), could resolve the stars in the previous question?

8. • The photographic equipment on a telescope is replaced by a CCD. If the photographic plate records 5 percent of the light reaching it, and the CCD records 90 percent, how much time will the new system take to collect as much information as the old detector recorded in a 1-hour exposure?

9. • The Moon lies about 380,000 km away. To what distances do the angular resolutions of *SST* ($3''$), *HST* ($0.05''$), and a radio interferometer ($0.001''$) correspond at that distance?

10. • The Andromeda galaxy lies about 2.5 million light-years away. To what distances do the angular resolutions of *SST* ($3''$), *HST* ($0.05''$), and a radio interferometer ($0.001''$) correspond at that distance?

11. • Based on collecting areas, how much more sensitive would you expect the Arecibo telescope (Figure 5.24) to be, compared with the large telescope in Figure 5.23?

12. • What is the equivalent single-mirror diameter of a telescope constructed from two separate 10-m mirrors? Four separate 8-m mirrors?

13. • Estimate the angular resolutions of (a) a radio interferometer with a 5000-km baseline, operating at a frequency of 5 GHz, and (b) an infrared interferometer with a baseline of 50 m, operating at a wavelength of 1 $\mu$m.

14. •• During a particular observation, the *Chandra* telescope detects photons with an average energy of 1.5 keV ∞ *(More Precisely 4-1)*, at a rate of 0.001 per second. What is the total power received, in watts?

15. •• The resolution of the *Chandra* telescope is about $1''$ for photons of energy 1 keV. The effective collecting diameter of the mirror assembly is 1.2 m. Based on these numbers, is *Chandra* diffraction limited?

*The Companion Website at www.aw-bc.com/chaisson provides algorithmically generated versions of each chapter's Problems along with additional quizzes, an Animations & Videos gallery, an Images gallery, an interactive Glossary, and a full eBook.*

# PART 2

# OUR PLANETARY SYSTEM

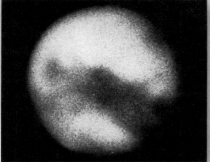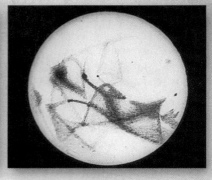

Early photo and sketch of Mars *(Lowell Observatory)*

Percival Lowell *(Lowell Observatory)*

The year 1877 was an important one in the study of the planet Mars. The Red Planet came unusually close to Earth, affording astronomers an especially good view. Of particular note was the discovery, by U.S. Naval Observatory astronomer Asaph Hall, of the two moons circling Mars. But most exciting was the report of the Italian astronomer Giovanni Schiaparelli on his observations of a network of linear markings that he termed *canali*. His work fueled the widespread idea more than a century ago that intelligent life exists on Mars.

In Italian, this word usually means "grooves" or "channels," but it can also mean "canals." Schiaparelli probably did not intend to imply that the *canali* were anything other than natural, but the word was translated into English as "canals," suggesting that the grooves had been constructed by intelligent beings. As often happens—then as now—the world's press (especially in the United States) sensationalized Schiaparelli's observations, and some astronomers began drawing elaborate maps of Mars, showing oases and lakes where canals met in desert areas.

Percival Lowell (1855–1916), a successful Boston businessman (and brother of the poet Amy Lowell and Harvard president Abbott Lawrence Lowell), became so fascinated by these reports that he abandoned his business and purchased a clear-sky site at Flagstaff, Arizona, where he built a private observatory. He devoted his fortune and energies until the day he died to achieving a better understanding of the Martian "canals." In doing so, he championed the idea that intelligent inhabitants of a drying (and dying) Mars had constructed a planetwide system of canals to transport water from the polar ice caps to the arid equatorial deserts.

Lowell was no slouch. He had earlier served as a distinguished diplomat and wrote extensively about the Far East. Later in life, he made an elaborate mathematical study of the orbit of Uranus, predicting the presence of an unseen body beyond Neptune—which was eventually found and named Pluto by Lowell Observatory's Clyde Tombaugh in 1930. And Lowell offered support to young astronomers, including Vesto Slipher, whose pioneering research on the recession of the galaxies helped found modern cosmology. Even so, Lowell is best remembered for his passionate belief in advanced Martian civilizations.

Mars today, via *Hubble (STScI)*

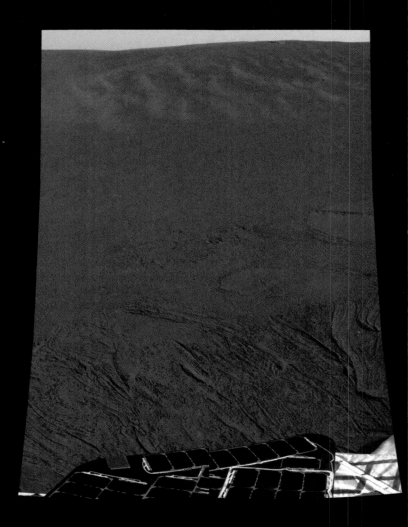

Equatorial Mars, via *Opportunity Rover (JPL)*

Panoramic view of Mars, via *Opportunity Rover (JPL)*

Today, we know that Mars is cool, dry, and probably lifeless; it certainly houses no intelligent beings and almost certainly never has. The Martian valleys and channels photographed by robot spacecraft in the 1970s are far too small to be the *canali* that Schiaparelli, Lowell, and others thought they saw on Mars. And the most recent armada of spacecraft—including the European *Mars Express* orbiter and the *U.S. Mars Reconnaissance Orbiter* and *Spirit* and *Opportunity* landers—that arrived at Mars in the early years of the 21st century reconfirm there is no liquid water there now.

The entire episode of Lowell's "canals" represents a classic case in which well-intentioned observers, perhaps obsessed with the notion of life on other worlds, let their personal opinions and prejudices seriously affect their analysis of reasonable data. The pair of globes of Mars opposite show how surface features, which were probably genuinely observed by astronomers a century ago, might have been imagined to be connected. The figure on the left of the pair is a photograph of how Mars actually looked in the best telescopes at the end of 19th century. The sketch at its right is an interpretation, done at the height of the canal hoopla, of that same view. The human eye, under physiological stress, tends to connect dimly observed, yet distinctly separated, features, causing old maps of Mars to have been as much a work of art as of science. Humans saw patterns and canals where none in fact existed—as noted by today's higher quality image of Mars above.

The chronicle of the Martian canals illustrates how the scientific method demands that we acquire new data to sort out sense from nonsense, fact from fiction. Rather than simply believing the claims about the Martian canals, scientists sought further observations to test Lowell's hypothesis. Eventually, improved observations, climaxing in several robotic missions to the Red Planet more than a century after all the fuss began, totally disproved the existence of artificial canals. It often takes time, but the scientific method does lead to progress in understanding reality.

The engineering feats of the modern space age allow us to probe many of the diverse worlds of the solar system in great detail. Here, we see the spectacular ring system around the planet Saturn, as imaged by the Cassini robot spacecraft that cruised by in 2005. Actually. this image is a mosaic of many smaller photos, which comprise the largest, most detailed, global view of Saturn's and its rings ever made. The color is also true, meaning that the pastels shown in this visible image are the closest approximation of Saturn's natural colors—just as our human eyes would see the planet up close. (JPL) ▶

# THE SOLAR SYSTEM

## An Introduction to Comparative Planetology

I n less than a single generation, we have learned more about the solar system—the Sun and everything that orbits it—than in all the centuries that went before. By studying the eight major planets, their moons, and the countless fragments of material that orbit in interplanetary space, astronomers have gained a richer outlook on our own home in space.

Space missions have visited all the planets of the solar system, extending astronomers' reach from the Earth-like inner planets to the giant gaseous worlds orbiting far from the Sun. Instruments aboard unmanned robots have taken close-up photographs of the planets and their moons and in some cases have made on-site measurements. Astronomers have come to realize that all solar system objects, large and small, have vital roles to play in furthering our understanding of our cosmic neighborhood.

## LEARNING GOALS

*Studying this chapter will enable you to*

1 Discuss the importance of comparative planetology to solar system studies.
2 Describe the overall scale and structure of the solar system.
3 Summarize the basic differences between the terrestrial and the jovian planets.
4 Identify and describe the major non-planetary components of the solar system.
5 Describe some of the spacecraft missions that have contributed significantly to our knowledge of the solar system.
6 Outline the theory of solar system formation that accounts for the overall properties of our planetary system.

 Visit www.aw-bc.com/chaisson for additional images, animations, quizzes, and eBook for this chapter.

## 6.1 An Inventory of the Solar System

The Greeks and other astronomers of old were aware of the Moon, the stars, and five planets—Mercury, Venus, Mars, Jupiter, and Saturn—in the night sky. ∞ (Sec. 2.2) They also knew of two other types of heavenly objects that were clearly neither stars nor planets. *Comets* appear as long, wispy strands of light in the night sky that remain visible for periods of up to several weeks and then slowly fade from view. *Meteors*, or "shooting stars," are sudden bright streaks of light that flash across the sky, usually vanishing less than a second after they first appear. These transient phenomena must have been familiar to ancient astronomers, but their role in the "big picture" of the solar system was not understood until much later.

### Discovering our Planetary System

Human knowledge of the basic content of the solar system remained largely unchanged from ancient times until the early 17th century, when the invention of the telescope made more detailed observations possible. Galileo Galilei was the first to capitalize on this new technology. (His simple telescope is shown in Figure 6.1.) Galileo's discovery of the phases of Venus and of four moons orbiting Jupiter early in the 17th century helped change forever humankind's vision of the universe. ∞ (Sec. 2.4)

As technological advances continued, knowledge of the solar system improved rapidly. Astronomers began

discovering objects invisible to the unaided human eye. By the end of the 19th century, astronomers had found Saturn's rings (1659), the planets Uranus (1781) and Neptune (1846), many planetary moons, and the first *asteroids*— "minor planets" orbiting the Sun, mostly in a broad band (called the *asteroid belt*) lying between Mars and Jupiter. Ceres, the largest asteroid and the first to be sighted, was discovered in 1801. A large telescope of mid-19th-century vintage is shown in Figure 6.2.

The 20th century brought continued improvements in optical telescopes. Thousands more asteroids were discovered, along with three more planetary ring systems, dozens of moons, and the first *Kuiper belt objects*, orbiting beyond Neptune. The century also saw the rise of both nonoptical—especially radio and infrared—astronomy and spacecraft exploration, each of which has made vitally important contributions to the field of planetary science.

The latter part of the 20th century also saw an entirely new avenue for planetary exploration—space flight. Astronauts have carried out experiments on the Moon (see Figure 6.3), and numerous unmanned probes have left Earth and traveled to all of the other planets. Figure 6.4 shows a view from the *Spirit* robot prospecting on the Martian surface in 2005, its solar panels most evident in this panoramic view from inside a shallow crater.

As currently explored, our **solar system** is known to contain one star (the Sun), eight planets, 165 moons (at last count) orbiting those planets, eight *asteroids* and more than 100 *Kuiper belt objects* larger than 300 km (200 miles) in diameter, tens of thousands of smaller (but well-studied) asteroids, myriad *comets* a few kilometers in diameter, and

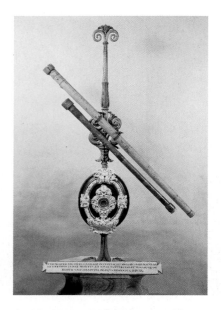

▲ FIGURE 6.1 **Early Telescope** The refracting telescope with which Galileo made his first observations was simple, but its influence on astronomy was immeasurable. (*Museo della Scienza; Scala/Art Resource, NY*)

▲ FIGURE 6.2 **Nineteenth-Century Telescope** By the mid-19th century, telescopes had improved enormously in both size and quality. Shown here is the Newtonian reflector built and used by Irish nobleman and amateur astronomer the Earl of Rosse. For 75 years, this 72-inch-diameter instrument was the largest telescope on Earth. (*Birr Scientific & Heritage Foundation*)

Our goal will be to develop (starting here, and concluding in Chapter 15) a comprehensive theory of the origin and evolution of our planetary system—a theory that explains all, or at least most, of the solar system's observed properties. We will seek to answer basic questions such as Why did planet X evolve in one way, while planet Y turned out completely different? and Why are the planets' orbits so orderly when their individual properties are not? In addressing these issues, we will find many similarities and common features among planets. However, each planet will also present new questions and afford unique insights into the ways planets work.

As we unravel the origin of our solar system, we may hope to learn something about planetary systems beyond our own. Since the mid-1990s, astronomers have detected more than 200 *extrasolar planets*—planets orbiting stars other than our own Sun. Many new planets are discovered each year (see Chapter 15), and our observations of them provide critical tests of modern theories of planet formation. Before the discovery of extrasolar planets, those theories had necessarily been based on observations of only our own solar system. Now astronomers have a whole new set of "proving grounds" in which to compare theory with reality.

Curiously, current data suggest that many of the newly discovered systems have properties rather different from those of our own, adding fuel to the long-standing debate among astronomers on the prevalence of planets like Earth and the possible existence of life as we know it elsewhere in the universe. It will be some time before astronomers can make definitive statements about the existence (or nonexistence) of planetary systems like our own.

▲ FIGURE 6.3 **Lunar Exploration** An *Apollo* astronaut does some lunar geology—prospecting near a huge boulder in the Mare Serenitatis during the final manned mission to the Moon in 1972. *(NASA)*

countless *meteoroids* less than 100 m across. The list will undoubtedly grow as we continue to explore our cosmic neighborhood.

## Comparative Planetology

As we proceed through the solar system in the next few chapters, we will seek to understand how each planet compares with our own and what each contributes to our knowledge of the solar system as a whole. We will use the powerful perspective of **comparative planetology**—comparing and contrasting the properties of the diverse worlds we encounter—to understand better the conditions under which planets form and evolve. Comparative planetology will be our indispensable guide as we proceed through the coming chapters.

### CONCEPT CHECK

✔ In what ways might observations of extrasolar planets help us understand our own solar system?

▲ FIGURE 6.4 *Spirit* **on Mars** The Mars rover *Spirit* took hundreds of images to create this true-color, 360° panorama of the Martian horizon from within Gusev crater. The robot, whose tracks into the shallow basin can be seen at right center, then measured the chemistry and mineralogy of soils and rocky outcrops. *(JPL)*

## 6.2 Measuring the Planets

Table 6.1 lists some basic orbital and physical properties of the eight planets, with a few other well-known solar-system objects (the Sun, the Moon, Pluto, an asteroid, and a comet) included for comparison. Note that the Sun, with more than a thousand times the mass of the next most massive object (the planet Jupiter), is clearly the dominant member of the solar system. In fact, the Sun contains about 99.9 percent of all solar system material. The planets—including our own—are insignificant in comparison.

Most of the quantities listed in Table 6.1 can be determined using methods described in Chapters 1 and 2. Here we present a brief summary of the properties listed in Table 6.1 and the techniques used to measure them:

- The *distance* of each planet from the Sun is known from Kepler's laws once the scale of the solar system is set by radar ranging on Venus. ∞ (Sec. 2.6)

- A planet's sidereal *orbital period* (that is, relative to the stars) can be measured from repeated observations of its location on the sky, so long as Earth's own motion around the Sun is properly taken into account. ∞ (Sec. 2.5)

- A planet's *radius* is found by measuring the angular size of planet—the angle from one side to the other as we view it on the sky—and then applying elementary geometry. ∞ (*More Precisely 1-3*)

- The *masses* of planets with moons may be calculated by applying Newton's laws of motion and gravity, just by observing the moons' orbits around the planets. ∞ (*More Precisely 2-3*) The *sizes* of those orbits, like the sizes of the planets themselves, are determined by geometry.

- The masses of Mercury and Venus (as well as those of our own Moon and the asteroid Ceres) are a little harder to determine accurately, because these bodies have no natural satellites of their own. Nevertheless, it is possible to measure their masses by careful observations of their gravitational influence on other planets or nearby bodies. Mercury and Venus produce small, but measurable, effects on each other's orbits, as well as on that of Earth. The Moon also causes small "wobbles" in Earth's motion as the two bodies orbit their common center of mass.

- These techniques for determining mass were available to astronomers well over a century ago. Today, the masses of most of the objects listed in Table 6.1 have been accurately measured through their gravitational interaction with artificial satellites and space probes launched from Earth. Only in the case of Ceres is the mass still poorly known, mainly because that asteroid's gravity is so weak.

| TABLE 6.1 **Properties of Some Solar System Objects** | | | | | | | | |
|---|---|---|---|---|---|---|---|---|
| Object | Orbital Semimajor Axis (AU) | Orbital Period (Earth Years) | Mass (Earth Masses ) | Radius (Earth Radii) | Number of Known Satellites | Rotation Period * (days) | Average Density (kg/m³) | (g/cm³) |
| Mercury | 0.39 | 0.24 | 0.055 | 0.38 | 0 | 59 | 5400 | 5.4 |
| Venus | 0.72 | 0.62 | 0.82 | 0.95 | 0 | −243 | 5200 | 5.2 |
| Earth | 1.0 | 1.0 | 1.0 | 1.0 | 1 | 1.0 | 5500 | 5.5 |
| Moon | — | — | 0.012 | 0.27 | — | 27.3 | 3300 | 3.3 |
| Mars | 1.52 | 1.9 | 0.11 | 0.53 | 2 | 1.0 | 3900 | 3.9 |
| Ceres (asteroid) | 2.8 | 4.7 | 0.00015 | 0.073 | 0 | 0.38 | 2700 | 2.7 |
| Jupiter | 5.2 | 11.9 | 318 | 11.2 | 63 | 0.41 | 1300 | 1.3 |
| Saturn | 9.5 | 29.4 | 95 | 9.5 | 56 | 0.44 | 700 | 0.7 |
| Uranus | 19.2 | 84 | 15 | 4.0 | 27 | −0.72 | 1300 | 1.3 |
| Neptune | 30.1 | 164 | 17 | 3.9 | 13 | 0.67 | 1600 | 1.6 |
| Pluto (Kuiper belt object) | 39.5 | 248 | 0.002 | 0.2 | 3 | −6.4 | 2100 | 2.1 |
| Hale–Bopp (comet) | 180 | 2400 | $1.0 \times 10^{-9}$ | 0.004 | — | 0.47 | 100 | 0.1 |
| Sun | — | — | 332,000 | 109 | — | 25.8 | 1400 | 1.4 |

*A negative rotation period indicates retrograde (backward) rotation relative to the sense in which all planets orbit the Sun.*

- A planet's *rotation period* may, in principle, be determined simply by watching surface features alternately appear and disappear as the planet rotates. However, with most planets this is difficult to do, as their surfaces are hard to see or may even be nonexistent. Mercury's surface features are hard to distinguish; the surface of Venus is completely obscured by clouds; and Jupiter, Saturn, Uranus, and Neptune have no solid surfaces at all—their atmospheres simply thicken and eventually become liquid as we descend deeper and deeper below the visible clouds. We will describe the methods used to measure these planets' rotation periods in later chapters.

- The final two columns in Table 6.1 list the *average density* of each object. **Density** is a measure of the "compactness" of matter. Average density is computed by dividing an object's mass (in kilograms, say) by its volume (in cubic meters, for instance). For example, we can easily compute Earth's average density. Earth's mass, as determined from observations of the Moon's orbit, is approximately $6.0 \times 10^{24}$ kg. ∞ *(More Precisely 2-3)* Earth's radius $R$ is roughly 6400 km, so its volume is $\frac{4}{3}\pi R^3 \approx 1.1 \times 10^{12}$ km$^3$, or $1.1 \times 10^{21}$ m$^3$. ∞ *(Sec. 1.7)* Dividing Earth's mass by its volume, we obtain an average density of approximately 5500 kg/m$^3$.

*On average*, then, there are about 5500 kilograms of Earth matter in every cubic meter of Earth volume. For comparison, the density of ordinary water is 1000 kg/m$^3$, rocks on Earth's surface have densities in the range 2000–3000 kg/m$^3$, and iron has a density of some 8000 kg/m$^3$. Earth's atmosphere (at sea level) has a density of only a few kilograms per cubic meter. Because many working astronomers are more familiar with the CGS (centimeter–gram–second) unit of density (grams per cubic centimeter, abbreviated g/cm$^3$, where 1 kg/m$^3$ = 1000 g/cm$^3$), Table 6.1 lists density in both SI and CGS units. *More Precisely 6-1* makes these methods a little more concrete by applying them to observations of the planet Jupiter.

## CONCEPT CHECK

✔ How do astronomers go about determining the bulk properties (i.e., masses, radii, and densities) of distant planets?

## 6.3 The Overall Layout of the Solar System

By earthly standards, the solar system is immense. The distance from the Sun to the Kuiper belt outside Neptune's orbit is about 50 AU, more than a million times Earth's radius and roughly 20,000 times the distance from Earth to the Moon (Figure 6.5). Yet, despite the solar system's vast extent, the planets all lie very close to the Sun, astronomically speaking. Even the diameter of the

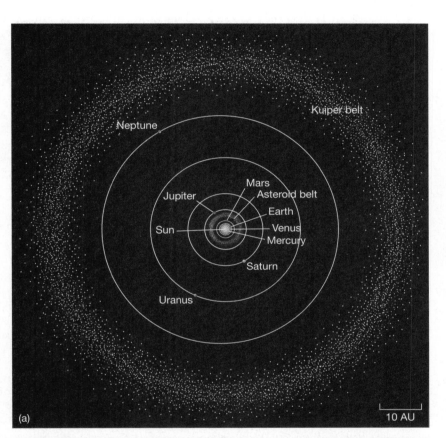

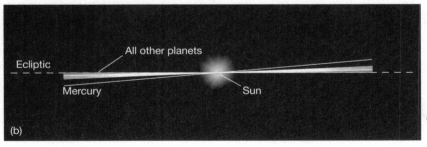

▲ FIGURE 6.5 **Solar System** Major bodies of the solar system: the Sun, planets, and asteroids. Part (a) shows a face-on view from directly above the ecliptic plane; part (b) shows the inclinations of the orbits in an edge-on view. Except for Mercury, the orbits of the planets are almost circular and lie nearly in the same plane. As we move outward from the Sun, the distance between adjacent orbits increases. The entire solar system spans nearly 100 AU, which is roughly the diameter of the Kuiper belt.

## MORE PRECISELY 6-1

### Computing Planetary Properties

Let's look a little more closely at some of the methods used to determine the physical properties of a planet. The accompanying figure shows the planet Jupiter and Europa, one of its inner moons. The orbits of both Jupiter and Earth have been measured very precisely so, at any instant, the distance between the two planets is accurately known. However, since Jupiter's orbital semi-major axis is 5.2 AU and that of Earth is 1 AU, for definiteness in what follows we will simply assume a distance of 4.2 AU, or about 630,000,000 km. ∞ (Sec. 2.6) This distance corresponds roughly to the point of closest approach between the two planets.

Europa

47"

(NASA)

R  I  V  U  X  G

Now that we know the distance to Jupiter, let's compute two other important planetary properties—size and mass.

**EXAMPLE 1** Determining the size is a matter of geometry. ∞ (More Precisely 1-3) As indicated in the figure, Jupiter's angular diameter at a distance of 4.2 AU from Earth is 46.8 arc seconds. (Recall that 1 arc second is 1/3600 of a degree.) Since the distance is known, it follows that the planet's actual diameter is

$$\text{diameter} = \text{distance} \times \frac{\text{angular diameter}}{57.3°}$$

$$= (4.2 \times 150{,}000{,}000 \text{ km}) \times \frac{(46.8/3600)°}{57.3°}$$

$$= 143{,}000 \text{ km,}$$

so Jupiter's radius is half the diameter, or 71,500 km.

**EXAMPLE 2** Jupiter's mass can be determined by application of Newton's laws of motion and gravity. ∞ (More Precisely 2-3) Jupiter's moon Europa is observed to orbit the planet with an orbital period $P = 3.55$ days. The orbit is circular, with an angular radius of 3.66 arc minutes, as seen from Earth. Converting this quantity to a distance in kilometers, as above, we find that the radius of Europa's orbit around Jupiter is $r = 671{,}000$ km. The moon thus travels a distance equal to the circumference of its orbit, $2\pi r = 4.22$ million km, in a time $P = 3.55$ days $= 307{,}000$ s. The orbital speed of Europa is $V = 2\pi r/P = 13.7$ km/s. Applying Newton's laws, we then determine Jupiter's mass to be

$$M = \frac{rV^2}{G} = 1.9 \times 10^{27} \text{ kg.}$$

**EXAMPLE 3** Knowing Jupiter's mass and radius, we can compute a further planetary property—its *density*. Dividing Jupiter's mass $M$ by the volume of a sphere of radius $R$, namely, $\frac{4}{3}\pi R^3$, we find

$$\text{density} = \frac{M}{\frac{4}{3}\pi R^3} = \frac{1.9 \times 10^{27} \text{ kg}}{1.5 \times 10^{24} \text{ m}^3} = 1240 \text{ kg/m}^3.$$

This value for the density differs from the figure listed in Table 6.1 because Jupiter is, in reality, not perfectly spherical. It is *flattened* somewhat at the poles, so our expression for the volume is not quite correct. When we take the flattening into account, the actual volume is a little lower than the number given here, and the density is correspondingly higher.

We have determined several important physical properties of Jupiter by combining observations made from Earth with our knowledge of simple geometry and Newton's laws of motion. In fact, Jupiter is perhaps the easiest planet to study in this way, as it is big, is relatively close, and has several easy-to-see satellites. Nevertheless, these fundamental techniques are applicable throughout the solar system—indeed, throughout the entire universe. Prior to the Space Age, they formed the basis for virtually all astronomical measurements of planetary properties.

---

Kuiper belt is just 1/1000 of a light-year, whereas the next nearest star is several light-years distant.

The planet closest to the Sun is Mercury. Moving outward, we encounter, in turn, Venus, Earth, Mars, Jupiter, Saturn, Uranus, and Neptune. In Chapter 2, we saw the basic properties of the planets' orbits. Their paths are all ellipses, with the Sun at (or very near) one focus. ∞ (Sec. 2.5) With the exception of the innermost

world, Mercury, most planets move on low-eccentricity orbits. Accordingly, we can reasonably think of most planets' orbits as circles centered on the Sun. Note that the planetary orbits are not evenly spaced, becoming farther and farther apart as we move outward from the Sun.

All the planets orbit the Sun counterclockwise as seen from above Earth's North Pole and in nearly the same

▲ **FIGURE 6.6 Planetary Alignment** This image shows six planets—Mercury, Venus, Mars, Jupiter, Saturn, and Earth—during a planetary alignment in April 2002. Because the planets all orbit in nearly the same plane, it is possible for them all to appear (by chance) in the same region of the sky, as seen from Earth. The Sun and new Moon are just below the horizon. As usual, the popular press contained many sensationalized predictions of catastrophes that would occur during this rare astronomical event. Also as usual, none came true. (*J. Lodriguss*)

plane as Earth (the plane of the ecliptic). Mercury deviates somewhat from the latter statement: Its orbital planes lie at 7° to the ecliptic. Still, as illustrated in the figure, we can think of the solar system as being quite flat—its "thickness" perpendicular to the plane of the ecliptic is a tiny fraction of its diameter. If we were to view the planets' orbits from a vantage point in the plane of the ecliptic about 50 AU from the Sun, none of the planets' orbits would be noticeably tilted. Figure 6.6 is a photograph of the planets Mercury, Venus, Mars, Jupiter, and Saturn taken during a chance planetary alignment in April 2002. These five planets can (occasionally) be found in the same region of the sky, in large part because their orbits lie nearly in the same plane in space.

CONCEPT CHECK
✔ In what sense is the solar system "flat"?

## 6.4 Terrestrial and Jovian Planets

On large scales, the solar system presents us with a sense of orderly motion. The planets move nearly in a plane, on almost concentric (and nearly circular) elliptical paths, in the same direction around the Sun, at steadily increasing orbital intervals. Although the individual details of the planets are much less regular, their overall properties allow a natural division into two broad classes.

### Planetary Properties

Figure 6.7 compares the planets with one another and with the Sun. A clear distinction can be drawn between the inner and the outer members of our planetary system based on densities and other physical properties. The inner planets—Mercury, Venus, Earth, and Mars—are

◀ **FIGURE 6.7 Sun and Planets** Relative sizes of the planets and our Sun, drawn to scale. Notice how much larger the jovian planets are than Earth and the other terrestrial planets and how much larger still is the Sun.

small, dense, and *solid*. The outer worlds—Jupiter, Saturn, Uranus, and Neptune—are large, of low density, and *gaseous*.

Because the physical and chemical properties of Mercury, Venus, and Mars are somewhat similar to Earth's, the four innermost planets are called the **terrestrial planets.** (The word *terrestrial* derives from the Latin word *terra*, meaning "land" or "earth.") The four terrestrial planets all lie within about 1.5 AU of the Sun. All are small and of relatively low mass—Earth is the largest and most massive of the four—and all have a generally *rocky* composition and solid surfaces.

Jupiter, Saturn, Uranus, and Neptune are all similar to one another chemically and physically, and very different from the terrestrial worlds. They are labeled the **jovian planets,** after Jupiter, the largest member of the group. (The word *jovian* comes from *Jove*, another name for the Roman god Jupiter.) The jovian worlds all orbit far from the Sun. They are all much larger than the inner planets and quite different from them in both composition and structure. They have no solid surfaces, and their outer layers are composed predominantly of the light gases *hydrogen* and *helium*.

Table 6.2 compares and contrasts some key properties of these two planetary classes.

## Differences Among the Terrestrial Planets

Despite their similarities in orbits and composition, the terrestrial planets have many important differences, too. We list a few here, as prelude to the more detailed accounts presented in Chapters 7–10:

- All four terrestrial planets have *atmospheres*, but the atmospheres are about as dissimilar as we could imagine, ranging from a near vacuum on Mercury to a hot, dense inferno on Venus.

- Earth alone has *oxygen* in its atmosphere and *liquid water* on its surface.

- *Surface conditions* on the four planets are quite distinct from one another, ranging from barren, heavily cratered terrain on Mercury to widespread volcanic activity on Venus.

- Earth and Mars *spin* at roughly the same rate—one rotation every 24 (Earth) hours—but Mercury and Venus both take months to rotate just once, and Venus rotates in the opposite sense from the others.

- Earth and Mars have *moons*, but Mercury and Venus do not.

- Earth and Mercury have measurable *magnetic fields*, of very different strengths, whereas Venus and Mars have none.

| TABLE 6.2 | Comparison of the Terrestrial and Jovian Planets |
| --- | --- |
| **Terrestrial Planets** | **Jovian Planets** |
| close to the Sun | far from the Sun |
| closely spaced orbits | widely spaced orbits |
| small masses | large masses |
| small radii | large radii |
| predominantly rocky | predominantly gaseous |
| solid surface | no solid surface |
| high density | low density |
| slower rotation | faster rotation |
| weak magnetic fields | strong magnetic fields |
| few moons | many moons |
| no rings | many rings |

Comparing the average densities of the terrestrial planets allows us to say something more about their overall *compositions*. However, before making the comparison, we must take into account how the weight of overlying layers compresses the interiors of the planets to different extents. When we do this, we find that the *uncompressed densities* of the terrestrial worlds—the densities they would have in the absence of any compression due to their own gravity—decrease as we move outward from the Sun: 5300, 4400, 4400, and 3800 kg/m$^3$ for Mercury, Venus, Earth, and Mars, respectively. The amount of compression is greatest for the most massive planets, Earth and Venus, and much less for Mercury and Mars. Partly on the basis of these figures, planetary scientists conclude that Earth and Venus are quite similar in overall composition. Mercury's higher density implies that it contains a higher proportion of some dense material—most likely nickel or iron. The lower density of Mars probably means that it is deficient in that same material.

Finding the common threads in the evolution of these four diverse worlds is no simple task! The goal of comparative planetology is to understand how four planets with broadly similar overall physical properties came to differ so much in detail.

## Terrestrial–Jovian Comparison

For all their differences, the terrestrial worlds still seem similar compared with the jovian planets. Perhaps the simplest way to express the major differences between the terrestrial and jovian worlds is to say that the jovian planets are everything the terrestrial planets are not. We will discuss the jovian planets in more detail in Chapters 11–13.

We highlight here some of the key differences between the terrestrial and jovian worlds:

- The terrestrial worlds lie close together, near the Sun; the jovian worlds are widely spaced through the outer solar system.

- The terrestrial worlds are small, dense, and rocky; the jovian worlds are large and gaseous, containing huge amounts of hydrogen and helium (the lightest elements), which are rare on the inner planets.

- The terrestrial worlds have solid surfaces; the jovian worlds have none (their dense atmospheres thicken with depth, eventually merging with their liquid interiors).

- The terrestrial worlds have weak magnetic fields, if any; the jovian worlds all have strong magnetic fields.

- All four jovian worlds are thought to contain large, dense "terrestrial" cores some 10 to 15 times the mass of Earth. These cores account for an increasing fraction of each planet's total mass as we move outward from the Sun.

- The terrestrial worlds have only three *moons* among them; the jovian worlds have many moons each, no two of them alike and none of them like our own.

- Furthermore, all the jovian planets have *rings*, a feature unknown on the terrestrial planets.

The job of the planetary scientist is made all the more complicated by the fact that the same theory that accounts for the jovian planets must also naturally explain the terrestrial ones, as well as the similarities and differences within each class. Section 6.7 presents an outline of the theory that does just that. Later, in Chapter 15, we will return to this theory in more depth and see how its predictions fare when faced with detailed observations of our planetary system and others. ∞ (Sec. 1.2)

CONCEPT CHECK
✔ Why do astronomers draw such a clear distinction between the inner and the outer planets?

# 6.5 Interplanetary Matter

In the vast space among the eight known major planets move countless chunks of rock and ice, mostly small, some quite large. All orbit the Sun, many on highly eccentric paths. This final component of the solar system is the collection of **interplanetary matter**—cosmic "debris" ranging in size from the relatively large *asteroids* and members of the *Kuiper belt*, through the smaller *comets* and even smaller *meteoroids*, down to the smallest grains of *interplanetary dust* that litter our cosmic environment (see Chapter 14).

The dust arises when larger bodies collide and break apart into smaller pieces that, in turn, collide again and are slowly ground into microscopic fragments, which eventually settle into the Sun or are swept away by the *solar wind*, a stream of energetic charged particles that continually flows outward from the Sun and pervades the entire solar system. The dust is quite difficult to detect in visible light, but infrared studies reveal that interplanetary space contains surprisingly large amounts of it. Our solar system is an extremely good vacuum by terrestrial standards, but positively dirty by the standards of interstellar or intergalactic space.

Asteroids (Figure 6.8a) and meteoroids are generally rocky in composition, somewhat like the outer layers of the terrestrial planets. The distinction between the two is simply a matter of size: Anything larger than 100 m in diameter (corresponding to a mass of about 10,000 tons) is conventionally termed an asteroid; anything smaller is a meteoroid. Their total mass is much less than that of Earth's Moon, so these objects play no important role in the present-day workings of the planets or their moons. Yet they are of crucial importance to our studies, for they provide the keys to answering some very fundamental questions about our planetary environment and what the solar system was like soon after its birth. Many of these bodies are made of material that has hardly evolved since the early days of the solar system. In addition, they often conveniently deliver themselves right to our doorstep, in the form of meteorites (the name we give them if they happen to survive the plunge through Earth's atmosphere and find their way to the ground—see Section 14.4), allowing us to study them in detail without having to fetch them from space.

Comets are quite distinct from the other small bodies in the solar system. They are generally icy rather than rocky in composition (although they do contain some rocky material) and typically have diameters in the 1–10-km range. They are quite similar in chemical makeup to some of the icy moons of the outer planets. Even more so than the asteroids and meteoroids, comets represent truly ancient material—the vast majority probably have not changed in any significant way since their formation long ago along with the rest of the solar system (see Chapter 15). Comets striking Earth's atmosphere do not reach the surface intact, so, until relatively recently (see below), astronomers had no actual samples of cometary material. However, comets do vaporize and emit radiation as their highly elongated orbits take them near the Sun (see Figure 6.8b), so scientists have long been able to determine a comet's makeup by spectroscopic study of the radiation it gives off as it is destroyed. ∞ (Sec. 4.2)

In 2004, NASA's *Stardust* probe made a close encounter with a comet passing through the inner solar system, and returned to Earth in 2006 carrying the first-ever samples of cometary dust. As discussed in more detail in Chapter 14, analysis of the dust confirms earlier inferences about comet composition based on the spectroscopic data,

▶ FIGURE 6.8 **Asteroid and Comet**
(a) Asteroids, like meteoroids, are generally composed of rocky material. This asteroid, Eros, is about 34 km long. It was photographed by the *Near Earth Asteroid Rendezvous* spacecraft that actually landed on the asteroid in 2001. (b) Comet Hale–Bopp, seen as it approached the Sun in 1997. Most comets are composed largely of ice and so tend to be relatively fragile. The comet's vaporized gas and dust form the tail, here extending away from the Sun for nearly a quarter of the way across the sky. (c) Pluto—seen here at center with its 3 known moons—is one of the largest members of the Kuiper belt. Formerly classified as a major planet, it was demoted during a heated debate among astronomers in 2006. (*JHU/APL; J. Lodriguss; NASA*)

(a)

(b)

and also provides new insights into other compounds less easily vaporized, and hence hard to detect by spectroscopic means.

Finally, beyond the outermost jovian planet, Neptune, lies the Kuiper belt—an "outer asteroid belt" consisting of icy, cometlike bodies ranging in size from a fraction of a kilometer to more than 1000 km in diameter. Because of their small sizes and great distances, most known Kuiper belt objects have been discovered only recently, mainly since the mid-1990s. However, the best known member of this class—Pluto (Figure 6.8c)—has been known for much longer, and is now the subject of controversy among astronomers. Pluto was originally classified as a planet following its discovery in 1930. However, it simply doesn't fit into the classification just described—in both mass and composition, it has much more in common with the icy jovian moons than with any terrestrial or jovian world. As a result, many astronomers questioned its "planetary" designation, an opinion that became stronger as the number of known Kuiper belt objects steadily increased.

The controversy was deepened by the detection of several Kuiper belt objects comparable in size to Pluto, culminating in 2003 with the discovery of a *larger* body, now aptly called Eris after the Greek goddess of discord, orbiting even farther from the Sun, beyond the Kuiper belt itself. In 2006, the International Astronomical Union, which oversees the rules for classifications in astronomy, decided that Pluto should no longer be classified as a major planet, but rather as a *dwarf planet*. It is the largest object in the Kuiper belt, but is no longer regarded as a planet on a par with the terrestrial or jovian worlds.

CONCEPT CHECK

✔ Why are astronomers so interested in interplanetary matter?

(c)

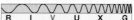
R    I    V    U    X    G

# 6.6 Spacecraft Exploration of the Solar System

Since the 1960s, dozens of unmanned space missions have traveled throughout the solar system. All of the planets have been visited and probed at close range, and spacecraft have visited numerous comets and asteroids. The first landing on an asteroid (by the *NEAR* spacecraft) occurred in February 2001 (see Figure 6.8a); to date, three probes have rendezvoused with comets passing through the inner solar system (see Section 14.2). The impact of these missions on our understanding of our planetary system has been nothing short of revolutionary.

In the next eight chapters, we will see many examples of the marvelous images radioed back to Earth by these missions, and we will discover how they fit into our modern picture of the solar system. Here, we focus on just a few of these remarkable technological achievements; Table 6.3 summarizes the various successful planetary missions highlighted in this section.

## The *Mariner 10* Flybys of Mercury

A "flyby," in NASA parlance, is any space mission in which a probe passes relatively close to a planet—within a few planetary radii, say—but does not go into orbit around it. In 1974, the U.S. spacecraft *Mariner 10* came within 10,000 km of the surface of Mercury, sending back high-resolution images of the planet. The photographs, which showed surface features as small as 150 m across, dramatically increased our knowledge of the planet. For the first time, we saw Mercury as a heavily cratered world, in many ways reminiscent of our own Moon.

*Mariner 10* was launched from Earth in November 1973 and was placed in an eccentric 176-day orbit about the Sun, aided by a gravitational assist (see *Discovery 6-1*) from the planet Venus (Figure 6.9). In that orbit, *Mariner 10*'s nearest point to the Sun (perihelion) is close to Mercury's path, and its farthest point away (aphelion) lies between the orbits of Venus and Earth. (∞ *More Precisely 2-1*) The 176-day period is exactly 2 Mercury years, so the spacecraft revisits Mercury roughly every 6 months. However, only on the first three encounters—in March 1974, September 1974, and March 1975—did the spacecraft return data. After that, the craft's supply of maneuvering fuel was exhausted; the craft still orbits the Sun today, but out of control and silent. In total, over 4000 photographs, covering about 45 percent of the planet's surface, were radioed back to Earth during the mission's active lifetime. The remaining 55 percent of Mercury is still unexplored.

No new missions have been sent to Mercury since *Mariner 10*. NASA plans a return in 2011, when the *Messenger* probe will be placed in orbit around the planet to

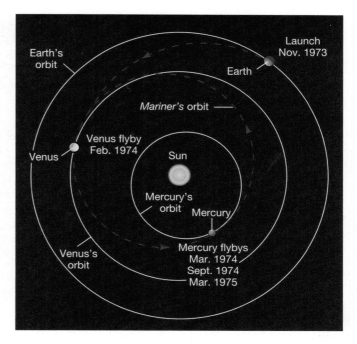

▲ **FIGURE 6.9 Mission to Mercury** The path of the *Mariner 10* probe to the planet Mercury included a gravitational boost from Venus. The spacecraft returned data from March 1974 until March 1975, providing astronomers with a wealth of information on Mercury.

map its entire surface at much higher resolution than was possible with *Mariner*. *Messenger* was launched in 2004 and will reach its final orbit after one gravity-assist from Earth (in 2005), two from Venus (in October 2006 and June 2007), and three from Mercury itself (in January and October 2008, and September 2009). Even on its first close pass of Mercury in early 2008 *Messenger* should already expand our knowledge of the planet significantly beyond the *Mariner 10* data. The European Space Agency also plans to place a spacecraft in orbit around Mercury at roughly the same time as the *Messenger* mission.

## Exploration of Venus

In all, some 20 spacecraft have visited Venus since the 1970s, far more than have spied on any other planet in the solar system. The Soviet space program took the lead in exploring Venus's atmosphere and surface, whereas American, and most recently European, spacecraft have performed extensive radar mapping of the planet from orbit.

The Soviet *Venera* (derived from the Russian word for Venus) program began in the mid-1960s, and the Soviet *Venera 4* through *Venera 12* probes parachuted into the planet's atmosphere between 1967 and 1978. The early

**TABLE 6.3    Some Missions to the Other Planets**

| Target Planet | Year(s) at Planet | Project | Launched by | Type of Mission | Scientific Achievements or Goals, and other Comments |
|---|---|---|---|---|---|
| Mercury | 1974–1975 | Mariner 10 | U.S. | flyby | Photographed 45 percent of the planet's surface; still orbiting the Sun |
| Venus | 1967 | Venera 4 | U.S.S.R. | atmospheric probe | First probe to enter the planet's atmosphere |
| | 1970 | Venera 7 | U.S.S.R. | lander | First landing on the planet's surface |
| | 1978–1992 | Pioneer Venus | U.S. | orbiter | Radar mapping of the entire planet |
| | 1983 | Venera 15, 16 | U.S.S.R. | orbiter | Radar mapping of northern hemisphere |
| | 1990–1994 | Magellan | U.S. | atmospheric probe, orbiter | High-resolution radar mapping of the entire planet |
| | 2006– | Venus Express | European Space Agency | orbiter | Studies of the dynamics and chemistry of the atmosphere of Venus; volcanism and other surface studies |
| Mars | 1965 | Mariner 4 | U.S. | flyby | Photographs showed cratered surface |
| | 1969 | Mariner 6, 7 | U.S. | flyby | More photographs of the planet's surface |
| | 1971 | Mariner 9 | U.S. | orbiter | First complete survey of the surface; revealed complex terrain and evidence of past geological activity |
| | 1976–1982 | Viking 1, 2 | U.S. | orbiter, lander | Detailed surface maps, implying geological and climatic change; first surface landings, first atmospheric and soil measurements; search for life |
| | 1997 | Mars Pathfinder | U.S. | lander | First Mars rover, local geological survey |
| | 1997–2006 | Mars Global Surveyor | U.S. | orbiter | High-resolution surface mapping, remote analysis of surface composition |
| | 2001– | Mars Odyssey | U.S. | orbiter | Remote surface chemical analysis, search for subsurface water, measurement of radiation levels |
| | 2003– | Mars Express | European Space Agency | atmospheric probe, orbiter | Studies of Martian atmosphere and geology; search for water and evidence of life |
| | 2004– | Mars Exploration Rover | U.S. | two landers | Assessment of likelihood that life arose on Mars, measurement of Martian climate and detailed geological surveys near the landing sites |
| | 2006– | Mars Reconnaissance Orbiter | U.S. | orbiter | History of water on Mars, detailed imaging of small-scale surface features |
| Jupiter | 1973 | Pioneer 10 | U.S. | flyby | First mission to the outer planets |
| | 1974 | Pioneer 11 | U.S. | flyby | Detailed close-up images; gravity assist from Jupiter to reach Saturn |
| | 1979 | Voyager 1 | U.S. | flyby | Detailed observations of planet and moons |
| | 1979 | Voyager 2 | U.S. | flyby | Continued reconnaissance of the Jovian system |
| | 1995–2003 | Galileo | U.S. | atmospheric probe, orbiter | Atmospheric studies; long-term precision measurements of the planet's moon system |
| Saturn | 1979 | Pioneer 11 | U.S. | flyby | First close-up observations of Saturn; now leaving the solar system |
| | 1981 | Voyager 1 | U.S. | flyby | Observations of planet, rings and moons; close-up measurements of the moon Titan |
| | 1982 | Voyager 2 | U.S. | flyby | Observations of planet and moons |
| | 2004– | Cassini-Huygens | U.S., European Space Agency | atmospheric probe, orbiter | Atmospheric studies; repeated and detailed measurements of the planet's moons |
| Uranus | 1986 | Voyager 2 | U.S. | flyby | Observations of planet and moons; "Grand Tour" of the outer planets |
| Neptune | 1989 | Voyager 2 | U.S. | flyby | Observations of planet and moons; now leaving the solar system |

spacecraft were destroyed by enormous atmospheric pressures before reaching the surface, but in 1970, *Venera 7* (Figure 6.10) became the first spacecraft to soft-land on the planet. During the 23 minutes it survived on the surface, it radioed back information on the planet's atmospheric pressure and temperature. Subsequent *Venera* landers transmitted photographs of the surface back to Earth and analyzed the atmosphere and the soil. None survived for more than an hour in the planet's hot, dense atmosphere. The data they sent back make up the entirety of our direct knowledge of Venus's surface.

The 1978 U.S. *Pioneer Venus* mission placed an orbiter at an altitude of some 150 km above Venus's surface and dispatched a "multiprobe" consisting of five separate instrument packages into the planet's atmosphere. During its hour-long descent to the surface, the probe returned information on the variation in density, temperature, and chemical composition with altitude in the atmosphere. The orbiter's radar produced images of most of the planet's surface.

The most recent U.S. mission was the *Magellan* probe (shown in Figure 6.11), which entered orbit around Venus in August 1990. Its radar imaging system could distinguish objects as small as 120 m across. Between 1991 and 1994, the probe mapped 98 percent of the surface of Venus with unprecedented clarity and made detailed measurements of the planet's gravity, rendering all previous data virtually obsolete. The mission ended in October 1994 with a (planned) plunge into the planet's dense atmosphere, sending back one final stream of high-quality data. Many theories of the processes shaping the planet's surface had to be radically altered or abandoned completely because of *Magellan*'s data.

▲ FIGURE 6.11 *Magellan* **Orbiter** The U.S. *Magellan* spacecraft was launched from the space shuttle *Atlantis* (at bottom) in May 1989 on a mission to explore the planet Venus. The large radio antenna at the top was used both for mapping the surface of Venus and for communicating with Earth. Contrast the relatively delicate structure of this craft, designed to operate in space, with the much more bulky design of the Venus lander shown in Figure 6.10. *(NASA)*

In April 2006 the European Space Agency placed the *Venus Express* probe in orbit around the planet. During its planned 2-year mission, *Venus Express* studied the atmosphere and surface of Venus, as well as the interaction between the planet's atmosphere and the *solar wind*—the stream of charged particles escaping from the Sun. As discussed in Chapter 9, the principal goals were to understand the structure, composition, and circulation of the planet's atmosphere.

## Exploration of Mars

Both NASA and the Soviet (now Russian) Space Agency have Mars exploration programs that began in the 1960s. However, the Soviet effort was plagued by a string of technical problems, along with a liberal measure of plain bad luck. As a result, almost all of the detailed planetary data we have on Mars has come from unmanned U.S. probes.

The first spacecraft to reach the Red Planet was *Mariner 4*, which flew by Mars in July 1965. The images sent back by the craft showed large numbers of craters caused by impacts of meteoroids with the planet's surface, instead of the Earth-like terrain some scientists had expected to find. Flybys in 1969 by *Mariner 6* and *Mariner 7* confirmed these findings, leading to the conclusion that Mars was a geologically dead planet having a heavily

▲ FIGURE 6.10 **Venus Lander** One of the Soviet *Venera* landers that reached the surface of Venus. The design was essentially similar for all the surface missions. Note the heavily armored construction, necessary to withstand the harsh conditions of high temperature and crushing atmospheric pressure on the planet's surface. *(Sovfoto/Eastfoto)*

cratered, old surface. This conclusion was soon reversed by the arrival in 1971 of the *Mariner 9* orbiter. The craft mapped the entire Martian surface at a resolution of about 1 km, and it rapidly became clear that here was a world far more complex than the dead planet imagined only a year or two previously. *Mariner 9's* maps revealed vast plains, volcanoes, drainage channels, and canyons. All these features were completely unexpected, given the data provided by the earlier missions. These new findings paved the way for the next step: actual landings on the planet's surface.

The two spacecraft of the U.S. *Viking* mission arrived at Mars in mid-1976. *Viking 1* and *Viking 2* each consisted of two parts. An orbiter mapped the surface at a resolution of about 100 m (about the same as the resolution achieved by *Magellan* for Venus), and a lander (see Figure 6.12) descended to the surface and performed a wide array of geological and biological experiments. *Viking 1* touched down on Mars on July 20, 1976. *Viking 2* arrived in September of the same year. By any standards, the *Viking* mission was a complete success: The orbiters and landers returned a wealth of long-term data on the Martian surface and atmosphere. *Viking 2* stopped transmitting data in April 1980. *Viking 1* continued to operate until November 1982.

The relative positions of Mars and Earth in their respective orbits make it most favorable to launch a spacecraft from Earth roughly every 26 months (the so-called *synodic period* of Mars relative to Earth—see *More Precisely 9-1*). It then takes roughly 8 to 9 months for the craft to arrive at Mars (Figure 6.13). The first U.S. probe to reach Mars since *Viking*—the *Mars Global Surveyor* orbiter—was launched from Earth in 1996 and arrived at Mars in late 1997, scanning the planet with cameras and other sensors and charting the Martian landscape at high resolution. Originally scheduled to end in 2002, the mission finally concluded in late 2006 when the spacecraft became unresponsive to commands issued from Earth.

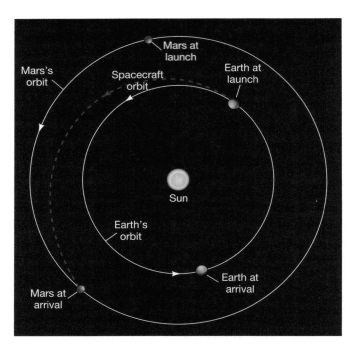

▲ **FIGURE 6.13 Mission to Mars** A typical orbital path from Earth to Mars. A spacecraft orbiting in the path shown can take anywhere from 8 to 9 months to make the trip, depending on precisely where Mars happens to be in its orbit, which is illustrated here as circular, but which, in reality, is somewhat eccentric.

*Mars Global Surveyor* was followed (and in fact overtaken) by *Mars Pathfinder*, which arrived at Mars in late June 1997. On July 4, *Pathfinder* parachuted an instrument package to the Martian surface. Near the ground, the parachute fell away and huge air bags deployed, enabling the robot to bounce softly to a safe landing. Side panels opened, and out came a small six-wheeled minirover, called *Sojourner* (Figure 6.14). During its 3-month lifetime, the lander took measurements of the Martian atmosphere and atmospheric dust, while *Sojourner* roamed the Martian countryside at a rate of a few meters per day, carrying out chemical analyses of the soil and rocks within about 50 m of the parent craft. In addition, over 16,000 images of the region were returned to Earth.

Mars exploration has now moved into high gear, with an ambitious series of missions spanning the first decade of the 21st century. A new U.S. orbiter, called *Mars Odyssey*, reached Mars in October 2001. Its sensors are designed to probe the chemical makeup of the Martian surface and look for possible water ice below. In the summer of 2003, no fewer than three Mars-bound missions left Earth. The European Space Agency launched *Mars Express*, consisting of an orbiter and a lander, and NASA launched *Mars Exploration Rover*, comprising two *Sojourner*-like landers, named *Spirit* and *Opportunity*.

*Mars Express* reached the planet in December 2003. Its orbiter worked perfectly and began transmitting data back to Earth, but its surface probe apparently crash-landed, as

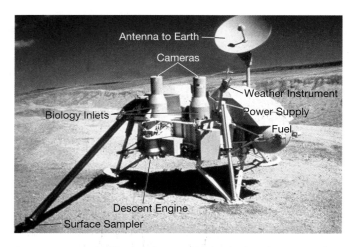

▲ **FIGURE 6.12 *Viking* Lander** A *Viking* lander, here being tested in the Mojave Desert prior to launch. For an idea of the scale, the reach of its extended arm at left is about 1 m. *(NASA)*

North Knob

Booboo

Sojourner Rover

1.7 meters

Yogi

Rover Tracks

Lamb

Hard Stop

Shaggy

Scooby-Doo

Lumpy

Airbags

Airbags

Solar Panel

◀ FIGURE 6.14 **Martian Lander** The Mars rover *Sojourner* was a small self-propelled vehicle used to explore the Martian surface. It is seen at left center using an x-ray spectrometer to determine the chemical composition of a large Martian rock, nicknamed Yogi by mission scientists. For scale, the rock is about 1 m tall and lies some 10 m from *Sojourner's* mother ship, *Pathfinder* (foreground). *(NASA)*

no signals were received from it. A month later, the two NASA probes successfully soft-landed by bouncing across the surface until they rolled to a stop. One ended up, as planned, in a huge crater that might be a dried-up lake; the other rolled into an equatorial depression that surprisingly displayed evidence for subsurface layering. Both craft were designed for a 2-month stay but, as of early 2007, they are entering their third year of service, searching for signs of water on Mars.

The most recent mission to Mars is NASA's *Mars Reconnaissance Orbiter*, which was launched in 2005 and arrived at the planet in May 2006, beginning its scientific work in November of that year. The main mission goals are to study the climate and geology of Mars and to determine whether life ever arose there.

NASA also has far-reaching plans for manned missions to Mars. Indeed, portions of the three most recent Mars missions are aimed specifically at determining the feasibility of human exploration of the planet. However, the enormous expense (and danger) of such an undertaking, coupled with the opinion of many astronomers that unmanned missions are economically and scientifically preferable to manned missions, makes the future of these projects uncertain at best.

## Missions to the Outer Planets

Two pairs of U.S. spacecraft launched in the 1970s—*Pioneer* and *Voyager*—revolutionized our knowledge of Jupiter and the jovian planets. *Pioneer 10* and *Pioneer 11* were launched in March 1972 and April 1973, respectively, and arrived at Jupiter in December 1973 and December 1974.

The *Pioneer* spacecraft took many photographs and made numerous scientific discoveries. Their orbital

trajectories also allowed them to observe the polar regions of Jupiter in much greater detail than later missions would achieve. In addition to their many scientific accomplishments, the *Pioneer* craft also played an important role as "scouts" for the later *Voyager* missions. The *Pioneer* series demonstrated that spacecraft could travel the long route from Earth to Jupiter without colliding with debris in the solar system. They also discovered—and survived—the perils of Jupiter's extensive radiation belts (somewhat like the Van Allen radiation belts that surround Earth—see Chapter 7—but on a much larger scale). In addition, *Pioneer 11* used Jupiter's gravity to propel it along the same trajectory to Saturn that the *Voyager* controllers planned for *Voyager 2*'s visit to Saturn's rings.

The two *Voyager* spacecraft (see Figure 6.15) left Earth in 1977 and reached Jupiter in March (*Voyager 1*) and July (*Voyager 2*) of 1979 to study the planet and its major satellites in detail. Each craft carried sophisticated equipment to investigate the planet's magnetic field, as well as radio, visible-light, and infrared sensors to analyze its reflected and emitted radiation. Both *Voyager 1* and *Voyager 2* used Jupiter's gravity to send them on to Saturn. *Voyager 1* was programmed to visit Titan, Saturn's largest moon, and so did not come close enough to the planet to receive a gravity-assisted boost to Uranus. However, *Voyager 2* went on to visit both Uranus and Neptune in a spectacularly successful "Grand Tour" of the outer planets. The data returned by the two craft are still being analyzed today. Like *Pioneer 11*, the two *Voyager* craft are now headed out of the solar system, still sending data as they race toward interstellar space. The figure in *Discovery 6-1* shows the past and present trajectories of the *Voyager* spacecraft.

The most recent mission to Jupiter is the U.S. *Galileo* probe, launched by NASA in 1989. The craft arrived at its

# DISCOVERY 6-1

## Gravitational "Slingshots"

*Celestial mechanics* is the study of the motions of gravitationally interacting objects, such as planets and stars, applying Newton's laws of motion to understand the intricate movements of astronomical bodies. ∞ (Sec. 2.8) Computerized celestial mechanics lets astronomers calculate planetary orbits to high precision, taking the planets' small gravitational influences on one another into account. Even before the computer age, the discovery of one of the outermost planets, Neptune, came about almost entirely through studies of the distortions of Uranus's orbit that were caused by Neptune's gravity.

Celestial mechanics is also an essential tool for scientists and engineers who wish to navigate manned and unmanned spacecraft throughout the solar system. Robot probes can now be sent on stunningly accurate trajectories, expressed in the trade with such slang phrases as "sinking a corner shot on a billion-kilometer pool table." Near-flawless rocket launches, aided by occasional midcourse changes in flight paths, now enable interplanetary navigators to steer remotely controlled spacecraft through an imaginary "window" of space just a few kilometers wide and a billion kilometers away.

However, sending a spacecraft to another planet requires a lot of energy—often more than can be conveniently provided by a rocket launched from Earth or safely transported in a shuttle for launch from orbit. Faced with these limitations, mission scientists often use their knowledge of celestial mechanics to carry out "slingshot" maneuvers, which can boost an interplanetary probe into a more energetic orbit and also aid navigation toward the target, all at no additional cost!

The accompanying figure illustrates a gravitational slingshot, or *gravity assist*, in action. A spacecraft approaches a planet, passes close by, and then escapes along a new trajectory. Obviously, the spacecraft's *direction* of motion is changed

by the encounter. Less obviously, the spacecraft's *speed* is also altered as the planet's gravity propels the spacecraft in the direction of the planet's motion. By a careful choice of incoming trajectory, the craft can speed up (by passing "behind" the planet, as shown) *or* slow down (by passing in front), by as much as twice the planet's orbital speed. Of course, there is no "free lunch"—the spacecraft gains energy from, or loses it to, the planet's motion, causing the planet's own orbit to change ever so slightly. However, since planets are so much more massive than spacecraft, the effect on the planet is insignificant.

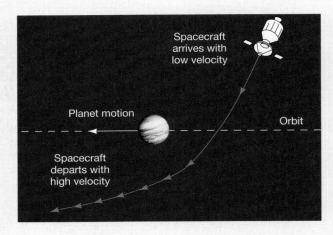

Such a slingshot maneuver has been used many times in missions to both the inner and the outer planets, as illustrated in the second figure, which shows the trajectories of the *Voyager* spacecraft through the outer solar system. The gravitational pulls of these giant worlds whipped the craft around at each visitation, enabling flight controllers to get considerable extra "mileage" out of the probes. *Voyager 1* is now high above

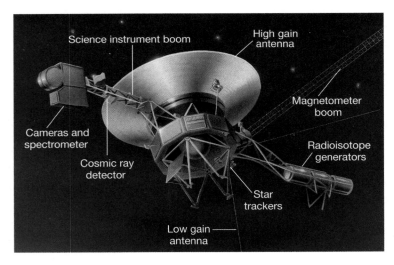

▲ FIGURE 6.15 *Voyager* Mission The two *Voyager* spacecraft that swung by several of the outer planets were identically constructed. Their main features are shown here. (*NASA*)

target in 1995 after a rather roundabout route involving a gravity assist from Venus and two from Earth itself. The mission consisted of an orbiter and an atmospheric probe. The probe descended into Jupiter's atmosphere in December 1995, slowed by a heat shield and a parachute, taking measurements and performing chemical analyses as it went. The orbiter executed a complex series of gravity-assisted maneuvers through Jupiter's system of moons, returning to some already studied by *Voyager* and visiting others for the first time. Some of *Galileo*'s main findings are described in Chapter 11.

The *Galileo* program was originally scheduled to last until December 1997, but NASA extended its lifetime for 6 more years to obtain even more detailed data on Jupiter's inner moons. The mission finally ended in September 2003. With the fuel supply dwindling, *Galileo*'s controllers elected to steer the craft directly into Jupiter, rather than

the plane of the solar system, having been deflected up and out following its encounter with Saturn. *Voyager 2* continued on for a "Grand Tour" of the four jovian planets. It is now outside the orbit of Pluto, in the Kuiper belt.

More recently, the *Cassini* mission to Saturn, which was launched in 1997 and arrived at its target in 2004, received *four* gravitational assists en route—two from Venus, one from Earth, and one from Saturn. Once in the Saturn system,

*Cassini* used the gravity of Saturn and its moons to propel it through a complex series of maneuvers designed to bring it close to all the major moons, as well as to the planet itself. Every encounter with a moon had a slingshot effect—sometimes accelerating and sometimes slowing the probe, but each time moving it into a different orbit—and every one of these effects was carefully calculated long before *Cassini* ever left Earth.

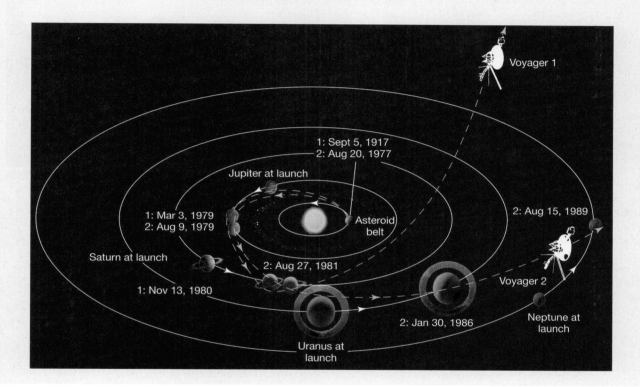

run any risk that it might collide with, and possibly contaminate, Jupiter's moon Europa, which (thanks largely to data returned by *Galileo*) is now a leading candidate in the search for extraterrestrial life. NASA has ambitious (but as yet unfunded) plans for a return to the Jupiter system, possibly as soon as 2010.

In October 1997, NASA launched the *Cassini-Huygens* mission to Saturn (Figure 6.16a). The launch from Cape Canaveral sparked controversy because of fears that the craft's plutonium power source might contaminate parts of our planet following an accident either during launch or during a subsequent gravity assist from Earth in 1999—one of four assists needed for the craft to reach Saturn's distant orbit. On arrival at its destination in 2004, it dispatched the *Huygens* probe built by the European Space Agency into the atmosphere of Titan, Saturn's largest moon. The *Cassini* orbiter then proceeded to orbit among the planet's moons, much as *Galileo* did at Jupiter. Both the probe and the orbiter have returned spectacular images and detailed

data about Saturn and its moons (Figure 6.16b), resolving some outstanding questions raised by *Voyager* about the Saturn system, but posing many new ones, too.

## 6.7    How Did the Solar System Form?

During the past four decades, interplanetary probes have vastly increased our knowledge of the solar system, and their data form the foundation for much of the discussion in the next eight chapters. However, we can understand the overall organization of our planetary system—the basic properties presented in Sections 6.3 and 6.4—without dwelling on these details. Indeed, some key elements of the modern theory of planetary formation predate the Space Age by many years. We present here (in simplified form) a brief version of the "standard" view of how the solar system came into being. This

(a)

(b)

| R | I | V | U | X | G |

▲ **FIGURE 6.16** *Cassini* **Mission to Saturn** (a) The *Cassini* spacecraft, shown here under construction prior to launch in 1997. For scale, note the engineers at bottom. (b) After reaching Saturn in 2004, *Cassini*'s cameras returned a steady stream of spectacular images, such as this one that shows a partial view of the planet's rings, a doughnut-shaped storm near the south pole, and two of its moons near top and bottom. (*NASA; ESA*)

picture will underlie much of our upcoming discussion of the planets, their moons, and the contents of the vast spaces between them.

## Nebular Contraction

One of the earliest heliocentric models of solar system formation may be traced back to the 17th-century French philosopher René Descartes. Imagine a large cloud of interstellar dust and gas (called a *nebula*) a light-year or so across. Now suppose that, due to some external influence, such as a collision with another interstellar cloud or perhaps the explosion of a nearby star, the nebula starts to contract under the influence of its own gravity. As it contracts, it becomes denser and hotter, eventually forming a star—the Sun—at its center (see Chapter 19). Descartes suggested that, while the Sun was forming in the cloud's hot core, the planets and their moons formed in the cloud's cooler outer regions. In other words, planets are by-products of the process of star formation.

In 1796, the French mathematician–astronomer Pierre Simon de Laplace developed Descartes' ideas in a more quantitative way. He showed mathematically that conservation of angular momentum (see *More Precisely 6-2*) demands that our hypothetical nebula spin

faster as it contracts. A decrease in the size of a rotating mass must be balanced by an increase in its rotational speed. The latter, in turn, causes the nebula's *shape* to change as it collapses. Centrifugal forces (due to rotation) tend to oppose the contraction in directions perpendicular to the rotation axis, with the result that the nebula collapses most rapidly along that axis. As shown in Figure 6.17, the fragment eventually flattens into a pancake-shaped primitive solar system. This swirling mass destined to become our solar system is usually referred to as the **solar nebula.**

If we now simply suppose that the planets formed out of this spinning material, then we can already understand the origin of much of the large-scale architecture observed in our planetary system today, such as the circularity of the planets' orbits and the fact that they move in the same sense in nearly the same plane (Figure 6.17c). The planets inherited all these properties from the rotating disk in which they were born. The idea that the planets formed from such a disk is called the **nebular theory.**

## Planetary Condensation

The variant of the nebular theory currently favored by most astronomers is known as the **condensation theory.** Again, we defer the details to Chapter 15, but in essence

## Angular Momentum

Most celestial objects rotate. Planets, moons, stars, and galaxies all have some **angular momentum,** which we can define as the tendency of a body to keep spinning or moving in a circle. Angular momentum is as important a property of an object as its mass or its energy.

Consider first a simpler motion—*linear momentum,* which is defined as the product of an object's mass and its velocity:

$$\text{linear momentum} = \text{mass} \times \text{velocity}$$

Linear momentum is the tendency of an object to keep moving in a straight line in the absence of external forces. Picture a truck and a bicycle rolling equally fast down a street. Each has some linear momentum, but you would obviously find it easier to stop the less massive bicycle. Although the two vehicles have the same speed, the truck has more momentum. We see, then, that the linear momentum of an object depends on the mass of the object. It also depends on the speed of the object: If two bicycles were rolling down the street at different speeds, the slower one could be stopped more easily.

Angular momentum is an analogous property of objects that are rotating or revolving. It is a measure of the object's tendency to keep spinning, or, equivalently, of how much effort must be expended to stop the object from spinning. However, in addition to mass and (angular) speed, angular momentum also depends on the way in which an object's mass is distributed.

Intuitively, we know that the more massive an object, or the larger it is, or the faster it spins, the harder it is to stop. In fact, angular momentum depends on the object's *mass, rotation rate* (measured in, say, revolutions per second), and *radius,* in a very specific way:

$$\text{angular momentum} \propto \text{mass} \times \text{rotation rate} \times \text{radius}^2.$$

(Recall that the symbol $\propto$ means "is proportional to"; the constant of proportionality depends on the details of how the object's mass is distributed.)

According to Newton's laws of motion, both types of momentum—linear and angular—must be *conserved* at all times. ∞ (Sec. 2.8) In other words, both linear and angular momentum must remain *constant* before, during, and after a physical change in any object (so long as no external forces act on the object). For example, as illustrated in the first figure, if a spherical object having some spin begins to contract, the previous relationship demands that it spin faster so that the product mass × angular speed × radius² remains constant. The sphere's mass does not change during the contraction, yet the size of the object clearly decreases. Its rotation speed must therefore increase in order to keep the total angular momentum unchanged. This constancy is referred to as **conservation of angular momentum.**

Figure skaters use the principle of conservation of angular momentum, too. They spin faster by drawing in their arms (as shown in the second pair of figures) and slow down by extending them. Here, the mass of the human body remains the same, but its lateral size changes, causing the body's rotation speed to increase or decrease, as the case may be, to keep its angular momentum unchanged.

(Orban/Corbis/Sygma)

**EXAMPLE 1** Suppose the sphere has radius 1 m and starts off rotating at 1 revolution per minute. It then contracts to one-tenth its initial size. Conservation of angular momentum entails that the sphere's final angular speed $A$ must satisfy the relationship

$$\text{mass} \times A \times (0.1\,\text{m})^2 = \text{mass} \times (1\,\text{rev/min}) \times (1\,\text{m})^2.$$

The mass is the same on either side of the equation and therefore cancels, so we find that $A = (1\,\text{rev/min}) \times (1\,\text{m})^2/(0.1\,\text{m})^2 = 100$ rev/min, or about 1.7 rev/s.

**EXAMPLE 2** Now suppose that the "sphere" is a large interstellar gas cloud that is about to collapse and form the solar nebula. Initially, let's imagine that it has a diameter of 1 light-year and that it rotates very slowly—once every 10 million years. Assuming that the cloud's mass stays constant, its rotation rate must *increase* to conserve angular momentum as the radius decreases. By the time it has collapsed to a diameter of 100 AU, the cloud has shrunk by a factor of (1 light-year/100 AU) $\approx$ 630. Conservation of angular momentum then implies that the cloud's (average) spin rate has increased by a factor of $630^2 \approx 400,000$, to roughly 1 revolution every 25 years—about the orbital period of Saturn.

Incidentally, the law of conservation of angular momentum also applies to planetary orbits (where the radius is now the distance from the planet to the Sun). In fact, Kepler's second law *is* just conservation of angular momentum, expressed another way. ∞ (Sec. 2.5)

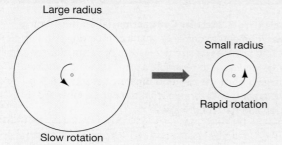

Large radius

Slow rotation

Small radius

Rapid rotation

the condensation theory adds to the nebular picture an additional basic fact about conditions in the disk while planets were forming. The Sun was forming at the disk's center, releasing huge amounts of energy as it contracted and grew hotter (see Chapter 19), *heating* the solar nebula in the process.

Planets could form only out of material that could solidify, or *condense*, from the nebular gas at any given location. Close to the Sun, the fierce heat meant that only *rocky* and *metallic* material could condense out, and the composition of the inner planets and asteroids reflects that fact. At greater distances, the temperature was lower, and *icy* material could form too. The far greater abundance of light elements in the solar nebula led to the jovian planets and the icy outer comets, moons, and Kuiper belt objects found in the outer solar system.

Astronomers are fairly confident that the solar nebula formed a disk, because similar disks have been observed (or inferred) around other stars. Figure 6.18(a) shows visible-light images of the region around a star called Beta Pictoris, lying about 60 light-years from the Sun. When the light from Beta Pictoris itself is suppressed and the resulting image enhanced by a computer, a faint disk of warm matter (viewed almost edge-on here) can be seen. This particular disk is roughly 1000 AU across—about 10 times the diameter of the Kuiper belt. Astronomers think that Beta Pictoris is a very young star, perhaps only 20 million years old and that we are witnessing it pass through an evolutionary stage similar to the one our own Sun underwent long ago. Figure 6.18(b) shows an artist's conception of the disk.

Theory indicates that the initial fragments of preplanetary material collided and coalesced to form larger bodies, ultimately leading to the solar system we see today. In the following chapters we will present and discuss the evidence for the condensation theory of solar system formation, including this important prediction that the early solar system was a very violent place, dominated by collisions and mergers between growing planets and smaller bodies.

In Chapter 15, having studied our own planetary system, we will assess how the condensation theory holds up in the face of detailed observational data. Then we will turn to planetary systems beyond our own—more than 150 of which are currently known—and examine what it means for the prospects of finding Earth-like planets, and maybe even life (Chapter 28), elsewhere in the Galaxy.

## CONCEPT CHECK

✔ How does the condensation theory account for the circular, planar orbits of the planets and the broad differences between terrestrial and the jovian worlds?

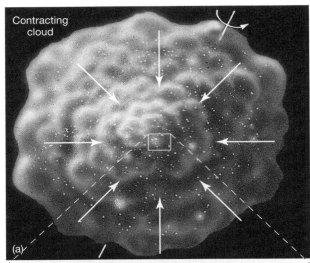

Contracting cloud

(a)

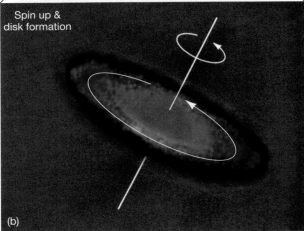

Spin up & disk formation

(b)

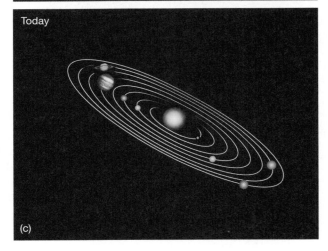

Today

(c)

▲ FIGURE 6.17 **Nebular Contraction** (a) Conservation of angular momentum demands that a contracting, rotating cloud must spin faster as its size decreases. (b) Eventually, the primitive solar system came to resemble a giant spinning pancake. The large blob at the center ultimately became the Sun. (c) The planets that formed from the nebula inherited its rotation and flattened shape.

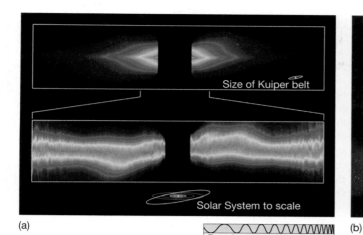

(a)    (b)

▲ **FIGURE 6.18** **Beta Pictoris** (a) A computer-enhanced view of a disk of warm matter surrounding the star Beta Pictoris. Both images in (a) display data taken at visible wavelengths, but are presented here in false color to accentuate the details; the bottom image is a close-up of the inner part of the disk, whose warp is possibly caused by the gravitational pull of unseen companions. In both images, the overwhelmingly bright central star has been covered to let us see the much fainter disk surrounding it. The disk is nearly edge-on to our line of sight. For scale, the dimension of the Kuiper belt (100 AU) has been drawn adjacent to the images. (b) An artist's conception of the disk of clumped matter, showing the warm disk with a young star at the center and several comet-sized or larger bodies already forming. The colors are thought to be accurate: At the outer edges of the disk, the temperature is low and the color is dull red. Progressing inward, the colors brighten and shift to a more yellowish tint as the temperature increases. Mottled dust is seen throughout—such protoplanetary regions are probably very "dirty." *(NASA; D. Berry)*

# CHAPTER REVIEW

## Summary

❶ The **solar system (p. 146)** consists of the Sun and everything that orbits it, including the eight major planets, the moons that orbit them, and the many small bodies found in interplanetary space. The science of **comparative planetology (p. 147)** compares and contrasts the properties of the diverse bodies found in the solar system and elsewhere in order to understand better the conditions under which planets form and develop.

❷ The major planets orbit the Sun in the same sense—counterclockwise as viewed from above Earth's North Pole—on roughly circular orbits that lie close to the plane of the ecliptic. The orbit of the innermost planet, Mercury, is the most eccentric and has the greatest orbital inclination. The spacing between planetary orbits increases as we move outward from the Sun. The diameter of Neptune's orbit is roughly 60 AU.

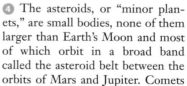

❸ The average **density (p. 149)** of a planet is obtained by dividing the planet's total mass by its volume. The innermost four planets in the solar system—Mercury, Venus, Earth, and Mars—have average densities comparable to Earth's and are generally rocky in composition. They are called the **terrestrial planets (p. 152).** The outer **jovian planets (p. 152)**—Jupiter, Saturn, Uranus, and Neptune—have much lower densities than the terrestrial worlds and are made up mostly of gaseous or liquid hydrogen and helium. Compared with the terrestrial worlds, the jovian planets are larger and more massive, rotate more rapidly, and have stronger magnetic fields. In addition, the jovian planets all have ring systems and many moons orbiting them.

❹ The asteroids, or "minor planets," are small bodies, none of them larger than Earth's Moon and most of which orbit in a broad band called the asteroid belt between the orbits of Mars and Jupiter. Comets are chunks of ice found chiefly in the outer solar system. Their importance to planetary astronomy lies in the fact that they are thought to be "leftover" material from the formation of the solar system and therefore contain clues to the very earliest stages of its development. The Kuiper belt is a band of icy bodies orbiting beyond the orbit of Neptune.

❺ All the major planets have been visited by unmanned space probes. Spacecraft have landed on Venus and Mars.

In many cases, to reach their destinations, the spacecraft were set on trajectories that included "gravitational assists" from one or more planets. The information returned by these probes has been vital to furthering astronomers' understanding of the solar system.

**6** According to the **nebular theory (p. 162)** of the formation of the solar system, a large cloud of dust and gas—the **solar nebula (p. 162)**—began to collapse under its own gravity. As it did so, it began to spin faster, to conserve angular momentum, eventually forming a disk out of which the planets arose. Planets formed in that rotating disk.

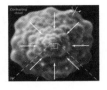

**7** The **condensation theory (p. 162)** builds on the nebular theory by considering the effects of solar heating on the planet formation process. At any given location, the temperature in the solar nebula determined which materials could condense out and hence also determined the composition of any planets forming there. The terrestrial planets are rocky because they formed in the hot inner regions of the solar nebula, near the Sun, where only rocky and metallic materials condensed out. Farther out, the nebula was cooler, and ices could also form, ultimately leading to the observed differences in composition between the inner and outer solar system.

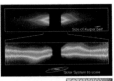

## Review and Discussion

1. Name and describe all the different types of objects found in the solar system. Give one distinguishing characteristic of each. Include a mention of interplanetary space.

2. What is comparative planetology? Why is it useful? What is its ultimate goal?

3. Why is it necessary to know the distance to a planet in order to determine the planet's mass?

4. List some ways in which the solar system is an "orderly" place.

5. Name some "disorderly" characteristics of the solar system.

6. Which are the terrestrial planets? Why are they given that name?

7. Which are the jovian planets? Why are they given that name?

8. Name three important differences between the terrestrial planets and the jovian planets.

9. Compare the properties of Pluto given in Table 6.1 with the properties of the terrestrial and jovian planets presented in Table 6.2. What do you conclude regarding the classification of Pluto as either a terrestrial or jovian planet?

10. Why are asteroids and meteoroids important to planetary scientists?

11. Comets generally vaporize upon striking Earth's atmosphere. How, then, do we know their composition?

12. Why has our knowledge of the solar system increased greatly in recent years?

13. How and why do scientists use gravity assists to propel spacecraft through the solar system?

14. Which planets have been visited by spacecraft from Earth? On which ones have spacecraft actually landed?

15. Why do you think *Galileo* and *Cassini* took such circuitous routes to Jupiter and Saturn, while *Pioneer* and *Voyager* did not?

16. How do you think NASA's new policy of building less complex, smaller, and cheaper spacecraft—with shorter times between design and launch—will affect the future exploration of the outer planets? Will missions like *Galileo* and *Cassini* be possible in the future?

17. What is the key ingredient in the modern condensation theory of the solar system's origin that was missing or unknown in the nebular theory?

18. Give three examples of how the condensation theory explains the observed features of the present-day solar system.

19. Why do you think the jovian planets are so much more massive than the terrestrial planets?

20. How did the temperature structure of the solar nebula determine planetary composition?

## Conceptual Self-Test: True or False/Multiple Choice

1. Most planets orbit the Sun in nearly the same plane as Earth does.

2. The largest planets also have the largest densities.

3. The total mass of all the planets is much less than the mass of the Sun.

4. The jovian planets all rotate more rapidly than Earth.

5. Saturn is the largest and most massive planet in the solar system.

6. Asteroids are similar in overall composition to the terrestrial planets.

7. Comets have compositions similar to the icy moons of the jovian planets.

8. All planets have moons.

9. Both *Voyager* missions used gravity assists to visit all four jovian planets.

10. Interstellar dust plays a key role in the formation of a planetary system.

11. A planet's mass can most easily be determined by measuring the planet's (a) moon's orbits; (b) angular diameter; (c) position in the sky; (d) orbital speed around the Sun.

12. If we were to construct an accurate scale model of the solar system on a football field with the Sun at one end and Neptune at the other, the planet closest to the center of the field would be (a) Earth; (b) Jupiter; (c) Saturn; (d) Uranus.

13. The inner planets tend to have (a) fewer moons; (b) faster rotation rates; (c) stronger magnetic fields; (d) higher gravity than the outer planets have.

14. The planets that have rings also tend to have (a) solid surfaces; (b) many moons; (c) slow rotation rates; (d) weak gravitational fields.

15. A solar system object of rocky composition and comparable in size to a small city is most likely (a) a meteoroid; (b) a comet; (c) an asteroid; (d) a planet.

16. The asteroids are mostly (a) found between Mars and Jupiter; (b) just like other planets, only younger; (c) just like other planets, only smaller; (d) found at the very edge of our solar system.

17. The *Sojourner* Mars rover, part of the Mars *Pathfinder* mission in 1997, was able to travel over an area about the size of (a) a soccer field; (b) a small U.S. city; (c) a very large U.S. city; (d) a small U.S. state.

18. To travel from Earth to the planet Neptune at more than 30,000 mph, the *Voyager 2* spacecraft took nearly (a) a year; (b) a decade; (c) three decades; (d) a century.

19. In the leading theory of solar system formation, the planets (a) were ejected from the Sun following a close encounter with another star; (b) formed from the same flattened, swirling gas cloud that formed the Sun; (c) are much younger than the Sun; (d) are much older than the Sun.

20. The solar system is differentiated because (a) all the heavy elements in the outer solar system have sunk to the center; (b) all the light elements in the inner solar system became part of the Sun; (c) all the light elements in the inner solar system were carried off in the form of comets; (d) only rocky and metallic particles could form close to the Sun.

## Problems

*Algorithmic versions of these Problems are available in the Practice Problems module of the Companion Website.*
*The number of dots preceding each Problem indicates its approximate level of difficulty.*

1. ●●●  Use the data given in Table 6.1 to calculate the angular diameter of Saturn when it lies 9 AU from Earth. Saturn's moon Titan is observed to orbit 3.1' from the planet. What is Titan's orbital period?

2. ●  Use Newton's law of gravity to compute your weight (a) on Earth, (b) on Mars, (c) on the asteroid Ceres, and (d) on Jupiter (neglecting temporarily the absence of a solid surface on this planet!). ∞ (Sec. 2.7)

3. ●  Only planets Mercury and Mars have orbits that deviate significantly from circles. Calculate the perihelion and aphelion distances of these planets from the Sun. ∞ *(More Precisely 2-1)*

4. ●●●  At closest approach, the planet Neptune lies roughly 29.1 AU from Earth. At that distance, Neptune's angular diameter is 2.3" Its moon Triton moves in a circular orbit with an angular diameter of 33.6" and a period of 5.9 days. Use these data to compute the radius, mass, and density of Neptune, and compare your results with the figures given in Table 6.1.

5. ●  Suppose the average mass of each of the 7000 asteroids in the solar system is about $10^{17}$ kg. Compare the total mass of all asteroids with the mass of Earth.

6. ●  Assuming a roughly spherical shape and a density of 3000 kg/m$^3$, estimate the diameter of an asteroid having the average mass given in the previous question.

7. ●●  A *short-period* comet is conventionally defined as a comet having an orbital period of less than 200 years. What is the maximum possible aphelion distance for a short-period comet with a perihelion of 0.5 AU? Where does this place the comet relative to the outer planets?

8. ●  How many times has *Mariner 10* now orbited the Sun?

9. ●●●  A spacecraft has an orbit that just grazes Earth's orbit at perihelion and the orbit of Mars at aphelion. What are the orbital eccentricity and semimajor axis of the orbit? How long does it take to go from Earth to Mars? (The orbit given is the so-called *minimum-energy orbit* for a craft leaving Earth and reaching Mars. Assume circular planetary orbits for simplicity.)

10. ●●  The asteroid Icarus has a perihelion distance of 0.2 AU and an orbital eccentricity of 0.7. What is Icarus's aphelion distance from the Sun?

11. ●●●  Earth and Mars were at closest approach in September 2003. The first *Mars Exploration Rover* was launched in June 2003 and arrived at Mars in January 2004. Sketch the orbits of the two planets and the trajectory of the spacecraft. Be sure to indicate the location of Mars at launch and of Earth when the spacecraft reached Mars. (For simplicity, neglect the eccentricity of Mars's orbit in your sketch.)

12. ●●  How long would it take for a radio signal to complete the round-trip between Earth and Saturn? Assume that Saturn is at its closest point to Earth. How far would a spacecraft orbiting the planet in a circular orbit of radius 100,000 km travel in that time? Do you think that mission control could maneuver the spacecraft in real time—that is, control all its functions directly from Earth?

13. ●●  An interstellar cloud fragment 0.2 light-year in diameter is rotating at a rate of one revolution per million years. It now begins to collapse. Assuming that the mass remains constant, estimate the cloud's rotation period when it has shrunk to (a) the size of the solar nebula, 100 AU across, and (b) the size of Earth's orbit, 2 AU across.

14. ●●  By what factor would Earth's rotational angular momentum change if the planet's spin rate were to double? By what factor would Earth's orbital angular momentum change if the planet's distance from the Sun were to double (assuming that the orbit remained circular)? The orbital angular momentum of a planet in a circular orbit is simply the product of the planet's mass, orbital speed, and distance from the Sun.

15. ●●  Consider a planet growing as material falls onto it from the solar nebula. As the planet grows, its density remains roughly constant. Does the force of gravity at the surface of the planet increase, decrease, or stay the same? Specifically, what would happen to the surface gravity and escape speed as the radius of the planet doubled? Give reasons for your answer.

*The Companion Website at www.aw-bc.com/chaisson provides algorithmically generated versions of each chapter's Problems, along with additional quizzes, an Animations & Videos gallery, an Images gallery, an interactive Glossary, and a full eBook.*

*Photographs like this one, showing Earth hovering in space like a "blue marble," help us appreciate our place in the universe. The air, water, land, and life comprise a complex, interactive system that constantly changes. Scientists are now trying to understand not only the details of regional events, such as volcanoes, earthquakes, and weather, but also to learn more about the global changes affecting the whole planet. This image is a mosaic of many photographs taken by the GOES-7 environmental satellite. Note the hurricane in the Gulf of Mexico. (NOAA/NASA)* ▶

# EARTH

## Our Home in Space

E arth is the best-studied terrestrial planet. From the matter of our world sprang life, intelligence, culture, and all the technology we now use to explore the cosmos. We ourselves are "Earthstuff" as much as are rocks, trees, and air. Now, as humanity begins to explore the solar system, we can draw on our knowledge of Earth to aid our understanding of the other planets.

By cataloging Earth's properties and attempting to explain them, we set the stage for our comparative study of the solar system. Every piece of information we glean about the structure and history of our own world may play a vital role in helping us understand the planetary system in which we live. If we are to appreciate the universe, we must first come to know our own planet. Our study of astronomy begins at home.

## LEARNING GOALS

*Studying this chapter will enable you to*

1 Summarize the physical properties of planet Earth.
2 Explain how Earth's atmosphere helps to heat us, as well as protect us.
3 Outline our current model of Earth's interior and describe some of the experimental techniques used to establish the model.
4 Summarize the evidence for the phenomenon of "continental drift" and discuss the physical processes that drive it.
5 Discuss the nature and origin of Earth's magnetosphere.
6 Describe how both the Moon and the Sun influence Earth's surface and affect our planet's spin.

 Visit www.aw-bc.com/chaisson for additional images, animations, quizzes, and eBook for this chapter.

## 7.1   Overall Structure of Planet Earth

The Earth Data box on p. 172 lists some of Earth's physical and orbital properties in detail. The data shown are determined with techniques that are conceptually similar to those presented in Chapter 6: using simple geometry to determine Earth's radius, the orbit of the Moon to measure our planet's mass, and so on. ∞ (Sec. 6.2) Throughout the body of this text, we will use rounded-off numbers whenever possible, taking our planet's mass and radius to be $6.0 \times 10^{24}$ kg and 6400 km, respectively.

Dividing mass by volume, we find that Earth's average density is around 5500 kg/m$^3$. This simple calculation allows us to make a very important deduction about the interior of our planet. From direct measurements, we know that the water that makes up much of Earth's surface has a density of 1000 kg/m$^3$, and the rock beneath us on the continents, as well as on the seafloor, has a density in the range from 2000 to 4000 kg/m$^3$. We can immediately conclude that, because the surface layers have densities much less than the average, much denser material must lie deeper, under the surface. Hence, we should expect that much of Earth's interior is made up of very dense matter, far more compact than the densest continental rocks on the surface.

On the basis of measurements made in many different ways—using aircraft in the atmosphere, satellites in orbit, gauges on the land, submarines in the ocean, and drilling gear below the rocky crust—scientists have built up the following overall picture of our planet: As indicated in Figure 7.1, Earth may be divided into six main regions. In Earth's interior, a thick **mantle** surrounds a smaller, two-part **core.** At the surface, we have (1) a relatively thin **crust,** comprising the solid continents and the seafloor, and (2) the **hydrosphere,** which contains the liquid oceans and accounts for some 70 percent of our planet's total surface area. An **atmosphere** of air lies just above the surface. At much greater altitudes, a zone of charged particles trapped by Earth's magnetic field forms the **magnetosphere.** Virtually all our planet's mass is contained within the surface and the interior. The gaseous atmosphere and the magnetosphere contribute hardly anything—less than 0.1 percent—to the total.

## 7.2   Earth's Atmosphere

From a human perspective, probably the most important aspect of Earth's atmosphere is that we can breathe it. Air is a mixture of gases, the most common of which are nitrogen (N$_2$, 78 percent by volume), oxygen (O$_2$, 21 percent), argon (Ar, 0.9 percent), and carbon dioxide (CO$_2$, 0.03 percent). The amount of water vapor (H$_2$O) varies from 0.1 to 3 percent, depending on location and climate. The presence of a large amount of oxygen makes our atmosphere unique in the solar system, and the presence of even trace amounts of water and carbon dioxide play vital roles in the workings of our planet.

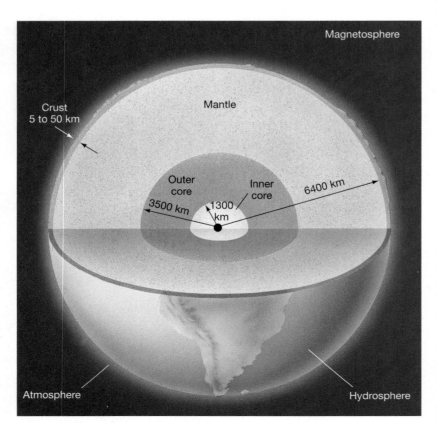

◄ FIGURE 7.1   **The Main Regions of Planet Earth** At the center lies our planet's solid inner core, about 2600 km in diameter, and surrounding this is a liquid outer core, some 7000 km across. Most of the rest of Earth's 13,000-km interior is taken up by the mantle, which is topped by a thin crust only a few tens of kilometers thick. The liquid portions of Earth's surface make up the hydrosphere. Above the hydrosphere and solid crust lies the atmosphere, most of it within 50 km of the surface. Earth's outermost region is the magnetosphere, extending thousands of kilometers into space.

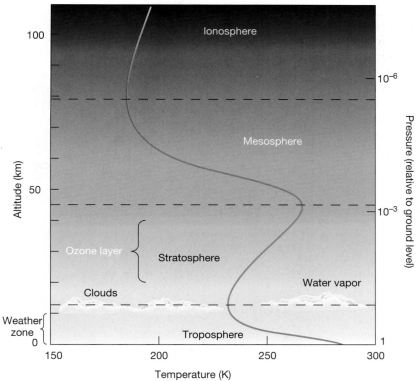

◀ **FIGURE 7.2 Earth's Atmosphere** Diagram of Earth's atmosphere showing the changes in temperature (blue curve, bottom axis) and pressure (right-hand axis) from the planet's surface to the bottom of the ionosphere. Pressure decreases steadily with increasing altitude, but the temperature may fall or rise, depending on height above the ground.

by solar ultraviolet radiation. Note how the temperature gradient (decreasing or increasing with altitude) changes from one atmospheric region to the next.

Atmospheric density decreases steadily with increasing altitude, and as the right-hand vertical axis in Figure 7.2 shows, so does pressure. Climbing even a modest mountain—4 or 5 km high, say—clearly demonstrates the thinning of the air in the troposphere. Climbers must wear oxygen masks when scaling the tallest peaks on Earth.

The troposphere is the region of Earth's (or any other planet's) atmosphere where *convection* occurs, driven by the heat of Earth's warm surface. **Convection** is the constant upwelling of warm air and the concurrent downward flow of cooler air to take its place, a process that physically transfers heat from a lower (hotter) to a higher (cooler) level. In Figure 7.3, part of Earth's surface is heated by the Sun. The air immediately above the warmed surface is heated, expands a little, and becomes less dense. As a result, the hot air becomes buoyant and starts to rise. At higher altitudes, the opposite effect occurs: The air gradually cools, grows denser, and

## Atmospheric Structure

Figure 7.2 shows a cross section of our planet's atmosphere. Compared with Earth's overall dimensions, the extent of the atmosphere is not great. Half of it lies within 5 km of the surface, and all but 1 percent is found below 30 km. The portion of the atmosphere below about 12 km is called the **troposphere.** Above it, extending up to an altitude of 40 to 50 km, lies the **stratosphere.** Between 50 and 80 km from the surface is the **mesosphere.** Above about 80 km, in the **ionosphere,** the atmosphere is kept partly ionized

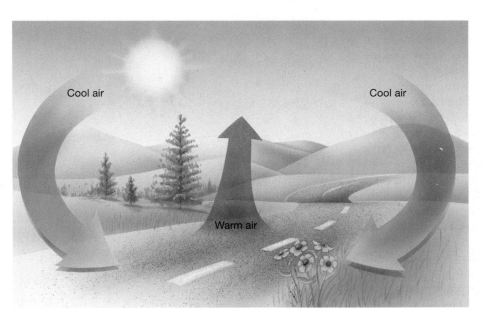

◀ **FIGURE 7.3 Convection** Convection occurs whenever cool matter overlies warm matter. The resulting circulation currents are familiar to us as the winds in Earth's atmosphere, caused by the solar-heated ground. Hot air rises, cools, and falls repeatedly back to Earth. Eventually, steady circulation patterns are established and maintained, provided that the source of heat (the Sun, in the case of Earth) remains intact.

sinks back to the ground. Cool air at the surface rushes in to replace the hot buoyant air. In this way, a circulation pattern is established. These *convection cells* of rising and falling air not only contribute to atmospheric heating, but also are responsible for surface winds. The constant churning motion in convection cells is responsible for all the weather we experience.

Atmospheric convection can also create clear-air turbulence—the bumpiness we sometimes experience on aircraft flights. Ascending and descending parcels of air, especially below fluffy clouds (themselves the result of convective processes, when water vapor condenses out at the cool tops of convection cells), can cause a choppy ride. For this reason, passenger aircraft tend to fly above most of the turbulence, at the top of the troposphere or in the lower stratosphere, where the atmosphere is stable and the air is calm.

Above about 100 km, in the ionosphere, the atmosphere is significantly ionized by the high-energy portion of the Sun's radiation spectrum, which breaks down molecules into atoms and atoms into ions. The degree of ionization increases with altitude. The presence of many free electrons makes this region of the upper atmosphere a good conductor of electricity, and the conductivity renders the ionosphere highly reflective to certain radio wavelengths. ∞ (Sec. 3.3) The reason that AM radio stations can be heard well beyond the horizon is that their signals bounce off the ionosphere before reaching a receiver. FM signals cannot be received from stations over the horizon, however, because the ionosphere is transparent to the somewhat shorter wavelengths of radio waves in the FM band.

## Atmospheric Ozone

Within the stratosphere is the **ozone layer,** where, at an altitude of around 25 km, incoming solar ultraviolet radiation is absorbed by atmospheric ozone and nitrogen. (Ozone [$O_3$] consists of three oxygen atoms combined into a single molecule. Ultraviolet radiation breaks ozone down, forming molecular oxygen [$O_2$] again.)

The ozone layer is one of the insulating spheres that serve to shield life on Earth from the harsh realities of outer space. Not so long ago, scientists judged space to be hostile to advanced life-forms because of what is missing out there: breathable air and a warm environment. Now most scientists regard outer space as harsh because of what is *present* out there: fierce radiation and energetic particles, both of which are injurious to human health. Without the protection of the ozone layer, advanced life (at least on Earth's surface) would be at best unlikely and at worst impossible.

Human technology has reached the point where it has begun to produce measurable—and possibly permanent—changes to our planet. One particularly undesirable by-product of our ingenuity is a group of chemicals known as

## PLANETARY DATA

### EARTH

*(NOAA/NASA)*

| | |
|---|---|
| Orbital semimajor axis ∞ (Sec 2.5) | 1.00 AU 149.6 million km |
| Orbital eccentricity | 0.017 |
| Perihelion ∞ (*More Precisely 2-1*) | 0.98 AU 147.1 million km |
| Aphelion | 1.02 AU 152.1 million km |
| Mean orbital speed | 29.79 km/s |
| Sidereal orbital period* | 1.000038 tropical years |
| Orbital inclination to the ecliptic | 0.01° |
| Mass | 5.976 × 10²⁴ kg |
| Equatorial radius | 6378 km |
| Mean density | 5520 kg/m³ |
| Surface gravity** | 9.80 m/s² |
| Escape speed | 11.2 km/s |
| Sidereal rotation period | 0.9973 solar day |
| Axial tilt*** | 23.45° |
| Magnetic axis tilt relative to rotation axis | 11.5° |
| Mean surface temperature | 290 K |
| Number of moons | 1 |

*Relative to the distant stars.*
**Acceleration due to gravity at Earth's surface.*
***Angle between Earth's axis of rotation and perpendicular to Earth's orbital plane (the plane of the ecliptic).*

chlorofluorocarbons (CFCs), relatively simple compounds once widely used for a variety of purposes—propellant in aerosol cans, solvents in dry-cleaning products, and coolant in air conditioners and refrigerators. In the 1970s, it was discovered that, instead of quickly breaking down after use, as had previously been thought, CFCs accumulate in the atmosphere and are carried high into the stratosphere by convection. There they are broken down by sunlight, releasing chlorine, which quickly reacts with ozone, turning it into oxygen. In chemical terms, the chlorine is said to act as a *catalyst*—it is not consumed in the reaction, so it survives to react with many more ozone molecules. A single chlorine atom can destroy up to 100,000 ozone molecules before being removed by other, less frequent chemical reactions.

Thus, even a small amount of CFCs is extraordinarily efficient at destroying atmospheric ozone, and the net result of CFC emission is a substantial increase in ultraviolet radiation levels at Earth's surface, with detrimental effects to most living organisms. Figure 7.4 shows a vast ozone "hole" over the Antarctic. The hole is a region where atmospheric circulation and low temperatures conspire each Antarctic spring to create a vast circumpolar cloud of ice crystals that act to promote the ozone-destroying reactions, resulting in ozone levels about 50 percent below

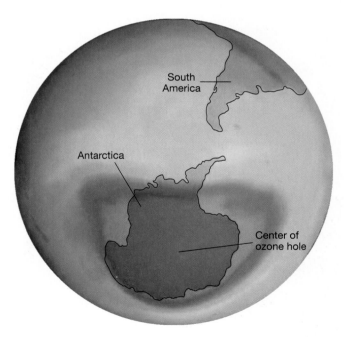

▲ **FIGURE 7.4 Antarctic Ozone Hole** This composite image constructed from satellite observations shows (in pink) a huge "hole" in the ozone layer over the Antarctic continent. The hole is a region where climatic conditions and human-made chemicals combine to rob our atmosphere of its protective ozone blanket. The depth and area of the hole have grown significantly since the hole was discovered in the 1980s. Its maximum size is now larger than North America.

normal for the region. Ozone depletion is not confined to the Antarctic, although the effect is greatest there. Smaller holes have been observed in the Arctic, and occasional ozone depletions of up to 20 percent have been reported at lower northern latitudes.

In the late 1980s, when the effects of CFCs on the atmosphere were realized, the world moved rapidly to curtail their production and use, with the goal of phasing them out entirely by 2030. Substantial cuts have already been made, and the agreement to do so has become a model of international cooperation. Still, scientists think that, even if all remaining CFC emissions were to stop today, it would nonetheless take several decades for CFCs to leave the atmosphere completely.

## Surface Heating

Much of the Sun's radiation manages to penetrate Earth's atmosphere, eventually reaching the ground. (See *More Precisely 7-1* for more on how the atmosphere affects incoming sunlight before it reaches the surface.) Most of this energy takes the form of visible and near-infrared radiation—ordinary sunlight. ∞ (Sec. 3.3) Essentially all of the solar radiation that is not absorbed by or reflected from clouds in the upper atmosphere is absorbed by Earth's surface. The result is that our planet's surface and most objects on it heat up considerably during the day. Earth cannot absorb this solar energy indefinitely, however. If it did, the surface would soon become hot enough to melt, and life on our planet would not exist.

As it heats up, Earth's surface reradiates much of the absorbed energy. This reemitted radiation follows the blackbody curve discussed in Chapter 3. ∞ (Sec. 3.4) As the surface temperature rises, the amount of energy radiated increases rapidly, in accordance with Stefan's law. Eventually, Earth radiates as much energy back into space as it receives from the Sun, and a stable balance is struck. In the absence of any complicating effects, this balance would be achieved at an average surface temperature of about 250 K (−23°C). Wien's law tells us that, at that temperature, most of the reemitted energy is in the form of infrared (heat) radiation.

But there *are* complications. Infrared radiation is partially blocked by Earth's atmosphere, primarily because of the presence of molecules of water vapor and carbon dioxide, which absorb very efficiently in the infrared portion of the spectrum. Even though these two gases are only trace constituents of our atmosphere, they manage to absorb a large fraction of all the infrared radiation emitted from the surface. Consequently, only some of that radiation escapes back into space. The remainder is trapped within our atmosphere, causing the temperature to increase.

This partial trapping of solar radiation is known as the **greenhouse effect.** The name comes from the fact that a similar process operates in a greenhouse, where sunlight passes relatively unhindered through glass panes, but

# MORE PRECISELY 7-1

## Why Is the Sky Blue?

Is the sky blue because it reflects the color of the ocean, or is the ocean blue because it reflects the color of the surrounding sky? The answer is the latter, and the reason has to do with the way that light is *scattered* by air molecules and minute dust particles. By *scattering*, we mean the process by which radiation is absorbed and then reradiated by the material through which it passes.

As sunlight passes through our atmosphere, it is scattered by gas molecules in the air. The British physicist Lord Rayleigh first investigated this phenomenon about a century ago, and today it bears his name—it is known as *Rayleigh scattering*. The process turns out to be highly sensitive to the wavelength of the light involved.

Rayleigh found that blue light is much more easily scattered than red light, basically because the wavelength of blue light (400 nm) is closer to the size of air molecules than the wavelength of red light (700 nm). He went on to prove mathematically, on the basis of the laws of electromagnetism, that the amount of scattering is inversely proportional to the *fourth* power of the wavelength:

$$\text{scattering by molecules} \propto \frac{1}{\text{wavelength}^4}.$$

Rayleigh's formula applies to scattering by particles (such as molecules) that are smaller than the wavelength of the light involved. Larger particles, such as dust, also preferentially scatter blue light, but by an amount that depends only inversely on the wavelength:

$$\text{scattering by dust} \propto \frac{1}{\text{wavelength}}.$$

---

**EXAMPLE 1** Let's compare the relative scattering of blue (400 nm) and red (700 nm) light by atmospheric molecules and dust. For Rayleigh scattering, blue light is scattered $(700/400)^4 \approx 9.4$ times more efficiently than red light. That is, blue photons are almost 10 times more likely to be scattered out of a beam of sunlight (taken out of the forward beam and redirected to the side) than are red photons. For scattering by dust, the corresponding factor is $(700/400) = 1.75$—not as big a differential, but still enough to have a large effect when the air happens to be particularly dirty.

---

When the Sun is high in the sky, the blue component of incoming sunlight is scattered much more than any other color component. Thus, some blue light is removed from the line of sight between us and the Sun and may scatter many times in the atmosphere before eventually entering our eyes, as shown in the first figure. Red or yellow light is scattered relatively little and arrives at our eyes predominantly along the line of sight to the Sun. The net effect is that the Sun is "reddened" slightly, because of the removal of blue light, whereas the sky away from the Sun appears blue. In outer space, where there is no atmosphere, there is no Rayleigh scattering of sunlight, and the sky is black (although, as we will see in Chapter 18, light from distant stars is reddened in precisely the same way as it passes through clouds of interstellar gas and dust).

At dawn or dusk, with the Sun near the horizon, sunlight must pass through much more atmosphere before reaching our eyes—so much so, in fact, that the blue component of the Sun's light is almost entirely scattered out of the line of sight, and even the red component is diminished in intensity. Accordingly, the Sun itself appears orange—a combination of its normal yellow color and a reddishness caused by the subtraction of virtually all of the blue end of the spectrum—and dimmer than at noon.

At the end of a particularly dusty day (second figure), when weather conditions or human activities during the daytime hours have raised excess particles into the air, short-wavelength Rayleigh scattering can be so heavy that the Sun appears brilliantly red. Reddening is often especially evident when we look at the westerly "sinking" summer Sun over the ocean, where seawater molecules have evaporated into the air, or during the weeks and months after an active volcano has released huge quantities of gas and dust particles into the air—as was the case in North America when the Philippine volcano Mount Pinatubo erupted in 1991.

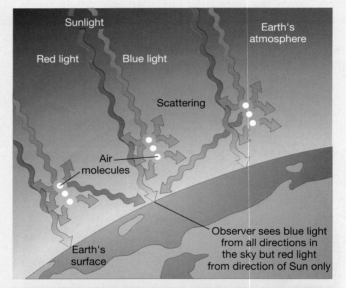

(NCAR)

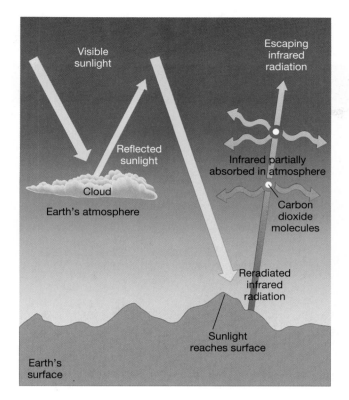

◄ **FIGURE 7.5 Greenhouse Effect** Sunlight that is not reflected by clouds reaches Earth's surface, warming it up. Infrared radiation reradiated from the surface is partially absorbed by the carbon dioxide (and also water vapor, not shown here) in the atmosphere, causing the overall surface temperature to rise.

global temperature increases of several kelvins over the next half century—enough to cause dramatic, and possibly catastrophic, changes in Earth's climate, ranging from rising sea levels to the accelerated spread of disease.

## Origin of Earth's Atmosphere

Why is our atmosphere made up of its present constituents? Why is it not composed entirely of nitrogen, say, or of carbon dioxide, like the atmospheres of Venus and Mars? The origin and development of Earth's atmosphere was a fairly complex and lengthy process.

When Earth first formed, any *primary atmosphere* it might have had would have consisted of the gases most common in the early solar system: hydrogen, helium, methane, ammonia, and water vapor—a far cry from the atmosphere we enjoy today. Almost all this low-density material, and especially any hydrogen or helium, escaped into space during the first half-billion or so years after Earth was formed. (For more information on how planets retain or lose their atmospheres, see *More Precisely 8-1.*)

Subsequently, Earth developed a *secondary atmosphere*, which was *outgassed* (expelled) from the planet's interior as a result of volcanic activity. Volcanic gases are rich in water vapor, methane, carbon dioxide, sulfur dioxide, and compounds containing nitrogen (such as nitrogen gas, ammonia, and nitric oxide). Solar ultraviolet radiation split the lighter, hydrogen-rich gases into their component atoms, allowing the hydrogen to escape and liberating much of the nitrogen from its bonds with other elements. As Earth's surface temperature fell, the water vapor condensed and oceans formed. Much of the carbon dioxide and sulfur dioxide became dissolved in the oceans or combined with surface rocks. Oxygen is such a reactive gas that any free oxygen that appeared at early times was removed as quickly as it formed. An atmosphere consisting largely of nitrogen slowly appeared.

The final major development in the story of our planet's atmosphere is known so far to have occurred only on Earth. *Life* appeared in the oceans more than 3.5 billion years ago, and organisms eventually began to produce atmospheric oxygen. The ozone layer formed, shielding the surface from the Sun's harmful radiation. Eventually, life spread to the land and flourished. The fact that oxygen is a major constituent of the present-day atmosphere is a direct consequence of the evolution of life on Earth.

much of the infrared radiation reemitted by the plants is blocked by the glass and cannot get out. Consequently, the interior of the greenhouse heats up, and flowers, fruits, and vegetables can grow even on cold wintry days.* The radiative processes that determine the temperature of Earth's atmosphere are illustrated in Figure 7.5. Earth's greenhouse effect makes our planet almost 40 K (40°C) hotter than would otherwise be the case.

The magnitude of the greenhouse effect is highly sensitive to the concentration of so-called **greenhouse gases** (that is, gases that absorb infrared radiation efficiently) in the atmosphere. Carbon dioxide and water vapor are the most important of these, although other atmospheric gases (such as methane) also contribute. The amount of carbon dioxide in Earth's atmosphere is increasing, largely as a result of the burning of fossil fuels (principally oil and coal) in the industrialized and developing worlds. Carbon dioxide levels have increased by over 20 percent in the last century, and they are continuing to rise at a present rate of 4 percent per decade. *Discovery 7-1* discusses the causes and some possible consequences of rising carbon dioxide levels in Earth's atmosphere.

In Chapter 9, we will see how a runaway increase in carbon dioxide levels in the atmosphere of the planet Venus radically altered conditions on its surface, causing its temperature to rise to over 700 K. Although no one is predicting that Earth's temperature will ever reach that of Venus, many scientists now think that our planet's increase in carbon dioxide levels, if left unchecked, may result in

*Note that, although this process does contribute to warming the interior of a greenhouse, it is not the most important effect. A greenhouse works mainly because its glass panes prevent convection from carrying heat up and away from the interior. Nevertheless, the name "greenhouse effect due to Earth's atmosphere has stuck.*

## CONCEPT CHECK

✔ Why is the greenhouse effect important for life on Earth?

# DISCOVERY 7-1

## The Greenhouse Effect and Global Warming

We saw in the text how greenhouse gases in Earth's atmosphere—notably, water vapor and carbon dioxide ($CO_2$)—tend to trap heat leaving the surface, raising our planet's temperature by several tens of degrees Celsius. In and of itself, this *greenhouse effect* is not a bad thing—in fact, it is the reason that water exists in the liquid state on Earth's surface, and thus it is crucial to the existence and survival of life on our planet (see Chapter 28). However, if atmospheric greenhouse gas levels rise unchecked, the consequences could be catastrophic.

Since the Industrial Revolution in the 18th century, and particularly over the past few decades, human activities on Earth have steadily raised the level of carbon dioxide in our atmosphere. Fossil fuels (coal, oil, and gas), still the dominant energy source of modern industry, all release $CO_2$ when burned. At the same time, the extensive forests that once covered much of our planet are being systematically destroyed to make room for human expansion. Forests play an important role in this situation because vegetation absorbs carbon dioxide, thus providing a natural control mechanism for atmospheric $CO_2$. Deforestation therefore also tends to increase the amount of greenhouse gases in Earth's atmosphere. The first figure shows atmospheric $CO_2$ levels over the past thousand years. Note the dramatic increase during the past two centuries.

*Global warming* is the slow rise in Earth's surface temperature caused by the increased greenhouse effect resulting from higher levels of atmospheric carbon dioxide. As shown in the second figure, average global temperatures have risen by about 0.5° C during the past century. This may not seem like much, but climate models predict that, if $CO_2$ levels continue to rise, a further increase of as much as 5° C is possible by the end of the 21st century. Such a rise would be enough to cause serious climate change on a global scale. Among the possible (some would say likely) consequences of such a temperature increase are the following phenomena.

- Melting of glaciers and the polar ice caps, leading to a rise in sea level of up to a meter by the year 2100, with the potential for widespread coastal flooding.

- Longer and more extreme periods of severe weather—heat waves, droughts and wildfires—yet with more precipitation (rain and snow) between them.

- Crop failures as Earth's temperate zones move toward the poles.

- Expansion of deserts in heavily populated equatorial regions.

- Increased numbers of mosquitoes and other pests spreading tropical diseases into unprotected populations.

The fossil record shows that major climatic changes have occurred many times before in Earth's history, but never at the rate predicted by these dire warnings. The unprecedented speed of the forecasted events may well be too rapid for many species (and some human societies) to survive.

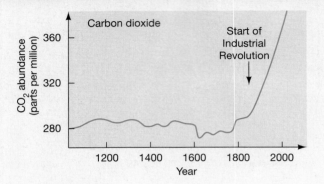

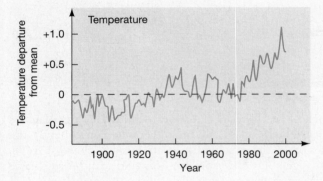

In Section 7.2, we described the danger to Earth's ozone layer posed by CFCs—another product of modern technology with unexpected global consequences—and saw how, once their environmental impact was identified, rapid steps were taken to curb their use. A concerted international response to global warming has been much slower in coming. Most scientists see the human-enhanced greenhouse effect as a real threat to Earth's climate, and they urge prompt and deep reductions in $CO_2$ emissions, along with steps to slow and ultimately reverse deforestation.

Some, however—particularly those connected with the industries most responsible for the production of greenhouse gases—argue that Earth's long-term response to increased greenhouse emissions is too complex for simple conclusions to be drawn and that immediate action is unnecessary. They suggest that the current temperature trend may be part of some much longer cycle or that natural environmental factors may in time stabilize, or even reduce, the level of $CO_2$ in the atmosphere without human intervention. In part because of these objections, international agreements to limit carbon emissions have failed to win approval in the United States (currently the largest producer of greenhouse gases), and thus have had only limited success.

Given the stakes, it is perhaps not surprising that these debates have become far more political than scientific in tone—not at all like the deliberative scientific method presented elsewhere in this text! The basic observations and much of the basic science are generally not seriously questioned, but the interpretation, long-term consequences, and proper response are all hotly debated. Separating the two sometimes is not easy, but the outcome may be of vital importance to life on Earth.

## 7.3 Earth's Interior

Although we reside on Earth, we cannot easily probe our planet's interior. Drilling gear can penetrate rock only so far before breaking. No substance used for drilling—even diamond, the hardest known material—can withstand the pressure below a depth of about 10 km. That's rather shallow compared with Earth's 6400-km radius. Fortunately, geologists have developed other techniques that indirectly probe the deep recesses of our planet.

### Seismic Waves

A sudden dislocation of rocky material near Earth's surface—an *earthquake*—causes the entire planet to vibrate a little. Earth rings like a giant bell. These vibrations are not random, however. They are systematic waves, called **seismic waves** (after the Greek word for "earthquake"), that move outward from the site of the quake. Like all waves, they carry information. This information can be detected and recorded with sensitive equipment—a *seismograph*—designed to monitor Earth tremors.

Decades of earthquake research have demonstrated the existence of many kinds of seismic waves. Two are of particular importance to the study of Earth's internal structure. First to arrive at a monitoring site after a distant earthquake are the primary waves, or *P-waves*. These are *pressure* waves, a little like ordinary sound waves in air, that alternately expand and compress the medium (the core or mantle) through which they move. Seismic P-waves usually travel at speeds ranging from 5 to 6 km/s and can travel through both liquids and solids. Some time later (the actual delay depends on the distance from the earthquake site), secondary waves, or *S-waves*, arrive. These are *shear* waves. Unlike P-waves, which vibrate the material through which they pass back and forth along the direction of travel of the wave, S-waves cause side-to-side motion, more like waves in a guitar string. The two types of waves are illustrated in Figure 7.6. S-waves normally travel through Earth's interior at 3 to 4 km/s; however, they cannot travel through liquid, which absorbs them.

The speeds of both P- and S-waves depend on the density of the matter through which the waves are traveling. Consequently, if we can measure the time taken for the waves to move from the site of an earthquake to one or more monitoring stations on Earth's surface, we can determine the density of matter in the interior. Figure 7.7 illustrates some P- and S-wave paths away from the site of an earthquake. Seismographs located around the world measure the times of arrival, as well as the strengths, of the seismic waves. Both observations contain much useful information—about the earthquake itself and about Earth's interior through which the waves pass. Notice that the waves do not travel in straight lines through the planet. Because the wave velocity increases with depth, waves that travel deeper tend to overtake those on slightly

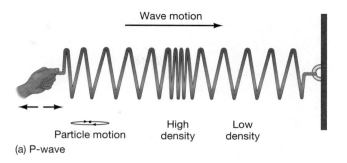

(a) P-wave

Wave motion

Particle motion

High density

Low density

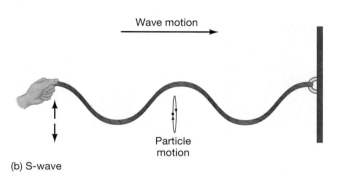

(b) S-wave

Wave motion

Particle motion

▲ **FIGURE 7.6 P and S Waves** (a) A pressure (P) wave traveling through Earth's interior causes material to vibrate in a direction parallel to the direction of motion of the wave. Material is alternately compressed and expanded. (b) A shear (S) wave produces motion perpendicular to the direction in which the wave travels, pushing material from side to side. Also shown is the motion of one typical particle: In case (a), the particle oscillates forward and backward about its initial position. In (b), the particle moves up and down.

shallower paths, and the waves bend as they move through the interior.

A particularly important result emerged after numerous quakes were monitored several decades ago: Seismic stations on the side of Earth opposite a quake never detect S-waves—these waves are blocked by material within Earth's interior. Furthermore, although P-waves always arrive at stations diametrically opposite the quake, parts of Earth's surface receive almost none (see Figure 7.7). Most geologists think that S-waves are absorbed by a liquid core at Earth's center and that P-waves are refracted at the core boundary, much as light is refracted by a lens. The result is the S- and P-wave "shadow zones" we observe. The fact that every earthquake exhibits these shadow zones is the best evidence that the core of our planet is hot enough to be liquid. *Discovery 7-2* presents some more seismic data on the structure of Earth's core.

The sizes of the shadow zones depend on the radius of the core, and careful analysis of the seismic data yield a core radius of about 3500 km. In fact, very faint P-waves *are* observed in the P-wave shadow zone indicated in Figure 7.7. These are thought to be reflected off the surface of a solid **inner core,** of radius 1300 km, lying at the center of the liquid **outer core.**

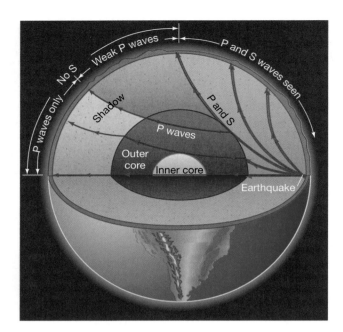

▲ **FIGURE 7.7 Seismic Waves** Earthquakes generate pressure (P, or primary) and shear (S, or secondary) waves that can be detected at seismographic stations around the world. The waves bend while moving through Earth's interior because of the variation in density and temperature within our planet. S-waves (colored red) are not detected by stations "shadowed" by the liquid core of Earth. P-waves (colored green) do reach the side of Earth opposite the earthquake, but their interaction with Earth's core produces another shadow zone, where almost no P-waves are seen.

## Modeling Earth's Interior

Because earthquakes occur often and at widespread places across the globe, geologists have accumulated a large amount of data about shadow zones and seismic-wave properties. They have used these data, along with direct knowledge of surface rocks, to build mathematical models of Earth's interior. Our knowledge of the deepest recesses of our planet is based almost entirely on modeling and indirect observation. We will find many more examples of this powerful combination throughout the text.

Figure 7.8 presents a model that most scientists accept. According to this model, Earth's outer core is surrounded by a thick mantle and topped with a thin crust. The mantle is about 3000 km thick and accounts for the bulk (80 percent) of our planet's volume. The crust has an average thickness of only 15 km—a little less (around 8 km) under the oceans and somewhat more (20–50 km) under the continents. The average density of crust material is around 3000 kg/m$^3$. Density and temperature both increase with depth. Specifically, from Earth's surface to its very center, the density increases from roughly 3000 kg/m$^3$ to a little more than 12,000 kg/m$^3$, and the temperature rises from just under 300 K to well over 5000 K. Much of the mantle has a density

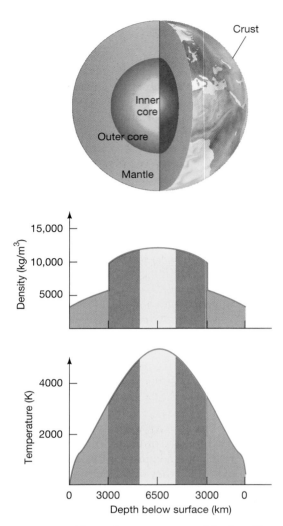

▲ **FIGURE 7.8 Earth's Interior** Computer models of Earth's interior imply that the density and temperature vary considerably through the mantle and the core. Note the sharp density change between Earth's core and mantle.

midway between the densities of the core and crust: about 5000 kg/m$^3$.

The high central density suggests to geologists that the inner parts of Earth must be rich in nickel and iron. Under the heavy pressure of the overlying layers, these metals (whose densities under surface conditions are around 8000 kg/m$^3$) can be compressed to the high densities predicted by the model. The sharp increase in density at the mantle–core boundary results from the difference in composition between the two regions. The mantle is composed of dense, but *rocky*, material—compounds of silicon and oxygen. The core consists primarily of even denser *metallic* elements. There is no similar jump in density or temperature at the inner core boundary—the material there simply changes from the liquid to the solid state.

# DISCOVERY 7-2

## Earth's "Rapidly" Spinning Core

Seismic waves racing out from the site of earthquakes allow geologists to map the interior makeup of planet Earth (Section 7.3). Recent analyses of old earthquake data suggest that Earth's inner core is spinning slightly *faster* than the rest of our planet. The whole planet rotates in the same direction (west to east), but the inner core takes about two-thirds of a second less to complete its daily rotation than the bulk of the matter surrounding it—which means that it gains a quarter turn each century.

Geologists arrived at this rather surprising result by carefully timing seismic P-waves moving through Earth's interior. Actually, the scientists used historical data acquired over the past few decades, mainly from 38 earthquakes that occurred in the south Atlantic Ocean whose seismic vibrations had to pierce the inner core on their way to monitoring stations in Alaska. As discussed in the text, only P-waves can make it through the liquid outer core, to be detected on the opposite side of Earth.

For many years, geologists have known that P-waves travel through the inner core about 3 to 4 percent faster along north–south pathways than along those close to the (east–west) plane of the equator. This difference probably results from the orientation of the iron atoms in the solid inner core; the atoms there must be aligned, much like the regular arrangement of atoms found in a crystal. Where the atomic alignment parallels the motion of the P-waves, those P-waves travel just a little faster—a bit like going "with the grain" rather than against it.

The new finding is that the route of the fastest seismic waves through Earth's deep interior is gradually shifting eastward, implying that the inner core has some motion slightly different from the rest of the planet. In the accompanying figure, the red region is the solid inner core, the orange region the liquid outer core. The dashed lines passing through the core depict the paths of the fastest seismic waves from 1900 to 1996. The data indicate that the axis of the "fast track" for the P-waves is shifting by about 1° per year relative to the crust above. That axis now intersects Earth's surface in the Arctic Ocean northeast of Siberia; some 30 years ago, it was 33° farther west. Gradual as it may

be, this shift amounts to about 20 km at the edge of the inner core—around a million times faster than the typical rates (centimeters per year) at which the continents creep across Earth's surface.

These discoveries may help us understand both how the inner core is continuing to grow by condensing from the molten outer core as Earth cools and how these two regions interact to produce Earth's magnetic field. As noted in Section 7.5, the dynamo theory of magnetism requires rotating, electrically conducting matter within Earth's interior. Detailed modeling of Earth's interior had also predicted that a rapidly moving jet stream of partly molten matter—a "mushy zone"—is probably established at the base of the outer core, causing the inner core to "superrotate"—rotate faster than the overlying layers. The shift in the axis of the fastest seismic waves indicates that this differential rotation pattern does indeed exist.

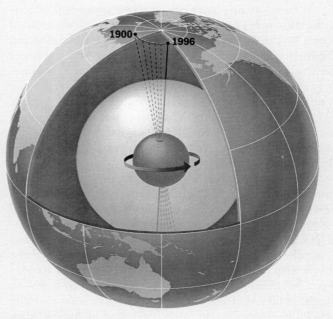

*(D. Rohr, Columbia Univ.)*

The model suggests that the core must be a mixture of nickel, iron, and some other lighter element, possibly sulfur. Without direct observations, it is difficult to be absolutely certain of the light component's identity. All geologists agree that much of the core must be liquid. The existence of the shadow zone demands this (and, as we will see, our current explanation of Earth's magnetic field relies on it). However, despite the high temperature, the pressure near the center—about 4 million times the atmospheric pressure at Earth's surface—is high enough to force the material there into the solid state.

Because geologists have been unable to drill deeper than about 10 km, no experiment has yet recovered a sample of Earth's mantle. However, we are not entirely ignorant of the mantle's properties. In a *volcano*, hot lava upwells from below the crust, bringing a little of the mantle to us and providing some inkling of Earth's interior. Observations of the chemical and physical properties of newly emerged lava are generally consistent with the model sketched in Figure 7.8.

The composition of the upper mantle is probably quite similar to the iron–magnesium–silicate mixtures known as *basalt.* You may have seen some dark gray basaltic rocks scattered across Earth's surface, especially near volcanoes. Basalt is formed as material from the mantle upwells from Earth's interior as lava, cools, and then solidifies. With a density between 3000 kg/m$^3$ and 3300 kg/m$^3$, basalt contrasts with the lighter *granite* (density 2700–3000 kg/m$^3$) that constitutes much of the rest of Earth's crust. Granite is richer than basalt in the light elements silicon and aluminum, which explains why the surface continents do not sink into the interior. Their low-density composition lets the crust "float" atop the denser matter of the mantle and core below.

## Differentiation

Earth, then, is not a homogeneous ball of rock. Instead, it has a layered structure, with a low-density crust at the surface, an intermediate-density material in the mantle, and a high-density core. Such variation in density and composition is known as **differentiation.**

Why isn't our planet just one big, rocky ball of uniform density? The answer appears to be that much of Earth was *molten* at some time in the past. As a result, the higher density matter sank to the core and the lower density material was displaced toward the surface. A remnant of this ancient heating exists today: Earth's central temperature is nearly equal to the surface temperature of the Sun. What processes were responsible for heating the entire planet to that extent? To answer this question, we must try to visualize the past, as sketched in Figure 7.9.

According to current models of solar system formation (see Chapter 15), when Earth formed 4.6 billion years ago, it did so by capturing material from its surroundings, growing in mass as it swept up "preplanetary" chunks of matter in its vicinity. As the young planet grew, its gravitational field strengthened and the speed with which newly captured matter struck its surface increased. This process generated a lot of heat—so much, in fact, that Earth may already have been partially or wholly molten by the time it reached its present size.

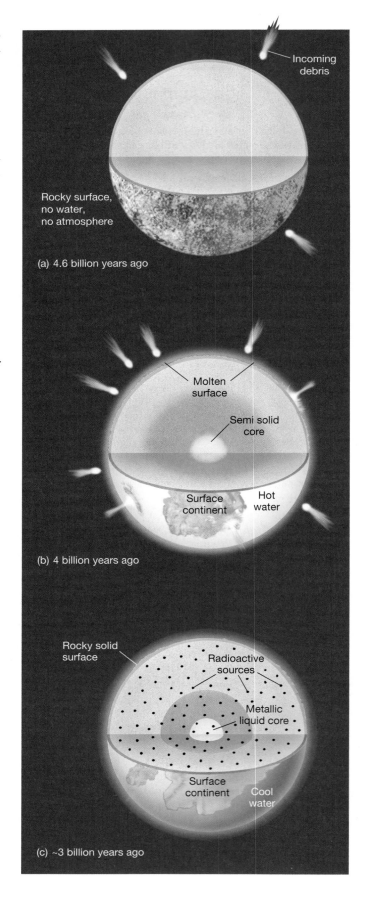

**(a) 4.6 billion years ago**

**(b) 4 billion years ago**

**(c) ~3 billion years ago**

▶ **FIGURE 7.9 Earth's Differentiation** Earth's interior changed considerably throughout its early history. (a) At its origin, 4.6 billion years ago, the Earth was probably already partly molten owing to debris bombardment and continued gravitational infall in its formative stage. (b) A second period of heavy bombardment, at about 4 billion years ago, likely caused its cooling surface layers to again become completely molten to a depth of tens of kilometers. (c) Early on especially, yet continuing to lesser extent to the present, radioactive heating from within has caused much of Earth's interior to liquefy, allowing its heavy metals to sink to the core while its lighter-weight rocks floated to the surface.

As Earth began to differentiate and heavy material sank to the center, even more gravitational energy was released, and the interior temperature must have increased still further.

Later, Earth continued to be bombarded with debris left over from the formation process. At its peak some 4 billion years ago, this secondary bombardment was probably intense enough to keep the surface molten, but only down to a depth of a few tens of kilometers. Erosion by wind and water has long since removed all trace of this early period from the surface of Earth, but the Moon still bears visible scars of the onslaught.

A second important process for heating Earth soon after its formation was **radioactivity**—the release of energy by certain rare heavy elements, such as uranium, thorium, and plutonium (see *More Precisely 7-2*). These elements release energy and heat their surroundings as their complex heavy nuclei decay (break up) into simpler lighter ones. Though the energy produced by the decay of a single radioactive atom is tiny, Earth contained a lot of radioactive atoms, and a lot of time was available. Rock is such a poor conductor of heat that the energy would have taken a very long time to reach the surface and leak away into space, so the heat built up in the interior, adding to the energy left there by Earth's formation.

Provided that enough radioactive elements were originally spread throughout the primitive Earth, rather like raisins in a cake, the entire planet—from crust to core—could have melted and remained molten for about a billion years. That's a long time by human standards, but not so long in the cosmic scheme of things. Measurements of the ages of some surface rocks indicate that Earth's crust finally began to solidify roughly 700 million years after it originally formed. Radioactive heating did not stop at that point, of course; it continued even after Earth's surface cooled and solidified. But radioactive decay works in only one direction, always producing lighter elements from heavier ones. Once gone, the heavy and rare radioactive elements cannot be replenished.

So the early source of heat diminished with time, allowing the planet to cool over the past 4 billion years. In this process, Earth has cooled from the outside in, much like a hot potato, since regions closest to the surface can most easily unload their excess heat into space. In that way, the surface developed a solid crust, and the differentiated interior attained the layered structure now implied by seismic studies.

## CONCEPT CHECK

✔ How might our knowledge of Earth's interior be changed if our planet were geologically inactive, with no volcanos or earthquakes?

## 7.4 Surface Activity

Earth is geologically alive today. Its interior seethes and its surface constantly changes. Figure 7.10 shows some examples of two kinds of surface geological activity: a volcano, whereby molten rock and hot ash upwell through fissures or cracks in the surface, and (the aftermath of) an earthquake, which occurs when the crust suddenly dislodges under great pressure. Catastrophic volcanoes and earthquakes are relatively rare events these days, but geological studies of rocks, lava, and other surface features imply that surface activity must have been more frequent, and probably more violent, long ago.

### Continental Drift

Many traces of past geological events are scattered across our globe. Erosion by wind and water has wiped away much of the evidence for ancient activity, but modern exploration has documented the sites of most of the recent activity, such as earthquakes and volcanic eruptions. Figure 7.11 is a map of the currently active areas of our planet. The red dots represent sites of volcanism or earthquakes. Nearly all these sites have experienced surface activity within the last century, some of them suffering much damage and the loss of many lives.

The intriguing aspect of Figure 7.11 is that the active sites are not spread evenly across our planet. Instead, they trace well-defined lines of activity, where crustal rocks dislodge (as in earthquakes) or mantle material upwells (as in volcanoes). In the mid-1960s, scientists realized that these lines are really the outlines of gigantic "plates," or slabs of Earth's surface.* Most startling of all, the plates are slowly moving—literally drifting around the surface of our planet. These plate motions have created the surface mountains, oceanic trenches, and other large-scale features across the face of planet Earth. In fact, plate motions have shaped the continents themselves. The process is popularly known as "continental drift." The technical term for the study of plate movement and its causes is **plate tectonics.** The major plates of the world are marked on Figure 7.11.

Taken together, the plates make up Earth's **lithosphere,** which contains both the crust and a small part of the upper mantle. The lithosphere is the portion of Earth that undergoes tectonic activity. The semisolid part of the mantle over which the lithosphere slides is known as the **asthenosphere.** The relationships between these regions of Earth are shown in Figure 7.12.

*Not all volcanoes are associated with plate boundaries. The Hawaiian Islands, located near the center of the Pacific plate, are a case in point. They are associated with a "hot spot" in Earth's upper mantle that melts the crust above it. Over millions of years, the motion of the Pacific plate across the underlying hot spot has resulted in a chain of volcanic islands.

(a)     (b)     (c)

▲ FIGURE 7.10 **Geological Activity** (a) An active volcano on Kilauea in Hawaii. Kilauea seems to be a virtually ongoing eruption. (b) Other, more sudden eruptions, such as that of Mount St. Helens in Washington State on May 18, 1980, are rare catastrophic events that can release more energy than the detonation of a thousand nuclear bombs. (c) The aftermath of an earthquake that claimed more than 5000 lives and caused billions of dollars' worth of damage in Kobe, Japan, in January 1995. (*P. Chesley/Getty Images, Inc.; USGS; H. Yamaguchi/Sygma*)

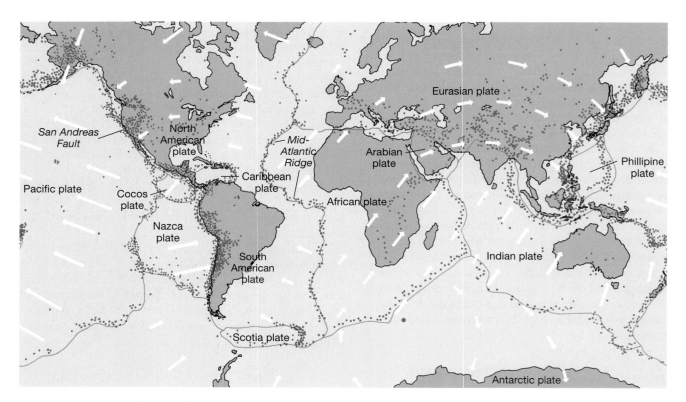

▲ FIGURE 7.11 **Global Plates** Red dots represent active sites where major volcanoes or earthquakes have occurred in the 20th century. Taken together, the sites outline vast "plates," indicated in dark blue, that drift around on the surface of our planet. The white arrows show the general directions and speeds of the plate motions.

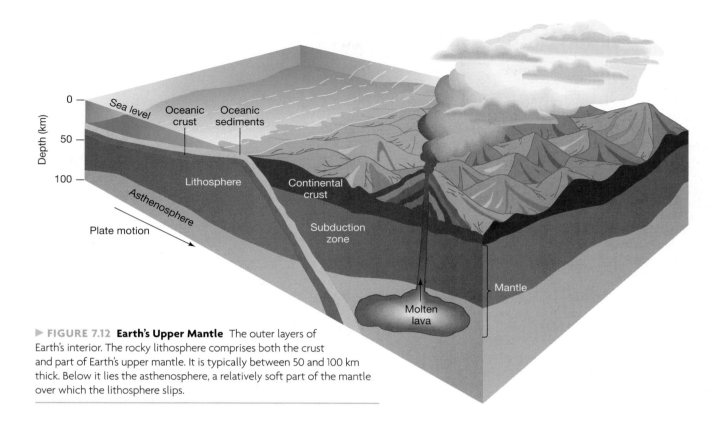

▶ **FIGURE 7.12 Earth's Upper Mantle** The outer layers of Earth's interior. The rocky lithosphere comprises both the crust and part of Earth's upper mantle. It is typically between 50 and 100 km thick. Below it lies the asthenosphere, a relatively soft part of the mantle over which the lithosphere slips.

The idea of continental drift was first suggested in 1912 by a German meteorologist named Alfred Wegener, who pointed out the remarkable geographic fit between the continents on either side of the Atlantic ocean. Note in Figure 7.11 how the Brazilian coast (on the easternmost part of South America) meshes nicely with the indented Ivory Coast along west Africa. In fact, most of the continental landmasses in the Southern Hemisphere fit together remarkably well. Following the arrows in the figure backwards, we can see that the fits are in fact roughly consistent with the present locations of the plates involved. The fit appears not to be as good in the Northern Hemisphere, but it improves markedly if we consider the entire continental shelves (the continental borders, which are under water) instead of just the portions that happen to stick up above sea level.

Few took Wegener's ideas seriously at the time, in part because there was no known mechanism that could drive the plates' motions. Nearly all scientists thought it preposterous that large segments of rocky crust could be drifting across the surface of our planet. These skeptical views persisted for more than half a century, when the accumulation of data in support of continental drift became overwhelming. Similar-looking fossils were found on opposite sides of the Atlantic Ocean at just the locations where the continents "fit together," and studies of the seafloor near the center of the Atlantic (to be discussed in more detail in a moment) strongly suggested

the formation of new crust as plates separated. Today, backed by evidence, Wegener's "crazy" theory forms the foundation for all geological studies of our planet's outer layers. ∞ (Sec. 1.2)

The plates are not simply slowing to a stop after some ancient initial movements. Rather, they are still drifting today, although at an extremely slow rate. Typically, the speeds of the plates amount to only a few centimeters per year—about the same rate as your fingernails grow. Still, this is well within the measuring capabilities of modern equipment. Curiously, one of the best ways of monitoring plate motion on a global scale is by making accurate observations of very distant astronomical objects. Quasars (see Chapter 25), lying many hundreds of millions of light-years from Earth, never show any measurable apparent motion on the sky stemming from their own motion in space. Thus, any apparent change in their position (after correction for Earth's motion) can be interpreted as arising from the motion of the telescope—that is, of the continental plate on which it is located!

On smaller scales, laser-ranging and other techniques now routinely track the relative motion of plates in many areas, such as California, where advance warning of earthquake activity is at a premium. During the course of Earth history, each plate has had plenty of time to move large distances, even at its sluggish pace. For example, a drift rate of only 2 cm per year can cause two continents (e.g., Europe and North America) to separate by

# MORE PRECISELY 7-2

## Radioactive Dating

In Chapter 4, we saw that atoms are made up of electrons and nuclei and that nuclei are composed of protons and neutrons. ∞ (Sec. 4.2) The number of protons in a nucleus determines which element it represents. However, the number of neutrons can vary. In fact, most elements can exist in several *isotopic* forms, all containing the same number of protons, but different numbers of neutrons in their nuclei. The particular nuclei we have encountered so far—the most common forms of hydrogen, helium, carbon, and iron—are all *stable*. For example, left alone, a carbon-12 nucleus, consisting of six protons and six neutrons, will remain unchanged forever. It will not break up into smaller pieces, nor will it turn into anything else.

Not all nuclei are stable, however. Many nuclei—such as carbon-14 (containing 6 protons and 8 neutrons), thorium-232 (90 protons, 142 neutrons), uranium-235 (92 protons, 143 neutrons), uranium-238 (92 protons, 146 neutrons), and plutonium-241 (94 protons, 147 neutrons)—are inherently *unstable*. Left alone, they will eventually break up into lighter "daughter" nuclei, emitting some elementary particles and releasing some energy in the process. The change happens spontaneously, without any external influence. This instability is known as **radioactivity.** The energy released by the disintegration of the radioactive elements just listed is the basis for nuclear fission reactors (and atomic bombs).

Unstable heavy nuclei achieve greater stability by disintegrating into lighter nuclei, but they do not do so immediately. Each type of "parent" nucleus takes a characteristic amount of time to decay. The **half-life** is the name given to the time required for half of a sample of parent nuclei to disintegrate. Notice that this is really a statement of probability. We cannot say *which* nuclei of a given element will decay in any given half-life interval; we can say only that half of them are expected to do so. If a given sample of material has half-life $T$, then we can write down a simple expression for the amount of material remaining after time $t$:

$$\text{fraction of material remaining} = (1/2)^{t/T}.$$

Thus, if we start with a billion radioactive nuclei embedded in a sample of rock, a half-billion nuclei will remain after one half-life, a quarter-billion after two half-lives, and so on. The first figure illustrates the decline in the number of parent nuclei as a function of time.

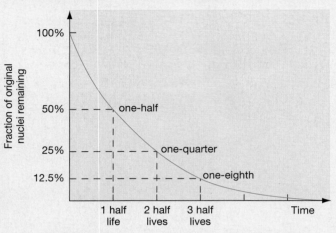

Every radioactive isotope has its own half-life, and most of their half-lives are now well known from studies conducted since the 1950s. For example, the half-life of uranium-235 is 713 million years, and that of uranium-238 is 4.5 billion years. Some radioactive elements decay much more rapidly, others much more slowly, but these two types of uranium are particularly important to geologists because their half-lives are comparable to the age of the solar system. The second figure illustrates the half-lives and decay reactions for four unstable heavy nuclei.

The decay of unstable radioactive nuclei into more stable *daughter* nuclei provides us with a useful tool for measuring the ages of any rocks we can get our hands on. The first step is to measure the amount of stable nuclei of a given kind (e.g., lead-206, which results from the decay of uranium-238). This amount is then compared with the amount of remaining unstable parent nuclei (in this case, uranium-238) from which the daughter nuclei descended. Knowing the rate (or half-life) at which the disintegration occurs, the age of the rock then follows directly. If half of the parent nuclei of some element have decayed, so that the number of daughter nuclei equals the number of parents, the age of the rock must be equal to the half-life of the radioactive nucleus studied. Similarly, if only a quarter of the parent nuclei remain (three times as many daughters as parents), the rock's age is twice the half-life of that element, and so on.

In practice, ages can be determined by these means to within an accuracy of a few percent. The most ancient rocks on Earth are dated at 3.9 billion years old. These rare specimens have been found in Greenland and Labrador.

---

some 4000 km over the course of 200 million years. That may be a long time by human standards, but it represents only about 5 percent of the age of Earth.

A common misconception is that the plates are the continents themselves. Some plates are indeed made mostly of continental landmasses, but other plates are made of a continent plus a large part of an ocean. For example, the Indian plate includes all of India, much of the Indian Ocean, and all of Australia and its surrounding south seas (see Figure 7.11). Still other plates are mostly ocean. The seafloor itself is a slowly drifting plate, and the oceanic water merely fills in the depressions between continents. The southeastern portion of the Pacific Ocean, called the Nazca plate, contains no landmass at all. For the most part, the continents are just passengers riding on much larger plates.

**EXAMPLE 1** Suppose that careful chemical analysis of a sample of rock reveals that, for every nucleus of uranium-238 remaining in the sample, there is 0.41 of a lead-206 nucleus. If we assume that there was no lead-206 initially present, and hence that every lead-206 nucleus is the decay product of a nucleus of uranium-238, we can easily calculate the fraction of uranium-238 nuclei remaining. The answer is

$$\text{fraction of uranium-238} = \frac{1}{1 + 0.41}$$

$$= 0.71 \approx \sqrt{\frac{1}{2}} = \left(\frac{1}{2}\right)^{0.5}.$$

From this equation, it follows that the elapsed time must be 0.5 times the half-life of uranium-238, or 2.25 billion years. If we were to repeat the analysis with uranium-235 and lead-207, we would expect to find consistent results within the measurement errors: 2.25 billion years is 2250/713 = 3.2 uranium-235 half-lives, so only $(1/2)^{3.2} \approx 11$ percent of any uranium-235 should remain—daughter lead-207 nuclei should outnumber parent uranium-235 nuclei by more than eight to one.

The radioactive-dating technique rests on the assumption that the rock has remained *solid* while the radioactive decays have been going on. If the rock melts, there is no particular reason to expect the daughter nuclei to remain in the same locations their parents had occupied, and the whole method fails. Thus, radioactive dating indicates the time that has elapsed since the last time the rock in question solidified. Hence, the 3.9-billion-year value represents only a portion—a lower limit—of the true age of our planet. It does not measure the duration of Earth's molten existence.

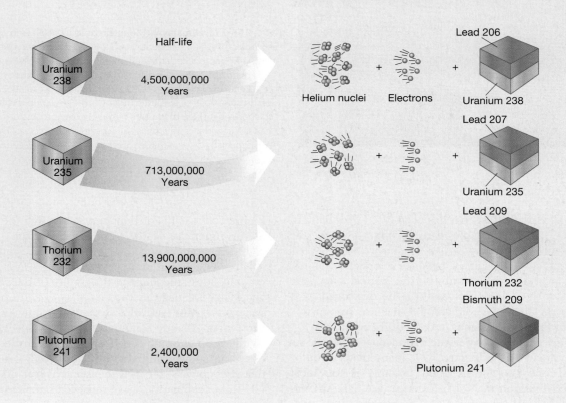

## Effects of Plate Motion

As the plates drift around, we might expect collisions to be routine. Indeed, plates do collide, but unlike two automobiles that collide and then stop, the surface plates are driven by enormous forces. They do not stop easily. Instead, they just keep crunching into one another, reshaping the landscape and causing violent seismic activity.

Figure 7.13(a) shows a collision currently occurring between two continental landmasses: The subcontinent of India, on the prow of the northward-moving Indian plate, is crashing into the landmass of Asia, located on the Eurasian plate (see Figure 7.11). The resulting folds of rocky crust create mountains—in this case, the snow-covered Himalayan mountain range at the upper right. A peak like Mount Everest (Figure 7.13b) represents

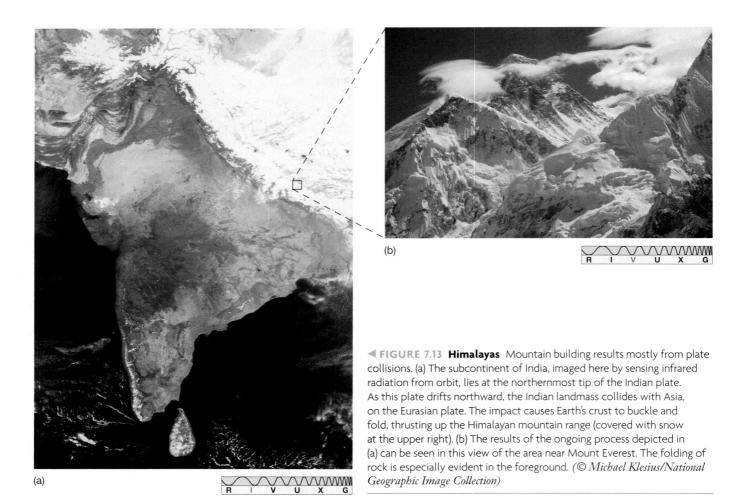

(a)

R I V U X G

(b)

R I V U X G

◀ **FIGURE 7.13 Himalayas** Mountain building results mostly from plate collisions. (a) The subcontinent of India, imaged here by sensing infrared radiation from orbit, lies at the northernmost tip of the Indian plate. As this plate drifts northward, the Indian landmass collides with Asia, on the Eurasian plate. The impact causes Earth's crust to buckle and fold, thrusting up the Himalayan mountain range (covered with snow at the upper right). (b) The results of the ongoing process depicted in (a) can be seen in this view of the area near Mount Everest. The folding of rock is especially evident in the foreground. (© *Michael Klesius/National Geographic Image Collection*)

a portion of Earth's crust that has been lifted over 8800 m by the slow, but inexorable, force produced when one plate plows into another.

Not all colliding plates produce mountain ranges. At other locations, called **subduction zones,** one plate slides under the other, ultimately to be destroyed as it sinks into the mantle. Subduction zones are responsible for most of the deep trenches in the world's oceans.

Nor do all plates experience head-on collisions. As noted by the arrows of Figure 7.11, many plates slide or shear past one another. A good example is the most famous active region in North America: the San Andreas Fault in California (Figure 7.14). The site of much earthquake activity, this fault marks the boundary where the Pacific and North American plates are rubbing past each other. The motion of these two plates, like that of moving parts in a poorly oiled machine, is neither steady nor smooth. The sudden jerks that occur when they do move against each other are often strong enough to cause major earthquakes.

At still other locations, the plates are moving apart. As they recede, new material from the mantle wells up between them, forming **midocean ridges.** Notice in

Figure 7.11 the major boundary separating the North and South American plates from the Eurasian and African plates, marked by the thin strip down the middle of the Atlantic Ocean. Discovered after World War II by oceanographic ships studying the geography of the seafloor, this giant fault is called the *Mid-Atlantic Ridge*. It extends, like a seam on a giant baseball, all the way from Scandinavia in the North Atlantic to the latitude of Cape Horn at the southern tip of South America. The entire ridge is a region of seismic and volcanic activity, but the only major part of it that rises above sea level is the island of Iceland.

Robot submarines have retrieved samples of the ocean floor at a variety of locations on either side of the Mid-Atlantic Ridge, and the ages of the samples have been measured by means of radioactive dating techniques. As depicted in Figure 7.15, the ocean floor closest to the ridge is relatively young, whereas material farther away, on either side, is older—exactly as we would expect if hot molten matter is upwelling and solidifying as the plates on either side drift apart. The Atlantic Ocean has apparently been growing in this way for the past 200 million years, the age of the oldest rocks found on any part of the Atlantic seafloor.

R I V U X G

▲ FIGURE 7.14 **Californian Fault** A small portion of the San Andreas fault in California. The fault is the result of the North American and Pacific plates sliding past one another. The Pacific plate, which includes a large slice of the California coast, is drifting to the northwest relative to the North American plate. The motions of the two plates are indicated by arrows whose lengths are proportional to the plate speeds. *(U.S. Department of the Interior)*

ocean floor has preserved within it a record of Earth's magnetism during past times, rather like a tape recording. Figure 7.16 is a diagram of a small portion of the Atlantic Ocean floor near Iceland. Earth's current magnetism is oriented in the familiar north–south fashion, and when samples of ocean floor close to the ridge are examined, the iron deposits are oriented just as expected: north–south. This material is the "young" basalt that upwelled and cooled fairly recently. However, samples retrieved farther from the ridge, corresponding to older material that upwelled long ago, are often magnetized with the *opposite* orientation. As we move away from the ridge, the imprinted magnetic field flips back and forth, more or less regularly and symmetrically on either side of the ridge.

The leading explanation of these different magnetic orientations is that they were caused by reversals in Earth's magnetic field that occurred as the plates drifted away from the central ridge. Working backward, we can use the fossil magnetic field to infer the past positions of the plates, as well as the orientation of Earth's magnetism. In addition to providing strong support for the idea of seafloor spreading, these measurements, when taken in conjunction with the data on the age of the seafloor, allow us to time our planet's magnetic reversals. On average, Earth's magnetic field reverses itself roughly every half-million years. Current theory suggests that such reversals are part of the way in which all planetary magnetic fields are generated. As we will see in Chapter 16, a similar phenomenon (with a reversal time of approximately 11 years) is also observed on the Sun.

## What Drives the Plates?

What process is responsible for the enormous forces that drag plates apart in some locations and ram them together in others? The answer is probably convection—the same physical process we encountered earlier in our study of the atmosphere. Figure 7.17 is a cross-sectional diagram of the top few hundred kilometers of our planet's interior. It depicts roughly the region in and around a midocean

Other studies of the Mid-Atlantic Ridge have yielded important information about Earth's magnetic field. As hot material (carrying traces of iron) from the mantle emerges from cracks in the oceanic ridges and solidifies, it becomes slightly magnetized, retaining an imprint of Earth's magnetic field *at the time it cooled*. Thus, the

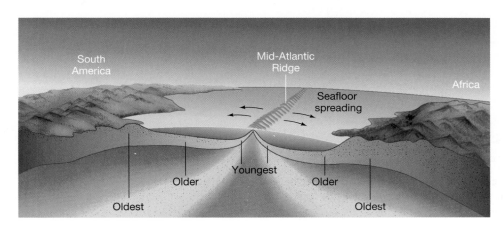

◀ FIGURE 7.15 **Seafloor Spreading** Samples of ocean floor retrieved by oceanographic vessels are youngest close to the Mid-Atlantic Ridge and progressively older farther away.

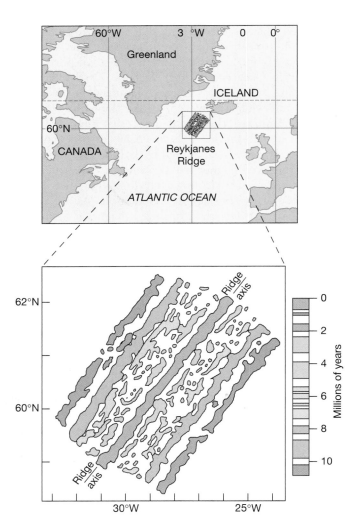

◀ FIGURE 7.16 **Magnetic Reversals** Samples of basalt retrieved from the ocean floor often reveal Earth's magnetism to have been oriented oppositely from the current north–south magnetic field. This simplified diagram shows the ages of some of the regions in the vicinity of the Mid-Atlantic Ridge (see Figure 7.11), together with the direction of the fossil magnetic field. The colored areas have the current orientation; they are separated by regions of reversed magnetic polarity.

ridge. There, the ocean floor is covered with a layer of sediment—dirt, sand, and dead sea organisms that have fallen through the seawater for millions of years. Below the sediment lies about 10 km of granite, the low-density rock that makes up the crust. Deeper still lies the upper mantle, whose temperature increases with depth. Below the base of the lithosphere, at a depth of perhaps 50 km, the temperature is sufficiently high that the mantle is soft enough to flow very slowly, although it is not molten. This region is the asthenosphere.

The setting is a perfect one for convection—warm matter underlying cool matter. The warm mantle rock rises, just as hot air rises in our atmosphere. Sometimes, the rock squeezes up through cracks in the granite crust. Every so often, such a fissure may open in the midst of a continental landmass, producing a volcano such as Mount St. Helens (see Figure 7.10b) or possibly a geyser like those at Yellowstone National Park. However, most such cracks are on the ocean floor. The Mid-Atlantic Ridge is a prime example.

Not all the rising warm rock in the upper mantle can squeeze through cracks and fissures. Some warm rock cools and falls back down to lower levels. In this way, large circulation patterns become established within the upper mantle, as depicted in Figure 7.17. Riding atop these

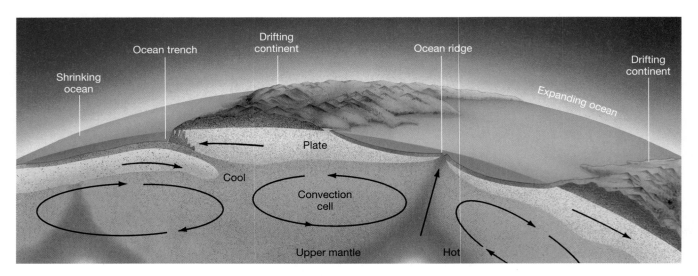

▲ FIGURE 7.17 **Plate Drift** The motion of Earth's tectonic plates is probably caused by convection—in this case, giant circulation patterns in the upper mantle that drag the plates across the surface.

convection patterns are the plates. The circulation is extraordinarily sluggish. Semisolid rock takes millions of years to complete one convection cycle. Although the details are far from certain and remain controversial, many researchers suspect that the large-scale circulation patterns near plate boundaries drive the motions of the plates.

This constant recycling of plate material provides a natural explanation for the **rock cycle**—the process by which surface rock on our planet is continuously redistributed and transformed from one type into another. Deep below the surface, in the asthenosphere, temperatures are high enough that mantle rock exists in the form of molten *magma*. When this material cools and hardens, it forms *igneous rocks*. (Granite and basalt are familiar examples.) Igneous rocks are associated with volcanic activity (in which the magma is called *lava*) and spreading regions such as the Mid-Atlantic Ridge, where magma emerges as two plates separate. The weathering and erosion of surface rocks produce sandy grains that are deposited as sediments and may eventually become compacted into *sedimentary rocks* such as sandstone and shale. Subsequently, at high temperatures or pressures, igneous or sedimentary rocks may be physically or chemically transformed into *metamorphic rocks* (e.g., marble and slate). Such conditions occur as plates collide and form mountain ranges or as a plate dives deep into a subduction zone.

## Past Continental Drift

Figure 7.18 illustrates how all the continents nearly fit together like pieces of a puzzle. Geologists think that sometime in the past a single gargantuan landmass dominated our planet. This ancestral supercontinent, known as *Pangaea* (meaning "all lands"), is shown in Figure 7.18(a). The rest of the planet was presumably covered with water. The present locations of the continents, along with measurements of their current drift rates, suggest that Pangaea was the major land feature on Earth approximately 200 million years ago. Dinosaurs, which were then the dominant form of life, could have sauntered from Russia to Texas via Boston without getting their feet wet. Pangaea explains the geographical and fossil evidence (cited earlier) that first led scientists to the idea of continental drift. The other frames in Figure 7.18 show how Pangaea split apart, its separate pieces drifting across Earth's surface, eventually becoming the familiar continents we know today.

There is nothing particularly special about a time 200 million years in the past. We do not suppose that Pangaea remained intact for some 4 billion years after the crust first formed, only to break up so suddenly and so recently. It is much more likely that Pangaea itself came into existence after an earlier period during which other plates, carrying widely separated continental masses, were driven together by tectonic forces, merging

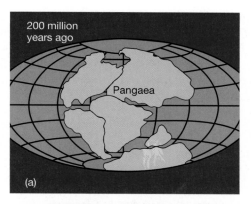

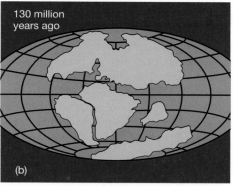

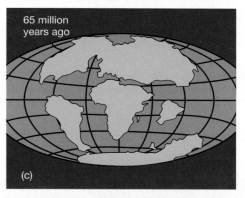

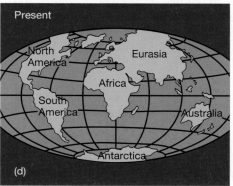

▲ FIGURE 7.18 **Pangaea** Given the current estimated drift rates and directions of the plates, we can trace their movements back into the past. About 200 million years ago, they would have been at the approximate positions shown in (a). The continents' current positions are shown in (d).

their landmasses to form a single supercontinent. It is quite possible that there has been a long series of "Pangaeas" stretching back in time over much of Earth's history, as tectonic forces have *continually* formed, destroyed, and re-formed our planet's landmasses. There probably will be many more.

CONCEPT CHECK

✔ Describe the causes, and some consequences, of plate tectonics on Earth.

## 7.5    Earth's Magnetosphere

Simply put, the *magnetosphere* is the region around a planet that is influenced by that planet's magnetic field. Discovered by artificial satellites launched in the late 1950s and sketched in Figure 7.19, Earth's magnetosphere extends far above the atmosphere, completely surrounding our planet. Close to Earth, the magnetic field is similar in overall structure to the field of a gigantic bar magnet (Figure 7.20). The *magnetic field lines*, which indicate the strength and direction of the field at any point in space, run from south to north, as indicated by the blue arrowheads in these figures.

The north and south *magnetic poles*, where the magnetic field lines intersect Earth's surface vertically, are roughly aligned with Earth's spin axis. Neither pole is fixed relative to our planet, however—both drift at a rate of some 10 km per year—nor are the poles symmetrically placed. At present, Earth's magnetic north pole lies in northern Canada, at a latitude of about 80°N, almost due north of the center of North America; the magnetic south

pole lies at a latitude of about 60°S, just off the coast of Antarctica south of Adelaide, Australia.

Earth's magnetosphere contains two doughnut-shaped zones of high-energy charged particles, one located about 3000 km, and the other 20,000 km, above Earth's surface. These zones are named the **Van Allen belts,** after the American physicist whose instruments on board one of the first artificial satellites initially detected them. We call them "belts" because they are most pronounced near Earth's equator and because they completely surround the planet. Figure 7.20 shows how these invisible regions envelop Earth, except near the North and South Poles.

The particles that make up the Van Allen belts originate in the solar wind—the steady stream of charged particles flowing from the Sun. ∞ (Sec. 6.5) Traveling through space, neutral particles and electromagnetic radiation are unaffected by Earth's magnetism, but electrically charged particles are strongly influenced. As illustrated in the inset to Figure 7.20, a magnetic field exerts a force on a moving charged particle, causing the particle to spiral around the magnetic field lines. In this way, charged particles—mainly electrons and protons—from the solar wind can become trapped by Earth's magnetism. Earth's magnetic field exerts electromagnetic control over these particles, herding them into the Van Allen belts. The outer belt contains mostly electrons; the much heavier protons accumulate in the inner belt.

We could never survive unprotected in the Van Allen belts. Unlike the lower atmosphere, on which humans and other life-forms rely for warmth and protection, much of the magnetosphere is subject to intense bombardment by large numbers of high-velocity, and potentially very harmful, charged particles. Colliding violently with an unprotected human body, these particles would deposit

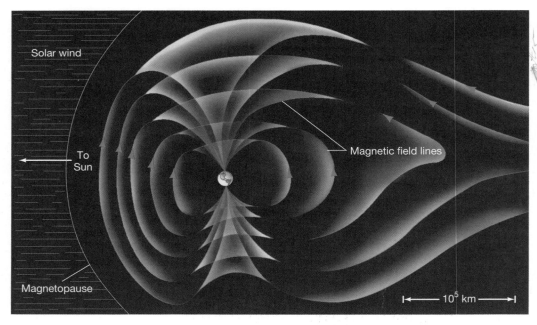

◄ **FIGURE 7.19  Earth's Magnetosphere** The magnetosphere is the region surrounding a planet wherein particles from the solar wind are trapped by the planet's magnetic field. Shown here is Earth's magnetosphere, with blue arrowheads on the field lines indicating the direction in which a compass needle would point. Far from Earth, the magnetosphere is greatly distorted by the solar wind, with a long "tail" extending from the nighttime side of Earth (here, at right) far into space. The magnetopause is the boundary of the magnetosphere in the sunward direction.

Solar wind

To Sun

Magnetic field lines

Magnetopause

|◄——— $10^5$ km ———►|

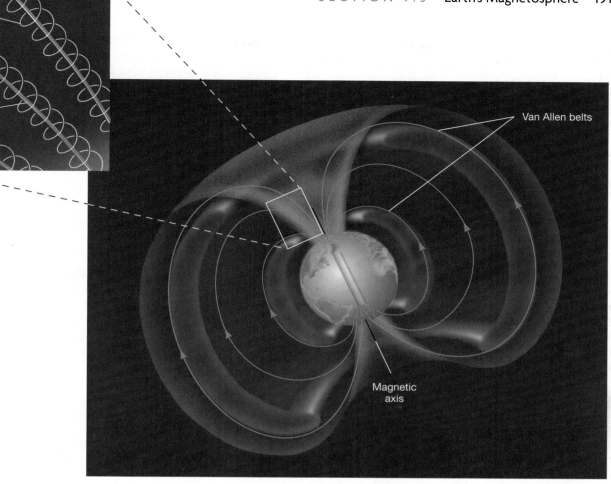

**▲ FIGURE 7.20 Van Allen Belts** Earth's magnetic field resembles somewhat the field of an enormous bar magnet buried deep inside our planet (but offset slightly from Earth's axis of rotation). High above Earth's atmosphere, the magnetosphere (light blue-green area) contains two doughnut-shaped regions (greyish areas) of magnetically trapped charged particles. These are the Van Allen belts. The inset illustrates how charged particles in a magnetic field spiral around the field lines. The convergence of the field lines near Earth's magnetic poles causes the particles to be reflected back toward the other pole. Thus, the charged particles tend to become "trapped" by Earth's magnetic field.

large amounts of energy wherever they made contact, causing severe damage to living organisms. Without sufficient shielding on the *Apollo* spacecraft, for example, the astronauts might not have survived the passage through the magnetosphere on their journey to the Moon.

Particles from the Van Allen belts often escape from the magnetosphere near Earth's north and south magnetic poles, where the field lines intersect the atmosphere. Their collisions with air molecules create a spectacular light show called an **aurora** (plural *aurorae*; Figure 7.21). This colorful display results when atmospheric molecules, excited upon collision with the charged particles, fall back to their ground states and emit visible light. Many different colors are produced because each type of atom or molecule can take one of several possible paths as it returns to its ground state. ∞ (Sec. 4.3) Aurorae are most brilliant at high latitudes, especially inside the Arctic and

Antarctic circles. In the north, the spectacle is called the *aurora borealis*, or *northern lights*. In the south, it is called the *aurora australis*, or *southern lights*.

Occasionally, particularly after a storm on the Sun (see Chapter 16), the Van Allen belts can become distorted by the solar wind and overloaded with many more particles than normal, allowing some particles to escape prematurely and at lower latitudes. For example, in North America, the aurora borealis is normally seen with any regularity only in northern Canada and Alaska. However, at times of greatest solar activity, the display has occasionally been seen as far south as the southern United States.

As is evident from Figure 7.19, Earth's magnetosphere is not symmetrical. Satellite mapping reveals that it is quite distorted, forming a teardrop-shaped cavity. On the sunlit (daytime) side of Earth, the magnetosphere is compressed by the flow of high-energy particles in the solar wind. The boundary between the magnetosphere

(a)

(b)

R   I   V   U   X   G

▲ FIGURE 7.21 **Aurorae** (a) A colorful aurora flashes rapidly across the sky, resembling huge windblown curtains glowing in the dark. Aurorae result from the emission of light radiation after magnetospheric particles collide with atmospheric molecules. The colors are produced as excited ions, atoms, and molecules recombine and cascade back to their ground states. (b) An aurora high above Earth, as photographed from a space shuttle (visible at left). *(NCAR; NASA)*

and this flow, known as the *magnetopause*, is found at about 10 Earth radii from our planet. On the side opposite the Sun, the field lines are extended away from Earth, with a long tail often reaching beyond the orbit of the Moon.

What is the origin of the magnetosphere and the Van Allen belts within it? Despite the artistic license in Figure 7.20, Earth's magnetism is not really the result of a huge bar magnet lying within our planet. In fact, geophysicists think that Earth's magnetic field is not a "permanent" part of our planet at all. Instead, it is thought to be continuously generated within the outer core and to exist only because Earth is rotating. As in the dynamos that run industrial machines, Earth's magnetism is produced by the spinning, electrically conducting, liquid metal core deep within our planet. The theory that explains planetary (and other) magnetic fields in terms of rotating, conducting material flowing in the planet's interior is known as **dynamo theory**. Both rapid rotation *and* a conducting liquid core are needed for such a mechanism to work. This connection between internal structure and magnetism is very important for studies of the other planets in the solar system: We can tell a lot about a planet's interior simply by measuring its magnetic field.

Earth's magnetic field plays an important role in controlling many of the potentially destructive charged particles that venture near our planet. Without the magnetosphere, Earth's atmosphere—and perhaps the surface, too—would be bombarded by harmful particles, possibly damaging many forms of life on our planet. Some researchers have even suggested that, had the magnetosphere not existed in the first place, life might never have arisen at all on our planet.

CONCEPT CHECK
✔ What does the existence of a planetary magnetic field tell us about a planet's interior?

## 7.6 The Tides

Earth is unique among the planets in that it has large quantities of liquid water on its surface. Approximately three-quarters of Earth's surface is covered by water, to an average depth of about 3.6 km. Only 2 percent of the water is contained within lakes, rivers, clouds, and glaciers. The remaining 98 percent is in the oceans, forming the *hydrosphere*.

Most people are familiar with the daily fluctuation in ocean level known as the **tides.** At most coastal locations on Earth, there are two low tides and two high tides each day. The "height" of the tides—the magnitude of the variation in sea level—can range from a few centimeters to many meters, depending on the location on Earth and the time of year. The height of a typical tide on the open ocean is about a meter, but if this tide is funneled into a narrow opening such as the mouth of a river, it can become much higher. For example, at the Bay of Fundy, on the U.S.–Canada border between Maine and New Brunswick, the high tide can reach nearly 20 m (approximately 60 feet, or the height of a six-story building) above the low-tide level. An enormous amount of energy is contained in the daily motion of the oceans. This energy is constantly eroding and reshaping our planet's coastlines. In some locations, it has been harnessed as a source of electrical power for human activities.

## Gravitational Deformation

What causes the tides? A clue comes from the observation that they exhibit daily, monthly, and yearly cycles. In fact, the tides are a direct result of the gravitational influence of the Moon and the Sun on Earth. We have already seen how gravity keeps Earth and the Moon in orbit about each other, and both in orbit around the Sun. ∞ (Sec. 2.7) For simplicity, let's first consider just the interaction between Earth and the Moon.

Recall that the strength of the gravitational force depends on the distance separating any two objects. Thus, the Moon's gravitational attraction is greater on the side of Earth that faces the Moon than on the opposite side, some 12,800 km (Earth's diameter) farther away. This difference in the gravitational force is small—only about 3 percent—but it produces a noticeable effect—a **tidal bulge.** As illustrated in Figure 7.22, Earth becomes slightly elongated, with the long axis of the distortion pointing toward the Moon.

Earth's oceans undergo the greatest deformation, because liquid can most easily move around on our planet's surface. (A bulge *is* actually raised in the solid material of Earth, but it is about a hundred times smaller than the oceanic bulge.) Thus, the ocean becomes a little deeper in some places (along the line joining Earth to the Moon) and shallower in others (perpendicular to this line). The daily tides we experience result as Earth rotates beneath this deformation.

The variation in the Moon's gravity across Earth is an example of a *differential force*, or **tidal force.** The average gravitational force between two bodies determines their orbit around one another. However, the *tidal* force, superimposed on that average, tends to deform the bodies. The tidal influence of one body on another diminishes very rapidly with increasing distance—in fact, as the inverse *cube* of the separation. For example, if the distance from Earth to the Moon were to double, the tides resulting from the Moon's gravity would decrease by a factor of eight. This rapid decline with increasing distance means that one object has to be very close or very massive in order to have a significant tidal effect on another. #19

We will see many situations in this book where tidal forces are critically important in understanding astronomical phenomena. We still use the word *tidal* in these other contexts, even though we are not discussing oceanic tides and, possibly, not even planets at all. In general astronomical use, the term refers to the deforming effect that the gravity of one body has on another.

Notice in Figure 7.22 that the side of Earth opposite the Moon also exhibits a tidal bulge. The different gravitational pulls—greatest on that part of Earth closest to the Moon, weaker at Earth's center, and weakest of all on Earth's opposite side—cause average tides on opposite sides of our planet to be approximately equal in height. On the side nearer the Moon, the ocean water is pulled slightly toward the Moon. On the opposite side, the ocean water is left behind as Earth is pulled closer to the Moon. Thus, high tide occurs *twice*, not once, each day at any given location. #19

Both the Moon and the Sun exert tidal forces on our planet. Thus, instead of one tidal bulge, there are actually two—one pointing toward the Moon, the other toward the Sun. Even though the Sun is 375 times farther away from Earth than is the Moon, the Sun's mass is so much greater (by a factor of 27 million) that its tidal influence is still significant—about half that of the Moon. The interaction between them accounts for the changes in the height of the tides over the course of a month or a year. When Earth, the Moon, and the Sun are roughly lined up (Figure 7.23a), the gravitational effects reinforce one another, so the highest tides are generally found at times of new and full moons. These tides are known as *spring tides*. When the Earth–Moon line is perpendicular to the Earth–Sun line (at the first and third quarters; Figure 7.23b), the daily tides are smallest. These are termed *neap tides*.

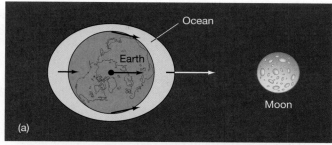

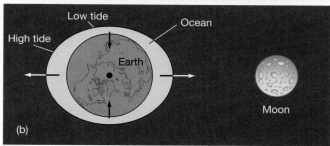

▲ **FIGURE 7.22 Lunar Tides** This exaggerated illustration shows how the Moon induces tides on both the near and far sides of Earth. The lengths of the arrows indicate the relative strengths of the Moon's gravitational pull on various parts of Earth. (a) The lunar gravitational forces acting on several locations on and inside Earth. The force is greatest on the side nearest the Moon and smallest on the opposite side. (b) The *differences* between the lunar forces experienced at the locations shown in part (a) and the force exerted by the Moon on Earth's center. Closest to the Moon, the relative force is toward the Moon because the Moon's gravitational pull is stronger at the surface than it is at the center. However, on the opposite side of Earth, the force at the surface is weaker than at the center, so the relative force is away from the Moon. The arrows thus represent the force with which the Moon tends to either pull matter away from or squeeze it toward Earth's center. Closest to the Moon, the oceans tend to be pulled away from Earth; on the far side, Earth tends to be pulled away from the oceans. The result is a tidal bulge.

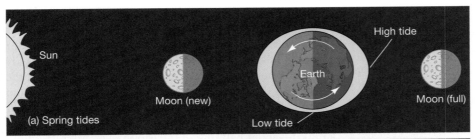

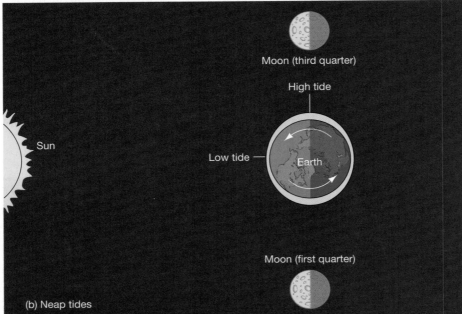

◀ FIGURE 7.23 **Solar and Lunar Tides**
The combined effects of the Sun and the
Moon produce variations in the high and
low tides. (a) When the Moon is either
full or new, Earth, Moon, and Sun are
approximately aligned, and the tidal
bulges raised in Earth's oceans by the
Moon and the Sun reinforce one another.
(b) At first- or third-quarter Moon, the
tidal effects of the Moon and the Sun
partially cancel each other, and the tides
are smallest. Because the Moon's tidal
effect is greater than that of the Sun
(since the Moon is much closer to us), the
net bulge points toward the Moon.

## Earth's Slowing Rotation

Earth rotates once on its axis (relative to the stars) in
$23^h 56^m$—one sidereal day. However, we know from fossil
measurements that Earth's rotation is gradually slowing
down, causing the length of the day to increase by about
1.5 milliseconds (ms) every century—not much on the
scale of a human lifetime, but over millions of years, this
steady slowing of Earth's spin adds up. At this rate, half a
billion years ago, the day was just over 22 hours long and
the year contained 397 days.

A number of natural biological clocks lead us to the
conclusion that Earth's spin rate is decreasing. For example, each day a growth mark is deposited on a certain type
of coral in the reefs off the Bahamas. These growth marks
are similar to the annual rings found in tree trunks, except
that in the case of coral, the marks are made daily, in
response to the day–night cycle of solar illumination.
However, they also show yearly variations as the coral's
growth responds to Earth's seasonal changes, allowing
us to perceive annual cycles. Coral growing today shows
365 marks per year, but ancient coral shows many more
growth deposits per year. Fossilized reefs that are five hundred million years old contain coral with nearly 400 deposits per year of growth.

Why is Earth's spin slowing? The main reason is the
tidal effect of the Moon. In reality, the tidal bulge raised

in Earth by the Moon does *not* point directly at the Moon,
as was shown in Figure 7.21. Instead, because of the
effects of friction, both between the crust and the oceans
and within Earth itself, Earth's rotation tends to drag the
tidal bulge around with it, causing the bulge to be
displaced by a small angle from the Earth–Moon line, in
the same direction as Earth's spin (Figure 7.24). The net
effect of the Moon's gravitational pull on this slightly
offset bulge is to *reduce* our planet's rotation rate. At the
same time, the Moon is spiraling slowly away from Earth,
increasing its average distance from our planet by about
4 cm per year.

This process will continue until Earth rotates on its
axis at exactly the same rate as the Moon orbits Earth. At
that time, the Moon will always be above the same point
on Earth and will no longer lag behind the bulge it
raises. Earth's rotation period will be 47 of our present
days, and the distance to the Moon will be 550,000 km
(about 43 percent greater than at present). However, this
will take a very long time—many billions of years—
to occur.

## CONCEPT CHECK

✔ In what ways do tidal forces differ from the familiar
inverse-square force of gravity?

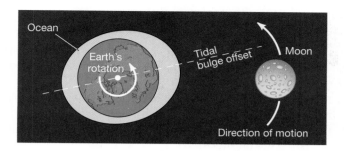

◄ **FIGURE 7.24** **Tidal Bulge** The tidal bulge raised in Earth by the Moon does not point directly at the Moon. Instead, because of the effects of friction, the bulge points slightly ahead of the Moon, in the direction of Earth's rotation. (The magnitude of the effect is greatly exaggerated in this diagram.) Because the Moon's gravitational pull on the near-side part of the bulge is greater than the pull on the far side, the overall effect is to decrease Earth's rotation rate.

# CHAPTER REVIEW

## Summary

① The six main regions of Earth are (from inside to outside) a central metallic **core (p. 170)**, which is surrounded by a thick rocky **mantle (p. 170)**, topped with a thin **crust (p. 170)**. The liquid oceans on our planet's surface make up the **hydrosphere (p. 170)**. Above the surface is the **atmosphere (p. 170)**, which is composed primarily of nitrogen and oxygen and thins rapidly with altitude. Surface winds and weather in the **troposphere (p. 171)**, the lowest region of Earth's atmosphere are caused by **convection (p. 171)**, the process by which heat is moved from one place to another by the upwelling or downflow of a fluid, such as air or water. Higher above the atmosphere lies the **magnetosphere (p. 170)**, where charged particles from the Sun are trapped by Earth's magnetic field.

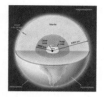

② At high altitudes, in the **ionosphere (p. 171)**, the atmosphere is kept ionized by the absorption of high-energy radiation and particles from the Sun. In the **stratosphere (p. 171)**, just above the troposphere, lies the **ozone layer (p. 172)**, where incoming solar ultraviolet radiation is absorbed. Both the ionosphere and the ozone layer help protect us from dangerous radiation from space. The **greenhouse effect (p. 173)** is the absorption and trapping of infrared radiation emitted by Earth's surface by atmospheric gases (primarily carbon dioxide and water vapor). It makes our planet's surface some 40 K warmer than would otherwise be the case. The air we breathe is not Earth's original atmosphere. That atmosphere was outgassed from our planet's interior by volcanoes and was then altered by solar radiation and, finally, by the emergence of life.

③ We study Earth's interior by observing how **seismic waves (p. 177)**, produced by earthquakes just below Earth's surface, travel through the mantle. We can also study the upper mantle by analyzing the material brought to the surface when a volcano erupts. Earth's center is dense and extremely hot. The planet's iron core consists of a solid **inner core (p. 177)** surrounded by a liquid **outer core (p. 177)**. The process by which heavy material sinks to the center of a planet lighter material rises to the surface is called **differentiation (p. 180)**. Earth's differentiation

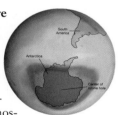

implies that our planet must have been at least partially molten in the past. One way in which this could have occurred is by the heat released during Earth's formation and subsequent bombardment by material from interplanetary space. Another possibility is the energy released by the decay of **radioactive (p. 181)** elements present in the material from which Earth formed.

④ Earth's surface is made up of about a dozen enormous slabs, or plates. The slow movement of these plates across the surface is called continental drift or **plate tectonics (p. 181)**. Earthquakes, volcanism, and mountain building are associated with plate boundaries, where

plates may collide, move apart, or rub against one another. The motion of the plates is thought to be driven by convection in Earth's mantle. The rocky upper layer of Earth that makes up the plates is the **lithosphere (p. 181)**. The semisolid region in the upper mantle over which the plates slide is called the **asthenosphere (p. 181)**. The constant recycling and transformation of crust material as plates separate, collide, and sink into the mantle is called the **rock cycle (p. 189)**. Evidence for past plate motion can be found in the geographical fit of continents, in the fossil record, and in the ages and magnetism of surface rocks.

⑤ Earth's magnetic field extends far beyond the surface of our planet. Charged particles from the solar wind are trapped by Earth's magnetic field lines to form the **Van Allen belts (p. 190)** that surround our planet. When particles

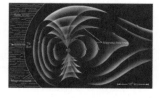

from the Van Allen belts hit Earth's atmosphere, they heat and ionize the atoms there, causing them to glow in an **aurora (p. 191)**. According to **dynamo theory (p. 192)**, planetary magnetic fields are produced by the motion of rapidly rotating, electrically conducting fluid (such as molten iron) in the planet's core.

⑥ The daily **tides (p. 192)** in Earth's oceans are caused by the gravitational effect of the Moon and the Sun, which raise **tidal bulges (p. 193)** in

the hydrosphere. The tidal effect of the Moon is almost twice that of the Sun. The size of the tides depends on the orientations of the Sun and the Moon relative to Earth. A differential gravitational force is always called a **tidal force (p. 193)**, even when no oceans or planets are involved. The tidal interaction between Earth and the Moon is causing Earth's spin to slow.

## Review and Discussion

1. By comparison with Earth's average density, what do the densities of the water and rocks in Earth's crust tell us about Earth's interior?

2. What is Rayleigh scattering? What is its most noticeable effect for us on Earth?

3. How do geologists use earthquakes to obtain information about Earth's interior?

4. Compare and contrast P-waves and S-waves, and explain how they are useful to geologists.

5. What is the greenhouse effect, and what effect does it have on Earth's surface temperature?

6. Give two reasons geologists think that part of Earth's core is liquid.

7. What clue to our planet's history does Earth's differentiation provide?

8. What is convection? What effect does it have on (a) Earth's atmosphere? (b) Earth's interior?

9. How does radioactivity allow us to estimate Earth's age?

10. How did radioactive decay heat Earth early in its history? When did this heating end?

11. What process is responsible for the surface mountains, oceanic trenches, and other large-scale features on Earth's surface?

12. Discuss how distant quasars, lying hundreds of millions of light-years from Earth, are used to monitor the motion of Earth's tectonic plates.

13. What conditions are needed to create a dynamo in Earth's interior? What effect does this dynamo have?

14. Give a brief description of Earth's magnetosphere, and tell how it was discovered.

15. How does Earth's magnetosphere protect us from the harsh realities of interplanetary space?

16. How do we know that Earth's magnetic field has undergone reversals in the past? How do you think Earth's magnetic field reversals might have affected the evolution of life on our planet?

17. Explain how the Moon produces tides in Earth's oceans.

18. If the Moon had oceans like Earth's, what would the tidal effect be like there? How many high and low tides would there be during a "day"? How would the variations in height compare with those on Earth?

19. If Earth had no moon, do you think we would know anything about tidal forces?

20. Is the greenhouse effect operating in Earth's atmosphere helpful or harmful? Give examples. What are the consequences of an enhanced greenhouse effect?

## Conceptual Self-Test: True or False/Multiple Choice

1. The troposphere is the part of the atmosphere in which convection occurs.

2. The oxygen in Earth's atmosphere is the result of the appearance of life on our planet.

3. Sunlight is absorbed by Earth's surface and then reemitted in the form of ultraviolet radiation.

4. Geologists obtain most of their information about Earth's mantle by drilling into our planet's interior.

5. Earth's core temperature is comparable to the surface temperature of the Sun.

6. When plates collide, they simply come to rest and fuse together.

7. Earth's magnetic field is the result of our planet's large, permanently magnetized iron core.

8. An aurora occurs when trapped electrons and protons in the magnetosphere collide with the upper atmosphere.

9. There is one high tide and one low tide per day at any given coastal location on Earth.

10. Tides are caused by the differences in the gravitational pulls of the planets from one side of Earth to the other.

11. If you were making a scale model of Earth representing our planet by a 12-inch basketball, the inner core would be about the size of (a) a $\frac{1}{2}$-inch ball bearing; (b) a 2-inch golf ball; (c) a 4-inch tangerine; (d) a 7-inch grapefruit.

12. Earth's average density is about the same as that of (a) a glass of water; (b) a heavy iron meteorite; (c) an ice cube; (d) a chunk of black volcanic rock.

13. According to Figure 7.2 ("Earth's Atmosphere"), commercial jet airplanes flying at 10 km are in (a) the troposphere; (b) the stratosphere; (c) the ozone; (d) the mesosphere.

14. If there were significantly more greenhouse gases, such as $CO_2$, in Earth's atmosphere, then (a) the ozone hole would close; (b) the ozone hole would get larger; (c) Earth's average temperature would change; (d) plants would grow faster than animals could eat them.

15. If seismometers registered P- and S-waves everywhere on the Moon, they would suggest that the Moon had (a) the same layered structure as Earth; (b) no molten core; (c) no moonquakes; (d) the same density throughout.

16. The deepest that geologists have drilled into Earth is about the same as (a) the height of the Statue of Liberty; (b) the altitude most commercial jet airplanes fly; (c) the distance between New York and Los Angeles; (d) the distance between the United States and China.

17. Due to plate tectonics, the width of the Atlantic Ocean is separating at a rate about the same as the growth of (a) grass; (b) human hair; (c) human fingernails; (d) dust in a typical home.

18. At Earth's geographic North Pole, a magnetic compass needle would point (approximately) (a) toward Alaska; (b) toward Kansas City; (c) toward Paris; (d) straight down.

19. If Earth had no Moon, then tides would (a) not occur; (b) occur more often and with more intensity; (c) still occur, but not really be measurable; (d) occur with the same frequency, but would not be as strong.

20. Which of the following statements is true? Because of the tides, (a) Earth's rotation rate is increasing; (b) the Moon is spiraling away from Earth; (c) Earth will eventually drift away from the Sun; (d) earthquake activity is increasing.

# Problems

 *Algorithmic versions of these Problems are available in the Practice Problems module of the Companion Website. The number of dots preceding each Problem indicates its approximate level of difficulty.*

1. • Verify that Earth's orbital perihelion and aphelion, mean orbital speed, surface gravity, and escape speed are correct as listed in the Earth Data box on p. 172.

2. • What would Earth's surface gravity and escape speed be if the entire planet had a density equal to that of the crust, say, 3000 kg/m³?

3. • Approximating Earth's atmosphere as a layer of gas 7.5 km thick, with uniform density 1.3 kg/m³, calculate the total mass of the atmosphere. Compare your result with Earth's mass.

4. ••• As discussed in the text, without the greenhouse effect, Earth's average surface temperature would be about 250 K. With the greenhouse effect, it is some 40 K higher. Use this information and Stefan's law to calculate the fraction of infrared radiation leaving Earth's surface that is absorbed by greenhouse gases in the atmosphere. ∞ (Sec. 3.4)

5. •• Most of Earth's ice is found in Antarctica, where permanent ice caps cover approximately 0.5 percent of Earth's total surface area and are 3 km thick, on average. Earth's oceans cover roughly 71 percent of our planet, to an average depth of 3.6 km. Assuming that water and ice have roughly the same density, estimate by how much sea level would rise if global warming were to cause the Antarctic ice caps to melt.

6. • Following an earthquake, how long would it take a P-wave, moving in a straight line with a speed of 5 km/s, to reach Earth's opposite side?

7. • On the basis of the data presented in the text, estimate the fractions of Earth's volume represented by (a) the inner core, (b) the outer core, (c) the mantle, and (d) the crust.

8. • At 3 cm/yr, how long would it take a typical plate to traverse the present width of the Atlantic Ocean, about 6000 km?

9. • In a certain sample of rock, it is found that 25 percent of uranium-238 nuclei have decayed into lead-206. On the basis of the data given in *More Precisely 7-2*, estimate the age of the rock sample.

10. •• A second sample of rock is found to contain three times as many lead-207 nuclei as uranium-235 nuclei. On the basis of the data given in *More Precisely 7-2*, what ratio of uranium-238 to lead-206 nuclei would you expect?

11. ••• Astronauts in orbit are weightless because they are falling freely in Earth's gravitational field, but they are still subject to tidal forces. Calculate the relative acceleration due to tidal forces of two masses in low Earth orbit, placed 1 m apart along a line extending radially outward from Earth's center. Compare this acceleration with the acceleration due to gravity at Earth's surface.

12. •• You are standing on Earth's surface during a total eclipse, and both the Moon and the Sun are directly overhead. By what fraction is your weight changed due to their combined tidal gravitational force?

13. •• You are standing on Earth's surface, and the full Moon is directly overhead. By what fraction is your weight decreased due to the combination of the Sun's and the Moon's tidal gravitational forces?

14. ••• The planet Jupiter exerts a strong tidal force on its innermost moon, Io. Compare Jupiter's tidal force on an object on Io's surface with the gravitational force on the body due to Io's own gravity. (For definiteness, perform the calculation for a 1-kg mass resting on Io's surface on the line joining the center of the planet to the center of the moon.) Io orbits at a distance of 422,000 km. Its mass and radius are $9.0 \times 10^{22}$ kg and 1800 km, respectively. Jupiter's mass is $1.9 \times 10^{27}$ kg.

15. •• Compare the magnitude of the tidal gravitational force on Earth due to Jupiter with that due to the Moon. Assume an Earth–Jupiter distance of 4.2 AU. On the basis of your answer, do you think that the tidal stresses caused by a "cosmic convergence"—a chance alignment of the four jovian planets so that they all appear from Earth to be in exactly the same direction in the sky—would have any noticeable effect on our planet?

*The Companion Website at www.aw-bc.com/chaisson provides algorithmically generated versions of each chapter's Problems, along with additional quizzes, an Animations & Videos gallery, an Images gallery, an interactive Glossary, and a full eBook.*

CHAPTER **8**

*America's manned exploration of the Moon was arguably the greatest engineering feat of the twentieth century. indeed one of the greatest of all time. Nine missions were launched to the Moon. a dozen astronauts were landed, and all returned safely to Earth. The Apollo program ended in 1972, as quickly as it had begun a decade before—largely because of political posturing at the height of the Cold War. Here, an Apollo 15 astronaut near Mount Hadley (see also Figure 8.22) is adjusting some instruments for testing the soil. (NASA) ▶*

# THE MOON AND MERCURY

## Scorched and Battered Worlds

The Moon is Earth's only natural satellite. Mercury, the smallest terrestrial world, is the planet closest to the Sun. Despite their different environments, these two bodies have many similarities—indeed, at first glance, you might even mistake one for the other. Both have heavily cratered, ancient surfaces, littered with boulders and pulverized dust. Both lack atmospheres to moderate day-to-night variations in solar heating and experience wild temperature swings as a result. Both are geologically dead.

In short, the Moon and Mercury differ greatly from Earth, but it is precisely those differences that make these desolate worlds so interesting to planetary scientists. Why is the Moon so unlike our own planet, despite its nearness to us, and why does planet Mercury apparently have so much more in common with Earth's Moon than with Earth itself? In this chapter, we explore the properties of these two worlds as we begin our comparative study of the planets and moons that make up our solar system.

## LEARNING GOALS

*Studying this chapter will enable you to*

1 Specify the general characteristics of the Moon and Mercury, and compare them with those of Earth.
2 Describe the surface features of the Moon and Mercury, and recount how those two bodies were formed by events early in their history.
3 Explain how the Moon's rotation is influenced by its orbit around Earth and Mercury's by its orbit around the Sun.
4 Explain how observations of cratering can be used to estimate the age of a body's surface.
5 Describe the evidence for ancient volcanism on the Moon and Mercury.
6 Compare the Moon's interior structure with that of Mercury.
7 Summarize the leading theory of the formation of the Moon.
8 Discuss how astronomers have pieced together the story of the Moon's evolution, and compare its evolutionary history with that of Mercury.

 Visit www.aw-bc.com/chaisson for additional images, animations, quizzes, and eBook for this chapter.

# 8.1 Orbital Properties

We begin our study of the Moon and Mercury by examining their orbits. This knowledge will, in turn, aid us in determining and explaining the other properties of these worlds. Detailed orbital and physical data are presented in the Moon Data box (p. 202) and the Mercury Data box (p. 220).

## The Moon

Parallax methods, described in Chapter 1, can provide us with quite accurate measurements of the distance to the Moon, using Earth's diameter as a baseline. ∞ (Sec. 1.7) Radar ranging yields more accurate distances. The Moon is much closer than any of the planets, and the radar echo bounced off the Moon's surface is strong. A radio telescope receives the echo after a round trip of 2.56 seconds. Dividing this time by 2 and multiplying it by the speed of light (300,000 km/s) gives us a distance of 384,000 km. (The actual distance at any particular time depends on the Moon's location in its slightly elliptical orbit around Earth.)

Current laser-ranging technology, using reflectors placed on the lunar surface by *Apollo* astronauts (see *Discovery 8-1*) to reflect laser beams fired from Earth, allows astronomers to measure the round-trip time with submicrosecond accuracy. Repeated measurements have allowed astronomers to determine the Moon's orbit to within a few centimeters. This precision is necessary for programming unmanned spacecraft to land successfully on the lunar surface.

## Mercury

Viewed from Earth, Mercury never strays far from the Sun. As illustrated in Figure 8.1(a), the planet's 0.4-AU orbital semimajor axis means that its angular distance from the Sun never exceeds 28°. Consequently, the planet is visible to the naked eye only when the Sun's light is blotted out—just before dawn or just after sunset (or, much less frequently, during a total solar eclipse)—and it is not possible to follow Mercury through a full cycle of phases. In fact, although Mercury was well known to ancient astronomers, they originally believed that this companion to the Sun was two different objects, and the connection between the planet's morning and evening appearances took some time to establish. However, later Greek astronomers were certainly aware that the "two planets" were really different alignments of a single body. Figure 8.1(b), a photograph taken just after sunset, shows Mercury above the western horizon, along with three other planets and the Moon.

Because Earth rotates at a rate of 15° per hour, Mercury is visible for at most 2 hours on any given night, even under the most favorable circumstances. For most

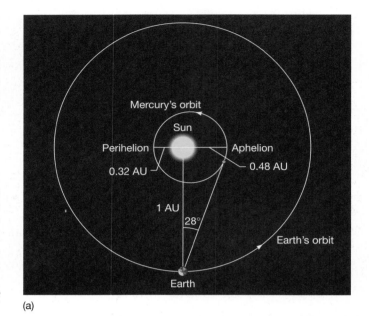

(a)

(b)

R I V U X G

▲ **FIGURE 8.1** **Evening Sky** (a) Mercury's orbit has a semimajor axis of just 0.4 AU, so the planet can never be farther than 28° from the Sun, as seen from Earth. Mercury's eccentric orbit means that this maximum separation is achieved only for the special configuration shown here, in which the Earth–Sun line is perpendicular to the long axis of Mercury's orbit and Mercury is near aphelion (its greatest distance from the Sun). (b) Four planets, together with the Moon, are visible in this photograph taken shortly after sunset. To the right of the Moon (top left) is the brightest planet, Venus. A little farther to the right is Mars, with the star Regulus just below and to its left. At the lower right, at the edge of the Sun's glare, are Jupiter and Mercury. (The Moon appears round rather than crescent shaped because the "dark" portion of its disk is indirectly illuminated by sunlight reflected from Earth. This "earthshine," relatively faint to the naked eye, is exaggerated in the overexposed photographic image.) (*J. Sanford/Photo Researchers, Inc.*)

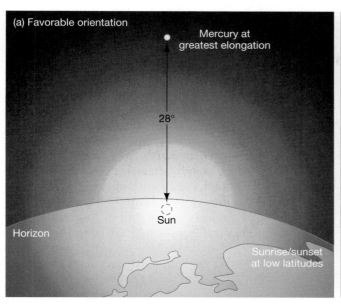

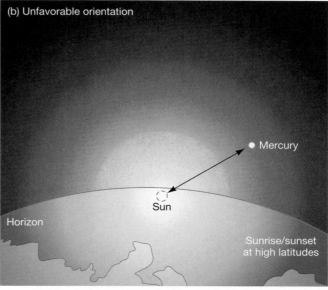

(a) Favorable orientation

Mercury at greatest elongation

28°

Horizon

Sun

Sunrise/sunset at low latitudes

(b) Unfavorable orientation

Mercury

Horizon

Sun

Sunrise/sunset at high latitudes

▲ FIGURE 8.2 **Visibility of Mercury** Favorable and unfavorable orientations of Mercury's orbit result from different orientations of Earth and different locations of the observer, both of which affect the angle between the horizon and Mercury's orbital plane. (a) At greatest elongation, Mercury lies some 28° from the Sun (see Figure 8.1). The planet is most easily visible when it also lies high above the horizon. (b) At the most unfavorable orientations, Mercury is close to both the Sun and the horizon.

observers at most times of the year, Mercury is considerably less than 28° above the horizon, so it is generally visible for a much shorter period (see Figure 8.2). Nowadays, large telescopes can filter out the Sun's glare and observe Mercury even during the daytime, when the planet is higher in the sky and atmospheric effects are reduced. (The amount of air that the light from the planet has to traverse before reaching our telescope decreases as the height of the planet above the horizon increases.) In fact, some of the best views of Mercury

have been obtained in this way. The naked-eye or amateur astronomer is generally limited to nighttime observations, however.

In all cases, it becomes progressively more difficult to view Mercury the closer (in the sky) its orbit takes it to the Sun. The best images of the planet therefore show a "half Mercury," close to its maximum angular separation from the Sun, or *maximum elongation*, as illustrated in Figure 8.3. (A planet's *elongation* is just its angular distance from the Sun, as seen from Earth.)

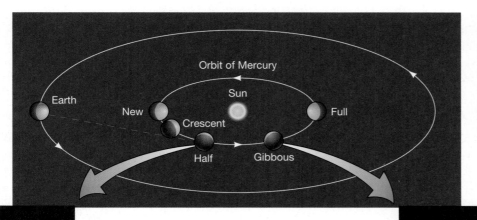

◀ FIGURE 8.3 **Phases of Mercury** Some views of Mercury at different points along its orbit. The best images of the planet are taken when it is at its maximum elongation (greatest apparent distance from the Sun) and show a "half Mercury" (cf. Figure 2.12a). (*R. Beebe*)

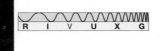

# MOON DATA

| | |
|---|---|
| Orbital semimajor axis* | 384,000 km |
| Orbital eccentricity | 0.055 |
| Perigee** | 363,000 km |
| Apogee† | 406,000 km |
| Mean orbital speed | 1.02 km/s |
| Sidereal orbital period | 27.3 Earth days*** |
| Synodic orbital period | 29.5 Earth days*** |
| Orbital inclination to the ecliptic | 5.2° |
| Greatest angular diameter, as seen from Earth' | 32.9 |
| Mass | 7.35 × 10²² kg 0.012 (Earth = 1) |
| Equatorial radius | 1738 km 0.27 (Earth = 1) |
| Mean density | 3340 kg/m³ 0.61 (Earth = 1) |
| Surface gravity | 1.62 m/s² 0.17 (Earth = 1) |
| Escape speed | 2.38 km/s |
| Sidereal rotation period | 27.3 Earth days*** |
| Axial tilt | 6.7° |
| Surface magnetic field | no detectable global field |
| Surface temperature | 100–400 K |

*Orbit around Earth.
**Closest approach to Earth.
***1 Earth (mean solar) day = 24 hours
†Greatest distance from Earth.

## 8.2 Physical Properties

From Earth, the Moon's angular diameter is about 0.5°. Knowing that and the distance to the Moon, we can easily calculate our satellite's true size, as discussed in Chapter 1. ∞ (*More Precisely 1-3*) The Moon's radius is about 1700 km, roughly one-fourth that of Earth. More precise measurements yield a lunar radius of 1738 km. We can determine Mercury's radius by similar reasoning. At its closest approach to Earth, at a distance of about 0.52 AU, Mercury's angular diameter is measured to be 13″ (arc seconds), implying a radius of about 2450 km, or 0.38 of Earth's radius. More accurate measurements by unmanned space probes yield a result of 2440 km.

As mentioned in Chapter 6, even before the Space Age, the masses of both the Moon and Mercury were already quite well known from studies of their effects on Earth's orbit. ∞ (Sec. 6.2) The mass of the Moon is $7.3 \times 10^{22}$ kg, approximately one-eightieth (0.012) the mass of Earth. The mass of Mercury is $3.3 \times 10^{23}$ kg—about 0.055 Earth mass.

The Moon's average density of 3300 kg/m³ contrasts with the average Earth value of about 5500 kg/m³, suggesting that the Moon contains fewer heavy elements (such as iron) than Earth does. In contrast, despite its many other similarities to the Moon, Mercury's mean density is 5400 kg/m³, only slightly less than that of Earth. Assuming that surface rocks on Mercury are of similar density to surface rocks on Earth and the Moon, we are led to the conclusion that the interior of Mercury must contain a lot of high-density material, most probably iron. In fact, since Mercury is considerably less massive than Earth, its interior is squeezed less by the weight of overlying material, so Mercury's iron core must actually contain a much larger fraction of the planet's mass than does our own planet's core. ∞ (Sec. 6.2)

Because the Moon and Mercury are so much less massive than Earth, their gravitational fields are also weaker. The force of gravity on the lunar surface is only about one-sixth that on Earth; Mercury's surface gravity is a little stronger—about 0.4 times Earth's. Thus, an astronaut weighing 180 lb on Earth would weigh a mere 30 lb on the Moon and 72 lb on Mercury. Those bulky space suits used by the *Apollo* astronauts on the Moon were not nearly as heavy as they appeared!

Astronomers have never observed any appreciable atmosphere on the Moon or Mercury, either spectroscopically from Earth or during close approaches by spacecraft. This is a direct consequence of these bodies' weak gravitational fields, as discussed in *More Precisely 8-1* (p. 210). Simply put, a massive object has a better chance of retaining an atmosphere, because the more massive an object is, the larger is the speed needed for atoms or molecules to escape from the object's gravitational pull. The Moon's escape speed is only 2.4 km/s, compared with 11.2 km/s for Earth; Mercury's escape speed is 4.2 km/s. Any primary atmospheres these worlds had initially, or secondary atmospheres that appeared later, are gone forever. ∞ (Sec. 7.2)

During its flybys of Mercury in 1974 and 1975, the U.S. space probe *Mariner 10* found traces of what was at first thought to be an atmosphere on the planet. ∞ (Sec. 6.6) However, this gas is now known to be temporarily trapped hydrogen and helium "stolen" from the solar wind by the planet's gravity. Mercury captures this gas and holds it for just a few weeks. Both the Moon and Mercury have extremely tenuous (less than a trillionth the density of Earth's atmosphere) envelopes of sodium and potassium. Scientists think that these atoms are torn out of the surface rocks following impacts with high-energy particles in the solar wind; they do not constitute a true atmosphere in any sense. Thus, neither the Moon nor Mercury has any protection against the harsh environment of interplanetary space. This fact is crucial in understanding their surface evolution and present-day appearance.

Lacking the moderating influence of an atmosphere, both the Moon and Mercury are characterized by wide variations in surface temperature. Noontime temperatures at the Moon's equator can reach 400 K, well above the boiling point of water. Because of its proximity to the Sun, Mercury's daytime temperature is even higher—radio observations of the planet's thermal emissions indicate that it can reach 700 K. ∞ (Sec. 3.4) But at night or in the shade, temperatures on both worlds fall to about 100 K, well below water's freezing point. Mercury's 600-K temperature range is the largest of any planet or moon in the solar system.

CONCEPT CHECK

✔ Why do the Moon and Mercury have no significant atmospheres, unlike Earth?

# 8.3 Surface Features on the Moon and Mercury

## Lunar Terrain

The first observers to point their telescopes at the Moon—most notable among them Galileo Galilei—saw large dark areas resembling (they thought) Earth's oceans. They also saw light-colored areas resembling the continents. Both types of regions are clear in Figure 8.4, a *mosaic* (a composite image constructed from many individual photographs) of the full Moon. The light and dark surface features are also evident to the naked eye, creating the face of the familiar "man in the Moon."

Today we know that the dark areas are not oceans, but extensive flat areas that resulted from lava flows during a much earlier period of the Moon's evolution. Nevertheless, they are still called **maria,** a Latin word meaning "seas" (singular: *mare*). There are 14 maria, all roughly circular. The largest of them (Mare Imbrium) is about 1100 km in diameter. The lighter areas, originally dubbed *terrae*, from the Latin word for "land," are now known to

▲ FIGURE 8.4 **Full Moon, Near Side** A photographic mosaic of the full Moon, north pole at the top. Because the Moon emits no visible radiation of its own, we can see it only by the reflected light of the Sun. Some prominent maria are labeled. (*UC/Lick Observatory*)

be elevated several kilometers above the maria. Accordingly, they are usually called the lunar **highlands.**

The smallest lunar features we can distinguish with the naked eye are about 200 km across. Telescopic observations further resolve the surface into numerous bowl-shaped depressions, or **craters** (after the Greek word for "bowl"). Most craters apparently formed eons ago, primarily as the result of meteoritic impact. In Figures 8.5(a) and (b), craters are particularly clear near the *terminator* (the line that separates day from night on the surface), where the Sun is low in the sky and casts long shadows that enable us to distinguish quite small surface details.

Due to the blurring effects of our atmosphere, the smallest lunar objects that telescopes on Earth's surface can resolve are about 1 km across (see Figure 8.5c). Much more detailed photographs have been taken by orbiting spacecraft and, of course, by visiting astronauts. Figure 8.6 is a view of some lunar craters taken from an orbiting spacecraft, showing features as small as 500 m across. Craters are found everywhere on the Moon's surface, although they are much more prevalent in the highlands. They come in all sizes—the largest are hundreds of kilometers in diameter; the smallest are microscopic.

Based on studies of lunar rock brought back to Earth by *Apollo* astronauts and unmanned Soviet landers, geologists have identified important differences in both *composition* and *age* between the highlands and the maria. The highlands are made largely of rocks rich in aluminum, making them lighter in color and lower in density (2900 kg/m³) than the material

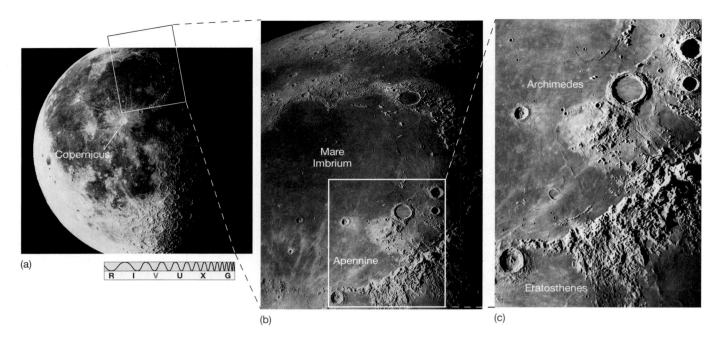

(a)

(b)

(c)

▲ **FIGURE 8.5** **Moon, Close Up** (a) The Moon near third quarter. Surface features are much more visible near the *terminator,* the line separating light from dark, where sunlight strikes at a sharp angle and shadows highlight the landscape. (b) Magnified view of a region near the terminator, as seen from Earth through a large telescope. The central dark area is Mare Imbrium, ringed at the bottom by the Apennine mountains. (c) Enlargement of a portion of (b). The smallest craters visible here have diameters of about 2 km, about twice the size of the Barringer crater on Earth shown in Figure 8.19. *(UC/Lick Observatory; Palomar)*

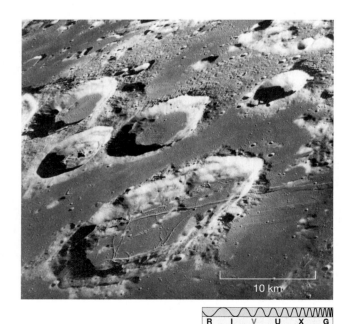

▲ **FIGURE 8.6** **Moon from Apollo** The Moon, as seen from the *Apollo 8* orbiter during the first human circumnavigation of our satellite in 1968. Craters ranging in size from 50 km to 500 m (also the width of the long fault lines) can be seen. *(NASA)*

in the maria, which contains more iron, giving it a darker color and greater density (3300 kg/m$^3$). Loosely speaking, the highlands represent the Moon's crust, whereas the maria are made of mantle material. Maria rock is quite similar to terrestrial basalt, and geologists think that it arose on the Moon much as basalt did on Earth, from the upwelling of molten material through the crust. ∞ (Sec. 7.3) Radioactive dating indicates ages of 4 to 4.4 billion years for highland rocks and from 3.2 to 3.9 billion years for those from the maria. ∞ (*More Precisely 7-2*)

All of the Moon's significant surface features have names. The 14 maria bear fanciful Latin names—Mare Imbrium ("Sea of Showers"), Mare Nubium ("Sea of Clouds"), Mare Nectaris ("Sea of Nectar"), and so on. Most mountain ranges in the highlands bear the names of terrestrial mountain ranges—the Alps, the Carpathians, the Apennines, the Pyrenees, and so on. Most of the craters are named after great scientists or philosophers, such as Plato, Aristotle, Eratosthenes, and Copernicus.

Because the Moon rotates once on its axis in exactly the same time it takes to complete one orbit around Earth, the Moon has a "near" side, which is always visible from Earth, and a "far" side, which never is (see Section 8.4). To the surprise of most astronomers, when the far side of the Moon was mapped, first by Soviet and later by U.S. spacecraft (see *Discovery 8-1*), no major maria were found there. The lunar far side (Figure 8.7) is composed almost entirely of highlands. This fact has great bearing on our theory of how the Moon's surface terrain came into being, for it implies that the processes involved could *not* have been entirely internal in nature. Earth's presence must somehow have played a role.

CONCEPT CHECK

✔ Describe three important ways in which the lunar maria differ from the highlands.

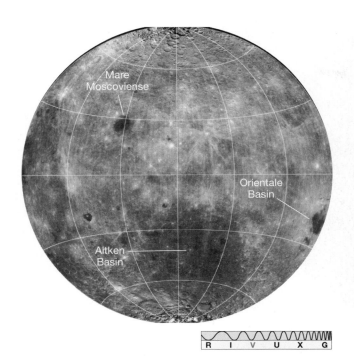

▲ FIGURE 8.7 **Full Moon, Far Side** The far side of the Moon, as seen by the *Clementine* military spacecraft. The large, dark region at center bottom outlines the South Pole–Aitken Basin, the largest and deepest impact basin known in the solar system. Only a few small maria exist on the far side. (*U.S. Dept. of Defense*)

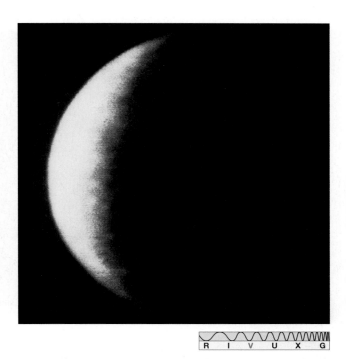

▲ FIGURE 8.8 **Mercury** Photograph of Mercury taken from Earth with a large ground-based optical telescope. Only a few faint surface features are discernible. (*Palomar Observatory/Caltech*)

## The Surface of Mercury

Mercury is difficult to observe from Earth because of Mercury's closeness to the Sun. Even with a fairly large telescope, we see it only as a slightly pinkish disk. Figure 8.8 is one of the few photographs of Mercury taken from Earth that shows any evidence of surface markings. Astronomers could only speculate about the faint, dark markings in the days before *Mariner 10*'s arrival. We now know that these markings are much like those seen by an observer gazing casually at Earth's Moon. The largest ground-based telescopes can resolve surface features on Mercury about as well as we can perceive features on the Moon with our unaided eyes.

In 1974, *Mariner 10* approached within 10,000 km of the surface of Mercury, sending back high-resolution images of the planet. ∞ (Sec. 6.6) These photographs, which showed surface features as small as 150 m across, revolutionized our knowledge of the planet. Figures 8.9(a) and (b) show views of Mercury taken by *Mariner 10* from a distance of about 200,000 km.

As discussed in Chapter 6, *Mariner 10* did not go into orbit around Mercury, but instead was placed in a some-what eccentric orbit around the Sun that brought it close to the planet every 176 days—exactly 2 Mercury years. ∞ (Sec. 6.6) However, the peculiar combination of Mercury's orbital period and rotation rate (discussed in more detail in Sec. 8.4) meant that *Mariner* saw the *same*

face of the planet at each approach. As a result, less than half of the planet's surface has been mapped. Together, the two mosaics in Figure 8.9 cover the known surface of Mercury. No similar photographs exist of the hemisphere that happened to be in shadow during each encounter.

Figure 8.10 shows a higher resolution photograph of the planet from a distance of 20,000 km. The similarities to the Moon are striking. We see no sign of clouds, rivers, dust storms, or other aspects of weather. Much of the cratered surface bears a strong resemblance to the Moon's highlands. Mercury, however, shows few extensive lava flow regions akin to the lunar maria.

## 8.4 Rotation Rates

The spins of both the Moon and Mercury are strongly influenced by their proximity to their parent bodies—Earth and the Sun, respectively. By studying the processes responsible for the rotation rates observed today, astronomers learn about the role of tidal forces in shaping the details of the solar system.

### The Rotation of the Moon

As mentioned earlier, the Moon's rotation period is precisely equal to its period of revolution about Earth—27.3 days—so the Moon keeps the same side facing Earth at all times (see Figure 8.11). To an astronaut standing on the Moon's near-side surface, Earth would appear almost stationary in the sky (although our planet's daily rotation would be clearly

(a)

(b)

◀ **FIGURE 8.9  Mercury, Up Close**  (a) Mercury is imaged here as a mosaic of photographs—a composite image constructed from many individual images—taken by the *Mariner 10* spacecraft in the mid-1970s during its approach to the planet. At the time, the spacecraft was 200,000 km away from Mercury. (b) *Mariner 10*'s view of Mercury as it sped away from the planet after each encounter. Again, the spacecraft was about 200,000 km away when the photographs making up this mosaic were taken. (*NASA*)

R  I  V  U  X  G

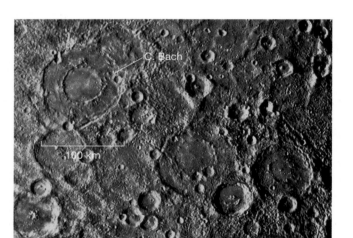

C. Bach

100 km

R  I  V  U  X  G

▲ **FIGURE 8.10  Mercury, Very Close**  Another photograph of Mercury by *Mariner 10*, this time from about 20,000 km above the planet's surface. The double-ringed crater at the upper left, named C. Bach and about 100 km across, exemplifies how many of the large craters on Mercury tend to form double, rather than single, rings. The reason is not yet understood. (*NASA*)

about 20 times greater than that on Earth, and the Moon's tidal bulge is correspondingly larger.

In Chapter 7, we saw how lunar tidal forces are causing Earth's spin to slow and how, as a result, Earth will eventually rotate on its axis at the same rate as the Moon revolves around Earth.  ∞ (Sec. 7.6) Earth's rotation will not become synchronous with the Earth–Moon orbital period for hundreds of billions of years. In the case of the Moon, however, the process has already gone to completion. The Moon's much larger tidal deformation caused it to evolve into a synchronous orbit long ago, and the Moon is said to have become *tidally locked* to Earth. Most of the moons in the solar system are similarly locked by the tidal fields of their parent planets.

Actually, the size of the lunar bulge is too great to be produced by Earth's present-day tidal influence. The explanation seems to be that, long ago, the distance from Earth to the Moon may have been as little as two-thirds of its current value, or about 250,000 km. Earth's tidal force on the Moon would then have been more than three times greater than it is today and could have accounted for the Moon's elongated shape. The resulting distortion could have "set" when the Moon solidified, thus surviving to the present day, and at the same time accelerating the synchronization of the Moon's orbit.

## Measurement of Mercury's Spin

In principle, the ability to discern surface features on Mercury should allow us to measure its rotation rate simply by watching the motion of a particular region around the planet. In the mid-19th century, an Italian astronomer named Giovanni Schiaparelli did just that. He concluded that Mercury always keeps one side facing the Sun, much as our Moon perpetually presents only one

evident). This condition, in which the spin of one body is precisely equal to (or *synchronized* with) its revolution around another body, is known as a **synchronous orbit.**

The fact that the Moon is in a synchronous orbit around Earth is no accident. It is an inevitable consequence of the gravitational interaction between those two bodies. Just as the Moon raises tides on Earth, Earth also produces a tidal bulge in the Moon. Indeed, because Earth is so much more massive, the tidal force on the Moon is

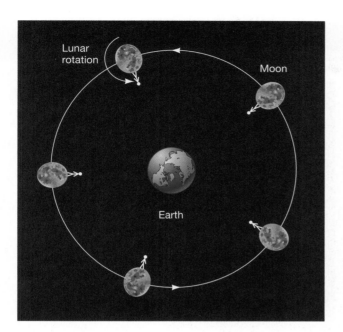

▲ FIGURE 8.11 **The Moon's Synchronous Rotation** As the Moon orbits Earth, it keeps one face permanently pointed toward our planet. To the astronaut shown here, Earth is always directly overhead. In fact, the Moon is slightly elongated in shape owing to Earth's tidal pull on it, with its long axis perpetually pointing toward Earth. (The elongation is highly exaggerated in this diagram.)

face to Earth. The explanation suggested for this supposed synchronous rotation was the same as that for the Moon: The tidal bulge raised in Mercury by the Sun had modified the planet's rotation rate until the bulge always pointed directly at the Sun. Although the surface features could not be seen clearly, the combination of Schiaparelli's observations and a plausible physical explanation was enough to convince most astronomers, and the belief that Mercury rotates synchronously with its revolution about the Sun (i.e., once every 88 Earth days) persisted for almost half a century.

In 1965, astronomers making radar observations of Mercury from the Arecibo radio telescope in Puerto Rico (see Figure 5.24) discovered that this long-held view was in error. The technique they used is illustrated in Figure 8.12, which shows a radar signal reflecting from the surface of a hypothetical planet. Let's imagine, for the purpose of this discussion, that the pulse of outgoing radiation is of a single frequency. The returning pulse bounced off the planet is very much weaker than the outgoing signal. Beyond this change, the reflected signal can be modified in two important ways. First, the signal as a whole may be redshifted or blueshifted as a consequence of the Doppler effect, depending on the overall radial velocity of the planet with respect to Earth. ∞ (Sec. 3.5) We will assume for simplicity that this velocity is zero, so that, on average, the frequency of the reflected signal is the same as that of the outgoing beam.

Second, if the planet is rotating, the radiation reflected from the side of the planet moving toward us returns at a slightly higher frequency than the radiation reflected from the receding side. (Think of the two hemispheres as being separate sources of radiation and moving at slightly different velocities, one toward us and one away.) The effect is very similar to the rotational line broadening discussed in Chapter 4 (see Figure 4.18), except that in this case the radiation we are measuring was not emitted by the planet, but only reflected from its surface. ∞ (Sec. 4.5) What we see in the reflected signal is a spread of frequencies on either side of the original frequency. By measuring the extent of that spread, we can determine the planet's rotational speed.

In this way, the Arecibo researchers found that the rotation period of Mercury is not 88 days, as had previously been thought, but 59 days, exactly two-thirds of the planet's orbital period. Because there are exactly three rotations for every two revolutions, we say that there is a 3:2 *spin–orbit resonance* in Mercury's motion. In this context, the term **resonance** just means that two characteristic times—here, Mercury's day and year—are related to each other in a simple way. An even simpler example of a spin–orbit resonance is the Moon's orbit around Earth. In that case, the rotation is synchronous with the revolution, and the resonance is said to be 1:1.

Figure 8.13 illustrates some implications of Mercury's curious rotation for a hypothetical inhabitant of the planet. Mercury's solar day—the time from noon to noon, say—is 2 Mercury years long! The Sun stays "up" in the black Mercury sky for almost 3 Earth months at a time, after which follow nearly 3 Earth months of darkness. At any given point in its orbit, Mercury presents the same face to the Sun, not every time it revolves, but *every other* time.

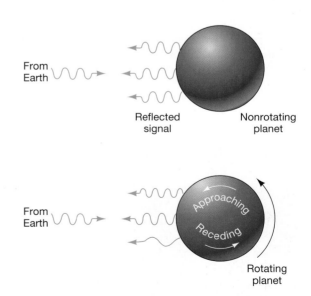

▲ FIGURE 8.12 **Planetary Radar** A radar beam (blue waves) reflected from a rotating planet yields information about both the planet's line-of-sight motion and its rotation rate (see Figure 4.18).

# DISCOVERY 8-1

## Lunar Exploration

The Space Age began in earnest on October 4, 1957, with the launch of the Soviet satellite *Sputnik 1*. Thirteen months later, on January 4, 1959, the Soviet *Luna 1*, the first human-made craft to escape Earth's gravity, passed the Moon. *Luna 2* crash-landed on the surface in September of that year, and *Luna 3* returned the first pictures of the far side a month later. The long-running *Luna* series established a clear Soviet lead in the early "space race" and returned volumes of detailed information about the Moon's surface. Several of the *Luna* missions landed and returned surface material to Earth.

The U.S. lunar exploration program got off to a rocky start. The first six attempts in the *Ranger* series, between 1961 and 1964, failed to accomplish their objective of just hitting the Moon. The last three were successful, however. *Ranger 7* collided with the lunar surface (as intended) on June 28, 1964. Five U.S. *Lunar Orbiter* spacecraft, launched in 1966 and 1967, were successfully placed in orbit around the Moon, and they relayed high-resolution images of much of the lunar surface back to Earth. Between 1966 and 1968, seven *Surveyor* missions soft-landed on the Moon and performed detailed analyses of the surface.

Many of these unmanned U.S. missions were performed in support of the manned *Apollo* program. On May 25, 1961, at a time when the U.S. space program was in great disarray, President John F. Kennedy declared that the United States would "send a man to the Moon and return him safely to Earth" before the end of the decade, and the *Apollo* program was born. On July 20, 1969, less than 12 years after *Sputnik* and only 8 years after the statement of the program's goal, *Apollo 11* commander Neil Armstrong became the first human to set foot on the Moon, in Mare Tranquilitatis (the Sea of Tranquility). Three-and-a-half years later, on December 14, 1972, scientist–astronaut Harrison Schmitt, of *Apollo 17*, was the last.

The astronauts who traveled in pairs to the lunar surface in each lunar lander (shown in the first photograph) performed numerous geological and other scientific studies on the surface.

(NASA)

The later landers brought with them a "lunar rover"—a small golf cart–sized vehicle that greatly expanded the area the astronauts could cover. Probably the most important single aspect of the *Apollo* program was the collection of samples of surface rock from various locations on the Moon. In all, some 382 kg of material was returned to Earth. Chemical analysis and radioactive dating of these samples revolutionized our understanding of the Moon's surface history. No amount of Earth-based observations could have achieved the same results.

Each *Apollo* lander left behind a nuclear-powered package of scientific instruments called the *Apollo Lunar Surface Experiments Package* (*ALSEP*, second photograph) to monitor the solar wind, measure heat flow in the Moon's interior, and, perhaps most important, record lunar seismic activity. With several *ALSEP*s on the surface, scientists could determine the location of "moonquakes" by triangulation and map the

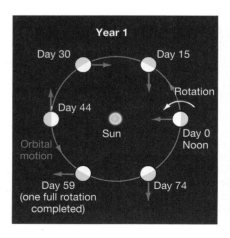

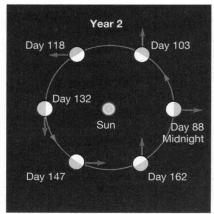

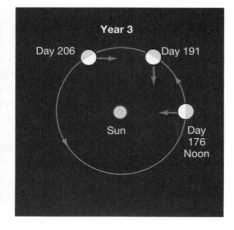

(NASA)

Moon's inner structure, obtaining information critical to our understanding of the Moon's evolution.

By any standards, the *Apollo* program was a spectacular success. It represents a towering achievement of the human race. The project's goals were met on schedule and within budget, and our knowledge of the Moon, Earth, and the solar system increased enormously. But the "Age of *Apollo*" was short lived. Public interest quickly waned. Over half a billion people breathlessly watched on television as Neil Armstrong set foot on the Moon, yet barely 3 years later, when the program was abruptly canceled for largely political (rather than scientific, technological, or economic) reasons, the landings had become so routine that they no longer excited the interest of the American public. Unmanned space science moved away from the Moon and

toward the other planets, and the manned space program foundered. Perhaps one of the most amazing—and saddest—aspects of the *Apollo* program is that only now, some three decades later, is the U.S. (and perhaps also China) gearing up for new crewed missions to the Moon in the coming decade.

In 1994, the small U.S. military satellite *Clementine* was placed in lunar orbit, to perform a detailed survey of the lunar surface. In January 1998, NASA returned to the Moon for the first time in a quarter century with the launch of *Lunar Prospector*, another small satellite on a 1-year mission to study the Moon's structure and origins. As discussed in more detail in *Discovery 8-2*, both missions were successful and have amply demonstrated the wealth of information that can be obtained by low-budget spacecraft. Following the spectacular end of the *Lunar Prospector* mission in 1999 (see Section 8.5), there are currently no active probes in lunar orbit. After a brief encounter with humankind, the Moon is—for now, at least—once again a lifeless, unchanging world.

Plans do exist to establish permanent human colonies on the Moon, both for commercial ventures, such as mining, and for scientific research. The possible discovery of water on the lunar surface (Section 8.5) alleviates one major logistical problem associated with such an undertaking. In 2006 NASA announced a new program to reestablish its lunar exploration program, now in concert with a possible manned mission to Mars. This program may also include the construction of large optical, radio, and other telescopes on the lunar surface. Such instruments, could be built larger than Earth-based devices and would benefit from perfect seeing and no light pollution.

Many astronomers are skeptical, arguing that the enormous cost of such facilities would outweigh the benefits they might offer compared to Earth-based and orbiting observatories. Others argue that the boost to space science from such a high-profile undertaking would easily justify the cost and that the existence of a suite of permanent, multiwavelength lunar observatories would be of enormous benefit to the field. At present, it remains to be seen whether the political will and economic resources exist to make this dream a reality.

ANIMATION/VIDEO   First Step on the Moon

## Explanation of Mercury's Rotation

Mercury's 3:2 spin–orbit resonance did not occur by chance. What mechanism establishes and maintains it? In the case of the Moon orbiting Earth, the 1:1 resonance is the result of tidal forces. In essence, the lunar rotation

◀ FIGURE 8.13 **Mercury's Rotation** Mercury's orbital and rotational motions combine to produce a day that is 2 Mercury years long. The red arrow represents an observer standing on the surface of the planet. At day 0 (center right in Year 1 drawing), it is noon for our observer and the Sun is directly overhead. By the time Mercury has completed one full orbit around the Sun and moved from day 0 to day 88, it has rotated on its axis exactly 1.5 times, so that it is now midnight at the observer's location. After another complete orbit, it is noon once again on day 176 (center right in Year 3 drawing). The eccentricity of Mercury's orbit is not shown in this simplified diagram.

period, which probably started off much shorter than its present value, has lengthened so that the tidal bulge created by Earth is fixed relative to the body of the Moon. Tidal forces (this time due to the Sun) are also responsible for Mercury's 3:2 resonance, but in a much more subtle way.

Mercury cannot settle into a 1:1 resonance because its orbit around the Sun is quite eccentric. By Kepler's second law, Mercury's orbital speed is greatest at perihelion (closest approach to the Sun) and least at aphelion (greatest distance from the Sun). ∞ (*More Precisely 2-1*) A moment's thought shows that, because of these variations in the planet's orbital speed, there is no way that the planet (rotating at a constant rate) can remain in a synchronous orbit. If its rotation were synchronous near perihelion, it would be too rapid at aphelion, and synchronism at aphelion would produce too slow a rotation at perihelion.

## MORE PRECISELY 8-1

### Why Air Sticks Around

Why do some planets and moons have atmospheres, while others do not, and what determines the composition of the atmosphere if one exists? Why does a layer of air, made up mostly of nitrogen and oxygen, lie just above Earth's surface? After all, experience shows that most gas naturally expands to fill all the volume available. Perfume in a room, fumes from a poorly running engine, and steam from a teakettle all disperse rapidly until we can hardly sense them. Why doesn't our planet's atmosphere similarly disperse by floating away into space?

The answer is that *gravity* holds it down. Earth's gravitational field exerts a pull on all the atoms and molecules in our atmosphere, preventing them from escaping. However, gravity is not the only influence acting, for if it were, all of Earth's air would have fallen to the surface long ago. *Heat*—the rapid random motion of the molecules in a gas—competes with gravity to keep the atmosphere buoyant. Let's explore this competition between gravity and heat in a little more detail.

All gas molecules are in constant random motion. The temperature of any gas is a direct measure of this motion: The hotter the gas, the faster the molecules are moving. ∞ (*More Precisely 3-1*) The Sun continuously supplies heat to our planet's atmosphere, and the resulting rapid movement of heated molecules produces *pressure*, which tends to oppose the force of gravity, preventing our atmosphere from collapsing under its own weight.

An important measure of the strength of a body's gravity is the body's *escape speed*—the speed needed for any object to escape forever from its surface. ∞ (Sec. 2.8) This speed increases with increased mass or decreased radius of the parent body (often a moon or a planet). In convenient (Earth) units, it can be expressed as

escape speed (in km/s)

$$= 11.2 \sqrt{\frac{\text{mass of body (in Earth masses)}}{\text{radius of body (in Earth radii)}}}.$$

(See the Earth Data box on p. 172.) Thus, Earth's escape speed is $11.2\sqrt{1/1} = 11.2$ km/s. If the *mass* of the parent body is quadrupled, the escape speed doubles. If the parent body's *radius* quadruples, then the escape speed is halved. In other words, you need high speed to escape the gravitational attraction of a very massive or very small body, but you can escape from a less massive or larger body at lower speeds.

To determine whether a planet will retain an atmosphere, we must compare the planet's escape speed with the *molecular speed*, which is the average speed of the gas particles making up the planet's atmosphere. This speed actually depends not only on the temperature of the gas, but also on the mass of the individual molecules—the hotter the gas or the smaller the molecular mass, the higher is the average speed of the molecules:

average molecular speed (in km/s)

$$= 0.157 \sqrt{\frac{\text{gas temperature } (K)}{\text{molecular mass (hydrogen atom masses)}}}.$$

Thus, increasing the absolute temperature of a sample of gas by a factor of four—for example, from 100 K to 400 K—doubles the average speed of its constituent molecules, and, at a given temperature, molecules of hydrogen ($H_2$: molecular mass = 2) in air move, on average, four times faster than molecules of oxygen ($O_2$: molecular mass = 32), which are 16 times heavier.

**EXAMPLE 1** For nitrogen ($N_2$: molecular mass = 28) and oxygen ($O_2$: molecular mass = 32) in Earth's atmosphere, where the temperature near the surface is nearly 300 K, the preceding formula yields the following average molecular speeds:

nitrogen:  speed $= 0.157$ km/s $\times \sqrt{\dfrac{300}{28}} = 0.51$ km/s;

oxygen:  speed $= 0.157$ km/s $\times \sqrt{\dfrac{300}{32}} = 0.48$ km/s.

Tidal forces always act so as to synchronize the rotation rate with the instantaneous orbital speed, but such synchronization cannot be maintained over Mercury's entire orbit. What happens? The answer is found when we realize that tidal effects diminish very rapidly with increasing distance. The tidal forces acting on Mercury at perihelion are much greater than those at aphelion, so perihelion "won" the struggle to determine the rotation rate. In the 3:2 resonance, Mercury's orbital and rotational motion are almost exactly synchronous *at perihelion*, so that particular rotation rate was naturally "picked out" by the Sun's tidal influence on the planet. Notice that even though Mercury rotates through only 180° between

one perihelion and the next (see Figure 8.13), the appearance of the tidal bulge is the *same* each time around.

Resonances such as these occur quite frequently in the solar system. Many additional examples can be found in the motion of the planets, their moons and rings, as well as in orbits of many asteroids and Kuiper belt objects. The rotation of Mercury is one of the simplest non-synchronous resonances known. Many resonances are much more complex. These intricate interactions are responsible for much of the fine detail observed in the motion of our planetary system.

The Sun's tidal influence also causes Mercury's rotation axis to be exactly perpendicular to its orbital plane.

These speeds are far smaller than the 11.2 km/s needed for a molecule to escape into space. As a result, Earth is able to retain its nitrogen–oxygen atmosphere. On the whole, *our planet's gravity simply has more influence than the heat of our atmosphere.*

In reality, the situation is a little more complicated than a simple comparison of speeds. Atmospheric molecules can gain or lose speed by bumping into one another or by colliding with objects near the ground. Thus, although we can characterize a gas by its average molecular speed, the molecules do not *all* move at the same speed, as illustrated in the accompanying figure. A tiny fraction of the molecules in any gas have speeds much greater than average—one molecule in two million has a speed more than three times the average, and one in $10^{16}$ exceeds the average by more than a factor of five. This means that at any instant, *some* molecules are moving fast enough to escape, even when the average molecular speed is much less than the escape speed. The result is that all planetary atmospheres slowly leak away into space.

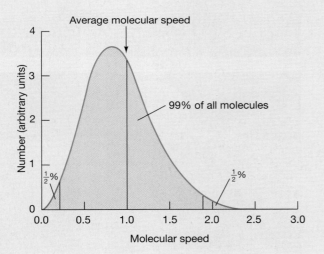

Average molecular speed

99% of all molecules

$\frac{1}{2}$%

$\frac{1}{2}$%

Number (arbitrary units)

Molecular speed

Don't be alarmed—the leakage is usually very gradual! As a rule of thumb, if the escape speed from a planet exceeds the average speed of a given type of molecule by a factor of six or more, then molecules of that type will not have escaped from the planet's atmosphere in significant quantities in the 4.6 billion years since the solar system formed. Conversely, if the escape speed is less than six times the average speed of molecules of a given type, then most of them will have escaped by now, and we should not expect to find them in the atmosphere.

For air on Earth, the mean molecular speeds of oxygen and nitrogen that we just computed are comfortably below one-sixth of the escape speed. However, if the Moon originally had an Earth-like atmosphere, that lunar atmosphere would have been heated by the Sun to much the same temperature as Earth's air today, so the average molecular speed would have been about 0.5 km/s. Because the Moon's escape speed is only $11.2\sqrt{0.012/0.27} = 2.4$ km/s—less than six times the average molecular speed—any original lunar atmosphere long ago dispersed into interplanetary space. Mercury's escape speed is $11.2\sqrt{0.055/0.38} = 4.2$ km/s However, its peak surface temperature is around 700 K, corresponding to an average molecular speed for nitrogen or oxygen of about 0.8 km/s, more than one-sixth of the escape speed, so there has been ample time for those gases to escape.

---

**EXAMPLE 2** We can use the foregoing arguments to understand some aspects of atmospheric *composition.* Hydrogen molecules ($H_2$: molecular mass = 2) move, on average, at about 1.9 km/s in Earth's atmosphere at sea level, so they have had time to escape since our planet formed ($6 \times 1.9$ km/s = 11.4 km/s, which is greater than Earth's 11.2-km/s escape speed). Consequently, we find very little hydrogen in Earth's atmosphere today. However, on the planet Jupiter, with a lower temperature (about 100 K), the speed of hydrogen molecules is correspondingly slower—about 1.1 km/s. At the same time, Jupiter's escape speed is 60 km/s, over five times higher than Earth's. For those reasons, Jupiter has retained its hydrogen—in fact, hydrogen is the dominant ingredient of Jupiter's atmosphere.

---

As a result, and because of Mercury's eccentric orbit and the spin–orbit resonance, some points on the surface get much hotter than others. In particular, the two (diametrically opposite) points on the equator where the Sun is directly overhead at perihelion get hottest of all. They are called the *hot longitudes.* The peak temperature of 700 K mentioned earlier occurs at noon at those two locations. At the *warm longitudes,* where the Sun is directly overhead at aphelion, the peak temperature is about 150 K cooler—a mere 550 K.

By contrast, the Sun is always on the horizon as seen from the planet's poles, so temperatures there never reach the sizzling levels of the equatorial regions. Earth-

based radar studies carried out during the 1990s suggest that Mercury's polar temperatures may be as low as 125 K and that, despite the planet's scorched equator, the poles may be covered with extensive sheets of water ice. (See Section 8.5 for similar findings regarding the Moon.)

CONCEPT CHECK

✔ How has gravity influenced the rotation rates of the Moon and Mercury?

## 8.5 Lunar Cratering and Surface Composition

On Earth, the combined actions of wind and water erode our planet's surface and reshape its appearance almost daily. Coupled with the never-ending motion of Earth's surface plates, the result is that most of the ancient history of our planet's surface is lost to us. The Moon, in contrast, has no air, no water, no plate tectonics, and no ongoing volcanic or seismic activity. Consequently, features dating back almost to its formation are still visible today.

### Meteoritic Impacts

The primary agent of change on the lunar surface is interplanetary debris, in the form of *meteoroids*. This material, much of it rocky or metallic in composition, is strewn throughout the solar system, orbiting the Sun in interplanetary space, perhaps for billions of years, until it happens to collide with some planet or moon. ∞ (Sec. 6.5) On Earth, most meteoroids burn up in the atmosphere, producing the streaks of light known as *meteors*, or "shooting stars." But the Moon, without an atmosphere, has no protection against this onslaught. Large and small meteoroids zoom in and collide with the surface, sometimes producing huge craters. Over billions of years, these collisions have scarred, cratered, and sculpted the lunar landscape. Craters are still being formed today—even as you read this—all across the surface of the Moon.

Meteoroids generally strike the Moon at speeds of several kilometers per second. At these speeds, even a small piece of matter carries an enormous amount of energy. For example, a 1-kg object hitting the Moon's surface at 10 km/s releases as much energy as the detonation of 10 kg of TNT! As illustrated in Figure 8.14, the impact of a meteoroid with the surface causes sudden and tremendous pressures to build up, heating the normally brittle rock and deforming the ground like heated plastic. The ensuing explosion pushes previously flat layers of rock up and out, forming a crater.

The diameter of the eventual crater is typically 10 times that of the incoming meteoroid; the depth of the crater is about twice the meteoroid's diameter. Thus, our 1-kg meteoroid, measuring perhaps 10 cm across, would produce a crater about 1 m in diameter and 20 cm deep. Shock waves from the impact pulverize the lunar surface to a depth many times that of the crater itself. Numerous rock samples brought back by the *Apollo* astronauts show patterns of repeated shattering and melting—direct evidence of the violent shock waves and high temperatures produced in meteoritic impacts. The material thrown out by the explosion

▶ **FIGURE 8.14 Meteoroid Impact** Several stages in the formation of a crater by meteoritic impact. (a) A meteoroid strikes the surface, releasing a large amount of energy. (b, c) The resulting explosion ejects material from the impact site and sends shock waves through the underlying surface. (d) Eventually, a characteristic crater surrounded by a blanket of ejected material results.

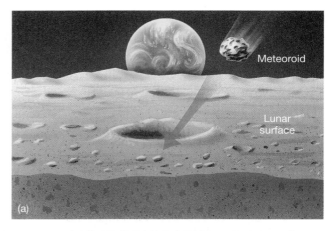

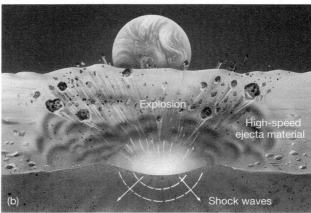

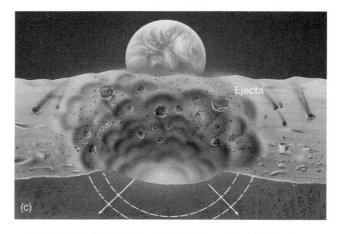

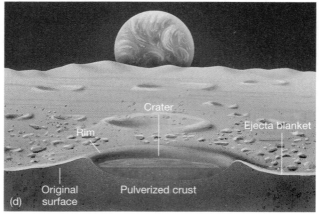

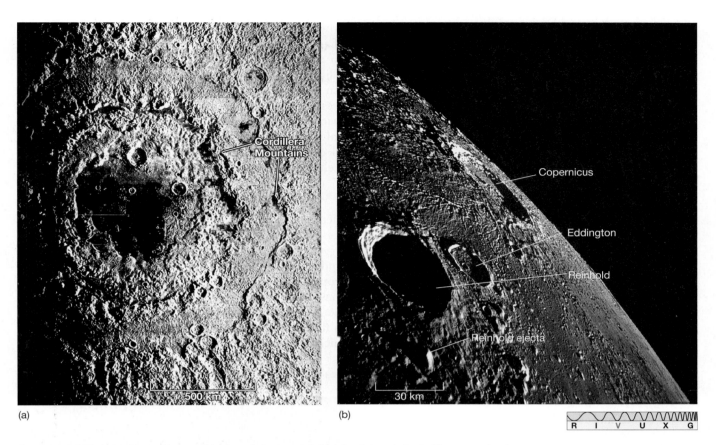

▲ FIGURE 8.15 **Large Lunar Craters** (a) A large lunar crater, called the Orientale Basin. The meteorite that produced this crater thrust up much surrounding matter, which can be seen as concentric rings of cliffs called the Cordillera Mountains. The outermost ring is nearly 1000 km in diameter. Notice the smaller, sharper, younger craters that have impacted this ancient basin in more recent times. (b) Two smaller craters called Reinhold and Eddington sit amid the secondary cratering resulting from the impact that created the 90-km-wide Copernicus crater (near the horizon) about a billion years ago. The ejecta blanket from crater Reinhold, 40 km across and in the foreground, can be seen clearly. The view was obtained by looking northeast from the lunar module during the *Apollo 12* mission. *(NASA)*

surrounds the crater in a layer called an *ejecta blanket*. The ejected debris ranges in size from fine dust to large boulders. Figure 8.15(a) shows the result of one particularly large meteoritic impact on the Moon. As shown in Figure 8.15(b), the larger pieces of ejecta may themselves form secondary craters.

In addition to the bombardment by meteoroids with masses of a gram or more, a steady "rain" of *micrometeoroids* (debris with masses ranging from a few micrograms up to about 1 gram) also eats away at the structure of the lunar surface. Some examples can be seen in Figure 8.16, a photomicrograph (a photograph taken through a microscope) of some glassy "beads" brought back to Earth by *Apollo* astronauts. The beads themselves were formed during the explosion following the impact of a meteoroid, when surface rock was melted, ejected, and rapidly cooled. Note how several of the beads also display fresh miniature craters caused by micrometeoroids that struck the beads after they had cooled and solidified.

In fact, the *rate* of cratering decreases rapidly with the size of the crater—fresh large craters are scarce, but small

craters are common. The reason for this is simple: There just aren't very many large chunks of debris in interplanetary space, so their collisions with the Moon are rare. At present average rates, one new 10-km (diameter) lunar crater is formed roughly every 10 million years, a new meter-sized crater is created about once a month, and centimeter-sized craters are formed every few minutes.

## Cratering History of the Moon

Astronomers can use the known ages (from radioactive dating) of Moon rocks to estimate the rate of cratering in the past. One very important result of this work is the discovery that the Moon was subjected to an extended period of intense meteoritic bombardment roughly 4 billion years ago. Indeed, this is a key piece of evidence supporting the condensation theory of solar system formation. ∞ (Sec. 6.7)

As we have seen, the heavily cratered highlands are older than the less-cratered maria, but the difference in cratering is not simply a matter of exposure time. Astronomers

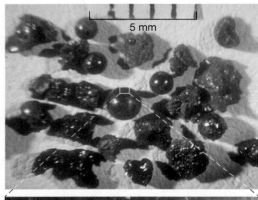

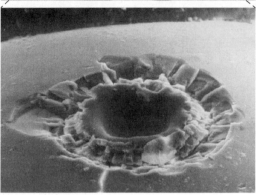

▲ FIGURE 8.16 **Microcraters** Craters of all sizes litter the lunar landscape. Some shown here, embedded in glassy beads retrieved by *Apollo* astronauts, measure only 0.01 mm across. (The scale at the top is in millimeters.) The beads themselves were formed during the explosion following a meteoroid impact, when surface rock was melted, ejected, and rapidly cooled. *(NASA)*

now think that the Moon, and presumably the entire inner solar system, experienced a sudden drop in meteoritic bombardment about 3.9 billion years ago. The highlands solidified and received most of their craters before that time, whereas the maria solidified afterward. The rate of cratering has remained relatively low ever since.

The great basins that comprise the maria are thought to have been created during the final stages of the heavy bombardment, between about 4.1 and 3.9 billion years ago. Subsequent volcanic activity filled the craters with lava, ultimately creating the formations we see today as the lava turned into solid rock. In a sense, then, the maria *are* oceans—ancient seas of molten lava, now solidified.

Not all these great craters became flooded with lava, however. One of the youngest craters is the Orientale Basin (Figure 8.15a), which formed about 3.9 billion years ago. This crater did not undergo much subsequent volcanism, and we can recognize its structure as an impact crater rather than as another mare. Similar "unflooded" basins are seen on the lunar far side (Figure 8.7).

Apart from meteorites found on Earth, the Moon is the only solar system object for which we have accurate age measurements, from radioactive dating of samples returned to Earth. However, studies of lunar cratering

provide astronomers with an important alternative means of estimating ages in the solar system. By counting craters on a planet, moon, or asteroid and using the Moon to calibrate the numbers, an approximate age for the surface can be obtained. In fact, this is how most of the ages presented in the next few chapters are determined. Note that, as with radioactive dating, the technique measures only the time since the surface in question last solidified—all cratering is erased and the clock is reset if the rock melts. ∞ *(More Precisely 7-2)*

## Lunar Dust

Meteoroid collisions with the Moon are the main cause of the layer of pulverized ejecta—also called lunar dust, or *regolith* (meaning "fine rocky layer")—that covers the lunar landscape to an average depth of about 20 m. This microscopic dust has a typical particle size of about 0.01 mm. In consistency, it is rather like talcum powder or ready-mix dry mortar. Figure 8.17 shows an *Apollo* astronaut's boot prints in the regolith, which is thinnest on the maria (10 m) and thickest on the highlands (over 100 m deep in places).

The constant barrage from space results in a slow, but steady, erosion of the lunar surface. The soft edges of the

▲ FIGURE 8.17 **Regolith** The lunar soil, or regolith, is a layer of powdery dust covering the lunar surface to a depth of roughly 20 m. Note the bootprints in the foreground of the *Apollo* astronaut, seen here adjusting some instruments for testing the composition of soil near Mount Hadley. The astronaut's weight has compacted the regolith to a depth of a few centimeters. Even so, these boot prints will probably survive for more than a million years. *(NASA)*

R I V U X G

FIGURE 8.18 **Lunar Surface** The lunar surface is not entirely changeless. Despite the complete lack of wind and water on the airless Moon, the surface has still eroded a little under the constant "rain" of meteoroids, especially micrometeoroids. Note the soft edges of the craters visible in the foreground of this image. In the absence of erosion, these features would be as jagged and angular today as they were when they formed. (The twin tracks were made by the *Apollo* lunar rover.) (*NASA*)

craters visible in the foreground of Figure 8.18 are the result of this process. In the absence of erosion, those features would still be as jagged and angular today as they were just after they formed. Instead, the steady buildup of dust due to innumerable impacts has smoothed their outlines and will probably erase them completely in about 100 million years.

From the known dependence of the cratering rate on the size of a crater, planetary scientists can calculate how many small craters they would expect to find, given the numbers of large craters actually observed. When they make this calculation, they find a shortage of craters less than about 20 m deep. These "missing" craters have been filled in by erosion over the lifetime of the Moon. This gives us a very rough estimate of the average erosion rate: about 5 m per billion years, or roughly 1/10,000 the rate on Earth.

The current lunar erosion rate is very low because meteoritic bombardment on the Moon is a much less effective erosive agent than are wind and water on Earth. For comparison, the Barringer Meteor Crater (Figure 8.19) in the Arizona desert, one of the largest meteoroid craters on Earth, is only 25,000 years old, but has already undergone noticeable erosion. It will probably disappear completely in just a few million years, quite a short time geologically. If a crater that size had formed on the Moon even 4 billion years ago, it would still be plainly visible today. Even the shallow boot prints shown in Figure 8.17 are likely to remain intact for several million years.

## Lunar Ice?

In contrast to Earth's soil, the lunar regolith contains no organic matter like that produced by biological organisms. No life whatsoever exists on the Moon. Nor were any

fossils found in *Apollo* samples. Lunar rocks are barren of life and apparently always have been. NASA was so confident of this fact that the astronauts were not even quarantined on their return from the last few *Apollo* landings. Furthermore, all the lunar samples returned by the U.S. and Soviet Moon programs were bone dry—they didn't even contain minerals having water molecules locked within their crystal structure. Terrestrial rocks, by contrast, are almost always 1 or 2 percent water. The main reasons for this lack of water are the Moon's lack of an atmosphere and the high (up to 400 K) daytime temperatures found over most of the lunar surface.

Some regions of the Moon *are* thought to contain water, however—in the form of ice. As early as the 1960s, some scientists had considered the theoretical possibility that ice might be found near the lunar poles. Since the Sun never rises more than a few degrees above the horizon, as seen from the Moon's polar regions, temperatures on the permanently shaded floors of craters near the poles never exceed about 100 K. Consequently, those scientists theorized, any water ice there could have remained permanently frozen since the very early days of the solar system, never melting or vaporizing and hence never escaping into space.

In November 1996, mission controllers of the *Clementine* spacecraft (see *Discovery 8-2*) reported that radar echoes captured by *Clementine* from an old, deep crater near the lunar south pole suggested deposits of low-density material, probably water ice, at a depth of a few meters. In early March 1998, NASA announced that sensitive equipment on board the *Lunar Prospector* mission had confirmed *Clementine*'s findings and in fact had detected large amounts of ice—possibly totaling trillions of tons—at both lunar poles. At first, it appeared that the ice was mainly in the form of tiny crystals mixed with the lunar regolith, spread over many tens of thousands of square kilometers of deeply shadowed crater floors. However, subsequent analysis of the data suggests that much of the ice may exist in the form of smaller, but more concentrated, "lakes" of nearly pure material lying perhaps half a meter below the surface.

The *Lunar Prospector* discovery of lunar ice was indirect; the instruments on board the spacecraft actually detected the presence of hydrogen (H), whose existence was taken as evidence of water ($H_2O$). In an attempt to gain more direct information about lunar ice, NASA scientists decided to end the *Lunar Prospector* mission in a very spectacular way. As the spacecraft neared the end of its lifetime, it was di-

◄ FIGURE 8.19 **Barringer Crater** The Barringer Meteor Crater, near Winslow, Arizona, is 1.2 km in diameter and 0.2 km deep. (Note the access road at right for scale.) Geologists think that a large meteoroid whacked Earth and formed this crater about 25,000 years ago. The meteoroid was probably about 50 m across and likely weighed around 200,000 tons. The inset shows a closeup of one of the interior walls of the crater. *(U.S. Geological Survey)*

rected to crash into one of the deep craters in which the ice was suspected to hide. Figure 8.20 shows the intended site of the impact and the trajectory to be taken by the satellite as it approached the surface. The hope was that the *Hubble Space Telescope* and ground-based telescopes on Earth might detect spectroscopic signatures of water vapor released by the impact. No water vapor was seen, although mission planners had stressed in advance that there were so many uncertainties involved in the effort that the probability of success was low—less than 10 percent. Thus, no conclusion can be drawn from the fact that water was not directly observed. Lunar ice remains a strong possibility, but its existence has not yet been definitively proven.

Assuming that it does exist, where did all this ice come from? Most likely, it was brought to the lunar surface by

meteoroids and comets. (We will see in Chapter 15 that this is the likely origin of Earth's water, too.) Any ice that survived the impact would have been scattered across the surface. Over most of the Moon, that ice would have rapidly vaporized and escaped, but in the deep basins near the poles, it survived and built up over time. Whatever its origin, the polar ice may be a crucial component of any serious attempt at human colonization of the Moon: The anticipated cost of transporting a kilogram of water from Earth to the Moon is between $2,000 and $20,000, prompting one *Clementine* scientist to describe the lunar ice deposits as "possibly the most valuable piece of real estate in the solar system."

(a)

R  I  V  U  X  G

◄ FIGURE 8.20 **Prospector Impact** (a) The intended impact site for the *Lunar Prospector* spacecraft was a deep crater close to the Moon's south pole. The purpose of the impact was to release water vapor for spectroscopic study by telescopes on or near Earth. (b) The trajectory of the spacecraft was designed to have the craft hit near the crater floor. *(U.S. Dept. of Defense)*

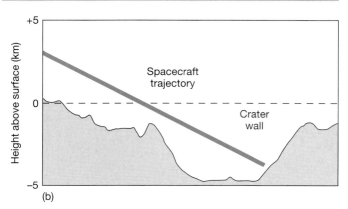

(b)

## Lunar Volcanism

Only a few decades ago, debate raged in scientific circles about the origin of lunar craters, with most scientists of the opinion that the craters were the result of volcanic activity. We now know that almost all lunar craters are actually meteoritic in origin. However, a few apparently are not. Figure 8.21 shows an intriguing alignment of several craters in a *crater-chain* pattern so straight that it is highly unlikely to have been produced by the random collision of meteoroids with the surface. Instead, the chain probably marks the location of a subsurface fault—a place where cracking or shearing of the surface once allowed molten matter to well up from below. As the lava cooled, it formed a solid "dome" above each fissure. Subsequently, the underlying lava receded and the centers of the domes collapsed, forming the craters we see today. Similar features have been observed on Venus by the orbiting *Magellan* probe (see Chapter 9).

Many other examples of lunar volcanism are known, both in telescopic observations from Earth and in the close-up photographs taken during the *Apollo* missions. Figure 8.22 shows a volcanic **rille,** a ditch where molten lava once flowed. There is good evidence for surface volcanism early in the Moon's history, and volcanism explains the presence of the lava that formed the maria. However, whatever volcanic activity once existed on the Moon ended long ago. The measured ages for rock samples returned from the Moon are all

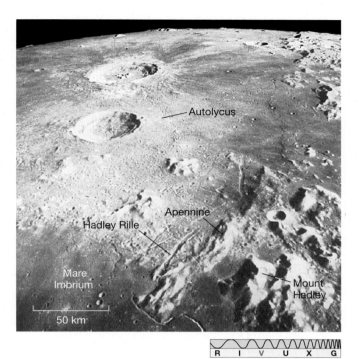

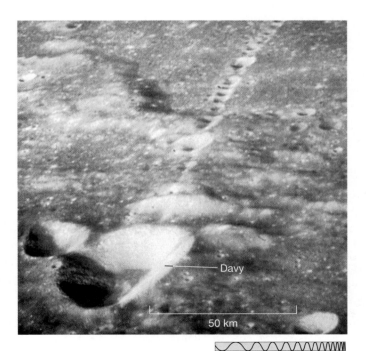

▲ FIGURE 8.21 **Crater Chain** This "chain" of well-ordered craters was photographed by an *Apollo 14* astronaut. The largest crater, called Davy, is located on the western edge of Mare Nubium. The entire field of view measures about 100 km across. (*NASA*)

▲ FIGURE 8.22 **Lunar Volcanism** A volcanic rille, photographed from the *Apollo 15* spacecraft orbiting the Moon, can be seen clearly here (bottom and center) winding its way through one of the maria. Called Hadley Rille, this system of valleys runs along the base of the Apennine Mountains (lower right) at the edge of the Mare Imbrium (to the left). Autolycus, the large crater closest to the center, spans 40 km. The shadow-sided, most prominent peak at lower right, Mount Hadley, rises almost 5 km high. (*NASA*)

greater than 3 billion years. (Recall from *More Precisely* 7-2 that the radioactivity clock starts "ticking" when the rock solidifies.) Apparently, the maria solidified over 3 billion years ago, and the Moon has been dormant ever since.

CONCEPT CHECK

✔ How has meteoritic bombardment affected the surface of the Moon?

## 8.6 The Surface of Mercury

Like craters on the Moon, almost all craters on Mercury are the result of meteoritic bombardment. However, Mercury's craters are less densely packed than their lunar counterparts, and there are extensive **intercrater plains.** The crater walls are generally not as high as those on the Moon, and the ejected material appears to have landed closer to the impact site exactly as we would expect on the basis of Mercury's stronger surface gravity.

One likely explanation for Mercury's relative lack of craters is that the older craters were filled in by volcanic

## The Moon on a Shoestring

Since the early 1990s, the watchword for unmanned exploration of space has been "smaller, faster, cheaper." Unlike previous generations of space missions, the emphasis now is on creating small-scale, lightweight systems that can be designed and built rapidly and cheaply, affording mission planners much greater flexibility, both in designing follow-up missions and in quickly changing mission parameters as circumstances warrant. Two lunar exploration satellites—*Clementine* and *Lunar Prospector*—have demonstrated the power of this approach. Both have returned high-quality science for total costs of about $70 million each, a small fraction of the cost of most other planetary spacecraft, which normally carry price tags of hundreds of millions, even billions, of dollars.

The *Clementine* satellite was sent to the Moon in 1994 by the U.S. Defense Department, largely to test some new sensing devices developed for the ballistic missile defense program. This was the first lunar mission by any nation since the crew of *Apollo 17* left the Moon in 1972. *Clementine* originally was a code word for a military-classified space project known as the Deep-Space Program Science Experiment. The vehicle and its onboard suite of instruments were designed to test the feasibility of miniaturizing a complex spacecraft, its engineering subsys-tems, and its sophisticated sensors for use in deep space. The spacecraft's total mass was less than 150 kg. *Clementine*'s technical design was a product of the Strategic Defense Initiative—the Star Wars program—now known as the Ballistic Missile Defense Organization. Its target—the Moon—was of no interest to the military, other than being a convenient, known object in the cosmic neighborhood.

The lunar portion of the mission was a spectacular success, although a follow-up mission to map an asteroid had to be canceled after a computer malfunction caused an onboard thruster to fire until it had used up all its fuel, leaving the spacecraft spinning out of control. *Clementine* made the first digital global map of the Moon, at very high resolution. In 2 months of operation, its sensors took over 2.5 million images with a clarity at least 10 times better than NASA's most sophisticated planetary camera in the 1990s—the one aboard the *Galileo* mission to Jupiter. *Clementine* was able to obtain global coverage of the Moon at visible, ultraviolet, and infrared wavelengths. In addition, the small craft carried lidar devices (the visible-light equivalent of radar), able to pulse the lunar surface and listen for echoes. In all, more than 50 advanced lightweight technologies were demonstrated on this powerful dwarf spacecraft.

The accompanying figure is a mosaic of about 1500 images centered on the Moon's south pole. The bottom half is part of the near side of the Moon, as seen from Earth; the top half is the far side that we never see from home. For scale, the double-ringed crater at the upper left, called Schroedinger, has an outer diameter of 320 km. (Note that the circular image is misleading—only a portion of the Moon is shown here.) The dark region at the pole is the center of an old, permanently shadowed depression called Aitken Basin. Its rim (outlined with a red dashed line) spans some 2000 km; the basin itself averages 10 km deep. This huge depression is the largest impact basin known in the solar system.

NASA's *Lunar Prospector* was launched in January 1998. Designed and built in just 22 months and weighing 295 kg fully fueled, the spacecraft orbited the Moon for 18 months, probing the lunar surface and interior with an array of onboard instruments: A *gamma-ray spectrometer* mapped the abundances of certain elements on the Moon's surface; two sensitive *magnetometers* probed the Moon's extremely weak magnetic field; a *neutron spectrometer* searched for water ice by detecting the element hydrogen on the Moon's surface; an *alpha-particle spectrometer* searched for particles emitted by radioactive gases leaking out of the lunar interior; and a *Doppler gravity experiment* made detailed measurements of the Moon's gravitational field by carefully monitoring small fluctuations in the craft's orbital velocity. *Lunar Prospector* did not carry a camera. The spacecraft's instrument package was designed in part to complement the imaging and radar capabilities of *Clementine*.

By far the most publicized aspect of the *Lunar Prospector* mission was its search for water ice at the lunar poles. Radar observations made by *Clementine* of the deep depression at the Moon's south pole (at the center of the image) had suggested deposits of water ice at a depth of a few meters. *Lunar Prospector*'s neutron spectrometer confirmed the presence of large amounts of hydrogen (and, presumably, therefore, water) there, although, as discussed in more detail in the text, a follow-up attempt to detect water directly was unsuccessful.

The data returned by *Clementine* and *Lunar Prospector* have allowed scientists to construct much more accurate models of the Moon's interior and to probe its past history with greater precision. But beyond the scientific results, important as they are, these spacecraft may have done something even more far reaching: By demonstrating that major scientific findings can come from low-cost, fast-turnaround missions, they may have changed forever the way planetary scientists explore the solar system.

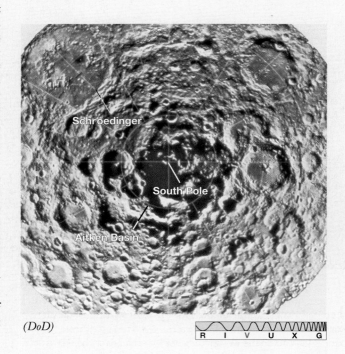

*(DoD)*

R I V U X G

activity, in much the same way as the Moon's maria filled in older craters as they formed. Most geologists think that volcanism did occur in Mercury's past, obscuring the old craters. However, the intercrater plains do not look much like maria—they are much lighter in color and not as flat. Although the details of how Mercury's landscape came to look the way it does remain unexplained, the apparent absence of rilles or other obvious features associated with very large-scale lava flows, along with the light color of the lava-flooded regions, suggests that Mercury's volcanic past was different from the Moon's.

Mercury has at least one type of surface feature not found on the Moon. Figure 8.23 shows a **scarp,** or cliff, on the surface that does not appear to be the result of volcanic or any other familiar geological activity. The scarp cuts across several craters, which indicates that whatever produced it occurred *after* most of the meteoritic bombardment was over. Mercury shows no evidence of crustal motions like plate tectonics on Earth—no fault lines, spreading sites, or indications of plate collisions are seen. ∞ (Sec. 7.4) The scarps, of which several are known from the *Mariner* images, probably formed when the planet's interior cooled and shrank long ago, much as wrinkles form on the skin of an old, shrunken apple. On the basis of the

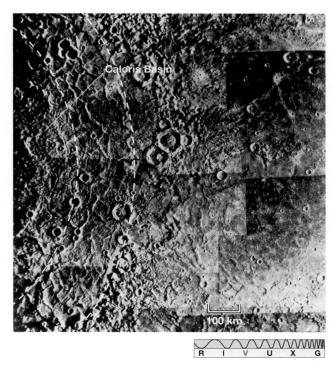

▲ FIGURE 8.24 **Mercury's Basin** Mercury's most prominent geological feature—the Caloris Basin—measures about 1400 km across and is ringed by concentric mountain ranges (dashed lines) that reach more than 3 km high in places. This huge circular basin, only half of which can be seen (at left) in this *Mariner 10* photo, is similar in size to the Moon's Mare Imbrium and spans more than half of Mercury's radius. *(NASA)*

amount of cratering observed in the surrounding terrain (as discussed in the previous section), astronomers estimate that the scarps probably formed about 4 billion years ago.

Figure 8.24 shows what may have been a result of the last great geological event in the history of Mercury: an immense bull's-eye crater called the Caloris Basin, formed eons ago by the impact of a large asteroid. (The basin is so called because it lies in Mercury's "hot longitudes"—see Section 8.3—close to the planet's equator; *calor* is the Latin word for "heat.") Because of the orientation of the planet during *Mariner 10*'s flybys, only half of the basin was visible. The center of the crater is off the left-hand side of the photograph. Compare this basin with the Orientale Basin on the Moon (Figure 8.15a). The impact-crater structures are quite similar, but even here there is a mystery: The patterns visible on the Caloris floor are unlike any seen on the Moon. Their origin, like the composition of the floor itself, is unknown.

So large was the impact that created the Caloris Basin that it apparently sent strong seismic waves reverberating throughout the entire planet. On the opposite side of Mercury from Caloris, there is a region of oddly rippled and wavy surface features, often referred to as *weird* (or *jumbled*) terrain. Scientists theorize that this terrain was produced when seismic waves from the Caloris impact

▲ FIGURE 8.23 **Mercury's Surface** Discovery Scarp on Mercury's surface, as photographed by *Mariner 10*. This cliff, or compressional feature, seems to have formed when the planet's crust cooled and shrank early in its history, causing a crease in the surface. Running diagonally across the center of the frame, the scarp is several hundred kilometers long and up to 3 km high in places. *(NASA)*

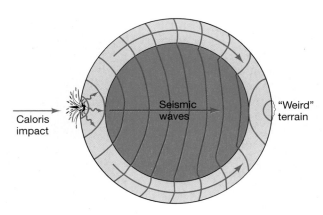

▲ FIGURE 8.25 **Weird Terrain** The refocusing of seismic waves after the Caloris Basin impact may have created the weird terrain on the opposite side of the planet.

traveled around the planet and converged on the diametrically opposite point, causing large-scale disruption of the surface there, as illustrated in Figure 8.25.

CONCEPT CHECK

✔ How do scarps on Mercury differ from geological faults on Earth?

## 8.7 Interiors

In Chapter 7 we saw how geologists combine bulk measurements of Earth's density, gravity, and magnetic field with seismic studies and mathematical models to build up a detailed model of the planet's interior. ∞(Secs. 7.1, 7.3, 7.5) Planetary scientists attempt to do much the same with the Moon and Mercury, but since less detailed data are available, the conclusions are correspondingly less precise.

### The Moon

The Moon's average density, about 3300 kg/m³, is similar to the measured density of lunar surface rock, virtually eliminating any chance that the Moon has a large, massive, and very dense nickel–iron core like that of Earth. In fact, the low density implies that the entire Moon is actually deficient in iron and other heavy metals compared with their abundance on our planet.

There is no evidence for any large-scale lunar magnetic field. *Lunar Prospector* detected some very weak surface magnetic fields—less than a thousandth of Earth's field—apparently associated with some large impact basins, but these are not thought to be related to conditions in the lunar core. As we saw in Chapter 7, researchers think that planetary magnetism requires a rapidly rotating liquid metal core, like Earth's. ∞ (Sec. 7.5) Thus, the absence of a lunar magnetic field could be a consequence of the Moon's slow rotation, the absence of a liquid core, or both.

## PLANETARY DATA

### MERCURY

(NASA)

| | |
|---|---|
| Orbital semimajor axis | 0.39 AU |
| | 57.9 million km |
| Orbital eccentricity | 0.206 |
| Perihelion | 0.31 AU |
| | 46 million km |
| Aphelion | 0.47 AU |
| | 69.8 million km |
| Mean orbital speed | 47.9 km/s |
| Sidereal orbital period | 88.0 Earth days* |
| | 0.241 tropical years |
| Synodic orbital period** | 115.9 Earth days* |
| Orbital inclination to the ecliptic | 7.00° |
| Greatest angular diameter, as seen from Earth | 13″ |
| Mass | $3.30 \times 10^{23}$ kg |
| | 0.055 (Earth = 1) |
| Equatorial radius | 2440 km |
| | 0.38 (Earth = 1) |
| Mean density | 5430 kg/m³ |
| | 0.98 (Earth = 1) |
| Surface gravity | 3.70 m/s² |
| | 0.38 (Earth = 1) |
| Escape speed | 4.2 km/s |
| Sidereal rotation period | 58.6 solar days |
| Axial tilt | 0.0° |
| Surface magnetic field | 0.011 (Earth = 1) |
| Magnetic axis tilt relative to rotation axis | <10° |
| Mean surface temperature | 100–700 K |
| Number of moons | 0 |

*1 Earth (mean solar) day = 24 hours
**The planet's apparent orbital period, taking Earth's own motion into account: specifically, the mean time between one closest approach to Earth and the next. ∞ (More Precisely 9-1)

Data from the gravity experiment aboard *Lunar Prospector*, combined with measurements made by the probe's magnetometers as the Moon passed through Earth's magnetic "tail" (see Figure 7.19), imply that the Moon may have a small iron core perhaps 300 km in radius. Near the center, the temperature may be as low as 1500 K, too cool to melt rock. However, seismic data collected by sensitive equipment left on the surface by *Apollo* astronauts (see *Discovery 8-1*) suggest that the inner parts of the core may be at least partially molten, implying a somewhat higher temperature. Our knowledge of the Moon's deep interior is still quite limited.

Based on a combination of seismic data, gravitational and magnetic measurements, and a good deal of mathematical modeling resting on assumptions about the Moon's interior composition, Figure 8.26 presents a schematic diagram of the Moon's interior structure. The central core is surrounded by a roughly 400-km-thick inner mantle of semisolid rock having properties similar to Earth's asthenosphere. ∞ (Sec. 7.4) Above these regions lies an outer mantle of solid rock, some 900–950 km thick, topped by a 60- to 150-km crust (considerably thicker than that of Earth). Together, these layers constitute the Moon's lithosphere. Outside the core, the mantle seems to be of almost uniform density, although it is chemically differentiated (i.e., its chemical properties change from the deep interior to near the surface). The crust material, which forms the

lunar highlands, is lighter than the mantle, which is similar in composition to the lunar maria.

The crust on the lunar far side is *thicker* than that on the side facing Earth. If we assume that lava takes the line of least resistance in getting to the surface, then we can readily understand why the far side of the Moon has no large maria: Volcanic activity did not occur on the far side simply because the crust was too thick to allow it to occur there.

But *why* is the far-side crust thicker? The answer is probably related to Earth's gravitational pull. Just as heavier material tends to sink to the center of Earth, the denser lunar mantle tended to sink below the lighter crust in Earth's gravitational field. The effect of this tendency was that the crust and the mantle became slightly off center with respect to each other. The mantle was pulled a little closer to Earth, while the crust moved slightly away. Thus, the crust became thinner on the near side and thicker on the far side.

### Mercury

Mercury's magnetic field, discovered by *Mariner 10*, is about a hundredth that of Earth. Actually, the discovery that Mercury has any magnetic field at all came as a surprise to planetary scientists. Having detected no magnetic field in the Moon (and, in fact, none in Venus or Mars, either), they had expected Mercury to have no measurable magnetism. Certainly, Mercury does not rotate rapidly, and it may lack a liquid metal core, yet a magnetic field undeniably surrounds it. Although weak, the field is strong enough to deflect the solar wind and create a small magnetosphere around the planet.

Scientists have no clear understanding of the origin of Mercury's magnetic field. If it is produced by ongoing dynamo action, as in Earth, then Mercury's core must be at least partially molten, and radar observations from Earth reported in 2007 appear to confirm this possibility. ∞ (Sec. 7.5) However, the absence of any recent surface geological activity suggests that the outer layers are solid to a considerable depth, as on the Moon. If the field is being generated dynamically, then Mercury's slow rotation may at least account for the field's weakness. Alternatively, Mercury's current weak magnetism may simply be the remnant of an extinct dynamo—much of the planet's iron core may have solidified long ago, but still bears a permanent magnetic imprint of the past. The models are inconclusive on this issue, and no spacecraft is scheduled to revisit Mercury until the first *Messenger* flyby in 2008. ∞ (Sec. 6.6)

Mercury's magnetic field and large average density together imply that the planet is differentiated. Even without the luxury of seismographs on the surface, we can infer that most of its interior must be dominated

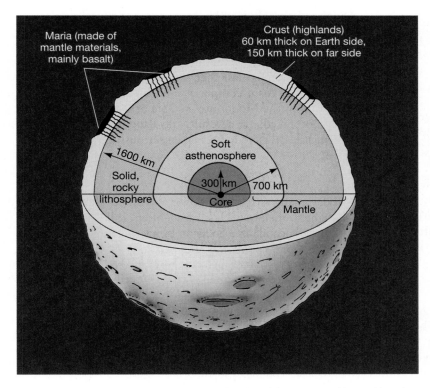

▲ FIGURE 8.26 **Lunar Interior** Cutaway diagram of the Moon. Unlike Earth's rocky lithosphere, the Moon's is very thick—nearly 1000 km. Below the lithosphere is the inner mantle, or lunar asthenosphere, a semisolid layer similar to the upper regions of Earth's mantle. At the center lies the core, which may be partly molten.

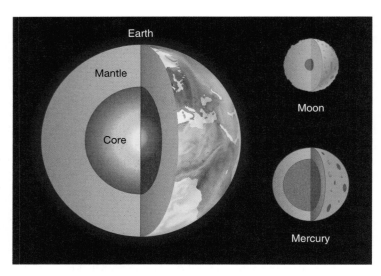

▲ FIGURE 8.27 **Terrestrial Interiors** The internal structures of Earth, the Moon, and Mercury, drawn to the same scale. Note how large a fraction of Mercury's interior is the planet's core.

by a large, heavy, iron-rich core with a radius of perhaps 1800 km. Probably a less-dense lunarlike mantle lies above this core, to a depth of about 500 to 600 km. Thus, about 40 percent of the volume of Mercury, or 60 percent of its mass, is contained in its iron core. The ratio of core volume to total planet volume is greater for Mercury than for any other object in the solar system. Figure 8.27 illustrates the relative sizes and internal structures of Earth, the Moon, and Mercury.

CONCEPT CHECK
✔ Why would we not expect strong magnetic fields on the Moon or Mercury?

## 8.8    The Origin of the Moon

Over the years, many theories have been advanced to account for the origin of the Moon. However, both the similarities *and* the differences between the Moon and Earth conspire to confound many promising attempts to explain the Moon's existence.

### Theories of Lunar Formation

One theory (the *sister*, or *coformation*, theory) suggests that the Moon formed as a separate object near Earth in much the same way as our own planet formed—the "blob" of material that eventually coalesced into Earth gave rise to the Moon at about the same time. The two objects thus formed as a double-planet system, each revolving about a common center of mass. Although once favored by many astronomers, this idea suffers from a major flaw: The Moon differs in both density and composition from Earth,

making it hard to understand how both could have originated from the same preplanetary material.

A second theory (the *capture* theory) maintains that the Moon formed far from Earth and was later captured by it. In this way, the density and composition of the two objects need not be similar, for the Moon presumably materialized in a quite different region of the early solar system. The objection to this theory is that the Moon's capture would be an extraordinarily difficult event; it might even be an impossible one. Why? Because the mass of our Moon is so large relative to that of Earth. It is not that our Moon is the largest natural satellite in the solar system, but it is unusually large compared with its parent planet. Mathematical modeling suggests that it is quite implausible that Earth and the Moon could have interacted in just the right way for the Moon to have been captured during a close encounter sometime in the past. Furthermore, although there are indeed significant differences in composition between our world and its companion, there are also many similarities—particularly between the mantles of the two bodies—that make it unlikely that they formed entirely independently of one another.

A third, older, theory (the *daughter*, or fission, theory) speculates that the Moon originated out of Earth itself. The Pacific Ocean basin has often been mentioned as the place from which protolunar matter may have been torn—the result, perhaps, of the rapid spin of a young, molten Earth. Indeed, there are some chemical similarities between the matter in the Moon's outer mantle and that in Earth's Pacific basin. However, this theory offers no solution to the fundamental mystery of how Earth could have been spinning so fast that it ejected an object as large as our Moon. Also, computer simulations indicate that the ejection of the Moon into a stable orbit simply would not have occurred. As a result, the daughter theory, in this form at least, is no longer taken seriously.

### The Impact Theory

Today, many astronomers favor a hybrid of the capture and daughter themes. This idea—often called the **impact theory**—postulates a collision by a large, Mars-sized object with a youthful and molten Earth. Such collisions may have been quite frequent in the early solar system. ∞ (Sec. 6.7) The collision presumed by the impact theory would have been more a glancing blow than a direct impact. The matter dislodged from our planet then reassembled to form the Moon.

Computer simulations of such a catastrophic event show that most of the bits and pieces of splattered Earth could have coalesced into a stable orbit. Figure 8.28 shows some of the stages of one such calculation. If Earth had already formed an iron core by the time the collision occurred, then the Moon would indeed have ended up with a composition similar to that of Earth's mantle. During the collision, any

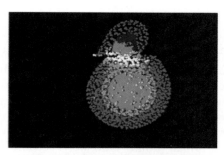

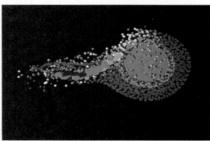

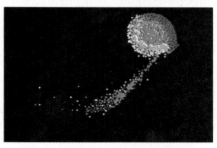

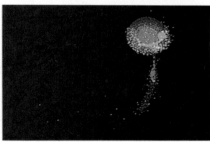

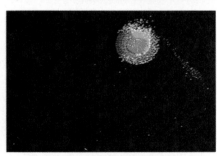

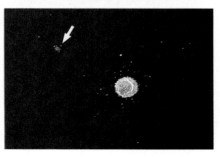

iron core in the colliding object itself would have been left behind in Earth, eventually to become part of Earth's core. Thus, both the Moon's overall similarity to that of Earth's mantle and its lack of a dense central core are naturally explained.

Over the past two decades, planetary scientists have come to realize that collisions like this probably played important roles in the formation of all the terrestrial planets (see Chapter 15). Because of the randomness inherent in such events, as well as the Moon's unique status as the only large satellite in the inner solar system, it seems that the Moon may not provide a particularly useful model for studies of the other moons in the solar system. Instead, as we will see, a moon's properties depend greatly on the characteristics of its parent planet.

Nevertheless, the quest to understand the origin of the Moon highlights the interplay between theory and observation that characterizes modern science. ∞ (Sec. 1.2) Detailed data from generations of unmanned and manned lunar missions have allowed astronomers to discriminate between competing theories of the formation of the Moon, discarding some and modifying others. At the same time, the condensation theory of solar system formation provides a natural context in which the currently favored impact theory can occur. ∞ (Sec. 6.7) Indeed, without the idea that planets formed by collisions of smaller bodies, such an impact might well have been viewed as so improbable that the theory would never have gained ground.

Finally, do not think that every last detail of the Moon's formation is understood or agreed upon by experts. That is far from the case. Some important aspects of the Moon's physical and chemical makeup are still inadequately explained—for example, the degree to which the Moon melted during its formation and whether current models are actually consistent with the observed lunar composition. The impact theory may well not be the last word on the subject. Still, past experience of the scientific method gives us confidence that the many twists and turns still to come will in the end lead us to a more complete understanding of our nearest neighbor in space.

CONCEPT CHECK

✔ How does the currently favored theory of the Moon's origin account for the Moon's lack of heavy materials compared with Earth's relative abundance and for the similarity in composition between the lunar crust and that of Earth?

◀ FIGURE 8.28 Moon Formation This sequence shows a simulated collision between Earth and an object the size of Mars. The sequence proceeds from top to bottom and zooms out dramatically. The arrow in the final frame shows the newly formed Moon. Red and blue colors represent rocky and metallic regions, respectively, and the direction of motion of the blue material in frames 2, 3, 4, and 5 is toward Earth. Note how most of the impactor's metallic core becomes part of Earth, leaving the Moon composed mainly of rocky material. (W. Benz)

## 8.9 Evolutionary History of the Moon and Mercury

Given all the data, can we construct reasonably consistent histories of the Moon and Mercury? The answer seems to be yes. Many specifics are still debated, but a broad consensus exists. Planned future missions to both bodies will continue to test and refine the pictures presented below.

### The Moon

The Moon formed about 4.6 billion years ago (see Chapter 15). The approximate age of the oldest rocks discovered in the lunar highlands is 4.4 billion years, so we know that at least part of the crust must already have solidified by that time and survived to the present. At its formation, the Moon was already depleted in heavy metals compared with Earth. Examine Figure 8.29 while studying the details that follow.

During the earliest phases of the Moon's existence— roughly the first half billion years or so—meteoritic bombardment must have been frequent enough to heat and remelt most of the *surface* layers of the Moon, perhaps to a depth of 400 km in places. The early solar system was surely populated with lots of interplanetary matter, much of it in the form of boulder-sized fragments that were capable of generating large amounts of energy upon colliding with planets and their moons. But the intense heat derived from such collisions could not have penetrated very far into the lunar interior: Rock simply does not conduct heat well.

This situation resembles the surface melting we suspect occurred on Earth from meteoritic impacts during the first billion years or so. But the Moon is much less massive than Earth and did not contain enough radioactive elements to heat it much further. Radioactivity probably heated the Moon a little, but not sufficiently to transform it from a warm, semisolid object to a completely liquid one. The chemical differentiation now inferred in the Moon's interior must have occurred during this period. If the Moon has a small iron core, that core also formed at this time.

About 3.9 billion years ago, around the time that Earth's crust solidified, the heaviest phase of the meteoritic bombardment ceased. The Moon was left with a solid crust, which would ultimately become the highlands, dented with numerous large basins, soon to flood with lava and become the maria (Figure 8.29a). Between 3.9 and 3.2 billion years ago, lunar volcanism filled the maria with the basaltic material we see today. The age of the youngest maria—3.2 billion years—indicates the time when the volcanic activity subsided. The maria are the sites of the last extensive lava flows on the Moon, over 3 billion years ago. Their smoothness, compared with the older, more rugged highlands, disguises their great age.

Small objects cool more rapidly than large ones because their interior is closer to the surface, on average. Being so small, the Moon rapidly lost its internal heat to space. As a consequence, it cooled much faster than Earth. As the Moon cooled, the volcanic activity ended and the thickness of the solid surface layer increased. With the exception of a few meters of surface erosion from eons of meteoritic bombardment (Figure 8.29c), the lunar landscape has remained more or less structurally frozen for the past 3 billion years. The Moon is dead now, and it has been dead for a long time.

### Mercury

Like the Moon, Mercury seems to have been a geologically dead world for much of the past 4 billion years. On both the Moon and Mercury, the absence of ongoing geological activity is a consequence of a thick, solid mantle that prevents volcanism or tectonic motion. Because of the

(a) 4 billion years ago

(b) 3 billion years ago

(c) Today

▲ FIGURE 8.29 **Lunar Evolution** Paintings of the Moon (a) about 4 billion years ago, after much of the meteoritic bombardment had subsided and the surface had solidified somewhat; (b) about 3 billion years ago, after molten lava had made its way up through surface fissures to fill the low-lying impact basins and create the smooth maria; and (c) today, with much of the originally smooth maria now heavily pitted with craters formed at various times within the past 3 billion years. (*U.S. Geological Survey*)

*Apollo* program, the Moon's early history is much better understood than Mercury's, which remains somewhat speculative. Indeed, what we do know about Mercury's history is gleaned mostly through comparison with the Moon.

When Mercury formed some 4.6 billion years ago, it was already depleted of lighter, rocky material. We will see later that this was largely a consequence of its location in the hot inner regions of the early solar system, although it is possible that a collision stripped away some of the planet's light mantle. During the next half-billion years, Mercury melted and differentiated, like the other terrestrial worlds. It suffered the same intense meteoritic bombardment as the Moon. Being more massive than the Moon, Mercury cooled more slowly, so its crust was thinner and volcanic activity more common at early times. More craters were erased, resulting in the intercrater plains found by *Mariner 10*.

As Mercury's large iron core formed and then cooled, the planet began to shrink, compressing the crust. This compression produced the scarps seen on Mercury's surface and may have prematurely terminated volcanic activity by squeezing shut the cracks and fissures on the surface. Thus, the extensive volcanic outflows that formed the lunar maria did not take place on Mercury. Despite its larger mass and greater internal temperature, Mercury has probably been geologically inactive even longer than the Moon.

# CHAPTER REVIEW

## Summary

**1** The Moon orbits Earth; Mercury is the closest planet to the Sun. Both the Moon and Mercury are airless, virtually unchanging worlds that exhibit extremes in temperature. Mercury has  no permanent atmosphere, although it does have a thin envelope of gas temporarily trapped from the solar wind. Both bodies are smaller and less massive than Earth, and have weaker gravities. The absence of atmospheric blankets result in hot dayside temperatures and cold nightside temperatures on the Moon and Mercury. Sunlight strikes the polar regions of both the Moon and Mercury at such an oblique angle that temperatures there are very low, with the result that both bodies may have significant amounts of water ice near their poles.

**2** The main surface features on the Moon are the dark **maria (p. 203)** and the lighter colored **highlands (p. 203).** Highland rocks are less dense than rocks from the maria and are thought to represent the Moon's crust. Maria rocks are thought to have originated in the lunar mantle. The surfaces of both the Moon and Mercury are covered with **craters**  **(p. 203)** of all sizes, caused by meteoroids striking from space. Lunar dust, called regolith, is made mostly of pulverized lunar rock, mixed with a small amount of material from impacting meteorites.

**3** The tidal interaction between Earth and the Moon is responsible for the Moon's **synchronous orbit (p. 206),** in which the same side of the Moon always faces our planet. The large lunar equatorial bulge probably indicates that the Moon once rotated more rapidly and orbited closer to Earth. Mercury's rotation rate is strongly  influenced by the tidal effect of the Sun. Because of Mercury's eccentric orbit, the planet rotates not synchronously, but exactly

three times for every two orbits around the Sun. The condition in which a body's rotation rate is simply related to its orbital period around some other body is known as spin–orbit **resonance (p. 207).**

**4** Meteoritic impacts are the main source of erosion on the surfaces of both the Moon and Mercury. The lunar highlands are older than the maria and are much more heavily cratered. The rate at which craters are formed decreases 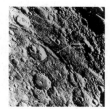 rapidly with increasing crater size. By measuring the ages of lunar rocks returned to Earth by *Apollo* astronauts, astronomers have deduced the rate of cratering in the past. They then use the amount of cratering to deduce the ages of regions on the Moon (and elsewhere) from which surface samples are unavailable.

**5** Evidence for past volcanic activity on the Moon is found in the form of crater chains and solidified lava channels called **rilles (p. 217).** Mercury's surface features bear a striking similarity to those of the Moon. The planet is heavily cratered, much like the lunar highlands. Among the differences  between Mercury and the Moon are Mercury's lack of lunarlike maria, its extensive **intercrater plains (p. 217),** and the great cracks, or **scarps (p. 219),** in its crust. The plains were caused by extensive lava flows early in Mercury's history. The scarps were apparently formed when the planet's core cooled and shrank, causing the surface to crack. Mercury has a large impact crater called the Caloris Basin, whose diameter is comparable to the radius of the planet. The impact that formed the crater apparently sent violent shock waves around the entire planet, buckling the crust on the opposite side.

**6** The Moon's average density is not much greater than that of its surface rocks, probably because the Moon cooled more rapidly than the larger Earth and solidified sooner, so there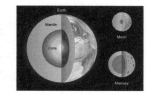

was less time for differentiation to occur, although the Moon likely has a small iron-rich core. The lunar crust is too thick and the mantle too cool for plate tectonics to occur. Mercury's average density is considerably greater—similar to that of Earth—implying that Mercury contains a large high-density core, probably composed primarily of iron. The Moon has no measurable large-scale magnetic field, a consequence of its slow rotation and lack of a molten metallic core. Mercury's weak magnetic field seems to have been "frozen in" long ago, when the planet's iron core solidified.

**7** The most likely explanation for the formation of the Moon is that the newly formed Earth was struck by a large (Mars-sized) object. Part of the colliding body remained behind as part of our planet. The rest ended up in orbit as the Moon.

**8** The absence of a lunar atmosphere and any present-day lunar volcanic activity are both consequences of the Moon's small size. Lunar gravity is too weak to retain any gases, and lunar volcanism was stifled by the Moon's cooling mantle shortly after extensive lava flows formed the maria more than 3 billion years ago.

The crust on the far side of the Moon is substantially thicker than the crust on the near side. As a result, there are almost no maria on the lunar far side. Mercury's evolutionary path was similar to that of the Moon for half a billion years after they both formed. Mercury's volcanic period probably ended before that of the Moon.

## Review and Discussion

1. How is the distance to the Moon most accurately measured?

2. Why is Mercury seldom seen with the naked eye?

3. Why did early astronomers think that Mercury was two separate planets?

4. Employ the concept of escape speed to explain why the Moon and Mercury have no significant atmospheres.

5. In what sense are the lunar maria "seas"?

6. Why is the surface of Mercury often compared with that of the Moon? List two similarities and two differences between the surfaces of Mercury and the Moon.

7. What does it mean to say that the Moon is in a synchronous orbit around Earth? How did the Moon come to be in such an orbit?

8. What does it mean to say that Mercury has a 3:2 spin–orbit resonance? Why didn't Mercury settle into a 1:1 spin–orbit resonance with the Sun, as the Moon did with Earth?

9. What is a scarp? How are scarps thought to have formed? Why do scientists think that the scarps on Mercury formed after most meteoritic bombardment ended?

10. What is the primary source of erosion on the Moon? Why is the average rate of lunar erosion so much less than on Earth?

11. What evidence do we have for ice on the Moon?

12. Name two pieces of evidence indicating that the lunar highlands are older than the maria.

13. In contrast with Earth, the Moon and Mercury undergo extremes in temperature. Why?

14. How is Mercury's evolutionary history like that of the Moon? How is it different?

15. Describe the theory of the Moon's origin favored by many astronomers.

16. Because the Moon always keeps one face toward Earth, an observer on the moon's near side would see Earth appear almost stationary in the lunar sky. Still, Earth would change its appearance as the Moon orbited Earth. How would Earth's appearance change?

17. The best place to aim a telescope or binoculars on the Moon is along the terminator line—the line between the Moon's light and dark hemispheres. Why? If you were standing on the lunar terminator, where would the Sun be in your sky? What time of day would it be if you were standing on Earth's terminator line?

18. Where on the Moon would be the best place from which to make astronomical observations? What would be this location's advantage over locations on Earth?

19. Explain why Mercury is never seen overhead at midnight in Earth's sky.

20. How is the varying thickness of the lunar crust related to the presence or absence of maria on the Moon?

## Conceptual Self-Test: True or False/Multiple Choice

1. Laser ranging can determine the distance to the Moon to an accuracy of a few centimeters.

2. Mercury can sometimes be seen at midnight.

3. Mercury's solar day is longer than its solar year.

4. The most accurate method for determining the distance to the Moon is by parallax.

5. Mercury's daytime temperature is higher than the Moon's because Mercury is more massive than the Moon.

6. Craters on the Moon and Mercury are primarily the result of volcanic activity.

7. There is no volcanic activity today on the surface of the Moon.

8. Although daytime temperatures on the Moon and Mercury are very high, it may still be possible for those two bodies to have large amounts of water ice at their poles.

9. The lunar maria's dark, dense rock originally was part of the lunar mantle.

10. The most likely scenario for the formation of the Moon is a collision between Earth and another planet-sized body.

11. Compared with the diameter of Earth's Moon, the diameter of Mercury is (a) larger; (b) smaller; (c) nearly the same.

12. In relation to the density of Earth's Moon, Mercury's density suggests that the planet (a) has an interior structure similar to that of the Moon; (b) has a dense metal core; (c) has a stronger magnetic field than the Moon; (d) is younger than the Moon.

13. Compared with the phases of Earth's Moon, Mercury goes from new phase to full phase (a) faster; (b) more slowly; (c) in about the same time.

14. Compared with the surface of Mercury, the surface of Earth's Moon has significantly (a) bigger craters; (b) more atmosphere; (c) more maria; (d) deeper craters.

15. Every two times Earth's Moon rotates on its axis, it orbits Earth (a) less than twice; (b) exactly two times; (c) more than twice; (d) three times.

16. Planets and moons showing the most craters have (a) the oldest surfaces; (b) been hit by meteors the most times; (c) the strongest gravity; (d) molten cores.

17. Compared with the Moon, Mercury has (a) a much smaller core; (b) a much larger core; (c) a similar-sized core.

18. The most likely theory of the formation of Earth's Moon is that it (a) was formed by the gravitational capture of a large asteroid; (b) formed simultaneously with Earth's formation; (c) was created from a collision scooping out the Pacific Ocean; (d) formed from a collision of Earth with a Mars-sized object.

19. Mercury, being smaller than Mars, probably cooled and solidified (a) faster, because it is smaller; (b) slower, because it is closer to the Sun; (c) in about the same time, because space is generally cold.

20. On the scale of the 5-billion-year age of the solar system, the Moon is (a) about the same age as Earth; (b) much younger than Earth; (c) much older than Earth.

## Problems

 *Algorithmic versions of these Problems are available in the Practice Problems module of the Companion Website.*
*The number of dots preceding each Problem indicates its approximate level of difficulty.*

1. • How long does a radar signal take to travel from Earth to Mercury and back when Mercury is at its closest point to Earth?

2. • The Moon's mass is one-eightieth that of Earth, and the lunar radius is one-fourth Earth's radius. On the basis of these figures, calculate the total weight on the Moon of a 100-kg astronaut with a 50-kg space suit and backpack, relative to his weight on Earth.

3. • What would be the same astronaut's weight on Mercury?

4. • Based on the data presented in the Moon Data box (p. 202), verify the values given for the Moon's perigee (minimum distance from Earth) and apogee (maximum distance from Earth), and estimate the Moon's minimum and maximum angular diameter, as seen from Earth. Compare these values with the angular diameter of the Sun (of actual diameter 1.4 million km), as seen from a distance of 1 AU.

5. • What is the angular diameter of the Sun, as seen from Mercury, at perihelion? At aphelion?

6. • The *Hubble Space Telescope* has a resolution of about 0.05″. What is the size of the smallest feature it can distinguish on the surface of the Moon (distance = 380,000 km)? On Mercury, at closest approach to Earth?

7. • What was the orbital period of the *Apollo 11* command module, orbiting 10 km above the lunar surface?

8. •• Compare the gravitational tidal acceleration of the Sun on Mercury (at perihelion; solar mass = $2 \times 10^{30}$ kg) with the tidal effect of Earth on the Moon (at perigee). ∞ (Sec. 7.6)

9. ••• Mercury's average orbital speed around the Sun is 47.9 km/s. Use Kepler's second law to calculate Mercury's speed (a) at perihelion and (b) at aphelion. ∞ (Sec. 2.5, *More Precisely 2-1*) Convert these speeds to angular speeds (in degrees per day), and compare them with Mercury's 6.1°-per-day rotation rate.

10. •• What would the lengths of a sidereal and a solar day on Mercury be if the planet were in a 4:3 spin–orbit resonance instead of the 3:2 resonance actually observed?

11. •• Assume that a planet will have lost its initial atmosphere by the present time if the average molecular speed exceeds one-sixth of the escape speed (see *More Precisely 8-1*). What would Mercury's mass have to be in order for it to still have a nitrogen atmosphere? The molecular weight of nitrogen is 28.

12. •• With the same assumptions as in the previous question, estimate the minimum molecular mass that might still be found in Mercury's atmosphere.

13. •• Using the rate given in the text for the formation of 10-km craters on the Moon, estimate how long would be needed for the entire Moon to be covered with new craters of that size. How much higher must the cratering rate have been in the past to cover the entire lunar surface with such craters in the 4.6 billion years since the Moon formed?

14. •• Repeat the previous question, for meter-sized craters.

15. • Using the data given in the text, calculate how long erosion would take to obliterate (a) the boot print in Figure 8.17, (b) the Barringer Meteor Crater in Figure 8.19, (c) lunar crater Reinhold in Figure 8.15(b).

*The Companion Website at www.aw-bc.com/chaisson provides algorithmically generated versions of each chapter's Problems, along with additional quizzes, an Animations & Videos gallery, an Images gallery, an interactive Glossary, and a full eBook.*

# CHAPTER 9

*Often called Earth's sister planet, Venus is nothing like Earth. When it comes to surface temperature, it's hot enough there (730 K) to melt lead. We now know that Venus's climate, like Earth's, has varied over time—largely the result of geological activity and atmospheric change. What we do not know well is why Venus became so very much hotter than Earth—or if Earth could someday heat up similarly. Here, this global view of the surface of Venus was created when the Magellan spacecraft's radar data were mapped onto a computer-simulated globe. (JPL) ▶*

# VENUS

## Earth's Sister Planet

Venus seems almost a carbon copy of our own world. The two planets are similar in size, density, and chemical composition. They orbit at comparable distances from the Sun. At formation, they must have been almost indistinguishable from one another. Yet they are now about as different as two terrestrial planets can be. Whereas Earth is a vibrant world, teeming with life, Venus is an uninhabitable inferno, with a dense, hot atmosphere of carbon dioxide, lacking any trace of oxygen or water.

Somewhere along their respective evolutionary paths, Venus and Earth diverged, and diverged radically. How did this occur? What were the factors leading to Venus's present condition? Why are Venus's surface, atmosphere, and interior so different from Earth's? In answering these questions, we will discover that a planet's environment, as well as its composition, can play a critical role in determining its future.

## LEARNING GOALS

*Studying this chapter will enable you to*

1 Summarize Venus's general orbital and physical properties.
2 Describe the characteristics of Venus's atmosphere and contrast it with that of Earth.
3 Compare the large-scale surface features and geology of Venus with those of Earth and the Moon.
4 Discuss the evidence for ongoing volcanic activity on Venus.
5 Explain why the greenhouse effect has produced conditions on Venus very different from those on Earth.
6 Describe Venus's magnetic field and internal structure.

Ⓦ Visit www.aw-bc.com/chaisson for additional images, animations, quizzes, and eBook for this chapter.

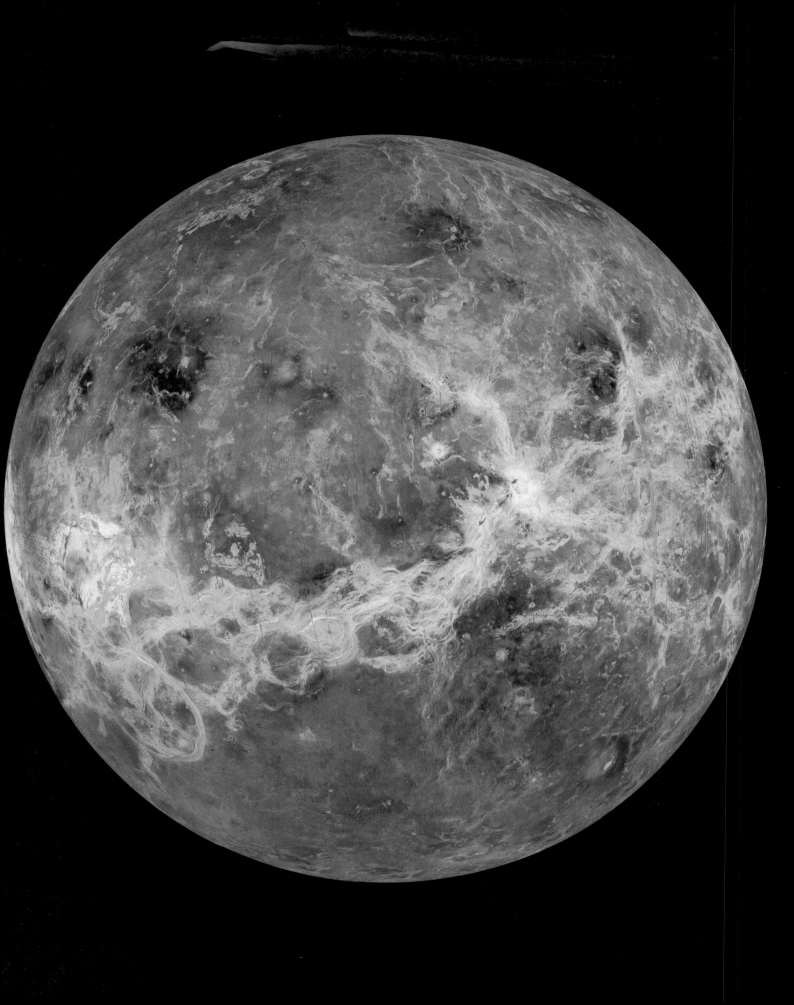

# 9.1 Orbital Properties

Venus is the second planet from the Sun. Its orbit lies within Earth's, so Venus, like Mercury, is always found fairly close to the Sun in the sky—our sister planet is never seen more than 47° from the Sun. Given Earth's rotation rate of 15° per hour, this means that Venus is visible above the horizon for at most 3 hours before the Sun rises or after it sets. Because we can see Venus from Earth only just before sunrise or just after sunset, the planet is often called the "morning star" or the "evening star," depending on where it happens to be in its orbit. Figure 9.1 shows Venus in the western sky just after sunset. The Venus Data box on p. 231 lists some of the planet's orbital and physical properties.

Venus is the third-brightest object in the entire sky (after the Sun and the Moon). It appears more than 10 times brighter than the brightest star, Sirius. You can see Venus even in the daytime if you know just where to look. On a moonless night away from city lights, Venus casts a faint shadow. The planet's brightness stems from the fact that Venus is highly reflective. Nearly 70 percent of the sunlight reaching Venus is reflected back into space. (Compare this percentage with roughly 10 percent in the case of Mercury and the Moon.) Most of the sunlight is reflected from clouds high in the planet's atmosphere.

We might expect Venus to appear brightest when it is "full"—that is, when we can see the entire sunlit side. However, because Venus orbits between Earth and the Sun, Venus is full when it is at its greatest distance from us—1.7 AU away on the other side of the Sun, as illustrated in Figure 9.2. Recall from Chapter 2 that this alignment is known as *superior conjunction*, where the term "conjunction" simply indicates that two objects are close together in the sky. ∞ (Sec. 2.2)

When Venus is closest to us, the planet is in the new phase, lying between Earth and the Sun (at *inferior conjunction*), and we again can't see it, because now the sunlit side faces away from us; only a thin ring of sunlight, caused by refraction in Venus's atmosphere, surrounds the planet. As Venus moves away from inferior conjunction, more and more of it becomes visible, but

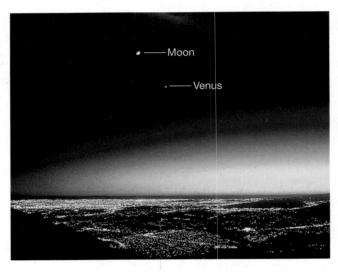

▲ FIGURE 9.1 **Venus at Sunset**  The Moon and Venus in the western sky just after sunset. Venus clearly outshines even the brightest stars in the sky. (*J. Schad/Photo Researchers, Inc.*)

its distance from us also continues to increase. Venus's maximum brightness, as seen from Earth, actually occurs about 36 days before or after its closest approach to our planet. At that time, Venus is about 39° from the Sun and 0.47 AU from Earth, and we see it as a rather fat crescent.

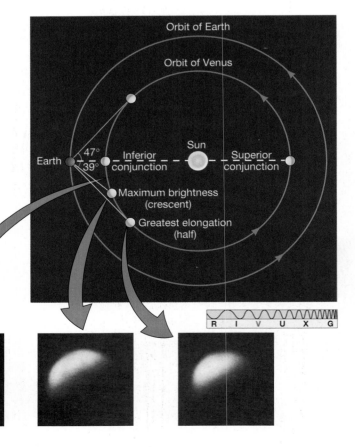

▶ FIGURE 9.2 **Venus's Brightness**  Venus appears full when it is at its greatest distance from Earth, on the opposite side of the Sun from us (superior conjunction). As its distance decreases, less and less of its sunlit side becomes visible. When closest to Earth, it lies between us and the Sun (inferior conjunction), so we cannot see the sunlit side of the planet at all. Venus appears brightest when it is about 39° from the Sun. (Compare Figure 2.12.) (Insets: *UC/Lick Observatory*)

## 9.2 Physical Properties

### Radius, Mass, and Density

We can determine Venus's radius from simple geometry, just as we did for Mercury and the Moon. ∞ (Sec. 8.2) At closest approach, when Venus is only 0.28 AU from us, its angular diameter is 64″. From this observation, we can determine the planet's radius to be about 6000 km. More accurate measurements from spacecraft give a value of 6052 km, or 0.95 Earth radii.

Like Mercury, Venus has no moon. Before the Space Age, astronomers calculated its mass by indirect means—through studies of its small gravitational effect on the orbits of the other planets, especially Earth. Now that spacecraft have orbited the planet, we know Venus's mass very accurately from measurements of its gravitational pull: Venus has a mass of $4.9 \times 10^{24}$ kg, or 0.82 the mass of Earth.

From its mass and radius, we find that Venus's average density is 5200 kg/m$^3$. As far as these bulk properties are concerned, then, Venus seems similar to Earth. If the planet's overall composition were similar to Earth's as well, we could then reasonably conclude that Venus's internal structure and evolution were basically Earth-like. We will review what evidence there is on this subject later in the chapter.

### Rotation Rate

The same clouds whose reflectivity makes Venus so easy to see in the night sky also make it impossible for us to discern any surface features on the planet, at least in visible light. As a result, until the advent of suitable radar techniques in the 1960s, astronomers did not know the rotation period of Venus. Even when viewed through a large optical telescope, the planet's cloud cover shows few features, and attempts to determine Venus's period of rotation by observing the cloud layer were frustrated by the rapidly changing nature of the clouds themselves. Some astronomers argued for a 25-day period, while others favored a 24-hour cycle.

Controversy raged until, to the surprise of all, radar observers announced that the Doppler broadening of their returned echoes implied a sluggish 243-day rotation period! ∞ (Sec. 8.3) Furthermore, Venus's spin was found to be *retrograde*—that is, in a sense opposite that of Earth and most other solar system objects and opposite that of Venus's orbital motion.

Planetary astronomers define "north" and "south" for each planet in the solar system by the convention that planets *always* rotate from west to east. With this definition, Venus's retrograde spin means that the planet's north pole lies *below* the plane of the ecliptic, unlike any of the other terrestrial worlds. Venus's axial tilt—the angle between its equatorial and orbital planes—is 177.4° (compared with 23.5° in the case of Earth). However, astronomical images of solar system objects conventionally place objects lying above the ecliptic at the top of the frame. Thus, with the

## PLANETARY DATA

### VENUS

| | |
|---|---|
| Orbital semimajor axis | 0.72 AU 108.2 million km |
| Orbital eccentricity | 0.007 |
| Perihelion | 0.72 AU 107.5 million km |
| Aphelion | 0.73 AU 108.9 million km |
| Mean orbital speed | 35.0 km/s |
| Sidereal orbital period | 224.7 Earth days*<br>0.615 tropical year |
| Synodic orbital period | 583.9 Earth days* |
| Orbital inclination to the ecliptic | 3.39° |
| Greatest angular diameter, as seen from Earth | 64″ |
| Mass | $4.87 \times 10^{24}$ kg<br>0.82 (Earth = 1) |
| Equatorial radius | 6052 km<br>0.95 (Earth = 1) |
| Mean density | 5240 kg/m$^3$<br>0.95 (Earth = 1) |
| Surface gravity | 8.87 m/s$^2$<br>0.91 (Earth = 1) |
| Escape speed | 10.4 km/s |
| Sidereal rotation period | −243.0 Earth days**<br>(retrograde) |
| Axial tilt | 177.4° |
| Surface magnetic field | <0.001 (Earth = 1) |
| Magnetic axis tilt relative to rotation axis | — |
| Mean surface temperature | 730 K |
| Number of moons | 0 |

*1 Earth (mean solar) day = 24 hours
**A negative sign denotes retrograde rotation.

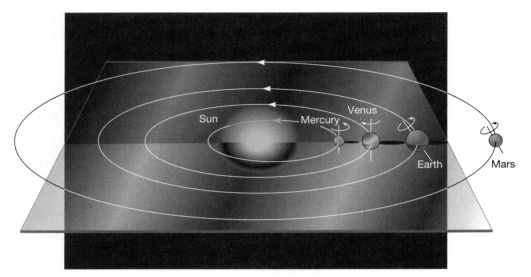

◀ FIGURE 9.3 **Terrestrial Planets' Spins** The inner planets of the solar system—Mercury, Venus, Earth, and Mars—display widely differing rotational properties. Although all orbit the Sun in the same direction and in nearly the same plane, Mercury's rotation is slow and prograde (in the same sense as its orbital motion), that of Venus is slow and retrograde, and those of Earth and Mars are fast and prograde. Venus rotates clockwise as seen from above the plane of the ecliptic, but Mercury, Earth, and Mars all spin counterclockwise. This is a perspective view, roughly halfway between a flat edge-on view and a direct overhead view.

preceding definition of north and south, all the images of Venus shown in this chapter have the *south* pole at the top.

Figure 9.3 illustrates Venus's retrograde rotation and compares it with the rotation of its neighbors Mercury, Earth, and Mars. Because of the planet's slow retrograde rotation, its solar day (from noon to noon) is quite different from its sidereal rotation period of 243 Earth days (the time for one "true" rotation relative to the stars). ∞ (Sec. 1.4) In fact, as illustrated in Figure 9.4, one Venus day is a little more than half a Venus year (225 Earth days).

Why is Venus rotating "backward" and why so slowly? At present, the best explanation planetary scientists can offer is that early in Venus's evolution, the planet was struck by a

large body, much like the one that may have hit Earth and formed the Moon, and that impact was sufficient to reduce the planet's spin almost to zero. ∞ (Sec. 8.8) Whatever its cause, the planet's rotation poses practical problems for Earth-bound observers. It turns out that Venus rotates almost exactly five times between one closest approach to Earth and the next. As a result, *Venus always presents nearly the same face to Earth at closest approach.* This means that observations of the planet's surface cover one side—the one facing us at closest approach—much more thoroughly than the other side, which we can see only when the planet is close to its maximum distance from Earth.

This nearly perfect 5:1 resonance between Venus's rotation and orbital motion is reminiscent of the Moon's synchronous orbit around Earth and Mercury's 3:2 spin–orbit resonance with the Sun. ∞ (Sec. 8.4) However, no known interaction between Earth and Venus can account for such an odd state of affairs. Earth's tidal effect on Venus is tiny and is much less than the Sun's tidal effect in any case.

Furthermore, the key word in the preceding paragraph is *nearly*. A resonance, if it existed, would require that the number of rotations per relative orbit be *exactly* five. The discrepancy amounts to less than 3 hours in 584 days (Venus's *synodic period* relative to Earth—see *More Precisely 9-1*), but it appears to be real, and if that is so, then no true resonance exists. Astronomers hate to appeal to coincidence to explain their observations, but the case of Venus's rotation appears to be just that. For now, we are simply compelled to accept this strange coincidence without explanation.

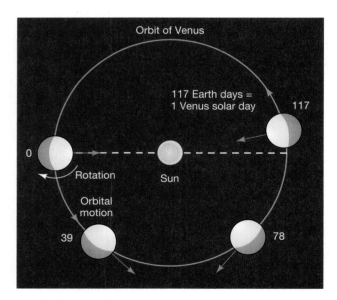

▲ FIGURE 9.4 **Venus's Solar Day** Venus's orbit and retrograde rotation combine to produce a solar day on Venus equal to 117 Earth days, or slightly more than half a Venus year. The red arrows represent a fixed location, or an observer standing, on the planet's surface. The numbers in the figure mark time in Earth days.

CONCEPT CHECK

✔ What is peculiar about Venus's rotation, and why does Venus rotate that way?

# MORE PRECISELY 9-1

## Synodic Periods and Solar Days

Recall from Chapter 1 that a body's *sidereal* orbital (or rotational) period is the time taken for the body to complete one orbit (or rotation) relative to the "fixed" stars. ∞ (Sec. 1.4) In many cases, however, we are more interested in how things look from our vantage point on Earth. The **synodic orbital period** is defined to be the time taken for the body to return to the same configuration relative to the Sun, *taking Earth's own motion into account.* Examples are the time from one full Moon to the next or between successive inferior conjunctions (times of closest approach) of Venus. ∞ (Sec. 1.6)

Similarly, a planet's **solar day**—the time from noon to noon—in general differs from its sidereal day because of the planet's motion around the Sun. In the case of Earth, the difference between a sidereal and a solar day is relatively small, as is the difference between a sidereal and a synodic month. ∞ (Secs. 1.4, 1.6) However, as we have seen, Mercury's sidereal and solar days differ significantly, as do the sidereal and synodic periods of all the planets. ∞ (Sec. 8.4) Let's take a moment to look at this topic in a little more detail.

Calculating the synodic period is easy. With Earth's sidereal orbital period (365.26 solar days) denoted by $P_E$ and that of the body in question by $P$, it follows that the *angular speeds* of Earth and the other body (in degrees per day) are, respectively, $360°/P_E$ and $360°/P$, since there are 360 degrees in one complete revolution. Thus, as illustrated in the accompanying figure, the rate at which the body "outruns" Earth is $360°/P - 360°/P_E$. The body's synodic period $S$ is, by definition, the time taken for the body to "lap" Earth (i.e., outstrip it by 360°), so it follows that

$$\frac{1}{S} = \frac{1}{P} - \frac{1}{P_E},$$

where $S$ is positive for the interior planets Mercury and Venus (having $P$ less than $P_E$, as in the figure) and negative for the exterior planets Mars, Jupiter, and so on (meaning that Earth overtakes them).

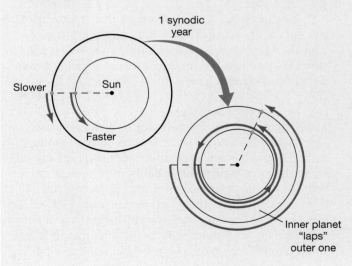

1 synodic year

Slower

Sun

Faster

Inner planet "laps" outer one

**EXAMPLE 1** The Moon has $P = 27.3$ days, so its synodic period $S$ is given by

$$\frac{1}{S} = \frac{1}{27.3} - \frac{1}{365.26} = \frac{1}{29.5}.$$

Hence, $S = 29.5$ days. Similarly, for Venus, $P = 224.7$ days, yielding the synodic period $S = 583.9$ days mentioned in the text. During this time, Earth orbits the Sun 1.6 times, and Venus 2.6, so the actual planetary orbits are a little more complicated than those sketched in the figure. For Mars, we have $P = 687$ days, so $S = -780$ days, accounting for the roughly 2-year interval between launch windows for space probes to that planet, as mentioned in Chapter 6. ∞ (Sec. 6.6) (The negative sign simply means that Earth overtakes Mars, which therefore appears to move backwards relative to Earth.)

Now let's consider how the competition between the planet's sidereal rotation period $R$ and its sidereal orbital period $P$ determines the length of its solar day. ∞ (Fig. 1.13) Reasoning similar to the preceding leads to the following expression for the planetary solar day $D$:

$$\frac{1}{D} = \frac{1}{R} - \frac{1}{P}.$$

For example, Earth has $R = 0.9973$ solar day and $P = P_E = 365.26$ solar days, which gives $D = 1$ solar day, as expected. The difference between Earth's solar and sidereal days is small because Earth rotates on its axis much more rapidly than it revolves around the Sun. For Mercury and Venus, however, the rotation and revolution periods are comparable, and consequently the difference between solar and sidereal days is much greater.

**EXAMPLE 2** Using the data from the Mercury Data box, p. 220, we find that, for Mercury, $R = 58.6$ (Earth) solar days and $P = 88.0$ solar days, so $D = 176.0$ solar days, or 2 sidereal years. ∞ (Sec. 8.4) Venus has $R = -243.0$ solar days and $P = 224.7$ solar days, so

$$\frac{1}{D} = \frac{-1}{243.0} - \frac{1}{224.7} = -\frac{1}{116.7}.$$

Thus, $D = -116.7$ solar days. (The negative signs, as usual, indicate retrograde rotation: The calculation implies that the Sun would move "backwards" through Venus's sky if someone on the surface could see it.) The extreme example would be a planet in a synchronous orbit around the Sun (or the Moon orbiting Earth), with $R = P$. In that case, $D$ would be infinite—the day would last forever, and the Sun would never set!

Sidereal periods are "physical," in the sense that they are the quantities to which the laws of Kepler and Newton refer. However, from the point of view of timekeeping, skywatching, or scheduling a mission to another planet, synodic periods, which tell us when a satellite returns to the same phase, or when a planet "returns to the same position" in the night sky, are just as important!

## 9.3 Long-Distance Observations of Venus

Because Venus, of all the other planets, most nearly matches Earth in size, mass, and density, and because its orbit is closest to us, it is often called Earth's sister planet. But unlike Earth, Venus has a dense atmosphere and thick clouds that are opaque to visible radiation, making its surface completely invisible from the outside at optical wavelengths. Figure 9.5 shows one of the best photographs of Venus taken with a large telescope on Earth. The planet presents an almost featureless white-yellow disk, although it shows occasional hints of cloud circulation.

Atmospheric patterns on Venus are much more evident when the planet is examined with equipment capable of detecting ultraviolet radiation. Some of Venus's atmospheric constituents absorb this high-frequency radiation, greatly increasing the cloud contrast. Figure 9.6(a) shows an ultraviolet image taken in 1979 by the U.S. *Pioneer Venus* spacecraft at a distance of 200,000 km from the planet's surface; Figure 9.6(b) shows a recent (2006) mosaic of infrared images from the European *Venus Express* orbiter, whose cameras partially penetrate the planet's thick haze. The large, fast-moving cloud patterns resemble Earth's high-altitude jet stream more than the great whirls characteristic of Earth's low-altitude clouds. The upper deck of clouds on Venus move at almost 400 km/h, encircling the planet in just 4 days—much faster than the planet itself rotates!

(a)

R  I  V  U  X  G

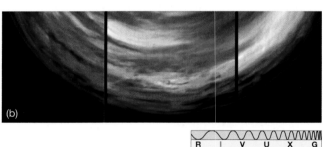

(b)

R  I  V  U  X  G

▲ FIGURE 9.6 **Venus, Up Close** (a) Venus as photographed by the *Pioneer* spacecraft's cameras 200,000 km away from the planet. This image was made by capturing solar ultraviolet radiation reflected from the planet's clouds, which are probably composed mostly of sulfuric acid droplets, much like the highly corrosive acid in a car battery. (b) Venus in the infrared, as seen by *Venus Express* on approach to the planet. The longer infrared wavelength allows us to "see" deeper into Venus's clouds. *(NASA; ESA)*

R  I  V  U  X  G

▲ FIGURE 9.5 **Venus** This photograph, taken from Earth, shows Venus with its creamy yellow mask of clouds. No surface detail can be seen, because the clouds completely obscure our view of whatever lies beneath them. *(AURA)*

Early spectroscopic studies of sunlight reflected from Venus's clouds revealed the presence of large amounts of carbon dioxide, but provided little evidence for any other atmospheric gases. Until the 1950s, astronomers generally believed that observational difficulties alone prevented them from seeing other atmospheric components. The hope lingered that Venus's clouds were actually predominantly water vapor, like those on Earth, and that below the cloud cover Venus might be a habitable planet similar to our own. Indeed, in the 1930s, scientists had measured the temperature of the atmosphere spectroscopically at about 240 K, not much different from the temperature of our own upper atmosphere. ∞ (Sec. 4.5, 7.2) Calculations of the planet's surface temperature—taking into account the cloud cover and Venus's proximity to the Sun, and assuming an atmosphere much like our own—suggested that Venus should have a surface temperature only 10 or 20 degrees higher than Earth's.

These hopes for an Earth-like Venus were dashed in 1956, when radio observations of the planet were used to measure its thermal energy emission. Unlike visible light, radio waves easily penetrate the cloud layer—and they gave the first indication of conditions on or near the surface: The radiation emitted by the planet has a blackbody spectrum characteristic of a temperature near 730 K! ∞ (Sec. 3.4) Almost overnight, the popular conception of Venus changed from that of a lush tropical jungle to an arid, uninhabitable desert.

Radar observations of the surface of Venus are routinely carried out from Earth with the Arecibo radio telescope. ∞ (Sec. 5.5) With careful signal processing, this instrument can achieve a resolution of a few kilometers, but it can adequately cover only a fraction (roughly 25 percent) of the planet. The telescope's view of Venus is limited by the planet's peculiar near resonance described in the previous section (which means that only one side of the planet can be studied) and also because radar reflections from regions near the "edge" of the planet are hard to obtain. However, the Arecibo data can usefully be combined with information received from probes orbiting Venus to build up a detailed picture of the planet's surface. Only with the arrival of the *Magellan* probe were more accurate data obtained.

## CONCEPT CHECK

✔ Why did early studies of Venus lead astronomers to such an inaccurate picture of the planet's surface conditions?

## 9.4  The Surface of Venus

Although the planet's clouds are thick and the terrain below them totally shrouded, we are by no means ignorant of Venus's surface. Detailed radar observations have been made both from Earth and from the *Venera, Pioneer Venus,* and *Magellan* spacecraft. ∞ (Sec. 6.6) Analysis of the radar echoes yields a map of the planet's surface. Except for the last two figures, all the views of Venus in this section are "radar-graphs" (as opposed to photographs) created in this way.

As Figure 9.7(a) illustrates, the early maps of Venus suffered from poor resolution; however, more recent probes—especially *Magellan*—have provided much sharper views. As in all the *Magellan* images, the light areas in Figure 9.7(b) represent regions where the surface is rough

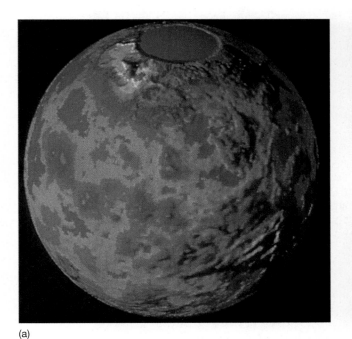

(a)

(b)

R  I  V  U  X  G

▲ FIGURE 9.7  **Venus Mosaics** (a) This image of the surface of Venus was made by a radar transmitter and receiver on board the *Pioneer* spacecraft, which is still in orbit about the planet, but is now inoperative. The two continent-sized landmasses are named Ishtar Terra (upper left) and Aphrodite (lower right). Colors represent altitude: blue is lowest, red highest. The spatial resolution is about 25 km. (b) A planetwide mosaic of *Magellan* images, colored in roughly the same way as part (a). The largest "continent" on Venus, Aphrodite Terra, is the yellow dragon-shaped area across the center of this image. See also the full-page, chapter-opening photo. *(NASA)*

and efficiently scatters *Magellan*'s sideways-looking radar beam back to the detector. Smooth areas tend to reflect the beam off into space instead and so appear dark. The strength of the returned signal as *Magellan* passed by thus results in a map of the planet's surface.

## Large-Scale Topography

Figure 9.8(a) shows basically the same *Pioneer Venus* data of Venus as Figure 9.7, except that this figure has been flattened out into a more conventional map. The altitude of the surface relative to the average radius of the planet is indicated by the use of color, with white representing the highest elevations and blue the lowest. (Note that the blue has nothing to do with oceans, nor does white indicate snow-capped mountains!) Figure 9.8(b) shows a map of Earth to the same scale and at the same spatial resolution. Some of Venus's main features are labeled in Figure 9.8(c).

The surface of Venus appears to be relatively smooth, resembling rolling plains with modest highlands and lowlands. Two continent-sized features, called Ishtar Terra and Aphrodite Terra (named after the Babylonian and

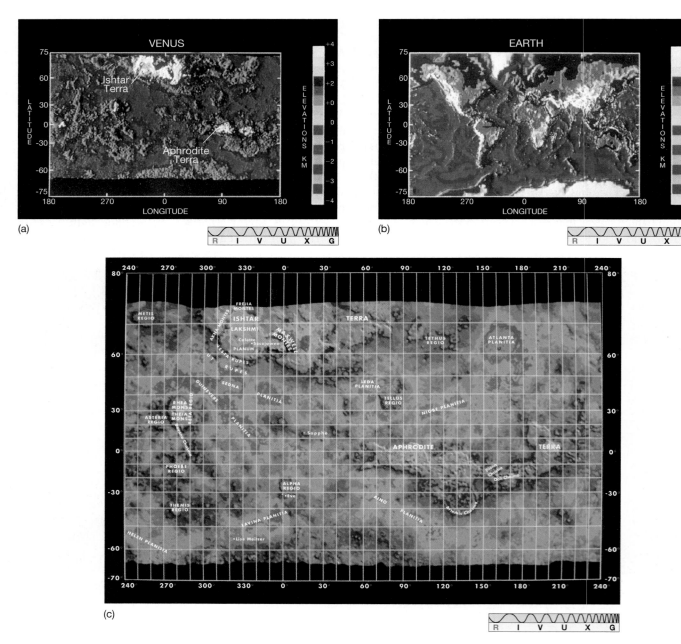

(a)

(b)

(c)

▲ **FIGURE 9.8 Venus Maps** (a) Radar map of the surface of Venus, based on *Pioneer Venus* data. Color represents elevation, with white the highest areas and blue the lowest. (b) A similar map of Earth, at the same spatial resolution. (c) Another version of (a), with major surface features labeled. Compare with Figure 9.7, and notice how the projection exaggerates the size of surface features near the poles. (*NASA*)

Greek counterparts, respectively, of Venus, the Roman goddess of love), adorn the landscape and contain mountains comparable in height to those on Earth. The elevated "continents" occupy only 8 percent of Venus's total surface area. For comparison, continents on Earth make up about 25 percent of the surface. The remainder of Venus's surface is classified as lowlands (27 percent) or rolling plains (65 percent), although there is probably little geological difference between the two terrains.

Note that, although Earth's tectonic plate boundaries are evident in Figure 9.8(b), no similar features can be seen in Figure 9.8(a). ∞ (Sec. 7.4) There simply appears to be no large-scale plate tectonics on Venus. Ishtar Terra ("Land of Ishtar") lies in the southern high latitudes (at the *tops* of Figures 9.7a and 9.8a—recall our earlier discussion of Venus's retrograde rotation). The projection used in Figure 9.8 makes Ishtar Terra appear larger than it really is—it is actually about the same size as Australia. This landmass is dominated by a great plateau known as Lakshmi Planum (Figure 9.9), some 1500 km across at its widest point and ringed by mountain ranges, including the Maxwell Montes range, which contains the highest peak on the planet, rising some 14 km above the level of Venus's deepest surface depressions. Again for comparison, the highest point on Earth (the summit of Mount Everest) lies about 20 km above the deepest section of Earth's ocean floor (Challenger Deep, at the bottom of the Marianas Trench on the eastern edge of the Philippines plate).

Figure 9.9(a) shows a large-scale *Venera* image of Lakshmi Planum, at a resolution of about 2 km. The

"wrinkles" are actually chains of mountains, hundreds of kilometers long and tens of kilometers apart. The red area immediately to the right of the plain is Maxwell Montes. On the western (right-hand) slope of the Maxwell range lies a great crater, called Cleopatra, about 100 km across. Figure 9.9(b) shows a *Magellan* image of Cleopatra, which was originally thought to be volcanic in origin. Close-up views of the crater's structure, however, have led planetary scientists to conclude that the crater is meteoritic in origin, although some volcanic activity was apparently associated with its formation when the colliding body temporarily breached the planet's crust. Notice the dark (smooth) lava flow emerging from within the inner ring and cutting across the outer rim at the upper right.

It is now conventional to name features on Venus after famous women—Aphrodite, Ishtar, Cleopatra, and so on. However, the early nonfemale names (e.g., Maxwell Montes, named after the Scottish physicist James Clerk Maxwell) predating this convention have stuck, and they are unlikely to change. Venus's other continent-sized formation, Aphrodite Terra, is located on the planet's equator and is comparable in size to Africa. Before *Magellan*'s arrival, some researchers had speculated that Aphrodite Terra might have been the site of something akin to seafloor spreading at the Mid-Atlantic ridge on Earth—a region where two lithospheric plates moved apart and molten rock rose to the surface in the gap between them, forming an extended ridge. ∞ (Sec. 7.4) With the low-resolution data then available, the issue could not be settled at the time.

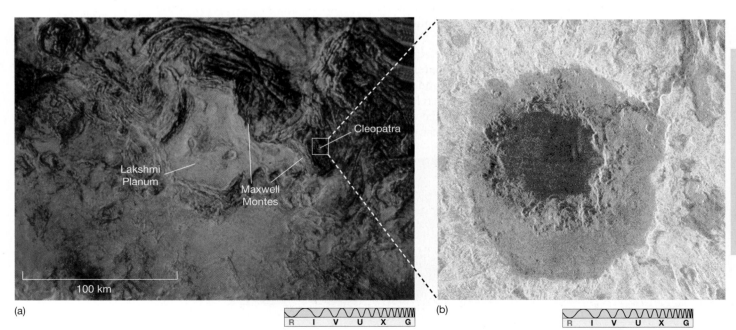

(a)

Lakshmi Planum

Maxwell Montes

Cleopatra

100 km

R I V U X G

(b)

R I V U X G

ANIMATION/VIDEO  Flight Over Sif Mons Volcano

▲ FIGURE 9.9 **Ishtar Terra** (a) A *Venera* orbiter image of a plateau known as Lakshmi Planum in Ishtar Terra. The Maxwell Montes mountain range (red) lies on the western margin of the plain, near the right-hand edge of the image. A meteor crater named Cleopatra is visible on the western slope of the Maxwell range. Note the two larger craters in the center of the plain itself. (b) A *Magellan* image of Cleopatra showing a double-ringed structure that identifies the feature to geologists as an impact crater. *(NASA)*

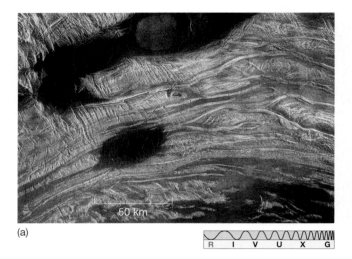

(a)

R I V U X G

The *Magellan* images now seem to rule out even this small-scale tectonic activity, and the Aphrodite region gives no indication of spreading. Figure 9.10(a) shows a portion of Aphrodite Terra called Ovda Regio. The crust appears buckled and fractured, with ridges running in two distinct directions across the image, suggesting that large compressive forces are distorting the crust. There seem to have been repeated periods of extensive lava flows. The dark regions are probably solidified lava flows. Some narrow lava channels, akin to rilles on the Moon, also appear. ∞ (Sec. 8.5) Such lava channels appear to be quite common on Venus. Unlike lunar rilles, however, they can be extremely long—hundreds or even thousands of kilometers (Figure 9.10b). These lava "rivers" often have lava "deltas" at their mouths, where they deposited their contents into the surrounding plains.

Figure 9.11 shows a series of angular cracks in the crust, thought to have formed when lava welled up from a deep fissure, flooded the surrounding area, and then retreated below the planet's surface. As the molten lava withdrew, the thin, new crust of solidified material collapsed under its own weight, forming the cracks we now see. Even

◀ **FIGURE 9.10**
**Aphrodite Terra** (a) A *Magellan* image of Ovda Regio, part of Aphrodite Terra. The intersecting ridges indicate repeated compression and buckling of the surface. The dark areas represent regions that have been flooded by lava upwelling from cracks like those shown in Figure 9.11. (b) This lava channel in Venus's south polar region, known as Lada Terra, extends for nearly 200 km. *(NASA)*

(b)

R I V U X G

taking into account the differences in temperature and composition between Venus's crust and Earth's, this terrain is not at all what we would expect at a spreading site similar to the Mid-Atlantic Ridge. ∞ (Sec. 7.4) Although there is no evidence for plate tectonics on Venus, it is likely that the stresses in the crust that led to the large mountain ranges were caused by convective motion within Venus's mantle—the same basic process that drives Earth's plates. Lakshmi Planum, for example, is probably the result of a "plume" of upwelling mantle material that raised and buckled the planet's surface.

*Venus Express*, currently (as of 2007) orbiting the planet, does not carry instruments capable of imaging the surface. However, its infrared sensors are designed to make precise temperature measurements at various levels in the atmosphere, including ground level. During 2006 the orbiter began mapping out Venus's surface temperature. Interestingly, since there is little

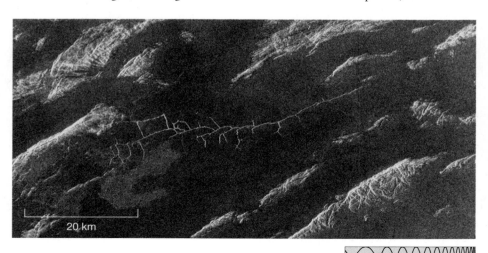

R I V U X G

◀ **FIGURE 9.11 Lava Flows** These cracks in Venus's surface, detected by *Magellan* in another part of Aphrodite Terra, have allowed lava to reach the surface and flood the surrounding terrain. The dark regions are smooth lava flows. The network of fissures visible here is about 50 km long. *(NASA)*

day–night variation in the planet's sweltering heat, temperature correlates well with altitude, and the temperature maps can be combined with the *Magellan* data to learn more about surface conditions. Ultimately, the *Venus Express* team hope to create maps of sufficient precision that they can detect hot spots on the surface, associated with volcanism or tectonic plumes, that are not evident in the *Magellan* maps.

## Volcanism and Cratering

On Earth, the principal agent of long-term, planetwide surface change is plate tectonics, driven by convection in our planet's mantle. ∞ (Secs. 7.3, 7.4) Volcanic and seismic activity are predominantly (although not exclusively) associated with plate boundaries. On Venus, without global plate tectonics, large-scale recycling of the crust by plate motion is not a factor in changing the planet's surface. Nevertheless, many areas of Venus have extensive volcanic features.

Most volcanoes on the planet are of the type known as **shield volcanoes.** Two large shield volcanoes, called Sif Mons and Gula Mons, are shown in Figure 9.12. Shield volcanoes, such as the Hawaiian Islands on Earth, are not associated with plate boundaries. Instead, they form when lava wells up through a "hot spot" in the crust and are built up over long periods of time by successive eruptions and lava flows. A characteristic of shield volcanoes is the formation of a *caldera*, or crater, at the summit when the underlying lava withdraws and the surface collapses. The distribution of volcanoes over the surface of Venus appears random—quite different from the distribution on Earth, where volcanic activity clearly traces out plate boundaries (see Figure 7.9)—consistent with the view that plate tectonics is absent on Venus.

More volcanic features are visible in Figure 9.13, which shows a series of seven

▶ **FIGURE 9.12 Volcanism on Venus** (a) Two larger volcanoes, known as Sif Mons (left) and Gula Mons, appear in this *Magellan* image. Color indicates height above a nominal planetary radius of 6052 km and ranges from purple (1 km, the level of the surrounding plain) to orange (corresponding to an altitude of about 4 km). The two volcanic calderas at the summits are about 100 km across. (b) A computer-generated view of Sif Mons, as seen from ground level. (c) Gula Mons, as seen from ground level. In (b) and (c), the colors are based on data returned from Soviet landers, and the vertical scales have been greatly exaggerated (by about a factor of 40), so these mountains look much taller relative to their widths than they actually are; Venus is actually a remarkably flat place. *(NASA)*

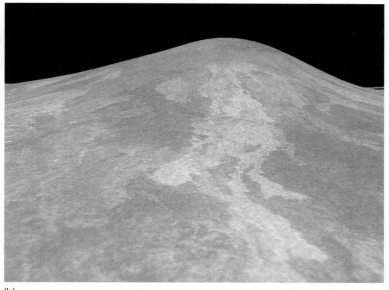
(b)

R I V U X G

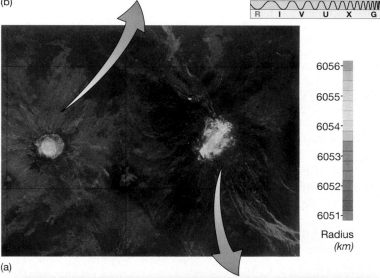
(a)

6056
6055
6054
6053
6052
6051

Radius
*(km)*

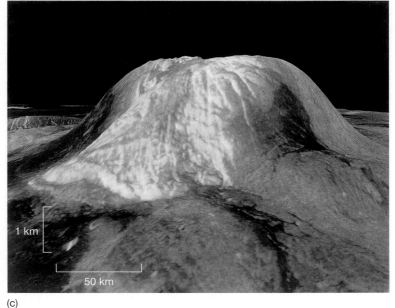
1 km

50 km

(c)

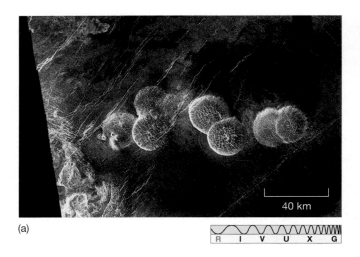

(a)

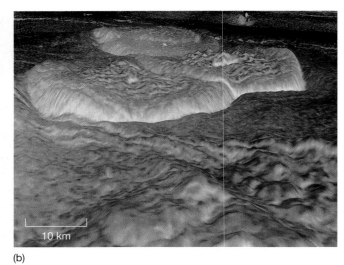

(b)

▲ FIGURE 9.13 **Lava Dome** (a) These dome-shaped structures resulted when viscous molten rock bulged out of the ground and then retreated, leaving behind a thin, solid crust that subsequently cracked and subsided. *Magellan* found features like these in several locations on Venus. (b) A three-dimensional representation of four of the domes. This computer-generated view is looking toward the right from near the center of the image in part (a). Colors in (b) are based on data returned by Soviet *Venera* landers. *(NASA)*

pancake-shaped **lava domes,** each about 25 km across. They probably formed when lava oozed out of the surface, formed the dome, and then withdrew, leaving the crust to crack and subside. Lava domes such as these are found in numerous locations on Venus. The largest volcanic structures on the planet are huge, roughly circular regions known as **coronae** (singular: *corona*). A large corona, called Aine, can be seen in Figure 9.14,

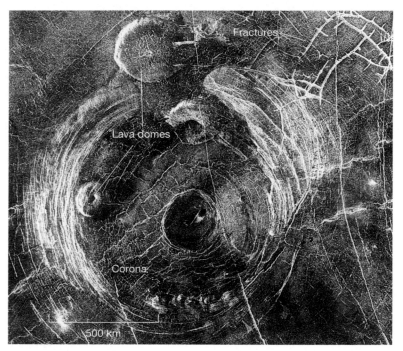

another large-scale mosaic of *Magellan* images. Coronae are unique to Venus. They appear to have been caused by upwelling mantle material, perhaps similar to the uplift that resulted in Lakshmi Planum but on a somewhat smaller scale. They generally have volcanoes both in and around them, and closer inspection of the rims usually shows evidence for extensive lava flows into the plains below.

There is overwhelming evidence for past surface activity on Venus. Has this activity now stopped, or is it still going on? Two pieces of indirect evidence suggest that volcanism continues today. First, the level of sulfur dioxide above Venus's clouds shows large and fairly frequent fluctuations. It is quite possible that these variations result from volcanic eruptions on the surface. If so, volcanism may be the primary cause of Venus's thick cloud cover. Second, both the *Pioneer Venus* and the *Venera* orbiters observed bursts of radio energy from Aphrodite and other regions of the planet's surface. The bursts are similar to those produced by lightning discharges that often occur in the plumes of

◀ FIGURE 9.14 **Venus Corona** This corona, called Aine, lies in the plains south of Aphrodite Terra and is about 300 km across. Coronae probably result from upwelling mantle material, causing the surface to bulge outward. Note the pancake-shaped lava domes at top, the many fractures in the crust around the corona, and the large impact craters with their surrounding white (rough) ejecta blankets that stud the region. *(NASA)*

erupting volcanoes on Earth, again suggesting ongoing activity. However, while these pieces of evidence are quite persuasive, they are still only circumstantial. No "smoking gun" (or to be more precise, erupting volcano) has yet been seen, so the case for active volcanism is not yet complete.

Not all the craters on Venus are volcanic in origin: Some, like Cleopatra (Figure 9.9b), were formed by meteoritic impact. Large impact craters on Venus are generally circular, but those less than about 15 km in diameter can be quite asymmetric in appearance. Figure 9.15(a) shows a *Magellan* image of a relatively small meteoritic impact crater, about 10 km across, in Venus's southern hemisphere. Geologists think that the light-colored region is the ejecta blanket—material ejected from the crater following the impact. The odd shape may be the result of a large meteoroid's breaking up just before impact into pieces that hit the surface near one another. Making craters such as these seems to be a fairly common fate for medium-sized bodies (1 km or so in diameter) that plow through Venus's dense atmosphere. Figure 9.15(b) shows the largest known impact feature on Venus: the 280-km-diameter crater called Mead. Its double-ringed structure is in many ways similar to the Moon's Mare Orientale (Figure 8.15a). Numerous impact craters (identifiable by their ejecta blankets) can also be discerned in Figure 9.14.

Venus's atmosphere is sufficiently thick that small meteoroids do not reach the ground, so there are no impact craters smaller than 2–3 km across, and atmospheric effects probably also account for the relative scarcity of impact craters less than 25 km in diameter. Contrast this with Earth, where even 10-m-sized craters are formed quite frequently (every few years) by small meteoroids striking the ground. (We don't see huge numbers of such craters on Earth because they are eroded away by wind and water quite rapidly—within a few tens of thousands of years.) On average, the number of large-diameter craters on Venus's surface per square kilometer is only about one-tenth that in the lunar maria. Applying similar crater-age estimates to Venus as we do to Earth and the Moon suggests that much of the surface of Venus is quite young—less than a billion years old, and perhaps as little as 200 or 300 million years in some places, such as the region shown in Figure 9.12. ∞ (Sec. 8.5)

Overall, the long-term degree of volcanism on Venus seems to be comparable to, but not as great as, that on Earth. However, planetary scientists think that the two planets differ in both the frequency and the severity of volcanic eruptions. On Earth, near-continuous volcanic activity at plate boundaries provides a natural "release valve," allowing energy from the interior to escape steadily through the surface in many small-scale volcanic events. On Venus, on the other hand, with no plate tectonics there is no such release mechanism, and heat from the planet's interior tends to build up in the upper mantle. Although small-scale volcanoes may still form and erupt from time to time, most of this pent-up energy seems to be released catastrophically in planetwide volcanic eruptions every few hundred million years.

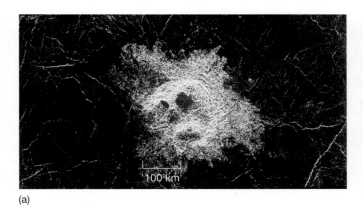

(a)

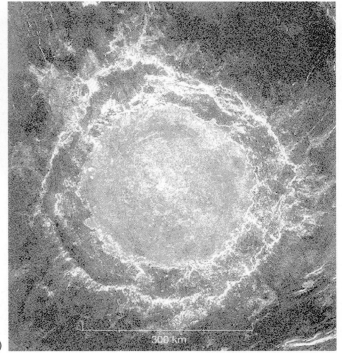

(b)

▲ FIGURE 9.15  **Impact Craters on Venus**  (a) A *Magellan* image of an apparent multiple-impact crater in Venus's southern hemisphere. The irregular shape of the light-colored ejecta seems to be the result of a meteoroid that fragmented just prior to impact. The dark regions in the crater may be pools of solidified lava associated with individual fragments. (b) Venus's largest crater, named Mead after anthropologist Margaret Mead, is about 280 km across. Bright (rough) regions clearly show its double-ringed structure. (*NASA*)

R  I  V  U  X  G

(a)

(b)

R I V U X G

◀ FIGURE 9.16 **Venus In Situ** (a) The first direct view of the surface of Venus, radioed back to Earth from the Soviet *Venera 9* spacecraft, which made a soft landing on the planet in 1975. The amount of sunlight penetrating Venus's cloud cover is about the same as that reaching Earth's surface on a heavily overcast day. (b) Another view of Venus, in true color, from *Venera 14.* Flat rocks like those visible in part (a) are seen among many smaller rocks and even fine soil on the surface. This landing site is not far from the *Venera 9* site shown in (a). The peculiar filtering effects of whatever light does penetrate the clouds make Venus's air and ground appear peach colored—in reality, they are most likely gray, like rocks on Earth. (*Russian Space Agency*)

Although erosion by the planet's atmosphere may play some part in obliterating surface features, the main erosive agent on Venus is volcanism, which appears to have "resurfaced" much of the planet roughly 500 million years ago.

### Data from the Soviet Landers

The 1975 soft landings of the Soviet *Venera 9* and *Venera 10* spacecraft directly established that Venus's surface is dry and dusty. Figure 9.16(a) shows one of the first photographs of the surface of Venus radioed back to Earth. Each craft lasted only about an hour before overheating, their electronic circuitry literally melting in this planetary oven. Typical rocks in the photo measure about 50 cm by 20 cm across—a little like flagstones on Earth. Sharp-edged and slablike, these rocks show little evidence of erosion. Apparently, they are quite young rocks, again supporting the idea of ongoing surface activity of some kind on Venus.

Later *Venera* missions took more detailed photographs, as shown in Figure 9.16(b). The presence of small rocks and finer material indicates the effects of erosive processes. These later missions also performed simple chemical analyses of the surface of Venus. The samples studied by *Venera 13* and *Venera 14* were predominantly basaltic in nature, again implying a volcanic past. However, not all the rocks were found to be basaltic: The *Venera 17* and *Venera 18* landers also found surface material resembling terrestrial granite, probably (as on Earth) part of the planet's ancient crust.

### CONCEPT CHECK

✔ Are volcanoes on Venus associated mainly with the movement of tectonic plates, as on Earth?

## 9.5 The Atmosphere of Venus

Measurements made by the *Venera* and *Pioneer Venus* spacecraft have allowed astronomers to paint a fairly detailed picture of Venus's atmosphere. ∞ (Sec. 6.6) The planet's hot, dense, carbon dioxide atmosphere contrasts sharply with that of Earth, even though, as we will see, the two may have had comparable beginnings.

### Atmospheric Structure

Figure 9.17 shows the variation of temperature and pressure with height. (Compare this figure with Figure 7.2, which gives similar information for Earth). The atmosphere of Venus is about 90 times more massive than Earth's, and it extends to a much greater height above the surface. On Earth, 90 percent of the atmosphere lies within about

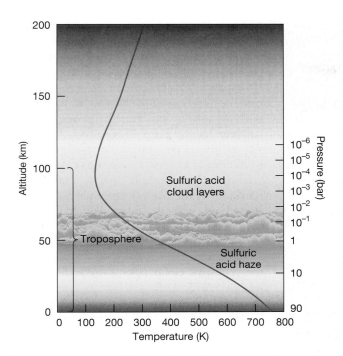

**▲ FIGURE 9.17 Venus's Atmosphere** The structure of the atmosphere of Venus, as determined by U.S. and Soviet probes. (One bar is the atmospheric pressure at sea level on Earth.)

10 km of sea level. On Venus, the 90 percent level is found at an altitude of 50 km instead. The surface temperature and pressure of Venus's atmosphere are much greater than Earth's. However, the temperature drops more rapidly with altitude, and the upper atmosphere of Venus is actually colder than our own.

Venus's troposphere extends up to an altitude of nearly 100 km. The reflective clouds that block our view of the surface lie between 50 and 70 km above the surface. Data from the *Pioneer Venus* multiprobe indicate that the clouds may actually be separated into three distinct layers within that altitude range. Below the clouds, extending down to an altitude of some 30 km, is a layer of haze. Below 30 km, the air is clear. Above the clouds, a high-speed "jet stream" blows from west to east at about 300–400 km/h, fastest at the equator and slowest at the poles. This high-altitude flow is responsible for the rapidly moving cloud patterns seen in ultraviolet light.

Figure 9.18 shows a sequence of three ultraviolet images of Venus in which the variations in the cloud patterns can be seen. Note the characteristic V-shaped appearance of the clouds—a consequence of the fact that, despite their slightly lower speeds, the winds near the poles have a shorter distance to travel in circling the planet and so are always forging ahead of winds at the equator. Near the surface, the dense atmosphere moves more sluggishly—indeed, the fluid flow bears more resemblance to that in Earth's oceans than to the flow in Earth's air. Surface wind speeds on Venus are typically less than 2 m/s (roughly 4 mph).

One of *Venus Express's* main missions is to study atmospheric circulation on Venus. In 2006 the orbiter returned a series of intriguing images of the planet's south pole, showing a **polar vortex** of swirling winds there (Figure 9.19). Polar vortices are well known to atmospheric scientists. Although they may look like giant hurricanes, they are not storms in the usual sense. They are relatively stable, long-lived flows circling the polar regions. They are expected in any body (planet or moon) with an atmosphere, although the details depend on the

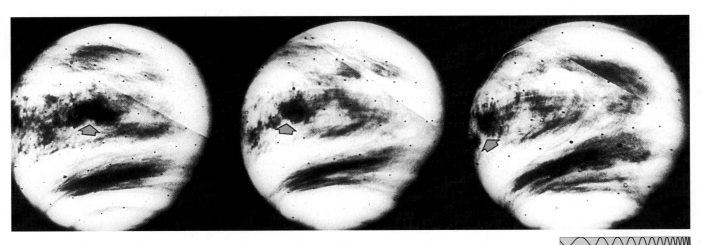

**▲ FIGURE 9.18 Atmospheric Circulation** Three ultraviolet views of Venus, taken by the *Pioneer Venus* orbiter, showing the changing cloud patterns in the planet's upper atmosphere. The wind flow is from right to left (or clockwise from above), in the direction opposite the sideways "V" in the clouds. Notice the motion of the dark region marked by the blue arrow. Venus's retrograde rotation means that north is at the bottom of these images and west to the right. The time difference between the left and right photographs is about 20 hours. *(NASA)*

R  I  V  U  X  G

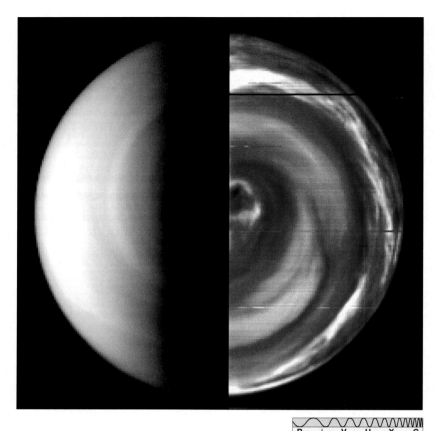

R   I   V   U   X   G

▲ **FIGURE 9.19 Venus Polar Vortex** The left side of this composite image shows Venus during the day when sunlight reflects from its cloud tops. By contrast, the false-color image at right is a nighttime view of radiation arising from deeper layers within, emphasizing the dynamic swirls and vortices of its lower-atmospheric cloud structures. *(ESA)*

properties of the atmosphere and the body's rotation rate. Earth's south polar vortex, for example, plays an important role in confining and concentrating the gases responsible for our planet's antarctic ozone hole. ∞ (Sec. 7.2)

NASA's *Pioneer Venus* orbiter discovered the planet's north polar vortex in 1978, and when *Venus Express* reached Venus in 2006 the search for the southern vortex was one of its top priorities. The Venus vortices present a puzzle to scientists because of their peculiar "double-lobed" structure (see right side of Figure 9.19), which is unique to Venus. This structure is not well understood. By making repeated observations of the southern vortex and watching how it changes in time, scientists hope to understand the forces driving it and gain clues to the global circulation of Venus's atmosphere.

## Atmospheric Composition

Carbon dioxide ($CO_2$) is the dominant component of Venus's atmosphere, accounting for 96.5 percent of it by volume. Almost all of the remaining 3.5 percent is nitrogen ($N_2$).

Trace amounts of other gases, such as water vapor, carbon monoxide, sulfur dioxide, and argon, are also present. This composition is clearly radically different from that of Earth's atmosphere. The absence of oxygen is perhaps not surprising, given the absence of life. (Recall our discussion of Earth's atmosphere in Chapter 7.) ∞ (Sec. 7.2) However, there is no sign of the large amount of water vapor we would expect to find if a volume of water equivalent to Earth's oceans had evaporated and remained in the planet's atmosphere. If Venus started off with an Earth-like composition, then something has happened to its water—Venus is now a very dry planet.

For a long time, the chemical makeup of the reflective cloud layer surrounding Venus was unknown. At first, scientists assumed that the clouds were water vapor or ice, as on Earth, but the reflectivity of the clouds at different wavelengths didn't match that of water ice. Later infrared observations carried out in the 1970s showed that the clouds (or at least the top layer of clouds) are actually composed of *sulfuric acid*, created by reactions between water and sulfur dioxide. Sulfur dioxide is an excellent absorber of ultraviolet radiation and could be responsible for many of the cloud patterns seen in ultraviolet light. Spacecraft observations confirmed the presence of all three compounds in the atmosphere and also indicated that there may be particles of sulfur suspended in and near the cloud layers, which may account for Venus's characteristic yellowish hue.

## The Greenhouse Effect on Venus

Given the distance of Venus from the Sun, the planet was not expected to be such a pressure cooker. As mentioned earlier, calculations based on Venus's orbit and reflectivity indicated a temperature not much different from Earth's, and early measurements of the cloud temperatures seemed to concur. Certainly, scientists reasoned, Venus could be no hotter than the sunward side of Mercury, and it should probably be much cooler. This reasoning was obviously seriously in error.

Why is Venus's atmosphere so hot? And if, as we think, Venus started off like Earth, why is it now so different? The answer to the first question is fairly easy: Given the present composition of its atmosphere, Venus is hot because of the greenhouse effect. Recall from our discussion in Chapter 7 that "greenhouse gases" in Earth's atmosphere—particularly water vapor and carbon dioxide—serve to trap heat from the Sun. ∞ (Sec. 7.2) By inhibiting the escape of infrared radiation reradiated from Earth's surface, these gases increase the planet's equilibrium temperature, in much the same way as an extra blanket keeps you warm on a cold night. Continuing the analogy a little further, the more blankets you place on the bed, the warmer you will become. Similarly, the more greenhouse gases there are in the atmosphere, the hotter the surface will be.

The same effect occurs naturally on Venus, whose dense atmosphere is made up almost entirely of a primary greenhouse gas, carbon dioxide. As illustrated schematically in Figure 9.20, the thick carbon dioxide blanket absorbs nearly 99 percent of all the infrared radiation released from the surface of Venus and is the immediate cause of the planet's sweltering 730 K surface temperature. Furthermore, the temperature is nearly as high at the poles as at the equator, and there is not much difference between the temperatures on the day and night sides. The circulation of the atmosphere spreads energy efficiently around the planet, making it impossible to escape the blazing heat, even during the planet's 2-month-long night.

## The Runaway Greenhouse Effect

But *why* is Venus's atmosphere so different from Earth's? Let's assume that the two planets started off with basically similar compositions. Why, then, is there so much carbon dioxide in the atmosphere of Venus, and why is the planet's atmosphere so dense? To address these questions, we must consider the processes that created the atmospheres of the terrestrial planets and then determined their evolution. In fact, we can turn the question around and ask instead, "Why is there so *little* carbon dioxide in Earth's atmosphere compared with that of Venus?"

Earth's atmosphere has evolved greatly since it first appeared. Our planet's secondary atmosphere was outgassed from the interior by volcanic activity 4 billion years ago. ∞ (Sec. 7.2) Since then, it has been reprocessed, in part by living organisms, into its present form. On Venus, the initial stages probably took place in more or less the same way, so that at some time in the past, Venus might well have had an atmosphere similar to the primitive secondary atmosphere on Earth, containing water, carbon dioxide, sulfur dioxide,

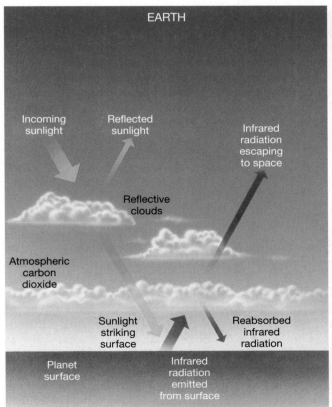
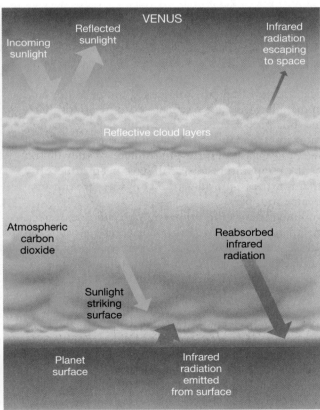

▲ FIGURE 9.20  **Greenhouse Effect on Earth and Venus**  Because Venus's atmosphere is much deeper and denser than Earth's, a much smaller fraction of the infrared radiation leaving the planet's surface escapes into space. The result is a much stronger greenhouse effect than on Earth and a correspondingly hotter planet. The outgoing infrared radiation is not absorbed at a single point in the atmosphere; instead, absorption occurs at all atmospheric levels. (The arrows indicate only that absorption occurs, not that it occurs at one specific level; the arrow thickness is proportional to the amount of radiation moving in and out.)

and nitrogen-rich compounds. What happened on Venus to cause such a major divergence from subsequent events on our own planet?

On Earth, sunlight split the nitrogen-rich compounds, releasing nitrogen into the air. Meanwhile, the water condensed into oceans, and much of the carbon dioxide and sulfur dioxide eventually became dissolved in them. Most of the remaining carbon dioxide combined with surface rocks. Thus, much of the secondary outgassed atmosphere quickly became part of the surface of the planet. If all the dissolved or chemically combined carbon dioxide were released back into Earth's present-day atmosphere, its new composition would be 98 percent carbon dioxide and 2 percent nitrogen, and it would have a pressure about 70 times its current value. In other words, apart from the presence of oxygen (which appeared on Earth only after the development of life) and water (the absence of which on Venus will be explained shortly), Earth's atmosphere would be a lot like that of Venus! The real difference between Earth and Venus, then, is that Venus's greenhouse gases never left the atmosphere, the way they did on Earth.

When Venus's secondary atmosphere appeared, the temperature was higher than on Earth, simply because Venus is closer to the Sun. However, the Sun was probably somewhat dimmer then (see Chapter 22)—perhaps only half its present brightness—so there is some uncertainty as to exactly how much hotter than Earth Venus actually was. If the temperature was already so high that no oceans condensed, the outgassed water vapor and carbon dioxide would have remained in the atmosphere, and the full greenhouse effect would have gone into operation immediately. If oceans did form and most of the greenhouse gases left the atmosphere, as they did on Earth,* the temperature must still have been sufficiently high that a process known as the **runaway greenhouse effect** came into play.

To understand the runaway greenhouse effect, imagine that we took Earth from its present orbit and placed it in Venus's orbit, some 30 percent closer to the Sun. At that distance from the Sun, the amount of sunlight striking Earth's surface would be about twice its present level, so the planet would warm up. More water would evaporate from the oceans, leading to an increase in atmospheric water vapor. At the same time, the ability of both the oceans and surface rocks to hold carbon dioxide would diminish, allowing more carbon dioxide to enter the atmosphere. As a result, greenhouse heating would increase, and the planet would warm still further, leading to a further increase in atmospheric greenhouse gases, and so on. Once started, the process would "run away,"

*In fact, careful study of the Magellan images reveals no sign of ancient seashores or ocean basins, nor evidence of erosion by rivers on Venus. However, it is unclear whether such features would have survived the heavy volcanism known to have occurred in the planet's more recent past.

eventually leading to the complete evaporation of the oceans, restoring all the original greenhouse gases to the atmosphere. Although the details are quite complex, basically the same thing would have happened on Venus long ago, ultimately resulting in the planetary inferno we see today.

The presence of atmospheric water vapor meant that the greenhouse effect on Venus was even more extreme in the past. By intensifying the blanketing effect of the carbon dioxide, the water vapor helped the surface of Venus reach temperatures perhaps twice as hot as current temperatures. At the high temperatures of the past, the water vapor was able to rise high into the planet's upper atmosphere—so high that it was broken up by solar ultraviolet radiation into its components, hydrogen and oxygen. The light hydrogen rapidly escaped, the reactive oxygen quickly combined with other atmospheric gases, and all water on Venus was lost forever. This is the reason that Venus lacks water today.

Although it is highly unlikely that global warming will ever send Earth down the path taken by Venus, this episode highlights the relative fragility of the planetary environment. ∞ *(Discovery 7-1)* No one knows how close to the Sun Earth could have formed before a runaway greenhouse effect would have occurred. But in comparing our planet with Venus, we have come to understand that there is an orbital limit, presumably between 0.7 and 1.0 AU, inside of which Earth would have suffered a similar catastrophic runaway. We must consider this "greenhouse limit" when we assess the likelihood that planets harboring life formed elsewhere in our Galaxy (see Chapter 28).

CONCEPT CHECK

✔ If Venus had formed at Earth's distance from the Sun, what might its climate be like today?

## 9.6 Venus's Magnetic Field and Internal Structure

In 1962, *Mariner 2* flew by Venus, carrying, among other instruments, magnetometers to measure the strength of the planet's magnetic field. None was detected, and subsequent Soviet and U.S. missions, carrying more sensitive detectors, have confirmed this finding. *Pioneer Venus* did detect a weak "induced" magnetic field produced by the interaction between the planet's upper atmosphere and the solar wind, but Venus apparently has no intrinsic magnetic field of its own.

Venus, with an average density similar to that of Earth, probably has a similar overall composition and a partially molten iron-rich core. The lack of any significant magnetic field on Venus, then, is almost surely the result of the planet's extremely slow rotation and consequent lack of dynamo action. ∞ (Sec. 7.5) Having no magnetosphere, Venus has no protection from the solar wind. The planet's upper atmosphere is continually bombarded by high-energy particles from the Sun, keeping the topmost layers permanently ionized. However, the great thickness of the atmosphere prevents any of these particles from reaching the surface.

None of the *Venera* landers carried seismic equipment, so no direct measurements of the planet's interior have been made, and theoretical models of the interior have little hard data to constrain them. However, to many geologists, the surface of Venus resembles that of the young Earth, at an age of perhaps a billion years. At that time, volcanic activity had begun, but the crust was still relatively thin and the convective processes in the mantle that drive plate tectonic motion were not yet fully established. Measurements of the planet's gravitational field suggest that Venus lacks an asthenosphere, the semisolid part of the upper mantle over which Earth's lithosphere slides. ∞ (Sec. 7.4)

Why has Venus remained in that immature state and not developed plate tectonics as Earth did? That question remains to be answered. Some planetary geologists have speculated that the high surface temperature has inhibited Venus's evolution by slowing the planet's cooling. Possibly, the high surface temperature has made the crust too soft for Earth-style plates to develop—or perhaps the high temperature and soft crust have led to more volcanism, tapping the energy that might otherwise go into convective motion. It may also be that the presence of water plays an important role in lubricating convection in the mantle and plate motion, so that arid Venus could never have evolved along the same path as Earth.

## CONCEPT CHECK

✔ If the interior of Venus is quite Earth-like, and Venus has a molten iron core, why doesn't the planet have a magnetic field as Earth does?

# CHAPTER REVIEW

## Summary

❶ The interior orbit of Venus with respect to Earth's means that Venus never strays far from the Sun in the sky. Because of its highly reflective cloud cover, Venus is brighter than any star in the sky, as seen from Earth. The planet's mass and radius are similar to those of Earth. The planet's rotation is slow and retrograde, most likely because of a collision between Venus and some other solar system body during the late stages of the planet's formation.

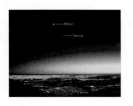

❷ The extremely thick atmosphere of Venus is nearly opaque to visible radiation, making the planet's surface invisible at optical wavelengths from the outside. Venus's atmosphere is nearly 100 times denser than Earth's and consists mainly of carbon dioxide. The temperature of the upper atmosphere is much like that of Earth's upper atmosphere, but the surface temperature of Venus is a searing 730 K. The planet's high-level winds circulate rapidly around the planet, and there are peculiarly shaped **polar vortices (p. 243)** at both poles.

❸ Venus's surface has been thoroughly mapped by radar from Earth-based radio telescopes and orbiting satellites. The planet's surface is mostly smooth, resembling rolling plains with modest highlands and lowlands. Two elevated continent-sized regions are called Ishtar Terra and Aphrodite Terra. There is no evidence

for plate tectonic activity as on Earth. Features called **coronae (p. 240)** are thought to have been caused by an upwelling of mantle material. For unknown reasons, the upwelling never developed into full convective motion. The surface of the planet appears to be relatively young, resurfaced by volcanism within the past few hundred

million years. Many **lava domes (p. 240)** and **shield volcanoes (p. 239)** were found by the *Magellan* orbiter on Venus's surface, but none of the volcanoes has yet proved to be currently active.

❹ Some craters on Venus are due to meteoritic impact, but the majority appear to be volcanic in origin. The evidence for currently active volcanoes on Venus includes surface features resembling those produced in Earthly volcanism, fluctuating levels of sulfur dioxide in Venus's atmosphere, and bursts

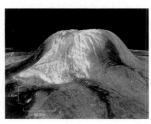

of radio energy similar to those produced by lightning discharges that often occur in the plumes of erupting volcanoes on Earth. However, no actual eruptions have been seen. Soviet spacecraft that landed on Venus photographed surface rocks with sharp edges and a slablike character. Some rocks appear predominantly basaltic in nature, implying a volcanic past, others resemble terrestrial granite and are part of the planet's ancient crust.

⑤ Although Venus and Earth may have started off with fairly similar surface conditions, their atmospheres are now very different. The total mass of Venus's atmosphere is about 90 times greater than 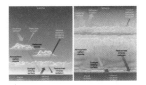 Earth's. The greenhouse effect caused by the large amount of carbon dioxide in Venus's atmosphere is the cause of the planet's current high temperatures. Almost all the water vapor and carbon dioxide initially present in Earth's early atmosphere quickly became part of the oceans or surface rocks. Because

Venus orbits closer to the Sun than does Earth, surface temperatures on Venus were initially higher, and the planet's greenhouse gases never left the atmosphere. The **runaway greenhouse effect (p. 246)** caused all the planet's greenhouse gases—carbon dioxide and water vapor—to end up in the atmosphere, leading to the extreme conditions we observe today.

⑥ Venus has no detectable magnetic field, almost certainly because the planet's rotation is too slow for any appreciable dynamo effect to have developed. To some planetary geologists, Venus's interior structure suggests that of the young Earth, before convection became established in the mantle.

## Review and Discussion

1. Why does Venus appear so bright to the eye? Upon what factors does the planet's brightness depend?

2. Explain why Venus is always found in the same general part of the sky as the Sun.

3. Why do astronomers think that the near resonance between Venus's rotation and revolution, as seen from Earth, is not a true resonance?

4. Describe one observational problem associated with Venus's near resonance of rotation and revolution.

5. What is our current best explanation of Venus's slow, retrograde spin?

6. If you were standing on Venus, how would Earth look?

7. How did radio observations of Venus made in the 1950s change our conception of the planet?

8. What did ultraviolet images returned by *Pioneer Venus* show about the planet's high-level clouds?

9. Name three ways in which the atmosphere of Venus differs from that of Earth.

10. What are the main constituents of Venus's atmosphere? What are clouds in the upper atmosphere made of?

11. What component of Venus's atmosphere causes the planet to be so hot? Explain why there is so much of this gas in the atmosphere of Venus, compared with its presence in Earth's

atmosphere. What happened to all the water that Venus must have had when the planet formed?

12. Earth and Venus are nearly alike in size and density. What primary fact caused one planet to evolve as an oasis for life, while the other became a dry and inhospitable inferno?

13. If Venus had formed at Earth's distance from the Sun, what do you imagine its climate would be like today? Why do you think so?

14. How do the "continents" of Venus differ from Earth's continents?

15. How are the impact craters of Venus different from those found on other bodies in the solar system?

16. What evidence exists that volcanism of various types has changed the surface of Venus?

17. What is the evidence for active volcanoes on Venus?

18. Given that Venus, like Earth, probably has a partially molten iron-rich core, why doesn't Venus also have a magnetic field?

19. Do you think there might be life on Venus? Explain your answer.

20. Do you think that Earth is in any danger of being subject to a runaway greenhouse effect like that on Venus?

## Conceptual Self-Test: True or False/Multiple Choice

1. Venus has a retrograde orbit around the Sun.

2. Venus is brightest when it is in its full phase.

3. The entire surface of Venus has been mapped by means of radar observations from Earth.

4. Venus's atmosphere is quite similar to that of Earth.

5. Venus has roughly the same temperature at its equator as at its poles.

6. Images from Magellan confirm that Venus has tectonic activity like that on Earth.

7. Evidence of lava flows is common on the surface of Venus.

8. Venus's mass has been accurately measured by orbiting spacecraft.

9. Most craters on the surface of Venus are the result of volcanism.

10. The greenhouse effect in the early atmosphere of Venus was most likely intensified by the presence of water vapor.

11. Venus is never seen at midnight because (a) it is closer to the Sun than is Earth; (b) it will be in its new phase then; (c) it is visible only at sunset; (d) it will be at superior conjunction.

12. Venus's permanent retrograde rotation about its axis results in the planet (a) always rising in the western sky; (b) orbiting the Sun in the opposite direction from Earth; (c) having its north pole below the plane of the ecliptic; (d) being brighter than any other planet.

13. Compared with Earth, Venus is (a) much smaller; (b) much larger; (c) about the same size.

14. Venus's surface is permanently obscured by clouds. As a result, the surface has been studied primarily by (a) robotic

landers (b) orbiting satellites using radar; (c) spectroscopy; (d) radar signals from Earth.

**15.** Compared with Earth, Venus has a level of plate tectonic activity that is (a) much more rapid; (b) virtually non-existent; (c) about the same.

**16.** Venus's atmosphere (a) has almost the same chemical composition as Earth's; (b) shows very high levels of humidity; (c) is composed mostly of carbon dioxide; (d) is predominantly made of acid droplets.

**17.** Compared with Earth's atmosphere, most of Venus's atmosphere is (a) compressed much closer to the surface; (b) spread out much farther from the surface; (c) similar in extent and structure.

**18.** Venus's atmospheric temperature is (a) about the same as Earth's; (b) cooler than temperatures on the planet Mercury; (c) hotter than temperatures on Mercury; (d) high due to the presence of sulfuric acid.

**19.** Carbon dioxide on Venus (a) is all in the atmosphere; (b) was absorbed in surface water and has evaporated into space; (c) has dissolved in the atmospheric acid; (d) is integrated into the surface rocks.

**20.** Venus lacks a planetary magnetic field because (a) it rotates very slowly; (b) it does not have a molten core; (c) there are no plate tectonics on the planet; (d) the core contains little or no iron.

## Problems

 *Algorithmic versions of these Problems are available in the Practice Problems module of the Companion Website. The number of dots preceding each Problem indicates its approximate level of difficulty.*

**1.** •• Using the data given in the text, calculate Venus's angular diameter, as seen by an observer on Earth, when the planet is (a) at its brightest, (b) at greatest elongation, and (c) at the most distant point in its orbit.

**2.** •• Seen from Earth, through how many degrees per night (relative to the stars) does Venus move around the time of inferior conjunction (closest approach to Earth)?

**3.** • How long does a radar signal take to travel from Earth to Venus and back when Venus is brightest? Compare this time with the round-trip time when Venus is at its closest point to Earth.

**4.** •• What would be the length of a solar day on Venus if the planet's rotation were prograde rather than retrograde?

**5.** ••• Calculate the mean angular orbital speeds of Venus and Mercury, in degrees per day. (For simplicity, imagine that the planets have circular orbits.) On the basis of these speeds, how long does it take for Mercury to "lap" Venus—that is, to complete exactly one extra revolution around the Sun? This length of time is the synodic period of Mercury, as seen from Venus.

**6.** • Draw a diagram showing the positions of Earth and Venus in their orbits over the course of one Venus synodic year (584 Earth days), starting at the point of closest approach of the planets. Mark the locations of the planets at intervals of 73 days, and indicate with an arrow (as in Figure 9.4) the orientation of some point on Venus's surface at each instant.

**7.** •• Compare the magnitude of the tidal gravitational acceleration on Venus (the difference between the accelerations at center and surface) due to Earth at closest approach with Venus's surface gravitational acceleration. Assume circular orbits for both planets. Repeat the question for the tidal acceleration on Venus due to the Sun. What do you conclude about the possibility that Venus's near resonance with Earth is the result of Earth's tidal influence?

**8.** • What is the size of the smallest feature that could be distinguished on the surface of Venus (at closest approach) by the Arecibo radio telescope at an angular resolution of 1″?

**9.** • Could an infrared telescope with an angular resolution of 0.1″ distinguish impact craters on the surface of Venus?

**10.** • When used as a radar instrument, the Arecibo installation can distinguish echoes received as little as $10^{-5}$ s apart. To what distance does this correspond? That distance is the effective resolution of Arecibo in making radar observations of Venus.

**11.** • *Pioneer Venus* observed high-level clouds moving around Venus's equator in 4 days. What was their speed in km/h? In mph?

**12.** • Approximating Venus's atmosphere as a layer of gas 50 km thick, with uniform density 21 kg/m$^3$, calculate the total mass of the atmosphere. Compare your answer with the mass of Earth's atmosphere (Chapter 7, Problem 3) and with the mass of Venus.

**13.** • According to Stefan's law (see Section 3.4), how much more radiation—per square meter, say—is emitted by Venus's surface at 730 K than is emitted by Earth's surface at 300 K?

**14.** ••• In the absence of any greenhouse effect, Venus's average surface temperature, like Earth's, would be about 250 K. In fact, it is about 730 K. Use this information and Stefan's law to estimate the fraction of infrared radiation leaving Venus's surface that is absorbed by carbon dioxide in the planet's atmosphere.

**15.** •• Calculate the orbital period of the *Magellan* spacecraft, moving around Venus on an elliptical orbit with a minimum altitude of 294 km and a maximum altitude of 8543 km above the planet's surface. In 1993, the spacecraft's orbit was changed to have minimum and maximum altitudes of 180 km and 541 km, respectively. What was the new period?

*The Companion Website at www.aw-bc.com/chaisson provides algorithmically generated versions of each chapter's Problems, along with additional quizzes, an Animations & Videos gallery, an Images gallery, an interactive Glossary, and a full eBook.*

# MARS

## A Near Miss for Life?

Named by the ancient Romans for their bloody god of war, Mars is for many people the most intriguing of all celestial objects. Over the years, it has inspired speculation that life—perhaps intelligent and possibly hostile—may exist there. With the dawn of the Space Age, those notions had to be abandoned. Visits by robot spacecraft have revealed no signs of life of any sort on Mars, even at the microbial level.

Still, the planet's properties are close enough to those of Earth that Mars is even now widely regarded as the second most hospitable environment for the appearance of life in the solar system, after Earth itself. At about the same time as Earth's "twin," Venus, was evolving into a searing inferno, the Mars of long ago may have had running water and blue skies. If life ever arose there, however, it must be long extinct. The Mars of today appears to be a dry, dead world.

## LEARNING GOALS

*Studying this chapter will enable you to*

1  Summarize the general orbital and physical properties of Mars.
2  Describe the observational evidence for seasonal changes on Mars.
3  Compare the surface features and geology of Mars with those of the Moon and Earth, and account for these characteristics in terms of Martian history.
4  Discuss the evidence that Mars once had a much denser atmosphere and running water on its surface.
5  Explain where that ancient water on Mars may be found today.
6  Compare the atmosphere of Mars with those of Earth and Venus, and explain why the evolutionary histories of these three worlds diverged so sharply.
7  Discuss what is known of the internal structure of Mars.
8  Describe the characteristics of the Martian moons, and explain their probable origin.

 Visit www.aw-bc.com/chaisson for additional images, animations, quizzes, and eBook for this chapter.

# 10.1    Orbital Properties

Mars is the fourth planet from the Sun and the outermost of the four terrestrial worlds in the solar system. It lies outside Earth's orbit, as illustrated in Figure 10.1(a), which shows the orbits of both planets drawn to scale. Because of its superior (exterior) orbit, Mars ranges in our sky from a position that appears close to the Sun (*conjunction*, with Earth and Mars at points A in the figure) to one on the opposite side of the sky from the Sun (*opposition*, at points B). ∞ (Sec. 2.2) Contrast this orbit with Mercury's and Venus's inferior orbits, which ensure that we never see those planets far from the Sun in the nighttime sky. ∞ (Secs. 8.1, 9.1) From our earthly viewpoint, Mars appears to traverse a great circle in the sky, keeping close to the ecliptic and occasionally executing retrograde loops. ∞ (Sec. 2.2) The Mars Data box on p. 253 lists some detailed orbital and physical data having to do with that planet.

Mars's orbital eccentricity is 0.093, much larger than that of most other planets—only the innermost planet, Mercury, has a more elongated orbit. Because of this relatively large eccentricity, Mars's perihelion distance from the Sun—1.38 AU (207 million km)—is substantially smaller than its aphelion distance—1.67 AU (249 million km)—resulting in a large variation in the amount of sunlight striking the planet over the course of its year. In fact, the intensity of sunlight on the Martian surface is almost 45 percent greater when the planet is at perihelion than when it is at aphelion. As we will see, this has a substantial effect on the Martian climate.

Mars is largest and brightest in the night sky at opposition, when Earth lies between Mars and the Sun (location B in Figure 10.1a). If this happens to occur near Martian perihelion, the two planets can come as close as 0.37 AU (56 million km). The separation is less than 0.38 AU because Earth's orbit is slightly eccentric and our planet actually lies about 1.01 AU from the Sun when such an opposition occurs. (As indicated in Figure 10.1, Earth reaches aphelion in early July, whereas opposition at Martian perihelion happens in late August.) The angular size of Mars under these most favorable circumstances is about 25″. Ground-based observations of the planet at those times can distinguish surface features as small as 100 km across—about the same resolution as the unaided human eye can achieve when viewing the Moon. Notice that, at this resolution, the extensive "canal systems" that once fueled such rampant speculation about life on Mars (see the Part 2 Opener on p. 142) could not possibly

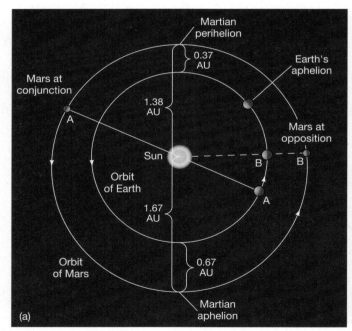

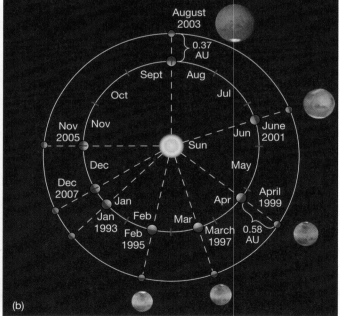

▲ FIGURE 10.1  **Mars Orbit** (a) The orbit of Mars compared with that of Earth. Observe that Mars's orbit is noticeably offcenter, unlike Earth's, whose eccentricity is barely perceptible here. When the planets are on opposite sides of the Sun, as at the points marked A, Mars is said to be at conjunction. The planets are at their closest at opposition, when Earth and Mars are aligned and on the same side of the Sun, as at the points marked B. (Note that to get from point A to point B, Earth must travel for nearly 13 months—all the way around its orbit and then some.) (b) Several oppositions of Mars, including the particularly favorable (close) configuration of August 2003 and the unfavorable oppositions of February 1995 and March 1997. The five images of Mars were taken over the course of eight years with the *Hubble* telescope. Note how the angular size of the planet at opposition varies with distance from Earth.

# PLANETARY DATA

## MARS

| | |
|---|---|
| Orbital semimajor axis | 1.52 AU |
| | 227.9 million km |
| Orbital eccentricity | 0.093 |
| Perihelion | 1.38 AU |
| | 206.6 million km |
| Aphelion | 1.67 AU |
| | 249.2 million km |
| Mean orbital speed | 24.1 km/s |
| Sidereal orbital period | 686.9 Earth days* |
| | 1.881 tropical years |
| Synodic orbital period | 779.9 Earth days* |
| Orbital inclination to the ecliptic | 1.85° |
| Greatest angular diameter, as seen from Earth | 24.5″ |
| Mass | $6.42 \times 10^{23}$ kg |
| | 0.11 (Earth = 1) |
| Equatorial radius | 3394 km |
| | 0.53 (Earth = 1) |
| Mean density | 3930 kg/m³ |
| | 0.71 (Earth = 1) |
| Surface gravity | 3.72 m/s² |
| | 0.38 (Earth = 1) |
| Escape speed | 5.0 km/s |
| Sidereal rotation period | 1.026 Earth days |
| | 24ʰ 37ᵐ |
| Axial tilt | 23.98° |
| Surface magnetic field | approximately 1/800 Earth = 1 |
| Magnetic axis tilt relative to rotation axis | ——— |
| Mean surface temperature | 210 K |
| | (range 150–310 K) |
| Number of moons | 2 |

*1 Earth (mean solar) day = 24 hours

actually have been observed. They were strictly a figment of the human imagination.

Figure 10.1(b), shows the dates and configurations of eight successive Martian oppositions between January 1993 and December 2007. Oppositions occur at roughly 780-day intervals—the synodic period of Mars relative to Earth—with corrections for the fact that, in accordance with Kepler's second law, the planets do not move at constant speeds around their orbits. ∞ (Sec. 2.5, *More Precisely 9-1*) Oppositions near Martian perihelion are less frequent, occurring roughly once every 15 years. The most recent such event, on August 28, 2003, was one of the closest ever, with Mars just 0.373 AU from Earth, affording unprecedented observing conditions for amateurs and professionals alike. On average, the two planets come within 0.38 AU of one another only about three times per century.

Although Mars is quite bright and easily seen at opposition, the planet is still considerably fainter than Venus. This faintness results from a combination of three factors. First, Mars is more than twice as far from the Sun as is Venus, so each square meter on the Martian surface receives less than one-quarter the amount of sunlight that strikes each square meter on Venus. Second, the surface area of Mars is only about 30 percent that of Venus, so there are fewer square meters to intercept the sunlight. Finally, Mars is much less reflective than Venus—only about 15 percent of the sunlight striking the planet is reflected back into space, compared with nearly 70 percent in the case of Venus. Still, at its brightest, Mars is brighter than any star. Its characteristic red color, visible even to the naked eye, makes the planet easily identifiable in the night sky.

## CONCEPT CHECK

✔ Why do the closest views of Mars from Earth occur roughly only once every 15 years?

## 10.2 Physical Properties

As with Mercury and Venus, we can determine the radius of Mars by means of simple geometry. From the data given earlier for the planet's size and distance, we obtain a radius of about 3400 km. More accurate measurements give a result of 3394 km, or 0.53 Earth radii.

Unlike Mercury and Venus, Mars has two small moons in orbit around it, both visible (through telescopes) from Earth. Named Phobos (Fear) and Deimos (Panic) for the sons of Ares (the Greek name for the war god known to the Romans as Mars) and Aphrodite (the Greek name for Venus, goddess of love), these moons are little more than large rocks trapped by the planet's gravity. We will return to their individual properties at the end of the chapter. The larger of the two, Phobos, orbits at a distance of just 9378 km from the center of the planet once every

459 minutes. Applying the modified version of Kepler's third law (which states that the square of a moon's orbital period is proportional to the cube of its orbital semimajor axis divided by the mass of the planet it orbits), we find that the mass of Mars is $6.4 \times 10^{23}$ kg, or 0.11 times that of Earth. ∞ (Sec. 2.8) Naturally, the orbit of Deimos yields the same result.

From the planet's mass and radius, it follows that the average density of Mars is 3900 kg/m$^3$, only slightly greater than that of the Moon. If we assume that Martian surface rocks are similar to those on the other terrestrial planets, this average density suggests the existence of a substantial higher density core within the planet. Planetary scientists suspect that this core is composed largely of iron sulfide (a compound about twice as dense as Martian surface rock) and has a diameter of about 2500 km.

Surface markings easily seen on Mars allow astronomers to track the planet's rotation. Mars rotates once on its axis every 24.6 hours. One Martian day is thus similar in length to one Earth day. The planet's equator is inclined to the orbital plane at an angle of 24.0°, again similar to Earth's inclination of 23.5°. Thus, as Mars orbits the Sun, we find both daily and seasonal cycles, just as on Earth. In the case of Mars, however, the seasons are complicated somewhat by variations in solar heating due to the planet's eccentric orbit—southern summer occurs around the time of Martian perihelion and so is significantly warmer than summer in the north.

## 10.3  Long-Distance Observations of Mars

At opposition, when Mars is closest to us and most easily observed, we see it as full, so the Sun's light strikes the surface almost vertically, casting few shadows and preventing us from seeing any topographic detail, such as craters or mountains. Even through a large telescope, Mars appears only as a reddish disk, with some light and dark patches and prominent polar caps. These surface features undergo slow seasonal changes over the course of a Martian year. We saw in Chapter 1 how the inclination of Earth's axis produces similar seasonal changes. ∞ (Sec. 1.4) Figure 10.2 shows some of the best images of Mars ever made from Earth or Earth orbit, along with a photograph taken by one of the U.S. *Viking* spacecraft en route to the planet.

When the planet is viewed from Earth, the most obvious Martian surface features are the bright polar caps (see Figure 10.2a), growing and diminishing according to the seasons and almost disappearing at the time of the Martian summer. The dark surface features on Mars also change from season to season, although their variability probably has little to do with the melting of the polar ice caps. To the more fanciful observers around the start of the 20th century, these changes suggested the seasonal growth of vegetation on the planet. It was but a small step from seeing polar ice caps and speculating about teeming

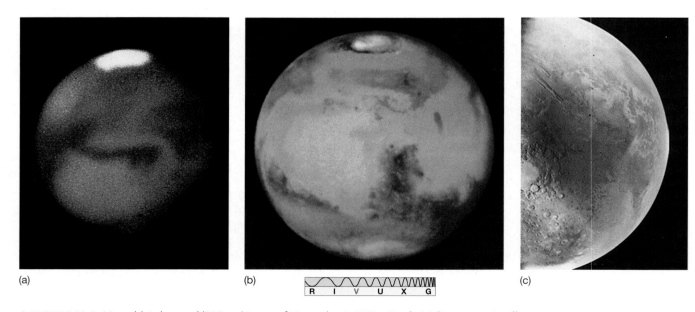

(a)                    (b)    R  I  V  U  X  G                    (c)

▲ **FIGURE 10.2  Mars** (a) A deep-red (800-nm) image of Mars, taken in 1991 at Pic-du-Midi, an exceptionally clear site in the French Alps. One of the planet's polar caps appears at the top and a few other surface markings are evident in this ground-based telescopic view. (b) A visible-light *Hubble Space Telescope* image of Mars, taken while the planet was near opposition in 2003. (c) A view of Mars taken from a *Viking* spacecraft during its approach in 1976. The planet's surface features can be seen clearly at a level of detail completely invisible from Earth. *(CNRS; NASA)*

ANIMATION/VIDEO    *Hubble* View of Mars

vegetation to imagining a planet harboring intelligent life, perhaps not unlike us.

But those speculations and imaginings were not to be confirmed. Spectroscopic observations from Earth and from Earth orbit revealed that the changing caps are mostly frozen carbon dioxide (i.e., dry ice), not water ice, as at Earth's North and South Poles. ∞ (Sec. 4.4) The polar caps do contain water, but it remains permanently frozen, and the dark markings seen in Figure 10.2, once thought (by some) to be part of a network of canals dug by Martians for irrigation purposes, are actually highly cratered and eroded areas around which surface dust occasionally blows. From a distance, the repeated covering and uncovering of these landmarks gives the impression of surface variability, but it's only the thin dust cover that changes.

The powdery Martian surface dust is borne aloft by strong winds that often reach hurricane proportions (hundreds of kilometers per hour). In fact, when the U.S. *Mariner 9* spacecraft went into orbit around Mars in 1971, a planetwide dust storm obscured the entire landscape. Had the craft been on a flyby mission (for a quick look) instead of an orbiting mission (for a longer view), its visit would have been a failure. Fortunately, the storm subsided, enabling the craft to radio home detailed information about the planet's surface.

CONCEPT CHECK

✔ Does Mars have seasons like those on Earth?

## 10.4 The Martian Surface

Maps of the surface of Mars returned by orbiting spacecraft show a wide range of geological features. Mars has huge volcanoes, deep canyons, vast dune fields, and many other geological wonders. Orbiters have performed large-scale surveys of much of the planet's surface, and lander data have complemented these planetwide studies with detailed information on (so far) three specific sites. ∞ (Sec. 6.6) The current focus of Martian exploration is the ongoing search for water on or below the planet's surface.

### Large-Scale Topography

Figure 10.3 shows a planetwide mosaic of thousands of images taken in the 1970s by the *Viking* orbiters. The images show some of the planet's topographic features in true color. More recently, *Mars Global Surveyor* has mapped out the Martian surface to an accuracy of a few meters, using an instrument called a *laser altimeter*, which analyzes pulses of laser light to measure the distance between the spacecraft and the planet's surface. Figure 10.4 shows some results of those measurements, projected onto a Martian globe. Figure 10.5 flattens things out into a more conventional map and marks some prominent surface features.

A striking feature of the terrain of Mars is the marked difference between the northern and southern hemispheres. The northern hemisphere is made up largely of rolling volcanic plains not unlike the lunar maria—indeed, this similarity was key to their identification as lava-flow features. Much larger than their counterparts on Earth or the Moon, these extensive lava plains were formed by

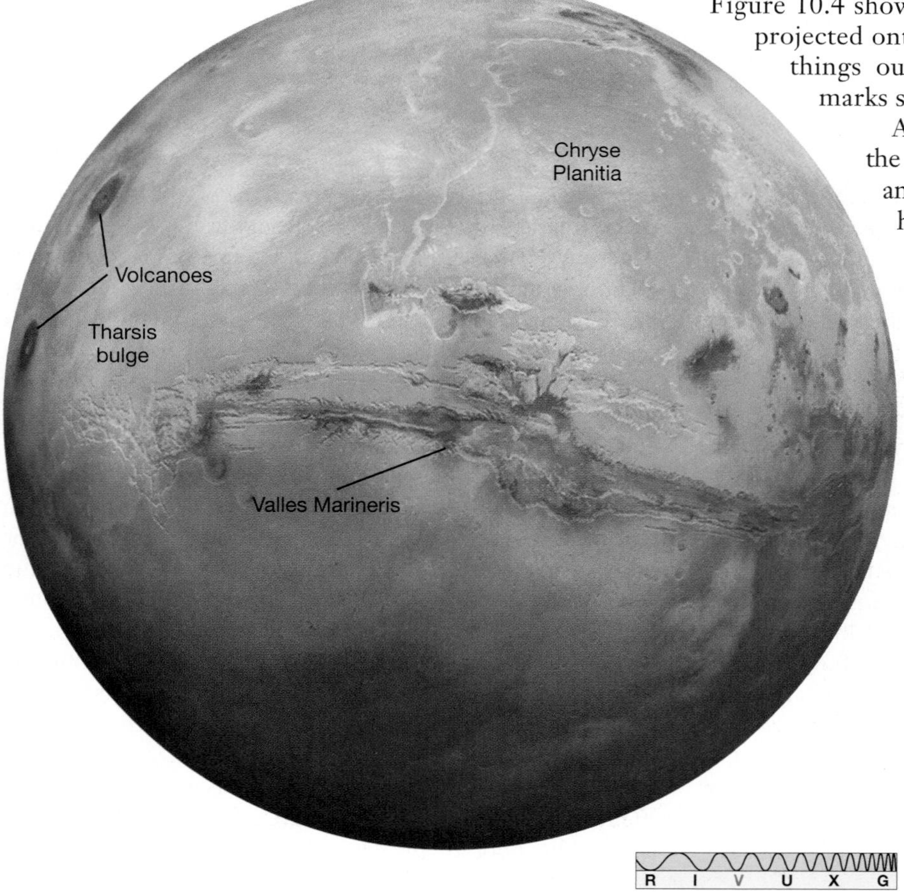


◄ **FIGURE 10.3 Tharsis** A mosaic of Mars based on images from a *Viking* spacecraft in orbit around the planet. Some 5000 km across, the Tharsis region bulges out from Mars's equatorial zone, rising to a height of about 10 km. The two large volcanoes on the left mark the approximate peak of the Tharsis bulge. One of the plains flanking this bulge, Chryse Planitia, is toward the right. Dominating the center of the field of view is a vast "canyon" known as Valles Marineris—the Mariner Valley. *(NASA)*

ANIMATION/VIDEO Flight Over Tharsis

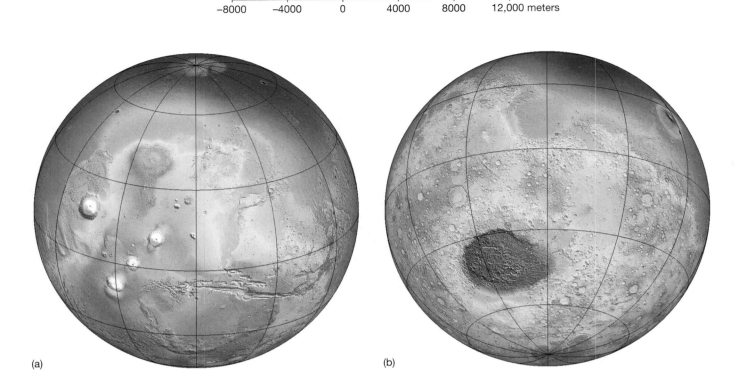

−8000    −4000    0    4000    8000    12,000 meters

(a)    (b)

▲ FIGURE 10.4 **Mars Globes** Two computer-generated globes of planet Mars, based on detailed measurements made by *Mars Global Surveyor.* Color represents height above (or below) the mean planetary radius, ranging from dark blue (−8 km), through green, yellow and red (+4 km), to white (over 8 km in altitude), as given on the scale above the images. Frame (a) shows roughly the same hemisphere as that in Figure 10.3, containing the Tharsis region of Mars. Frame (b) shows the planet's other hemisphere, dominated by the giant Hellas impact basin. *(NASA)*

eruptions involving enormous volumes of material. They are strewn with blocks of volcanic rock, as well as with boulders blasted out of impact areas by infalling meteoroids. (The Martian atmosphere is too thin to offer much resistance to incoming debris.) The southern hemisphere consists of heavily cratered highlands lying some 5 km above the level of the lowland north. Most of the dark regions visible from Earth are mountainous regions in the south. Figure 10.6 contrasts typical terrains in the two hemispheres.

The northern plains are much less cratered than the southern highlands. On the basis of the arguments presented in Chapter 8, this smoother surface suggests that the northern surface is younger. ∞ (Sec. 8.5) Its age is perhaps 3 billion years, compared with 4 billion in the south. In places, the boundary between the southern highlands and the northern plains is quite sharp—the surface level can drop by as much as 4 km in a horizontal distance of 100 km or so. Most scientists assume that the southern terrain is the original crust of the planet. How most of the northern hemisphere could have been lowered in elevation and subsequently flooded with lava remains a mystery.

The major geological feature on the planet is the Tharsis bulge (marked in Figure 10.3). Roughly the size of North America, Tharsis lies on the Martian equator and rises some 10 km higher than the rest of the Martian surface. To its east lies Chryse Planitia (the "Plains of Gold"), to the west a region called Isidis Planitia (the "Plains of Isis," an Egyptian goddess). These features are wide depressions, hundreds of kilometers across and up to 3 km deep. If we wished to extend the idea of "continents" from Earth and Venus to Mars, we would conclude that Tharsis is the only continent on the Martian surface. However, as on Venus, there is no sign of plate tectonics on Mars—the absence of fault lines or other evidence of plate motion tells geologists that the "continent" of Tharsis is not drifting as its Earthly counterparts are. ∞ (Sec. 7.4) Tharsis appears to be even less heavily cratered than the northern plains, making it the youngest region on the planet, an estimated 2 to 3 billion years old.

Almost diametrically opposite Tharsis, in the southern highlands, lies the Hellas Basin, which, paradoxically, contains the *lowest* point on Mars. (Hellas is clearly visible in Figure 10.4 and is labeled in Figure 10.5.) Some 3000 km across, the floor of the basin lies nearly 9 km below the basin's rim and over 6 km below the average level of the

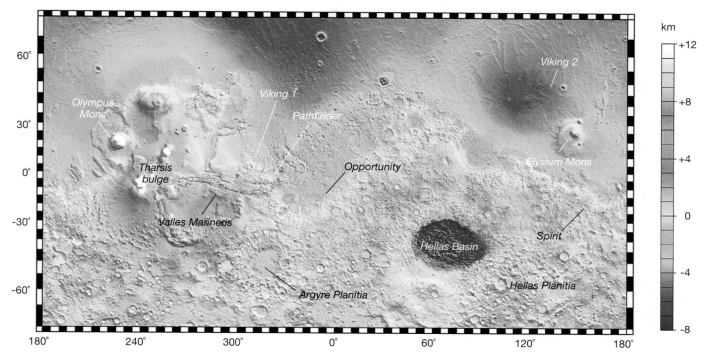

▲ **FIGURE 10.5 Mars Map** The *Mars Global Surveyor* data of Figure 10.4, now displayed as a flat map. The Mariner Valley (see also Figures 10.4 and 10.7) can be seen at left, whereas the opposite side of the planet is dominated by the giant Hellas impact basin. Note the great difference in elevation between the northern and southern hemispheres. Some surface features are labeled, as are the *Viking, Pathfinder,* and *Exploration Rover* robot landing sites. *(NASA)*

planet's surface. Its shape and structure identify the Hellas Basin as an impact feature, similar in many ways to the south pole–Aitken Basin on the far side of Earth's Moon. ∞ (*Discovery 8-2*) The formation of the Hellas Basin must have caused a major redistribution of the young Martian crust—perhaps even enough to account for a substantial portion of the highlands around it, according to some researchers. The basin's heavily cratered floor indicates that the impact occurred very early on in Martian history—some 4 billion years ago—during the heavy bombardment that accompanied the formation of the terrestrial planets. ∞ (Secs. 6.7, 8.5)

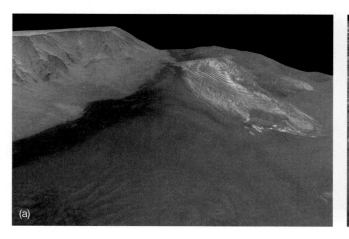
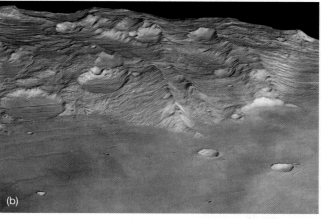

▲ **FIGURE 10.6 Mars Up Close** (a) The northern hemisphere of Mars, like this one near Chryse Planitia, consists of rolling, volcanic plains. (b) The southern Martian highlands, like this one in Hellas Planitia, are heavily cratered. Both of these *Mars Express* photographs are in true color and show roughly the same scale, nearly 100 km across. *(ESA)*

R I V U X G

### The Martian "Grand Canyon"

A particularly prominent feature associated with the Tharsis bulge is a great "canyon" known as Valles Marineris (the Mariner Valley). Shown in its entirety in Figure 10.3 and in more detail in Figure 10.7, this feature is not really a canyon in the terrestrial sense, because running water played no part in its formation. Planetary astronomers theorize that it was formed by the same crustal forces that caused the entire Tharsis region to bulge outward, making the surface split and crack. The resulting cracks, called *tectonic fractures*, are found all around the Tharsis bulge. Valles Marineris is the largest of them. Cratering studies suggest that the cracks are at least 2 billion years old; age estimates for Valles Marineris range up to 3.5 billion years. ∞ (Sec. 8.5) Similar (but much smaller) cracks, originating from similar causes, have been found in the Aphrodite Terra region of Venus. ∞ (Sec. 9.4)

Valles Marineris runs for almost 4000 km along the Martian equator, about one-fifth of the way around the planet. At its widest, it is some 120 km across, and it is as deep as 7 km in places. Like many Martian surface features, it simply dwarfs Earthly competition. The Grand Canyon in Arizona would easily fit into one of its side "tributary" cracks. Valles Marineris is so large that it can even be seen from Earth—in fact, it was one of the few "canals" observed by 19th-century astronomers that actually corresponded to a real feature on the planet's surface. (It was known as the Coprates canal.) We must reemphasize, however, that this Martian feature was not constructed by intelligent beings, nor was it carved by a river, nor is it a result of Martian plate tectonics. For some reason, the crustal forces that formed it never developed into full-fledged plate motion as exists on Earth.

### Volcanism on Mars

Mars contains the largest known volcanoes in the solar system. Three very large volcanoes are found on the Tharsis bulge, two of them visible on the left-hand side of Figure 10.3. The largest volcano of all is Olympus Mons (Figure 10.8), northwest of Tharsis, lying just over the left (western) horizon of Figure 10.3. This volcano measures some 700 km in diameter at its base—only slightly smaller than the state of Texas—and rises to a height of 25 km above the surrounding plains. The caldera, or crater, at its summit, measures 80 km across. The other three large volcanoes are a little smaller—a "mere" 18 km high—and lie near the top of the bulge.

Like Maxwell Mons on Venus, none of these volcanoes is associated with plate motion on Mars—as just mentioned, there is none. Instead, they are shield volcanoes, sitting atop a hot spot in the underlying Martian mantle. ∞ (Sec. 9.4)

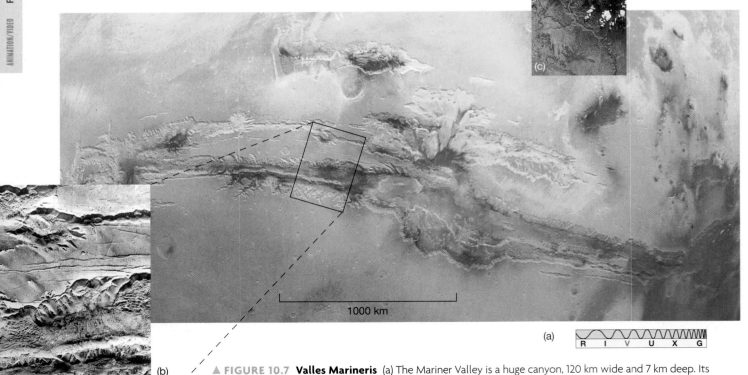

1000 km

(a)

R  I  V  U  X  G

(b)

(c)

▲ **FIGURE 10.7 Valles Marineris** (a) The Mariner Valley is a huge canyon, 120 km wide and 7 km deep. Its length is about 4000 km, or nearly the full breadth of the continental United States. (b) A close-up view shows the complexity of the valley walls and dry tributaries. (c) A comparison, to scale, with Earth's Grand Canyon, which is a mere 20 km wide and 2 km deep, suggests just how big the Mariner Valley is. (*NASA*)

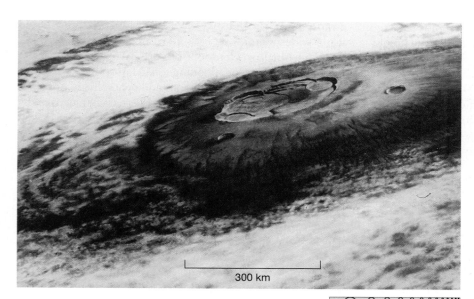

◀ FIGURE 10.8 **Olympus Mons** The largest volcano known on Mars or anywhere else in the solar system, Olympus Mons is nearly three times taller than Mount Everest on Earth, measuring about 700 km across its base and rising 25 km high at its peak. This Martian mountain seems currently inactive and may have been extinct for at least several hundred million years. By comparison, the largest volcano on Earth, Hawaii's Mauna Loa, measures a mere 120 km across and peaks just 9 km above the Pacific Ocean floor. *(NASA)*

300 km

R I V U X G

All four show distinctive lava channels and other flow features similar to those found on shield volcanoes on Earth. *Viking* and *Mars Global Surveyor* images of the Martian surface reveal many hundreds of volcanoes. Most of the largest are associated with the Tharsis bulge, but many smaller volcanoes are found in the northern plains.

The great height of Martian volcanoes is a direct consequence of the planet's low surface gravity. (See the Mars Data box on p. 253.) As lava flows and spreads to form a shield volcano, its eventual height depends on the new mountain's ability to support its own weight. The lower the gravity, the less is the weight and the higher is the mountain. It is no accident that Maxwell Mons on Venus and the Hawaiian shield volcanoes on Earth rise to roughly the same height (about 10 km) above their respective bases—Earth and Venus have similar surface gravity. Mars's surface gravity is only 40 percent that of Earth, so volcanoes rise roughly 2.5 times as high.

Are the Martian shield volcanoes still active? Scientists have found no direct evidence for recent or ongoing eruptions. However, if these volcanoes have been around since the Tharsis uplift (as the formation of the Tharsis bulge is known) and were active as recently as 100 million years ago (an age estimate based on the extent of impact cratering on their slopes), some of them may still be at least intermittently active. Millions of years, though, may pass between eruptions.

### Impact Cratering

The *Mariner* spacecraft found that the surfaces of Mars and its two moons are pitted with impact craters formed by meteoroids falling in from space. On Mars, as on Venus, there is a lack of small impact craters, less than roughly 5 km in diameter. ∞ (Sec. 9.4) This time, though, the explanation is not that such craters do not form—small meteoroids

have no trouble penetrating the thin Martian atmosphere. Instead, thin or not, the atmosphere is an efficient erosive agent, transporting dust from place to place and erasing small impact craters faster than they can form.

Overall, erosion on Mars is about 100 times slower than on Earth, but still far faster than on the Moon or Venus. For comparison, a 1-km diameter crater might survive for 100 million years on Mars. On Earth, it would be gone in a million years or so, but it would remain intact for tens of billions of years on the Moon before being obliterated by meteoritic erosion. On Venus, the crater would most likely survive until the next large-scale volcanic resurfacing event. ∞ (Sec. 9.4)

As on the Moon and Venus, the extent of large impact cratering (craters too big to have been filled in by erosion since they formed) serves as an age indicator for the Martian surface. ∞ (Sec. 8.5) The ages quoted earlier, ranging from 4 billion years for the southern highlands to a few hundred million years in the youngest volcanic areas, were obtained in this way.

### CONCEPT CHECK

✔ How do we know that the northern Martian lowlands are younger than the southern highlands?

## 10.5 Water on Mars

Mars today is apparently dry and desolate, yet astronomers have strong evidence that that was not always the case. Unlike Earth, where water is abundant, and Venus, which (as we saw in Chapter 9) has been devoid of water for billions of years, Mars offers intriguing hints that it may once have harbored liquid water on its surface. ∞ (Sec. 9.5) The relatively low rate of surface erosion on Mars means that many surface features formed billions of years ago are still detectable, providing astronomers with a unique opportunity—in principle, at least—to probe the presence of water on that planet over the entire span of Martian history.

Because water is such a vital ingredient to the development of life on Earth (see Chapter 28), its presence on Mars has important implications for life there too. Let's take a closer look at conditions on the Martian surface since the planet formed.

## Evidence for Past Running Water

Although the great surface cracks in the Tharsis region are not really canyons and were not formed by running water, photographic evidence reveals that liquid water once existed in great quantity on the surface of Mars. Two types of flow feature are seen: the *runoff channels* and the *outflow channels.*

The **runoff channels** (one of which is shown in Figure 10.9a) are found in the southern highlands. They are extensive systems—sometimes hundreds of kilometers in total length—of interconnecting, twisting channels that seem to merge into larger, wider channels. They bear a strong resemblance to river systems on Earth, and geologists think that this is just what they are—the dried-up beds of long-gone rivers that once carried rainfall on Mars from the mountains down into the valleys. To most scientists, these runoff channels speak of a time 4 billion years ago (the age of the Martian highlands), when the

atmosphere was thicker, the surface warmer, and liquid water widespread.

The **outflow channels** (Figure 10.10a) are probably relics of catastrophic flooding on Mars long ago. They appear only in equatorial regions and generally do not form the extensive interconnected networks that characterize the runoff channels. Instead, they are probably the paths taken by huge volumes of water draining from the southern highlands into the northern plains. The onrushing water arising from these flash floods probably also formed the odd teardrop-shaped "islands" (resembling the miniature versions seen in the wet sand of our beaches at low tide) that have been found on the plains close to the ends of the outflow channels (Figure 10.10b). Judging from the width and depth of the channels, the flow rates must have been truly enormous—perhaps as much as a hundred times greater than the $10^5$ tons per second carried by the Amazon River, the largest river system on Earth. Flooding shaped the outflow channels about 3 billion years ago, about the same time as the northern volcanic plains formed.

Were the runoff channels river systems like those on Earth, part of a planetwide *water* cycle, in which rain fell, forming rivers that drained into lakes or oceans, which in turn evaporated to form clouds and more rain? Or were the runoff channels formed by periodically melting underground ice, much like the outflow channels but on a gentler scale, with no associated lakes or other extended bodies of water? Astronomers have debated this issue for years, but, based on the accumulated data from the current crop of Mars probes, the evidence now seems to favor the former view. Many planetary scientists think that Mars may have enjoyed an extended early period during which rivers, lakes, and perhaps even oceans adorned its surface.

Figure 10.11 is a 2003 *Mars Global Surveyor* image showing what mission specialists think may be a *delta*—a fan-shaped network of channels and sediment deposits where a river once flowed into a larger body of water, in this case a lake filling a crater in the southern highlands. Other researchers go even further, suggesting that the data provide evidence for large open expanses of water on the early Martian surface. Figure 10.12(a) is a computer-generated view of the Martian north polar region, based on *Mars Global Surveyor* images, showing the extent of what may have been an ancient ocean covering much of the northern lowlands. The Hellas basin (Figure 10.5) is another possible candidate for an ancient Martian sea.

The idea of a warm, wet early Mars has been controversial. Proponents

▲ **FIGURE 10.9 Martian Channel** (a) This runoff channel on Mars is about 400 km long and up to 5 km wide in places. (b) The Red River running from Shreveport, Louisiana, to the Mississippi River. The two differ mainly in that there is currently no liquid water in this, or any other, Martian valley. *(ESA; NASA)*

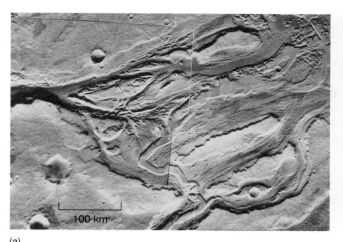

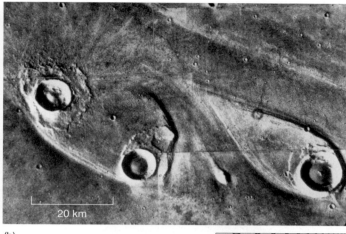

(a)                                                              (b)

R  I  V  U  X  G

▲ FIGURE 10.10  **Martian Outflow**  (a) An outflow channel near the Martian equator bears witness to a catastrophic flood that occurred about 3 billion years ago. (b) The onrushing water that carved out the outflow channels was responsible for forming these oddly shaped "islands" as the flow encountered obstacles—impact craters—in its path. Each "island" is about 40 km long. *(NASA)*

point to features such as the terraced "beaches" shown in Figure 10.12(b), which might conceivably have been left behind as a lake or ocean evaporated and the shoreline receded. But detractors maintain that the terraces could also have been created by geological activity, perhaps related to the tectonic forces that depressed the northern hemisphere far below the level of the south, in which case they have nothing whatever to do with Martian water. Furthermore, *Mars Global Surveyor* data released in 2003 seem to indicate that the Martian surface contains too few *carbonate* rock layers—compounds containing carbon and oxygen that should have been formed in abundance in ancient oceans. Their absence would support the picture of a cold, dry Mars that never experienced the extended mild period required to form lakes and oceans.

However, as more data have been obtained, the evidence in favor of ancient lakes or seas has strengthened. The *Mars Express* orbiter has detected "hydrated" chemical compounds in surface rocks over broad swaths of the planet, strongly suggesting that those regions were wet for extended periods of time. In addition, as discussed below, direct chemical analyses made by the most recent *NASA* landers indicate that their landing sites also experienced long periods in the past during which liquid water existed on the surface.

## Subsurface Ice

As far as we can tell, there is no liquid water on the Martian surface today. However, the detailed appearance of Martian impact craters provides an important piece of information about conditions just below the planet's surface. The ejecta blankets surrounding many Martian

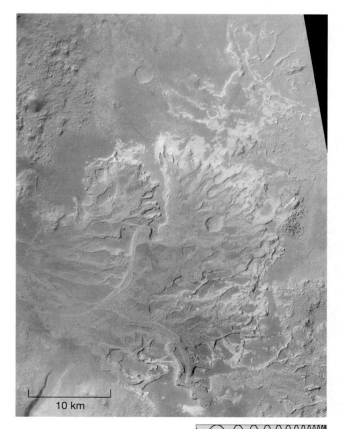

◀ FIGURE 10.11  **Martian River Delta**  Did this fan-shaped region of twisted streams form as a river flowed into a larger sea? If it did, the *Mars Global Surveyor* image supports the idea that Mars once had large bodies of liquid water on its surface. Not all scientists agree with this interpretation, however. *(NASA)*

R  I  V  U  X  G

craters look quite different from their lunar counterparts. Figure 10.13 compares the Copernicus crater on the Moon with the (fairly typical) crater Yuty on Mars. The material surrounding the lunar crater is just what one would expect from an explosion ejecting a large volume of dust, soil, and boulders. However, the ejecta blanket on Mars gives the distinct impression of a liquid that has splashed or flowed out of the crater. Geologists think that this *fluidized ejecta* crater indicates that a layer of **permafrost,** or water ice, lies just a few meters under the surface. The explosive impact heated and liquefied the ice, resulting in the fluid appearance of the ejecta.

More direct evidence for subsurface ice was obtained in 2002, when a gamma-ray spectrometer aboard the *Mars Odyssey* orbiter detected extensive deposits of water ice crystals (actually, the hydrogen they contain) mixed with the Martian surface layers at high latitudes (more than 50° north and south of the equator). In some locations ice appears to compose as much as 50 percent by volume of the planet's soil. The instrument was similar in design to the one carried by *Lunar Prospector,* which searched for (and found) ice crystals in the regolith near the lunar poles. ∞ (Sec. 8.5, *Discovery 8-2*) Radar aboard *Mars Express* has confirmed these results and also suggests deep deposits of ice extending hundreds of meters below the surface in many locations.

Ironically, one site where ice has been found had in fact been the target of NASA's *Mars Polar Lander* probe, which disappeared in 1999. Unfortunately, none of the five successful Martian landers to date has put down in a region where ice has been detected. ∞ (Sec. 6.6)

Prior to the arrival of *Mars Global Surveyor* in 2000, astronomers thought that all the water below the Martian surface existed in the form of ice. However, since then, *Surveyor* mission scientists have reported the discovery of numerous small-scale "gullies" in Martian cliffs and crater walls that apparently were carved by running water in the

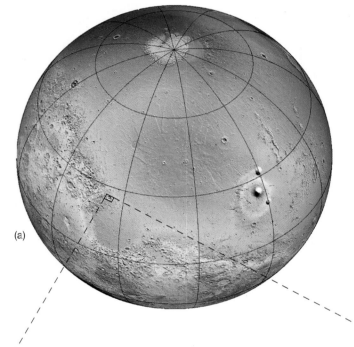

(a)

◀ **FIGURE 10.12 Ancient Ocean?** (a) A possible ancient Martian ocean once might have spanned the polar regions. The blue regions in this computer-generated map actually indicate depth below the average radius of the planet, thus they also outline the extent of the ocean inferred from *Mars Global Surveyor* data. (The color elevation scale is nearly the same as in Figure 10.4.) (b) This high-resolution image shows tentative evidence for erosion by standing water in the floor of Holden Crater, about 140 km across. *(NASA)*

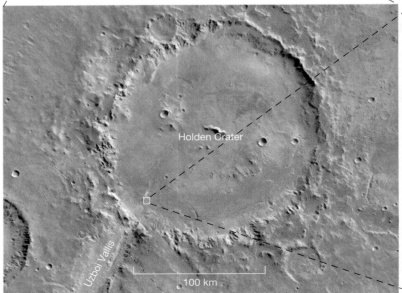

Holden Crater

Uzboi Vallis

100 km

(b)

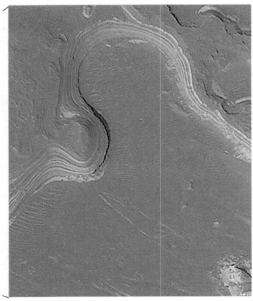

R   I   V   U   X   G

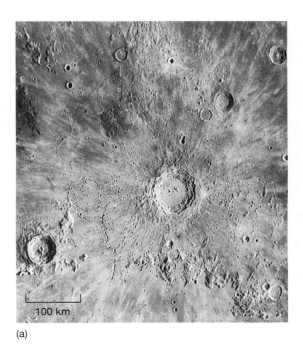

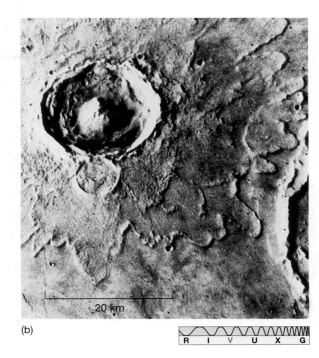

R I V U X G

▲ FIGURE 10.13 **Crater Comparison** (a) The large lunar impact crater Copernicus is typical of those found on Earth's Moon. Its ejecta blanket appears to be composed of dry, powdery material. (b) The ejecta from Mars's crater Yuty (18 km in diameter) evidently was once liquid. This type of crater is sometimes called a "splosh" crater. *(NASA)*

relatively recent past. These features are too small to have been resolved by the *Viking* cameras. Figure 10.14(a) shows one such gully, found in the inner rim of a Martian impact crater in the southern highlands. Its structure has many similarities to the channels carved by flash floods on Earth.

The ages of these intriguing Martian features are uncertain and might be as great as a million years in some cases. However, the *Surveyor* team speculate that some of them may still be active today and that liquid water might exist in some regions of Mars at depths of less than 500 m. Some scientists dispute this interpretation, arguing that the "fluid" responsible for the gullies could have been solid (granular) or even liquid carbon dioxide, expelled under great pressure from the Martian crust. Others point to features such as that shown in Figure 10.14(b), a very similar looking gully found in an impact crater in the Arctic—perhaps the closest we can come on Earth to replicating the harsh conditions on Mars—which formed when subsurface ice became exposed to sunlight and melted. Perhaps the same occurred on Mars, in which case, even if the gullies were created by flowing water, that does not necessarily imply liquid water below the surface.

Recent data from *Surveyor*'s cameras have deepened the mystery of the Martian surface flows. Figure 10.15 shows two images, taken 6 years apart, of an unnamed impact crater in the southern highlands. The white streak in the second image is thought (by some) to be a frozen mudslide, where liquid water briefly flowed down the inside of the crater wall, carrying rocky debris with it, then froze on the chilly Martian surface. At the time of writing, its composition is uncertain, but one thing is clear—whatever it is, it formed *recently*, demonstrating that the production of this, and presumably other, similar surface features is an ongoing process. It seems that a lot more study, perhaps even a human visit, will be needed before this issue is settled.

## The Martian Polar Caps

We have already noted that the Martian polar caps are composed predominantly of carbon dioxide frost—dry ice—and show seasonal variations. Each cap in fact consists of two distinct parts—the **seasonal cap,** which grows and shrinks each year, and the **residual cap,** which remains permanently frozen. At maximum size, in southern midwinter, the southern seasonal cap is some 4000 km across. Half a Martian year later, the northern cap is at its largest, reaching a diameter of roughly 3000 km. The two seasonal polar caps do not have the same maximum size because of the eccentricity of Mars's orbit around the Sun. During southern winter, Mars is considerably farther from the Sun than half a year later, in northern winter. Thus the southern winter season is longer and colder than that of the north, and the polar cap grows correspondingly larger.

The seasonal caps are composed entirely of carbon dioxide. Their temperatures are never greater than about

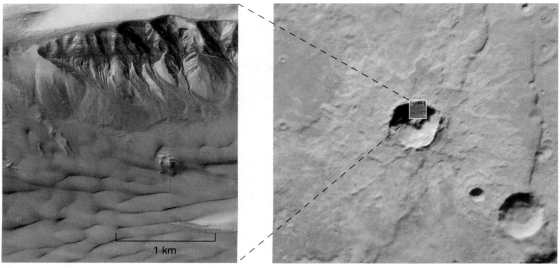

(a)

(b)

R   I   V   U   X   G

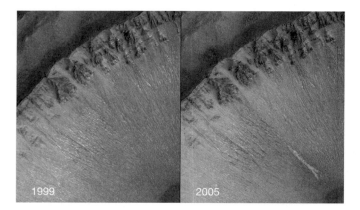

1999     2005

▲ **FIGURE 10.15 Recent Martian Outflow** This comparison between two *Mars Global Surveyor* images, taken in 1999 and 2005, of a Martian impact crater shows that something—the white streak (lower right), possibly water—flowed across the surface during that 6-year period. If so, then the flows responsible for the gullies in Figure 10.14 are not just a thing of the past—the activity is ongoing. *(NASA)*

▲ FIGURE 10.14 **Running Water on Mars?** (a) This high-resolution *Mars Global Surveyor* view (left) of a crater wall (right) near the Mariner Valley shows evidence of "gullies" apparently formed by running water in the relatively recent past. (b) Similar gullies in the Haughton impact crater on the Arctic's Devon Island formed when underground ice temporarily melted, causing streams of water to flow on the surface; *(Gordon Osinski/Canadian Space Agency; NASA)*

150 K (−120°C), the point at which dry ice can form. During the Martian summer, when sunlight striking a cap is most intense, carbon dioxide evaporates into the atmosphere, and the cap shrinks. In the winter, atmospheric carbon dioxide refreezes, and the cap reforms. As the caps grow and shrink, they cause substantial variations (up to 30 percent) in the Martian atmospheric pressure—a large fraction of the planet's atmosphere freezes out and evaporates again each year. From studies of these atmospheric fluctuations, scientists can estimate the amount of carbon dioxide in the seasonal polar caps. The maximum thickness of the seasonal caps is thought to be about 1 m.

The residual caps (Figure 10.16) are smaller and brighter than the seasonal caps and show an even more marked north–south asymmetry. The southern residual cap is about 350 km across and, like the seasonal caps, is probably made mostly of carbon dioxide, although it also contains some water ice. Its temperature remains below 150 K at all times. Its composition, long suspected by theorists, was finally established only in 2004, by spectroscopic imaging observations made by the *Mars Express* orbiter.

The northern residual cap is much larger—about 1000 km across—and warmer, with a temperature that can exceed 200 K in northern summertime. Planetary scientists think that the northern residual cap is made mostly of water ice, an opinion strengthened by spectroscopic observations that show an increase in the concentration of water vapor above the north pole in northern summer as some small fraction of its water ice evaporates in the Sun's

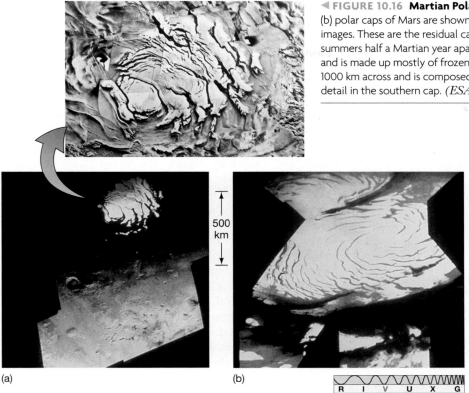

◄ FIGURE 10.16 **Martian Polar Caps** The southern (a) and northern (b) polar caps of Mars are shown to scale in these mosaics of *Mariner 9* images. These are the residual caps, seen here during their respective summers half a Martian year apart. The southern cap is some 350 km across and is made up mostly of frozen carbon dioxide. The northern cap is about 1000 km across and is composed mostly of water ice. The inset shows greater detail in the southern cap. *(ESA; NASA)*

(a)  (b)

R I V U X G

heat. (Note that, in this terminology, Earth's polar caps are both residual, and are composed entirely of water ice.) The thickness of the Martian caps is uncertain, but it is likely that, as on Earth, they represent a significant storehouse for water on the planet.

Why is there such a temperature difference (at least 50 K) between the two residual polar caps, and why is the northern cap warmer, despite the fact that the planet's northern hemisphere is generally cooler than the south (see Section 10.2)? The reason is not fully understood, but it seems to be related to the giant dust storms that envelop the planet during southern summer. These storms, which last for a quarter of a Martian year (about six Earth months), tend to blow the dust from the warmer south into the cooler northern hemisphere. The northern ice cap becomes dusty and less reflective. As a result, it absorbs more sunlight and warms up.

## Climate Change on Mars

Aside from the gullies mentioned earlier, which are suggestive but by no means conclusive, astronomers have no direct evidence for liquid water anywhere on the surface of Mars today, and the amount of water vapor in the Martian atmosphere is tiny. Yet even setting aside the unproven hints of ancient oceans, the extent of the outflow channels clearly implies that a huge total volume of liquid water must have existed on Mars in the distant past. Where did all that water go? Some of it may have entered the atmosphere and escaped into space, but, as best we can tell,

most of it today is locked in the permafrost layer under the Martian surface, with a little more contained in the polar caps.

Planetary scientists trace the history of water on Mars along the following broad lines. Early on, conditions on the planet were much warmer—perhaps even Earth-like—and liquid water was widespread, forming the runoff channels as rainfall drained into river valleys. Roughly 4 billion years ago, for reasons discussed below, climatic conditions changed and the water began to freeze, forming the permafrost and drying out the river beds. Mars remained frozen for about a billion years, until volcanic (or perhaps some other) activity heated large regions of the surface, melting the subsurface ice and causing flash floods that created the outflow channels.

Subsequently, volcanic activity subsided, the water refroze, and Mars once again became a dry world. The present level of water vapor in the Martian atmosphere is the maximum possible, given the atmosphere's present density and temperature. Estimates of the total amount of water stored as permafrost and in the polar caps are uncertain, but it is thought that if all the water on Mars were to become liquid, it would cover the surface to a depth of roughly 10 meters.

## The View from the Martian Landers

Remote sensing—taking images and other measurements from orbit—has been vitally important to our understanding of Mars, but in some cases there is just no substitute for a close-up look. To date, five U.S spacecraft have successfully landed on the Martian surface (see Table 6.3). Their landing sites, marked on Figure 10.5, spanned a variety of Martian terrains. Their goals included detailed geological and chemical analysis of Martian surface rocks, the search for life, and the search for water.

*Viking 1* landed in Chryse Planitia, a broad depression to the east of Tharsis. The view that greeted its cameras (Figure 10.17) was a windswept, gently rolling, rather desolate plain, littered with rocks of all sizes, not unlike a high desert on Earth. The surface rocks visible in

R I V U X G

▲ FIGURE 10.17 *Viking 1* This is the view from the *Viking 1* spacecraft now parked on the surface of Mars. The fine-grained soil and the reddish rock-strewn terrain stretching toward the horizon contains substantial amounts of iron ore; the surface of Mars is literally rusting away. The sky is a pale pink color, the result of airborne dust. *(NASA)*

Figure 10.17 are probably part of the ejecta blanket of a nearby impact crater. *Viking 2* landed somewhat farther north, in a region of Mars called Utopia, chosen in part because mission planners anticipated greater seasonal climatic variations there. The plain on which *Viking 2* landed was flat and featureless. From space, the landing site appeared smooth and dusty. In fact, the surface turned out to be very rocky, even rockier than the Chryse site, and without the dust layer the mission directors had expected (Figure 10.18). The views that the two landers recorded may turn out to be quite typical of the low-latitude northern plains.

The *Viking* landers performed numerous chemical analyses of the Martian regolith. One important finding

of these studies was the high iron content of the planet's surface. Chemical reactions between the iron-rich surface soil and free oxygen in the atmosphere is responsible for the iron oxide ("rust") that gives Mars its characteristic color. Although the surface layers are rich in iron relative to Earth's surface, the overall abundance is similar to Earth's average iron content. On Earth, much of the iron has differentiated to the center. Chemical differentiation does not appear to have been nearly so complete on Mars.

The next successful mission to the Martian surface was *Mars Pathfinder.* During the unexpectedly long lifetime of its mission in 1997 (it lasted almost 3 months instead of the anticipated 1), the lander performed measurements of the Martian atmosphere and atmospheric dust while its robot rover *Sojourner* (visible in Figure 6.14) carried out chemical analyses of the soil and rocks within about 50 m of the parent craft. ∞ (Sec. 6.6) In addition, more than 16,000 images of the region were returned to Earth. *Sojourner* lander found that the soil at its landing site was similar to that found by the *Viking* landers. However, analyses of nearby rocks revealed a chemical makeup different from that of the Martian meteorites found on Earth (see *Discovery 10-1*).

The landing site for the *Pathfinder* mission had been carefully chosen to lie near the mouth of an outflow channel, and the size distribution and composition of the many rocks and boulders surrounding the lander were consistent with their having been deposited there by flood waters. In addition, the presence of numerous rounded pebbles strongly suggested the erosive action of running water at some time in the past.

The most recent visitors to the Martian surface are the twin landers *Spirit* and *Opportunity* of the *Mars Exploration Rover* mission. ∞ (Sec. 6.6) During their operating lifetimes on the planet—well into their third years at the time of writing, again far exceeding the 3-month "prime mission" expectations of mission planners—these two robots have made extensive chemical and geological studies of rocks within a few kilometers of their landing sites on opposite sides of the planet, with the primary goal of finding evidence for liquid water on the surface at some time in the past. Their findings provide the best evidence

(a)

◄ **FIGURE 10.18** *Viking 2* Another view of the Martian surface, this one rock strewn and flat, as seen through the camera aboard the *Viking 2* robot that soft-landed on the northern Utopian plains. The discarded canister is about 20 cm long. The 0.5-m scars in the dirt were made by the robot's shovel. (*NASA*)

indicator of past water, and measurements made by the rover do indeed suggest that the rocks near the landing site have been alternately underwater and dry for extended periods of time, possibly as a shallow lake alternately filled and evaporated repeatedly over the course of Martian history. If life ever did exist at this site, the sorts of rocks found there might have preserved a fossil record very well, but *Opportunity* was not equipped to carry out such studies.

## CONCEPT CHECK

✔ Where has all the Martian water gone?

yet for standing water on the ancient Martian surface and have changed the minds of many skeptical scientists on the subject of water on Mars.

*Spirit*'s landing site was rocky and similar in many ways to the terrains encountered by earlier landers, although closer study by the science team reveals that most of the rocks in the lander's vicinity do appear to have been extensively altered by water long ago. Halfway around the planet, however, *Opportunity* appears to have hit the jackpot in its quest, finding itself surrounded by rocks showing every chemical and geological indication of having been very wet—possibly immersed in salt water—in the past (Figure 10.19).

*Opportunity*'s landing site had been chosen in part because the *Mars Odyssey* orbiter had detected a compound called hematite in the surface rocks there, a possible

(b)

▼ **FIGURE 10.19** *Opportunity* **Rover** (a) A panoramic view of the terrain near where NASA's *Opportunity* rover landed on Mars in 2004. This is Endurance crater, roughly 130 m across. (b) A (false color) close-up of a clump of rocks at the bottom of Endurance crater, where *Opportunity* found evidence for extensive liquid water in Mars's past. (*NASA*)

# DISCOVERY 10-1

## Life on Mars?

Even before the *Viking* missions reached Mars in 1976, most astronomers had abandoned hope of finding life there. Scientists knew that Mars had no large-scale canal systems, no surface water, almost no oxygen in its atmosphere, and no seasonal vegetation changes. The current absence of liquid water on Mars especially dims the chances for life there now. However, running water and, possibly, a dense atmosphere in the past may have created conditions suitable for the emergence of life long ago. (See Chapter 28 for a fuller discussion of what constitutes "life," scientifically speaking, and why water plays such an important role.)

In the hope that some form of microbial life—perhaps bacteria or other microscopic organisms—might have survived to the present day, the *Viking* landers carried out experiments designed to detect biological activity. The accompanying pair of photographs shows the Martian surface before and after the robot arm of one of the landers dug a shallow trench to scoop up soil samples. The arm is visible in the first frame.

All three *Viking* biological experiments assumed some basic similarity between hypothetical Martian bacteria and those found on Earth. A *gas-exchange experiment* offered a nutrient broth to any residents of a sample of Martian soil and looked for gases that would signal metabolic activity. A *labeled-release experiment* added compounds containing radioactive carbon to the soil and then waited for results signaling that Martian organisms had either eaten or inhaled the carbon. Finally, a *pyrolitic-release experiment* added radioactively tagged carbon dioxide to a sample of Martian soil and atmosphere, waited a while, and then removed the gas and tested the soil (by heating it) for signs that something had absorbed the tagged gas.

Initially, all three experiments appeared to be giving positive signals! However, subsequent careful studies showed that all the results could be explained by inorganic (i.e., nonliving) chemical reactions. Thus, at present, we have no irrefutable evidence for even microbial life on the Martian surface. Most scientists think that the *Viking* robots detected peculiar reactions that mimicked the basic chemistry of living organisms in some ways, but they did not detect life itself.

A criticism of the *Viking* experiments is that they searched only for life now living. Today, Mars seems locked in an ice age—the kind of numbing cold that would prohibit sustained life as we know it. If bacterial life did arise on an Earth-like early Mars, however, then we might be able to find its fossilized remains preserved on or near the Martian surface. Surprisingly, one place to look for life on Mars is right here on Earth. Scientists think that some meteorites found on Earth's surface come from the Moon and from Mars. These meteorites were apparently blasted off these bodies long ago during an impact of some sort, thrown into space, and eventually trapped by Earth's gravity, ultimately to fall to the ground. The most fascinating of the rocks are surely those from the Red Planet—for one of them may harbor fossil evidence for past life on Mars!

The next figure shows ALH84001, a blackened 2-kg meteorite about 17 cm across, found in 1984 in Antarctica. On the basis of estimates of the cosmic-ray exposure it received before reaching Earth, the rock is thought to have been blasted off Mars about 16 million years ago. Looking at this specimen through a microscope (inset), scientists can see rounded orange-brown "globules" of carbonate minerals on the rock's shiny crust. Because carbonates form only in the presence of water, the presence of these globules suggests that carbon dioxide gas and liquid water existed near ground level at some point in Mars's history, a conclusion that planetary scientists had drawn earlier from studies of *Viking*'s orbital images of valleys apparently carved by water when the Martian climate was wetter and warmer.

In a widely viewed press conference in Washington, DC, in 1996, a group of scientists argued, on the basis of all the data accumulated from studies of ALH84001, that they had discovered fossilized evidence for life on Mars. The key pieces of evidence they presented for primitive Martian life were as follows: (1) Bacteria on Earth can produce structures similar to the globules shown in the inset. (2) The meteorite contains traces of *polycyclic aromatic hydrocarbons*—a tongue-twisting name for a class of complex organic molecules (usually abbreviated PAHs) that, although not directly involved in known biological cycles on Earth, occur among the decay products of plants and other organisms. (3) High-powered electron microscopes show that ALH84001 contains tiny teardrop-shaped crystals of magnetite and iron sulfide embedded in places where the carbonate has dissolved. On Earth, bacteria are known to manufacture similar chemical crystals. (4) On very small scales, elongated and egg-shaped structures are seen within the carbonate globules. The researchers interpret these minute structures as fossils of primitive organisms.

The photomicrograph in the second figure shows this fourth, and most controversial, piece of evidence—curved, rodlike structures that resemble bacteria on Earth. Scale is

(NASA)

R I V U X G

(Before)          (After)

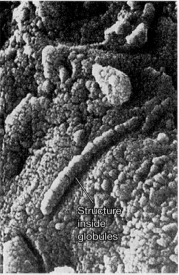

(NASA)

crucial here, however. The structures are only about 0.5 $\mu$m across, 30 times smaller than ancient bacterial cells found fossilized on Earth. Furthermore, several key tests have not yet been done, such as cutting through the suspected fossilized tubes to search for evidence of cell walls, semipermeable membranes, or any internal cavities where body fluids would have resided. Nor has anyone yet found in ALH84001 any amino acids, the basic building blocks of life as we know it (see Section 28.1).

These results remain highly controversial. Many experts do not agree that life has been found on Mars—not even fossilized life. Skeptics maintain that all the evidence could be the result of chemical reactions not requiring any kind of biology. Carbonate compounds are common in all areas of chemistry; PAHs are found in many lifeless places (glacial ice, asteroid-belt meteorites, interstellar clouds, and even the exhaust fumes of automobiles); bacteria are not needed to produce crystals; and it remains unclear whether the tiny tubular structures shown are animal, vegetable, or merely mineral. In addition, there is the huge problem of contamination—after all, ALH84001 was found on Earth and apparently sat in the Antarctic ice fields for 13,000 years before being picked up by meteorite hunters.

During 1999, the team released a new analysis of a second meteorite, named Nakhla (shown in the third figure), discovered in the Sahara Desert in 1911 and also thought to have come from Mars. Again, the scientists reported evidence for microbial life, in the form of clusters of minute spheres and ovals found within tiny clay-filled cracks deep inside the meteorite, having similarities in size, shape, and arrangement to known bacteria on Earth. Since Nakhla is a volcanic basalt rock that solidified about 1.3 billion years ago (as opposed to 4 billion years ago for ALH84001), the new work suggests that life might have spanned the entire history of Mars. If so, then life might still be present there today—but the opponents remain largely unconvinced.

As things now stand, it's a matter of interpretation—at the frontiers of science, issues are usually not as clear-cut as we would hope. The scientific method requires that we continually test and retest the competing theories to try to determine the truth. Only additional analysis and new data—perhaps in the form of samples returned directly from the Martian surface—will tell conclusively whether primitive Martian life existed long ago, although most workers in the field seem to have concluded that, taken as a whole, the results do not support the claim of ancient life on Mars. Still, even some skeptics concede that as much as 20 percent of the organic (carbon-based) molecules in ALH84001 could have originated on the Martian surface—although that is a far cry from proving the existence of life there.

Should the claim of life on Mars hold up against the weight of healthy skepticism in the scientific community, these findings may go down in history as one of the greatest scientific discoveries of all time. We are—or at least were—not alone in the universe! Maybe.

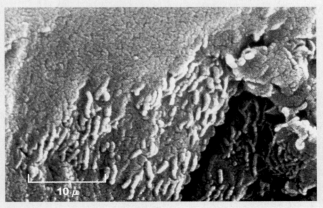

(NASA)

## 10.6 The Martian Atmosphere

Long before the arrival of the *Mariner* and *Viking* spacecraft, astronomers knew from Earth-based spectroscopy that the Martian atmosphere was quite thin and composed primarily of carbon dioxide. In 1964, *Mariner 4* confirmed these results, finding that the atmospheric pressure was only about 1/150 the pressure of Earth's atmosphere at sea level and that carbon dioxide made up at least 95 percent of the total atmosphere. With the arrival of *Viking*, more detailed measurements of the Martian atmosphere could be made. Its composition is now known to be 95.3 percent carbon dioxide, 2.7 percent nitrogen, 1.6 percent argon, 0.13 percent oxygen, 0.07 percent carbon monoxide, and about 0.03 percent water vapor. The level of water vapor is quite variable. Weather conditions encountered by *Mars Pathfinder* were quite similar to those found by *Viking 1*.

### Atmospheric Structure and Weather

As the *Viking* landers descended to the surface, they made measurements of the temperature and pressure at various heights. The results are shown in Figure 10.20. The Martian atmosphere contains a troposphere (the lowest-lying atmospheric zone, where convection and weather occur), which varies both from place to place and from season to

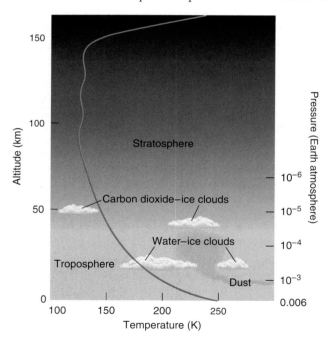

▲ **FIGURE 10.20 Martian Atmosphere** Structure of the Martian atmosphere, as determined by *Viking* and *Mars Global Surveyor*. The troposphere, which rises to an altitude of about 30 km in the daytime, occasionally contains clouds of water ice or, more frequently, dust during the planetwide dust storms that occur each year. Above the troposphere lies the stratosphere. Note the absence of a higher temperature zone in the stratosphere, indicating the absence of an ozone layer.

season. ∞ (Sec. 7.2) The variability of the troposphere arises from the variability of the Martian surface temperature. At noon in the summertime, surface temperatures may reach 300 K. Atmospheric convection is strong, and the top of the troposphere can reach an altitude of 30 km. At night, the atmosphere retains little heat, and the temperature can drop by as much as 100 K. Convection then ceases and the troposphere vanishes.

On average, surface temperatures on Mars are about 50 K cooler than on Earth. The low early-morning temperatures often produce water-ice fog in the Martian canyons (Figure 10.21). Higher in the atmosphere, in the stratosphere, temperatures are low enough for carbon dioxide to solidify, giving rise to a high-level layer of carbon dioxide clouds and haze.

For most of the year, there is little day-to-day variation in the Martian weather: The Sun rises, the surface warms up, and light winds blow until sunset, when the temperature drops again. Only in the southern summer does the daily routine change. Strong surface winds (without rain or snow) sweep up the dry dust, carry it high into the stratosphere, and eventually deposit it elsewhere on the planet. At its greatest fury, a Martian storm floods the atmosphere with dust, making the worst storm we could imagine on Earth's Sahara Desert seem inconsequential by comparison. The dust can remain airborne for months at a time. The blown dust forms systems of sand dunes similar in appearance to those found on Earth.

### Atmospheric Evolution

Although there is some superficial similarity in composition between the atmospheres of Mars and Venus, the two planets obviously have quite different atmospheric histories—Mars's "air" is over 10,000 times thinner than that on Venus. As with the other planets we have studied, we can ask *why* the Martian atmosphere is as it is.

Presumably, Mars acquired a secondary atmosphere *outgassed* from the planet's interior quite early in its history, just as the other terrestrial worlds did. ∞ (Sec. 7.2) Around 4 billion years ago, as indicated by the runoff channels in the highlands, Mars may have had a fairly dense atmosphere, complete with blue skies, oceans, and rain. Even taking into account the larger distance from Mars to the Sun and the fact that the Sun was about 30 percent less luminous 4 billion years ago (see Chapter 22), planetary scientists estimate that the greenhouse effect from a Martian atmosphere a few times denser than Earth's present atmosphere could have kept conditions fairly comfortable. A surface temperature above 0°C (the freezing point of water) seems quite possible.

Sometime during the next billion years, most of the Martian atmosphere disappeared. Possibly some of it was expelled by impacts with large bodies in the early solar system, and a substantial part may have leaked away into space because of the planet's weak gravity.

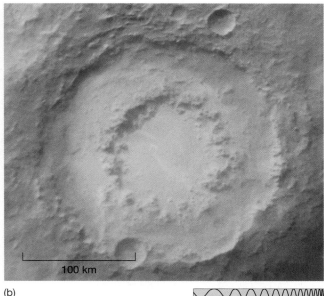

(a)

(b)

▲ FIGURE 10.21 **Fog in the Canyons** (a) As the Sun's light reaches and heats the canyon floor, it drives water vapor from the surface. When this vapor comes in contact with the colder air above the surface, it condenses again, and a temporary water-ice fog results, as seen here, near Mars's northern polar cap. (b) Fog also shrouds the floor of the 200-km-wide Lowell Crater, imaged here by *Mars Global Surveyor* in the autumn of 2000. (*NASA*)

∞ (*More Precisely 8-1*) However, most of the remainder probably became unstable and was lost in a kind of "reverse runaway greenhouse effect," as we now describe. The following scenario, summarized in Figure 10.22, is accepted by many planetary scientists, although not by all—those who discount the evidence presented earlier for liquid water and a thick early atmosphere on Mars obviously require no explanation for their absence today.

The key mechanism controlling the level of carbon dioxide in a terrestrial planet's atmosphere is the absorption of carbon dioxide into the rocks that make up the planet's crust. The presence of liquid water greatly accelerates this process—carbon dioxide dissolves in the liquid water of the planet's rivers and lakes (and oceans, if any), ultimately reacting with surface material to form carbonate rocks. The result—at least in the absence of any opposing effect—is a continual depletion of atmospheric carbon dioxide.

On Venus, as we saw in Chapter 9, an opposing effect did exist, as the familiar greenhouse effect ran away to high temperatures and pressures. ∞ (Sec. 9.5) Carbon dioxide *left* the surface and entered the atmosphere as the temperature rose, resulting in the extreme conditions we now find on that planet today.

On Earth, a very different process acts to counteract the absorption of carbon dioxide into the surface. *Plate tectonics* constantly recycles our planet's carbon dioxide, returning it to the atmosphere via volcanic activity. ∞ (Sec. 7.4) Eventually, these two competing processes come into balance, and the atmospheric

concentration of carbon dioxide tends to remain roughly constant.

However, neither of these opposing processes operated on Mars. The planet was too cool for the greenhouse effect to run away and, as we have seen, Mars's interior cooled faster than Earth's and the planet apparently never developed large-scale plate motion. Even taking into consideration the large volcanoes discussed earlier in this chapter, Mars has had on average far less volcanism than Earth does, so the processes depleting carbon dioxide have been much more effective than those replenishing it, creating a "one-way street" in which the level of atmospheric carbon dioxide steadily declined.

As the Martian carbon dioxide was consumed and its greenhouse effect diminished, the planet cooled, causing still more carbon dioxide to leave the atmosphere. The reason for this is just as described above—lower temperatures allowed more carbon dioxide to be absorbed in the surface layers. The result was a runaway to *lower* temperatures and *decreasing* levels of carbon dioxide in the atmosphere—just the opposite the sequence of events on Venus. Calculations show that much of the Martian atmospheric carbon dioxide could have been depleted in this way in a relatively short period of time, perhaps as quickly as a few hundred million years (Figure 10.23), although some of it might have been replenished by volcanic activity, possibly extending the "comfortable" lifetime of the planet to a half-billion years or so. Much of the debate about the presence of liquid water on the Martian surface revolves around the time it took for the planet's surface to freeze.

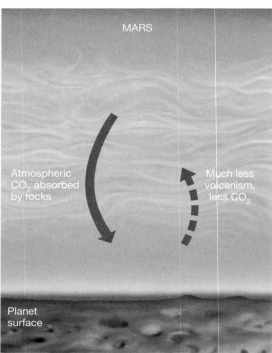

▲ FIGURE 10.22 **Atmospheric Change** Over time, carbon dioxide ($CO_2$) in a planet's atmosphere tends to be absorbed by surface rocks and water. On Earth, geological and volcanic activity returns $CO_2$ to the atmosphere, and a stable balance is struck. On Mars, the rate of volcanic activity is much lower, and the $CO_2$ returned was insufficient to replenish the atmosphere. As a result, the Martian atmosphere thinned and cooled, losing most of its $CO_2$ in as little as a few hundred million years.

As the temperature continued to fall, water froze out of the atmosphere, lowering still further the level of atmospheric greenhouse gases and accelerating the cooling. (Recall from Section 9.5 that water vapor also contributes to the greenhouse effect.) Eventually, even carbon dioxide began to freeze out, particularly at the poles, and Mars reached the frigid state we see today—a cold, dry planet with most of its original complement of atmospheric gases now residing in or under the barren surface.

CONCEPT CHECK
✔ What happened to the Martian atmosphere?

(a) Ancient Mars          (b) Today's Mars

◀ FIGURE 10.23 **Martian Evolution** (a) Artist's conception of Mars some several billion years ago, with a dwindling atmosphere and some lingering surface water. (b) A photo of Mars today. (*Kees Veenenbos*)

## 10.7 Martian Internal Structure

The *Viking* landers carried seismometers to probe the internal structure of Mars. However, one failed to work, and the other was unable to clearly distinguish seismic activity from the buffeting of the Martian wind. As a result, no seismic studies of the Martian interior have yet been carried out. On the basis of studies of the stresses that occurred during the Tharsis uplift, astronomers estimate the thickness of the crust to be about 100 km.

During its visit to Mars in 1965, *Mariner 4* detected no planetary magnetic field, and for many years the most that could be said about the Martian magnetic field was that its strength was no more than a few thousandths the strength of Earth's field (to the level of sensitivity of *Mariner*'s instruments). The *Viking* spacecraft were not designed to make magnetic measurements. In September 1997, *Mars Global Surveyor* detected a very weak Martian field, about 1/800 times that of Earth. However, this is probably a local anomaly, akin to the magnetic fluctuations detected by *Lunar Prospector* at certain locations on the surface of Earth's Moon, and not a global field. ∞ (Sec. 8.7)

Because Mars rotates rapidly, the absence of a global magnetic field is taken to mean that the planet's core is nonmetallic, nonliquid, or both. ∞ (Sec. 7.5) The small size of Mars indicates that any radioactive (or other internal) heating of its interior would have been less effective at heating and melting the planet than similar heating on Earth. The heat was able to reach the surface and escape more easily than on a larger planet such as Earth or Venus.

The evidence we noted earlier for ancient surface activity, especially volcanism, suggests that at least parts of the planet's interior must have melted and possibly differentiated at some time in the past. But the lack of current activity, the absence of any significant magnetic field, the relatively low density (3900 kg/m³), and an abnormally high abundance of iron at the surface all suggest that Mars never melted as extensively as did Earth. The latest data indicate that the Martian core has a diameter of about 2500 km, is composed largely of iron sulfide (a compound about twice as dense as surface rock), and is still at least partly molten.

The history of Mars appears to be that of a planet on which large-scale tectonic activity almost started, but was stifled by the planet's rapidly cooling outer layers. The large upwelling of material that formed the Tharsis bulge might have developed into full-fledged plate tectonic motion on a larger, warmer planet, but the Martian mantle became too rigid and the crust too thick for that to occur. Instead, the upwelling continued to fire volcanic activity, almost up to the present day, but, geologically, much of the planet apparently died 2 billion years ago.

### CONCEPT CHECK

✔ What is the principal reason for the lack of geological activity on Mars today?

## 10.8 The Moons of Mars

Unlike Earth's Moon, Mars's moons are tiny compared with their parent planet and orbit very close to it, relative to the planet's radius. Discovered by American astronomer Asaph Hall in 1877, the two Martian moons—Phobos ("fear") and Deimos ("panic")—are only a few tens of kilometers across. Their composition is quite unlike that of the planet. They are quite difficult to study from Earth because their proximity to Mars makes it hard to distinguish them from their much brighter parent. The *Mariner* and *Viking* orbiters, however, studied both in great detail.

As shown in Figure 10.24, Phobos and Deimos are both quite irregularly shaped and heavily cratered. The larger of the two is Phobos (Figure 10.24a), which is about 28 km long and 20 km wide and is dominated by an enormous 10-km-wide crater named Stickney (after Angelina Stickney, Asaph Hall's wife, who encouraged him to persevere in his observations). The smaller Deimos (Figure 10.24b) is only 16 km long by 10 km wide. Its largest crater is 2.3 km in diameter. The fact that both moons have quite dark surfaces, reflecting no more than 6 percent of the light falling on them, contributes to the difficulty in observing them from Earth.

Phobos and Deimos move in circular, equatorial orbits, and they rotate synchronously (i.e., they each keep the same face permanently turned toward the planet). These characteristics are direct consequences of the tidal influence of Mars. Both moons orbit Mars in the prograde sense—that is, in the same sense (counterclockwise, as seen from above the north celestial pole) as the planet orbits the Sun and rotates on its axis.

Phobos lies only 9378 km (less than three planetary radii) from the center of Mars and, as we saw earlier, has an orbital period of 7 hours and 39 minutes. This period is much less than a Martian day, so an observer standing on the Martian surface would see Phobos move "backward" across the Martian sky—that is, in a direction opposite that of the apparent daily motion of the Sun. Because the moon moves faster than the observer, it overtakes the planet's rotation, rising in the west and setting in the east, crossing the sky from horizon to horizon in about 5.5 hours. Deimos lies somewhat farther out, at 23,459 km, or slightly less than seven planetary radii, and orbits in 30 hours and 18 minutes. Because it completes its orbit in more than a Martian day, it moves "normally," as seen from the ground (i.e., from east to west), taking almost 3 days to traverse the sky.

Astronomers estimated the masses of the two moons on the basis of measurements of their gravitational effect on the *Viking* orbiters. The density of the Martian moons is around 2000 kg/m³, far less than that of any world we have yet encountered in our outward journey through the solar system. This is one reason that astronomers think it unlikely that Phobos and Deimos formed along with Mars. Instead, it is more probable that they are asteroids that were slowed

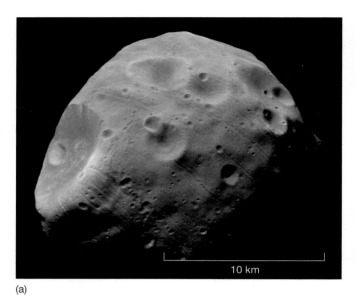

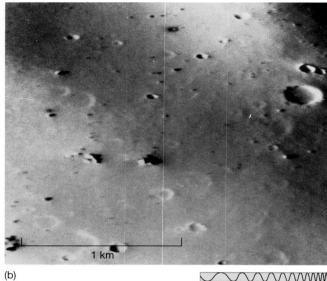

(a)

(b)

| R | I | V | U | X | G |

▲ FIGURE 10.24 **Martian Moons** (a) A *Mars Express* photograph of the potato-shaped Phobos, not much larger than Manhattan Island. The prominent crater (called Stickney) at left is about 10 km across. (b) Like Phobos, the smaller moon, Deimos, has a composition unlike that of Mars. Both moons are probably captured asteroids. This close-up photograph of Deimos was taken by a *Viking* orbiter. The field of view is only 2 km across, and most of the boulders shown are about the size of a house. *(ESA)*

and captured by the outer fringes of the early Martian atmosphere (which, as we have just seen, was probably much denser than the atmosphere today). It is even possible that they are remnants of a single object that broke up during capture. Phobos, on its low-altitude orbit, continues to interact with the planet's upper atmosphere. Its orbit is expected to decay, plunging the moon into the planet's surface in just a few tens of millions of years.

If they are indeed captured asteroids, Phobos and Deimos represent material left over from the earliest

stages of the solar system. Astronomers study them not to gain insight into Martian evolution, but rather because the moons contain information about the young solar system, before the major planets had formed.

## CONCEPT CHECK

✔ In what ways do Phobos and Deimos differ from Earth's Moon?

# CHAPTER REVIEW

## Summary

❶ Mars lies outside Earth's orbit and traverses the entire plane of the ecliptic, as seen from Earth. Mars is about half the radius and one-tenth the mass of Earth. It rotates at almost the same rate as Earth, and its axis of rotation is inclined to the ecliptic at almost the same angle as Earth's axis. Surface temperatures on Mars average about 50 K cooler than those on Earth. Otherwise, Martian weather is reminiscent of that on Earth, with dust storms, clouds, and fog.

❷ As a result of its axial tilt, Mars has daily and seasonal cycles much like those on our own planet, but they are more complex than those on Earth because of Mars's eccentric orbit. From Earth, the most obvious Martian surface features are the polar caps, which grow and diminish as the seasons change on Mars. The two polar

caps on Mars each consist of a **seasonal cap (p. 263)**, composed of carbon dioxide, which grows and shrinks, and a **residual cap (p. 263)**, of water ice, which remains permanently frozen. The appearance of the planet also changes because of seasonal dust storms that obscure its surface.

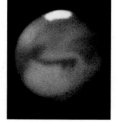

❸ The Martian surface has vast plains, huge volcanoes, and deep channels and canyons. Mars's major surface feature is the Tharsis bulge, located on the planet's equator. This feature may have been caused by a "plume" of upwelling material in the youthful Martian mantle. Associated with the bulge are Olympus Mons, the largest known volcano in the solar system, and a huge crack, called the Valles Marineris, in the planet's surface. The height of the Martian volcanoes is a direct consequence of Mars's low surface gravity. No evidence for recent or ongoing eruptions has been found. On the other side of Mars from Tharsis lies the Hellas basin, the site of a violent meteoritic impact early in the planet's history. There is a marked difference between the two Martian hemispheres. The northern hemisphere consists of rolling volcanic plains and lies several kilometers below the level of the heavily cratered southern hemisphere. The lack of craters in the north suggests that this region is younger. The cause of the north–south asymmetry is not known.

**4** There is strong evidence that Mars once had running water on its surface. **Runoff channels (p. 260)** are the remains of ancient Martian rivers, whereas **outflow channels (p. 260)** are the paths taken by flash floods that cascaded from the southern highlands into the northern plains. *Mars Global Surveyor* and *Mars Express* images also strongly suggest that liquid water once existed in great quantity on Mars, and the *Mars Exploration Rover* landers have returned direct evidence for a wet Martian past. The planet may have enjoyed a relatively brief, warm "Earth-like" phase early on in its evolution, with a thick atmosphere and rain, rivers, and lakes or even oceans.

**5** Today, much of the water on Mars is locked up in the polar caps and in the layer of permafrost lying under the Martian surface. *Viking* observations of fluidized ejecta surrounding impact craters indicated the presence of subsurface ice, and *Mars Odyssey* and *Mars Express* subsequently detected extensive ice deposits

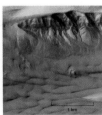

mixed with and lying under the planet's surface layers. *Mars Explorer* has found numerous gullies in crater walls that appear to have been formed by running water, and some flows are known to have occurred within the past few years. Whether these flows imply liquid water or ice below the surface remains unclear.

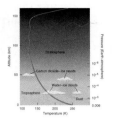

**6** Like the atmosphere of Venus, Mars's atmosphere is composed primarily of carbon dioxide. However, unlike Venus's atmosphere, the cool Martian atmosphere has a density less than 1 percent that of Earth's. Mars may once have had a dense atmosphere, but it was lost, partly to space and partly to surface rocks and subsurface **permafrost (p. 262)** and polar caps. Even today, the thin atmosphere is slowly leaking away.

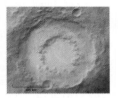

**7** Mars has an extremely weak magnetic field, which, together with the planet's rapid rotation, implies that its core is nonmetallic, nonliquid, or both. The lack of current volcanism, the absence of any significant magnetic field, the planet's relatively low density, and a high abundance of surface iron all suggest that Mars never melted and differentiated as extensively as did Earth. Convection in the Martian interior seems to have been stifled 2 billion years ago by the planet's rapidly cooling and solidifying mantle.

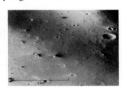

**8** The Martian moons Phobos and Deimos are probably asteroids captured by Mars early in its history. Their densities are far less than that of any planet in the inner solar system. These moons may be representative of conditions in the early solar system.

## Review and Discussion

1. Why is opposition the best time to see Mars from Earth?
2. Why are some Martian oppositions better than others for viewing Mars?
3. For a century, there was speculation that intelligent life had constructed irrigation canals on Mars. What did the "canals" turn out to be?
4. Imagine that you will be visiting the southern hemisphere of Mars during its summer. Describe the atmospheric conditions you might face.
5. Describe the two Martian polar caps, their seasonal and permanent composition, and the differences between them.
6. Why is Mars red?
7. Describe the major large-scale surface features of Mars.
8. Why were Martian volcanoes able to grow so large?
9. Why couldn't you breathe on Mars?
10. What is the evidence that water once flowed on Mars?
11. Is there liquid water on Mars today?
12. Is there water on Mars today, in any form?
13. Why do some scientists think Mars once had an extensive ocean? Where was it located?
14. How were the masses of Mars's moons measured, and what did these measurements tell us about their origin?
15. What do measurements of Martian magnetism tell us about the planet's interior?
16. What is the evidence that Mars never melted as extensively as did Earth?
17. How would Earth look from Mars?
18. If humans were sent to Mars to live, what environmental factors would have to be considered? What resources might Mars provide, and which would have to come from Earth?
19. Since Mars has an atmosphere, and it is composed mostly of a greenhouse gas, why isn't there a significant greenhouse effect to warm its surface?
20. Compare and contrast the evolution of the atmospheres of Mars, Venus, and Earth.

# Conceptual Self-Test: True or False/Multiple Choice

1. Seen from Mars, Earth would go through phases, just as Venus and Mercury do.

2. Because Mars has such a thin atmosphere, the planet has no significant surface winds.

3. Seasonal changes in the appearance of Mars are caused by vegetation on the surface.

4. Olympus Mons is the largest impact crater in the solar system.

5. There are many indications of past plate tectonics on Mars.

6. Valles Marineris is comparable in size to Earth's Grand Canyon.

7. The polar caps of Mars are composed entirely of frozen carbon dioxide.

8. The great height of Martian volcanoes is a direct result of the planet's low gravity.

9. NASA's *Opportunity* rover detected liquid water just under the Martian surface.

10. Water once flowed on the surface of Mars.

11. Compared with the Earth's orbit, the orbit of Mars (a) has the same eccentricity; (b) is more eccentric; (c) is less eccentric; (d) is smaller.

12. As seen from Earth, Mars exhibits a retrograde loop about once every (a) week; (b) 6 months; (c) 2 years; (d) decade.

13. Compared with Earth's diameter, the diameter of Mars is (a) significantly larger; (b) significantly smaller; (c) nearly the same size; (d) unknown.

14. The lengths of the seasons on Mars can be determined by observing the planet's (a) tilt; (b) eccentricity; (c) polar caps; (d) moons.

15. In terms of area, the extinct Martian volcano Olympus Mons is about the size of (a) Mt. Everest; (b) Colorado; (c) North America; (d) Earth's Moon.

16. Figure 10.5 ("Mars Map") clearly shows (a) surface water and ice at northern latitudes; (b) a giant canyon stretching all the way across the planet; (c) iron deposits in the mid-latitudes; (d) cratered terrain in the south.

17. The best evidence for the existence of liquid water on an ancient Mars is Figure (a) 10.12; (b) 10.14; (c) 10.15; (d) 10.17.

18. Compared with the atmosphere of Venus, the Martian atmosphere has (a) a significantly higher temperature; (b) significantly more carbon dioxide; (c) a significantly lower atmospheric pressure; (d) significantly more acidic compounds.

19. In comparison to the atmosphere of Venus, the vastly different atmospheric character of Mars is likely due to a/an (a) ineffective greenhouse effect; (b) reverse greenhouse effect; (c) absence of greenhouse gases that would hold in heat; (d) greater distance from the Sun.

20. The moons of Mars (a) are probably captured asteroids; (b) formed following a collision with Earth; (c) are the remnants of a larger moon; (d) formed simultaneouly with Mars.

# Problems

1. •• By calculating the rate at which Earth overtakes Mars in its orbit (see *More Precisely 9-1*), verify the value of Mars's synodic period given in the Mars Data box on p. 253.

2. • Calculate the minimum and maximum angular diameters of the Sun, as seen from Mars.

3. •• What is the maximum elongation of Earth, as seen from Mars? (For simplicity, assume circular orbits for both planets.)

4. • What will be the minimum size of a Martian surface feature resolvable during the 2003 opposition by an Earth-based telescope with an angular resolution of 0.05″?

5. •• Use the reasoning presented in Chapter 1 to calculate the difference in length between the mean Martian solar day and the Martian sidereal day. ∞ (Sec. 1.4)

6. • Verify that the surface gravity on Mars is 40 percent that of Earth.

7. • What would you weigh on Mars?

8. • How long would it take the wind in a Martian dust storm, moving at a speed of 150 km/h, to encircle the planet's equator?

9. • The mass of the Martian atmosphere is about 1/150 the mass of Earth's atmosphere and is composed mainly (95 percent) of carbon dioxide. Using the result of problem 3 in Chapter 7 to determine the mass of Earth's atmosphere, estimate the total mass of carbon dioxide in the atmosphere of Mars. Compare your answer with the mass of a seasonal polar cap, approximated as a circular sheet of frozen carbon dioxide ("dry ice," having a density of 1600 kg/m$^3$) of diameter 3000 km and thickness 1 m.

10. • Compare the mass of a seasonal polar cap (see previous question) with that of a residual cap of diameter 1000 km, thickness 1 km, and density 1000 kg/m$^3$.

11. • The Hellas impact basin is roughly circular, 3000 km across, and 6 km deep. Taking the Martian crust to have a density of 3000 kg/m$^3$, estimate how much mass was blasted off the Martian surface when the basin formed. Compare your answer with the present total mass of the Martian atmosphere. (See problem 9.)

12. •• The outflow channel shown in Figure 10.10 is about 10 km across and 100 m deep. If it carried 10$^7$ metric tons (10$^{10}$ kg) of water per second, as stated in the text, estimate the speed at which the water must have flowed.

13. •• Calculate the total mass of a uniform layer of water covering the entire Martian surface to a depth of 2 m. (See Section 10.5.) Compare your answer with the mass of Mars.

14. •• Using the data from Section 10.8, compute the time interval between each rising of the moons Phobos and Deimos and the next, as seen from the Martian surface.

15. •• Using the data given in the text, calculate the maximum angular sizes of Phobos and Deimos, as seen by an observer standing on the Martian surface directly under their orbits. Would a Martian observer ever see a total solar eclipse? (See problem 2.)

# 11

# JUPITER
## Giant of the Solar System

Beyond the orbit of Mars, the solar system is very different from our own backyard. The outer solar system presents us with a totally unfamiliar environment: huge gas balls, peculiar moons, complex ring systems, and a wide variety of physical and chemical phenomena, many of which are still only poorly understood. Although the jovian planets—Jupiter, Saturn, Uranus, and Neptune—differ from one another in many ways, we will find that they have much in common, too. As with the terrestrial planets, we will learn from their differences as well as from their similarities.

Our study of these alien places begins with the jovian planet closest to Earth: Jupiter, the largest planet in the solar system. In mass, composition, and internal structure, it offers a model for the other jovian worlds.

## LEARNING GOALS

*Studying this chapter will enable you to*

1 Specify the ways in which Jupiter differs from the terrestrial planets in its physical and orbital properties.

2 Discuss the processes responsible for the appearance of Jupiter's atmosphere.

3 Describe Jupiter's internal structure and composition, and explain how their properties are inferred from external measurements.

4 Summarize the characteristics of Jupiter's magnetosphere.

5 Discuss the orbital properties of the Galilean moons of Jupiter, and describe the appearance and physical properties of each moon.

6 Explain how tidal forces can produce enormous internal stresses in a jovian moon, and discuss some effects of those stresses.

 Visit www.aw-bc.com/chaisson for additional images, animations, quizzes, and eBook for this chapter.

## 11.1    Orbital and Physical Properties

Named after the most powerful god of the Roman pantheon, Jupiter is by far the largest planet in the solar system. Ancient astronomers could not have known the planet's true size, but their choice of names was apt. The Jupiter Data box on p. 282 presents some orbital and physical data on the planet.

### The View from Earth

Jupiter is the fifth planet from the Sun and the innermost jovian planet (Figure 11.1). It is the third-brightest object in the night sky (after the Moon and Venus), making it easy to locate and study. As in the case of Mars, Jupiter is brightest when it is near opposition. When this happens to occur close to perihelion, the planet can be up to 50″ across, and a lot of detail can be discerned through even a small telescope.

Figure 11.2(a) is a photograph of Jupiter, taken through a telescope on Earth. In contrast to the terrestrial worlds, Jupiter has many moons that vary greatly in size and other properties. The four largest, visible in this telescopic view (and, to a few people, with the naked eye), are known as the **Galilean moons,** after Galileo Galilei, who discovered them in 1610. ∞ (Sec. 2.4) Figure 11.2(b) is a *Hubble Space Telescope* image of Jupiter taken during the opposition of December 1990. Notice both the alternating light and dark bands that cross the planet parallel to its equator and also the large oval at the

lower right. These atmospheric features are quite unlike anything found on the inner planets. Figure 11.2(c) is an up-close, true-color image of Jupiter's north polar region, taken by the *Cassini* spacecraft as it passed the planet in 2001 en route to Saturn.

### Mass and Radius

Since astronomers have been able to study the motion of the Galilean moons for quite some time, Jupiter's mass has long been known to high accuracy. ∞ (*More Precisely 6-1*) It is $1.9 \times 10^{27}$ kg, or 318 Earth masses—more than twice the mass of all the other planets combined. In the broadest sense, our solar system is a two-object system with a lot of additional debris. Nonetheless, as massive as Jupiter is, it is still only 1/1000 the mass of the Sun.

Knowing Jupiter's distance and angular size, we can easily determine the planet's radius, which turns out to be 71,500 km, or 11.2 Earth radii. More dramatically stated, more than 1400 Earths would be needed to equal the volume of Jupiter. From the planet's size and mass, we derive a density of 1300 kg/m³ for Jupiter. Here (as if we needed it) is yet another indicator that Jupiter is radically different from the terrestrial worlds: It is clear that, whatever Jupiter's composition, it cannot possibly be made up of the same material as the inner planets. (Recall from Chapter 7 that Earth's average density is 5500 kg/m³). ∞ (Sec. 7.1)

In fact, theoretical studies of the planet's internal structure indicate that Jupiter must be composed

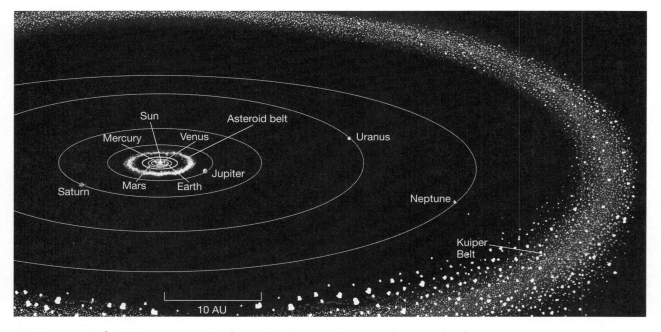

▲ **FIGURE 11.1  Solar System Perspective**  This is a variation on Figure 6.5—neither an overhead view or an edge-on view of our solar system, but an oblique view from a distant perspective—illustrating the jovian planets relative to their terrestrial cousins. Jupiter orbits at a distance of 5.2 AU from the Sun, outside the asteroid belt but well inside the Kuiper belt.

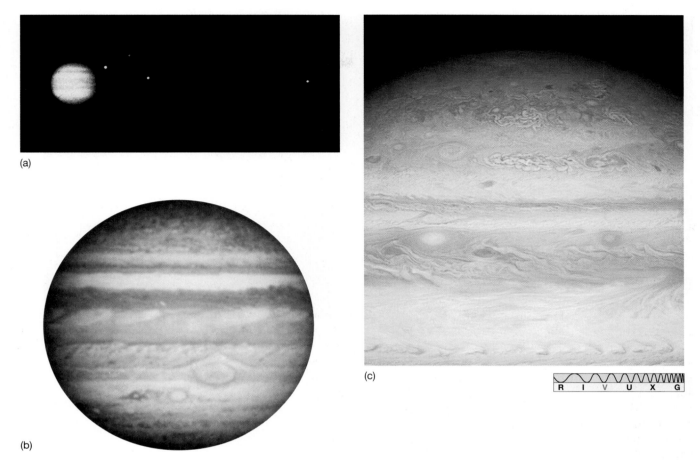

(a)

(b)

(c)

R I V U X G

▲ FIGURE 11.2 **Jupiter** (a) Photograph of Jupiter made through a ground-based telescope, showing the planet and several of its Galilean moons. (b) A *Hubble Space Telescope* image of Jupiter, in true color. Features as small as a few hundred kilometers across are resolved. (c) A *Cassini* spacecraft image of Jupiter, taken while the vehicle was on its way to Saturn, shows intricate clouds of different heights, thicknesses, and chemical composition. (See also the full-page opening photo for this chapter.) *(NASA; AURA)*

primarily of hydrogen and helium. The enormous pressure in the planet's interior due to Jupiter's strong gravity greatly compresses these light gases, whose densities on Earth (at room temperature and sea level) are 0.08 and 0.16 kg/m$^3$, respectively, producing the relatively high average density we observe.

## Rotation Rate

As with other planets, we can attempt to determine Jupiter's rotation rate simply by timing a surface feature as it moves around the planet. However, in the case of Jupiter (and, indeed, all the gaseous outer planets), there is a catch: Jupiter has no solid surface. All we see are the features of clouds in the planet's upper atmosphere. With no solid surface to "tie them down," different parts of Jupiter's atmosphere move independently of one another.

Visual observations and Doppler-shifted spectral lines indicate that the equatorial zones rotate a little faster (with

a period of 9$^h$50$^m$) than the higher latitudes (with a period of 9$^h$55$^m$). Jupiter thus exhibits **differential rotation**—the rotation rate is not constant from one location to another. Differential rotation is not possible in solid objects like the terrestrial planets, but it is normal for fluid bodies such as Jupiter.

Observations of Jupiter's magnetosphere provide a more meaningful measurement of the rotation period. The planet's magnetic field is strong and emits radiation at radio wavelengths as charged particles accelerate in response to Jupiter's magnetic field. Careful studies show a periodicity of 9$^h$55$^m$ at these radio wavelengths. We assume that this measurement matches the rotation of the planet's interior, where the magnetic field arises. ∞ (Sec. 7.5) Thus, Jupiter's interior rotates at the same rate as the clouds at the planet's poles. The equatorial zones rotate more rapidly.

A rotation period of 9$^h$55$^m$ is fast for such a large object. In fact, Jupiter has the fastest rotation rate of any planet in the solar system, and this rapid spin has altered

TUTORIAL  Jupiter—Differential Rotation

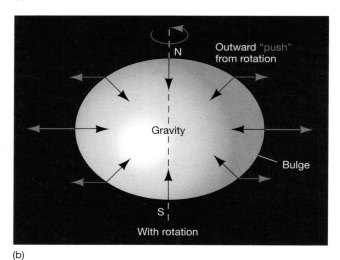

ANIMATION/VIDEO    Jupiter's Rotation

▲ FIGURE 11.3 **Rotational Flattening** All spinning objects tend to develop an equatorial bulge because rotation causes matter to push outward against the inward-pulling gravity. The size of the bulge depends on the mechanical strength of the matter and the rate of rotation. The inward-pointing arrows denote gravity, the outward arrows the "push" due to rotation.

Jupiter's shape. As illustrated in Figure 11.3, a spinning object tends to flatten and develop a bulge around its midsection. ∞ (Sec. 6.7) The more loosely the object's matter is bound together, or the faster it spins, the larger the bulge becomes. In objects like Jupiter, which are made up of gas or loosely packed matter, high spin rates can produce a quite pronounced bulge. Jupiter's equatorial radius (71,500 km) exceeds its polar radius (66,900 km) by about 6.5 percent.*

*Earth also bulges slightly at the equator because of rotation. However, our planet is much more rigid than Jupiter, and the effect is much smaller—the equatorial diameter is only about 40 km larger than the distance from pole to pole, a tiny difference compared with Earth's full diameter of nearly 13,000 km. Relative to its overall dimensions, Earth is smoother and more spherical than a billiard ball.

# PLANETARY DATA

## JUPITER

| Orbital semimajor axis | 5.20 AU |
| | 778.4 million km |
| Orbital eccentricity | 0.048 |
| Perihelion | 4.95 AU |
| | 740.7 million km |
| Aphelion | 5.46 AU |
| | 816.1 million km |
| Mean orbital speed | 13.1 km/s |
| Sidereal orbital period | 11.86 tropical years |
| Synodic orbital period | 398.88 Earth days* |
| Orbital inclination to the ecliptic | 1.31° |
| Greatest angular diameter, as seen from Earth | 50″ |
| Mass | $1.90 \times 10^{27}$ kg |
| | 317.8 (Earth = 1) |
| Equatorial radius | 71,492 km |
| | 11.21 (Earth = 1) |
| Mean density | 1330 kg/m³ |
| | 0.241 (Earth = 1) |
| Surface gravity (at cloud tops) | 24.8 m/s² |
| | 2.53 (Earth = 1) |
| Escape speed | 59.5 km/s |
| Sidereal rotation period | 0.41 Earth day |
| Axial tilt | 3.08° |
| Surface magnetic field | 13.89 (Earth = 1) |
| Magnetic axis tilt relative to rotation axis | 9.6° |
| Surface temperature | 124 K (at cloud tops) |
| Number of moons | 16 (more than 10 km in diameter) |
| | 63 (total) |

*1 Earth (mean solar) day = 24 hours

But there is more to the story of Jupiter's shape. Jupiter's observed equatorial bulge also tells us something important about the planet's deep interior. Careful calculations indicate that Jupiter would be *more* flattened than it actually is if its core were composed of hydrogen and helium alone. To account for the planet's observed shape, we must assume that Jupiter has a dense, compact core, probably of rocky composition, about 5–10 times the mass of Earth. This is one of the few pieces of data we have on Jupiter's internal structure.

### CONCEPT CHECK

✔ How do observations of a planet's magnetosphere allow astronomers to measure the rotation rate of the interior?

## 11.2  The Atmosphere of Jupiter

Jupiter is visually dominated by two features: a series of ever-changing atmospheric bands arranged parallel to the equator and an oval atmospheric blob called the **Great Red Spot**, or, often, just the "Red Spot." The bands of clouds, clearly visible in Figure 11.2, display many colors—pale yellows, light blues, deep browns, drab tans, and vivid reds, among others. Shown in more detail in Figure 11.4, a close-up photograph taken as *Voyager 1* sped past in 1979, the Red Spot is the largest of many features associated with Jupiter's weather. It seems to be a hurricane twice the size of planet Earth that has persisted for hundreds of years, the largest of numerous long-lived storm systems in the planet's atmosphere.

### Atmospheric Composition

Spectroscopic studies of sunlight reflected from Jupiter gave astronomers their first look at the planet's atmospheric composition. Radio, infrared, and ultraviolet observations provided more details later. The most abundant gas is molecular hydrogen ($H_2$, 86.1 percent by number of molecules), followed by helium (He, 13.8 percent). Together, these two gases make up over 99 percent of Jupiter's atmosphere. Small amounts of atmospheric methane ($CH_4$), ammonia ($NH_3$), and water vapor ($H_2O$) are also found. Researchers think that hydrogen and helium in those same proportions make up the bulk of the planet's interior as well.

The abundance of hydrogen and helium on Jupiter is a direct consequence of the planet's strong gravity. Unlike the gravitational pull of the terrestrial planets, the gravity of the much more massive jovian planets is powerful enough to have retained even hydrogen. ∞ (*More Precisely 8-1*) Little, if any, of Jupiter's original atmosphere has escaped since the planet formed 4.6 billion years ago.

### Atmospheric Bands

Astronomers generally describe Jupiter's banded appearance—and, to a lesser extent, the appearance of the other jovian worlds as well—as a series of bright **zones** and dark **belts** crossing the planet. These variations appear to be the result of *convective motion* in the planet's atmosphere. ∞ (Sec. 7.2) *Voyager* sensors indicated that the light-colored zones lie above upward-moving convective currents in Jupiter's atmosphere. The dark belts are regions

10,000 km

R  I  V  U  X  G

◀ FIGURE 11.4  **Jupiter's Red Spot** *Voyager 1* took this photograph of Jupiter's Great Red Spot (upper right) from a distance of about 100,000 km. Resolution is about 100 km. Note the complex turbulence to the left of both the Red Spot and the smaller white oval below it. (For scale, planet Earth has been superposed.) (*NASA*)

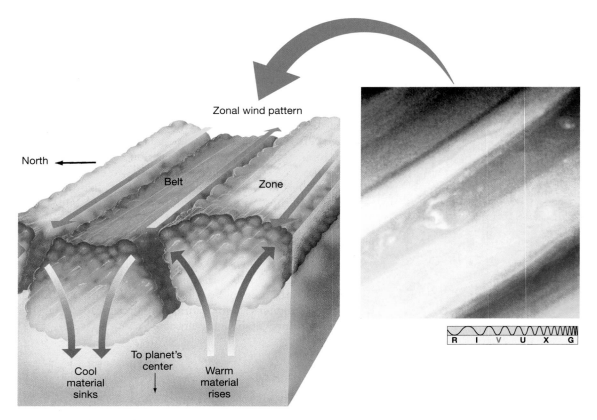

North

Zonal wind pattern

Belt

Zone

Cool
material
sinks

To planet's
center

Warm
material
rises

R  I  V  U  X  G

▲ **FIGURE 11.5** **Jupiter's Convection** The colored bands in Jupiter's atmosphere are associated with vertical convective motion. Upwelling warm gas results in zones of lighter color; the darker bands overlie regions of lower pressure where cooler gas sinks back down into the atmosphere. As on Earth, surface winds tend to blow from high- to low-pressure regions. Jupiter's rapid rotation channels those winds into an east–west flow pattern, as indicated by the three yellow-red arrows drawn atop the belts and zones. The inset is a *Voyager* photo of part of Jupiter's cloud layer, as seen from above, showing the planet's actual banded structure. *(NASA)*

representing the other part of the convection cycle, during which material is generally sinking downward, as illustrated schematically in Figure 11.5.

Thus, because of the upwelling material below them, the zones are regions of *high pressure*. The belts, conversely, are *low-pressure* regions. The belts and zones are Jupiter's equivalents of the familiar high- and low-pressure systems that cause our weather on Earth. A major difference between Jupiter and Earth is that Jupiter's rapid rotation has caused these systems to wrap all the way around the planet, instead of forming localized circulating storms, as on our own world.

Observations made by the *Cassini* mission in 2000 during its Jupiter flyby have challenged this standard view, suggesting instead that upward convection is actually confined to the belts. For now, planetary scientists have no clear resolution to the apparent contradiction between the *Voyager* and the *Cassini* findings.

Underlying the bands is an apparently very stable pattern of eastward and westward wind flow, known as Jupiter's **zonal flow.** Figure 11.5 illustrates how the wind direction alternates between adjacent bands as Jupiter's rotation deflects surface winds into eastward or westward

streams. The interaction between convective motion in Jupiter's atmosphere and the planet's rapid rotation channels the largest convective eddies into the observed zonal pattern. Smaller eddies—like the Red Spot—cause localized irregularities in the zonal flow.

The connection between Jupiter's belts and zones and the zonal flow pattern is evident in Figure 11.6, which shows the wind speed at different planetary latitudes, measured relative to the rotation of the planet's interior (determined from studies of Jupiter's magnetic field). As mentioned earlier, the equatorial regions of the atmosphere rotate faster than the planet; their average flow speed is some 85 m/s, or about 300 km/h, in the easterly direction. The speed of this equatorial flow is quite similar to that of the jet stream on Earth. At higher latitudes, there are alternating regions of westward and eastward flow, roughly symmetric about the equator, with the flow speed generally diminishing toward the poles. Near the poles, where the zonal flow disappears, the band structure vanishes also.

Because of the pressure difference between the two, the zones lie slightly higher in the atmosphere than do the belts. The associated temperature differences (the

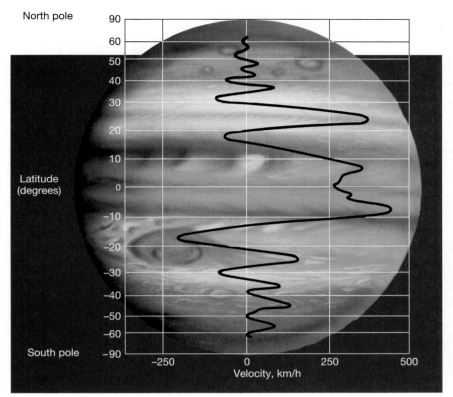

North pole

Latitude (degrees)

South pole

◄ **FIGURE 11.6 Zonal Flow** The wind speed in Jupiter's atmosphere, measured relative to the planet's internal rotation rate. Alternations in wind direction are associated with the atmospheric band structure.

planets, weather on Jupiter is the result of convection in the troposphere, so the clouds, which are associated with planetary weather systems, all lie at negative altitudes in the diagram. Just above the troposphere lies a thin, faint layer of haze created by *photochemical* reactions (reactions involving sunlight) similar to those which cause smog on Earth. The temperature at this level is about 110 K; it increases with altitude as the atmosphere absorbs solar ultraviolet radiation.

Jupiter's clouds are arranged in three main layers. Below the haze, at a depth of about 40 km (shown as −40 km in Figure 11.7), lies a layer of white, wispy clouds made up of ammonia ice. The temperature here is approximately 125–150 K; it increases quite rapidly with increasing depth. A few tens of kilometers below the ammonia clouds, the temperature is a little warmer—over 200 K—and the clouds are probably made up mostly of droplets or crystals of ammonium hydrosulfide, produced by reactions between ammonia and hydrogen sulfide in the planet's atmosphere. At deeper levels in the atmosphere, the ammonium hydrosulfide clouds give way to clouds of water ice or water vapor. This lowest cloud layer, which is not seen in visible-light images of Jupiter, lies some 80 km below the top of the troposphere.

Instead of being white (the color of ammonium hydrosulfide on Earth), Jupiter's middle cloud layer is tawny in color. This is the level at which atmospheric chemistry begins to play a role in determining Jupiter's appearance. Many planetary scientists think that molecules containing the element sulfur, and perhaps even sulfur itself, are important in influencing the cloud colors—particularly the reds, browns, and yellows, all colors associated with sulfur or its compounds. It is also possible that compounds containing the element phosphorus contribute to the coloration.

Deciphering the detailed causes of Jupiter's distinctive colors is a difficult task, however. The cloud chemistry is complex and highly sensitive to small changes in atmospheric conditions, such as pressure and temperature, as well as to chemical composition. The atmosphere is in incessant, churning motion, causing conditions to change from place to place and from hour to hour. In addition, the energy that powers the reactions comes in many different forms: the planet's own interior heat, solar ultraviolet

temperature increases as we descend into the atmosphere, as we will see next) and the resulting differences in chemical reactions are the basic reasons for the different colors of these jovian features. The zones and belts vary in both latitude and intensity during the year, although the general banded pattern remains. The variations are not seasonal in nature: Having a low-eccentricity orbit and a rotation axis almost exactly perpendicular to its orbital plane, Jupiter has no seasons. ∞ (Sec. 1.4) Instead, the annual changes appear to be the result of dynamic motion in the planet's atmosphere.

## Atmospheric Structure and Color

None of the atmospheric gases listed earlier can, by itself, account for Jupiter's observed coloration. For example, frozen ammonia and water vapor would simply produce white clouds, not the many colors actually seen. Scientists suspect that the colors of the clouds are the result of complex chemical processes occurring in the planet's turbulent upper atmosphere, although the details are still not fully understood. When we observe Jupiter's colors, we are actually looking down to many different depths in the planet's atmosphere.

Based on the best available data and mathematical models, Figure 11.7 is a cross-sectional diagram of Jupiter's atmosphere. Since the planet lacks a solid surface to use as a reference level for measuring altitude, the top of the troposphere is conventionally taken to lie at 0 km. As on all

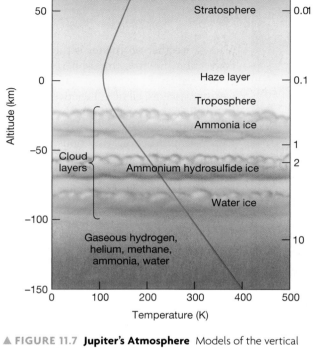

▲ FIGURE 11.8 *Galileo's* **Entry Site** The arrow on this image shows where the *Galileo* atmospheric probe plunged into Jupiter's cloud deck on December 7, 1995. The entry location was in Jupiter's equatorial zone and apparently almost devoid of upper-level clouds. Until its demise, the probe took numerous weather measurements, transmitting those signals to the orbiting mother ship, which then relayed them to Earth. (*NASA*)

▲ FIGURE 11.7 **Jupiter's Atmosphere** Models of the vertical structure of Jupiter's atmosphere suggest that the planet's clouds are arranged in three main layers, each with quite different colors and chemistry. The colors we see in photographs of the planet depend on the cloud cover. The white regions are the tops of the upper ammonia clouds. The yellows, reds, and browns are associated with the second cloud layer, which is composed of ammonium hydrosulfide ice. The lowest (bluish) cloud layer is water ice; however, the overlying layers are sufficiently thick that this level is not seen in visible light. The blue curve shows how Jupiter's atmospheric temperature depends on altitude. (For comparison with Earth, see Figure 7.2.)

radiation, aurorae in the planet's magnetosphere, and lightning discharges within the clouds themselves. All of these factors combine to keep a complete explanation of Jupiter's appearance beyond our present grasp.

The preceding description of Jupiter's atmosphere, based largely on *Voyager* data, was put to the test in December 1995, when the *Galileo* atmospheric probe arrived at the planet. ⚭ (Sec. 6.6) The probe survived for about an hour before being crushed by atmospheric pressure at an altitude of −150 km (i.e., right at the bottom of Figure 11.7). Overall, *Galileo's* findings on wind speed, temperature, and composition were in good agreement with the picture just presented. However, the probe's entry location was in Jupiter's equatorial zone and, as luck would have it, coincided with an atypical "hole" almost devoid of upper-level clouds (see Figure 11.8). The probe measured a temperature of 425 K at 150 km depth—a little higher than indicated in Figure 11.7, but consistent with the craft's

having entered a clearing in Jupiter's cloud decks, where convective heat can more readily rise (and thus be detected). The probe also measured a slightly lower than expected water content, but that, too, may be normal for the hot, windy regions near Jupiter's equator.

The experts were somewhat surprised by the depth to which Jupiter's winds continued. *Galileo's* probe measured high wind speeds throughout its descent into the clouds, not just at the cloud tops, implying that heat deep within the planet, rather than sunlight, drives Jupiter's weather patterns. Finally, complex organic molecules were sought, but not found. Some simple carbon-based molecules, such as ethane ($C_2H_6$), were detected by one of the onboard spectrometers, but nothing suggesting prebiotic compounds (molecules that could combine to form the building blocks of life—see Section 28.1) or bacteria floating in the atmosphere was found. That same instrument also detected traces of phosphine ($PH_3$), which may be a key coloring agent for Jupiter's clouds.

## Weather on Jupiter

In addition to the zonal flow pattern, Jupiter has many "small-scale" weather patterns. The Great Red Spot (Figure 11.4) is a prime example. It was first reported by

## A Cometary Impact

In July 1994 astronomers were granted a rare alternative means of studying Jupiter's atmosphere and interior—the collision of a comet (called Shoemaker–Levy 9, after its discoverers) with the planet!

When it was discovered in March 1993, comet Shoemaker–Levy 9 appeared to have a curious, "squashed" appearance. Higher resolution images (see the first accompanying figure) revealed that the comet was really made up of several pieces, the largest no more than 1 km across. All the pieces were following the same orbit, but they were spread out along the comet's path, like a string of pearls 1 million km long.

How could such an unusual object have originated? Tracing the orbit backward in time, researchers calculated that early in July 1992 the comet had approached within about 100,000 km of Jupiter. They realized that the objects shown in the figure were the fragments produced when a previously "normal" comet was captured by Jupiter and torn apart by its strong gravitational field. The data revealed an even more remarkable fact: On its next approach, roughly a year later, the comet would collide with Jupiter!

Between July 16 and July 22, 1994, fragments from Shoemaker–Levy 9 struck Jupiter's upper atmosphere, plowing into it at a speed of more than 60 km/s and causing a series of enormous explosions. Every major telescope on Earth, the *Hubble Space Telescope*, *Galileo* (which was only 1.5 AU from the planet at the time), and even *Voyager 2* were watching. Each impact created, for a period of a few minutes, a brilliant fireball hundreds

of kilometers across with a temperature of many thousands of kelvins. The largest of the fireballs were bigger than planet Earth. The energy released in each explosion was comparable to a billion terrestrial nuclear detonations, rivaling in violence the prehistoric impact suspected of causing the extinction of the dinosaurs on Earth 65 million years ago (see *Discovery 14-1*). One of the largest pieces of the comet, fragment G, produced the spectacular fireball shown in the second image.

The effects on the planet's atmosphere and the vibrations produced throughout Jupiter's interior were observable for days after the impact. The fallen material from the impacts spread slowly around Jupiter's bands and reached completely around the planet after five months. It took years for all the cometary matter to settle into Jupiter's interior.

As best we can determine, none of the cometary fragments breached the jovian clouds. Only *Galileo* had a direct view of the impacts on the back side of Jupiter, and in every case the explosions seemed to occur high in the atmosphere, above the uppermost cloud layer. Most of the dark material seen in the images is probably pieces of the comet rather than parts of Jupiter. Spectral lines from silicon, magnesium, and iron were detected in the aftermath of the collisions, and the presence of these metals might explain the dark material observed near some of the impact sites (third image). Water vapor was also detected spectroscopically, again apparently from the melted and vaporized comet—the composition of which resembled a "dirty snowball," just as astronomers had long predicted (see Section 14.2).

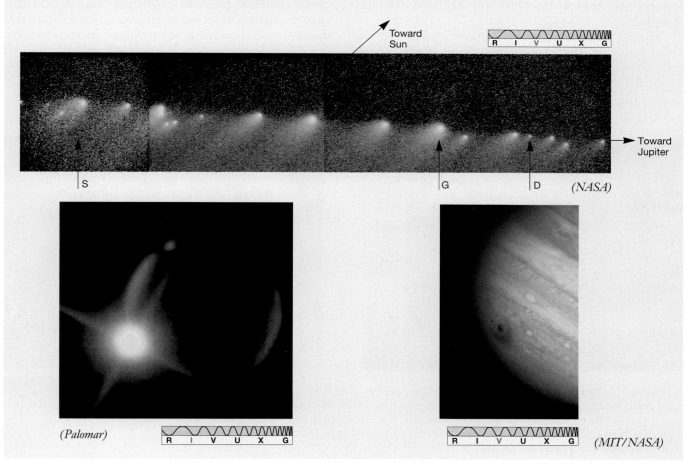

Toward Sun

Toward Jupiter

R I V U X G

S          G          D          (NASA)

(Palomar)          R I V U X G

R I V U X G          (MIT/NASA)

ANIMATION/VIDEO   Comet Impact with Jupiter

British scientist Robert Hooke in the mid-17th century, so we can be reasonably sure that it has existed continuously, in one form or another, for over 300 years. It may well be much older. *Voyager* observations showed the spot to be a region of swirling, circulating winds, rather like a whirlpool or a terrestrial hurricane—a persistent and vast atmospheric storm. The size of the spot varies, although it averages about twice the diameter of Earth. Its present dimensions are roughly 25,000 km by 15,000 km. The spot rotates around Jupiter at a rate similar to that of the planet's interior, perhaps suggesting that the roots of the Great Red Spot lie far below the atmosphere.

The origin of the spot's red color is uncertain, as is its source of energy, although it is generally supposed that the spot is somehow sustained by Jupiter's large-scale atmospheric motion. Repeated observations show that the gas flow around the spot is counterclockwise, with a period of about 6 days. Turbulent eddies form and drift away from its edge. The spot's center, however, remains quite tranquil in appearance, like the eye of a hurricane on Earth. The zonal motion north of the Great Red Spot is westward, whereas that to the south is eastward (see Figure 11.9), supporting the idea that the spot is confined and powered by the zonal flow. However, the details of how it is so confined are still a matter of conjecture. Computer simulations of the complex fluid dynamics of Jupiter's atmosphere are only now beginning to hint at answers.

Storms, which as a rule are much smaller than the Great Red Spot, may be quite common on Jupiter. Spacecraft photographs of the dark side of the planet reveal bright flashes resembling lightning. The *Voyager* mission

discovered many smaller light- and dark-colored spots that are also apparently circulating storm systems. Note the **white ovals** in Figures 11.4 and 11.9, south of the spot. Like the spot itself, they rotate counterclockwise. Their high cloud tops give them their color. These particular white ovals are known to be at least 40 years old. Figure 11.10 shows a **brown oval**, a "hole" in the clouds that allows us to look down into Jupiter's lower atmosphere. For unknown reasons, brown ovals appear only at latitudes around 20°N. Although not as long lived as the Great Red Spot, these systems can persist for many years or even decades.

Continuous monitoring of conditions on the outer planets has recently yielded important insights into the formation and evolution of large storm systems on the jovian worlds. In the late 1990s astronomers noted with interest the collision and merger of three relatively small white ovals in Jupiter's atmosphere (Figure 11.11a). For several years the resultant larger system remained a white oval, but in early 2006 it changed from white to brown to red, becoming in a matter of months a smaller version of the Great Red Spot! Figure 11.11(b) shows a *Hubble Space Telescope* image of the new red spot.

Scientists speculate that the red coloration of the Great Red Spot may be due to that storm's enormous size and strength, which lifts its cloud tops high above the surrounding clouds, where solar ultraviolet radiation causes chemical reactions producing the color. In that case, the reddening of the smaller spot may indicate that that storm is intensifying, and perhaps may even rival the Great Red Spot some day. Merger and growth may well be the mechanism by which large storms on the jovian planets form and strengthen.

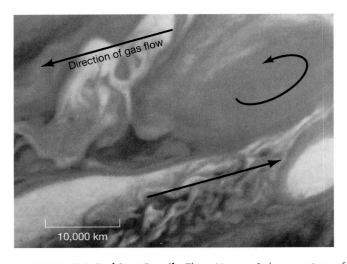

▲ FIGURE 11.9 **Red Spot Details** These *Voyager 2* close-up views of the Great Red Spot, taken 4 hours apart, show clearly the turbulent flow around its edges. The general direction of motion of the gas north of (above) the spot is westward (to the left), whereas gas south of the spot flows east. The spot itself rotates counterclockwise, suggesting that it is being "rolled" between the two oppositely directed flows. The colors have been exaggerated somewhat to enhance the contrast. (*NASA*)

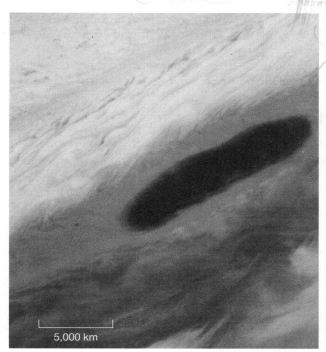

5,000 km

R  I  V  U  X  G

▲ FIGURE 11.10 **Brown Oval** This brown oval in Jupiter's northern hemisphere is actually a break in the upper cloud layer, allowing us to see deeper into the atmosphere, where the clouds are brown. The oval's length is approximately equal to Earth's diameter. *(NASA)*

Despite these many mysteries, we can offer at least a partial explanation for the longevity of storm systems on Jupiter. On Earth, a large storm, such as a hurricane, forms over the ocean and may survive for many days, but it dies quickly once it encounters land. Earth's continental landmasses disrupt the flow patterns that sustain the storm. Jupiter has no continents, so once a storm becomes established and reaches a size at which other storm systems cannot destroy it, apparently little affects it. The larger the system, the longer its lifetime.

CONCEPT CHECK

✔ List some similarities and differences between Jupiter's belts, zones, and spots, on the one hand, and weather systems on Earth, on the other.

## 11.3 Internal Structure

Much of our knowledge of Jupiter's interior comes from theoretical modeling. Indeed, apart from data gained following the collision of a comet with Jupiter in 1994 (see *Discovery 11-1*), we have very little direct evidence of the planet's internal properties. Planetary scientists use all available bulk data on the planet—mass, radius,

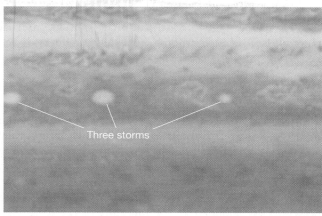

Three storms

(a)

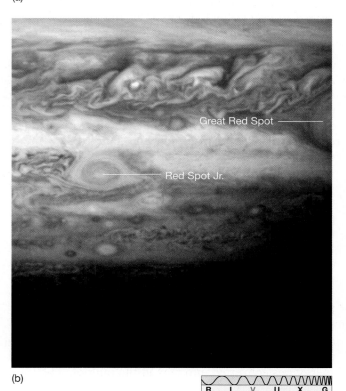

Great Red Spot

Red Spot Jr.

(b)

R  I  V  U  X  G

▲ FIGURE 11.11 **Red Spot Junior** (a) Between 1997 and 2000, astronomers watched as three white ovals in Jupiter's southern hemisphere merged to form a single large storm. Each oval, captured here by the *Cassini* spacecraft cameras, is about half the size of Earth. (b) In early 2006 the white oval turned red, producing a second red spot! The color change may indicate that the storm is intensifying. *(NASA)*

composition, rotation, temperature, etc.—to construct a model of the interior that agrees with observations. Modeling is an integral part of the scientific method, and our statements about Jupiter's structure are really statements about the model that best fits the observed facts. ∞ (Sec. 1.2) However, because the planet consists largely of hydrogen and helium—two simple gases whose physics we think we understand well—we can be fairly confident that Jupiter's internal structure is now understood.

## An Internal Energy Source

On the basis of Jupiter's distance from the Sun, astronomers had expected to find the temperature of the cloud tops to be around 105 K. At that temperature, they reasoned, Jupiter would radiate back into space exactly the same amount of energy as it received from the Sun. When radio and infrared observations were first made of the planet, however, astronomers found that its blackbody spectrum corresponded to a temperature of 125 K instead. Subsequent measurements, including those made by *Voyager* and *Galileo*, have verified that finding.

Although a difference of 20 K may seem small, recall from Chapter 3 that the energy emitted by a planet grows as the *fourth* power of the surface temperature (in Jupiter's case, the temperature of the cloud tops). ∞ (Sec. 3.4) A planet at 125 K therefore radiates $(125/105)^4$, or about twice as much energy as a planet at 105 K radiates. Put another way, Jupiter actually emits about twice as much energy as it receives from the Sun. Thus, unlike any of the terrestrial planets, *Jupiter must have its own internal source of heat.*

What is responsible for Jupiter's extra energy? It is not the decay of radioactive elements within the planet. That process must be occurring, as in Earth, but estimates of the total amount of energy released into Jupiter's interior are far below the levels needed to account for the temperature we measure. ∞ (Sec. 7.3) Nor is it nuclear fusion, the process that generates energy in the Sun. The temperature in Jupiter's interior, high as it is, is still far too low for that (see *Discovery 11-2*). Instead, astronomers theorize that the source of Jupiter's excess energy is the slow escape of gravitational energy released during the planet's formation. As the planet took shape, some of its gravitational energy was converted into heat in the interior. That heat

is still slowly leaking out through the planet's heavy atmospheric blanket, resulting in the excess emission we observe.

Despite the huge amounts of energy involved—Jupiter emits about $4 \times 10^{17}$ watts more energy than it receives from the Sun—the loss is slight compared with the planet's total energy. On the basis of the planet's mass and temperature, as well as the rate at which thermal energy is leaving the planet, astronomers calculate that the average temperature of the interior of Jupiter decreases by only about a millionth of a kelvin per year. ∞ (*More Precisely 3-1*)

## Jupiter's Deep Interior

Jupiter's clouds, with their complex chemistry, are probably less than 200 km thick. Below them, the temperature and pressure steadily increase as the atmosphere becomes the "interior" of the planet.

Both the temperature and the density of Jupiter's atmosphere increase with depth below the cloud cover. However, no "surface" of any kind exists anywhere inside. Instead, Jupiter's atmosphere just becomes denser and denser because of the pressure of the overlying layers. At a depth of a few thousand kilometers, the gas makes a gradual transition into the liquid state (see Figure 11.12). By a depth of about 20,000 km, the pressure is about 3 million times greater than atmospheric pressure on Earth. Under those conditions, the hot liquid hydrogen is compressed so much that it undergoes another transition, this time to a "metallic" state with properties in many ways similar to those of a liquid metal. Of particular importance for Jupiter's magnetic field (see Section 11.4) is that this metallic hydrogen is an excellent conductor of electricity.

As mentioned earlier, Jupiter's observed flattening requires that there be a relatively small (i.e., relatively small compared with the size of Jupiter), dense core at its center. On the basis of *Voyager* data, scientists once thought that the core might contain as much as 20 Earth masses of material. However, following *Galileo*'s arrival, it now appears that the core's mass could be as low as 5 Earth masses and perhaps even less. The precise

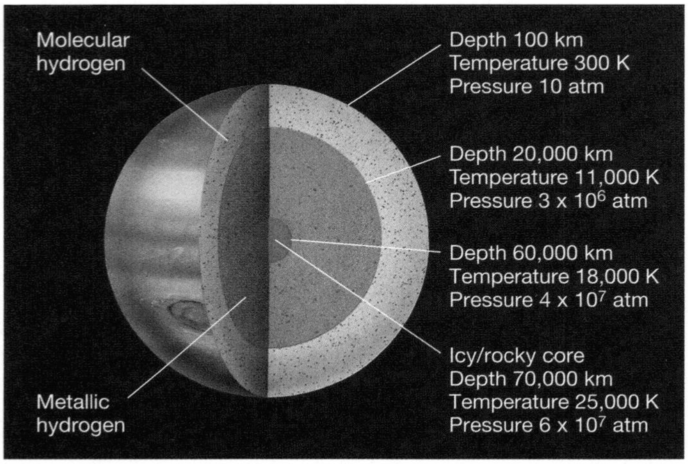

Molecular hydrogen

Metallic hydrogen

Depth 100 km
Temperature 300 K
Pressure 10 atm

Depth 20,000 km
Temperature 11,000 K
Pressure $3 \times 10^6$ atm

Depth 60,000 km
Temperature 18,000 K
Pressure $4 \times 10^7$ atm

Icy/rocky core
Depth 70,000 km
Temperature 25,000 K
Pressure $6 \times 10^7$ atm

◀ FIGURE 11.12 **Jupiter's Interior** Jupiter's internal structure, as deduced from *Voyager* measurements and theoretical modeling. The outer radius represents the top of the cloud layers, some 70,000 km from the planet's center. The density and temperature increase with depth, and the atmosphere gradually liquefies at a depth of a few thousand kilometers. Below a depth of 20,000 km, the hydrogen behaves like a liquid metal. At the center of the planet lies a large rocky core, somewhat terrestrial in composition, but much larger than any of the inner planets. Although the values are uncertain, the temperature and pressure at the center are probably about 25,000 K and 60 million (Earth) atmospheres, respectively.

# DISCOVERY 11-2

## Almost a Star?

Jupiter has a starlike composition—predominantly hydrogen and helium, with a trace of heavier elements. Did Jupiter ever come close to becoming a star itself? Might the solar system have formed as a double-star system? Probably not. Unlike a star, Jupiter is cold. Its central temperature is far too low to ignite the nuclear fires that power our Sun (see Section 16.6). Jupiter's mass would have to increase 80-fold before its central temperature would rise to the point where nuclear reactions could begin, converting Jupiter into a small, dim star.

Even so, it is interesting to note that, although Jupiter's present-day energy output is very small (by solar standards, at least), it must have been much greater in the distant past, while the planet was still contracting rapidly toward its present size. For a brief period—perhaps a few hundred million years—Jupiter might actually have been as bright as a faint star, although its brightness never came within a factor of 100 of the Sun's. Still, seen from Earth at that time, Jupiter would have been about 100 times brighter than the Moon!

What might have happened had our solar system formed as a double-star system? Conceivably, had Jupiter been massive enough, its radiation might have produced severe temperature fluctuations on all the planets, perhaps to the point of making life on Earth impossible. Even if Jupiter's brightness were too low to cause us any problems, its gravitational pull (which would be 1/12 that of the Sun if its mass were 80 times its present value) might have made the establishment of stable, roughly circular planetary orbits in the inner solar system an improbable event, again to the detriment of life on Earth.

Curiously, in recent years astronomers have come to realize that, had Jupiter been too *small*, that also could have adversely affected the chances for life on our planet! As we will see in Chapter 15, Jupiter played a crucial role in clearing debris from the outer solar system during and after the period when the planets formed. ∞ (Sec. 6.7) Had that not occurred, the meteoritic bombardment of our planet might have been too severe and too extended for complex life ever to have evolved. ∞ (Sec. 8.5) Many stars near the Sun are now known to have Jupiter-sized planets orbiting them. It seems that the size of the "Jupiter," or second-largest body, in a newborn planetary system may be a critical factor in determining the likelihood of the appearance of life there.

composition of the core is unknown, but planetary scientists think that it contains much denser materials than the rest of the planet.

Current best estimates indicate that the core consists of "rocky" materials, similar to those found on the terrestrial worlds. (Note that the term *rocky* here refers to the *chemical composition* of the core, not to its physical state. At the high temperatures and pressures found deep in the jovian interiors, the core material bears little resemblance to rocks found on Earth's surface.) In fact, it now appears that all four jovian planets contain similarly large rocky cores and that the formation of such a large "terrestrial" planetary core may be a necessary stage in the process of building up a gas giant (see Section 15.2).

Because of the enormous pressure at the center of Jupiter—approximately 60 million times that on Earth's surface, or 12 times that at Earth's center—the core must be compressed to a very high density (perhaps twice the density of Earth's core). The jovian core is probably not much more than 20,000 km in diameter (still big enough for Earth to fit inside, with plenty of room left over), and the central temperature may be as high as 25,000 K.

## CONCEPT CHECK

✔ How have astronomers determined the properties of Jupiter's core?

## 11.4 Jupiter's Magnetosphere

For decades, ground-based radio telescopes monitored radiation leaking from Jupiter's magnetosphere, but only when the *Pioneer* and *Voyager* spacecraft reconnoitered the planet in the mid-1970s did astronomers realize the full extent of its magnetic field. The *Galileo* probe spent many years orbiting within Jupiter's magnetosphere, returning a wealth of detailed information about its structure.

Jupiter, it turns out, is surrounded by a vast sea of energetic charged particles, mostly electrons and protons, somewhat similar to Earth's Van Allen belts, but much, much larger. The radio radiation detected on Earth is emitted when these particles are accelerated to very high speeds—close to the speed of light—by Jupiter's powerful magnetic field. This radiation is several thousand times more intense than that produced by Earth's magnetic field. The particles present a serious hazard to manned and unmanned space vehicles alike. Sensitive electronic equipment (not to mention even more sensitive human bodies) requires special protective shielding to operate for long in this hostile environment. *Galileo* was not expected to survive as long as it did.

Direct measurements from spacecraft show Jupiter's magnetosphere to be almost 30 million km across, roughly a million times more voluminous than Earth's magnetosphere and far larger than the entire Sun. As with Earth's, the size and shape of Jupiter's magnetosphere are determined by the interaction between the planet's magnetic field and the solar wind. Jupiter's magnetosphere has a long tail extending away from the Sun at least as far as Saturn's orbit (over 4 AU farther out from the Sun), as sketched in Figure 11.13. However, on the sunward side, the *magnetopause*—the boundary of Jupiter's magnetic influence on the solar wind—lies only 3 million km from the planet. Near Jupiter's surface, the magnetic field channels particles from the magnetosphere into the upper atmosphere, forming aurorae vastly larger and more energetic than those observed on Earth (Figure 11.14). ∞ (Sec. 7.5)

The outer magnetosphere of Jupiter appears to be quite unstable, sometimes deflating in response to "gusts" in the solar wind and then reexpanding as the wind subsides. In the inner magnetosphere, Jupiter's rapid rotation has forced most of the charged particles into a flat *current sheet*, lying on the planet's magnetic equator, quite unlike the Van Allen belts surrounding Earth. ∞ (Sec. 7.5) The portion of the magnetosphere

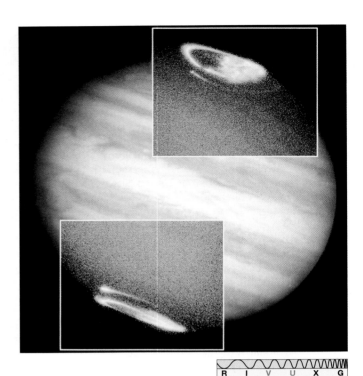

R I V U X G

▲ FIGURE 11.14 **Aurorae on Jupiter** Aurorae on Jupiter, as seen by the *Hubble Space Telescope*. The main image was taken in visible (true-color) light, but the two insets at the poles were taken in the ultraviolet part of the spectrum. The oval-shaped aurorae, extending hundreds of kilometers above Jupiter's surface, result from charged particles escaping the jovian magnetosphere and colliding with the atmosphere, causing the gas to glow. *(NASA)*

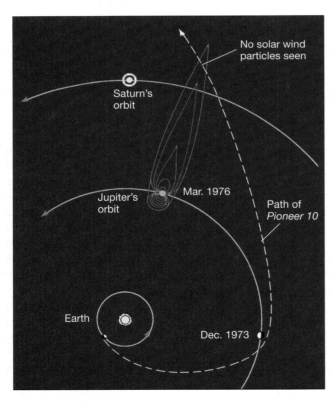

▲ FIGURE 11.13 *Pioneer 10* **Mission** The *Pioneer 10* spacecraft (a forerunner of the *Voyager* missions) did not detect any solar particles while moving far behind Jupiter in 1976. Accordingly, as sketched here, Jupiter's magnetosphere apparently extends beyond the orbit of Saturn.

close to Jupiter is sketched in Figure 11.15. Notice that the planet's magnetic axis is not exactly aligned with its rotation axis, but is inclined to it at an angle of approximately 10°. Jupiter's magnetic field happens to be oriented opposite Earth's, with field lines running from north to south, rather than south to north as in the case of our own planet (see Figure 7.19).

Both ground- and space-based observations of the radiation emitted from Jupiter's magnetosphere imply that the *intrinsic* strength of the planet's magnetic field is nearly 20,000 times greater than Earth's. The existence of such a strong field further supports our theoretical model of Jupiter's internal structure. The conducting liquid interior that is thought to make up most of the planet should combine with Jupiter's rapid rotation to produce a large dynamo effect and a strong magnetic field, just as are observed. ∞ (Sec. 7.5)

## CONCEPT CHECK

✔ Why is Jupiter's magnetosphere so much larger than Earth's?

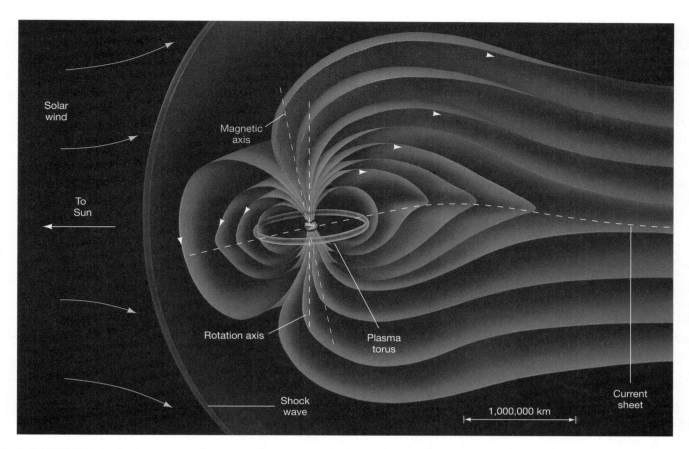

Solar
wind

Magnetic
axis

To
Sun

Rotation axis

Plasma
torus

Shock
wave

1,000,000 km

Current
sheet

▲ FIGURE 11.15 **Jupiter's Magnetosphere** Jupiter's inner magnetosphere is characterized by a flat current sheet consisting of charged particles squeezed into the magnetic equatorial plane by the planet's rapid rotation. The plasma torus, a ring of charged particles associated with the moon Io, is discussed in Section 11.5.

## 11.5 The Moons of Jupiter

As of early 2007, Jupiter's official satellite count stands at 63. Table 11.1 presents some properties of the 16 largest members of Jupiter's moon system—those with diameters of 10 km or more. The 47 small bodies not in the table are also the most recently detected of Jupiter's moons; all have been found since 1999 by systematic surveys (made from Earth) of the space around the giant planet. As discussed in *Discovery 11-3*, their sizes, orbits, and sheer growing numbers are causing some astronomers to reconsider just what the definition of *moon* should be.

The four largest satellites—the Galilean moons—are each comparable in size to Earth's Moon. ∞ (Sec. 2.4) Moving outward from Jupiter, the four are named Io, Europa, Ganymede, and Callisto, after the mythical attendants of the Roman god Jupiter. They move in nearly circular orbits about their parent planet. When the *Voyager 1* spacecraft passed close to the Galilean moons in 1979, it sent some remarkably detailed photographs back to Earth, allowing planetary scientists to discern fine surface features on each moon. ∞ (Sec. 6.6) More recently, in the

late 1990s, the *Galileo* mission expanded our knowledge of these small, but complex, worlds still further. We will consider the Galilean satellites in more detail momentarily.

Within the orbit of Io lie four small satellites, all but one discovered by *Voyager* cameras. The largest of the four, Amalthea, is less than 300 km across and is irregularly shaped. E. E. Barnard discovered it in 1892. Amalthea orbits at a distance of 181,000 km from Jupiter's center—only 110,000 km above the cloud tops. Its rotation, like that of most of Jupiter's satellites, is synchronous with its orbit because of Jupiter's strong tidal field. Amalthea rotates once per orbital period—every 11.7 hours.

Beyond the Galilean moons lie eight more small satellites, all discovered in the 20th century, but before the *Voyager* missions. They fall into two groups of four moons each. The moons in the inner group move in eccentric, inclined orbits, about 11 million km from the planet. The outer four moons lie about 22 million km from Jupiter. Their orbits, too, are fairly eccentric, but *retrograde*, moving in a sense opposite that of all the other moons' orbits (and Jupiter's rotation). It is very likely that each group represents a single body that was captured by Jupiter's strong

**TABLE 11.1   The Major Moons of Jupiter***

| Name | Distance from Jupiter (km) | (planetary radii) | Orbital Period (days) | Size (longest diameter, km) | Mass** (Earth Moon masses) | Density (kg/m³) | (g/cm³) |
|---|---|---|---|---|---|---|---|
| Metis | 128,000 | 1.79 | 0.29 | 40 | | | |
| Adrastea | 129,000 | 1.80 | 0.30 | 20 | | | |
| Amalthea | 181,000 | 2.54 | 0.50 | 260 | | | |
| Thebe | 222,000 | 3.10 | 0.67 | 100 | | | |
| Io | 422,000 | 5.90 | 1.77 | 3640 | 1.22 | 3500 | 3.5 |
| Europa | 671,000 | 9.38 | 3.55 | 3130 | 0.65 | 3000 | 3.0 |
| Ganymede | 1,070,000 | 15.0 | 7.15 | 5270 | 2.02 | 1900 | 1.9 |
| Callisto | 1,880,000 | 26.3 | 16.7 | 4800 | 1.46 | 1900 | 1.9 |
| Leda | 11,100,000 | 155 | 239 | 10 | | | |
| Himalia | 11,500,000 | 161 | 251 | 170 | | | |
| Lysithea | 11,700,000 | 164 | 259 | 24 | | | |
| Elara | 11,700,000 | 164 | 260 | 80 | | | |
| Ananke | 21,200,000 | 297 | −631[†] | 20 | | | |
| Carme | 22,600,000 | 316 | −692[†] | 30 | | | |
| Pasiphae | 23,500,000 | 329 | −735[†] | 36 | | | |
| Sinope | 23,700,000 | 332 | −758[†] | 28 | | | |

*Moons larger than 10 km in diameter. This table does not include the 45 recently discovered small moons described in the text. All of these small moons move on inclined, eccentric, mainly retrograde orbits some 10–25 million km from the planet.

**Mass of Earth's Moon = $7.4 \times 10^{22}$ kg = $3.9 \times 10^{-5}$ Jupiter masses.

[†]Retrograde orbit.

gravitational field long after the planet and its larger moons originally formed. Both bodies subsequently broke up, either during or after the capture, resulting in the two families of similar orbits we see today. The masses, and hence the densities, of these small worlds are unknown. However, their appearance and sizes suggest compositions more like asteroids or comets than their larger Galilean companions.

## The Galilean Moons: a Model of the Inner Solar System

Jupiter's Galilean moons have several interesting parallels with the terrestrial planets. Their orbits are direct (i.e., in the same sense as Jupiter's rotation), are roughly circular, and lie close to Jupiter's equatorial plane. Figure 11.16 shows the moons' orbits, with Jupiter to scale. Also shown in the figure are some representative orbits of the *Galileo* spacecraft, illustrating schematically how the probe used gravity assists from Galilean moons to maneuver through Jupiter's system of satellites. ∞ (*Discovery 6-1*)

The four Galilean moons range in size from slightly smaller than Earth's Moon (Europa) to slightly larger than Mercury (Ganymede). Figure 11.17 is a *Voyager 1* image

of Io and Europa, with Jupiter providing a spectacular backdrop. Figure 11.18 shows the four Galilean moons to scale.

The similarity to the inner solar system continues with the fact that the moons' densities decrease with increasing distance from Jupiter. ∞ (Sec. 6.4) Largely on the basis of detailed measurements made by *Galileo* of the moons' gravitational fields, together with mathematical models of the interiors, researchers have built up fairly detailed pictures of each moon's composition and internal structure (Figure 11.18). The innermost two Galilean moons, Io and Europa, have thick rocky mantles, possibly similar to the crusts of the terrestrial planets, surrounding iron–iron sulfide cores. Io's core accounts for about half that moon's total radius. Europa has a water–ice outer shell between 100 and 200 km thick. The two outer moons, Ganymede and Callisto, are clearly deficient in rocky materials. Lighter materials, such as water and ice, may account for as much as half of their total mass. Ganymede appears to have a relatively small metallic core topped by a rocky mantle and a thick icy outer shell. Callisto seems to be a largely undifferentiated mixture of rock and ice.

Many astronomers think that the formation of Jupiter and the Galilean satellites may in fact have mimicked, on

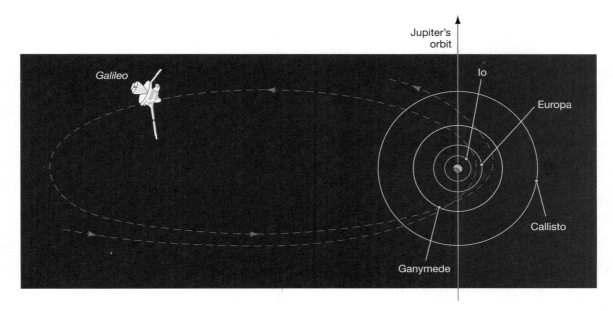

▲ FIGURE 11.16 *Galileo* at Jupiter The orbits of Jupiter's Galilean moons, to scale, as seen from above the planet's north pole. The orbits are prograde and circular and lie in the planet's equatorial plane. Also drawn are some orbits of the *Galileo* probe (dashed red), showing it swinging around the interior moons of Jupiter. Each time it reconnoitered a moon, the probe would enter a slightly different orbit, allowing it to see more detail during hundreds of close encounters throughout the mission. Here, overtaking Europa on the "inside track," *Galileo* is slowed by that moon's gravity, placing the spacecraft in a lower orbit. Subsequent encounters with this and other moons would further modify *Galileo*'s trajectory, allowing the vehicle to visit all of the inner moons in the Jupiter system repeatedly.

a small scale, the formation of the Sun and the inner planets. For that reason, studies of the Galilean moon system could provide us with valuable insight into the processes that created our own world. We will return to this parallel in Chapter 15. So interested were mission planners in

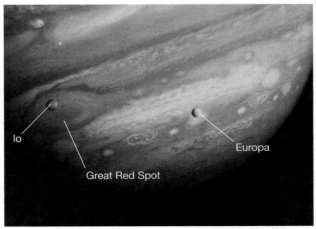

▲ FIGURE 11.17 Jupiter, Up Close *Voyager 1* took this photo of Jupiter with ruddy Io on the left and pearllike Europa toward the right. Note the scale of objects here: Both Io and Europa are comparable in size to our Moon, and the Red Spot is roughly twice as big as Earth. *(NASA)*

learning more about the Galilean moon system that the already highly successful *Galileo* mission was extended for 6 more years to allow for even more detailed study, particularly of Europa. ∞ (Sec. 6.6) The Galilean moons were scrutinized at resolutions as fine as a few meters during numerous extremely close passages by the spacecraft.

Not all the properties of the Galilean moons find analogs in the inner solar system, however. For example, all four Galilean satellites are locked into states of synchronous rotation by Jupiter's strong tidal field, so they all keep one face permanently pointing toward their parent planet. By contrast, of the terrestrial planets, only Mercury is strongly influenced by the Sun's tidal force, and even its orbit is not synchronous. ∞ (Sec. 8.4) Finally, inspection of Table 11.1 shows a remarkable coincidence in the orbital periods of the three inner Galilean moons: Their periods are almost exactly in the ratio 1:2:4 (and the fourth moon Callisto is not too far from being the "8" in the sequence). This configuration may be the result of a complex, but poorly understood, three-body (or perhaps even four-body) resonance in the Galilean moon system, something not found among the terrestrial worlds.

## Io: The Most Active Moon

Io, the densest of the Galilean moons, is the most geologically active object in the entire solar system. Its mass and radius are fairly similar to those of Earth's Moon,

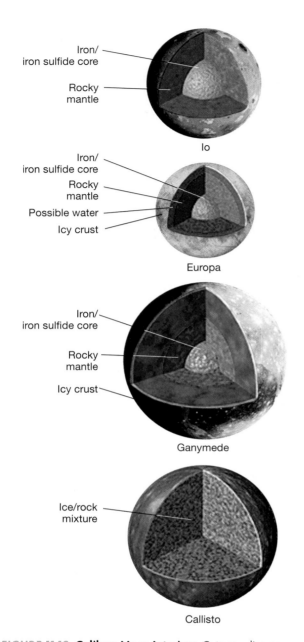

Io

Iron/iron sulfide core
Rocky mantle

Europa

Iron/iron sulfide core
Rocky mantle
Possible water
Icy crust

Ganymede

Iron/iron sulfide core
Rocky mantle
Icy crust

Callisto

Ice/rock mixture

▲ FIGURE 11.18 **Galilean Moon Interiors** Cutaway diagrams showing the interior structure of the four Galilean satellites. Moving outward from Io to Callisto, we see that the moons' densities steadily decrease as the composition shifts from rocky mantles and metallic cores in Io and Europa, to a thick icy crust and smaller core in Ganymede, to an almost uniform rock and ice mix in Callisto. Both Ganymede and Europa are thought to have layers of liquid water beneath their icy surfaces.

but there the resemblance ends. Shown in Figure 11.19, Io's surface is a collage of reds, yellows, and blackish browns—resembling a giant pizza in the minds of some startled *Voyager* scientists. As the spacecraft sped past Io, it made an outstanding discovery: Io has active volcanoes! *Voyager 1* photographed eight erupting volcanoes. Six were still erupting when *Voyager 2* passed by 4 months later.

By the time *Galileo* arrived in 1995, several of the eruptions observed by *Voyager* had subsided. However, many new ones were seen—in fact, *Galileo* found that Io's surface features can change significantly in as little as a few weeks. In all, more than 80 active volcanoes have been identified on Io. The largest, called Loki (on the far side of Figure 11.20), is larger than the state of Maryland and emits more energy than all of Earth's volcanoes combined.

The top right inset in Figure 11.20 shows a volcano called Prometheus ejecting matter at speeds of up to 2 km/s to an altitude of about 150 km. These high-speed gases are quite unlike the (relatively) sluggish ooze that emanates from Earth's volcanoes. According to *Galileo*'s instruments, lava temperatures on Io generally range from 650 to 900 K, with the higher end of the range implying that at least some of the volcanism is similar to that found on Earth. However, temperatures as high as 2000 K—far hotter than any earthly volcano—have been measured at some locations. Mission scientists speculate that these "superhot" volcanoes may be similar to those which occurred on Earth more than 3 billion years ago.

The orange color immediately surrounding the volcanoes most likely results from sulfur compounds in the ejected material. In stark contrast to the surfaces of the other Galilean moons, Io's surface is neither cratered nor streaked (the circular features visible in Figures 11.19 and 11.20 are volcanoes), but is instead exceptionally smooth, mostly varying in altitude by less than about 1 km, although some volcanoes are several kilometers high. The smoothness is apparently the result of molten matter that constantly fills in any "dents and cracks." This remarkable moon has the youngest surface of any known object in the solar system. Io also has a thin, temporary atmosphere made up primarily of sulfur dioxide, presumably the result of gases ejected by volcanic activity.

Io's volcanism has a major effect on Jupiter's magnetosphere. All the Galilean moons orbit within the magnetosphere and play some part in modifying its properties, but Io's influence is particularly marked. Although many of the charged particles in Jupiter's magnetosphere come from the solar wind, there is strong evidence that Io's volcanism is the primary source of heavy ions in the inner regions. Jupiter's magnetic field continually sweeps past Io, gathering up the particles its volcanoes spew into space and accelerating them to high speed. The result is the *Io plasma torus* (Figure 11.21; see also Figure 11.15), a doughnut-shaped region of energetic heavy ions that follows Io's orbital track, completely encircling Jupiter. (A plasma is a gas that has been heated to such high temperatures that all its atoms are ionized. A few neutral atoms have also been observed in the Io plasma torus.)

The plasma torus is quite easily detectable from Earth, but before *Voyager* its origin was unclear. *Galileo* made detailed studies of the plasma's dynamic and rapidly varying magnetic field. Spectroscopic analysis shows that sulfur is indeed one of the torus's major constituents, strongly

(a)

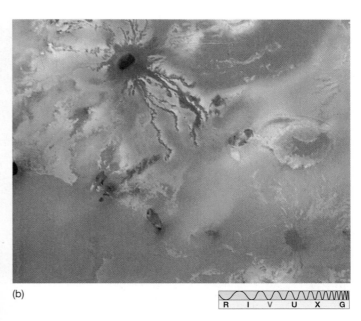

(b)

R I V U X G

▲ FIGURE 11.19 **Io** Jupiter's innermost moon, Io, is quite different in character from the other three Galilean satellites. Its surface is kept smooth and brightly colored by the moon's constant volcanism. The resolution of the *Galileo* photograph in (a) is about 7 km. In the more detailed *Voyager* image (b), features as small as 2 km across can be seen. (*NASA*)

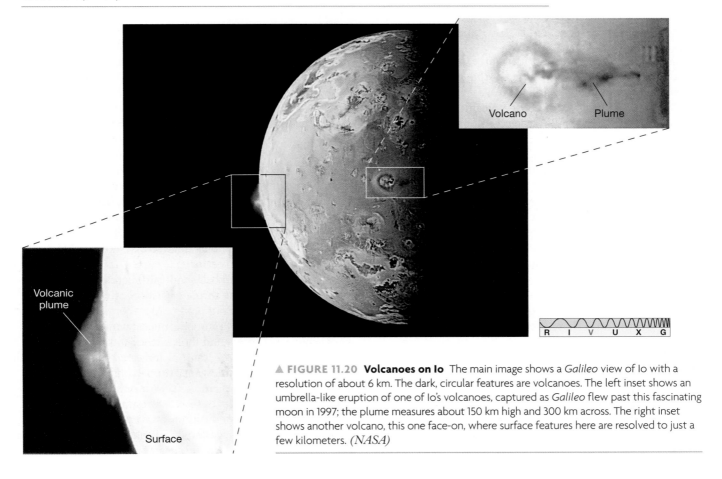

Volcano      Plume

Volcanic plume

Surface

R I V U X G

▲ FIGURE 11.20 **Volcanoes on Io** The main image shows a *Galileo* view of Io with a resolution of about 6 km. The dark, circular features are volcanoes. The left inset shows an umbrella-like eruption of one of Io's volcanoes, captured as *Galileo* flew past this fascinating moon in 1997; the plume measures about 150 km high and 300 km across. The right inset shows another volcano, this one face-on, where surface features here are resolved to just a few kilometers. (*NASA*)

ANIMATION/VIDEO    Io Cutaway

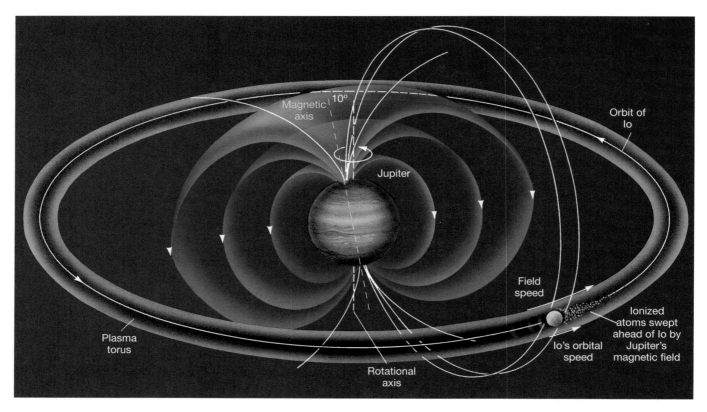

▲ **FIGURE 11.21 Io Plasma Torus** The torus is the result of material being ejected from Io's volcanoes and swept up by Jupiter's rapidly rotating magnetic field. Spectroscopic analysis indicates that the torus is made mainly of sodium and sulfur atoms and ions.

implicating Io's volcanoes as its source. As a hazard to spacecraft—manned or unmanned—the plasma torus is formidable, with lethal radiation levels.

What causes such astounding volcanic activity on Io? The moon is far too small to have geological activity like that on Earth. Io should be long dead, like our own Moon. At one time, some scientists suggested that Jupiter's magnetosphere might be the culprit: Perhaps the (then-unknown) processes creating the plasma torus were somehow also stressing the moon. We now know that this is not the case. The real source of Io's energy is *gravity*—Jupiter's gravity. Io orbits very close to Jupiter—only 422,000 km, or 5.9 Jupiter radii, from the center of the planet. As a result, Jupiter's huge gravitational field exerts strong tidal forces on the moon. If Io were the only satellite in the Jupiter system, it would long ago have come into a state of synchronous rotation with the planet, just as our own Moon has with Earth, for the reasons discussed in Chapter 8. ∞ (Sec. 8.4) In that case, Io would move in a perfectly circular orbit, with one face permanently turned toward Jupiter, and the tidal bulge would be stationary with respect to the moon.

But Io is not alone. As it orbits, it is constantly tugged by the gravity of its nearest large neighbor, Europa. The tugs are small and not enough to cause any great tidal effect, but they are sufficient to make Io's orbit slightly noncircular, preventing the moon from settling into a precisely synchronous state. The reason for this effect is exactly the same as in the case of Mercury, also as discussed in Chapter 8. ∞ (Sec. 8.4) In a noncircular orbit, the moon's speed varies from place to place as it revolves around its planet, but its rate of rotation on its axis remains constant. Thus, it cannot keep one face always turned toward Jupiter. Instead, as seen from Jupiter, Io rocks or "wobbles" slightly from side to side as it moves. The large (100-m) tidal bulge, however, always points directly toward Jupiter, so it moves back and forth across Io's surface as the moon wobbles. These conflicting forces result in enormous tidal stresses that continually flex and squeeze Io's interior.

Just as the repeated back-and-forth bending of a piece of wire can produce heat through friction, the ever-changing distortion of Io's interior constantly energizes the moon. This generation of large amounts of heat within Io ultimately causes huge jets of gas and molten rock to squirt out of the moon's surface. *Galileo's* sensors indicated extremely high temperatures in the outflowing material. It is likely that much of Io's interior is soft or molten, with only a relatively thin solid crust overlying it. Researchers estimate that the total amount of heat generated within Io as a result of tidal flexing is about 100 million megawatts. This phenomenon makes Io one of the most fascinating objects in our solar system.

# DISCOVERY 11-3

## Jupiter's Many Moons

The moons of Jupiter listed in Table 11.1 were all discovered long before *Galileo* reached the planet in 1995. The spacecraft focused on studying Jupiter and its inner moons and discovered no new moons itself, so the number of known satellites of Jupiter stayed fixed at 16. Since 1999, however, the number has exploded, reaching 63 at the time of writing and very likely to increase further. How did that happen?

Remarkably, this host of new moons was discovered not via close-up spacecraft exploration, but rather by painstaking long-distance observations made from Earth, using large ground-based telescopes and specially designed instruments and software to scan large areas of the sky for very faint objects. Steadily improving technology means that small bodies once far too faint for Earth-based telescopes to see are now being detected and cataloged almost routinely. The first pair of figures shows a typical discovery image sequence, wherein a faint and eminently undistinguished-looking moon (marked by the arrow) is revealed by its motion relative to the background stars. The two images were taken 40 minutes apart in 2003 by the Canada–France–Hawaii telescope on Mauna Kea.

These latest additions to Jupiter's family of satellites have characteristics comparable to those of the outermost eight moons listed in Table 11.1. All are very small—less than 10 km, and in some cases as little as 1 km, in diameter—and their masses are unknown. Their orbits are all moderately eccentric, with semimajor axes between 10 and 25 million km and inclined at 15–40° to Jupiter's equatorial plane. Most of the orbits are retrograde. Very likely, these newcomers have the same origin as do the eight outlying satellites just mentioned—space "junk" captured by Jupiter, probably long ago. The second figure shows the orbits of all the known moons of Jupiter, to scale. The orbits of the Galilean satellites can be seen at the center. All the other orbits are those of irregular, small moons that may very well not be original members of the Jupiter system.

While the origin of Jupiter's moons is an interesting puzzle and may have much to teach us about the early solar system, the rapidly escalating number of tiny bodies orbiting Jupiter and the other giant planets is beginning to pose a classification problem for astronomers. At the moment, any object—no matter how small—orbiting a planet is called a moon. This simple criterion works well when the list of known moons consists of a relatively small number of relatively large objects, hundreds or thousands of kilometers across. However, as observations improve and more and more kilometer-sized jovian moons are discovered, some researchers have begun suggesting that perhaps a size limit should be imposed to separate "real" moons (like the Galilean satellites) from captured cometary material. In this way, we might distinguish objects that formed along with Jupiter from material that formed elsewhere and that now—by chance—happens to orbit Jupiter.

For the present, though, no such limit exists, so we can expect to hear reports of a steadily increasing number of "moons" in the coming years.

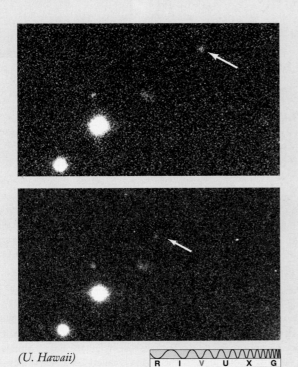

(U. Hawaii)

R I V U X G

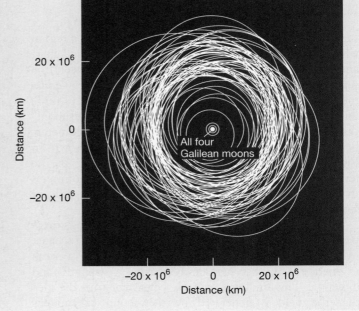

All four
Galilean moons

20 x 10⁶

0

−20 x 10⁶

Distance (km)

−20 x 10⁶   0   20 x 10⁶
Distance (km)

## Europa: Liquid Water Locked in Ice

Europa (Figure 11.22) is a world very different from Io. Lying outside Io's orbit, 671,000 km (9.4 Jupiter radii) from Jupiter, Europa showed relatively few craters on its surface in images taken by *Voyager*, suggesting geologic youth—perhaps just a few million years. Recent activity has erased any scars of ancient meteoritic impacts. The dark areas are rocky deposits that may have come from the moon's interior or that may have been swept up by Europa as it moved in its orbit. Europa's surface also displays a vast network of lines crisscrossing bright, clear fields of water ice. Some of these linear "bands," or fractures, extend halfway around the satellite and resemble, in some ways, the pressure ridges that develop in ice floes on Earth's polar oceans.

On the basis of *Voyager* images, planetary scientists had theorized that Europa might be completely covered by an ocean of liquid water with its top frozen at the low temperatures prevailing so far from the Sun. In this view, the cracks in the surface are attributed to the tidal influence of Jupiter and the gravitational pulls of the other Galilean satellites, although these forces are considerably weaker than those powering Io's violent volcanic activity. However, other researchers had contended that Europa's fractured surface was instead related to some form of tectonic activity—involving ice rather than rock. High-resolution *Galileo* observations now appear to strongly support the former idea. Figure 11.22(c) is a *Galileo* image of this weird moon, showing what look like icebergs—flat chunks of ice that have been broken apart, moved several kilometers, and reassembled, perhaps by the action of water currents below. Mission scientists estimate that Europa's surface ice may be several kilometers thick and that there may be a 100-km-deep liquid ocean below it.

Other detailed images of Europa's surface lend further support to this hypothesis. Figure 11.22(d) shows a region where Europa's icy crust appears to have been pulled apart and new material has filled in the gaps between the separating ice sheets. Elsewhere on the surface, *Galileo* found what appeared to be the icy equivalent of lava flows on Earth—regions where water apparently "erupted" through the surface and flowed for many kilometers before solidifying. The smooth "trenches" shown in Figure 11.22(d) strongly suggest local flooding of the terrain. The scarcity of impact craters on Europa implies that the processes responsible for these features did not stop long ago. Rather, they must be ongoing.

Further evidence comes from studies of Europa's magnetic field. Measurements made by *Galileo* on repeated flybys of the moon revealed that Europa has a weak magnetic field that constantly changes strength and direction. This finding is entirely consistent with the idea that the field is generated by the action of Jupiter's magnetism on a shell of electrically conducting fluid about 100 km below Europa's surface—in other words, the salty

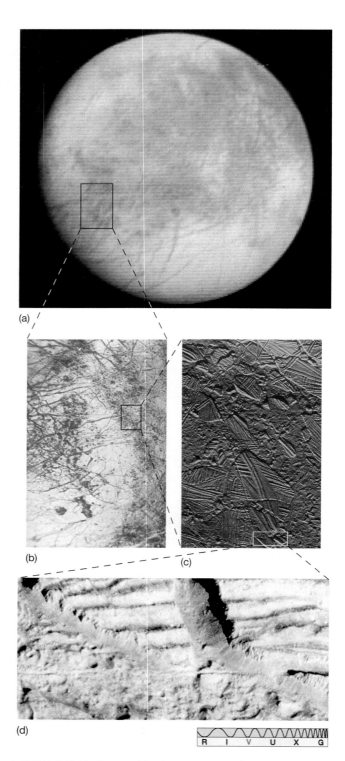

▲ FIGURE 11.22 **Europa** (a) *Voyager 1* mosaic of Europa. Resolution is about 5 km. (b) Europa's icy surface is only lightly cratered, indicating that some ongoing process must be obliterating impact craters soon after they form. The origin of the cracks crisscrossing the surface is uncertain. (c) At 5-m resolution, this image from the *Galileo* spacecraft shows a smooth yet tangled surface resembling the huge ice floes that cover Earth's polar regions. This region is called Conamara Chaos. (d) This detailed *Galileo* image shows "pulled apart" terrain that suggests liquid water upwelling from the interior and freezing, filling in the gaps between separating surface ice sheets. (*NASA*)

layer of liquid water suggested by the surface observations. *Galileo*'s observations convinced quite a few skeptical scientists of the reality of Europa's ocean.

The likelihood that Europa has an extensive layer of liquid water below its surface ice opens up many interesting avenues of speculation about the possible development of life there. In the rest of the solar system, only Earth has liquid water on or near its surface, and most scientists agree that water played a key role in the appearance of life on Earth (see Chapter 28). Europa may well contain more liquid water than exists on our entire planet! Of course, the existence of water does not *necessarily* imply the emergence of life: Europa, even in its liquid ocean, is still a hostile environment compared with Earth. Nevertheless, the possibility—even a remote one—of life on Europa was an important motivating factor in the decision to extend the *Galileo* mission for 6 more years.

## Ganymede and Callisto: Fraternal Twins

The two outermost Galilean moons are Ganymede (at 1.1 million km, or 15 planetary radii, from the center of Jupiter) and Callisto (at 1.9 million km, or 26 Jupiter radii). The density of each is only about 2000 kg/m$^3$, suggesting that they harbor substantial amounts of ice throughout and are not just covered by thin icy or snowy surfaces. Ganymede, shown in Figure 11.23, is the largest moon in the solar system, exceeding not only Earth's Moon, but also the planet Mercury in size. It has many impact craters on its surface and patterns of dark and light markings that are reminiscent of the highlands and maria on Earth's own Moon. In fact, Ganymede's history has many parallels with that of the Moon (with water ice replacing lunar rock). The large, dark region clearly visible in Figure 11.23(a) is called Galileo Regio.

As on the inner planets, we can estimate ages on Ganymede by counting craters. We learn that the darker regions, such as Galileo Regio, are the oldest parts of Ganymede's surface. These regions are the original icy surface of the moon, just as the ancient highlands on our own Moon are its original crust. The surface darkens with age as micrometeorite dust slowly covers it. The light-colored parts of Ganymede are much less heavily cratered, so they must be younger. They are Ganymede's "maria" and probably formed in a manner similar to the way that maria on the Moon were created. ∞ (Sec. 8.5) Intense meteoritic bombardment caused liquid water—Ganymede's counterpart to our own Moon's molten lava—to upwell from the interior and flood the affected regions before solidifying.

Not all of Ganymede's surface features follow the lunar analogy. Ganymede has a system of grooves and ridges (shown in Figure 11.23c) that may have resulted from crustal tectonic motion, much as Earth's surface undergoes mountain building and faulting at plate boundaries. ∞ (Sec. 7.4) Ganymede's large size indicates that its original radioactivity probably helped heat and differentiate its interior, after which

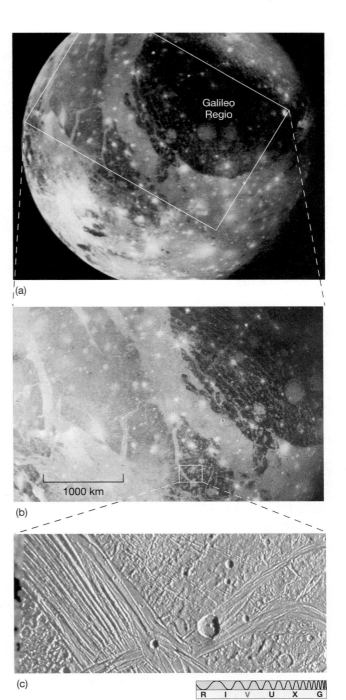

▲ **FIGURE 11.23 Ganymede** (a) and (b) *Voyager 2* images of Ganymede. The dark regions are the oldest parts of the moon's surface and probably represent its original icy crust. The largest dark region visible here, called Galileo Regio, spans some 3200 km. The lighter, younger regions are the result of flooding and freezing that occurred within a billion years or so of Ganymede's formation. The light-colored spots are recent impact craters. (c) Grooved terrain on Ganymede may have been caused by a process similar to plate tectonics on Earth. The area shown in this *Galileo* image spans about 50 km and reveals a multitude of ever-smaller ridges, valleys, and craters, right down to the resolution limit of the spacecraft's camera (about 300 m, three times the length of a football field). The image suggests erosion of some sort, possibly even caused by water. (*NASA*)

the moon cooled and the crust cracked. Ganymede seems to have had some early plate tectonic activity, but the process stopped about 3 billion years ago, when the cooling crust became too thick for such activity to continue. The *Galileo* data suggest that the surface of Ganymede may be older than was previously thought. With the improved resolution of that spacecraft's images (Figure 11.23c), some regions thought to have been smooth, and hence young, are now seen to be heavily splintered by fractures and thus probably very old.

In 1996, *Galileo* detected a weak magnetosphere surrounding Ganymede, making it the first moon in the solar system on which a magnetic field had been observed and implying that Ganymede has a modest iron-rich core. Ganymede's magnetic field is about 1 percent that of Earth. In December 2000, the magnetometer team reported fluctuations in the field strength similar to those near Europa, suggesting that Ganymede, too, may have liquid or perhaps "slushy" water under its surface. Recent observations of surface formations similar to those attributed to flowing water "lava" on Europa appear to support this view.

Callisto, shown in Figure 11.24, is in many ways similar in appearance to Ganymede, although it has more craters and fewer fault lines. Its most obvious feature is a huge series of concentric ridges surrounding each of two large basins. The larger of the two, on Callisto's Jupiter-facing side, is named Valhalla and measures some 3000 km across. It is

clearly visible in the figure. The ridges resemble the ripples made as a stone hits water, but on Callisto they probably resulted from a cataclysmic impact with an asteroid or comet. The upthrust ice was partially melted, but it resolidified quickly, before the ripples had a chance to subside.

Today, both the ridges and the rest of the crust are frigid ice and show no obvious signs of geological activity (such as the grooved terrain on Ganymede). Apparently, Callisto froze before plate tectonic or other activity could start. The density of impact craters on the Valhalla basin indicates that it formed long ago, perhaps 4 billion years in the past. Yet, even on this frozen world, there are hints from *Galileo*'s magnetometers that there might be a thin layer of water, or more likely slush, deep below the surface.

Ganymede's internal differentiation indicates that the moon was largely molten at some time in the past; Callisto is undifferentiated and hence apparently never melted. Researchers are uncertain why two such similar bodies should have evolved so differently. Complicating matters further is Ganymede's magnetic field and possible subsurface liquid water, which suggest that the moon's interior may still be relatively warm. If that is so, then Ganymede's heating and differentiation must have happened quite recently—less than a billion years ago, based on recent estimates of how rapidly the moon's heat escapes into space.

Scientists have no clear explanation for how Ganymede could have evolved in this manner. Heating by

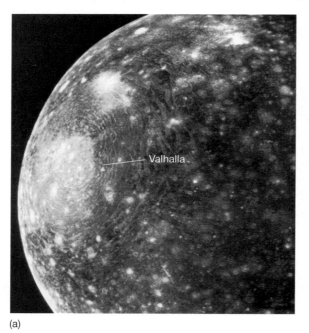

Valhalla

(a)

100 km

R  I  V  U  X  G

(b)

▲ FIGURE 11.24  **Callisto**  (a) Callisto, the outermost Galilean moon of Jupiter, is similar to Ganymede in overall composition, but is more heavily cratered. The large series of concentric ridges visible at left is known as Valhalla. Extending nearly 1500 km from the basin center, the ridges formed when "ripples" from a large meteoritic impact froze before they could disperse completely. Resolution in this *Voyager 2* image is about 10 km. (b) This higher-resolution *Galileo* image of Callisto's equatorial region, about 300 × 200 km in area, displays more clearly its heavy cratering. *(NASA)*

meteoritic bombardment ended too early, and radioactivity probably could not have provided enough energy at such a late time. ∞ (Sec. 7.3) Some astronomers speculate that interactions among the inner moons, possibly related to the 1:2:4 near resonance mentioned earlier, may have been responsible. These interactions might have caused Ganymede's orbit to change significantly about a billion years ago, and prior tidal heating by Jupiter could have helped melt the moon's interior.

---

### CONCEPT CHECKS

✔ What is the ultimate source of all the activity observed on Jupiter's Galilean satellites?

✔ Why are scientists so interested in the existence of liquid water on Europa and Ganymede?

---

## 11.6  Jupiter's Ring

Yet another remarkable finding of the 1979 *Voyager* missions was the discovery of a faint ring of matter encircling Jupiter in the plane of the planet's equator (see Figure 11.25). The ring lies roughly 50,000 km above the top cloud layer of the planet, inside the orbit of the innermost moon. A thin sheet of material may extend all the way down to Jupiter's cloud tops, but most of the ring is confined within a region only a few thousand kilometers across. The outer edge of the ring is quite sharply defined. In the direction perpendicular to the equatorial plane, the ring is only a few tens of kilometers

thick. The small, dark particles that make up the ring may be fragments chipped off by meteoritic impact from two small moons—Metis and Adrastea, discovered by *Voyager*—that lie very close to the ring itself.

Despite differences in appearance and structure, Jupiter's ring can perhaps be best understood by studying the most famous planetary ring system—that of Saturn—so we will postpone further discussion of ring properties until the next chapter.

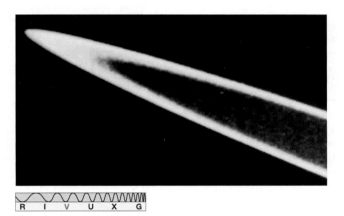

R  I  V  U  X  G

▲ **FIGURE 11.25  Jupiter's Ring** Jupiter's faint ring, as photographed (nearly edge-on) by *Voyager 2*. Made of dark fragments of rock and dust possibly chipped off the innermost moons by meteorites, the ring was unknown before the two *Voyager* spacecraft arrived at the planet. It lies in Jupiter's equatorial plane, only 50,000 km above the cloud tops. (*NASA*)

---

# CHAPTER REVIEW

## Summary

❶ Jupiter is the largest planet in the solar system. Its mass is more than twice the mass of all the other planets combined, although still only 1/1000 the mass of the Sun. Composed primarily of hydrogen and helium, Jupiter rotates rapidly, producing a pronounced equatorial bulge. The planet's flattened shape allows astronomers to infer the presence of a large rocky core in its interior. Unlike the terrestrial planets, Jupiter displays **differential rotation (p. 281)**: the planet has no solid surface, and the rotation rate varies from place to place in the atmosphere. Measurements of radio emission from Jupiter's magnetosphere provide a measure of the planet's interior rotation rate. Jupiter has many moons and a faint, dark ring extending down to the planet's cloud tops.

❷ Jupiter's atmosphere consists of three main cloud layers, forming bands of bright **zones (p. 283)** and darker **belts (p. 283)** crossing the planet parallel to the equator. The bands are the result of convection in Jupiter's interior and the planet's rapid rotation. The lighter zones are the tops of upwelling, warm

currents, and the darker belts are cooler regions where gas is sinking. Underlying them is a stable pattern of eastward or westward wind flow called a **zonal flow (p. 284)**. The wind direction alternates as we move north or south away from the equator. The colors we see are the result of chemical reactions, fueled by the planet's interior heat, solar ultraviolet radiation, auroral phenomena, and lightning. The main weather pattern on Jupiter is the **Great Red Spot (p. 283)**, an Earth-sized hurricane that has been raging for at least three centuries. Other, smaller weather systems—**white ovals (p. 288), brown ovals (p. 288),** and a recently formed new red spot—are also observed and can persist for decades.

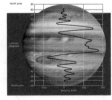

❸ Jupiter's atmosphere becomes hotter and denser with depth, eventually becoming liquid. Interior pressures are so high that the hydrogen is "metallic" in nature near the

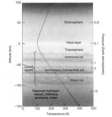

center. The planet has a large "terrestrial" core about 10 times the mass of Earth. Jupiter radiates about twice as much energy into space as it receives from the Sun. The source of this energy is most likely heat released into the planet's interior when Jupiter formed 4.6 billion years ago, now slowly leaking from the surface.

④ The magnetosphere of Jupiter is about a million times more voluminous than Earth's magnetosphere, and the planet has a long magnetic "tail" extending away from the Sun to at least the distance of Saturn's orbit. Energetic particles spiral around magnetic field lines, accelerated by Jupiter's rotating magnetic field, producing intense radio radiation.

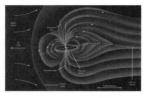

⑤ Jupiter and its system of moons resemble a small solar system. Sixty-three moons have been discovered so far. The outermost eight resemble asteroids and have retrograde orbits, suggesting that they may have been captured by Jupiter's gravity long after the planets and largest moons formed. Jupiter's four major moons are called the **Galilean moons (p. 280),** after their discoverer, Galileo Galilei. Their densities decrease with increasing distance from the planet. The innermost Galilean moon, Io, has active volcanoes and a smooth surface. Europa has a cracked, icy surface that probably conceals an ocean of liquid water, making the moon an interesting

candidate for life in the solar system. Ganymede and Callisto have ancient, heavily cratered surfaces. Ganymede, the largest moon in the solar system, shows evidence of past geological activity, but now appears to be unmoving rock and ice, although recent evidence suggests that it, too, may have subsurface liquid water. Callisto apparently froze before tectonic activity could start there.

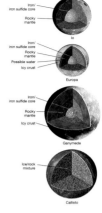

⑥ Io's volcanism is powered by the constant flexing of the moon by Jupiter's tidal forces. As Io orbits Jupiter, the moon "wobbles" because of the gravitational pull of Europa. The ever-changing distortion of its interior energizes Io, and geyserlike volcanoes keep its surface smooth with constant eruptions. Europa's fields of ice are nearly devoid of craters, but have extensive large-scale fractures, due most likely to the tidal influence of Jupiter, the gravitational

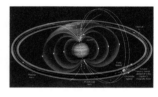

effects of the other Galilean satellites, and small-scale "chaos" caused by the action of the underlying ocean.

## Review and Discussion

1. In what sense does our solar system consist of only two important objects?

2. What is differential rotation, and how is it observed on Jupiter?

3. What does Jupiter's degree of flattening tell us about the planet's interior?

4. Describe some of the ways in which the *Voyager* mission changed our perception of Jupiter.

5. Describe some of the ways in which the *Galileo* mission changed our perception of Jupiter.

6. What is the Great Red Spot? What is known about the source of its energy?

7. What is the cause of the colors in Jupiter's atmosphere?

8. Why has Jupiter retained most of its original atmosphere?

9. Explain the theory that accounts for Jupiter's internal heat source.

10. What is Jupiter thought to be like beneath its clouds? Why do we think this?

11. What is responsible for Jupiter's enormous magnetic field?

12. In what sense are Jupiter and its moons like a miniature solar system?

13. How does the density of the Galilean moons vary with increasing distance from Jupiter? Is there a trend to this variation? If so, why?

14. What is the cause of Io's volcanic activity?

15. What evidence do we have for liquid water below Europa's surface?

16. Why do scientists think that Ganymede's interior may have been heated as recently as a billion years ago?

17. How does the amount of cratering vary among the Galilean moons? Does it depend on their location? If so, why?

18. Why is there speculation that the Galilean moon Europa might be an abode for life?

19. What might be the consequences of the discovery of life on Europa?

20. Water is relatively uncommon among the terrestrial planets. Is it common among the moons of Jupiter?

## Conceptual Self-Test: True or False/Multiple Choice

1. The solid surface of Jupiter lies just below the cloud layers that are visible from Earth.

2. Storms in Jupiter's atmosphere are generally much longer lived than storms on Earth.

3. The element helium plays an important role in producing the colors in Jupiter's atmosphere.

4. Jupiter is noticeably flattened due to its rapid rotation.

5. Jupiter emits more energy than it receives from the Sun.

6. Although often referred to as a gaseous planet, Jupiter is mostly liquid in its interior.

7. Because of Jupiter's strong magnetic field, the Galilean satellites rotate synchronously with their orbits around the planet.

8. Io is the only moon in the solar system with active volcanoes.

9. Scientists speculate that Europa may have liquid water below its frozen surface.

10. Ganymede shows evidence of ancient plate tectonics.

11. Compared with Earth's orbit, the orbit of Jupiter is approximately (a) half as large; (b) twice as large; (c) 5 times larger; (d) 10 times larger.

12. Compared with Earth's density, the density of Jupiter is (a) much greater; (b) much less; (c) about the same.

13. The main constituent of Jupiter's atmosphere is (a) hydrogen; (b) helium; (c) ammonia; (d) carbon dioxide.

14. Figure 11.6 ("Zonal Flow") shows that the most rapid westerly wind flows on Jupiter occur at (a) northern midlatitudes; (b) equatorial latitudes; (c) southern mid-latitudes; (d) polar latitudes.

15. According to Figure 11.7 ("Jupiter's Atmosphere"), if ammonia and ammonium hydrosulfide ice were transparent to visible light, Jupiter would appear (a) bluish; (b) red; (c) tawny brown; (d) exactly as it does now.

16. Jupiter's rocky core is (a) smaller than Earth's Moon; (b) comparable in size to Mars; (c) almost the same size as Venus; (d) larger than Earth.

17. Jupiter's magnetosphere extends far into space, stretching about (a) 1 AU; (b) 5 AU; (c) 10 AU; (d) 20 AU beyond the planet.

18. The moon of Jupiter most similar in size to Earth's Moon is (a) Io; (b) Europa; (c) Ganymede; (d) Callisto.

19. Io's surface appears very smooth because it (a) is continually resurfaced by volcanic activity; (b) is covered with ice; (c) has been shielded by Jupiter from meteorite impacts; (d) is liquid.

20. The Galilean moons of Jupiter are sometimes described as a miniature inner solar system because (a) there are the same number of Galilean moons as there are terrestrial planets; (b) the moons have generally "terrestrial" composition; (c) the moons' densities decrease with increasing distance from Jupiter; (d) the moons all move on circular, synchronous orbits.

## Problems

 *Algorithmic versions of these Problems are available in the Practice Problems module of the Companion Website.*
*The number of dots preceding each Problem indicates its approximate level of difficulty.*

1. • How does the force of gravity at Jupiter's cloud tops compare with the force of gravity at Earth's surface?

2. • What are the angular diameters of the orbits of Jupiter's four Galilean satellites, as seen from Earth at closest approach (assuming, for definiteness, that opposition occurs near perihelion)?

3. • Using the figures given in the text, calculate how long it takes Jupiter's equatorial winds to circle the planet, relative to the interior.

4. •• Calculate the rotational speed (in km/s) of a point on Jupiter's equator, at the level of the cloud tops. Compare it with the orbital speed just above the cloud tops.

5. • Given Jupiter's age and current atmospheric temperature, what is the smallest possible mass the planet could have and still have retained its hydrogen atmosphere?  ∞ *(More Precisely 8-1)*

6. • If Jupiter had been just massive enough to fuse hydrogen (see *Discovery 11-1*), calculate what the planet's gravitational force on Earth would have been at closest approach, relative to the gravitational pull of the Sun. Assume circular orbits. Also, estimate what the magnitude of the planet's tidal effect on our planet would have been, again relative to that of the Sun.

7. • Calculate the ratio of Jupiter's mass to the total mass of the Galilean moons. Compare your answer with the ratio of Earth's mass to that of the Moon.

8. ••• At what distance would a satellite orbit Jupiter in the time taken for Jupiter to rotate exactly once, so that the satellite would appear stationary above the planet?  ∞ *(More Precisely 2-3)*

9. • Illustrate the 1:2:4 "resonance" mentioned in the text by drawing a diagram showing the locations of Io, Europa, and Ganymede at various times over the course of one Ganymede orbit. Show the moons' locations at intervals of Io's orbital period.

10. •• Estimate the strength of Jupiter's gravitational tidal acceleration on Europa, and compare it with the moon's own surface gravity.  ∞ *(Sec. 7.6)* Compare your answer with the strength of Earth's gravitational tidal force on the Moon, relative to the Moon's own surface gravity.

11. •• Estimate the strength of Jupiter's gravitational tidal force on Europa, relative to the moon's own surface gravity.

12. •• Estimate the strength of Jupiter's gravitational tidal force on Ganymede, relative to the moon's own surface gravity.

13. •• Calculate the strength of Europa's gravitational pull on Io at closest approach, relative to Jupiter's gravitational attraction on Io.

14. •• What are the surface gravity and escape speed of Europa?

15. •• Compare the apparent sizes of the Galilean moons, as seen from Jupiter's cloud tops, with the angular diameter of the Sun as seen from Jupiter. Would you expect ever to see a total solar eclipse from Jupiter's cloud tops?

*The Companion Website at www.aw-bc.com/chaisson provides algorithmically generated versions of each chapter's Problems, along with additional quizzes, an Animations & Videos gallery, an Images gallery, an interactive Glossary, and a full eBook.*

CHAPTER

# 12

The number of known moons in the solar system increased rapidly during the late 1990s. Better telescopes enabled astronomers to take a more complete inventory of objects in our cosmic neighborhood. But the more closely we examine these alien worlds, the more complex they seem to be. In this puzzling image taken by the Cassini spacecraft in 2005, Saturn's tiny ice-covered moon Enceladus (only 500 km in diameter) shows evidence of a youthful terrain in the south where craters are mostly absent. The long blue streaks (about 1 km wide) are likely fractures in the ice through which gas escapes to form a thin but real atmosphere. (JPL) ▶

# SATURN
## Spectacular Rings and Mysterious Moons

S aturn is one of the most beautiful and enchanting of all astronomical objects. Its rings are a breathtaking sight when viewed through even a small telescope, and they are probably the planet's best-known feature. Aside from its famous rings, though, Saturn presents us with another example of a jovian planet, allowing us to explore further the properties of these giant gaseous worlds.

Saturn is in many ways similar to its larger neighbor, Jupiter, in terms of composition, size, and structure, although its lower mass and greater distance from the Sun mean that Saturn's atmospheric colors are far less pronounced than those on Jupiter, and its weather patterns, although just as violent, are much harder to see. On the other hand, Saturn's rings and moons differ greatly from those of Jupiter. The comparison between these worlds provides us with invaluable insight into the structure and evolution of all the jovian planets.

## LEARNING GOALS

*Studying this chapter will enable you to*

1  Summarize the orbital and physical properties of Saturn, and compare them with those of Jupiter.
2  Describe the composition and structure of Saturn's atmosphere and interior.
3  Explain why Saturn's internal heat source and magnetosphere differ from those of Jupiter.
4  Describe the structure and composition of Saturn's rings.
5  Define the Roche limit, and explain its relevance to the origin of Saturn's rings.
6  Summarize the general characteristics of Titan, and discuss the chemical processes in its atmosphere.
7  Discuss some of the orbital and geological properties of Saturn's smaller moons.

 Visit www.aw-bc.com/chaisson for additional images, animations, quizzes, and eBook for this chapter.

## 12.1   Orbital and Physical Properties

Saturn was the outermost planet known to ancient astronomers. Named after the father of Jupiter in Roman mythology, Saturn orbits the Sun at almost twice the distance of Jupiter. The planet's sidereal orbital period of 29.4 Earth years was the longest natural unit of time known to the ancient world. The Saturn Data box on p. 310 presents orbital and physical data on the planet.

### Overall Properties

At opposition, when most of the Earth- (or Earth-orbit) based photographs of Saturn were taken, the planet is at its brightest and can lie within 8 AU of Earth. However, its great distance from the Sun still makes Saturn considerably fainter than either Jupiter or Mars. Saturn ranks behind Jupiter, the inner planets, and several of the brightest stars in the sky in terms of apparent brightness.

As with Jupiter, Saturn's many moons allowed an accurate determination of the planet's mass long before the arrival of spacecraft from Earth. Saturn's mass is $5.7 \times 10^{26}$ kg, or 95 times the mass of Earth. ∞ (Sec. 6.1) Although less than one-third the mass of Jupiter, Saturn is still an enormous body, at least by terrestrial standards.

From Saturn's distance and angular size, the planet's radius—and hence its average density—quickly follow. ∞ (*More Precisely 6-1*) Saturn's equatorial radius is 60,000 km, or 9.5 Earth radii. The average density is 700 kg/m³—less than the density of water (1000 kg/m³). Here we have a planet that would float in the ocean—if Earth had one big enough! Saturn's low average density indicates that, like Jupiter, it is composed primarily of hydrogen and helium. Saturn's lower mass, however, results in lower interior pressure, so these gases are less compressed than in Jupiter's case.

### Rotation Rate

Saturn, like Jupiter, rotates very rapidly and differentially. ∞ (Sec. 11.1) The atmospheric rotation period, determined by tracking weather features observed in the planet's atmosphere, is $10^{h}14^{m}$ at the equator and roughly $10^{h}40^{m}$ at higher latitudes. However, the rotational period of the interior, obtained from *Cassini* measurements of magnetospheric outbursts, which should better trace the rotation of the planet's core, is $10^{h}46^{m}$, significantly longer than the surface values.

Curiously, the *Cassini* measurement is about 6 minutes longer than the corresponding result obtained (using similar means) by *Voyager* more than 20 years earlier. Scientists are uncertain as to the cause of this difference, although they do not think that Saturn's actual rotation rate has changed by as much as 1 percent during this relatively short time. Rather, it seems that the planet's magnetic field is not as good an indicator of interior

rotation as was previously thought. Saturn's axis of rotation is significantly tilted with respect to the planet's orbital plane—27°, similar to that of both Earth and Mars. Researchers think that the interaction between the planet's magnetic field and the moon Enceladus (see Section 12.5) tends to slow the field's rotation, accounting for the observed difference in periods.

Because of Saturn's lower density, this rapid rotation makes Saturn even more flattened than Jupiter. In fact, Saturn is the "flattest" planet in the solar system, with a polar radius of just 54,000 km, about 10 percent less than the planet's equatorial radius. Careful calculations show that this degree of flattening, large as it is, is less than would be expected for a planet composed of hydrogen and helium alone. Accordingly, astronomers think that Saturn also has a rocky core, perhaps as much as 15 times the mass of Earth, or 1.5 times the mass of Jupiter's core.

### Rings

Saturn's best-known feature is its spectacular *ring system*. Because the rings lie in the planet's equatorial plane, their appearance (as seen from Earth) changes as Saturn orbits the Sun, as shown in Figure 12.1. As Saturn moves along its orbit, the angles at which the rings are illuminated and at which we view them vary. When the planet's north or south pole is tipped toward the Sun, during Saturn's summer or winter, the highly reflective rings are at their brightest. During Saturn's spring and fall, the rings are close to being edge-on, both to the Sun and to us, so they seem to disappear altogether. The last "ring crossing" occurred in 1996; the next will take place in 2010.

One important deduction that we can make from this simple observation is that the rings are very thin. In fact, we now know that their thickness is only a few tens of meters, even though they are over 200,000 km in diameter.

### CONCEPT CHECK

✔ Why do some of the Earth-based images of Saturn in this chapter show the rings seen from above, whereas others show them from below?

## 12.2   Saturn's Atmosphere

Saturn is much less colorful than Jupiter. Figure 12.2 shows yellowish and tan cloud belts that parallel the equator, but these regions display less atmospheric structure than do the belts on Jupiter. No obvious large, long-lived "spots" or "ovals" adorn Saturn's cloud decks. Bands and storms do exist, but the color changes that distinguish them on Jupiter are largely absent on Saturn. This is the highest-resolution global image ever made of Saturn—a mosaic of

▶ FIGURE 12.1 **Ring Orientation** Over time, Saturn's rings change their appearance to terrestrial observers as the tilted ring plane orbits the Sun. At some times during Saturn's 29.5-year orbital period the rings seem to disappear altogether as Earth passes through their plane and we view them edge-on. Numbers along Saturn's orbit indicate the year. The roughly true-color images (inset) span a period of several years from the mid-1990s (bottom) to nearly the present (top), showing how the rings change from our perspective, from almost edge-on to more nearly face-on. See also Figure 12.2 for a closeup image of its tilted ring system relative to Earth. (*NASA*)

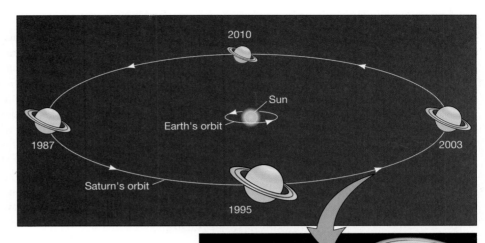

more than 100 photographs taken in true color—showing great subtlety in the structure of its cloud deck.

## Composition and Coloration

Astronomers first observed methane in the spectrum of sunlight reflected from Saturn in the 1930s, about the same time that it was discovered on Jupiter. However, it was not until the early 1960s, when more sensitive observations became possible, that ammonia was finally detected on Saturn. In the planet's cold upper atmosphere, most ammonia is in the solid or liquid form, with relatively little of it present as a gas to absorb sunlight and create spectral lines. Astronomers finally made the first accurate determinations of Saturn's hydrogen and helium content in the late 1960s. These Earth-based measurements were later confirmed with the arrival of the *Pioneer* and *Voyager* spacecraft in the 1970s.

Saturn's atmosphere consists of molecular hydrogen ($H_2$, 92.4 percent), helium (He, 7.4 percent), methane ($CH_4$, 0.2 percent), and ammonia ($NH_3$, 0.02 percent). As on Jupiter, hydrogen and helium dominate—these most abundant elements never escaped from Saturn's atmosphere because of the planet's large mass and low temperature (see *More Precisely 8-1*). However, the fraction of helium on Saturn is far less than is observed on Jupiter (where, as we saw, helium accounts for nearly 14 percent of the atmosphere) or in the Sun.

It is extremely unlikely that the processes that created the outer planets preferentially stripped Saturn of nearly half its helium or that the missing helium somehow escaped from the planet while the lighter hydrogen remained behind. Instead, astronomers think that, at some time in Saturn's past, the heavier helium began to sink toward the

◀ FIGURE 12.2 **Saturn** This spectacular image, acquired in 2005 by the *Cassini* spacecraft while approaching Saturn, is actually a mosaic of many images taken in true color. Note the subtle coloration of the planet and the detail in its rings. Resolution is 40 km. (*NASA*)

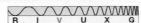

## PLANETARY DATA

### SATURN

| Orbital semimajor axis | 9.54 AU<br>1427 million km |
| --- | --- |
| Orbital eccentricity | 0.054 |
| Perihelion | 9.02 AU<br>1349 million km |
| Aphelion | 10.05 AU<br>1504 million km |
| Mean orbital speed | 9.65 km/s |
| Sidereal orbital period | 29.42 tropical years |
| Synodic orbital period | 378.09 Earth days* |
| Orbital inclination to the ecliptic | 2.49° |
| Greatest angular diameter,<br>as seen from Earth | 21″ |
| Mass | $5.68 \times 10^{26}$ kg<br>95.16 (Earth = 1) |
| Equatorial radius | 60,268 km<br>9.45 (Earth = 1) |
| Mean density | 687 kg/m³<br>0.125 (Earth = 1) |
| Surface gravity (at cloud tops) | 10.4 m/s²<br>1.07 (Earth = 1) |
| Escape speed | 35.5 km/s |
| Sidereal rotation period | 0.45 Earth day |
| Axial tilt | 26.73° |
| Surface magnetic field | 0.67 (Earth = 1) |
| Magnetic axis tilt relative<br>to rotation axis | 0.8° |
| Surface temperature | 97 K (at cloud tops) |
| Number of moons | 18 (more than<br>10 km in diameter)<br>56 (total) |

*1 Earth (mean solar) day = 24 hours

center of the planet, reducing its abundance in the outer layers and leaving them relatively rich in hydrogen. We will return to the reasons for this differentiation and its consequences in a moment.

Figure 12.3 illustrates Saturn's atmospheric structure (recall the corresponding diagram for Jupiter, Figure 11.7). In many respects, Saturn's atmosphere is quite similar to Jupiter's, except that the temperature is a little lower because of Saturn's greater distance from the Sun and because its clouds are somewhat thicker. Since Saturn, like Jupiter, lacks a solid surface, we take the top of the troposphere as our reference level and set it to 0 km. The top of the visible clouds lies about 50 km below this level. As on Jupiter, the clouds are arranged in three distinct layers, composed (in order of increasing depth) of ammonia, ammonium hydrosulfide, and water ice. Above the clouds lies a layer of haze formed by the action of sunlight on Saturn's upper atmosphere.

The total thickness of the three cloud layers in Saturn's atmosphere is roughly 200 km, compared with about 80 km on Jupiter, and each layer is itself somewhat thicker than its counterpart on Jupiter. The reason for this difference is Saturn's weaker gravity (due to its lower

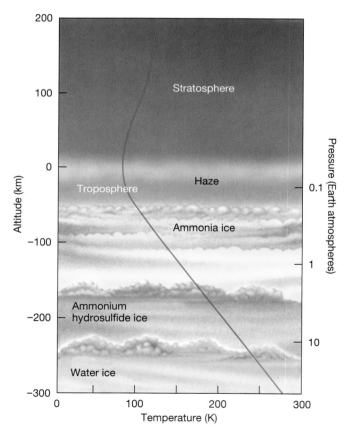

▲ FIGURE 12.3 **Saturn's Atmosphere** The vertical structure of Saturn's atmosphere contains several cloud layers, like Jupiter, but Saturn's weaker gravity results in thicker clouds and a more uniform appearance. The colors shown are intended to represent Saturn's visible-light appearance.

mass). At the haze level, Jupiter's gravitational field is nearly two-and-a-half times stronger than Saturn's, so Jupiter's atmosphere is pulled much more powerfully toward the center of the planet. Thus, Jupiter's atmosphere is compressed more than Saturn's, and the clouds are squeezed more closely together.

The colors of Saturn's cloud layers, as well as the planet's overall butterscotch hue, are due to the same basic cloud chemistry as on Jupiter. However, because Saturn's clouds are thicker, there are few holes and gaps in the top layer, so we rarely glimpse the more colorful levels below. Instead, we see different levels only in the topmost layer, which accounts for Saturn's rather uniform appearance.

## Weather

Saturn has atmospheric wind patterns that are in many ways reminiscent of those on Jupiter. There is an overall east–west zonal flow, which is apparently quite stable. Computer-enhanced images of the planet that bring out more cloud contrast (see Figure 12.4) clearly show the existence of bands, oval storm systems, and turbulent flow patterns looking very much like those seen on Jupiter. Scientists think that Saturn's bands and storms have essentially the same cause as does Jupiter's weather. Ultimately,

the large-scale flows and small-scale storm systems are powered by convective motion in Saturn's interior and by the planet's rapid rotation.

The zonal flow on Saturn is considerably faster than on Jupiter and shows fewer east–west alternations, as can be seen from Figure 12.5 (see also Figure 11.6). The equatorial eastward jet stream, which reaches a speed of about 400 km/h on Jupiter, moves at a brisk 1500 km/h on Saturn and extends to much higher latitudes. Not until latitudes 40° north and south of the equator are the first westward flows found. Latitude 40° north also marks the strongest bands on Saturn and the most obvious ovals and turbulent eddies. Astronomers still do not fully understand the reasons for the differences between Jupiter's and Saturn's flow patterns.

In September 1990, amateur astronomers detected a large white spot in Saturn's southern hemisphere, just below the equator. In November of that year, when the *Hubble Space Telescope* imaged the phenomenon in more detail, the spot had developed into a band of clouds completely encircling the planet's equator. Some of the *Hubble* images are shown in Figure 12.6(a). Astronomers suspect that the white coloration arose from crystals of ammonia ice formed when an upwelling plume of warm gas penetrated the cool upper cloud layers. Because the

▲ **FIGURE 12.4 Saturn's Cloud Structure** More structure is seen in Saturn's cloud cover when computer processing and artificial color are used to enhance the contrast of the image, as in these *Voyager* images of the entire gas ball and a smaller, magnified piece of it. (*NASA*)

R I V U X G

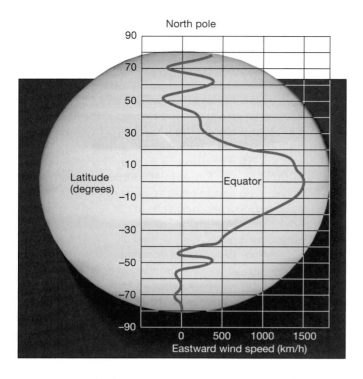

North pole

◀ FIGURE 12.5 **Saturn's Zonal Flow** Winds on Saturn reach speeds even greater than those on Jupiter. As on Jupiter, the visible bands appear to be associated with variations in wind speed.

crystals were freshly formed, they had not yet been affected by the chemical reactions that still color the planet's other clouds.

Such spots are relatively rare on Saturn. The previous one visible from Earth appeared in 1933, but it was much smaller than the 1990 system and much shorter lived, lasting for only a few weeks. Figure 12.6(b), taken in 1998, shows two somewhat smaller systems. The turbulent flow patterns seen around the 1990 white spot had many similarities to the flow around Jupiter's Great Red Spot. Scientists hope that routine observations of such temporary atmospheric phenomena on the outer worlds will enable them to gain greater insight into the dynamics of planetary atmospheres.

Since *Cassini*'s arrival at Saturn researchers have been able to study the planet's storm systems in much

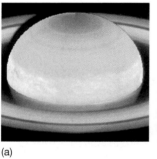
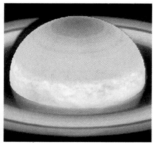
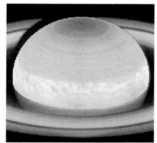
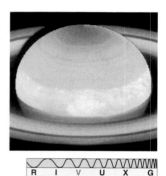

(a)

R I V U X G

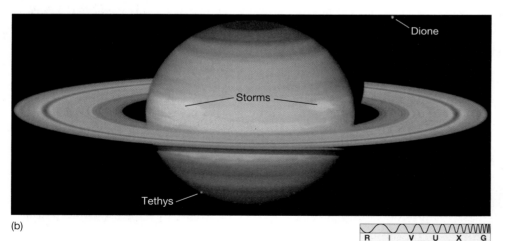

Dione

Storms

Tethys

(b)

R I V U X G

▲ FIGURE 12.6 **Saturn Storms** (a) Circulating and evolving cloud systems on Saturn, imaged by the *Hubble Space Telescope* at approximately 2-hour intervals. (b) This infrared image, displayed in false color and created for greater contrast by combining observations made at three different wavelengths—1.0 μm (blue), 1.8 μm (green), and 2.1 μm (red)—enhances the belt structure and storms near the equator. Blue coloration indicates regions where the atmosphere is relatively free of haze (see Figure 12.3); green and yellow indicate increasing levels of haze; red and orange indicate high-level clouds. Two small storm systems near the equator appear whitish. Two of Saturn's moons are also visible in the image, as noted. *(NASA)*

more detail. Figure 12.7 shows a particularly large and complex system that appeared in Saturn's southern hemisphere in late 2004. Dubbed the "Dragon Storm" for its twisting, convoluted shape, it is the bright orange region just above and to the right of center in this (false color) infrared image. The storm is thought to be Saturn's equivalent of a thunderstorm on Earth (although on a scale comparable to our entire planet). *Cassini*'s detectors measured strong bursts of radio waves associated with the storm; they are most likely produced by intense lightning discharges deep below the cloud tops. The lightning is thought to be powered by convection and precipitation (water and ammonia "rain"), just as on Earth, but the bursts are millions of times stronger than anything ever witnessed here at home.

Astronomers think that the storm is actually a long-lived phenomenon rooted deep in Saturn's atmosphere—perhaps a little like Jupiter's Great Red Spot—and normally completely hidden below the upper cloud layers. ∞ (Sec. 11.2) Only occasionally does the storm flare up, producing a bright plume visible from outside. The outburst shown in Figure 12.7 subsided within a year, but researchers suspect that a comparable outburst at roughly the same latitude in 2006 had the same underlying power source.

*Cassini* has also detected numerous small dark storms apparently associated with this and other large systems. These smaller storms seem to be "spun off" from the larger systems and subsequently merge with the planet's zonal flow, perhaps providing a means for energy to flow from Saturn's warm interior into the cold atmosphere. However, the "big picture" connecting these large- and small-scale storm systems on Saturn with the Red Spot, white ovals, and the recently discovered "Red Spot Junior" on Jupiter, remains to be worked out. ∞ (Sec. 11.2)

Figure 12.8 shows *Cassini*'s view of an Earth-sized *polar vortex* at Saturn's south pole. The existence of the vortex had been known for some years before this image was obtained in late 2006, but close-up imagery from *Cassini* has allowed scientists to study the system with unprecedented clarity. As mentioned in Chapter 9, such vortices are expected at the poles of any planet with an atmosphere, and *Venus Express* has mapped a similar vortex on Venus in great detail. ∞ (Sec. 9.5) However, the opportunity to compare polar vortices in two planets with such different atmospheres, orbits, and rotational properties as Venus and Saturn offers astronomers a unique exercise in comparative planetology.

CONCEPT CHECK

✔ Why are atmospheric features on Saturn generally less vivid than those on Jupiter?

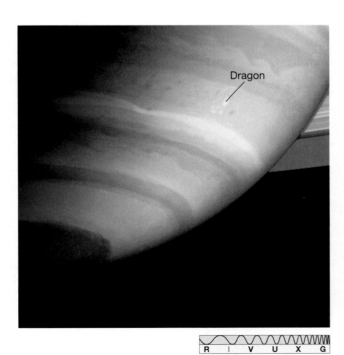

▲ FIGURE 12.7  **Saturn's "Dragon Storm"**  This false-color, near-infrared image from *Cassini* shows what is thought to be a vast thunderstorm on Saturn. Called the "Dragon Storm," it generated bursts of radio waves resembling the static created by lightning on Earth. (*NASA*)

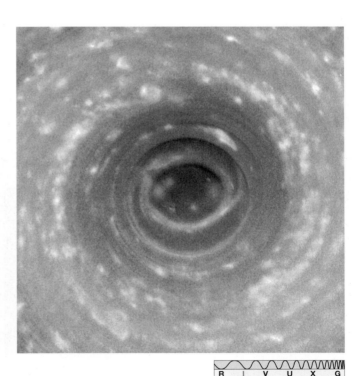

▲ FIGURE 12.8  **Saturn's Polar Vortex**  This infrared *Cassini* image shows a huge swirling polar vortex at Saturn's south pole. The swirling, hurricane-like storm has a well-developed eye wall and calm winds at its center. The size of this vortex is slightly larger than the entire Earth, and its winds are about twice that of a category 5 hurricane on Earth. (*NASA*)

## 12.3    Saturn's Interior and Magnetosphere

Figure 12.9 depicts Saturn's internal structure. (Compare with Figure 11.12 for the case of Jupiter.) The picture was pieced together by planetary scientists using the same tools—spacecraft observations and theoretical modeling—that they employed to infer Jupiter's inner workings. Saturn has the same basic internal parts as Jupiter, but their relative proportions are somewhat different: Saturn's metallic hydrogen layer is thinner and its core is larger. Because of its lower mass, Saturn has a less extreme core temperature, density, and pressure than does Jupiter. The central pressure is around one-fifth of Jupiter's—about 2 to 3 times the pressure at the center of Earth.

### Internal Heating

Infrared measurements indicate that Saturn's surface (i.e., cloud-top) temperature is 97 K, substantially higher than the temperature at which Saturn would reradiate all the energy it receives from the Sun. In fact, Saturn radiates almost three times more energy than it absorbs. Thus, Saturn, like Jupiter, has an internal energy source. ∞ (Sec. 11.3) But the explanation behind Jupiter's excess energy—that the planet has a large reservoir of heat left over from its formation—doesn't work for Saturn. Smaller than Jupiter, Saturn must have cooled more rapidly—rapidly enough that its original supply of energy was used up long ago. What, then, is happening inside Saturn to produce this extra heat?

The explanation for the origin of Saturn's extra heat also explains the mystery of the planet's apparent helium deficit. At the temperatures and high pressures found in Jupiter's interior, liquid helium *dissolves* in liquid hydrogen. Inside Saturn, where the internal temperature is lower, the helium doesn't dissolve so easily and tends to form

droplets instead. The phenomenon is familiar to cooks, who know that it is generally much easier to dissolve ingredients in hot liquids than in cold ones. Saturn probably started out with a fairly uniform solution of helium dissolved in hydrogen, but the helium tended to condense out of the surrounding hydrogen, much as water vapor condenses out of Earth's atmosphere to form a mist. The amount of helium condensation was greatest in the planet's cool outer layers, where the mist turned to rain about 2 billion years ago. A light shower of liquid helium has been falling through Saturn's interior ever since. This **helium precipitation** is responsible for depleting the outer layers of their helium content.

So we can account for the unusually low abundance of helium in Saturn's atmosphere: Much of it has rained down to lower levels. But what about the excess heating? The answer is simple: As the helium sinks toward the center, the planet's gravitational field compresses it and heats it up. The gravitational energy thus released is the source of Saturn's internal heat. In the distant future—in a billion years or so—the helium rain will stop, and Saturn will cool until its outermost layers radiate only as much energy as they receive from the Sun. When that happens, the temperature at Saturn's cloud tops will be 74 K. As Jupiter cools, it, too, may someday experience precipitate helium in its interior, causing its surface temperature to rise once again.

### Magnetospheric Activity

Saturn's electrically conducting interior and rapid rotation produce a strong magnetic field and an extensive magnetosphere. Probably because of the considerably smaller mass of Saturn's metallic hydrogen zone, the planet's basic magnetic field strength is only about one-twentieth that of Jupiter, or about a thousand times greater than that of Earth. The magnetic field at Saturn's cloud tops (roughly 10 Earth radii from the planet's center) is approximately the same as that at Earth's surface. *Voyager* measurements indicate that, unlike Jupiter's and Earth's magnetic axes, which are slightly tilted, Saturn's magnetic field is not inclined with respect to its axis of rotation. Saturn's magnetic field, like Jupiter's, is oriented opposite that of Earth: that is, an Earth compass needle would point toward Saturn's south pole rather than its north. ∞ (Sec. 11.4) Figure 12.10 shows an aurora on Saturn, imaged in 1998 by the *Hubble Space Telescope*.

Saturn's magnetosphere extends about 1 million km toward the Sun and is large enough to contain the planet's ring system and the innermost 16 small moons. Saturn's largest moon, Titan, orbits about 1.2 million km from the planet, so it is found sometimes just inside the outer magnetosphere and sometimes just outside, depending on the intensity of the solar wind (which tends to push the sunward side of the magnetosphere closer to the planet). Because no major moons lie deep within Saturn's magnetosphere, the details of its structure are different from those of Jupiter's magnetosphere. For example, there is

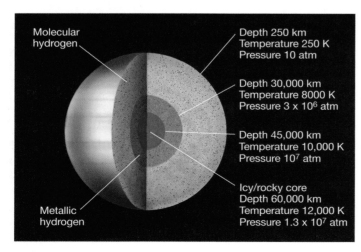

▲ **FIGURE 12.9 Saturn's Interior** Saturn's internal structure, as deduced from *Voyager* observations and computer modeling.

▲ **FIGURE 12.10 Aurora on Saturn** An ultraviolet camera aboard the *Hubble Space Telescope* recorded this image of a remarkably symmetrical (orange) aurora on Saturn during a solar storm in 1998. *(NASA)*

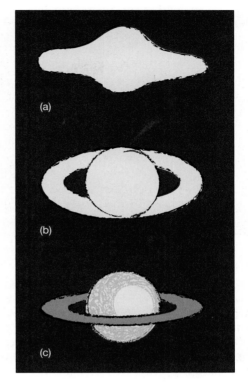

▲ **FIGURE 12.11 Sketches of Saturn's Rings** Three artist's re-renderings of sketches of Saturn's rings, made (a) by Galileo in 1610, (b) by Galileo in 1616, and (c) by Huygens in 1655.

no equivalent of the Io plasma torus. ∞ (Sec. 11.5) Like Jupiter, Saturn emits radio waves, but as luck would have it, they are reflected from Earth's ionosphere (they lie in the AM band) and were not detected until the *Voyager* craft approached the planet.

CONCEPT CHECK
✔ Where did Saturn's atmospheric helium go?

## 12.4 Saturn's Spectacular Ring System

The most obvious and well-known aspect of Saturn's appearance is, of course, its *planetary ring system.* Astronomers now know that all the jovian planets have rings, but Saturn's are by far the brightest, the most extensive, and the most beautiful. ∞ (Sec. 11.6)

### The View from Earth

Galileo saw Saturn's rings first in 1610, but he did not recognize what he saw as a planet with a ring. At the resolution of his small telescope, the rings looked like bumps on the planet, or perhaps (he speculated) parts of a triple planet of some sort. Figure 12.11(a) and (b)

shows two of Galileo's early sketches of Saturn. By 1616, Galileo had already realized that the "bumps" were not round, but rather elliptical in shape. In 1655, the Dutch astronomer Christian Huygens realized what the bumps were: a thin, flat ring, completely encircling the planet (Figure 12.11c).

In 1675, the French–Italian astronomer Giovanni Domenico Cassini discovered the first ring feature: a dark band about two-thirds of the way out from the inner edge. From Earth, the band looks like a gap in the ring (an observation that is not too far from the truth, although we now know that there is actually some ring material within it). This "gap" is named the **Cassini division,** in honor of its discoverer. Careful observations from Earth show that the inner "ring" is in reality also composed of two rings. From the outside in, the three rings are known somewhat prosaically as the **A, B,** and **C rings.** The Cassini division lies between the A and B rings. The much narrower **Encke gap,** some 300 km wide, is found in the outer part of the A ring. These ring features are marked on Figure 12.12. No finer ring details are visible from our Earthly vantage point. Of the three main rings, the B ring is brightest, followed by the somewhat fainter A ring, and then by the almost translucent C ring. A more complete list of ring properties appears in Table 12.1. (The D, E, F, and G rings listed in the table are discussed later in this section.)

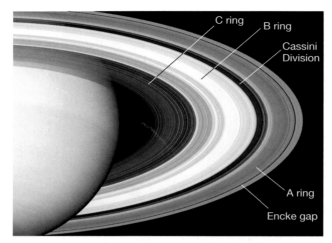

R I V U X G

▲ **FIGURE 12.12 Saturn's Rings** Much fine structure, especially in the rings, appears in this image of Saturn taken with the *Cassini* spacecraft in 2005. The main ring features long known from Earth-based observations—the A, B, and C rings, the Cassini Division, and the Encke gap—are marked and are shown here in false color in order to enhance contrast. The other rings listed in Table 12.1 are not visible here. *(NASA)*

## What Are Saturn's Rings?

A fairly obvious question—and one that perplexed the best scientists and mathematicians on Earth for almost two centuries—is "What are Saturn's rings made of?" By the middle of the 19th century, various dynamic and thermodynamic arguments had conclusively proved that the rings could be neither solid, liquid, nor gas!

| TABLE 12.1 | The Rings of Saturn | | | | |
|---|---|---|---|---|---|
| **Ring** | **Inner Radius** | | **Outer Radius** | | **Width** |
| | (km) | (planetary radii) | (km) | (planetary radii) | (km) |
| D | 67,000 | 1.11 | 74,700 | 1.24 | 7700 |
| C | 74,700 | 1.24 | 92,000 | 1.53 | 17,300 |
| B | 92,000 | 1.53 | 117,500 | 1.95 | 25,500 |
| Cassini Division | 117,500 | 1.95 | 122,300 | 2.03 | 4800 |
| A | 122,300 | 2.03 | 136,800 | 2.27 | 14,500 |
| Encke gap* | 133,400 | 2.22 | 133,700 | 2.22 | 300 |
| F | 140,300 | 2.33 | 140,400 | 2.33 | 100 |
| G | 165,800 | 2.75 | 173,800 | 2.89 | 8000 |
| E | 180,000 | 3.00 | 480,000 | 8.00 | 300,000 |

*\* The Encke gap lies within the A ring.*

What is left? In 1857, after showing that a solid ring would become unstable and break up, Scottish physicist James Clerk Maxwell suggested that the rings are composed of a great number of small particles, all independently orbiting Saturn, like so many tiny moons. That inspired speculation was verified in 1895, when Lick Observatory astronomers measured the Doppler shift of sunlight reflected from the rings and showed that the velocities thus determined were exactly what would be expected from separate particles moving in circular orbits in accordance with Newton's law of gravity. ∞ (Secs. 2.8, 3.5)

What sort of particles make up the rings? The fact that they reflect most (over 80 percent) of the sunlight striking them had long suggested to astronomers that they are made of ice, and infrared observations in the 1970s confirmed that water ice is indeed a prime constituent of the rings. Radar observations and later *Voyager* and *Cassini* studies of scattered sunlight showed that the diameters of the particles range from fractions of a millimeter to tens of meters, with most particles being about the size (and composition) of a large snowball on Earth.

We now know that the rings are truly thin—only 10–15 m thick, according to *Cassini* measurements. Stars can occasionally be seen through them, like automobile headlights penetrating a snowstorm. Why are the rings so thin? The answer seems to be that collisions between ring particles tend to keep them all moving in circular orbits in a single plane. Any particle that tries to stray from this orderly motion finds itself in an orbit that soon runs into other ring particles. Over long periods, the ensuing jostling serves to keep all of the particles moving in circular, planar orbits. The asymmetric gravitational field of Saturn (a result of its flattened shape) sees to it that the rings lie in the planet's equatorial plane.

## The Roche Limit

But why a ring of particles at all? What process produced the rings in the first place? To answer these questions, consider the fate of a small moon orbiting close to a massive planet such as Saturn. The moon is held together by internal forces—its own gravity, for example. As we bring our hypothetical moon closer to the planet, the tidal force on it increases. Recall from Chapter 7 that the effect of such a tidal force is to stretch the moon along the direction toward the planet—that is, to create a tidal bulge. Recall also that the tidal force increases rapidly with decreasing distance from the planet. ∞ (Sec. 7.6) As the moon is brought closer to the planet, it reaches a point where the tidal force tending to stretch it out becomes *greater* than the internal forces holding it together. At that point, the moon is torn apart by the planet's gravity, as shown in Figure 12.13. The pieces of the satellite then pursue their own individual orbits around that planet, eventually spreading all the way around it in the form of a ring.

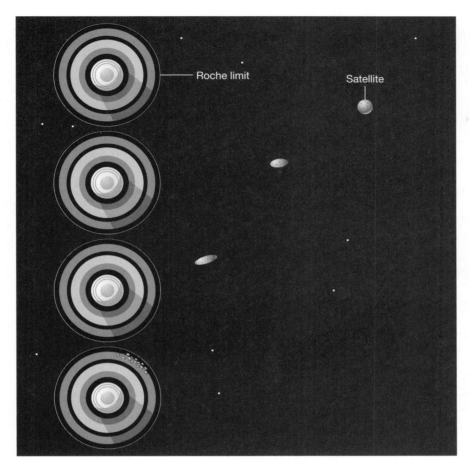

Roche limit

Satellite

◀ **FIGURE 12.13  Roche Limit** From top to bottom, these four frames illustrate how the tidal field of a planet first distorts (near top), and then destroys (at bottom), a moon that strays too close. Note that the distortion is exaggerated in the middle panels, and the moon's infall does not happen directly; rather, its breakup occurs over the course of many orbits.

even within the Roche limit because they are held together mostly by interatomic (electromagnetic) forces, not by gravity.

## The Rings in Detail

Thus, as the two *Voyager* probes approached Saturn in 1980 and 1981, scientists on Earth were fairly confident that they understood the nature of the rings. However, there were many surprises in store. The *Voyager* flybys changed forever our view of this spectacular region in our cosmic backyard, revealing the rings to be vastly more complex than astronomers had imagined. *Cassini*'s 4-year tour of the Saturn system a quarter century later allowed much more extended and detailed study of many of the phenomena discovered by *Voyager* and yielded many new insights into this fascinating system.

As the *Voyager* probes approached Saturn, it became obvious that the main rings were actually composed of tens of thousands of narrow **ringlets,** shown (as seen by *Cassini*) in Figure 12.15. Although *Voyager* cameras did find several new gaps in the rings, the ringlets in the figure are actually not separated from one another by empty space. Instead, detailed studies reveal that the rings contain concentric regions of alternating high and low concentrations of ring particles. The ringlets are the high-density peaks. According to theory, the gravitational influence of Saturn's inner moons and the mutual gravitational attraction of the ring particles enables waves of matter to form and move in the plane of the rings, rather like ripples on the surface of a pond. The wave crests wrap all the way around the rings, forming tightly wound spiral patterns called *spiral density waves* that resemble grooves in a huge celestial phonograph record.

Although the ringlets are the result of spiral waves in the rings, the true gaps are not. The narrower gaps—roughly 20 of them—are thought to be kept clear by the

For any given planet and any given moon, the critical distance inside of which the moon is destroyed is known as the *tidal stability limit*, or the **Roche limit,** after the 19th-century French mathematician Edouard Roche, who first calculated it. As a handy rule of thumb, if our hypothetical moon is held together by its own gravity and its average density is comparable to that of the parent planet (both reasonably good approximations for Saturn's larger moons), then the Roche limit is roughly 2.4 times the radius of the planet. Thus, for Saturn, no moon can survive within a distance of 144,000 km of the planet's center, about 7000 km beyond the outer edge of the A ring. The main (A, B, C, D, and F) rings of Saturn occupy the region inside Saturn's Roche limit.

These considerations apply equally well to the other jovian worlds. Figure 12.14 shows the location of the ring system of each jovian planet relative to the planet's Roche limit. Given the approximations in our assumptions, we can conclude that all the major planetary rings are found within the Roche limit of their parent planet. Notice that, strictly speaking, the calculation of this limit applies only to low-density moons massive enough for their own gravity to be the dominant force binding them together. Sufficiently small moons (less than 10 km or so in diameter) can survive

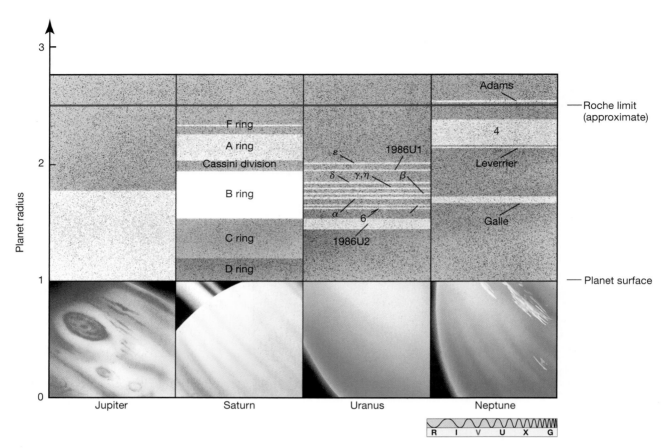

▲ FIGURE 12.14 **Jovian Ring Systems** The rings of Jupiter, Saturn, Uranus, and Neptune. All distances are expressed in planetary radii. The red line represents the Roche limit, and all the rings lie within (or very close to) this limit of their parent planets. Note that the red line is just an approximation—the details depend on the internal structure of the planet, and Neptune's Adams ring is actually consistent with Roche's calculations.

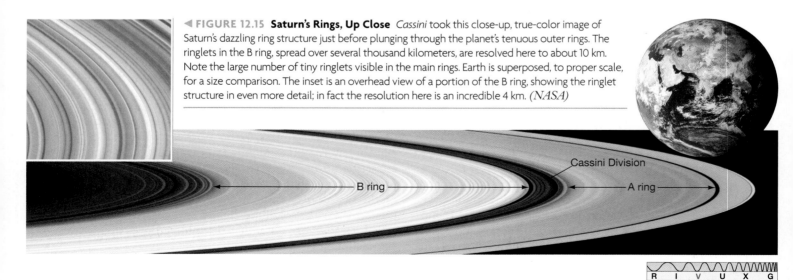

◀ FIGURE 12.15 **Saturn's Rings, Up Close** *Cassini* took this close-up, true-color image of Saturn's dazzling ring structure just before plunging through the planet's tenuous outer rings. The ringlets in the B ring, spread over several thousand kilometers, are resolved here to about 10 km. Note the large number of tiny ringlets visible in the main rings. Earth is superposed, to proper scale, for a size comparison. The inset is an overhead view of a portion of the B ring, showing the ringlet structure in even more detail; in fact the resolution here is an incredible 4 km. *(NASA)*

action of small *moonlets* embedded in them. These moonlets are larger (perhaps 10 or 20 km in diameter) than the largest true ring particles, and they simply "sweep up" ring material through collisions as they go. However, despite careful searches of *Voyager* and *Cassini* images, only two of these moonlets have so far been found. The 30-km-wide moon Pan, discovered in archival *Voyager* data in 1991, orbits in the Encke gap (marked in Figure 12.12). The 7-km Daphnis, imaged by *Cassini* in 2005, resides in the narrow Keeler gap, close to the outer edge of the A ring.

Astronomers have found indirect evidence for embedded moonlets in the form of the scalloped "wakes" they leave behind them in the rings, but no other direct sightings have occurred. Despite their elusiveness, moonlets are nevertheless regarded as the best explanation for the small gaps and associated fine structure in the rings. In 2006 *Cassini* found evidence for a new class of intermediate-sized (100-m diameter) moonlets, which may play an important role in clearing some small or partial gaps.

*Voyager 2* found a series of faint rings, now known collectively as the **D ring**, inside the inner edge of the C ring, stretching down almost to Saturn's cloud tops. The D ring contains relatively few particles and is so dark that it is completely invisible from Earth. Two other faint rings, discovered by *Pioneer 11* and *Voyager 1*, respectively, lie well outside the main ring structure. Both the **G ring** and the **E ring**, are faint and diffuse, more like Jupiter's ring than the main A, B, and C rings of the Saturn system. ∞ (Sec. 11.6) The E ring appears to be associated with volcanism on the moon Enceladus (see Section 12.5).

Figure 12.16 is a *Cassini* image showing the rings from a perspective never before seen—behind the planet looking back toward the eclipsed Sun. Just as diffuse airborne dust is most easily seen against the light streaming in through a sunlit window, the planet's faint rings show up clearly in this remarkable back-lit view. In addition, this image reveals several additional faint rings, some of them also associated with the orbits of various small moons.

The *Voyager 2* cameras revealed one other completely unexpected feature. A series of dark radial "spokes" formed on the B ring, moved around the planet for about one ring orbital period, and then disappeared (Figure 12.17). Careful scrutiny of these peculiar drifters showed that they were composed of very fine (micron-sized) dust hovering a few tens of meters *above* the plane of the rings. Scientists think that this dust was held in place by electromagnetic forces generated in the ring plane, perhaps resulting from collisions among particles there or interactions with the planet's magnetic field. The spokes faded as the ring revolved.

Astronomers had expected that the creation and dissolution of such spokes would be regular occurrences in the Saturn ring system, but during the first year of *Cassini*'s tour, none were seen. However, as with the planet's faint rings, lighting and viewing conditions seem to be critical to the spokes' visibility. As the planet's orientation changed relative to the Sun, spokes were finally observed in late 2005. Spoke activity is expected to be common for the remainder of the *Cassini* mission, and researchers hope that repeated observations will allow them to understand this peculiar phenomenon.

## Orbital Resonances and Shepherd Satellites

*Voyager* images showed that the largest gap in the rings, the Cassini division, is not completely empty of matter. In fact, as can be seen in Figure 12.15, the division contains a series of faint ringlets and gaps (and, presumably, embedded moonlets, too). The overall concentration of ring particles in the division as a whole is, however, much lower than in the A and B rings. The diffuse nature of the division causes it to appear bright in Figure 12.16. Although its small internal gaps probably result from embedded satellites, the division itself does not. Instead, it owes its existence to another solar-system *resonance*, this time involving particles orbiting in the division, on the one hand, and Saturn's innermost major moon, Mimas, on the other. ∞ (Sec. 8.4)

A ring particle moving in an orbit within the Cassini division has an orbital period exactly half that of Mimas. Particles in the division thus complete exactly two orbits around Saturn in the time taken for Mimas to orbit once—a configuration known as a 2:1 resonance. Applying Kepler's third law (recast to refer not to the planets, but to Saturn's moons), we can show that this 2:1 resonance with Mimas corresponds to a radius of 117,000 km, the inner edge of the division. ∞ (*More Precisely 2-3*)

The effect of this resonance is that particles in the division receive a gravitational tug from Mimas at exactly the same location in their orbit every other time around. Successive tugs reinforce one another, and the initially circular trajectories of the ring particles soon get stretched out into ellipses. In their new orbits, these particles collide with other particles and eventually find their way into new circular orbits at other radii. The net effect is that the number of ring particles in the Cassini division is greatly reduced.

Particles in "nonresonant" orbits (i.e., at radii whose orbital period is not simply related to the period of Mimas) also are acted upon by Mimas's gravitational pull. But the times when the force is greatest are spread uniformly around the orbit, and the tugs cancel out. It's a little like pushing a child on a swing: Pushing at the same point in the swing's motion each time produces much better results than do random shoves. Thus, Mimas (or any other moon) has a large effect on the ring at those radii at which a resonance exists and little or no effect elsewhere.

▲ **FIGURE 12.16 Back-Lit Rings** *Cassini* took this spectacular image of Saturn's rings as it passed through Saturn's shadow. Note how the back-lit main ring system differs from the front-lit view shown earlier in Figure 12.12. The normally hard to see F, G, and E outer rings are clearly visible (and marked) in this contrast-enhanced image. The inset shows the moon Enceladus orbiting within the E ring; its eruptions likely give rise to the ring's icy particles. *(NASA)*

We now know that resonances between ring particles and moons play an important role in shaping the fine structure of Saturn's rings. For example, the sharp outer edge of the A ring is thought to be governed by a 3:2 resonance with Mimas (three ring orbits in two Mimas orbital periods). Most theories of planetary rings predict that the ring system should spread out with time, basically because of collisions among ring particles. Instead, the A ring's outer edge is "patrolled" by a small satellite named Atlas, held in place by the gravity of Mimas, that prevents ring particles from diffusing away. Compare Tables 12.1 and 12.2 and see if you can identify other resonant connections between Saturn's moons or between the moons and the rings. (You should be able to find quite a few—Saturn's ring system is a complex place!)

Outside the A ring lies perhaps the strangest ring of all. The faint, narrow **F ring** (shown in Figure 12.18) was discovered by *Pioneer 11* in 1979, but its full complexity became evident only when *Voyager 1* took a closer look. Unlike the inner major rings, the F ring is narrow—less than a hundred kilometers wide. It lies just inside Saturn's Roche limit, separated from the A ring by about 3500 km. Its narrowness by itself is unusual, as is its slightly eccentric shape, but the F ring's oddest feature is its irregular, "kinked" structure, making it look as though it is made up of several separate strands braided together! This remarkable appearance sent dynamicists scrambling in search of an explanation. It now seems as though the ring's intricate structure, as well as its thinness, arise from the influence of two small moons, known as **shepherd satellites,** that orbit on either side of it (Figure 12.19).

These two small, dark satellites, each little more than 100 km in diameter, are called Prometheus and Pandora. They orbit about 1000 km on either side of the F ring, and their gravitational influence on the F-ring particles keeps the ring tightly confined in its narrow orbit. As illustrated in Figure 12.20(a), a ring particle overtaking the moon on an interior orbit (that is, closer to Saturn) loses energy due

▲ **FIGURE 12.17 Spokes in the Rings** Saturn's B ring showed a series of dark temporary "spokes" as *Voyager 2* flew by at a distance of about 4 million km. The spokes are caused by small particles suspended just above the ring plane. *Cassini*, too, has seen spokes, although (so far) they have not been as prominent as those seen by *Voyager* 25 years earlier. *(NASA)*

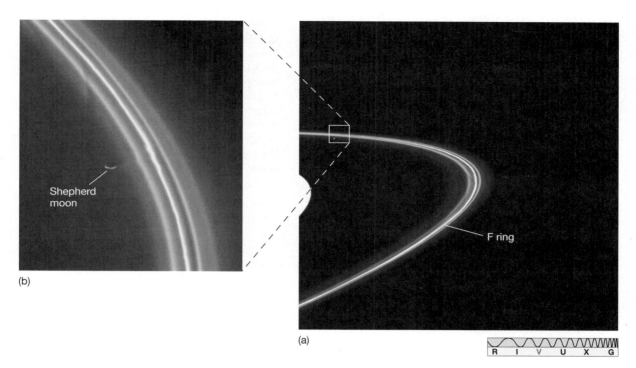

▲ FIGURE 12.18 **Shepherd Moon** (a) Saturn's narrow F ring appears to contain kinks and braids, making it unlike any of Saturn's other rings. Its thinness can be explained by the effects of two shepherd satellites that orbit near the ring—one a few hundred kilometers inside, the other a similar distance outside. (b) One of the potato-shaped shepherd satellites (Prometheus here roughly 100 km across) can be seen at the right of this enlarged view. *(NASA)*

to the moon's gravitational attraction and sinks inward. Conversely, a particle orbiting outside the moon is overtaken and forced outward. The combination of these effects from two moons (Figure 12.20b) can confine a narrow ring—any particle straying too far out of the F ring is gently guided back into the fold by one or the other of the shepherd moons. (The moon Atlas confines the A ring in a somewhat similar way.)

However, the details of how Prometheus and Pandora produce the braids in the F ring, and why the two moons are there at all, in such similar orbits, remain unclear. There is evidence that other eccentric rings found in the gaps in the A, B, and C rings may also result from the effects of shepherd moonlets. Beyond the F ring, the G ring apparently lacks ringlets and peculiar internal structure, but its relative narrowness and sharp edges suggest the presence of shepherd satellites, although so far none has been found. *Cassini* discovered that the G ring has regions of enhanced brightness, known as *arcs*, that seem to move in resonance with Mimas and may harbor a confining moon or moonlets, but no definitive sighting has yet been made.

◀ FIGURE 12.19 **F Ring Structure** Kinks, waves, and other substructure in the F ring can be seen in this high-resolution *Cassini* image. A back-and-forth dance of the shepherd moons (Pandora shown here) gravitationally sculpts scalloped-shaped clumps in the core of the rings. *(NASA)*

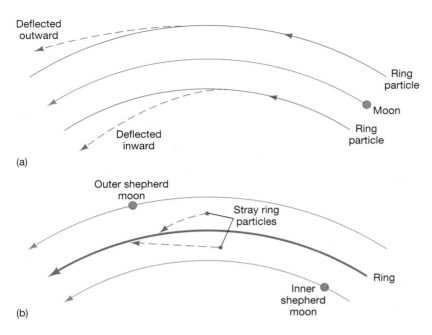

◀ FIGURE 12.20 **Moon–Ring Interaction** (a) Strange as it may seem, the net effect of the interactions between a moon and ring particles is that the moon tends to push those particles away from it. (b) The F-ring shepherd satellites operate by forcing errant F-ring particles back into the main ring. Each moon operates as in part (a), so that the ring is confined between the two moons. As a consequence of Newton's third law of motion, the satellites themselves slowly drift away from the ring (unless they, in turn, are held in place by interactions with other, larger moons).

## The Origin of the Rings

Two possible origins have been suggested for Saturn's rings. Astronomers estimate that the total mass of ring material is no more than $10^{15}$ tons—enough to make a satellite about 250 km in diameter. If such a satellite strayed inside Saturn's Roche limit or was destroyed (perhaps by a collision) near that radius, a ring could have resulted. An alternative view is that the rings represent material left over from Saturn's formation stage 4.6 billion years ago. In this scenario, Saturn's tidal field prevented any moon from forming inside the Roche limit, so the material has remained a ring ever since. Which view is correct?

All the dynamic activity observed in Saturn's rings suggests to many researchers that the rings must be quite young—perhaps no more than 50 million years old, or 100 times younger than the solar system. There is just too much going on for the ring to have remained stable for billions of years, so they probably aren't left over from the planet's formative stages. If this is so, then either the rings are continuously replenished, perhaps by fragments of Saturn's moons chipped off by meteorites, or they are the result of a relatively recent, possibly catastrophic, event in the planet's system—a small moon that may have been hit by a large comet or even by another moon.

Astronomers prefer not to invoke catastrophic events to explain specific phenomena, but the more we learn of the universe, the more we realize that catastrophe probably plays an important role. For now, the details of the formation of Saturn's ring system simply aren't known.

CONCEPT CHECK

✔ What do the Roche limit and orbital resonances have to do with planetary rings?

## 12.5   The Moons of Saturn

Saturn has the most extensive, and in many ways the most complex, system of natural satellites of all the planets. The planet's 18 largest moons are listed in Table 12.2. Observations of sunlight reflected from them suggests that most are covered with snow and ice. Many of them are probably made almost entirely of water ice. Even so, they are a curious and varied lot, and many aspects of their structure and history are still not well understood. Most of our detailed knowledge of these moons comes from the *Pioneer* and *Voyager* flybys in the late 1970s and early 1980s, and from *Cassini*, which is currently orbiting the planet. ∞ (Sec. 6.6)

The moons fall into three fairly natural groups. First, there are the many "small" moons—irregularly shaped chunks of ice, all less than 400 km across—that exhibit a bewildering variety of complex and fascinating motion. Second, there are six "medium-sized" moons—spherical bodies with diameters ranging from about 400 to 1500 km—that offer clues to the past and present state of the environment of Saturn while presenting many puzzles regarding their own appearance and history. Finally, there is Saturn's single "large" moon—Titan—which, at 5150 km in diameter, is the second-largest satellite in the solar system. (Jupiter's Ganymede is a little bigger.) Titan has an atmosphere denser than Earth's and (some scientists think) surface conditions that may be conducive to life. Notice, incidentally, that Jupiter has no "medium-sized" moons, as just defined: The Galilean satellites are large, like Titan, and all of Jupiter's other satellites are small—no more than 200 km in diameter.

As in the case of Jupiter, improving observational techniques in recent years have led to an explosion in the number of known moons orbiting Saturn—from just the 18 listed in Table 12.2 in 1995 to at least 56 as of early

2007. ∞ (*Discovery 11-3*) Like the many small satellites of Jupiter, these moons are all very faint (and hence small) and revolve around Saturn quite far from the planet on rather inclined, often retrograde, orbits, much as Saturn's other "small" moons do. Most likely, they are chunks of debris captured from interplanetary space after close encounters with Saturn.

## Titan: A Moon with an Atmosphere

The largest and most intriguing of all Saturn's moons, Titan, was discovered by Christian Huygens in 1655. Even through a large Earth-based optical telescope, this moon is visible only as a barely resolved reddish disk. However, long before the *Voyager* or *Cassini* missions, astronomers already knew from spectroscopic observations that the moon's reddish coloration was caused by something quite special—an atmosphere. So eager were mission planners to obtain a closer look that they programmed *Voyager 1* to pass very close to Titan, even though it meant that the spacecraft could not then use Saturn's gravity to continue on to

Uranus and Neptune. Instead, *Voyager 1* left the Saturn system on a path taking the craft out of the solar system well above the plane of the ecliptic. ∞ (*Discovery 6-1*)

A *Voyager 1* image of Titan is shown in Figure 12.21(a). Unfortunately, despite the spacecraft's close pass, the moon's surface remained a mystery. A thick, uniform layer of haze, similar to the photochemical smog (created by chemical reactions powered by light) found over many cities on Earth, that envelops the moon completely obscured the spacecraft's view. Still, *Voyager 1* was able to provide mission specialists with detailed atmospheric data. Figure 12.21(b) shows one of the best Earth-based views of the moon, taken in infrared light to penetrate the smog. A detailed, close-up look at Titan is a key part of the ongoing *Cassini* mission. As described in *Discovery 12-1*, the spacecraft's carefully choreographed trajectory includes a total of 44 visits to the moon by the end of the primary mission in 2008.

Titan's atmosphere is thicker and denser even than Earth's, and it is certainly far more substantial than that of any other moon. Prior to *Voyager 1*'s arrival in 1980, only methane and a few other simple hydrocarbons had

## TABLE 12.2   The Major Moons of Saturn*

| Name | Distance from Saturn (km) | Distance from Saturn (planetary radii) | Orbit Period (days) | Size (longest diameter, km) | Mass** (Earth moon masses) | Density (kg/m³) | Density (g/cm³) |
|------|------|------|------|------|------|------|------|
| Pan | 133,600 | 2.22 | 0.57 | 20 | $4.0 \times 10^{-8}$ | | |
| Atlas | 138,000 | 2.28 | 0.60 | 37 | | | |
| Prometheus | 139,000 | 2.31 | 0.61 | 148 | $1.9 \times 10^{-6}$ | | |
| Pandora | 142,000 | 2.35 | 0.63 | 110 | $1.8 \times 10^{-6}$ | | |
| Epimetheus | 151,000 | 2.51 | 0.69 | 138 | $7.5 \times 10^{-6}$ | | |
| Janus | 151,000 | 2.51 | 0.69 | 199 | $2.7 \times 10^{-5}$ | | |
| Mimas | 186,000 | 3.08 | 0.94 | 398 | 0.00051 | 1100 | 1.1 |
| Enceladus | 238,000 | 3.95 | 1.37 | 498 | 0.00099 | 1100 | 1.1 |
| Calypso | 295,000 | 4.89 | 1.89 | 30 | | | |
| Telesto | 295,000 | 4.89 | 1.89 | 30 | | | |
| Tethys | 295,000 | 4.89 | 1.89 | 1060 | 0.0085 | 1000 | 1.0 |
| Dione | 377,000 | 6.26 | 2.74 | 1120 | 0.014 | 1400 | 1.4 |
| Helene | 377,000 | 6.26 | 2.74 | 32 | | | |
| Rhea | 527,000 | 8.74 | 4.52 | 1530 | 0.032 | 1200 | 1.2 |
| Titan | 1,220,000 | 20.3 | 16.0 | 5150 | 1.83 | 1900 | 1.9 |
| Hyperion | 1,480,000 | 24.6 | 21.3 | 370 | | | |
| Iapetus | 3,560,000 | 59.1 | 79.3 | 1440 | 0.022 | 1000 | 1.0 |
| Phoebe | 13,000,000 | 215 | −550† | 230 | | | |

\* *Moons larger than 10 km in diameter only; does not include the 13 recently discovered small moons described in the text.*
\*\* *Mass of Earth's Moon = $7.4 \times 10^{22}$ kg = $1.3 \times 10^{-4}$ Saturn masses.*
† *Retrograde orbit.*

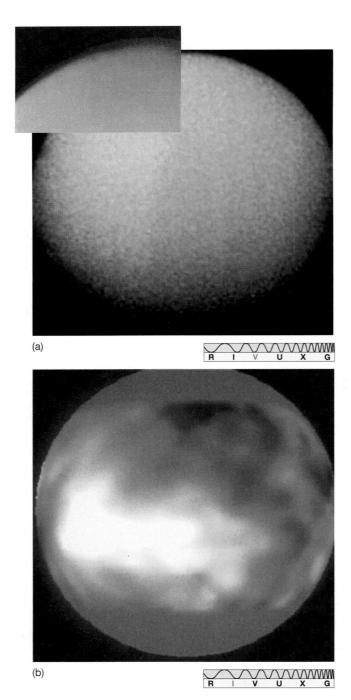

(a)

R I V U X G

(b)

R I V U X G

▲ FIGURE 12.21 **Titan** (a) Larger than the planet Mercury and roughly half the size of Earth, Titan, Saturn's largest moon, was photographed in visible light from only 4000 km away as *Voyager 1* passed by in 1980. Only Titan's upper cloud deck can be seen here. The inset shows a contrast-enhanced image of the haze layer (falsely colored blue). (b) In the infrared, as captured with the adaptive-optics system on the Canada–France–Hawaii telescope on Mauna Kea, some large-scale surface features can be seen. The bright regions are thought to be highlands, possibly covered with frozen methane. The brightest area is nearly 4000 km across—about the size of Australia. (*NASA; CFHT*)

been conclusively detected on Titan (*hydrocarbons* are molecules consisting solely of hydrogen and carbon atoms; methane, $CH_4$, is the simplest). Radio and infrared observations made by *Voyager 1* and *Cassini* showed that the atmosphere is actually made up mostly of nitrogen ($N_2$, roughly 90 percent) and argon (Ar, up to 10 percent), with a small percentage of methane and other trace gases.

Titan's atmosphere seems to act like a gigantic chemical factory. Powered by the energy of sunlight, the atmosphere is undergoing a complex series of chemical reactions that maintain steady (but trace) levels of hydrogen gas ($H_2$), the hydrocarbons ethane ($C_2H_6$) and propane ($C_3H_8$), and carbon monoxide (CO), ultimately resulting in the observed smog. The upper atmosphere is thick with aerosol haze (droplets so small that they remain suspended in the atmosphere), and the unseen surface appears to be covered with organic material that has settled down from the clouds.

Figure 12.22 shows the probable structure of Titan's atmosphere. The diagram was drawn following the *Voyager* flybys, but it is largely consistent with the later data returned by *Cassini*. Despite the moon's low mass (a little less than twice that of Earth's Moon), and hence its low surface *gravity* (one-seventh of Earth's), the atmospheric pressure at ground level is 60 percent greater than on Earth. Titan's atmosphere contains about 10 times more gas than Earth's atmosphere. Titan's surface temperature is a frigid 94 K, roughly what we would expect on the basis of that moon's distance from the Sun. At the temperatures typical of the lower atmosphere, methane and ethane may behave rather like water on Earth, raising the possibility of ethane snow and methane rain, fog, and even oceans! At higher levels in the atmosphere, the temperature rises, the result of photochemical absorption of solar radiation.

Because of Titan's weaker gravitational pull, the atmosphere extends some 10 times farther into space than does our own. The top of the main haze layer lies about 200 km above the surface, although there are additional layers, seen primarily through their absorption of ultraviolet radiation, at 300 km and 400 km. (See the insert in Figure 12.22). Below the haze the atmosphere is reasonably clear, although rather gloomy, because so little sunlight gets through. *Cassini* detected low-lying methane clouds at roughly the altitudes predicted by the model, although the clouds were less common than scientists had expected.

Why does Titan have such a thick atmosphere, when similar moons of Jupiter, such as Ganymede and Callisto, have none? The answer seems to be a direct result of Titan's greater distance from the Sun. The moons of Saturn formed at considerably lower temperature than did those of Jupiter. Such conditions would have enhanced the ability of the water ice that makes up the bulk of Titan's interior to absorb methane and ammonia, both of which were present in abundance at those early times. As a result, Titan was initially laden with much more methane and ammonia gas than was either Ganymede or Callisto.

# DISCOVERY 12-1

## Dancing Among Saturn's Moons

In Chapter 6 we described how mission planners routinely use "gravitational slingshots" to accelerate and redirect a spacecraft's trajectory as it speeds toward its target. ∞ *(Discovery 6-1)* Many of NASA's planetary missions have relied heavily on this expertise. ∞ *(Sec. 6.6)* The *Galileo* and *Cassini* probes also made heavy use of gravity assists to control the spacecrafts' intricate orbits within the moon systems of Jupiter and Saturn. Here we describe how *Cassini* has managed to make repeated close approaches to many of the moons in the Saturn system—the key to the extraordinary data now streaming back to scientists on Earth.

The first figure illustrates how a spacecraft can be moved onto a higher or lower orbit depending on the details of its interaction with a moon. Passing behind the moon, the spacecraft acquires some of the moon's energy and speeds up, moving farther from the planet. Conversely, passage in front of the moon slows the spacecraft down, reducing its orbital semimajor axis. In either case, the *direction* of the spacecraft's trajectory is changed, setting it on course for its next destination. In the case of *Cassini* and Saturn, the moon responsible for the lion's share of the close encounters is Titan—very convenient, as a close study of Titan is one of the mission's prime objectives.

By carefully combining a series of close encounters with Titan and other moons, *Cassini*'s controllers sent the probe throughout the moon system, allowing it to visit many of the moons multiple times. The second figure shows a possible trajectory computed before launch. Each orbit is the result of a carefully computed interaction with Titan or some other moon during the previous pass. In fact, this diagram differs in detail from *Cassini*'s actual flight path—small errors in the exact distance from each moon at one pass cause large changes in the next encounter, and the actual trajectory was revised several times during the mission.

*Cassini*'s primary mission is scheduled to end in June 2008. However, past experience with *Galileo* and many recent Mars probes suggests that the spacecraft will last much longer

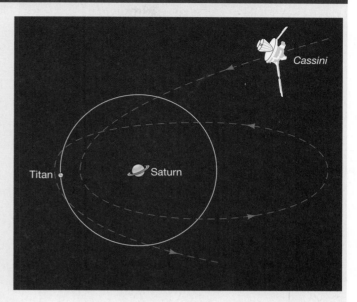

than that, and plans exist for an extended mission that will revisit moons of particular interest. Even as *Cassini*'s fuel supply dwindles, mission controllers should still be able to adjust its trajectory, using mainly Titan's gravity, to reach the desired parts of the Saturn system.

Eventually, though, the maneuvering fuel will be nearly gone and the mission will have to be brought to a close. Mission planners are undecided as to the best way to end the *Cassini* tour. They could place *Cassini* into a wide stable orbit, taking measurements for years to come until the probe's nuclear fuel cells are depleted. Or the probe could crash into Saturn, just as *Galileo* did with Jupiter. Even crash-landing on Titan has been considered, although that would risk polluting the moon with nuclear material. A final option is to send *Cassini* flying through the main rings of Saturn, taking fascinating photographs and amassing data until it "hits something big." Stay tuned!

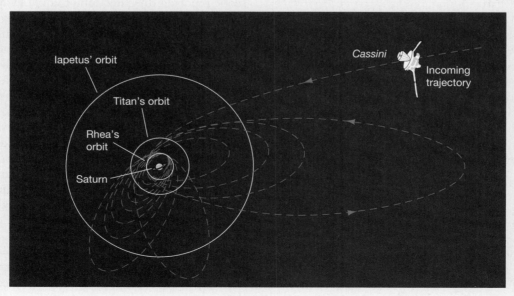

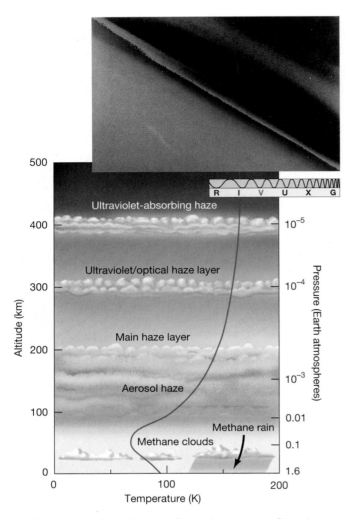

▲ **FIGURE 12.22 Titan's Atmosphere** The structure of Titan's atmosphere, as deduced from *Voyager 1* observations. The solid blue line represents temperature at different altitudes. The inset shows the haze layers in Titan's upper atmosphere, depicted in false-color green above Titan's orange surface in this *Voyager 1* image. *(NASA)*

As Titan's internal radioactivity warmed the moon, the ice released the trapped gases, forming a thick methane–ammonia atmosphere. Sunlight split the ammonia into hydrogen, which escaped into space, and nitrogen, which remained in the atmosphere. The methane, which is more tightly bound and so was less easily broken apart, survived intact. Together with argon outgassed from Titan's interior, these gases form the basis of the atmosphere we see on Titan today.

## Titan's Surface and Interior Structure

*Cassini's* observations have refined our models of Titan's atmosphere, but they have revolutionized our knowledge of the moon's surface and interior. Under the frigid conditions found on Titan, as on Ganymede and Callisto, water ice plays the role of rock on Earth, and liquid water the

role of lava. ∞ (Sec. 11.5) Before *Cassini* came on the scene, speculation about what might be found on Titan's surface ran the gamut from oceans of liquid methane or ethane to icy valleys laden with hydrocarbon sludge.

Figure 12.23 shows how *Cassini's* infrared instruments can penetrate the moon's atmosphere, revealing details of the surface. The image shows light and dark regions near the center of the field of view, thought to be icy plateaus, apparently smeared with hydrocarbon tar. Ridges and cracks on the moon's surface suggest that geological activity, in the form of "titanquakes," may be common. The rather blurred boundaries between the light and dark regions, the peculiar surface markings in the light-colored region, and the absence of extensive cratering suggest that some sort of erosion is occurring, perhaps as a result of wind or volcanic activity. Radar imaging reveals few large (10 to 100 km diameter) craters on the moon's surface, but there do not seem to be as many small craters as would be expected given Titan's location in Saturn's congested ring plane. The inset in the figure shows what may be an icy volcano, supporting the view that the moon's surface is geologically active.

In January 2005, the *Huygens* probe, transported to Saturn by *Cassini* and released 3 weeks earlier, arrived at Titan and parachuted through the thick atmosphere to the moon's surface (Figure 12.24a). Figure 12.24(b) shows an intriguing image radioed back from *Huygens* during its descent. Interpretations vary, but it appears to show a network of drainage channels leading to a shoreline. The probe landed on solid ground, and for the next hour transmitted images and instrument readings to *Cassini* as it passed overhead. The view from the surface (Figure 12.24c) reveals a hazy view of an icy landscape. The "rocks" in the foreground are a few centimeters across and show evidence of erosion by liquid of some sort. The nature of the "ocean" in Figure 12.24(b) remains uncertain—it may well consist of slush rather than liquid.

In 2003, radio astronomers using the Arecibo telescope reported the detection of liquid hydrocarbon lakes on Titan's surface. ∞ (Sec. 5.5) In early 2007, *Cassini* mission specialists confirmed this finding, presenting radar images showing numerous lakes, some many tens of kilometers in length, near Titan's north polar regions, currently the coolest part of the moon (see Figure 12.1, and note that Titan's tidal locking means that its rotation axis is parallel to that of Saturn). Figure 12.25 shows an example of these new images. As with the *Magellan* radar images of Venus, the darkest regions are extremely smooth, implying that they are composed of liquid; the shapes of the regions also strongly suggest liquid bodies. Given the surface temperature, the lakes are most likely composed of methane (which becomes liquid at temperatures below 110 K), possibly with ethane dissolved in it.

Scientists had long maintained that Titan's methane had to be maintained by liquid lakes on the surface or underground. These observations represent the first

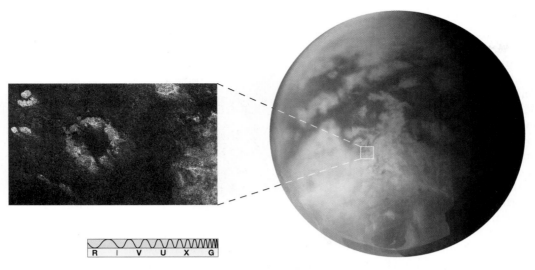

R I V U X G

▲ **FIGURE 12.23** **Titan Revealed** *Cassini's* infrared telescopes revealed this infrared, false-color view of Titan's surface in late 2004. The circular area near the center may be an old impact basin and the dark linear feature to its northwest perhaps mountain ranges caused by ancient tectonic activity. The inset shows a surface feature thought to be an icy volcano, further suggesting some geological activity on this icy moon's surface. Resolution of the larger image is 25 km; that of the inset is 10 times better. *(NASA/ESA)*

confirmation that such lakes actually exist, making Titan only the second body in the solar system (after Earth) known to have extensive liquid features on its surface. Titan's lakes are expected to vary with the seasons, and scientists hope that repeated observations on subsequent *Cassini* passes will show changes in the size and structure of these regions as winter on Titan gives way to spring.

Exploration of the lakes by *Cassini* and future missions may afford scientists the opportunity to study the kind of chemistry thought to have occurred billions of years ago on Earth—the prebiotic chemical reactions that eventually led to life on our own planet (see Chapter 28).

Astronomers had also long expected that Titan's internal composition and structure would be similar to those of

(a)

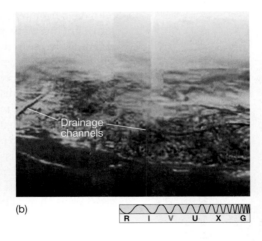

Drainage channels

(b)

R I V U X G

(c)

▲ **FIGURE 12.24** **The View from *Huygens*** (a) Artist's conception of the *Huygens* lander parachuting through Titan's thick atmosphere. (b) This photograph of the surface was taken from an altitude of 8 km as the probe descended. It shows a network of channels reminiscent of streams or rivers draining from the light-shaded uplifted terrain (at center) into a darker, low-lying region (at bottom). Resolution is about 20 m. (c) *Huygens's* view of its landing site, in approximately true color. The foreground rocklike objects are only a few centimeters across. *(D. Ducros; NASA/ESA)*

ANIMATION/VIDEO  *Huygens* Landing on Titan

327

Ganymede and Callisto, because all three moons have quite similar masses, radii, and, hence, average densities. (Titan's density is 1900 kg/m³. ∞ (Sec. 11.5) Titan contains a rocky core surrounded by a thick mantle of water ice. Each pass of *Cassini* allows scientists to probe the gravity of Titan, and repeated passes, coupled with knowledge of the properties of Titan's likely constituents, have resulted in the construction of some remarkably detailed models of the moon's interior. Figure 12.26 shows a recent such model. It indeed shows a rocky core and an icy mantle, but, intriguingly, also predicts the presence of a thick layer of liquid water a few tens of kilometers below the surface. Thus, Titan joins Europa, Ganymede, and Earth on the list of solar system objects containing large bodies of liquid water, with all that that implies for the prospects of life developing there. ∞ (Sec. 11.5)

## CONCEPT CHECK

✔ Why are planetary scientists so interested in Titan?

▲ **FIGURE 12.25 Titan's Lakes** In 2006, radar aboard *Cassini* detected numerous smooth regions (colored dark blue in this radio image), thought to be lakes of liquid methane, near Titan's north pole. The image is foreshortened since the radar system aboard the spacecraft was "pinging" signals over the horizon. The scene spans about 200 km vertically, with smallest details resolved to an amazing 0.5 km. *(NASA/ESA)*

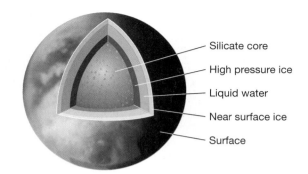

▲ **FIGURE 12.26 Titan's Interior** Based on measurements of Titan's gravitational field during numerous flybys, Titan's interior appears to be largely a rocky-silicate mix. Most intriguing is the subsurface layer of liquid water, similar to that hypothesized on Jupiter's Europa and Ganymede. *(NASA/ESA)*

## Saturn's Medium-Sized Moons

Saturn's complement of midsized moons consists (in order of increasing distance from the planet) of Mimas (at 3.1 planetary radii), Enceladus (4.0), Tethys (4.9), Dione (6.3), Rhea (8.7), and Iapetus (59.1). These moons are shown, to proper scale, in Figure 12.27. All six were known from Earth-based observations long before the Space Age. The inner five move on nearly circular trajectories, and all are tidally locked into synchronous rotation (so that one side always faces the planet) by Saturn's gravity. They therefore all have permanently "leading" and "trailing" faces as they move in their orbits, a fact that is important in understanding their often asymmetrical surface markings.

Unlike the densities of the Galilean satellites of Jupiter, the densities of these six moons do not show any correlation with distance from Saturn. The densities of Saturn's midsized moons are all between 1000 and 1400 kg/m³, implying that nearness to the central planetary heat source was a less important influence during their formation than it was in the Jupiter system. ∞ (Sec. 11.5) Scientists think that the midsized moons are composed largely of rock and water ice, like Titan. Their densities are lower than Titan's primarily because their smaller masses produce less compression of their interiors. All show heavy cratering, indicating the cluttered and violent planetary environment in the early solar system as fragments collided to form the outer planets and their moons. ∞ (Sec. 6.7)

The largest of the six, Rhea, has a mass only one-thirtieth that of Earth's Moon, and its icy surface is highly reflective and heavily cratered. At the low temperatures found on the surface of this moon, water ice is very hard and behaves rather like rock on the inner planets. For that reason, Rhea's surface craters look very much like craters on the Moon or Mercury. The density of craters is similar to that in the lunar highlands, indicating that the surface is old, and there is no evidence of extensive geological activity.

Prior to *Cassini*'s arrival, Rhea's main riddle was the presence of so-called *wispy terrain*—prominent light-colored streaks—on its trailing side (the right side of the image in Figure 12.27 and mapped in Figure 12.28a). The leading face, by contrast, shows no such markings, only craters. Astronomers thought that the wisps might have been caused by some event in the distant past during which water was somehow released from the interior and condensed on the surface. However, *Cassini* images reveal that the markings are in fact bright complexes of ice cliffs created by *tectonic fractures*, where stresses in the moon's icy interior as it cooled and solidified caused the surface layers to crack and buckle. ∞ (Sec. 10.4) Any similar features on the leading side have presumably been obliterated by cratering, which should be more frequent on the satellite's forward-facing surface.

Inside Rhea's orbit lie the orbits of Tethys and Dione. These two moons are comparable to each other in size and have masses somewhat less than half the mass of Rhea. Like Rhea, they have reflective surfaces that are heavily cratered, but each shows signs of surface activity, too. Dione's trailing face (at the left of the moon in Figure 12.27) has prominent bright streaks, which *Cassini* revealed to be ice cliffs, as just described on Rhea. As illustrated in more detail in Figure 12.28(b), the cliffs cut across many craters, showing them to be considerably more recent than the period of heaviest bombardment. ∞ (Sec. 8.5) Dione also has "maria" of sorts, where flooding appears to have obliterated the older craters. The cracks on Tethys (upper left in Figure 12.27) may also be tectonic fractures, or could possibly be the result of a violent impact early in the moon's history (a large impact basin lies on the far side of the moon in the view shown here).

The innermost, and smallest, medium-sized moon is Mimas. Despite its low mass—only 1 percent the mass of Rhea—its closeness to the rings causes resonant interactions with the ring particles, resulting most notably in the Cassini division, as we have already seen. Possibly

Mimas

Enceladus

Iapetus

Earth's
Moon

Rhea

Tethys

Dione

R I V U X G

▲ FIGURE 12.27 **Saturn's Mid-Sized Moons** Saturn's six medium-sized satellites, to scale, as seen by the *Cassini* spacecraft. All are heavily cratered and all are shown here in natural color. Iapetus has a ridge around its middle and shows a clear contrast between its light-colored (icy) surface (at top in this image) and its dark cratered hemisphere (at bottom). The light-colored wisps on Rhea are thought to be water and ice released from the moon's interior during some long-ago period of activity. Dione and Tethys show possible evidence of ancient geological activity. Enceladus appears to be volcanically active, presumably because of Saturn's tidal influence. Mimas's main surface feature is the large crater Herschel, plainly visible at the center in this image. For scale, part of Earth's Moon is shown at left. See also the full-page chapter opening photo for a closeup look at the fascinating moon, Enceladus. *(NASA)*

because of its proximity to the rings, Mimas is heavily cratered. The moon's chief surface feature is an enormous crater, called Herschel, on the leading face (at center in Figure 12.27). The diameter of this crater is almost one-third that of the moon itself. The impact that formed Herschel must have come very close to destroying Mimas completely. It is quite possible that the debris produced by such impacts is responsible for creating or maintaining the spectacular rings we see.

Enceladus orbits just outside Mimas. Its size, mass, composition, and orbit are so similar to those of Mimas that one might guess that the two moons would also be similar to each other in appearance and history. However, this is not so. Enceladus is so bright and shiny—it reflects virtually 100 percent of the sunlight falling on it—that astronomers think that its surface must be completely coated with fine crystals of pure ice, the icy "ash" of water "volcanoes" formed when liquid water emerges under pressure from the moon's interior.

*Voyager* found that the moon bears visible evidence of large-scale volcanic activity of some sort. Much of its surface is devoid of impact craters, which seem to have been erased by what look like lava flows, except that the "lava" is water, temporarily liquefied during recent internal upheavals and now frozen again. Arguing that the

processes involved may actually be more similar to the geothermal activity found in many volcanic regions on Earth, some astronomers prefer to describe these features as *geysers*, rather than volcanoes. Similar activity has been found on Neptune's moon Triton (see Section 13.5).

Long before *Cassini*'s arrival, the apparent association of Enceladus with the nearby thin cloud of small, reflective particles making up Saturn's E ring provided strong circumstantial evidence that the moon is responsible for the ring. The E ring is known to be densest near Enceladus. Calculations indicate that the ring is unstable because of the disruptive effects of the solar wind, supporting the view that volcanism on Enceladus continually supplies new particles to maintain the ring.

*Cassini* confirmed much of the speculation about Enceladus's internal activity and its connection with the E ring, finding evidence for every stage of the scenario just outlined. The probe detected icy jets emerging from geysers near the moon's south pole (see the chapter opening photo on page 307) and a transient water-vapor atmosphere surrounding the moon, densest around the south pole. *Cassini* also found a large increase in the density of E-ring particles near Enceladus as the atmosphere continually escapes from the moon's weak gravity. Thus the geysers steadily replenish the atmosphere as it escapes to replenish the E ring.

These remarkable findings still leave some questions unanswered. First, why is there so much activity on such a small moon? The answer remains unclear, but the best current explanation seems to be that the internal heating is the result of tidal stresses, much like those driving the volcanism on Io. ⚬ (Sec. 11.5) Enceladus appears to be locked into an orbital resonance with some of the other moons, causing its orbit to be slightly nonspherical. Even though Saturn's tidal force on Enceladus is only one-quarter of the force exerted by Jupiter on Io, the departure from perfect synchronism may be enough to cause the activity observed.

A second question raised by *Cassini* is, why is the activity concentrated on the *south* pole? Instead of being one of the coolest spots on the moon, as one might expect based on the small amount of sunlight reaching the surface there, the south pole of Enceladus is actually several

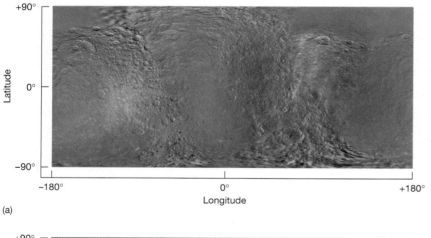

(a)

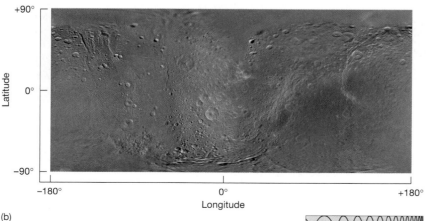

(b)

R I V U X G

◄ FIGURE 12.28  **Mid-Sized Moons, Close Up** Global maps of Rhea (a) and Dione (b), based on many images taken by *Cassini*. The suspected snow and ice cliffs are the whitish streaks throughout these flat maps. (*NASA*)

kelvins *warmer* than the equator! One possibility is that upwelling material at a warm spot on the moon may have caused the entire moon to "roll over" to place the low-density warm region on the rotation axis—that is, at the pole—in much the same way as a spinning bowling ball will tend to rotate in such a way as to place the (low-density) holes on the spin axis. Scientists await the next *Cassini* flyby in 2008 for more data to test this startling theory about this mysterious icy moon.

The outermost midsized moon is Iapetus. This moon orbits Saturn on a somewhat eccentric, inclined orbit with a semimajor axis of 3.6 million km. Its mass is about three-quarters that of Rhea. Iapetus is a two-faced moon. The dark, leading face (at the bottom in Figure 12.27) reflects only about 3 percent of the sunlight reaching it, whereas the icy trailing side reflects 50 percent. Spectroscopic studies of the dark regions seem to indicate that the material originates on Iapetus, in which case the moon is not simply sweeping up dark material as it orbits. Similar dark deposits seen elsewhere in the solar system are thought to be organic (containing carbon) in nature; they can be produced by the action of solar radiation on hydrocarbon (e.g. methane) ice. But how the dark markings can adorn only one side of Iapetus in that case is still unknown.

Iapetus's other prominent surface feature is a giant 20-km-high, 1400-km-long ridge spanning half of the moon's circumference. Discovered by *Cassini* in 2005, it is clearly visible cutting across the bottom third of the moon in Figure 12.27. It is unique in the solar system, and so far defies explanation.

## The Small Satellites

Finally we come to Saturn's dozen or so small moons. Their masses are poorly known (they are inferred mainly from their gravitational effects on the rings), but they are thought to be similar in composition to the small moons of Jupiter. The outermost small moons, Hyperion and Phoebe, were discovered in the 19th century, in 1848 and 1898, respectively. The others were first detected in the second half of the 20th century. Only the moons in or near the rings themselves were actually discovered by the *Voyager* spacecraft.

Just 10,000 km beyond the F ring lie the so-called *co-orbital satellites* Janus and Epimetheus. As the name implies, these two satellites "share" an orbit, but in a very strange way. At any given instant, both moons are in circular orbits about Saturn, but one of them has a slightly smaller orbital radius than the other. Each satellite obeys Kepler's laws, so the inner satellite orbits slightly faster than the outer one and slowly catches up to it. The inner moon takes about four Earth years to "lap" the outer one. As the inner satellite gains ground on the outer one, a strange thing happens: As illustrated in Figure 12.29, when the two get close enough to begin to feel each other's weak gravity, they switch orbits—the new inner moon (which used to be the outer one) begins to pull away from its companion, and the whole process begins again! No one knows how the co-orbital satellites came to be engaged in this curious dance. Possibly they are portions of a single moon that broke up, perhaps after a meteoritic impact, leaving the two pieces in almost the same orbit.

In fact, several of the other small moons also share orbits, this time with larger moons. Telesto and Calypso have orbits that are synchronized with the orbit of Tethys, so that the two smaller moons always remain fixed relative to the larger moon, lying precisely 60° ahead of and 60° behind it as it travels around Saturn (see Figure 12.30). The moon Helene is similarly tied to Dione. These 60° points are known as **Lagrangian points,** after the French mathematician Joseph Louis Lagrange, who first studied them.

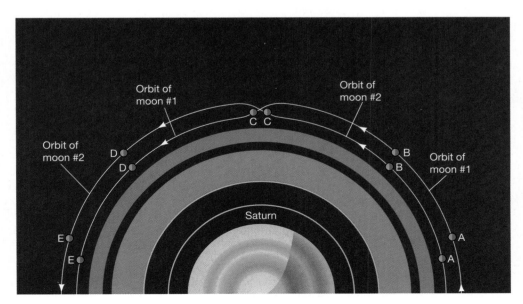

◄ FIGURE 12.29 **Orbit-Sharing Satellites** The peculiar motion of Saturn's co-orbital satellites Janus and Epimetheus, which play a never-ending game of tag as they move around the planet in their orbits. The labeled points represent the locations of the two moons at a few successive times. From A to C, moon 2 gains on moon 1. However, before it can overtake it, the two moons swap orbits, and moon 1 starts to pull ahead of moon 2 again, through points D and E. The whole process then repeats, apparently forever.

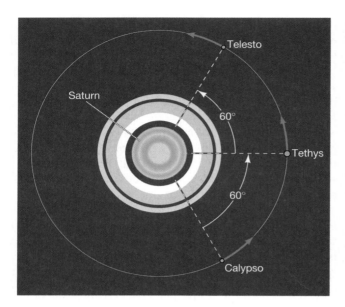

▲ **FIGURE 12.30 Synchronous Orbits** The orbits of the moons Telesto and Calypso are tied to the motion of the moon Tethys. The combined gravitational pulls of Saturn and Tethys keep the small moons exactly 60° ahead and behind the larger moon at all times, so all three moons share an orbit and never change their relative positions.

R   I   V   U   X   G

▲ **FIGURE 12.31 Hyperion** The small moon Hyperion is irregularly shaped and deeply pitted with dark-floored craters. Its spongy appearance is likely caused by dark material, trapped within the moon's many craters, that melt ever deeper into this low-density object. Its eccentric orbit and irregular shape means that Saturn's gravity causes it to tumble in a chaotic, unpredictable way. *(NASA)*

Later we will see further examples of this special 1:1 orbital resonance in the motion of some asteroids about the Sun, trapped in the Lagrangian points of Jupiter's orbit.

The strangest motion of all is that of the moon Hyperion (Figure 12.31), which orbits between Titan and Iapetus, at a distance of 1.5 million km from the planet. Unlike most of Saturn's moons, Hyperion has a rotation that is not synchronous with its orbital motion. Because of the gravitational effect of Titan, Hyperion's orbit is not circular, so synchronous rotation cannot occur. In response to the competing gravitational influences of Titan and Saturn, this irregularly shaped satellite constantly changes both its rotational speed and its axis of rotation, in a condition known as **chaotic rotation.** As Hyperion orbits Saturn, it tumbles apparently at random, never stopping and never repeating itself, in a completely unpredictable way.

Since the 1970s, the study of chaos on Earth has revealed new classes of unexpected behavior in even very simple systems. Hyperion is one of the few other places in the universe where this behavior has been unambiguously observed.

CONCEPT CHECK

✔ Why do Saturn's midsized moons show asymmetric surface markings?

# CHAPTER REVIEW

## Summary

**1** Saturn was the outermost planet known to ancient astronomers. Its rings and moons were not discovered until after the invention of the telescope. The rings lie in the planet's equatorial plane which is tilted at 27° with respect to the planet's orbit, so their appearance from Earth changes as Saturn orbits the Sun. Saturn is smaller than Jupiter, but still much larger than any of the terrestrial worlds. Like Jupiter, Saturn rotates rapidly, producing a pronounced flattening, and displays differential rotation. Strong radio emission from the planet's magnetosphere allows the rotation rate of the interior to be determined.

**2** As on Jupiter, weather systems are seen on Saturn, although they are less distinct. Large storms are occasionally seen. Saturn has weaker gravity and a more extended atmosphere than Jupiter. The planet's overall butterscotch hue is due to cloud chemistry similar to that occurring in Jupiter's atmosphere. Saturn, like Jupiter, has bands, ovals, and turbulent flow patterns powered by convective motion in the interior. *Cassini* imaged the planet's south polar vortex.

**3** Again like Jupiter, Saturn emits far more radiation into space than it receives from the Sun. Unlike Jupiter's, Saturn's excess energy emission is the result of **helium precipitation (p. 314)** in the planet's interior, where helium liquefies and forms droplets that then fall toward the center of the planet. This process is also responsible for Saturn's observed helium deficit. Saturn's interior is theoretically similar to that of Jupiter, but with a thinner layer of metallic hydrogen and a larger core. Its lower mass gives Saturn a less extreme core temperature, density, and pressure than Jupiter's core has. Saturn's conducting interior and rapid rotation produce a strong magnetic field and an extensive magnetosphere that contains the planet's ring system and many of the innermost moons.

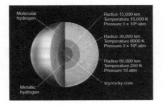

**4** From Earth, the main visible features of the rings are the **A, B,** and **C rings (p. 315)**, the **Cassini division (p. 315)**, and the **Encke gap (p. 315)**. The rings are made up of trillions of icy particles ranging in size from dust grains to boulders. Their total mass is comparable to that of a small moon. Both divisions are dark because they are almost empty of ring particles. The main rings contain tens of thousands of narrow **ringlets (p. 317)**. Interactions between the ring particles and the planet's inner moons are responsible for much of the fine structure observed. The narrow **F ring (p. 320)** lies just outside the A ring and has a kinked, braided structure,

apparently caused by two small **shepherd satellites (p. 320)** that orbit close to the ring and prevent it from breaking up. Beyond the F ring is the faint, narrow **G ring (p. 319)**, whose sharp edges and bright arcs suggest a shepherd moonlet, although none has been found. The faint **D ring (p. 319)**, lies between the C ring and Saturn's cloud layer. The diffuse **E ring (p. 319)** is associated with the moon Enceladus.

**5** The **Roche limit (p. 317)** of a planet is the distance within which the planet's tidal field would overwhelm the internal gravity of a moon, tearing it apart and forming a ring. All known planetary ring systems lie inside their parent planets' Roche limits. Planetary rings may have lifetimes of only a few tens of millions of years. If so, the fact that we see rings around all four jovian planets means that they must constantly be reformed or replenished, perhaps by material chipped off moons by meteoritic impact or by the tidal destruction of entire moons.

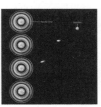

**6** Saturn's single large moon Titan is the second-largest moon in the solar system. Its thick atmosphere obscures the moon's surface and may be the site of complex cloud and surface chemistry. The moon's surface is so cold that water behaves like rock and liquid methane flows like water. Sensors aboard *Cassini* have allowed mission scientists to map the moon's surface for the first time, revealing evidence for ongoing erosion and volcanic activity. The *Huygens* probe landed on the icy surface and photographed what may be channels carved by flowing methane. The existence of Titan's atmosphere is a direct consequence of the cold conditions that prevailed at the time of the moon's formation.

**7** The medium-sized moons of Saturn are made up predominantly of rock and water ice. They show a wide variety of surface terrains, are heavily cratered, and are tidally locked into synchronous orbits by the planet's gravity. The innermost midsized moon Mimas exerts an influence over the structure of the rings. The Cassini division is the result of resonance between ring particles there and Mimas. The moon Iapetus has an equatorial ridge and a marked contrast between its leading and trailing faces, and Enceladus has a highly reflective appearance, the result of water "volcanoes" on its surface. Rhea and Dione have extensive ice cliffs on their surfaces, the result of cracking of the outer layers as the moons cooled. Saturn's small moons exhibit a wide variety of complex motion. Several moons "share" orbits, in some cases lying at the **Lagrangian points (p. 331)** 60° ahead of and 60° behind the orbit of a larger moon. The moon Hyperion tumbles in an unpredictable way as it orbits the planet.

## Review and Discussion

1. Seen from Earth, Saturn's rings sometimes appear broad and brilliant, but at other times seem to disappear. Why?

2. What is a ring crossing? When will the next one occur?

3. Why does Saturn have a less varied appearance than Jupiter?

4. What does Saturn's shape tell us about its deep interior?

5. Compare and contrast the atmospheres and weather systems of Saturn and Jupiter, and tell how the differences affect each planet's appearance.

6. Compare the thicknesses of Saturn's various layers (clouds, molecular hydrogen, metallic hydrogen, and core) with the equivalent layers in Jupiter. Why do the thicknesses differ?

7. What mechanism is responsible for the relative absence of helium in Saturn's atmosphere, compared with Jupiter's atmosphere?

8. Is Saturn as a whole deficient in helium relative to Jupiter?

9. When were Saturn's rings discovered? When did astronomers realize what they were?

10. What would happen to a satellite if it came too close to Saturn?

11. What evidence supports the idea that a relatively recent catastrophic event was responsible for Saturn's rings?

12. What effect does Mimas have on Saturn's rings?

13. What are shepherd satellites?

14. When *Voyager 1* passed Saturn in 1980, why didn't it see the surface of Titan, Saturn's largest moon?

15. Compare and contrast Titan with Jupiter's Galilean moons.

16. Why does Titan have a dense atmosphere, whereas other large moons in the solar system don't?

17. What is the evidence for geological activity on Enceladus?

18. What mystery is associated with Iapetus?

19. Describe the behavior of Saturn's co-orbital satellites.

20. Imagine what the sky would look like from Saturn's moon Hyperion. Would the Sun rise and set in the same way it does on Earth? How do you imagine Saturn might look?

## Conceptual Self-Test: True or False/Multiple Choice

1. Saturn probably does not have a rocky core.

2. Unlike the weather on Jupiter, no storm systems have ever been seen on Saturn.

3. Relative to Jupiter's atmosphere, Saturn's atmosphere is deficient in helium.

4. Saturn is the only planet with a ring system.

5. The composition of Saturn's ring particles is predominantly water ice.

6. Although Saturn's ring system is tens of thousands of kilometers wide, it is only a few tens of meters thick.

7. Saturn's rings exist because they lie within the planet's Roche limit.

8. Two small shepherd satellites are responsible for the unusually complex form of Saturn's F ring.

9. Titan's surface is obscured by thick clouds of ammonia ice.

10. Astronomers think that Titan's surface may be covered with water.

11. From Figure 12.1 ("Ring Orientation"), the next time Saturn's rings will appear roughly edge-on as seen from Earth will be around (a) 2004; (b) 2010; (c) 2020; (d) 2035.

12. Compared with the time it takes Jupiter to orbit the Sun once, the time it takes Saturn, which is twice as far away, to orbit the Sun is (a) significantly less than twice as long; (b) about twice as long; (c) significantly more than twice as long.

13. Saturn's cloud layers are much thicker than those of Jupiter because Saturn has (a) more moons; (b) lower density; (c) a weaker magnetic field; (d) weaker surface gravity.

14. According to Figure 12.5 ("Saturn's Zonal Flow"), the winds on Saturn are fastest at (a) the north pole; (b) 50° N latitude; (c) the equator; (d) 50° S latitude.

15. Saturn's icy–rocky core is roughly (a) half the mass of; (b) the same mass as; (c) twice as massive as; (d) 10 times more massive than planet Earth.

16. Of the following, which are most like the particles found in Saturn's rings? (a) house-sized rocky boulders; (b) grains of silicate sand; (c) asteroids from the asteroid belt; (d) fist-sized snowballs.

17. A moon placed at a planet's Roche limit will (a) change color; (b) break into smaller pieces; (c) develop a magnetic field; (d) flatten into a disk.

18. The atmospheric pressure at the surface of Titan is (a) less than; (b) about the same as; (c) about one-and-a-half times greater than; (d) about 16 times greater than the atmospheric pressure at Earth's surface.

19. A tidally locked moon of Saturn (a) always presents the same face to the planet; (b) does not rotate; (c) always stays above the same point on the planet's surface; (d) maintains a constant distance from all the other moons.

20. The moons Telesto and Calypso, orbiting at the Lagrangian points of Saturn and the moon Tethys (a) orbit twice as far from Saturn as does Tethys; (b) orbit closer to Saturn than does Tethys; (c) always stay the same distance apart; (d) always stay between Saturn and the Sun.

## Problems

*Algorithmic versions of these Problems are available in the Practice Problems module of the Companion Website.*
*The number of dots preceding each Problem indicates its approximate level of difficulty.*

1. • What is the angular diameter of Saturn's A ring, as seen from Earth at closest approach?

2. • What is the size of the smallest feature visible in Saturn's rings, as seen from Earth at closest approach with a resolution of 0.05″?

3. •• What would be the mass of Saturn if it were composed entirely of hydrogen at a density of 0.08 kg/m$^3$, the density of hydrogen at sea level on Earth? Assume for simplicity that Saturn is spherical. Compare your answer with Saturn's actual mass and with the mass of Earth.

4. • How long does it take for Saturn's equatorial flow, moving at 1500 km/h, to encircle the planet? Compare your answer with the wind-circulation time on Jupiter.

5. •• Use Stefan's law to calculate what the surface temperature on Saturn would be in the absence of any internal heat source if Saturn's actual surface temperature is 97 K and the planet radiates three times more energy than it receives from the Sun.

6. • On the basis of the data given in Sections 12.1 and 12.3 (Figure 12.9), estimate the average density of Saturn's core.

7. •• The text states that the total mass of material in Saturn's rings is about 10$^{15}$ tons (10$^{18}$ kg). Suppose the average ring particle is 6 cm in radius (the size of a large snowball) and has a density of 1000 kg/m$^3$. How many ring particles are there?

8. • What is the orbital speed of ring particles at the inner edge of the B ring, in km/s? Compare your answer with the speed of a satellite in low Earth orbit (500 km altitude, say). Why are these speeds so different?

9. • Show that Titan's surface gravity is about one-seventh of Earth's, as stated in the text. What is Titan's escape speed?

10. •• Assuming a spherical shape and a uniform density of 2000 kg/m$^3$, calculate how small an icy moon would have to be before a fastball pitched at 40 m/s (about 90 mph) could escape.

11. ••• Calculate the orbital radii of particles having the following properties: (a) a 3:1 orbital resonance with Tethys—that is, orbiting Saturn three times for every orbit of Tethys; (b) a 2:1 resonance with Mimas (two orbits for every orbit of Mimas); (c) a 3:2 resonance with Mimas (three orbits for every two of Mimas); (d) a 2:1 resonance with Dione.

12. •• Compare Saturn's tidal gravitational effect on Mimas with Mimas's own surface gravity.

13. •• Compare Saturn's tidal gravitational effect on Titan with Titan's own surface gravity. On the basis of these numbers and the corresponding numbers for Jupiter's Galilean moons, would you expect significant internal heating in Titan?

14. ••• Sunlight reflected back to Earth from a particle in Saturn's rings is Doppler shifted twice—first because of the relative motion between the source of the radiation (the Sun) and the ring particle and then again by the relative motion between the particle and the observer on Earth (see Section 3.5). As a result, if Earth, Saturn, and the Sun are roughly aligned (i.e., Saturn is near opposition), the observed Doppler shift corresponds to *twice* the particle's orbital speed. A certain solar spectral line, of wavelength 656.112 nm, is reflected from the rings and observed on Earth. If the rings happen to be seen almost edge-on, what is the line's observed wavelength in light reflected from (a) the approaching inner edge of the B ring? (b) the receding inner edge of the B ring? (c) the approaching outer edge of the A ring? and (d) the receding outer edge of the A ring?

15. ••• On the basis of the data given in the text, estimate the difference in orbital radii between Saturn's two co-orbital satellites.

*The Companion Website at www.aw-bc.com/chaisson provides algorithmically generated versions of each chapter's Problems, along with additional quizzes, an Animations & Videos gallery, an Images gallery, an interactive Glossary, and a full eBook.*

# 13

# URANUS AND NEPTUNE

## The Outer Worlds of the Solar System

The two outermost planets were unknown to the ancients and were discovered by telescopic observations: Uranus in 1781 and Neptune in 1846. Uranus and Neptune have similar bulk properties, so it is natural to consider them together; they are part of the jovian family of planets.

Yet as we study the properties of Uranus and Neptune in more detail, we find important differences between the outer and inner jovian worlds. The two outermost jovian planets are the smallest and least massive of the four, and their internal structures and the details of their atmospheric composition differ significantly from those of their larger jovian cousins. Their moons and rings, too, deviate from those found around Jupiter and Saturn. These differences are not mere anomalies; rather, they have much to tell us about the environment in which the outer planets formed and evolved.

## LEARNING GOALS

*Studying this chapter will enable you to*

1  Describe how both chance and calculation played major roles in the discoveries of the outer planets.
2  Summarize the similarities and differences between Uranus and Neptune, and compare these planets with the other two jovian worlds.
3  Describe what is known about the interiors of Uranus and Neptune.
4  Explain what the moons of the outer planets tell us about their past.
5  Contrast the rings of Uranus and Neptune with those of Jupiter and Saturn.

Visit www.aw-bc.com/chaisson for additional images, animations, quizzes, and eBook for this chapter.

# 13.1 The Discoveries of Uranus and Neptune

The two outermost planets, Uranus and Neptune, were unknown to ancient astronomers. Both were discovered after the dawn of the modern scientific age, and both discoveries are testaments to the power of two pillars of modern science—improved *technology* and mathematical *modeling.* ∞ (Sec. 1.2)

## Uranus

The planet Uranus was discovered by British astronomer William Herschel in 1781. Herschel was engaged in charting the faint stars in the sky when he came across an odd-looking object that he described as "a curious either nebulous star or perhaps a comet." Repeated observations showed that it was neither. The object appeared as a disk in Herschel's 6-inch telescope and moved relative to the stars, but it traveled too slowly to be a comet. Herschel soon realized that he had found the seventh planet in the solar system.

This was the first new planet discovered in well over 2000 years, and the event caused quite a stir at the time. The story goes that Herschel's first instinct was to name the new planet "Sidus Georgium" (Latin for "George's star"), after his king, George III of England. The world was saved from a planet named George by the wise advice of another astronomer, Johann Bode, who suggested instead that the tradition of using names from Greco–Roman mythology be continued and that the planet be named Uranus, after the father of Saturn.

Uranus is in fact just barely visible to the naked eye if you know exactly where to look. At opposition, it has a maximum angular diameter of 4.1″ and shines just above the unaided eye's threshold of visibility. It looks like a faint, undistinguished star. No wonder it went unnoticed by the ancients. Even today, few astronomers have seen it without a telescope.

Through a large Earth-based optical telescope (Figure 13.1), Uranus appears hardly more than a tiny pale greenish disk. With the flyby of *Voyager 2* in 1986, our knowledge of Uranus increased dramatically, although closeup images of the planet still showed virtually no surface detail (Figure 13.2). The apparently featureless atmosphere of Uranus contrasts sharply with the bands and spots visible on all the other jovian worlds. Some orbital and physical properties of Uranus are presented in the Uranus Data box on p. 342.

## Neptune

Following the discovery of Uranus, astronomers set about charting its orbit and quickly discovered a small discrepancy between the planet's predicted position and where they actually observed it. Try as they might, astronomers could not find an elliptical orbit that fit the

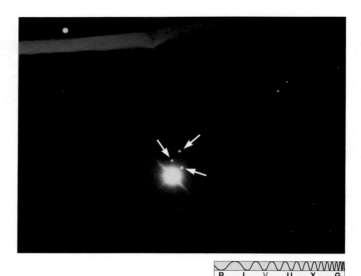

R  I  V  U  X  G

▲ **FIGURE 13.1 Uranus from Earth** Details are virtually invisible on photographs of Uranus made with large Earth-based telescopes. (Arrows point to three of the planet's moons.) *(UC/Lick Observatory)*

planet's trajectory to within the accuracy of their measurements. Half a century after Uranus's discovery, the discrepancy had grown to a quarter of an arc minute, far too big to be explained away as observational error.

The logical conclusion was that an unknown body must be exerting a gravitational force on Uranus—much

R  I  V  U  X  G

▲ **FIGURE 13.2 Uranus, Close Up** This image of Uranus, taken from a distance of about 1 million km, was sent back to Earth by the *Voyager 2* spacecraft as it whizzed past the giant planet at 10 times the speed of a rifle bullet. The image approximates the planet's true color, but shows virtually no detail in the nearly featureless upper atmosphere, except for a few wispy clouds in the northern hemisphere. *(NASA)*

weaker than that of the Sun, but still measurable. But what body could that be? Astronomers realized that there had to be *another* planet in the solar system perturbing Uranus's motion.

In the 1840s, two mathematicians independently solved the difficult problem of determining the new planet's mass and orbit. A British astronomer, John Adams, reached the solution in September 1845; in June of the following year, the French mathematician Urbain Leverrier came up with essentially the same answer. British astronomers seeking the new planet found nothing during the summer of 1846. In September, a German astronomer named Johann Galle began his own search from the Berlin Observatory, using a newly completed set of more accurate sky charts. He found the new planet within one or two degrees of the predicted position—on his first attempt. After some wrangling over names and credits, the new planet was named Neptune, and Adams and Leverrier (but not Galle!) are now jointly credited with its discovery.

Neptune's orbital and physical properties are listed in the Neptune Data box on p. 351. With an orbital period of 163.7 Earth years, Neptune has not yet completed one revolution since its discovery. Unlike Uranus, distant Neptune cannot be seen with the naked eye, although it can be seen with a small telescope—in fact, according to his notes, Galileo might actually have seen Neptune, although he had no idea what it really was at the time. Through a large telescope, Neptune appears as a bluish disk, with a maximum angular diameter of 2.4″ at opposition.

Figure 13.3 shows a long Earth-based exposure of Neptune and its largest moon, Triton. Neptune is so distant that surface features on the planet are virtually impossible to discern. Even under the best observational conditions, only a few markings can be seen. These features are suggestive of

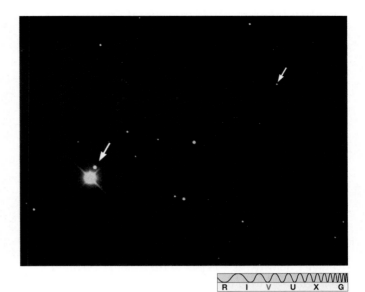

▲ FIGURE 13.3 **Neptune from Earth** Neptune and two of its moons, Triton (left arrow) and Nereid (right), imaged with a large Earth-based telescope. *(UC/Lick Observatory)*

multicolored cloud bands, with light bluish hues seeming to dominate. With *Voyager 2*'s arrival, much more detail emerged, as shown in Figure 13.4. Superficially, at least, Neptune resembles a blue-tinted Jupiter, with atmospheric bands and spots clearly evident.

## CONCEPT CHECK

✔ How did observations of the orbit of Uranus lead to the discovery of Neptune?

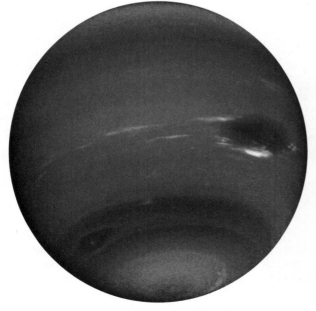

(a)

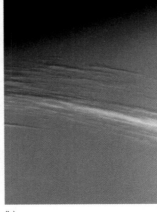

(b)

◄ FIGURE 13.4 **Neptune, Close Up** (a) Neptune as seen in natural color by *Voyager 2*, from a distance of roughly 1 million km. (b) A closer view, resolved to about 10 km, shows cloud streaks ranging in width from 50 km to 200 km. *(NASA)*

ANIMATION/VIDEO   Rotation of Neptune

## 13.2 Orbital and Physical Properties

With orbital semimajor axes of 19.2 and 30.1 AU, respectively, and orbital periods on the order of a century, Uranus and Neptune lie in the outer reaches of the Sun's planetary family. Their orbits lie just inside the Kuiper belt, for which (see Chapters 14 and 15) Uranus and Neptune are largely responsible.

Figure 13.5 shows Uranus and Neptune to scale, along with Earth for comparison. The two giant planets are quite similar to one another in their bulk properties. The radius of Uranus is 4.0 times that of Earth, that of Neptune 3.9 Earth radii. Their masses (first determined from terrestrial observations of their larger moons and later refined by *Voyager 2*) are 14.5 Earth masses for Uranus and 17.1 Earth masses for Neptune. Uranus's average density is 1300 kg/m$^3$, and Neptune's is 1600 kg/m$^3$. These densities imply that large rocky cores constitute a greater fraction of the planets' masses than do the cores of either Jupiter or Saturn. The cores themselves are probably comparable in size, mass, and composition to those of the two larger giants.

Like the other jovian planets, Uranus has a short rotation period. Earth-based observations of the Doppler shifts in spectral lines first indicated that Uranus's "day" was between 10 and 20 hours long. The precise value of the planet's rotation period—accurately determined when *Voyager 2* timed radio signals associated with Uranus's magnetosphere—is now known to be 17.2 hours. Again, as with Jupiter and Saturn, the planet's atmosphere rotates differentially. However, Uranus's atmosphere actually rotates *faster* at the poles (where the period is 14.2 hours) than near the equator (where the period is 16.5 hours).

Each planet in our solar system seems to have some outstanding peculiarity, and Uranus is no exception. Unlike all the other planets, whose spin axes are roughly perpendicular to the plane of the ecliptic, Uranus's axis of rotation lies almost within that plane—98° from the perpendicular, to be precise. (Because the north pole lies below the ecliptic plane, the rotation of Uranus, like that of Venus, is classified as retrograde.) We might say that, relative to the other planets, Uranus lies tipped over on its side. As a result, the north (spin) pole of Uranus, at some time in its orbit, points almost directly toward the Sun.* Half a Uranus year later, its south pole faces the Sun, as illustrated in Figure 13.6. When *Voyager 2* encountered the planet in 1986, the north pole happened to be pointing nearly at the Sun, so it was midsummer in the northern hemisphere.

The strange orientation of Uranus's rotation axis produces some extreme seasonal effects. Starting at the height of northern summer, when the north pole points closest to the Sun, an observer near that pole would find the Sun would never set. Rather, it would appear to move in a small circle in the sky around the planet's north celestial pole as the planet rotated, completing one circuit (counterclockwise) every 17 hours. ∞ (Sec. 1.3) Over time, as Uranus moved along its orbit and its rotation axis pointed farther and farther from the Sun, the circle would gradually increase in size, with the Sun dipping slightly lower in the sky each day. Eventually, the Sun would begin to set and rise again in a daily cycle, and the nights would grow progressively longer with each passing day. Twenty-one Earth years after the summer solstice, the autumnal equinox would occur, with day and night each 8.5 hours long.

The days would continue to shorten, until one day the Sun would fail to rise at all. The ensuing

*As in Chapter 9, we adopt the convention that a planet's rotation is always counterclockwise as seen from above the north pole (i.e., planets always rotate from west to east). ∞ (Sec. 9.2)

Jupiter

Saturn

Earth

Uranus

Neptune

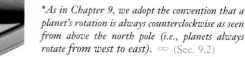

◀ FIGURE 13.5 **Jovian Planets** Jupiter, Saturn, Uranus, and Neptune, showing their relative sizes compared to Earth. Uranus and Neptune are quite similar in their bulk properties, each one probably having a core about 10 times more massive than Earth. Jupiter and Saturn are both much larger, but their rocky cores are probably comparable in mass to those of Uranus and Neptune. Note the very different atmospheric features of these five worlds. *(NASA)*

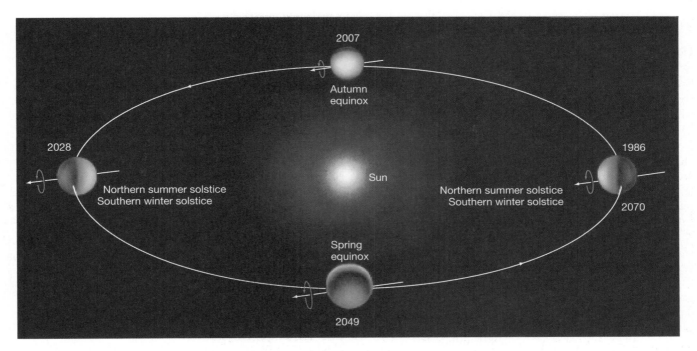

▲ FIGURE 13.6 **Seasons on Uranus** Because of Uranus's axial tilt of 98°, the planet experiences the most extreme seasons known in the solar system. The equatorial regions have two "summers" (warm seasons, around the times of the two equinoxes) and two "winters" (cool seasons, at the solstices) each year, and the poles are alternately plunged into darkness for 42 Earth years at a time.

period of total darkness would be equal in length to the earlier period of constant daylight, plunging the northern hemisphere into the depths of winter. Eventually, the Sun would rise again; the days would lengthen through the vernal equinox and beyond, and in time the observer would again experience a long summer of uninterrupted (though dim) sunshine. From the point of view of an observer on the equator, by contrast, summer and winter would be almost equally cold seasons, with the Sun never rising far above the horizon. Spring and fall would be the warmest times of year, with the Sun passing almost overhead each day.

No one knows why Uranus is tilted in this way—the other planets have rotation axes lying well out of the ecliptic plane. Some scientists have speculated that a catastrophic event, such as a glancing collision between the planet and another planet-sized body during the formative stages of the solar system, might have altered Uranus's spin axis. ∞ (Sec. 6.7) There is no direct evidence for such an occurrence, however, and no theory to tell us how we should seek to confirm it.

Neptune's clouds show more variety and contrast than do those of Uranus, and Earth-based astronomers studying them determined a rotation rate for Neptune even before *Voyager 2*'s flyby in 1989. The average rotation period of Neptune's atmosphere is 17.3 hours (quite similar to that of Uranus). Measurements of Neptune's radio emissions by *Voyager 2* showed that the magnetic field of the planet, and presumably also its interior, rotates once every 16.1 hours. Thus, Neptune is unique among the jovian worlds in that its atmosphere rotates *more slowly* than its interior. Neptune's axis of rotation is inclined 29.6° to a line perpendicular to the planet's orbital plane, quite similar to the 27° tilt of Saturn.

CONCEPT CHECK
✔ What is unusual about the rotation of Uranus?

## 13.3  The Atmospheres of Uranus and Neptune

### Composition

Spectroscopic studies of sunlight reflected from Uranus's and Neptune's dense clouds indicate that the two planets' outer atmospheres (the parts we actually measure spectroscopically) are quite similar to the atmospheres of Jupiter and Saturn. The most abundant element is molecular hydrogen ($H_2$, 84 percent), followed by helium (He, about 14 percent) and methane ($CH_4$), which is more abundant on Neptune (about 3 percent) than on Uranus (2 percent). Ammonia ($NH_3$), which plays such an important role in the Jupiter and Saturn systems, is not present in any significant quantity in the outermost jovian worlds.

## URANUS

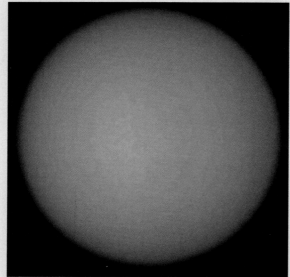

| | |
|---|---|
| Orbital semimajor axis | 19.19 AU |
| | 2871 million km |
| Orbital eccentricity | 0.047 |
| Perihelion | 18.29 AU |
| | 2736 million km |
| Aphelion | 20.10 AU |
| | 3006 million km |
| Mean orbital speed | 6.80 km/s |
| Sidereal orbital period | 83.75 tropical years |
| Synodic orbital period | 369.66 Earth days* |
| Orbital inclination to the ecliptic | 0.77° |
| Greatest angular diameter, as seen from Earth | 4.1″ |
| Mass | $8.68 \times 10^{25}$ kg |
| | 14.54 (Earth = 1) |
| Equatorial radius | 25,559 km |
| | 4.01 (Earth = 1) |
| Mean density | 1271 kg/m³ |
| | 0.230 (Earth = 1) |
| Surface gravity (at cloud tops) | 8.87 m/s² |
| | 0.91 (Earth = 1) |
| Escape speed | 21.3 km/s |
| Sidereal rotation period | −0.72 Earth day (retrograde) |
| Axial tilt | 97.92° |
| Surface magnetic field | 0.74 (Earth = 1) |
| Magnetic axis tilt relative to rotation axis | 58.6° |
| Mean surface temperature | 58 K |
| Number of moons | 27 (more than 10 km in diameter) |
| | 27 (total) |

*1 Earth (mean solar) day = 24 hours*

The abundances of gaseous ammonia and methane vary systematically among the jovian planets. Jupiter has much more gaseous ammonia than methane, but moving outward from the Sun, we find that the more distant planets have steadily decreasing amounts of ammonia and relatively greater amounts of methane. The reason for this variation is temperature. Ammonia gas freezes into ammonia ice crystals at about 70 K. This temperature is cooler than the cloud-top temperatures of Jupiter and Saturn, but warmer than those of Uranus (58 K) and Neptune (59 K). Thus, the outermost jovian planets have little or no *gaseous* ammonia in their atmospheres, so their spectra (which record atmospheric gases only) show only traces of ammonia.

The increasing amounts of methane are largely responsible for the outer jovian planets' blue coloration. Methane absorbs long-wavelength red light quite efficiently, so sunlight reflected from the planets' atmospheres is deficient in red and yellow photons and appears bluish-green or blue. As the concentration of methane increases, the reflected light should appear bluer—just the trend that is observed: Uranus, with less methane, looks bluish-green, whereas Neptune, with more, looks distinctly blue.

### Weather

*Voyager 2* detected just a few cloud features in Uranus's atmosphere (Figure 13.2), and even those became visible only after extensive computer enhancement. Figure 13.7 shows a series of *Hubble Space Telescope* views of the planet. Parts (a) through (c) are heavily processed optical images that show the progress of a pair of bright clouds around the planet. Part (d) shows a false-color, near-infrared rendition of Uranus. The colors in this image generally indicate the depth to which we can see into the atmosphere. Blue-green regions are clear atmospheric regions where astronomers can study conditions down to the lower cloud levels. Yellow-gray colors show sunlight reflecting from higher cloud layers or from atmospheric haze. Orange-red colors, such as the prominent "spots" on the south (right) edge of this image, indicate very high clouds, much like the wispy, white cirrus clouds often seen at high altitudes in Earth's atmosphere. Like cirrus clouds on Earth, these Uranian clouds are made up predominantly of ice crystals, formed in the planet's cold upper atmosphere.

Uranus apparently lacks any significant internal heat source, and because the planet has a low surface temperature, its clouds are found only at low-lying, warmer levels in the atmosphere. The absence of high-level clouds means that we must look deep into the planet's atmosphere to see any structure, so the bands and spots that characterize flow patterns on the other jovian worlds are largely "washed out" on Uranus by intervening stratospheric haze.

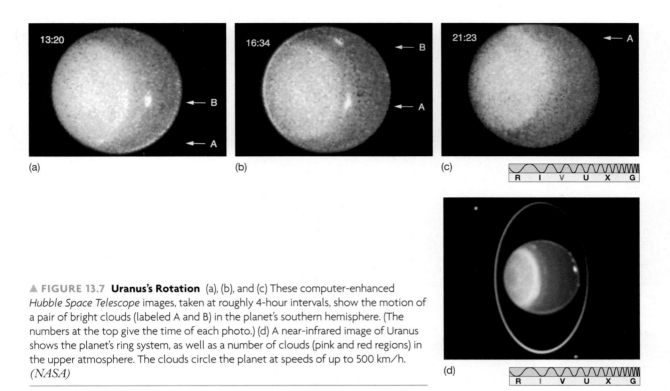

▲ FIGURE 13.7 **Uranus's Rotation** (a), (b), and (c) These computer-enhanced *Hubble Space Telescope* images, taken at roughly 4-hour intervals, show the motion of a pair of bright clouds (labeled A and B) in the planet's southern hemisphere. (The numbers at the top give the time of each photo.) (d) A near-infrared image of Uranus shows the planet's ring system, as well as a number of clouds (pink and red regions) in the upper atmosphere. The clouds circle the planet at speeds of up to 500 km/h. *(NASA)*

From computer-processed images such as those shown in Figure 13.7, astronomers have learned that Uranus's atmospheric clouds and flow patterns move around the planet in the same sense as the planet's rotation, with wind speeds ranging from 200 to 500 km/h. In fact, tracking these clouds allowed Uranus's differential rotation, mentioned earlier, to be measured. Despite the odd angle at which sunlight is currently striking the surface (recall that it is now late summer in the northern hemisphere), the planet's rapid rotation still channels the wind flow into bands reminiscent of those found on Jupiter and Saturn. Wind speeds are greater near the north pole, possibly because that part of the planet currently receives the greatest amount of solar heating. Even though the predominant wind flow is in the east–west direction, the atmosphere seems to be quite efficient at transporting energy from the heated north to the unheated southern hemisphere. Although much of the south is currently in total darkness, the temperature there is only a few kelvins less than in the north.

Neptune's clouds and band structure are much more easily seen than Uranus's. Although Neptune lies at a greater distance from the Sun, the planet's upper atmosphere is actually slightly warmer than that of Uranus. Like Jupiter and Saturn, but unlike Uranus, Neptune has an internal energy source—in fact, Neptune radiates 2.7 times more heat than it receives from the Sun. The cause of this heating is still uncertain. Some scientists have suggested that Neptune's excess methane has helped "insulate" the planet, tending to maintain its initially high internal temperature.

If that is so, then the source of Neptune's internal heat is the same as Jupiter's: energy left over from the planet's formation. ∞ (Sec. 11.3) The combination of extra heat and less haze may be responsible for the greater visibility of Neptune's atmospheric features (see Figure 13.8), as its cloud layers lie at higher levels in the atmosphere than do those of Uranus.

Neptune sports several storm systems similar in appearance to those seen on Jupiter (and assumed to be produced and sustained by the same basic processes). The largest such storm, known simply as the **Great Dark Spot,** is shown in Figure 13.8(a). ∞ (Sec. 11.2) Discovered by *Voyager 2* in 1989, the spot was about the size of Earth, was located near the planet's equator, and exhibited many of the same general characteristics as the Great Red Spot on Jupiter. The flow around it was counterclockwise, as with the Red Spot, and there appeared to be turbulence where the winds associated with the Great Dark Spot interacted with the zonal flow to its north and south. The flow around this and other dark spots may drive updrafts to high altitudes, where methane crystallizes out of the atmosphere to form high-lying cirrus clouds—those visible in Figure 13.8(a) lie some 50 km above the main cloud tops. Astronomers did not have long to study the Dark Spot's properties, however: As shown in Figure 13.8(b), when the *Hubble Space Telescope* viewed Neptune after the mid-1990s, the spot had vanished, although several new storms (bright spots) had appeared.

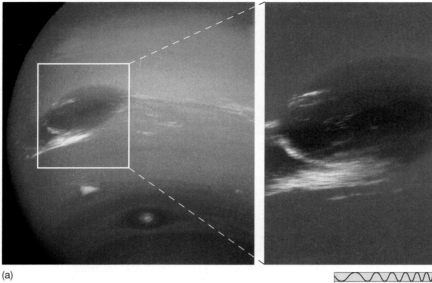

(a)

R I V U X G

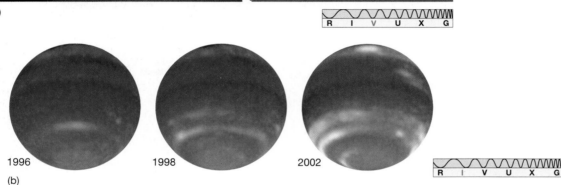

1996    1998    2002

(b)

R I V U X G

◀ FIGURE 13.8 **Neptune's Dark Spot** (a) Close-up views, taken by *Voyager 2* in 1989, of the Great Dark Spot of Neptune, a large storm system in the planet's atmosphere, possibly similar in structure to Jupiter's Great Red Spot. Resolution in the photo on the right is about 50 km; the entire dark spot is roughly the size of planet Earth. (b) These *Hubble Space Telescope* views of Neptune were taken years apart (as marked) when the planet was some 4.5 billion km from Earth. The cloud features (mostly methane ice crystals) are tinted pink here because they were imaged in the infrared, but they are really white in visible light. Note that the Great Dark Spot has disappeared in recent years. *(NASA)*

Infrared views such as those shown in Figure 13.8(b) reveal Neptune's dynamic weather patterns. The planet's weather can change in as little as a few rotation periods, and winds blow at speeds in excess of 1500 km/h—almost half the speed of sound in Neptune's upper atmosphere—with storms the size of Earth more the rule than the exception. The planet's stormy disposition is well established, but very difficult to understand. On Earth, weather systems are driven by the heat of the Sun. However, Neptune lies far from the Sun, in the outer solar system, and the Sun's heating effect is minuscule—nearly a thousand times less than at Earth. How can Neptune be so cold, yet so active?

Intriguingly, the three images of Neptune shown in Figure 13.8(b) reveal that the planet's southern hemisphere has brightened significantly over the period shown. Apparently, despite the Sun's faint heating, the planet is responding to the increase in solar energy as its southern half slowly moves from winter into spring.

CONCEPT CHECK

✔ Why are planetary scientists puzzled by the strong winds and rapidly changing storm systems on Neptune?

## 13.4 Magnetospheres and Internal Structure

*Voyager 2* found that both Uranus and Neptune have fairly strong internal magnetic fields—about a hundred times stronger than Earth's field and one-tenth as strong as Saturn's. However, because Uranus and Neptune are so much larger than Earth, the magnetic fields at the cloud tops—spread out over far larger volumes than is the field on Earth—are actually comparable in strength to Earth's field. Uranus and Neptune each have substantial magnetospheres, populated largely by electrons and protons either captured from the solar wind or created from ionized hydrogen gas escaping from the planets themselves.

When *Voyager 2* arrived at Uranus, it discovered that the planet's magnetic field was tilted at about 60° to the axis of rotation. On Earth, such a tilt would put the north magnetic pole somewhere in the Caribbean. Furthermore, on Uranus, the magnetic field lines are *not* centered on the planet. It is as though Uranus's field were due to a bar magnet that is tilted with respect to the planet's rotation axis and displaced from the center by about one-third the radius of the planet. Figure 13.9 shows the magnetic field structures of the four jovian planets, with Earth's also

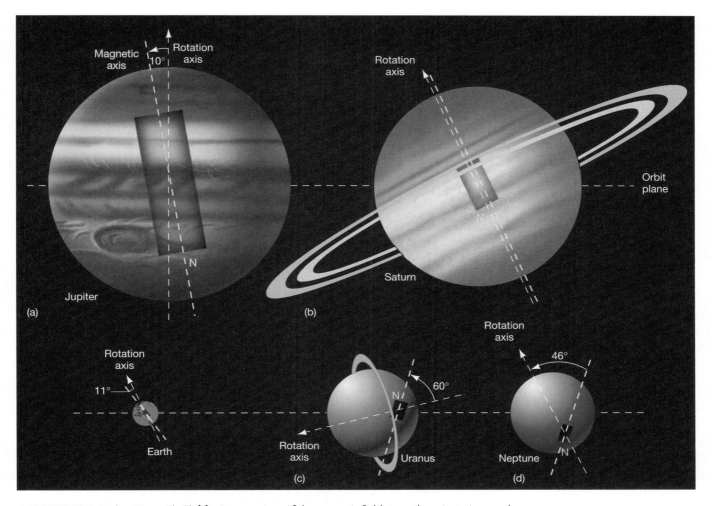

▲ FIGURE 13.9 **Jovian Magnetic Fields** A comparison of the magnetic field strengths, orientations, and offsets in the four jovian planets: (a) Jupiter, (b) Saturn, (c) Uranus, and (d) Neptune. The planets are drawn to scale, and in each case the magnetic field is represented as though it came from a bar magnet (a simplification, for purposes of illustration only). The size and location of each magnet represent the strength and orientation of the planetary field. Notice that the fields of Uranus and Neptune are significantly offset from the center of the planet and are much inclined to the planet's rotation axis. Earth's magnetic field is shown for comparison. One end of each magnet is marked N to indicate the polarity of Earth's field.

shown for comparison. The locations and orientations of the bar magnets represent the observed planetary fields, and the sizes of the bars indicate magnetic field strength.

Because dynamo theories generally predict that a planet's magnetic axis should be roughly aligned with its rotation axis—as on Earth, Jupiter, Saturn, and the Sun—the misalignment on Uranus suggested to some researchers that perhaps the planet's field had been caught in the act of reversing. ∞ (Sec. 7.5) Another possibility was that the oddly tilted field was in some way related to the planet's axial tilt—perhaps one catastrophic collision skewed both axes at the same time. Those ideas evaporated in 1989 when *Voyager 2* found that Neptune's field is also inclined to the planet's axis of rotation, at an angle of 46° (see Figure 13.9d), and also substantially offset from the center of the planet. It now appears that the internal

structures of Uranus and Neptune are different from those of Jupiter and Saturn, and this difference changes how the former planets' magnetic fields are generated.

Theoretical models indicate that Uranus and Neptune have rocky cores similar to those found in Jupiter and Saturn—about the size of Earth and perhaps 10 times more massive. However, the pressure outside the cores of Uranus and Neptune (unlike the pressure within Jupiter and Saturn) is too low to force hydrogen into the metallic state, so hydrogen stays in its molecular form all the way into the planets' cores. Astronomers theorize that deep below the cloud layers, Uranus and Neptune may have high-density, "slushy" interiors containing thick layers of water clouds. It is also possible that much of the planets' ammonia is dissolved in the hypothetical water, accounting for the absence of ammonia at higher levels. Such an

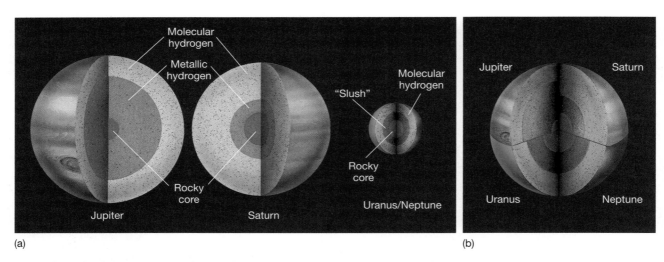

▲ FIGURE 13.10 **Jovian Interiors** A comparison of the interior structures of the four jovian planets. (a) The planets drawn to scale. (b) The relative proportions of the various internal zones.

ammonia solution would provide a thick, electrically conducting ionic layer that could conceivably explain the planets' misaligned magnetic fields if the circulating electrical currents that generate the fields occur mainly in regions far from the planets' centers and rotation axes.

At present, we simply don't know enough about the interiors of Uranus and Neptune to assess the correctness of this picture. Our current state of knowledge is summarized in Figure 13.10, which compares the internal structures of the four jovian worlds.

CONCEPT CHECK

✔ What is odd about the magnetic fields of Uranus and Neptune?

## 13.5 The Moon Systems of Uranus and Neptune

Like Jupiter and Saturn, both Uranus and Neptune have extensive moon systems, each consisting of a few large moons, long known from ground-based observations, and many smaller moonlets, discovered by *Voyager 2* or recently detected from Earth.

### Uranus's Moons

As of early 2007, some 27 moons are known to orbit Uranus. The properties of those more than 25 km in diameter are listed in Table 13.1.

William Herschel discovered and named Titania and Oberon, the two largest of Uranus's five major moons, in 1789. British astronomer William Lassell found Ariel and Umbriel, the next-largest moons, in 1851. Gerard Kuiper found Miranda, the smallest, in 1948. In order of increasing distance from the planet,

they are Miranda (at 5.1 planetary radii), Ariel (7.5), Umbriel (10.4), Titania (17.1), and Oberon (22.8). Ten smaller moons discovered by *Voyager 2* all lie inside the orbit of Miranda. Many of them are intimately related to the Uranian ring system. All of these moons revolve in the planet's skewed equatorial plane, almost perpendicular to the ecliptic, in circular, tidally locked orbits, sharing their parent's extreme seasons.

Of the remaining 12 moons, one, orbiting close to the planet, was found after careful reanalysis of *Voyager 2* images. All the rest were discovered via systematic ground-based searches made since 1997, with techniques similar to those which have been so successful in identifying new moons of Jupiter and Saturn. ∞ (*Discovery 11-3, Sec. 12.5*) These small bodies orbit far from Uranus, mostly on retrograde, highly inclined orbits. Like the outer moons of Jupiter and Saturn, and like Phobos and Deimos of Mars, each is thought to be interplanetary debris captured following a glancing encounter with the planet's atmosphere.

The five largest Uranian moons are similar in many respects to the six midsized moons of Saturn. ∞ (Sec. 12.5) Their densities lie in the range from 1100 to 1700 kg/m³, suggesting a composition of ice and rock, like Saturn's moons, and their diameters range from 1600 km for Titania and Oberon, to 1200 km for Umbriel and Ariel, to 480 km for Miranda. Uranus has no moons comparable to the Galilean satellites of Jupiter or to Saturn's single large moon, Titan. Figure 13.11 shows Uranus's five large moons to scale, along with Earth's Moon and Neptune's only midsized moon (named Proteus) for comparison.

The outermost of the five moons, Titania and Oberon, are heavily cratered and show little indication of geological activity. Their overall appearance (and quite possibly their history) is comparable to that of Saturn's

R  I  V  U  X  G

▲ FIGURE 13.11  **Moons of Uranus and Neptune**  The five largest moons of Uranus, and Proteus, the sole midsized moon of Neptune, are shown to scale, with part of Earth's Moon (also to scale) for comparison. In order of increasing distance from the planet, the Uranian moons are Miranda, Ariel, Umbriel, Titania, and Oberon. The smallest details visible on these moons are about 15 km across. The appearance, structure, and history of Titania and Oberon may be quite similar to those of Saturn's moon Rhea. Umbriel is one of the darkest bodies in the solar system, although it has a bright white spot on its sunward side. Ariel is similar in size, but has a brighter surface, with signs of past geological activity. *(NASA; Lick Observatory)*

moon Rhea, except that they lack Rhea's wispy streaks. Also, like all Uranian moons, they are considerably less reflective than Saturn's satellites, suggesting that their icy surfaces are quite dirty.

One possible reason for the lesser reflectivity may simply be that the planetary environment in the vicinity of Uranus and Neptune contains more small "sooty" particles than do the parts of the solar system that are closer to the Sun. An alternative explanation, now considered more likely by many planetary scientists, cites the effects of radiation and high-energy particles that strike the surfaces of these moons. The impacts tend to break up the molecules on the moons' surfaces, eventually leading to chemical reactions that slowly build up a layer of dark, organic material. This **radiation darkening** is thought to contribute to the generally darker coloration of many of the moons and rings in the outer solar system. In either case, the longer a moon has been inactive and

untouched by meteoritic impact, the darker its surface should be.

The darkest of the moons of Uranus is Umbriel. This moon displays little evidence of any past surface activity; its only mark of distinction is a bright spot about 30 km across, of unknown origin, in its northern hemisphere. By contrast, Ariel, similar in size to Umbriel, but closer to Uranus, does appear to have undergone some activity in the past. Ariel shows signs of resurfacing in places and exhibits surface cracks a little like those seen on another of Saturn's moons, Tethys. However, unlike Tethys, whose cracks are probably due to meteoritic impact, Ariel's activity likely occurred when internal forces and external tidal stresses (due to the gravitational pull of Uranus) distorted the moon and cracked its surface.

Strangest of all of Uranus's icy moons is Miranda, shown in Figure 13.12. Before the *Voyager 2* encounter, astronomers expected that Miranda would resemble

**TABLE 13.1   The Major Moons of Uranus***

| Name | Distance from Uranus (km) | (planetary radii) | Orbital Period (days) | Size (longest diameter, km) | Mass** (Earth Moon masses) | Density (kg/m³) | (g/cm³) |
|---|---|---|---|---|---|---|---|
| Cordelia | 49,800 | 1.95 | 0.34 | 26 | | | |
| Ophelia | 53,800 | 2.10 | 0.38 | 32 | | | |
| Bianca | 59,200 | 2.31 | 0.43 | 44 | | | |
| Cressida | 61,800 | 2.42 | 0.46 | 66 | | | |
| Desdemona | 62,700 | 2.45 | 0.47 | 58 | | | |
| Juliet | 64,400 | 2.52 | 0.49 | 84 | | | |
| Portia | 66,100 | 2.59 | 0.51 | 110 | | | |
| Rosalind | 69,900 | 2.74 | 0.56 | 58 | | | |
| Belinda | 75,300 | 2.94 | 0.62 | 68 | | | |
| Puck | 86,000 | 3.36 | 0.76 | 150 | | | |
| Miranda | 130,000 | 5.08 | 1.41 | 480 | 0.00090 | 1100 | 1.1 |
| Ariel | 191,000 | 7.48 | 2.52 | 1160 | 0.018 | 1600 | 1.6 |
| Umbriel | 266,000 | 10.4 | 4.14 | 1170 | 0.016 | 1400 | 1.4 |
| Titania | 436,000 | 17.1 | 8.71 | 1580 | 0.048 | 1700 | 1.7 |
| Oberon | 583,000 | 22.8 | 13.5 | 1520 | 0.041 | 1600 | 1.6 |
| Caliban (S/1997 U1) | 7,231,000 | 283 | −580[†] | 100 | | | |
| Sycorax (S/1997 U2) | 12,179,000 | 477 | −1290[†] | 190 | | | |
| Prospero (S/1999 U3) | 16,256,000 | 636 | −1980[†] | 30 | | | |
| Setebos (S/1999 U1) | 17,418,000 | 681 | −2230[†] | 30 | | | |

*For reasons of space, only moons more than 25 km in diameter are listed. We have omitted 8 moons with diameters between 10 and 25 km.*
** *Mass of Earth's Moon = $7.4 \times 10^{22}$ kg = $8.5 \times 10^{-4}$ Uranus mass.*
[†] *Retrograde orbit.*

Mimas, the moon of Saturn whose size and location it most closely approximates. However, instead of being a relatively uninteresting, cratered, geologically inactive world, Miranda displays a wide range of surface terrains, including ridges, valleys, large oval faults, and many other tortuous geological features.

To explain why Miranda seems to combine so many different types of surface features, some researchers have hypothesized that this baffling object has been catastrophically disrupted several times (by internal or external processes), with the pieces falling back together in a chaotic, jumbled way. Certainly, the frequency of large craters on the outer moons suggests that destructive impacts may once have been quite common in the Uranian system. It will be a long time, though, before we can obtain more detailed information to test this theory.

## Neptune's Moons

From Earth, we can see only two moons orbiting Neptune. William Lassell discovered Triton, the inner moon, in 1846. The moon Nereid was located by Gerard Kuiper in 1949. *Voyager 2* discovered six additional moons, all less than a few hundred kilometers across and all lying within Nereid's orbit. Five more small moons, on wide, eccentric orbits, have been discovered by ground-based surveys since 2002. The planet's 13 known moons are listed in Table 13.2. Proteus, Neptune's only midsized moon (by our previous definition) is shown in Figure 13.11. ∞ (Sec. 12.5)

In its moons, we find Neptune's contribution to our list of solar system peculiarities. Unlike the other jovian worlds, Neptune has no regular moon system—that is, no

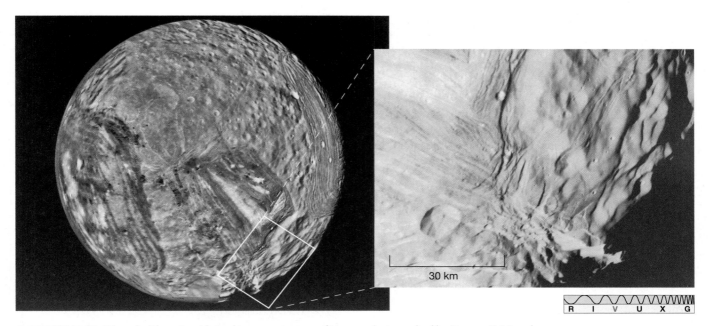

30 km

R I V U X G

▲ **FIGURE 13.12 Miranda** The asteroid-sized innermost moon of Uranus, photographed by *Voyager 2*. Miranda has a strange, fractured surface suggestive of a violent past, but the cause of the grooves and cracks is currently unknown. Resolution in the inset is a remarkable 2 km. The long "canyon" near the bottom of the inset is nearly 20 km deep. *(NASA)*

| TABLE 13.2 | The Moons of Neptune* | | | | | | |
|---|---|---|---|---|---|---|---|
| **Name** | **Distance from Neptune** (km) | (planetary radii) | **Orbital Period** (days) | **Size** (longest diameter, km) | **Mass**** (Earth Moon masses) | **Density** (kg/m³) | (g/cm³) |
| Naiad | 48,200 | 1.95 | 0.29 | 58 | | | |
| Thalassa | 50,100 | 2.02 | 0.31 | 80 | | | |
| Despina | 52,500 | 2.12 | 0.33 | 150 | | | |
| Galatea | 62,000 | 2.50 | 0.43 | 160 | | | |
| Larissa | 73,500 | 2.97 | 0.55 | 210 | | | |
| Proteus | 118,000 | 4.75 | 1.12 | 440 | | | |
| Triton | 355,000 | 14.3 | −5.88[†] | 2710 | 0.292 | 2100 | 2.1 |
| Nereid | 5,510,000 | 223 | 360 | 340 | 0.0000034 | 1200 | 1.2 |
| S/2002 N1 | 15,686,000 | 630 | −1870[†] | 60 | | | |
| S/2002 N2 | 22,337,000 | 900 | 2930 | 40 | | | |
| S/2002 N3 | 22,613,000 | 910 | 2980 | 40 | | | |
| S/2003 N1 | 46,738,000 | 1890 | −9140[†] | 40 | | | |
| S/2002 N4 | 47,280,000 | 1910 | −9010[†] | 60 | | | |

*All known moons of Neptune are currently estimated to be more than 10 km in diameter.*
**Mass of Earth's Moon = $7.4 \times 10^{22}$ kg = $7.3 \times 10^{-4}$ Neptune mass.**
[†] *Retrograde orbit.*

moons on roughly circular, equatorial, prograde orbits. The largest moon, Triton, is 2700 km in diameter and occupies a circular retrograde orbit 355,000 km (14.3 planetary radii) from the planet, inclined at about 20° to Neptune's equatorial plane. Triton is the only large moon in our solar system to have a retrograde orbit. The other moon visible from Earth, Nereid, is only 340 km across. This moon orbits Neptune in the prograde sense, but on an elongated trajectory that brings it as close as 1.4 million km to the planet and as far away as 9.7 million km. Nereid is probably similar in both size and composition to Neptune's small inner moons.

*Voyager 2* approached to within 24,000 km of Triton's surface, providing us with virtually all that we now know about that distant, icy world. Astronomers redetermined the moon's radius (correcting it downward by about 20 percent) and measured its mass for the first time. Along with Saturn's Titan and the four Galilean moons of Jupiter, Triton is one of the six large moons in the outer solar system. Triton is the smallest of them, with about half the mass of the next smallest, Jupiter's Europa.

Lying 4.5 billion km from the Sun, and with a fairly reflective surface, Triton has a surface temperature of just 37 K. It has a tenuous nitrogen atmosphere, perhaps a hundred thousand times thinner than Earth's, and a surface that most likely consists primarily of water ice. A *Voyager 2* mosaic of Triton's south polar region is shown in Figure 13.13. The moon's low temperatures produce a layer of nitrogen frost that forms and evaporates over the polar caps, a little like the carbon dioxide frost responsible for the seasonal caps on Mars. The frost is visible as the pinkish region at the right of the figure.

Overall, Triton exhibits a marked lack of cratering, presumably indicating that surface activity has obliterated the evidence of most impacts. There are many other signs of an active past. For example, Triton's face is scarred by large fissures similar to those seen on Ganymede, and Triton's odd cantaloupe-like terrain may indicate repeated faulting and deformation over the moon's lifetime. In addition, Triton has numerous frozen "lakes" of water ice (Figure 13.14), which may be volcanic in origin. The basic process may be similar to the water volcanism inferred (but not directly observed) on Saturn's moon Enceladus. ∞ (Sec. 12.5)

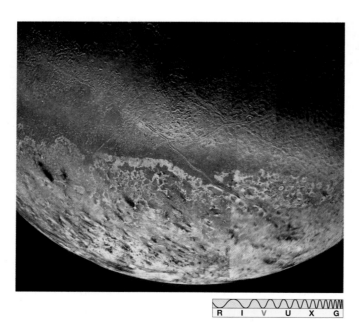

▲ FIGURE 13.13 **Triton** The south polar region of Triton, showing a variety of terrains ranging from deep ridges and gashes to what appear to be lakes of frozen water, all indicative of past surface activity. The pinkish region at lower right is nitrogen frost, forming the moon's polar cap. Resolution is about 4 km. The long black streaks at bottom left were probably formed by geysers of liquid nitrogen on the surface. (*NASA*)

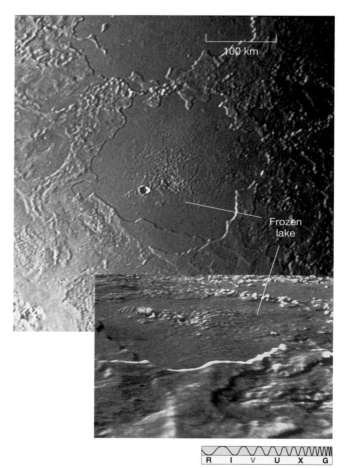

▲ FIGURE 13.14 **Water Ice on Triton** Scientists think that this roughly circular lakelike feature on Triton may have been caused by the eruption of an ice volcano. The water "lava" has since solidified, leaving a smooth surface. The absence of craters implies that this eruption was a relatively recent event. The frozen lake is about 200 km in diameter; its details are resolved to a superb 1 km. The inset is a computer-generated view along Triton's surface, illustrating the topography of the area. (*NASA*)

Triton's surface activity is not just a thing of the past. As *Voyager 2* passed the moon, its cameras detected two great jets of nitrogen gas erupting from below the surface and rising several kilometers above it. It is thought that these "geysers" form when liquid nitrogen below Triton's surface is heated and vaporized by some internal energy source or perhaps even by the Sun's feeble light. Vaporization produces high pressure, which forces the gas through fissures in the crust, creating the displays *Voyager 2* saw. Scientists conjecture that nitrogen geysers may be common on Triton and are perhaps responsible for much of the moon's thin atmosphere. The long black streaks at the bottom left of Figure 13.13 may have formed when geysers carried dark carbon-rich material from the moon's interior to the surface. Winds in Triton's thin atmosphere may also play a role in spreading the material over the surface.

The event or events that placed Triton on a retrograde orbit and Nereid on such an eccentric path are unknown, but they are the subject of considerable speculation. Triton's peculiar orbit and surface features suggest to some astronomers that the moon did not form as part of the Neptunian system, but instead was captured, perhaps not too long ago, astronomically speaking—maybe even as little as a few hundred million years. Other astronomers, basing their views on Triton's chemical composition, maintain that the moon formed "normally," but was later kicked into its abnormal orbit by some catastrophic event, such as an interaction with another, similar-sized body.

The surface deformations on Triton certainly suggest fairly violent and relatively recent events in the moon's past. However, they were most likely caused by the tidal stresses produced in Triton as Neptune's gravity made the moon's orbit more circular and synchronized its spin, and they give little indication of the processes responsible for the orbit.

Whatever its past, Triton's future is fairly clear. Because of its retrograde orbit, the tidal bulge Triton raises on Neptune tends to make the moon spiral *toward* the planet rather than away from it (as our Moon moves away from Earth). ∞ (Sec. 7.6) Thus, Triton is doomed to be torn apart by Neptune's tidal gravitational field, probably in no more than 100 million years or so, the time required for the moon's inward spiral to bring it inside Neptune's Roche limit. ∞ (Sec. 12.4) The shredded moon will form a new ring around the planet (see Figure 12.12). By that time, it is conceivable that Saturn's ring system may have disappeared, so Neptune will then be the only planet in the solar system with spectacular rings!

CONCEPT CHECK

✔ Why is Triton much less heavily cratered than the other moons of Uranus and Neptune?

## 13.6  The Rings of the Outermost Jovian Planets

All the jovian planets have rings. However, just as the ring system of Saturn differs greatly from that of Jupiter, the ring systems of Uranus and Neptune also differ both from one another and from those of the two larger jovian worlds.

### The Rings of Uranus

The ring system surrounding Uranus was discovered in 1977, when astronomers observed a **stellar occultation:** The rings passed in front of a bright star, momentarily dimming the star's light (Figure 13.15). Such an alignment happens a few times per decade and allows astronomers to measure planetary structures that are too small and faint to

be detected directly. The 1977 observation was actually aimed at studying the planet's atmosphere by watching how it absorbed starlight. However, 40 minutes before and after Uranus itself occulted (passed in front of) the star, the flickering starlight revealed the presence of a set of rings. The discovery was particularly exciting because, at the time, only Saturn was known to have rings. Jupiter's rings went unseen until *Voyager 1* arrived there in 1979, and those of Neptune were unambiguously detected only in 1989, by *Voyager 2*.

The ground-based observations revealed the presence of a total of nine thin rings around Uranus. The main rings, in order of increasing radius, are named Alpha, Beta, Gamma, Delta, and Epsilon, and they range from 44,000 to 51,000 km from the planet's center. All lie within the Roche limit of Uranus, which is about 62,000 km from the planet's center. A fainter ring, known as the Eta ring, lies between the Beta and

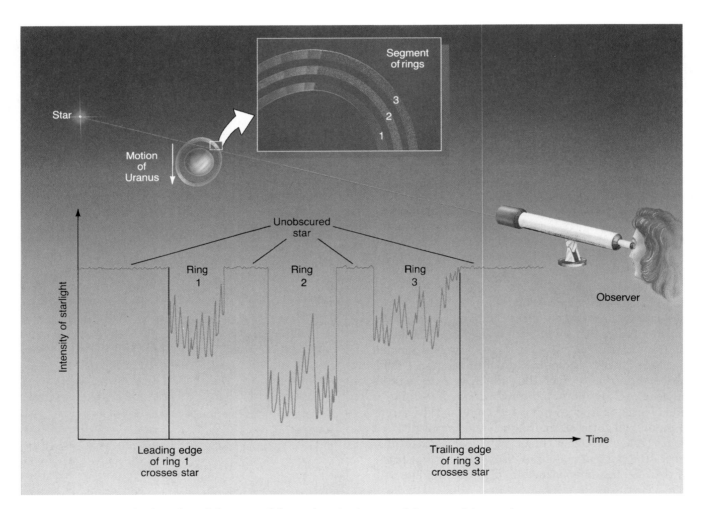

▲ FIGURE 13.15  **Occultation of Starlight**  By carefully watching the dimming of distant starlight as a planet crosses the line of sight, astronomers can infer fine details about that planet. The rings of Uranus were discovered with this technique.

Gamma rings, and three other faint rings, known as 4, 5, and 6, lie between the Alpha ring and the planet itself. In 1986, *Voyager 2* discovered two more even fainter rings, one between Delta and Epsilon and one between ring 6 and Uranus. The main rings are shown in Figure 13.16. More details on the rings are given in Table 13.3.

The rings of Uranus are quite different from those of Saturn. Whereas Saturn's rings are bright and wide, with relatively narrow gaps between them, the rings of Uranus are dark, narrow, and widely spaced. With the exception of the Epsilon ring and the diffuse innermost ring, the rings of Uranus are all less than about 10 km wide, and the spacing between them ranges from a few hundred to about a thousand kilometers. However, like Saturn's rings, all Uranus's rings are less than a few tens of meters

| **TABLE 13.3** | **The Rings of Uranus** | | | | |
|---|---|---|---|---|---|
| Ring | Inner Radius | | Outer Radius* | | Width |
| | (km) | (planetary radii) | (km) | (planetary radii) | (km) |
| 1986U2R | 37,000 | 1.45 | 39,500 | 1.55 | 2500 |
| 6 | 41,800 | 1.64 | | | 2 |
| 5 | 42,200 | 1.65 | | | 2 |
| 4 | 42,600 | 1.67 | | | 3 |
| Alpha | 44,700 | 1.75 | | | 4–10 |
| Beta | 45,700 | 1.79 | | | 5–11 |
| Eta | 47,200 | 1.83 | | | 2 |
| Gamma | 47,600 | 1.86 | | | 1–4 |
| Delta | 48,300 | 1.90 | | | 3–7 |
| 1986U1R | 50,000 | 1.96 | | | 2 |
| Epsilon | 51,200 | 2.00 | | | 20–100 |

\* *Most of Uranus's rings are so thin that there is little difference between their inner and outer radii.*

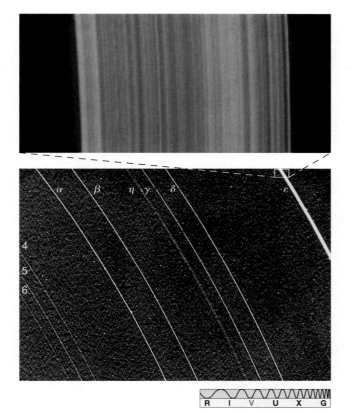

▲ FIGURE 13.16 **Uranus's Rings** The main rings of Uranus, as imaged by *Voyager 2*. All nine of the rings known before the spacecraft's arrival can be seen in this photo. From the inside out, they are labeled from 6 to Epsilon. Resolution is about 10 km, which is just about the width of most of these rings. The two rings discovered by *Voyager 2* are too faint to be seen here. The inset at top shows a close-up of the Epsilon ring, revealing some internal structure. The width of this ring averages 30 km; special image processing has magnified the resolution in the inset to about 100 m—the size of a football field. (*NASA*)

thick (that is, measured in the direction perpendicular to the ring plane).

The density of particles within Uranus's rings is comparable to that found in Saturn's A and B rings. The particles that make up Saturn's rings range in size from dust grains to large boulders, but in the case of Uranus, the particles show a much smaller spread—few, if any, are smaller than a centimeter or so in diameter. The ring particles are also considerably less reflective than Saturn's ring particles, possibly because they are covered with the same dark material as Uranus's moons. The Epsilon ring (shown in detail in the inset for Figure 13.16) exhibits properties a little like those of Saturn's F ring. It has a slight eccentricity of 0.008 and is of variable width, although no braids were found in it. It also appears to be composed of ringlets.

Like the F ring of Saturn, Uranus's narrow rings require *shepherd satellites* to keep them from diffusing. ∞ (Sec. 12.4) In fact, the theory of shepherd satellites was first worked out to explain the rings of Uranus, which had been detected by stellar occultation even before *Voyager 2*'s encounter with Saturn. Thus, the existence of the F ring did not come as quite such a surprise as it might have otherwise! Presumably, many of the small inner satellites of Uranus play some role in governing the appearance of the rings. *Voyager 2* detected Cordelia and Ophelia, the shepherds of the Epsilon ring (see Figure 13.17). Many other, undetected, shepherd satellites must also exist.

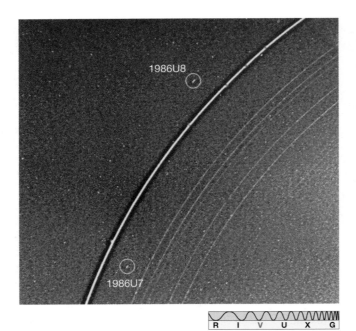

▲ **FIGURE 13.17 Uranian Shepherd Moons** These two small moons, named Cordelia (U7) and Ophelia (U8), were discovered by *Voyager 2* in 1986. They tend to shepherd Uranus's Epsilon ring, keeping it from diffusing. *(NASA)*

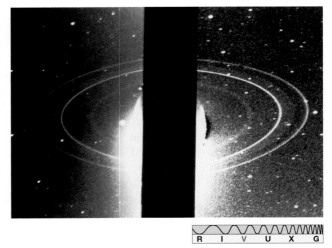

▲ **FIGURE 13.18 Neptune's Faint Rings** In this long-exposure image, Neptune (center) is heavily overexposed and has been artificially obscured (by an instrument) to make the rings easier to see. One of the two fainter rings lies between the inner bright ring (Leverrier) and the planet. The others lie between the Leverrier ring and the outer bright ring (known as the Adams ring). *(NASA)*

## The Rings of Neptune

As shown in Figure 13.18 and presented in more detail in Table 13.4, Neptune is surrounded by five dark rings. Three are quite narrow, like the rings of Uranus; the other two are broad and diffuse, more like Jupiter's ring. The dark coloration probably results from radiation darkening, as discussed earlier in the context of the moons of Uranus. All the rings lie within Neptune's Roche limit. The outermost (Adams) ring is noticeably clumped in places. From Earth, we see not a complete ring, but only partial arcs—the unseen parts of the ring are simply too thin (unclumped) to be detected. The connection between the rings and the planet's small inner satellites has not yet been firmly established, but many astronomers think that the clumping is caused by shepherd satellites.

Although all the jovian worlds have ring systems, the rings themselves differ widely from planet to planet. Is there some "standard" way in which rings form around a planet? Also, is there a standard manner in which ring systems evolve? Or do the processes of ring formation and evolution depend entirely on the particular planet in question? If, as now appears to be the case, ring systems are relatively short lived, their formation must be a fairly common event. Otherwise, we would not expect to find rings around all four jovian planets at once. There are many indications that the individual planetary environment plays an important role in determining a ring system's appearance and longevity. Although many aspects of ring formation and evolution are now understood, it must be admitted that no comprehensive theory yet exists.

### CONCEPT CHECK

✔ What does the Epsilon ring of Uranus have in common with the F ring of Saturn?

| Table 13.4 | The Rings of Neptune | | | | |
|---|---|---|---|---|---|
| **Ring** | **Inner Radius** | | **Outer Radius*** | | **Width** |
| | (km) | (planetary radii) | (km) | (planetary radii) | (km) |
| Galle (1989N3R) | 40,900 | 1.65 | 42,900 | 1.73 | 2000 |
| Leverrier (1989N2R) | 53,200 | 2.15 | | | 100 |
| Lassell (1989N4R)** | 53,200 | 2.15 | 57,200 | 2.31 | 4000 |
| Arago (1989N4R)** | 57,200 | 2.31 | | | 100 |
| Adams (1989N1R) | 62,900 | 2.54 | | | 50 |

*\* Three of Neptune's rings are so thin that there is little difference between their inner and outer radii.*
*\*\* Lassell and Arago were originally identified as a single ring.*

# CHAPTER REVIEW

## Summary

**1** The outer planets Uranus and Neptune were unknown to ancient astronomers. Uranus was discovered in the 18th century, by chance. Neptune was discovered after mathematical calculations of Uranus's slightly non-Keplerian orbit revealed the presence of an eighth planet. At opposition, Uranus is barely visible to the unaided eye. Through a telescope, the planet appears as a pale green disk. Neptune cannot be seen with the naked eye, but a telescope shows it as a tiny bluish disk. Today, we know the giant planets Uranus and Neptune mainly through data taken by *Voyager 2*.

**2** The masses of the outer planets are determined from measurements of their orbiting moons. The radii of Uranus and Neptune were relatively poorly known until the *Voyager 2* flybys in the 1980s. Uranus and Neptune have similar bulk properties; they are smaller, less massive, and denser than Jupiter or Saturn. For unknown reasons, Uranus's spin axis lies nearly in the plane of the ecliptic, leading to extreme seasonal variations in solar heating on the planet as it orbits the Sun. Surface features are barely discernible on Uranus, but computer-enhanced images from *Voyager 2* revealed atmospheric clouds and flow patterns moving beneath the planet's haze. Neptune, although farther away from us, has atmospheric features that are clearer because of warmer temperatures and less haze. The **Great Dark Spot (p. 343)** on Neptune had many similarities to Jupiter's Red Spot, but disappeared in 1994.

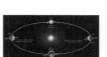

**3** The relatively high densities of Uranus and Neptune imply large, rocky cores making up greater fractions of the planets' masses than in either Jupiter or Saturn.

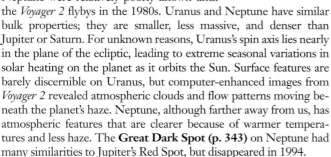

Unlike the other jovian planets, Uranus has no excess heat emission. The source of Neptune's excess energy, like that of Jupiter's, is most likely heat left over from the planet's formation. Both Uranus and Neptune have substantial magnetospheres. *Voyager 2* discovered that the magnetic fields of the two planets are tilted at large angles to the planets' rotation axes. The reason for these large tilts is not known.

**4** All but two of Uranus's moons revolve in the planet's equatorial plane, almost perpendicular to the ecliptic, in circular synchronous orbits. Like the moons of Saturn, the medium-sized moons of Uranus are made up predominantly of rock and water ice. Many of them are heavily cratered and in some cases must have come close to being destroyed by the meteoritic impacts whose craters we now see. The strange moon Miranda has geological features that suggest repeated violent impacts in the past. Neptune's large moon Triton has a fractured surface of water ice and a thin atmosphere of nitrogen, probably produced by nitrogen "geysers" on its surface. Triton is the only large moon in the solar system to have a retrograde orbit around its parent planet. This orbit is unstable and will eventually cause Triton to be torn apart by Neptune's gravity.

**5** Uranus has a series of dark, narrow rings, first detected from Earth by **stellar occultation (p. 352)**—their obscuration of the light received from background stars. Shepherd satellites are responsible for the rings' thinness. Neptune has three narrow rings like Uranus's and one broad ring like Jupiter's. The four were discovered by *Voyager 2*. The dark coloration of both the rings and the moons of the outer giant planets may be due to **radiation darkening (p. 347),** whereby exposure to solar high-energy radiation slowly causes a dark hydrocarbon layer to build up on a body's icy surface.

## Review and Discussion

1. How was Uranus discovered?

2. Why did astronomers suspect an eighth planet beyond Uranus?

3. How did astronomers determine where to look for Neptune?

4. How did Uranus come to be spinning "on its side"?

5. What is responsible for the overall colors of Uranus and Neptune?

6. Why are storms and other atmospheric features more easily seen on Neptune than on Uranus?

7. How are the interiors of Uranus and Neptune thought to differ from those of Jupiter and Saturn?

8. How do the magnetic fields of Uranus and Neptune compare with that of Earth?

9. Describe a day on Titania.

10. What is unique about Miranda? Give a possible explanation.

11. Why are the icy moons of the outer planets so dark?

12. How does Neptune's moon system differ from those of the other jovian worlds? What do these differences suggest about the origin of Neptune's moon system?

13. What causes Triton's geysers?

14. What is the predicted fate of Triton?

15. How were the rings of Uranus discovered?

16. How were the rings of Neptune discovered?

17. The rings of Uranus are dark, narrow, and widely spaced. Which of these properties makes them different from the rings of Saturn?

18. Why are the Uranian rings so narrow and sharply defined?

19. How do the rings of Neptune differ from those of Uranus and Saturn?

20. Why was the discovery of Uranus in 1781 so surprising? Might there be similar surprises in store for today's astronomers?

## Conceptual Self-Test: True or False/Multiple Choice

1. Since its discovery, Uranus has completed just two-and-a-half orbits of the Sun.

2. During the northern summer of Uranus, an observer near the north pole would observe the Sun high and almost stationary in the sky.

3. Uranus's rotation axis is almost parallel to the plane of the ecliptic.

4. *Voyager 2* observed nitrogen geysers on the surface of Neptune.

5. The magnetic fields of both Uranus and Neptune are highly tilted relative to their rotation axes and significantly offset from the planets' centers.

6. Triton's orbit is unusual because it is retrograde.

7. Uranus was discovered via its gravitational effect on Neptune.

8. The radius of Uranus was determined accurately by observing the planet pass in front of a background star during the late 1970s.

9. Triton is smaller than Earth's Moon.

10. Neptune has the most extensive ring system in the solar system.

11. The discovery of new planets mostly requires (a) complex calculations and large supercomputers; (b) the patient use of improving technology; (c) an astronomy degree from a large university; (d) pure luck.

12. Uranus was discovered about the same time as (a) Columbus reached North America; (b) the U.S. Declaration of Independence; (c) the American Civil War; (d) the Great Depression in the United States.

13. Compared with Uranus, the planet Neptune is (a) much smaller; (b) much larger; (c) roughly the same size; (d) tilted on its side.

14. The jovian planets with the largest diameters also tend to (a) have the slowest rotation rates; (b) move most slowly in their orbit around the Sun; (c) have the fewest moons; (d) have magnetic field axes most closely aligned with their axes of rotation.

15. The five largest moons of Uranus (a) all orbit in the ecliptic plane; (b) can never come between Uranus and the Sun; (c) all orbit directly above the planet's equator; (d) all have significantly eccentric orbits.

16. Moons that show few craters probably (a) are captured asteroids; (b) have been shielded from impacts by their host planet; (c) have had their smaller craters obliterated by larger impacts; (d) have warm interiors.

17. A gas giant planet orbiting a distant star would be expected to have (a) a ring system like that of Saturn; (b) a density less than water; (c) many large moons orbiting in different directions; (d) evidence for hydrogen in its spectrum.

18. Uranus's rings were discovered by the occultation of starlight, as shown in Figure 13.15 ("Occultation of Starlight"). If Uranus were moving more rapidly relative to Earth, the graph in that figure would appear (a) more compressed horizontally; (b) the same; (c) more stretched out horizontally.

19. The discovery of a moon orbiting a planet allows astronomers to measure (a) the planet's mass; (b) the moon's mass and density; (c) the planet's ring structure; (d) the planet's cratering history.

20. The solar system object most similar to Neptune is (a) Earth; (b) Jupiter; (c) Saturn; (d) Uranus.

# Problems

*Algorithmic versions of these Problems are available in the Practice Problems module of the Companion Website.*
*The number of dots preceding each Problem indicates its approximate level of difficulty.*

1. •• Calculate the time between successive closest approaches of (a) Saturn and Uranus and (b) Uranus and Neptune. For simplicity, assume circular orbits and calculate the time taken for the inner plate to "lap" the other one, as in problem 5 of Chapter 9. ∞ *(More Precisely 9-1)* Compare your answers with Neptune's sidereal orbital period.

2. • What is the gravitational force exerted on Uranus by Neptune, at closest approach? Compare your answer with the Sun's gravitational force on Uranus.

3. •• What is the angular diameter of the Sun, as seen from Uranus? Compare your answer with the angular diameter of Titania, as seen from the planet's cloud tops. Would you expect solar eclipses to occur on Uranus?

4. •• If the core of Uranus has a radius twice that of planet Earth and an average density of 8000 kg/m³, what is the mass of Uranus outside the core? What fraction of the planet's total mass is core?

5. • Estimate the speed of cloud A in Figure 13.7, assuming that it lies near the equator. Is your estimate consistent with the rotation speed of the planet?

6. • Estimate the strength of Neptune's gravitational tidal acceleration on Triton, relative to the moon's own surface gravity. Compare your answer with the corresponding ratio for Jupiter and Io. (See Chapter 11, Problem 10.)

7. • Add up the masses of all the moons of Uranus and Neptune. (Neglect the masses of the small moons—they contribute little to the result.) How does this sum compare with the mass of Earth's Moon?

8. ••• Astronomers on Earth are observing the occultation of a star by Uranus and its rings (see Figure 13.15). It so happens that the event is occurring when Uranus is at opposition, and the center of the planet appears to pass directly across the star. Assuming circular planetary orbits and, for simplicity, taking the rings to be face-on, calculate (a) how long the 90-km-wide Epsilon ring takes to cross the line of sight and (b) the time interval between the passage of the Alpha ring and that of the Epsilon ring. (*Hint:* The apparent motion of Uranus is due to a combination of both the planet's own motion and Earth's motion in their respective orbits around the Sun.)

9. •• On the basis of the earlier discussion of planetary atmospheres, would you expect Triton to have retained a nitrogen atmosphere? ∞ *(More Precisely 8-1)*

10. •• How long does it take for the inner shepherd moon in Figure 13.17 to "lap" the outer one?

11. ••• Neptune's apparent motion relative to the stars is actually due mainly to Earth's orbital motion. Given this fact, estimate the angle through which Neptune moves over the course of a week around the time of opposition.

12. • What would be your weight on Neptune? On Triton?

13. • How close is Triton to Neptune's Roche limit?

14. • From Wien's law, at what wavelength does Uranus's thermal emission peak? In what part of the electromagnetic spectrum does this wavelength lie? ∞ *(More Precisely 3-2)*

15. • What is the round-trip travel time of light from Earth to Neptune (at a distance of 30 AU)? How far would a spacecraft orbiting the planet at a speed of 0.5 km/s travel during that time?

*The Companion Website at www.aw-bc.com/chaisson provides algorithmically generated versions of each chapter's Problems, along with additional quizzes, an Animations & Videos gallery, an Images gallery, an interactive Glossary, and a full eBook.*

# 14

Only within the past couple of decades have scientists taken seriously the idea that life on Earth has been disrupted over the course of billions of years by asteroid and comet impacts. Here, by contrast, is the impact of a comet by a probe sent to collide with it. This image of the oblong, 6-km-long nucleus of Comet Tempel 1 was taken 1 minute after it obliterated a small projectile launched from the Deep Impact spacecraft that glided past in 2005. The objective was to gain a better understanding of comets and other nearby cosmic debris, especially given the external threat they represent to life on Earth. (JPL) ▶

# SOLAR SYSTEM DEBRIS

## Keys to Our Origin

According to the current definition, there are only eight planets in the solar system. But hundreds of thousands of other celestial bodies are also known to revolve around the Sun. These minor bodies—the asteroids, comets, Kuiper belt objects, and meteoroids—are small and of negligible mass compared with the planets and their major moons. On the basis of statistical deductions, astronomers estimate that there are more than a billion such objects still to be discovered. Yet each is a separate world, with its own story to tell about the early solar system.

These small bodies may seem to be only rocky and icy "debris," but more than the planets themselves, they hold a record of the formative stages of our planetary system. Many are nearly pristine, unevolved bodies with much to teach us about our local origins.

## LEARNING GOALS

*Studying this chapter will enable you to*

1   Describe the orbital properties of the major groups of asteroids.
2   Summarize the composition and physical properties of a typical asteroid.
3   Detail the composition and structure of a typical comet, and explain the formation and appearance of its tail.
4   Discuss the characteristics of cometary orbits and what they tell us about the probable origin of comets.
5   Describe the composition of the solar system beyond Neptune, and explain why astronomers no longer regard Pluto as a planet.
6   Distinguish among the terms *meteor, meteoroid,* and *meteorite.*
7   Summarize the orbital and physical properties of meteoroids, and explain what these properties suggest about the probable origin of meteoroids.

 Visit www.aw-bc.com/chaisson for additional images, animations, quizzes, and eBook for this chapter.

## 14.1 Asteroids

**Asteroids** are relatively small, predominantly rocky objects that revolve around the Sun. Their name literally means "starlike bodies," but asteroids are definitely not stars. They are too small even to be classified as planets. Astronomers often refer to them as "minor planets" or, sometimes, "planetoids."

Asteroids differ from planets in both their orbits and their size. As illustrated in Figure 14.1, they generally move on somewhat eccentric trajectories between Mars and Jupiter, unlike the almost circular paths of the major planets. Few asteroids are larger than 300 km in diameter, and most are far smaller—as small as a tenth of a kilometer across. The largest known asteroid, Ceres (inset to Figure 14.1), is just 1/10,000 the mass of Earth and measures only 940 km across. Taken together, the known asteroids amount to less than the mass of the Moon, so they do not contribute significantly to the total mass of the solar system.

### Orbital Properties

European astronomers discovered the first asteroids early in the 19th century as they searched the sky for an additional planet orbiting between Mars and Jupiter. Italian astronomer Giuseppe Piazzi was the first to discover an asteroid. He detected Ceres in 1801, and measured its orbital semimajor axis to be 2.8 AU. Within a few years, three more asteroids—Pallas (2.8 AU), Juno (2.7 AU), and Vesta (3.4 AU)—were discovered.

By the start of the 20th century, astronomers had cataloged several hundred asteroids with well-determined orbits. Now, at the beginning of the 21st century, the list has grown to over 75,000. The total number of known asteroids (including those whose orbits are not yet known with sufficient accuracy to make them "official") now exceeds 200,000. The vast majority of these bodies are found in a region of the solar system known as the **asteroid belt,** located between 2.1 and 3.3 AU from the Sun—roughly midway between the orbits of Mars (1.5 AU) and Jupiter (5.2 AU). All but one of the known asteroids revolve about the Sun in prograde orbits, in the same sense as the planets.

Such a compact concentration of asteroids in a well-defined belt suggests that they are either the fragments of a planet broken up long ago or primal rocks that never managed to accumulate into a genuine planet. On the basis of the best evidence currently available, and consistent with the theory of planet formation described in Chapter 15, researchers favor the latter view: There is far too little mass in the belt to constitute a planet, and the marked chemical differences among individual asteroids

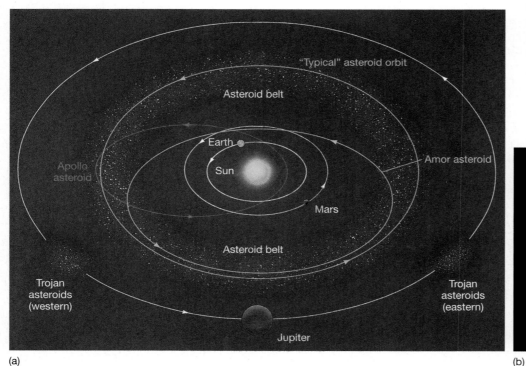

(a)                                                                      (b)

▲ **FIGURE 14.1** **Inner Solar System** (a) The main asteroid belt, along with the orbits of Earth, Mars, and Jupiter (drawn obliquely, that is neither face-on nor edge-on). Note the Trojan asteroids at two locations in Jupiter's orbit. Some Apollo (Earth-crossing) and Amor (Mars-crossing) orbits are shown. (We will learn more about these classes of asteroids later in the chapter.) (b) An ultraviolet image of the largest known asteroid, Ceres, as seen by the *Hubble* telescope. Little surface detail is evident, although image processing reveals what seems to be a large impact crater some 250 km across near the center of the frame. (*NASA/SWRI*)

strongly suggest that they could not all have originated in a single body. Instead, astronomers think that the strong gravitational field of Jupiter continuously disturbs the motions of these chunks of primitive matter, nudging and pulling at them and preventing them from aggregating into a planet.

## Physical Properties

With only a few exceptions (such as Ceres, shown in Figure 14.1), asteroids are too small to be resolved by Earth-based telescopes, so astronomers must rely on indirect methods to find their sizes, shapes, and composition. Consequently, only a few of their physical and chemical properties are accurately known. To the extent that astronomers can determine their compositions, asteroids have been found to differ not only from the eight known planets and their many moons, but also among themselves.

Asteroids are classified by their spectroscopic properties. The darkest, or least reflective, asteroids contain a large fraction of carbon in their makeup. These asteroids are known as *C-type* (or *carbonaceous*) asteroids. The more reflective *S-type* asteroids contain silicate, or rocky, material. Generally speaking, S-type asteroids predominate in the inner portions of the asteroid belt, and the fraction of C-type bodies steadily increases as we move outward. Overall, about 15 percent of all asteroids are S-type, 75 percent are C-type, and 10 percent are other types (mainly the *M-type* asteroids, containing large fractions of nickel and iron). Many planetary scientists think that the carbonaceous asteroids consist of very primitive material representative of the earliest stages of the solar system. Carbonaceous asteroids have not been subject to significant heating or undergone chemical evolution since they first formed 4.6 billion years ago.

In most cases, astronomers estimate the sizes of asteroids from the amount of sunlight they reflect and the amount of heat they radiate. These observations are difficult, but size measurements have been obtained in this way for more than 1000 asteroids. On rare occasions, astronomers witness an asteroid occulting a star, allowing them to determine the asteroid's size and shape with great accuracy. The largest asteroids are roughly spherical, but the smaller ones can be highly irregular.

The three largest asteroids—Ceres, Pallas, and Vesta—have diameters of 940 km, 580 km, and 540 km, respectively. Only two dozen or so asteroids are more than 200 km across, and most are much smaller. Almost assuredly, many hundreds of thousands more await discovery. However, observers estimate that they are mostly very small. Probably 99 percent of all asteroids larger than 100 km are known and cataloged, and at least 50 percent of asteroids larger than 10 km are accounted for. Although the vast majority of asteroids are probably less than a few kilometers across, most of the *mass* in the asteroid belt resides in objects greater than a few tens of kilometers in diameter.

Vesta is unique among asteroids in that, despite its small size, it appears to have undergone *volcanism* in its distant past. On the basis of their orbits and their overall spectral similarities to Vesta, numerous meteorites (Section 14.3) found on Earth are thought to have been chipped off that asteroid following collisions with other members of the asteroid belt. Remarkably, these meteorites have compositions similar to that of terrestrial basalt, indicating that they were subject to ancient volcanic activity. Why Vesta should exhibit such past activity, while the larger Ceres and Pallas do not, is not known. Possibly, Vesta was once part of a larger body that subsequently broke up. No other fragments of that body have yet been conclusively identified, although the spectrum and surface properties of a tiny 2-km-long asteroid discovered in 1999 suggests that it may once have been part of Vesta.

## Asteroid Observations from Space

The first close-up views of asteroids were provided by the Jupiter probe *Galileo*, which, on its rather roundabout path to the giant planet, passed twice through the asteroid belt, making close encounters with asteroid Gaspra in October 1991 and asteroid Ida in August 1993 (Figure 14.2). ∞ (Sec. 6.6) Both Gaspra and Ida are S-type asteroids. Technical problems limited the amount of data that could be sent back from the spacecraft during the flybys. Nevertheless, the images produced by *Galileo* showed far more detail than any photographs made from Earth.

Gaspra and Ida are irregularly shaped bodies with maximum diameters of about 20 km and 60 km, respectively. They are pitted with craters ranging in size from a few hundred meters to 2 km across and are covered with a layer of dust of variable thickness. Ida is much more heavily cratered than Gaspra, in part because it resides in a denser part of the asteroid belt. Also, scientists think that Ida has suffered more from the ravages of time. Ida is about a billion years old, far older than Gaspra, which is estimated to have an age of just 200 million years, based on the extent of cratering on its surface. Both asteroids are thought to be fragments of much larger objects that broke up into many smaller pieces following violent collisions long ago.

To the surprise of most mission scientists, closer inspection of the Ida image (Figure 14.2b) revealed the presence of a tiny moon, now named Dactyl, just 1.5 km across, orbiting the asteroid at a distance of about 90 km. A few such *binary asteroids* (two asteroids orbiting one another as they circle the Sun) had previously been observed from Earth. However, the binary systems known before the Ida flyby were all much larger than Ida—a moon the size of Dactyl cannot be detected from the ground. Scientists think that, given the relative congestion of the asteroid belt, collisions between asteroids may be quite common, providing a source of both interplanetary

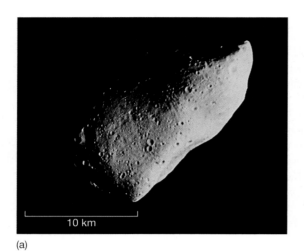

(a)                                                 (b)

▲ FIGURE 14.2 **Gaspra and Ida** (a) The S-type asteroid Gaspra, as seen from a distance of 1600 km by the space probe *Galileo* on its way to Jupiter. (b) The S-type asteroid Ida, photographed by *Galileo* from a distance of 3400 km. (Ida's moon, Dactyl, is visible at right.) The resolution in these photographs is about 100 m. True-color images show the surfaces of both bodies to be a fairly uniform shade of gray. Sensors on board the spacecraft indicated that the amount of infrared radiation absorbed by these surfaces varies from place to place, probably because of variations in the thickness of the dust layer blanketing them. *(NASA)*

dust and smaller asteroids and possibly deflecting one or both of the bodies involved onto eccentric, Earth-crossing orbits. The less-violent collisions may be responsible for the binary systems we see.

By studying the *Galileo* images, astronomers were able to obtain limited information on Dactyl's orbit around Ida and hence (using Newton's law of gravity—see *More Precisely 2-3*) to estimate Ida's mass at about $5–10 \times 10^{16}$ kg. This information in turn allowed them to measure Ida's density as 2200–2900 kg/m$^3$, a range consistent with the asteroid's rocky, S-type classification.

In June 1997, the *Near Earth Asteroid Rendezvous* (*NEAR*) spacecraft visited the C-type asteroid Mathilde on its way to the mission's main target: the S-type asteroid Eros. Shown in Figure 14.3, Mathilde is some 60 km across. By sensing its gravitational pull, *NEAR* measured Mathilde's mass to be about $10^{17}$ kg, implying a density of just 1400 kg/m$^3$. To account for this low density, scientists speculate that the asteroid's interior must be quite porous. Indeed, many smaller asteroids seem to be more like loosely bound "rubble piles" than pieces of solid rock. The interior's relatively soft consistency may also explain the unexpectedly large size of many of the craters observed on Mathilde's surface. A solid object would probably have shattered after an impact violent enough to cause such large craters. However, like crumple zones in a car, Mathilde's porous interior could have absorbed and dissipated the impactor's energy, allowing the asteroid to survive the event.

Upon arrival at Eros on February 14, 2000, *NEAR* (now renamed *NEAR–Shoemaker*) went into orbit around the asteroid. For 1 year, the spacecraft sent back high-resolution images of Eros (Figure 14.4) and made detailed measure-

ments of its size and shape (34 × 11 × 11 km), as well as its gravitational and magnetic fields, composition, and structure. The craft's various sensors revealed Eros to be a heavily cratered body of mass $7 \times 10^{15}$ kg and roughly uniform density around 2700 kg/m$^3$. The asteroid's interior seems to be solid rock—not rubble, as in the case of Mathilde—although it is extensively fractured due to innumerable

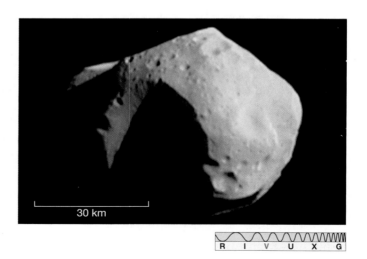

▲ FIGURE 14.3 **Asteroid Mathilde** The C-type asteroid Mathilde, imaged by the *NEAR* spacecraft en route to the near-Earth asteroid Eros. Mathilde measures some 60 × 50 km and rotates every 17.5 days. The largest craters visible in this image are about 20 km across—much larger than those seen on Gaspra or Ida. The reason may be this asteroid's low density (approximately 1400 kg/m$^3$) and rather soft composition. *(JHU/NASA)*

impacts in the past. The measurements are consistent with Eros being a primitive, unevolved sample of material from the early solar system. In February 2001, *NEAR–Shoemaker* landed on Eros, sending back a series of close-up images as the craft descended to the surface. Remarkably, despite its having no landing gear, the spacecraft survived the low-velocity impact and maintained radio contact with Earth for 16 more days before communication finally ceased.

Apart from Ida, Mathilde, and Eros, the masses of most asteroids are unknown. However, a few of the largest asteroids do have strong enough gravitational fields for their effects on their neighbors to be measured and their masses thereby determined to reasonable accuracy. Their computed densities are generally compatible with the rocky or carbonaceous compositions just described.

## CONCEPT CHECK

✔ Describe some basic similarities and differences between asteroids and the inner planets.

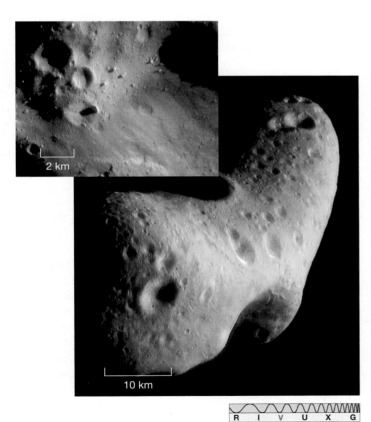

▲ FIGURE 14.4 **Asteroid Eros** A mosaic of detailed images of the asteroid Eros, as seen by the *NEAR* spacecraft (which actually landed on this asteroid). Craters of all sizes, ranging from 50 m (the resolution of the image) to 5 km, pit the surface. The inset shows a close-up image of a "young" section of the surface, where loose material from recent impacts has apparently filled in and erased all trace of older craters. (*JHU/NASA*)

2 km

10 km

R I V U X G

## Earth-Crossing Asteroids

The orbits of most asteroids have eccentricities lying in the range from 0.05 to 0.3, ensuring that they always remain between the orbits of Mars and Jupiter. Very few asteroids have eccentricities greater than 0.4. Those that do are of particular interest to us, however, as their paths may intersect Earth's orbit, leading to the possibility of a collision with our planet. These bodies are collectively known as **Earth-crossing asteroids.** Those stray asteroids having highly elliptical orbits or orbits that do not lie in the main asteroid belt, have probably been influenced by nearby Mars and, especially, Jupiter. The gravitational fields of those two planets can disturb normal asteroid orbits, deflecting them into the inner solar system. Earth-crossing asteroids are termed *Apollo asteroids* (after the first known Earth-crossing asteroid, Apollo) if their orbital semimajor axes exceed 1 AU and *Aten asteroids* otherwise. Asteroids whose orbits cross only the orbit of Mars are known as *Amor asteroids* (see Figure 14.1).

As of early 2007, more than 4400 Earth-crossing asteroids are known. Most have been discovered since the late 1990s, when systematic searches for such objects began. More than 800 Earth crossers are officially designated "potentially hazardous," meaning that they are more than about 150 m in diameter (three times the size of the impactor responsible for the Barringer crater shown in Figure 8.19) and move in orbits that could bring them within 0.05 AU (7.5 million km) of our planet.

From a human perspective, perhaps the most important consequence of the existence of Earth-crossers is the very real possibility of an actual collision with Earth. For example, the mile-wide Apollo asteroid Icarus (Figure 14.5) missed our planet by "only" 6 million km in 1968—a very close call by cosmic standards. More recently, in June 2002, the 100-m-diameter asteroid 2002 MN came much closer, missing us by a mere 120,000 km (less than one-third of the distance to the Moon). It was detected 3 days *after* it passed our planet! All told, between 1997 and 2007, more than 200 asteroids (that we know of!) passed within 0.05 AU of our planet. A similar number are expected to pass within this distance between 2007 and 2017.

None of the currently known potentially hazardous asteroids are expected to impact Earth during the next century—the closest predicted near miss will occur in April, 2029, when the 350-m asteroid 2004 MN4 will pass just 23,000 km above our planet's surface. That said, calculations imply that most Earth-crossing asteroids *will* in fact eventually collide with Earth. On average, during any given million-year period, our planet is struck by about three asteroids. Because Earth is largely covered with water, two of those impacts are likely to occur in the ocean and only one on land. Several dozen large land basins and eroded craters on our planet are suspected to

ANIMATION/VIDEO   Orbiting Eros, NEAR Descent, NEAR Landing

R I V U X G

▲ FIGURE 14.5 **Asteroid Icarus** The Earth-crossing asteroid Icarus has an orbit that passes within 0.2 AU of the Sun, well within Earth's orbit. Icarus occasionally comes close to Earth, making it one of the best-studied asteroids in the solar system. Its motion relative to the stars makes it appear as a streak (marked) in this long-exposure photograph. (*Palomar/Caltech*)

be sites of ancient asteroid collisions (see, for example, Figure 14.29, later in the chapter). The many large impact craters on the Moon, Venus, and Mars are direct evidence of similar events on other worlds (see also *Discovery 11-1*).

Most known Earth-crossing asteroids are relatively small—about 1 km in diameter (although one 10 km in diameter has been identified). Even so, a visit of even a kilometer-sized asteroid to Earth could be catastrophic by human standards. Such an object packs enough energy to devastate an area some 100 km in diameter. The explosive power would be equivalent to about a million 1-megaton nuclear bombs—a hundred times more than all the nuclear weapons currently in existence on Earth. A fatal blast wave (the shock from the explosion, spreading rapidly outward from the site of the impact) and a possible accompanying tsunami (tidal wave) from an ocean impact would doubtless affect a much larger area still. Should an asteroid hit our planet hard enough, it might even cause the extinction of entire species—indeed, many scientists think that the extinction of the dinosaurs was the result of just such an impact (see *Discovery 14-1*).

Some astronomers take the prospect of an asteroid impact sufficiently seriously that they maintain an "asteroid watch"—an effort to catalog and monitor all Earth-crossing asteroids in order to maximize our warning time of any impending collision. Several large, dedicated

telescopes now scan the skies for faint objects in our neighborhood. Currently, our options in the event that an impending impactor is seen are very limited—science fiction movies aside, we could neither destroy nor deflect an asteroid just a few days away from our planet. However, scientists are confident that, given enough warning, a small "nudge" from thrusters placed on the impactor's surface years ahead of the collision could shift its orbit by just enough to miss us.

## Orbital Resonances

Although most asteroids orbit in the main belt, between about 2 and 3 AU from the Sun, an additional class of asteroids, called the **Trojan asteroids,**\* orbits at the distance of Jupiter. Several hundred such asteroids are now known. They are locked into a 1:1 orbital resonance with Jupiter by that planet's strong gravity, just as some of the small moons of Saturn share orbits with the medium-sized moons Tethys and Dione, as described in Chapter 12. ∞ (Sec. 12.5)

Calculations first performed by the French mathematician Joseph Louis Lagrange in 1772 show that there are exactly five places in the solar system where a small body can orbit the Sun in synchrony with Jupiter, subject to the combined gravitational influence of both large bodies. (Lagrange in fact demonstrated that five such points exist for any planet.) These places are now known as the *Lagrangian points* of the planet's orbit. As illustrated in Figure 14.6, three of the points (referred to as $L_1$, $L_2$, and $L_3$) lie on the line joining Jupiter and the Sun (or its extension in either direction). The other two—$L_4$ and $L_5$—are located on Jupiter's orbit, exactly 60° ahead of and 60° behind the planet. All five Lagrangian points revolve around the Sun at the same rate as Jupiter.

In principle, an asteroid placed at any of the Lagrangian points will circle the Sun in lockstep with Jupiter, always maintaining the same position relative to the planet. However, the three Lagrangian points that are in line with Jupiter and the Sun are known to be *unstable*—a body displaced, however slightly, from any of those points will tend to drift slowly away from it, not back toward it. Since matter in the solar system is constantly subjected to small perturbations—by the planets, the asteroids, and even the solar wind—matter does not accumulate in these regions. Thus, no asteroids orbit near the $L_1$, $L_2$, or $L_3$ point of Jupiter's orbit.

This is not the case for the other two Lagrangian points, $L_4$ and $L_5$. They are both *stable*—matter placed near them tends to remain in their vicinity. Consequently, asteroids tend to accumulate near these points (see Figure 14.1).

\**The first few hundred asteroids discovered were named after characters in Greek mythology. The first asteroid found in this orbit was called Achilles. As more asteroids were found sharing this orbit, they became known as the Trojans.*

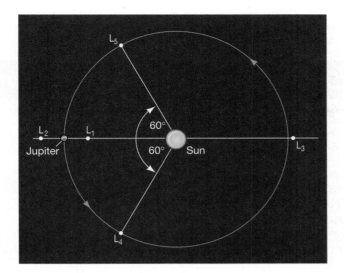

▲ FIGURE 14.6 **Lagrangian Points** The Lagrangian points of the Jupiter–Sun system, where a third body could orbit in synchrony with Jupiter on a circular trajectory. Only the $L_4$ and $L_5$ points are stable. They are the locations of the Trojan asteroids (see Figure 14.1).

For unknown reasons, Trojan asteroids tend to be found near Jupiter's leading ($L_4$, or *eastern*, as it lies to the east of Jupiter in the sky) Lagrangian point, rather than the trailing ($L_5$) point. Recently, a few small asteroids have been found similarly trapped in the Lagrangian points of Venus, Earth, and Mars.

The main asteroid belt also has structure—not so obvious as the Trojan orbits or the prominent gaps and ringlets in Saturn's ring system, but nevertheless of great dynamic significance. A graph of the number of asteroids having various orbital semimajor axes (Figure 14.7a) shows that there are several prominently underpopulated regions in the distribution. These "holes" are known as the **Kirkwood gaps,** after their discoverer, the 19th-century American astronomer Daniel Kirkwood.

The Trojan asteroids share an orbit with Jupiter, orbiting in 1:1 resonance with the planet. The Kirkwood

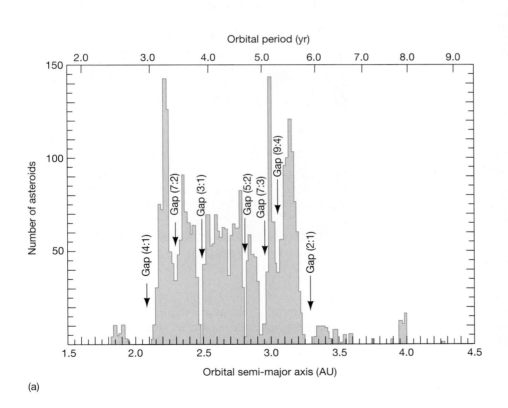

(a)

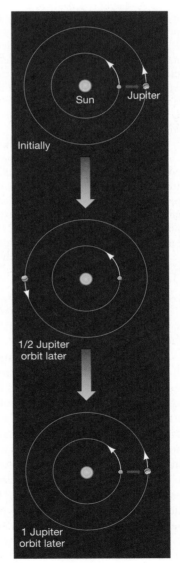

(b)

▲ FIGURE 14.7 **Kirkwood Gaps** (a) The distribution of asteroid semimajor axes shows some prominent gaps caused by resonances with Jupiter's orbital motion. Note, for example, the prominent gap at 3.3 AU, which corresponds to the 2:1 resonance—the orbital period is 5.9 years, exactly half that of Jupiter. (b) An asteroid in a 2:1 resonance with Jupiter receives a strong gravitational tug from the planet each time they are closest together (as in panels 1 and 3). Because the asteroid's period is precisely half that of Jupiter, the tugs come at exactly the same point in every other orbit, and their effects reinforce each other.

# DISCOVERY 14-1

## What Killed the Dinosaurs?

The name *dinosaur* derives from the Greek words *deinos* ("terrible") and *sauros* ("lizard"). Dinosaurs were no ordinary reptiles: In their prime, roughly 100 million years ago, the dinosaurs were the all-powerful rulers of Earth. Their fossilized remains have been uncovered on all the world's continents. Despite their dominance, according to the fossil record, these creatures vanished from Earth quite suddenly about 65 million years ago. What happened to them?

Until fairly recently, the prevailing view among paleontologists—scientists who study prehistoric life—was that dinosaurs were rather small-brained, cold-blooded creatures. In chilly climates, or even at night, the metabolisms of these huge reptiles would have become sluggish, making it difficult for them to move around and secure food. The suggestion was that they were poorly equipped to adapt to sudden changes in Earth's climate, so they eventually died out. However, a competing, and still controversial, view of dinosaurs has emerged: Recent fossil evidence suggests that many of these monsters may in fact have been warm-blooded and relatively fast-moving creatures—not at all the dull-witted, slow-moving giants of earlier conception. In any case, no species able to dominate Earth for more than 100 million years could have been too poorly equipped for survival. For comparison, humans have thus far dominated for a little over 2 million years.

If the dinosaurs didn't die out simply because of stupidity and inflexibility, then what happened to cause their sudden and complete disappearance? Many explanations have been offered for the extinction of the dinosaurs. Devastating plagues, magnetic field reversals, increased tectonic activity, severe climate changes, and supernova explosions have all been proposed. ∞ (Secs. 7.4, 7.5) In the 1980s, it was suggested that a huge extraterrestrial object collided with Earth 65 million years ago, and this is now (arguably) the leading explanation for the demise of the dinosaurs.

According to this idea, a 10- to 15-km-wide asteroid or comet struck Earth, releasing as much energy as 10 million

or more of the largest hydrogen bombs humans have ever constructed and kicking huge quantities of dust (including the pulverized remnants of the impactor itself) high into the atmosphere. (See the first figure.) The dust may have shrouded our planet for many years, virtually extinguishing the Sun's rays during that time. On the darkened surface, plants could not survive. The entire food chain was disrupted, and the dinosaurs, at the top of that chain, eventually became extinct.

Although we have no direct astronomical evidence to confirm or refute this idea, we can estimate the chances that a large asteroid or comet will strike Earth today on the basis of observations of the number of objects that are presently on Earth-crossing orbits. The second figure shows the likelihood of an impact as a function of the size of the colliding body. The horizontal scale indicates the energy released by the collision, measured in *megatons* of TNT. The megaton—$4.2 \times 10^{16}$ joules,

*(C. Butler/Astrostock-Sanford)*

---

gaps result from other, more complex, orbital resonances with Jupiter. For example, an asteroid with a semimajor axis of 3.3 AU would (by Kepler's third law) orbit the Sun in exactly half the time taken by Jupiter. ∞ (Sec. 2.5) The gap at 3.3 AU, then, corresponds to a 2:1 resonance. An asteroid at that particular resonance receives a regular, periodic tug from Jupiter at the same point in every other orbit (Figure 14.7b). The cumulative effect of all the tugs is to deflect the asteroid into an elongated orbit—one that crosses the orbit of Mars or Earth. Eventually, the asteroid collides with one of those two planets or comes close enough that it is pushed onto an entirely different trajectory. In this way, Jupiter's gravity creates the Kirkwood gaps, and some of the cleared-out asteroids become Apollo or Amor asteroids.

Notice that, although there are many similarities between this mechanism and the resonances that produce the gaps in Saturn's rings (see Chapter 12), there are differences, too. ∞ (Sec. 12.4) Unlike Saturn's rings, where eccentric orbits are rapidly circularized by collisions among ring particles, there are no *physical* gaps in the asteroid belt. The in-and-out motion of the belt asteroids as they travel in their eccentric orbits around the Sun means that no part of the belt is actually empty. Only when we look at semimajor axes (or, equivalently, at orbital *energies*) do the gaps become apparent.

## CONCEPT CHECK

✔ Why are astronomers so interested in Earth-crossing asteroids?

the explosive yield of a large nuclear warhead—is the only common terrestrial measure of energy adequate to describe the violence of these occurrences.

We see that 100-million-megaton events, like the planet-wide catastrophe that supposedly wiped out the dinosaurs, are very rare, occurring only once every 10 million years or so. However, smaller impacts, equivalent to "only" a few tens of kilotons of TNT (roughly equivalent to the bomb that destroyed Hiroshima in 1945), could happen every few years—and we may be long overdue for one. The most recent large impact was the Tunguska explosion in Siberia, in 1908, which packed a roughly 1-megaton punch (see Figure 14.30).

The main geological evidence supporting the theory that the dinosaurs' extinction was the result of an asteroid impact is a layer of clay enriched with the element iridium. The layer is found in 65-million-year-old rocky sediments all around our planet. Iridium on Earth's surface is rare, because most of it sank into our planet's interior long ago. The abundance of iridium in this one layer of clay is about 10 times greater than in other terrestrial rocks, but it matches closely the abundance of iridium found in meteorites (and, we assume, in asteroids and comets, too). The site of the catastrophic impact has also been tentatively identified as being near Chicxulub, in the Yucatán Peninsula in Mexico, where evidence of a heavily eroded, but not completely obliterated, crater of just the right size and age has been found.

The theory is not without its detractors, however. Perhaps predictably, the idea of catastrophic change on Earth being precipitated by events in interplanetary space was rapidly accepted by most astronomers, but it remains controversial among some paleontologists and geologists. Opponents argue that the amount of iridium in the clay layer varies greatly from place to place across the globe, and there is no complete explanation of why that should be so. Perhaps, they suggest, the iridium was produced by volcanoes, and has nothing to do with an extraterrestrial impact at all.

Still, in the 20 years since the idea was first suggested, the focus of the debate seems to have shifted. The reality of a major impact 65 million years ago has become widely accepted, and much of the argument now revolves around the question of whether that event actually caused the extinction of the dinosaurs or merely accelerated a process that was already underway. Either way, the realization that such catastrophic events can and do occur marks an important milestone in our understanding of evolution on our planet. This realization was bolstered by the Shoemaker–Levy 9 impact on Jupiter in 1994. ∞ (*Discovery 11-1*) In addition, there is growing evidence for even larger impacts in the more distant past, with yet more sweeping evolutionary consequences. As is often the case in science, the debate has evolved, sometimes erratically, as new data have been obtained, but a real measure of consensus has already been achieved, and important new insights into our planetary environment have been gained. ∞ (Sec. 1.2)

As a general rule, we can expect that global catastrophes are bad for the dominant species on a planet. As the dominant species on Earth, we are the ones who now stand to lose the most.

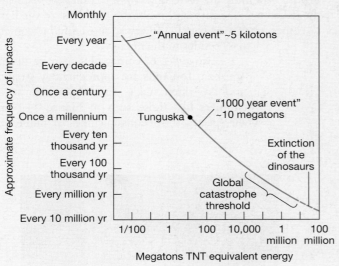

## 14.2 Comets

**Comets** are usually discovered as faint, fuzzy patches of light on the sky while they are still several astronomical units away from the Sun. Traveling in a highly elliptical orbit with the Sun at one focus (Figure 14.8), a comet brightens and develops an extended **tail** as it nears the Sun. (The name "comet" derives from the Greek word *kome*, meaning "hair.") As the comet departs the Sun's vicinity, its brightness and tail diminish until it once again becomes a faint point of light receding into the distance. Like the planets, comets emit no visible light of their own—they shine by reflected (or reemitted) sunlight. Each year, a few dozen are detected as they pass through the inner solar system. Many more must pass by unseen.

### Comet Appearance and Structure

The various parts of a typical comet are shown in Figure 14.9. Even through a large telescope, the **nucleus,** or main solid body, of a comet is no more than a minute point of light. A typical cometary nucleus is extremely small—only a few kilometers in diameter. During most of the comet's orbit, far from the Sun, only this frozen nucleus exists. When a comet comes within a few astronomical units of the Sun, however, its icy surface becomes too warm to remain stable. Part of the comet becomes gaseous and expands into space, forming a diffuse **coma** ("halo") of dust and evaporated gas around the nucleus. The process by which a solid (in this case ice) changes directly into a gas, without

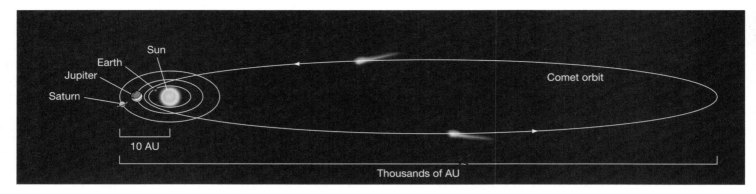

▲ FIGURE 14.8 **Distant Orbit** Comets move in highly eccentric paths that carry them far beyond the known planets.

first becoming liquid, is called *sublimation*. Frozen carbon dioxide (dry ice) on Earth provides a familiar example of this process. In space, sublimation is the rule, rather than the exception, for the behavior of ice when it is exposed to heat.

The coma becomes larger and brighter as the comet nears the Sun. At maximum size, the coma can measure 100,000 km in diameter—almost as large as Saturn or Jupiter. Engulfing the coma, an invisible **hydrogen envelope,** usually distorted by the solar wind, stretches across millions of kilometers of space. The comet's tail, which is most pronounced when the comet is closest to the Sun and the rate of sublimation from the nucleus is greatest, is much larger still, sometimes spanning as much as 1 AU. From Earth, only the coma and tail of a comet are visible to the naked eye. Despite the size of the tail, most of the light comes from the coma. However, most of the comet's mass resides in the nucleus.

Two types of comet tails may be distinguished. **Ion tails** are approximately straight and are often made up of glowing, linear streamers like those seen in Figure 14.10(a). Their spectra show emission lines of numerous

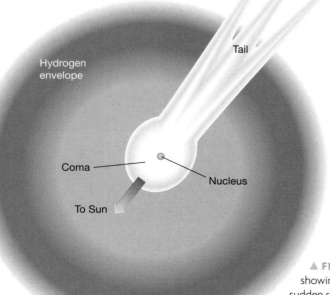

(a)

(b)

R I V U X G

▲ FIGURE 14.9 **Comet Structure** (a) Diagram of a typical comet, showing the nucleus, coma, hydrogen envelope, and tail. The tail is not a sudden streak in time across the sky, as in the case of meteors or fireworks. Instead, it travels along with the rest of the comet (as long as the comet is sufficiently close to the Sun for the tail to exist). Note how the invisible hydrogen envelope is usually larger than the visible extent of the comet; it is often even much larger than drawn here. (b) Halley's Comet in 1986, about one month before it rounded the Sun at perihelion. *(NOAO)*

ionized molecules—molecules that have lost some of their normal complement of electrons—including carbon monoxide, nitrogen, and water, among many others. ∞ (Sec. 4.4) **Dust tails** are usually broad, diffuse, and gently curved (Figure 14.10b). They are rich in microscopic dust particles that reflect sunlight, making the tail visible from afar. (Ion and dust tails are also referred to as *Type I* and *Type II* tails, respectively.)

Comets' tails are in all cases directed *away* from the Sun by the solar wind (the invisible stream of matter and radiation escaping the Sun). Consequently, as depicted in Figure 14.11, the tail always lies outside the comet's orbit and actually leads the comet during the portion of the orbit that is outbound from the Sun.

Ion tails and dust tails differ in shape because of the different responses of gas and dust to the forces acting in interplanetary space. Every tiny particle in space in our solar system—including those in comets' tails—follows an orbit determined by gravity and the solar wind. If gravity alone were acting, the particle would follow the same curved path as its parent comet, in accordance with Newton's laws of motion. ∞ (Sec. 2.7) If the solar wind were the only influence, the tail would be swept up by it and would trail radially outward from the Sun. Ion tails are much more strongly influenced by the solar wind than by the Sun's gravity, so those tails always point directly away from the Sun. The heavier dust particles have more of a tendency to follow the comet's orbit, giving rise to the slightly curved dust tails.

◄ **FIGURE 14.10  Comet Tails** (a) A comet with a primarily ion tail. Called comet Giacobini–Zinner and seen here in 1959, its coma measured 70,000 km across; its tail was well over 500,000 km long. (b) Photograph of a comet having both an ion tail (dark blue) and a dust tail (white blue), both marked in the inset, showing the gentle curvature and inherent fuzziness of the dust. This is comet Hale–Bopp in 1997. At the comet's closest approach to the Sun, its tail stretched nearly 40° across the sky. (*U.S. Naval Observatory; Aaron Horowitz/Corbis*)

(a)

Ion tail

Dust tail

(b)

R  I  V  U  X  G

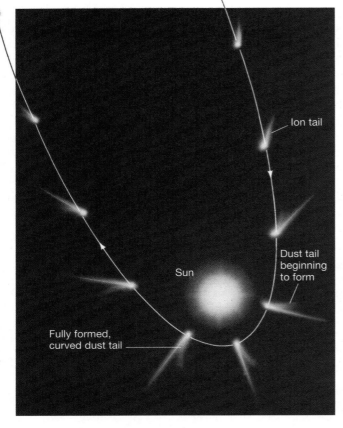

▲ FIGURE 14.11 **Comet Trajectory** As it approaches the Sun, a comet develops an ion tail, which is always directed away from the Sun. Closer in, a curved dust tail, also directed generally away from the Sun, may appear. Notice that although the ion tail always points directly away from the Sun on both the inbound and the outgoing portions of the orbit, the dust tail has a marked curvature and tends to lag behind the ion tail. (Compare this figure with a photo of a real comet, for example Figure 4.10.)

Arguably the most famous comet of all is Halley's comet. In 1705, the British astronomer Edmund Halley realized that the appearance of this comet in 1682 was not a one-time event. Basing his work on previous sightings of the comet, Halley calculated its path and found that the comet orbited the Sun with a period of 76 years. He predicted its reappearance in 1758. Halley's successful determination of the comet's trajectory and his prediction of its return was an early triumph of Newton's laws of motion and gravity. ∞ (Sec. 2.8) Although Halley did not live to see his calculations proved correct, the comet was named in his honor.

Once astronomers knew the comet's period, they traced the appearances of the comet backwards in time. Historical records from many ancient cultures show that Halley's comet has been observed at every one of its passages since 240 B.C. A spectacular show, the tail of Halley's comet can reach almost a full astronomical unit in length, stretching many tens of degrees across the sky. Figure 14.12(a) shows Halley's comet as seen from Earth in 1910. Its most recent appearance, in 1986 (Figure 14.12b and also Figure 14.9b), was not ideal for terrestrial viewing, as the perihelion happened to occur on roughly the opposite side of the Sun from Earth, but the comet was closely scrutinized by spacecraft (see below). The comet's orbit is shown in Figure 14.13. Its next scheduled visit to the inner solar system is in 2061.

## Physical Properties of Comets

The mass of a comet can occasionally be estimated by watching how the comet interacts with other solar system objects or by determining the size of the nucleus and assuming a density characteristic of icy composition. These methods yield typical cometary masses ranging from $10^{12}$ to $10^{16}$ kg, comparable to the masses of small asteroids.

A comet's mass decreases with time, because some material is lost each time the comet rounds the Sun as material evaporates from its surface. For comets that travel

▲ FIGURE 14.12 **Halley's Comet** (a) Halley's comet as it appeared in 1910. Top, on May 10, with a 30° tail, bottom, on May 12, with a 40° tail. (b) Halley on its return and photographed with higher resolution on March 14, 1986. *(Caltech; Mt. Stromlo and Siding Springs Observatories)*

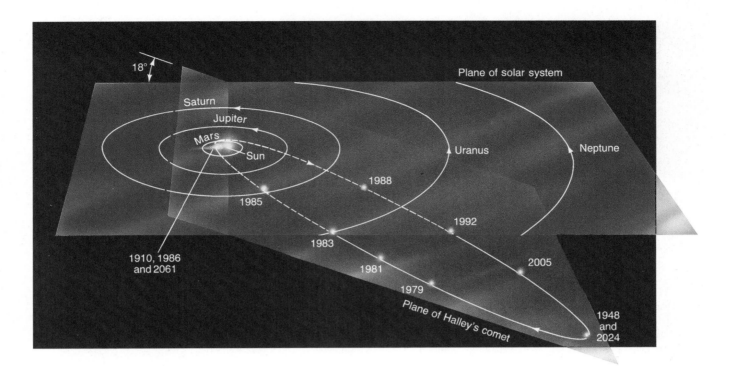

▲ FIGURE 14.13 **Halley's Orbit** Halley's comet has a smaller orbital path and a shorter period than most comets, but its orbital orientation is not typical of a short-period comet. Sometime in the past, this comet must have encountered a jovian planet (probably Jupiter itself), which threw it into a tighter orbit that extends not to the Oort cloud, but merely a little beyond Neptune. Edmund Halley applied Newton's law of gravity to predict this comet's return.

within an astronomical unit of the Sun, the evaporation rate can reach as high as $10^{30}$ molecules per second—about 30 tons of cometary material lost for every second the comet spends near the Sun (within Earth's orbit, say). Astronomers have estimated that this loss of material will destroy even a large comet, such as Halley or Hale–Bopp (Figure 14.10b), in just a few thousand orbits. Occasionally, as shown in Figure 14.14, the process can be much more violent, causing a comet suddenly to flare up in brightness and then rapidly fade as its nucleus disintegrates.

In seeking the physical makeup of a cometary body itself, astronomers are guided by the observation that comets have dust that reflects light and also certain gas that emits spectral lines of hydrogen, nitrogen, carbon, and oxygen. Even as the atoms, molecules, and dust particles boil off, creating the coma and tail, the nucleus itself remains a cold mixture of gas and dust, hardly more than a ball of loosely packed ice with a density of about 100 kg/m³ and a temperature of only a few tens of kelvins. Experts now consider cometary nuclei to be made up largely of dust particles trapped within a mixture of methane, ammonia, carbon dioxide, and ordinary water ice. (These constituents should be fairly familiar to you as the main components of most of the small moons in the outer solar system.) ∞ (Secs. 12.5, 13.5) Because of this composition, comets are often described as "dirty snowballs."

## Space Missions to Comets

So far, three comets have received close-up visits from human spacecraft. When Halley's comet rounded the Sun in 1986, a small armada of spacecraft launched by the (former) USSR, Japan, and a group of western European countries went to meet it. One of the Soviet craft, *Vega 2*, traveled through the comet's coma and came within 8000 km of the nucleus. Using positional knowledge of the comet gained from the Soviet craft encounter, the European *Giotto* spacecraft (named after the Italian artist who painted an image of Halley's comet not long after its appearance in the year 1301) was navigated to within 600 km of the nucleus. This was a daring trajectory, since at 70 km/s—the speed of the craft relative to the comet—a colliding dust particle becomes a devastating bullet. Debris did in fact damage *Giotto*'s camera, but not before it sent home a wealth of data. Figure 14.15 shows *Giotto*'s view of the comet's nucleus, along with a sketch of its structure.

Halley's nucleus is an irregular, potato-shaped object, somewhat larger than astronomers had estimated. Spacecraft measurements showed it to be 15 km long by as much as 10 km wide, and almost jet black—as dark as finely ground charcoal or soot. The solid nucleus was enveloped by a cloud of dust, which scattered light throughout the

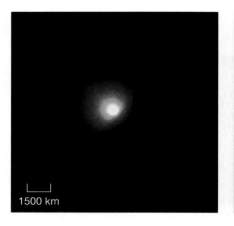

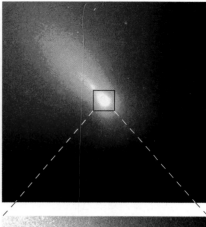

▲ FIGURE 14.14 **Comet Flare-Up** As a comet approaches the heat of the Sun, its icy material vaporizes, causing the spectacular tail we observe. Sometimes, as in this sequence of *Hubble Space Telescope* images captured on three consecutive days in July 2000, the vaporization can be explosive, causing a sudden flare-up in brightness. Shortly after these images were taken, the comet's brightness diminished rapidly as its nucleus disintegrated. This was comet C/1999 S4 LINEAR, named (in part) after its "discoverer," an automated Lincoln Lab telescope in New Mexico, conducting a search for Earth-crossing asteroids. The inset shows a dramatic shower of "minicomets" as the nucleus of the comet fragmented. (*NASA*)

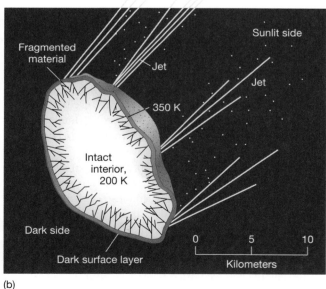

(a)

(b)

▲ FIGURE 14.15 **Halley, Up Close** (a) The European *Giotto* spacecraft resolved the nucleus of Halley's comet, showing it to be very dark, although heavy dust in the area obscured any surface features. Resolution here is about 50 m—half the length of a football field. At the time this image was made, in March 1986, the comet was within days of perihelion, and the Sun was toward the right. The brightest areas are jets of evaporated gas and dust spewing from the comet's nucleus. (b) A diagram of Halley's nucleus, showing its size, shape, jets, and other physical and chemical properties. (*ESA/Max Planck Institut*)

coma. Partly because of this scattering and partly because of dimming by the dust, none of the visiting spacecraft were able to discern much surface detail on the nucleus.

The spacecraft found direct evidence for several jets of matter streaming from the nucleus. Instead of evaporating uniformly from the whole surface to form the comet's coma and tail, gas and dust apparently vent from small areas on the sunlit side of Halley's nucleus. The force of these jets may be largely responsible for the comet's observed 53-hour rotation period. Like maneuvering rockets on a spacecraft, such jets can cause a comet to change its rotation rate and even to veer away from a perfectly elliptical orbit. Astronomers had hypothesized the existence of these nongravitational forces on the basis of slight deviations from Kepler's laws observed in some cometary trajectories. The Halley encounter was the first time astronomers actually saw the jets at work.

The past few years have seen two new missions to study comets at close range. In February 1999, NASA launched the *Stardust* mission, with the objective of collecting the first ever samples of cometary material and returning them to Earth. In January 2001, the spacecraft used a gravity assist from Earth to boost it onto a path to intercept comet P/Wild 2 ("Wild" is German, pronounced "Vilt"). The comet was chosen because it is a relative newcomer to the inner solar system, having been deflected onto its present orbit by an encounter with Jupiter in 1974. It therefore has not been subject to much solar heating or loss of mass by evaporation since it formed long ago.

In January 2004, *Stardust* approached within 200 km of the comet's nucleus (Figure 14.16a), collecting cometary particles in a specially designed foamlike "aerogel" detector (see Figure 14.16b). *Stardust* returned to Earth in January 2006, returning the debris to mission scientists, who are now studying the detailed physical, chemical, and even biological properties of a body that most probably has not changed significantly since our solar system formed. As shown in Figure 14.16(c), the aerogel performed flawlessly, providing researchers with the first-ever samples of cometary material. Among other findings, detailed chemical analysis of the samples has revealed evidence for nitrogen-rich organic material apparently formed in deep space and the unexpected presence of silicate materials that should only have formed at high temperatures, possibly challenging astronomers' current models of solar system formation (Chapter 15).

The second NASA mission had a considerably more violent end. On July 4, 2005, a 400-kg projectile from NASA's *Deep Impact* spacecraft crashed into comet Tempel 1 at more than 10 km/s (23,000 mph), blasting gas and debris from the comet's surface into interplanetary space, while the spacecraft itself watched from a safe distance of 500 km. Figure 14.17 is an artist's conception of the mothership's close view of the cratering at the moment of impact, and the full-page chapter-opening photo shows an actual image of the resulting explosion about 1 minute

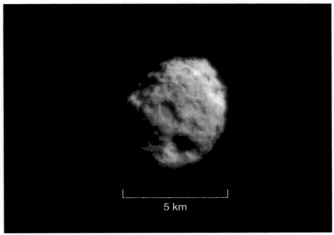

(a)

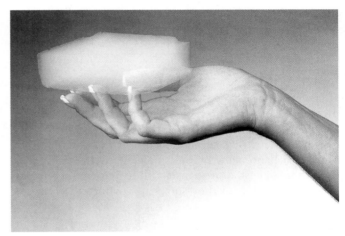

(b)

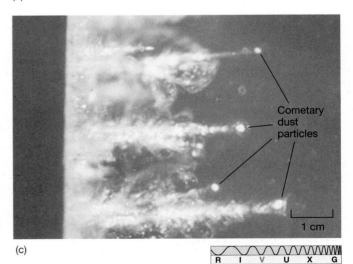

Cometary dust particles

1 cm

(c)

R I V U X G

▲ FIGURE 14.16 *Stardust* at Wild-2 (a) The *Stardust* spacecraft captured this image of comet Wild-2 in January 2004, just before the craft passed through the comet's coma. (b) Onboard is a detector made of a foamlike jello (called aerogel) that is 99.8% air, yet is strong enough to stop and store cometary dust particles as they hit the spacecraft. (c) Upon return of the craft to Earth in 2006, analysis began of the minute tracks in the aerogel, the ends of which contain captured comet dust fragments. *(NASA)*

▲ FIGURE 14.17　**Deep Impact**　This is an artist's rendering of the moment of impact of a projectile sent into Comet Tempel 1 by the *Deep Impact* mothership (depicted at right). For an actual look at what happened, see the chapter-opening photo on page 359. *(JPL)*

after impact. Spectroscopic analysis of the ejected gas has provided scientists with their clearest view yet of the internal composition of a comet, and hence of the primordial matter of the early solar system, confirming the presence of water ice and many organic molecules. Observations of the crater suggest a low-density "fluffy" internal composition, consistent with the "snowball" picture of cometary structure presented earlier.

## Comet Orbits

Comets that survive their close encounter with the Sun—some break up entirely—continue their outward journey to the edge of the solar system. Their highly elliptical orbits take many comets far beyond Pluto, perhaps even as far as 50,000 AU, where, in accord with Kepler's second law, they move more slowly and so spend most of their time. ∞ (Sec. 2.5) Most comets take hundreds of thousands, and some even take millions, of years to complete a single orbit around the Sun. These comets are known as *long-period comets*. However, a few *short-period comets*, conventionally defined as those having orbital periods of less than 200 years, return for another encounter within a relatively short time. According to Kepler's third law, short-period comets do not venture far beyond the distance of Neptune at aphelion.

Unlike the orbits of the other solar system objects we have studied so far, the orbits of comets are not necessarily confined to within a few degrees of the ecliptic plane. Short-period comets do tend to have prograde orbits lying close to the ecliptic, but long-period comets exhibit all inclinations and all orientations, both prograde and retrograde, roughly uniformly distributed in all directions from the Sun.

The short-period comets originate beyond the orbit of Neptune, in the **Kuiper belt** (named after Gerard

Kuiper, a pioneer in infrared and planetary astronomy). A little like the asteroids in the inner solar system, most Kuiper belt comets move in roughly circular orbits between about 30 and 50 AU from the Sun, never venturing inside the orbits of the jovian planets. Occasionally, however, a close encounter between two comets, or (more likely) the cumulative gravitational influence of one of the outer planets, "kicks" a Kuiper belt comet into an eccentric orbit that brings it into the inner solar system and into our view. The observed orbits of these comets reflect the flattened structure of the Kuiper belt.

What of the long-period comets? How do we account for their apparently random orbital orientations? Only a tiny portion of a typical long-period cometary orbit lies within the inner solar system, so it follows that, for every comet we see, there must be many more similar objects at great distances from the Sun. On these general grounds, many astronomers reason that there must be a huge "cloud" of comets far beyond the orbit of Pluto, completely surrounding the Sun. This region, which may contain trillions of comets, with a total mass comparable to the mass of the inner planets, is named the **Oort cloud**, after the Dutch astronomer Jan Oort, who first wrote (in the 1950s) of the possibility of such a vast reservoir of inactive, frozen comets orbiting far from the Sun. The Kuiper belt and the orbits of some typical Oort cloud comets are sketched in Figure 14.18.

The observed orbital properties of long-period comets have led researchers to conclude that the Oort cloud may be up to 100,000 AU in diameter. Like those of the Kuiper belt, however, most of the comets of the Oort cloud never come anywhere near the Sun. Indeed, Oort cloud comets rarely approach even the orbit of Pluto, let alone that of Earth. Only when the gravitational field of a passing star happens to deflect a comet into an extremely eccentric orbit that passes through the inner solar system do we actually get to see the comet.

Because the Oort cloud surrounds the Sun in all directions, instead of being confined near the plane of the ecliptic like the Kuiper belt, the long-period comets we see can come from any direction in the sky. Despite their great distances and long orbital periods, Oort cloud comets are still gravitationally bound to the Sun. Their orbits are governed by precisely the same laws of motion that control the planets' orbits.

## CONCEPT CHECK

✔ In terms of composition, how do comets differ from asteroids?

✔ In what sense are the comets we see *un*representative of comets in general?

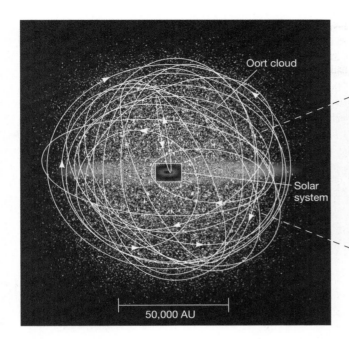

▲ FIGURE 14.18 **Comet Reservoirs** Diagram of the Oort cloud, showing a few cometary orbits. Most Oort cloud comets never come close to the Sun. Of all the orbits shown, only the most elongated ellipse represents a comet that will actually enter the solar system (which, on the scale of this drawing, is much smaller than the overlaid box at the center of the figure) and possibly become visible from Earth. (See also Figure 14.8.) (inset) The Kuiper belt, the source of short-period comets, whose orbits tend to hug the plane of the ecliptic.

## 14.3 Beyond Neptune

No one has ever observed any comets in the faraway Oort cloud—they are just too small and dim for us to see from Earth. But in the 1990s such faint objects began to be inventoried in the relatively nearby Kuiper belt, just beyond Neptune's orbit, some 30 to 50 AU from the Sun. They are collectively referred to as **Kuiper belt objects.** And as the search has broadened, other, even more distant bodies have been found. The generic term for any small body orbiting beyond Neptune—including members of the Kuiper belt—is **trans-Neptunian object.**

Ground-based telescopes have led the way in recent years in the painstaking effort to capture the meager amounts of sunlight reflected from these dark inhabitants of the outer solar system. However, one resident of the Kuiper belt has been known for decades—Pluto. Let's begin our study of this distant region by reviewing what is known about its most prominent member.

### The Serendipitous Discovery of Pluto

Around the end of the 19th century, observations of the orbits of Uranus and Neptune suggested that Neptune's influence was not sufficient to account for all the irregularities in Uranus's motion. Furthermore, it seemed that Neptune itself might be affected by some other unknown body—perhaps even another planet. Following their success in the discovery of Neptune, astronomers hoped to pinpoint the location of this new object by using similar techniques. One of the most ardent searchers was Percival Lowell, a capable, persistent observer and one of the best-known astronomers of

his day. (Recall that he was also the leading proponent of the theory that the "canals" on Mars were constructed by an intelligent race of Martians—see the Part 2 Opener on p. 142.)

Basing his investigation primarily on the motion of Uranus (Neptune's orbit was still relatively poorly determined at the time), and using techniques similar to those developed earlier in the search for Neptune by Adams and Leverrier, Lowell set about calculating where the supposed new body should be. He sought it, without success, during the decade preceding his death in 1916. Not until 14 years later did American astronomer Clyde Tombaugh, working with improved equipment and better photographic techniques at the Lowell Observatory, succeed in finding the new body, only 6° away from Lowell's predicted position. It was named Pluto, for the Roman god of the dead who presided over eternal darkness (and also because its first two letters and its astronomical symbol, ♇ are Lowell's initials). The discovery of Pluto was announced on March 13, 1930, Percival Lowell's birthday, and also the anniversary of Herschel's discovery of Uranus.

On the face of it, the discovery of Pluto looked like another spectacular success for celestial mechanics. However, it now appears that the supposed irregularities in the motions of Uranus and Neptune did not exist, and that the mass of Pluto, not measured accurately until the 1980s, is far too small to have caused them anyway. The discovery of Pluto owed much more to simple luck than to elegant mathematics!

## Pluto's Orbital and Physical Properties

Pluto's orbit is quite elongated, with an eccentricity of 0.25. It is also inclined at 17.2° to the plane of the ecliptic. Because of its substantial orbital eccentricity, Pluto's distance from the Sun varies considerably. At perihelion, it lies 29.7 AU (4.4 billion km) from the Sun, *inside* the orbit of Neptune. At aphelion, the distance is 49.3 AU (7.4 billion km), well outside Neptune's orbit. Pluto last passed perihelion in 1989 and remained inside Neptune's orbit until February 1999. Given Pluto's 248-year orbital period, this will not occur again until the middle of the 23rd century. The orbits of Neptune and Pluto are sketched in Figure 14.19.

Pluto's orbital period is exactly 1.5 times that of Neptune—in other words, the two bodies are locked in a 3:2 resonance (two orbits of Pluto for every three of Neptune) as they orbit the Sun. As a result, even though their orbits appear to cross, Pluto and Neptune are in no danger of colliding with each other. Because of the resonance and Pluto's tilted orbital plane, the distance between the two bodies at closest approach is actually about 17 AU (compare with Pluto's closest approach to Uranus of just 11 AU). As with other solar system resonances, Pluto's 3:2 synchronization with Neptune is not a matter of chance. As explained in Section 15.4, it is a direct consequence of the evolution of the outer solar system billions of years ago.

Pluto is so far away that little is known of its physical nature. Until the late 1970s, studies of sunlight reflected from its surface suggested a rotation period of just under a week, but measurements of its mass and radius were uncertain. All this changed in July 1978, when astronomers at the U.S. Naval Observatory discovered that Pluto has a companion. It was named Charon, after the mythical boatman who ferried the dead across the river Styx into Hades, Pluto's domain. The discovery photograph of Charon is shown in Figure 14.20(a). Charon is the small bump near the top of the image. Figure 14.20(b) shows a 2005 *Hubble Space Telescope* view that clearly resolves the two bodies and in addition shows two more small satellites orbiting the Pluto–Charon system.

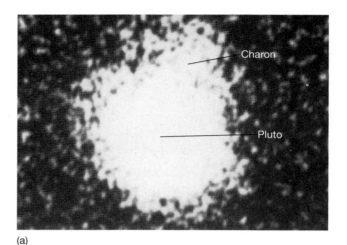

(a)

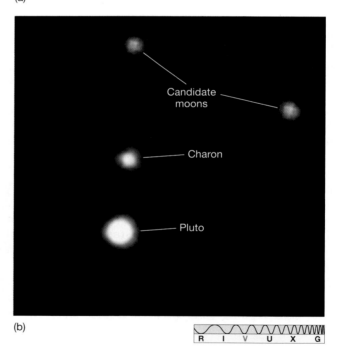

(b)

▲ **FIGURE 14.20 Pluto and Charon** (a) The photograph from which Pluto's moon, Charon, was discovered. The moon is the small blotch of light at the top right portion of the image. The larger blob of reflected sunlight is Pluto itself. (b) The Pluto–Charon system, shown to the same scale and better resolved than in part (a), as seen by the *Hubble Space Telescope*. The angular separation of the planet and its moon is about 0.9″. Two additional small moons have now been confirmed and given the names Nix and Hydra. (*U.S. Naval Observatory; NASA*)

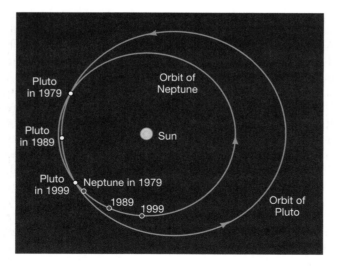

▲ **FIGURE 14.19 Neptune and Pluto** The orbits of Neptune and Pluto cross, although Pluto's orbital inclination and a 3:2 resonance prevent the planets from actually coming close to each other. Between 1979 and 1999, Pluto was actually inside Neptune's orbit.

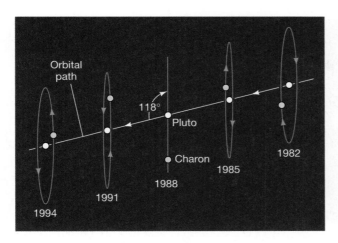

▲ **FIGURE 14.21 Pluto–Charon Eclipses** The orbital orientation of Charon produced a series of eclipses between 1985 and 1991. Observations of eclipses of Charon by Pluto and of Pluto by Charon have provided detailed information about the sizes and orbits of both bodies.

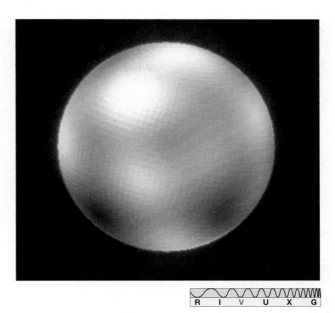

R I V U X G

▲ **FIGURE 14.22 Pluto** A three-dimensional surface map of Pluto—not a photograph, but rather a "modeled" view created by carefully combining 24 *Hubble Space Telescope* images with a mathematical description of the surface. *(NASA)*

The discovery of Charon permitted astronomers to measure the masses and radii of both bodies with great accuracy. Charon's orbit is inclined at an angle of 118° to the plane of Pluto's path around the Sun. By pure luck, over the 6-year period from 1985 to 1991 (less than 10 years after Charon's discovery), the two bodies happened to be oriented in such a way that viewers on Earth saw a series of *eclipses*, in which Pluto and Charon repeatedly passed in front of each other, as seen from our vantage point. Figure 14.21 sketches this orbital configuration. With more good fortune, these eclipses took place while Pluto was closest to the Sun, making for the best possible Earth-based observations.

Basing their calculations principally on the variations in reflected light as Pluto and Charon periodically hid each other, astronomers calculated that Pluto and Charon move in a circular, tidally locked orbit with a period of 6.4 days and a separation of 19,700 km, implying a mass for Pluto of 0.0021 Earth mass ($1.3 \times 10^{22}$ kg), far smaller than any previous estimate. *(More Precisely 2-3)* Pluto's diameter is 2270 km, about one-fifth the size of Earth. Charon is about 1300 km across. If both bodies have the same composition (probably a reasonable assumption), Charon's mass must be about one-sixth that of Pluto.

The masses and radii of Pluto and Charon imply average densities of 2100 kg/m³—just what we would expect for bodies of that size made up mostly of water ice, like the large moons of the outer planets. ∞ (Sec. 6.2) In fact, Pluto is very similar in both mass and radius to Neptune's large moon, Triton—which, as we have seen, is thought to be a captured Kuiper belt object. ∞ (Sec. 13.5) Spectroscopy reveals the presence of *frozen* methane as a major surface constituent of Pluto, implying

a temperature of no more than 50 K. Pluto may also have a thin methane atmosphere, associated with the methane ice on its surface.

Recent computer-generated maps (Figure 14.22) have begun to hint at surface features on Pluto and suggest that Charon may have bright polar caps, although their composition and nature are unknown. Some of the dark regions may be craters or impact basins, as on Earth's Moon. ∞ (Sec. 8.3) However, astronomers eager for a closer look will have to wait until 2015, when NASA's *New Horizons* mission to Pluto and the Kuiper belt, launched in 2006, is scheduled to reach its destination.

## Properties of Trans-Neptunian Objects

Most Kuiper belt objects are not nearly as well observed as Pluto, which happens to be the largest known member of the class and orbits near the inner edge of the belt, making it appear relatively bright as seen from Earth. Most trans-Neptunian objects—even the large ones—are very faint: Figure 14.23 shows some of the best available images (apart from those of Pluto and Charon) of some of these distant bodies. Astronomers generally have only limited information on them—a *segment of an orbit*, from which distance, semimajor axis, period, and eccentricity can be inferred, its *brightness*, which translates into an estimated diameter, and sometimes *brightness variations*, which may imply rotation or the presence of a companion.

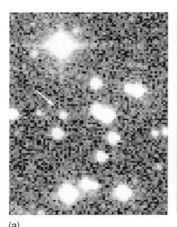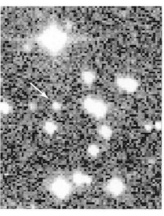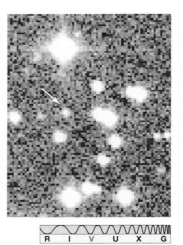

(a)

R I V U X G

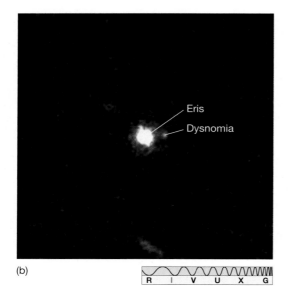

Eris
Dysnomia

(b)

R I V U X G

◀ FIGURE 14.23 **Kuiper Belt Object** (a) Some of the best available images of a Kuiper belt object. Known as Pholus, the object itself is the fuzzy blob (marked with an arrow) that changes position between one frame and the next. It may be almost 1000 km across and lies more than 40 AU from Earth. (b) The trans-Neptunian object Eris and its small moon Dysnomia (named after the Greek goddess of discord and her daughter of chaos and lawlessness) were recently imaged in the infrared at the Keck Observatory in Hawaii. *(LPL/Keck)*

native American creation god), discovered in 2002, is roughly 1200 km across—larger than the largest asteroid, Ceres, and more than half the size of Pluto. The (still unnamed) object 2005 FY$_9$, discovered in 2005, is some 2000 km in diameter. However, the final blow came with the discovery in 2005 of the (non-Kuiper belt) object Eris, whose diameter ( measured in 2006 by *HST*) is 2400 km—*larger* than Pluto.

The sizes of some of the biggest known trans-Neptunian objects are compared in Figure 14.24, and their orbits are sketched in Figure 14.25. These figures include yet another intriguing object called Sedna, discovered in 2003 and thought to have a diameter of about 1500 km. It is the farthest known object in the solar system. Its highly elliptical orbit takes it out to almost 1000 AU from the Sun—almost to the (theoretical) inner edge of the Oort cloud. There may well be more Pluto-sized (or larger) objects still out there, waiting to be discovered—the possibility has not been conclusively ruled out. Systematic faint surveys of a broad swath of the sky that includes the entire plane of the ecliptic (as are planned within the next decade) will be needed before any definitive statement can be made.

### The King of the Kuiper Belt

Even before the discovery of Eris, many astronomers had already concluded that Pluto was not a different type of object at all, but simply the largest known member of the Kuiper belt, playing much the same role in the Kuiper belt as Ceres does among the asteroids. Once it became clear that Eris was most likely larger than Pluto, pressure mounted to find a classification that reflected astronomers' new understanding of the outer solar system. In 2006 the International Astronomical Union

Nevertheless, as astronomers have refined their observational techniques, the number of known objects beyond Neptune has risen rapidly. As of early 2007, the count stands at just over 1000; most are in the Kuiper belt. Because they are so small and distant, researchers reason that only a tiny fraction of the total have so far been observed, and the total number of Kuiper belt objects larger than 100 km is estimated to be more than 100,000. If that is so, then the combined mass of all the debris in the Kuiper belt could well be hundreds of times larger than the mass of the inner asteroid belt (although still less than the mass of Earth).

Unfortunately for Pluto's planetary status, as the details were filled in and the numbers of known trans-Neptunian objects increased, it became more and more clear to astronomers that Pluto is not distinctly different from the other small bodies in the outer solar system, as had once been supposed. The Kuiper belt object Quaoar (pronounced "Kwah-o-ar," and named for a

▲ FIGURE 14.24 **Trans-Neptunian Objects** Some large trans-Neptunian objects, including Pluto and the largest known, called Eris, with part of Earth and the Moon added for scale. Most diameters are approximate, as they have been estimated from the object's observed brightness. (*NASA; Caltech*)

(IAU) adopted the first ever *definition* of a planet: A body is considered a planet if

1. it orbits the Sun,
2. it is massive enough that its own gravity has caused its shape to be approximately spherical, and
3. it has "cleared the neighborhood" around its orbit of other bodies.

"Clearing the neighborhood" means that the body has swept up (collided with) any debris whose orbit happens to intersect its path, or that its gravity has kicked most such debris to other parts of the solar system. ∞ (*Discovery 6-1*)

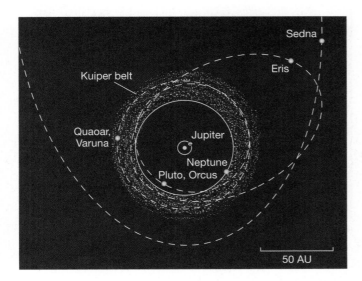

▲ FIGURE 14.25 **Orbits in the Outer Solar System** Orbits of some prominent residents of the outer solar system: Jupiter, Neptune, the Kuiper belt, Pluto, and several more trans-Neptunian objects. Sedna's orbit extends for 10 times the outermost distance to Eris.

Pluto certainly satisfies the first two criteria. However, part 3 of this definition *excludes* Pluto from planetary status. The wording is chosen specifically to ensure that a planet is massive enough to dominate its immediate neighborhood, and Pluto, which orbits within the congested Kuiper belt and is resonantly tied to Neptune's orbital motion, clearly does not meet that condition. (Indeed, as we will see in Chapter 15, the Kuiper belt was formed as Neptune and Uranus established themselves as planets by clearing their own orbital neighborhoods of interplanetary debris.) As if for consolation, so as not to completely strip Pluto of its title, the IAU invented a new category for bodies that satisfy criteria 1 and 2 but not 3—*dwarf planet.* Under these criteria, Pluto, Eris, and Ceres are now classified as dwarf planets. Several of the other bodies in Figure 14.24 may also be placed in this category once their properties are better determined.

The IAU decision has sparked controversy among astronomers. Some are unhappy at Pluto's demotion from the solar system "A list." They argue (probably correctly) that the new definition was concocted largely to exclude Pluto and Eris from planetary status and that criterion 3 in its current form is really too vague to be of much scientific value. Others applaud the redefinition as long-overdue recognition of Pluto's true identity as a large Kuiper belt object, but object to the new term "dwarf planet" as redundant and unnecessarily confusing.

The arguments over terminology are probably far from over, but it seems unlikely that future changes will restore Pluto to its former status. Simply put, it is not sufficiently different from the other known Kuiper belt objects to warrant inclusion in a different category. Few astronomers doubt that if Pluto were discovered today, its classification as a member (the largest yet, the headlines would say!) of the Kuiper belt would be assured.

Pluto's reclassification illustrates the way in which science evolves. Our conception of the cosmos has undergone many changes—some radical, others more gradual—since the time of Copernicus. ∞ (Sec. 2.3) We have seen several examples already, and we will see many more later in this book. As our understanding grows, our terminology and classifications change. The situation with Pluto has a close parallel in the discovery of the first asteroids in the early 19th century. They, too, were initially classified as planets—indeed, in the 1840s, leading astronomy texts listed no fewer than 11 planets in our solar system, including as numbers 5 through 8 the asteroids Vesta, Juno, Ceres, and Pallas. Within a couple of decades, however, the discovery of several dozen more asteroids had made it clear that these small bodies represented a whole new class of solar system objects, separate from the major planets, and the number of planets fell to 8 (including the then newly discovered Neptune).

Much of the observational work on the Kuiper belt began as a search for a 10th planet. It is ironic that the end result of these efforts has been a reduction in the number of "true" solar system planets back to eight!

CONCEPT CHECKS

✔ Why do astronomers no longer regard Pluto as a planet?

## 14.4 Meteoroids

On a clear night, it is possible to see a few *meteors*— "shooting stars"—every hour. A **meteor** is a sudden streak of light in the night sky caused by friction between air molecules in Earth's atmosphere and an incoming piece of interplanetary matter—an asteroid, a comet, or a **meteoroid.** The friction heats and excites the air molecules, which then emit light as they return to their ground states, producing the characteristic bright streak shown in Figure 14.26. Recall from Section 6.5 that the distinction between an asteroid and a meteoroid is simply a matter of size. Both are chunks of rocky interplanetary debris; meteoroids are conventionally taken to be less than 100 m in diameter.

Note that the brief flash that is a meteor is in no way similar to the broad, steady swath of light associated with a comet's tail. A meteor is a fleeting event in Earth's atmosphere, whereas a comet tail exists in deep space and can be visible in the sky for weeks or even months. Before encountering the atmosphere, the piece of debris causing a meteor was almost certainly a meteoroid, because these small interplanetary fragments are far more common than either asteroids or comets. Any piece of interplanetary debris that survives its fiery passage through our atmosphere and finds its way to the ground is called a **meteorite.**

(a)

(b)

R I V U X G

▲ FIGURE 14.26 **Meteor Trails** A bright streak called a meteor is produced when a fragment of interplanetary debris plunges into the atmosphere, heating the air to incandescence. (a) A small meteor photographed against stars and the Northern Lights provide a stunning background for a bright meteor trail. (b) These meteors (one with a red smoke trail) streaked across the sky during the height of the Leonid meteor storm of November 2001. (*P. Parviainen; J. Lodriguss*)

▶ **FIGURE 14.27 Meteor Showers** A meteoroid swarm associated with a given comet intersects Earth's orbit at specific locations, giving rise to meteor showers at certain fixed times of the year. A portion of the comet breaks up as it rounds the Sun, at the point marked 1. Fragments continue along the comet's orbit, gradually spreading out (points 2 and 3). The rate at which the debris disperses around the orbit is much slower than depicted here. It actually takes many orbits for the material to disperse as shown, but eventually the fragments extend all around the orbit, more or less uniformly. If the orbit happens to intersect Earth's, the result is a meteor shower each time Earth passes through the intersection (point 4).

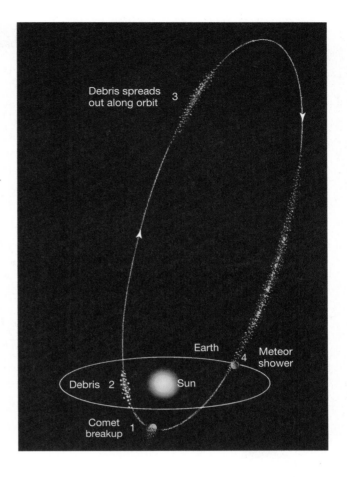

## Cometary Fragments

Smaller meteoroids are mainly the rocky remains of broken-up comets. Each time a comet passes near the Sun, some cometary fragments are dislodged from the main body. The fragments initially travel in a tightly knit group of dust or pebble-sized objects, called a **meteoroid swarm,** moving in nearly the same orbit as the parent comet. Over the course of time, the swarm gradually disperses along the orbit, and eventually the **micrometeoroids,** as these small meteoroids are known, become more or less smoothly spread all the way around the parent comet's orbit.

If Earth's orbit happens to intersect the orbit of such a young cluster of meteoroids, a spectacular **meteor shower** can result. Earth's motion takes our planet across a given comet's orbit at most twice a year (depending on the precise orbit of each body). Intersection occurs at the same time each year (see Figure 14.27), so the appearance of certain meteor showers is a regular and (fairly) predictable event.

Table 14.1 lists some prominent meteor showers, the dates they are visible from Earth, and the comets from which they are thought to originate. Meteor showers are usually named for their *radiant*, the constellation from whose direction they appear to come (Figure 14.28). For example, the Perseid shower is seen to emanate from the constellation Perseus. It can last for several days, but reaches maximum every year on the morning of August 12, when upward of 50 meteors per hour can be observed. Astronomers can use the speed and direction of a meteor's flight to compute the meteor's interplanetary trajectory. This is how certain meteoroid swarms have come to be identified with well-known comet orbits. For example, the Perseid shower shares the same orbit as comet 1862III (also known as comet Swift–Tuttle), the third comet discovered in the year 1862.

## Stray Asteroids

Larger meteoroids—more than a few centimeters in diameter—are usually *not* associated with swarms of cometary debris. Generally regarded as small bodies that

have strayed from the asteroid belt, possibly as the result of collisions with or between asteroids, these objects have produced most of the cratering on the surfaces of the Moon, Mercury, Venus, Mars, and some of the moons of the jovian planets. When these large meteoroids enter Earth's atmosphere with a typical velocity of nearly 20 km/s, they produce energetic shock waves, or "sonic booms," as well as bright streaks in the sky and dusty trails of discarded debris. Such large meteors are sometimes known as *fireballs*. The greater the speed of the incoming object, the hotter its surface becomes and the faster it burns up. A few large meteoroids enter the atmosphere at such high speed (about 75 km/s) that they either fragment or disperse entirely at high altitudes.

The more massive meteoroids (at least a ton in mass and a meter across) do make it to Earth's surface, producing a crater such as the kilometer-wide Barringer Crater shown in Figure 8.19. From the size of this crater, we can estimate that the meteoroid responsible for its formation must have had a mass of about 200,000 tons. Since only 25 tons of iron meteorite fragments have been found at the crash site, the remaining mass must have been scattered by the explosion at impact, broken down by subsequent erosion, or buried in the ground.

**TABLE 14.1   Some Prominent Meteor Showers**

| Morning of Maximum Activity | Name of Shower | Rough Hourly Count | Parent Comet |
|---|---|---|---|
| Jan. 3 | Quadrantid | 40 | — |
| Apr. 21 | Lyrid | 10 | 1861I (Thatcher) |
| May 4 | Eta Aquarid | 20 | Halley |
| June 30 | Beta Taurid | 25‡ | Encke |
| July 30 | Delta Aquarid | 20 | — |
| Aug. 11 | Perseid | 50 | 1862III (Swift–Tuttle) |
| Oct. 9 | Draconid | up to 500 | Giacobini–Zinner |
| Oct. 20 | Orionid | 30 | Halley |
| Nov. 7 | Taurid | 10 | Encke |
| Nov. 16 | Leonid | 12* | 1866I (Tuttle) |
| Dec. 13 | Geminid | 50 | 3200 (Phaeton)† |

*Every 33 years, as Earth passes through the densest region of this meteoroid swarm, we see intense showers that can exceed 1000 meteors per minute for brief periods. This intense activity is next expected to occur in 2032.
†Phaeton is actually an asteroid and shows no signs of cometary activity, but its orbit matches the meteoroid paths very well.
‡Meteor count peaks after sunrise.

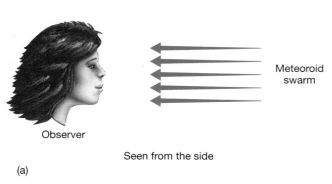

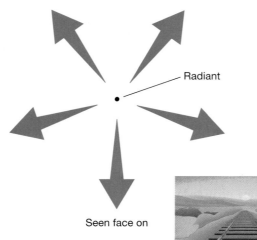

Currently, Earth is scarred with nearly 100 craters larger than 0.1 km in diameter. Most of these craters are so heavily eroded by weather and distorted by crustal activity that they can be identified only in satellite photography, as shown in Figure 14.29. Fortunately, such major collisions between Earth and large meteoroids are thought to be rare events now. Researchers estimate that, on average, they occur only once every few hundred thousand years (see *Discovery 14-1*).

The orbits of large meteorites that survive their plunge through Earth's atmosphere can be reconstructed in a manner similar to that used to determine the orbits of meteor showers. In most cases, their computed orbits do indeed intersect the asteroid belt, providing the strongest evidence we have that they were once part of the belt before being redirected, probably by a collision with another asteroid, into the Earth-crossing orbit that led to the impact with our planet. Not all meteoroids come from the asteroid belt, however: As we have already seen, some are known to have originated on the surface of Mars. ∞ *(Discovery 10-1)* In addition, detailed composition studies reveal that others most likely came from the Moon, blasted off the lunar surface by violent impacts long ago.

◀ FIGURE 14.28 **Radiant** (a) A group of meteoroids approaches an observer. All are moving in the same direction at the same speed. (b) From the observer's viewpoint, the trajectories of the meteoroids (and the meteor shower they produce) appear to spread out from a central point, called the radiant, in much the same way as parallel railroad tracks seem to converge in the distance (c).

▲ FIGURE 14.29 **Manicouagan Reservoir** This photograph, taken from orbit by the U.S. *Skylab* space station, shows the ancient impact basin that forms Quebec's Manicouagan Reservoir. A large meteorite landed there about 200 million years ago. The central floor of the crater rebounded after the impact, forming an elevated central peak. The lake, 70 km in diameter, now fills the resulting ring-shaped depression. (*NASA*)

▲ FIGURE 14.30 **Tunguska Debris** The Tunguska event of 1908 leveled trees over a vast area. Although the impact of the blast was tremendous and its sound audible for hundreds of kilometers, the Siberian site is so remote that little was known about the event until scientific expeditions arrived to study it many years later. (*Sovfoto/Eastfoto*)

Not all encounters of meteoroids with Earth result in an impact. One of the most recent meteoritic events occurred in central Siberia on June 30, 1908 (Figure 14.30). The presence of only a shallow depression, as well as a complete absence of fragments, implies that this Siberian intruder exploded several kilometers above the ground, leaving a blasted depression at ground level, but no well-formed crater. Recent calculations suggest that the object in question was a rocky meteoroid about 30 m across. The explosion, estimated to have been equal in energy to a 10-megaton nuclear detonation, was heard hundreds of kilometers away and produced measurable increases in atmospheric dust levels all across the Northern Hemisphere.

## Meteorite Properties

One feature that distinguishes small micrometeoroids, which burn up in Earth's atmosphere, from larger meteoroids, which reach the ground, is their composition—as evidenced by their strikingly different densities. The average density of meteoritic fireballs that are too small to reach the ground (but that can be captured by high-flying aircraft) is about 500–1000 kg/m$^3$. Such a low density is typical of comets, which are made of loosely packed ice and dust. In contrast, the meteorites that reach Earth's surface are often much denser—up to 5000 kg/m$^3$—suggesting a composition more like that of the asteroids. Meteorites like those shown in Figure 14.31 have received close scrutiny from planetary scientists—prior to the Space Age, they were the only type of extraterrestrial matter we could touch and examine in terrestrial laboratories.

Most meteorites are rocky in composition (Figure 14.32a), although a few percent are composed mainly of iron and nickel (Figure 14.32b). The basic composition of the rocky meteorites is much like that of the inner planets and the Moon, except that some of their lighter elements—such as hydrogen and oxygen—appear to have boiled away long ago when the bodies from which the meteorites originated were molten. Some meteorites show clear evidence of strong heating at some time in their past, most likely indicating that they originated on a larger body that either underwent some geological activity or was partially melted during the collision that liberated the fragments that eventually became the meteorites. Others show no such evidence and probably date from the formation of the solar system.

Most primitive of all are the *carbonaceous* meteorites, so called because of their relatively high carbon content. These meteorites are black or dark gray and may well be related to the carbon-rich C-type asteroids that populate the outer asteroid belt. (Similarly, the silicate-rich stony meteorites are probably associated with the inner S-type asteroids.) Many carbonaceous meteorites contain significant amounts of ice and other

▲ FIGURE 14.31 **Large Meteorites** (a) The world's second largest meteorite, the Ahnighito, on display at the American Museum of Natural History in New York, serves as a jungle gym for curious children. This 34-ton rock is so heavy that the Museum floor had to be specially reinforced to support its weight. (b) The Wabar meteorite, discovered in the Arabian desert. Although small fragments of the original meteor had been collected more than a century before, the 2000-kg main body was not found until 1965. *(Corbis-Blair; Jon Mandaville/Aramco World)*

volatile substances, and they are usually rich in organic molecules.

Finally, almost all meteorites are *old*. Direct radioactive dating shows most of them to be between 4.4 and 4.6 billion years old—roughly the age of the oldest lunar rocks. Meteorites, along with lunar rocks, comets, and Kuiper belt objects like Pluto, provide essential clues to the original state of matter in the solar neighborhood.

We will return to their critical role in our understanding of solar system formation in Chapter 15.

CONCEPT CHECK

✔ What are meteoroids, and why are they important to planetary scientists?

▲ FIGURE 14.32 **Meteorite Samples** (a) A stony (silicate) meteorite often has a dark fusion crust, created when its surface is melted by the tremendous heat generated during its passage through the atmosphere. The coin at the bottom is for scale. (b) Iron meteorites are much rarer than the stony variety and often contain some nickel as well. Most show characteristic crystalline patterns when their surfaces are cut, polished, and etched with acid. *(Science Graphics)*

# CHAPTER REVIEW

## Summary

**1** More than 100,000 **asteroids (p. 360)** have been cataloged. Most orbit in a broad band called the **asteroid belt (p. 360)** between the orbits of Mars and Jupiter. They are probably primal rocks that never 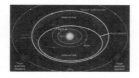 clumped together to form a planet. The **Trojan asteroids (p. 364)** share Jupiter's orbit, remaining 60° ahead of or behind that planet as it moves around the Sun. A few **Earth-crossing asteroids (p. 363)** have orbits that intersect Earth's orbit and will probably collide with our planet one day. The **Kirkwood gaps (p. 365)** in the main asteroid belt have been cleared by Jupiter's gravity.

**2** The largest asteroids are a few hundred kilometers across. Most are much smaller. The total mass of all asteroids combined is less than the mass of Earth's Moon. Asteroids are classified according to the properties of their reflected light:  Brighter, S-type (silicate) asteroids dominate the inner asteroid belt, whereas darker, C-type (carbonaceous) asteroids are more plentiful in the outer regions. The C-type asteroids are thought to have changed little since the solar system formed. Smaller asteroids tend to be irregular in shape and may have undergone violent collisions in the past.

**3** **Comets (p. 367)** are fragments of icy material that normally orbit far from the Sun. As a comet approaches the Sun, its surface ice begins to vaporize. We see the comet by the sunlight reflected from the dust and vapor released.  The **nucleus (p. 367)** of a comet may be only a few kilometers in diameter. It is surrounded by a **coma (p. 367)** of dust and gas and an extensive invisible **hydrogen envelope (p. 368)**. Stretching behind the comet is a long **tail (p. 367)**, formed by the interaction between the cometary material and the solar wind. The comet's **ion tail (p. 368)** consists of ionized gas particles and always points directly away from the Sun. The comet's **dust tail (p. 369)** is less affected by the solar wind and has a somewhat curved shape. Comets are icy, dusty bodies, sometimes called "dirty snowballs," that are thought to be leftover material unchanged since the formation of the solar system. Their masses are comparable to the masses of small asteroids.

**4** Unlike the orbits of most other bodies in the solar system, comets' orbits are often highly elongated and not confined to the ecliptic plane. Most comets are thought to reside in the **Oort cloud (p. 374)**, a vast "reservoir" of cometary 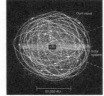 material, tens of thousands of astronomical units across, completely surrounding the Sun. A very small fraction of comets happen to have highly elliptical orbits that bring them into the inner solar system. Comets with orbital periods less than about 200 years are thought to originate not in the Oort cloud, but in the **Kuiper belt (p. 374)**, a broad band lying roughly in the plane of the ecliptic, beyond the orbit of Neptune. More than 1000 **Kuiper belt objects (p. 375)** are now known. Pluto is the largest member of the class.

**5** Pluto is the best-known body orbiting beyond Neptune. It was discovered in the 20th century after a laborious search for a planet that was supposedly affecting Uranus's orbital motion. However, we now know that  Pluto is far too small to have any detectable influence on Uranus's path. Pluto has a moon, Charon, whose mass is about one-sixth that of the planet itself. Studies of Charon's orbit around Pluto have allowed the masses and radii of both bodies to be accurately determined. Several bodies comparable in size to Pluto and Charon orbit beyond Neptune. At least one—Eris—is larger than Pluto. Neither Eris nor Pluto are currently classified as planets because their masses are too low to have cleared their orbital neighborhoods of other bodies.

**6** **Meteors (p. 380)**, or "shooting stars," are bright streaks of light that flash across the sky as **meteoroids (p. 380)**—pieces of interplanetary debris—enter Earth's atmosphere. If a meteoroid reaches the ground, it is called a **meteorite (p. 380)**.  Each time a comet rounds the Sun, some cometary material becomes dislodged, forming a **meteoroid swarm (p. 381)**— a group of small **micrometeoroids (p. 381)** following the comet's original orbit. If Earth happens to pass through the comet's orbit, a **meteor shower (p. 381)** occurs. Larger meteoroids are probably pieces of material chipped off asteroids following collisions in the asteroid belt.

**7** The major difference between meteoroids and asteroids is their size: The dividing line between them is conventionally taken to be 100 m. Meteorites are thought to be composed of the same material that makes up the asteroids, and the few orbits  that have been determined are consistent with an origin in the asteroid belt. Some meteorites show evidence of heating, but the oldest ones do not. Most meteorites are between 4.4 and 4.6 billion years old. Comets and stray asteroids are responsible for most of the cratering on the various worlds in the solar system. The most recent large impact on Earth occurred in 1908, when an asteroid apparently exploded several miles above Siberia.

## Review and Discussion

1. What are the Trojan, Apollo, and Amor asteroids?
2. How are asteroid masses measured?
3. How have the best photographs of asteroids been obtained?
4. What are the Kirkwood gaps? How did they form?
5. How do the C-type and S-type asteroids differ?
6. Are all asteroids found in the asteroid belt?
7. What are comets like when they are far from the Sun? What happens when they enter the inner solar system?
8. Where in the solar system do most comets reside?
9. Describe the various parts of a comet while it is near the Sun.
10. What are the typical ingredients of a comet nucleus?
11. How do we know what comets are made of?
12. What are some possible fates of comets?

13. Describe two ways in which a comet's orbit may change.
14. In what ways is the Kuiper belt similar to the asteroid belt? In what ways do they differ?
15. Why has the number of planets in the solar system recently decreased?
16. Explain the difference between a meteor, a meteoroid, and a meteorite.
17. What causes a meteor shower?
18. What do meteorites reveal about the age of the solar system?
19. Why can comets approach the Sun from any direction, but asteroids generally orbit close to the plane of the ecliptic?
20. What might be the consequences if a 10-km-diameter meteorite struck Earth today?

## Conceptual Self-Test: True or False/Multiple Choice

1. Asteroids, meteoroids, and comets are remnants of the early solar system.
2. Most asteroids move on roughly circular orbits between the orbits of Earth and Mars.
3. The Kirkwood gaps are two broad zones within the asteroid belt in which no asteroids are found.
4. Some comets travel in orbits that take them up to 50,000 AU from the Sun.
5. Cometary orbits always lie close to the plane of the ecliptic.
6. The Oort cloud is the large cloud of gas surrounding a comet while it is near the Sun.
7. Pluto is no longer the largest body known to orbit beyond Neptune.
8. Some meteorites found on Earth originally came from the Moon or Mars.
9. Comets are the sources of many meteor showers.
10. Astronomers have succeeded in tracing the orbits of some meteorites back into the asteroid belt.
11. According to Figure 14.1 ("Inner Solar System"), the asteroid groups with the smallest perihelion distances also tend to have orbits that (a) are slowest; (b) are nearly circular; (c) are most eccentric; (d) extend nearly to Jupiter.
12. Most main-belt asteroids are about the size of (a) the Moon; (b) North America; (c) a U.S. state; (d) a small U.S. city.

13. Spectroscopic studies indicate that the majority of asteroids contain large fractions of (a) carbon; (b) silicate rocks; (c) iron and nickel; (d) ice.
14. Trojan asteroids orbiting at Jupiter's Lagrangian points are located (a) far outside Jupiter's orbit; (b) close to Jupiter; (c) behind and in front of Jupiter, sharing its orbit; (d) between Mars and Jupiter.
15. The tails of a comet (a) point away from the Sun; (b) point opposite the direction of motion of the comet; (c) curve from right to left; (d) curve clockwise with the interplanetary magnetic field.
16. Compared with the orbits of the short-period comets, the orbits of long-period comets (a) tend to lie in the plane of the ecliptic; (b) look like short-period orbits, but are simply much larger; (c) are much less eccentric; (d) can come from all directions.
17. Kuiper belt objects are not regarded as planets because (a) they orbit too far from the Sun; (b) their masses are too low to clear other bodies from their orbital paths; (c) they are all irregular in shape; (d) they are predominantly icy in composition.
18. According to the chart in *Discovery 14-1*, an impact resulting in global catastrophe is expected to occur roughly once per (a) year; (b) century, (c) millennium, (d) million years.

19. A meteorite is a piece of interplanetary debris that (a) burns up in Earth's atmosphere; (b) misses Earth's surface; (c) glances off Earth's atmosphere; (d) survives the trip to the surface.

20. According to Table 14.1, the meteor shower that occurs closest to the autumnal equinox is the (a) Lyrids; (b) Beta Taurids; (c) Perseids; (d) Orionids.

## Problems

 *Algorithmic versions of these Problems are available in the Practice Problems module of the Companion Website. The number of dots preceding each Problem indicates its approximate level of difficulty.*

1. • (a) The asteroid Pallas has an average diameter of 520 km and a mass of $3.2 \times 10^{20}$ kg. How much would a 100-kg astronaut weigh there? (b) What is the asteroid's escape speed?

2. • You are standing on the surface of a spherical asteroid 10 km in diameter, of density 3000 kg/m$^3$. Could you throw a small rock fast enough that it escapes? Give the speed required in km/s and mph.

3. •• How large would the asteroid in the previous problem have to be for your weight to be 1 percent of your weight on Earth?

4. • Can you find a simple orbital resonance with Jupiter that can account for the small Kirkwood gap evident in Figure 14.7(a) at a semimajor axis around 2.7 AU?

5. •• The asteroid Icarus (Figure 14.5) has a perihelion of 0.19 AU and an orbital eccentricity of 0.83. Calculate the asteroid's orbital semimajor axis and aphelion distance from the Sun. Do these figures, by themselves, *necessarily* imply that Icarus will one day collide with Earth?

6. •• Calculate the minimum and maximum angular diameter of the Sun, as seen from Icarus (problem 5). How does your answer compare with the Sun's diameter as seen from Earth?

7. •• Using the data given in the text, estimate Dactyl's orbital period as it orbits Ida.

8. ••• (a) *NEAR's* initial orbit around Icarus had a periapsis (distance of closest approach) of about 100 km from the asteroid's center. If the mass of Eros is $6.7 \times 10^{15}$ kg and the orbit has an eccentricity of 0.3, calculate the spacecraft's orbital period. (b) Subsequently, *NEAR* moved into a closer orbit, with periapsis 14 km and apoapsis (greatest distance) 60 km. What was the new orbital period?

9. •• (a) Calculate the orbital period of a comet with a perihelion distance of 0.5 AU and aphelion in the Oort cloud, at a distance of 50,000 AU from the Sun. (b) A short-period comet has a perihelion distance of 1 AU and an orbital period of 125 years. What is its maximum distance from the Sun?

10. ••• As comet Hale–Bopp rounded the Sun, nongravitational forces changed its orbital period from 4200 years to 2400 years. By what factor did the comet's semimajor axis change? Given that the perihelion remained unchanged at 0.914 AU, calculate the old and new orbital eccentricities.

11. • Astronomers estimate that comet Hale–Bopp lost mass at an average rate of about 350,000 kg/s during the time it spent close to the Sun—a total of about 100 days. Estimate the total amount of mass lost and compare it with the comet's estimated mass of $5 \times 10^{15}$ kg.

12. •• It has been hypothesized that Earth is under continuous bombardment by house-sized "minicomets" with typical diameters of 10 m, at the rate of some 30,000 per day. Assuming spherical shapes and average densities of 100 kg/m$^3$, calculate the total mass of material reaching Earth each year. Compare the total mass received in the past 1 billion years (assuming that all rates were the same in the past) with the mass of Earth's oceans (see Chapter 7, problem 5).

13. • A particular comet has a total mass of $10^{13}$ kg, 95 percent of which is ice and dust. The remaining 5 percent is in the form of rocky fragments with an average mass of 100 g. How many meteoroids would you expect to find in the swarm formed by the breakup of this comet?

14. •• It is observed that the number of asteroids or meteoroids of a given diameter is roughly inversely proportional to the square of the diameter. Approximating the actual distribution of asteroids first as a single 1000-km body (e.g., Ceres), then as one hundred 100-km bodies, then as ten thousand 10-km asteroids, and so on, and assuming constant densities of 3000 kg/m$^3$, calculate the total mass (in units of Ceres's mass) in the form of 1000-km bodies, 100-km bodies, 10-km bodies, 1-km bodies, and 100-m bodies.

15. •• A meteoroid was created by a collision in the asteroid belt 2 billion years ago. Its greatest distance from the Sun is 3 AU, and its eccentricity is 0.8. How many times has the meteoroid crossed Earth's orbital radius?

*The Companion Website at www.aw-bc.com/chaisson provides algorithmically generated versions of each chapter's Problems, along with additional quizzes, an Animations & Videos gallery, an Images gallery, an interactive Glossary, and a full eBook.*

# 15

*The formation of our solar system was a long-ago event, with much of the matter of our primordial galactic cloud eventually either comprising the Sun and planets or ejected back into deep space. Now, some 4.5 billion years later, it is not easy to reconstruct what exactly did happen here. Astronomers therefore observe other young star systems, hoping to gain some insight about the origins of our own solar system. Here, the* Spitzer Space Telescope *has taken this infrared image of W5, with its towering pillars of cool gas and dust illuminated at their tips with light from warm embryonic stars. (SSC/JPL)* ▶

# THE FORMATION OF PLANETARY SYSTEMS
## The Solar System and Beyond

Having completed the chapters on the planets, you may be struck by the vast range of physical and chemical properties found in the solar system. The planets present a long list of interesting features and bizarre peculiarities, and the list grows even longer when we also consider their moons. Every object has its idiosyncrasies, some of them due to particular circumstances, others the result of planetary evolution. Each time a new discovery is made, we learn a little more about the properties and history of our planetary system. Still, our astronomical neighborhood might seem more like a great junkyard than a smoothly running planetary system.

Can we really make any sense of the collection of solar system matter? Are there underlying principles that unify the knowledge we have gained? And if there are, do they extend to planetary systems beyond our own? The answer, as we will see, is "Maybe. . . ."

## LEARNING GOALS

*Studying this chapter will enable you to*

1 List the major facts that any theory of solar system formation must explain and indicate how the leading theory accounts for them.

2 Explain how the terrestrial planets formed.

3 Say why planetary densities depend on distance from the Sun.

4 Discuss the leading theories for the formation of the jovian worlds.

5 Describe how comets and asteroids formed, and explain their role in determining planetary properties.

6 Outline the properties of known extrasolar planets.

7 Discuss how extrasolar planets fit in with current theories of solar system formation.

 Visit www.aw-bc.com/chaisson for additional images, animations, quizzes, and eBook for this chapter.

## 15.1   Modeling Planet Formation

The origin of the planets and their moons is a complex and as yet incompletely solved puzzle, although the basic outlines of the processes involved are becoming understood. ∞ (Sec. 6.7) Most of our knowledge of the solar system's formative stages has emerged from studies of interstellar gas clouds, fallen meteorites, and Earth's Moon, as well as of the various planets observed with ground-based telescopes and planetary space probes. Ironically, studies of Earth itself do not help much, because information about our planet's early stages eroded away long ago. Meteorites and comets provide perhaps the most useful information, for nearly all have preserved within them traces of solid and gaseous matter from the earliest times.

Until the mid-1990s, theories of the formation of planetary systems concentrated almost exclusively on our own solar system, for the very good reason that astronomers had no other examples of planetary systems against which to test their ideas. However, all that has now changed. As of mid-2007, we know of more than two hundred **extrasolar planets**—planets orbiting stars other than the Sun—to challenge our theories. And challenge them they do! As we will see, the other planetary systems discovered to date seem to have properties quite different from our own and may well require us to radically rethink our conception of how stars and planets form.

### Model Requirements

Although we now know of many extrasolar planetary systems, we currently have only limited information on each—little more than estimates of orbits and masses for the largest planets. Accordingly, we begin our study by outlining the comprehensive theory that accounts, in detail, for most of the observed properties of the planetary system we know best: the solar system. Later we will assess how our theory holds up in the face of the growing body of extrasolar data.

Any theory of the origin and architecture of our planetary system must adhere to the known facts. We know of 10 outstanding properties of our solar system as a whole:

1. *Each planet is relatively isolated in space.* The planets exist as independent bodies at progressively larger distances from the central Sun; they are not bunched together. In rough terms, each planet tends to be twice as far from the Sun as its next inward neighbor.

2. *The orbits of the planets are nearly circular.* In fact, with the exception of Mercury, which we will argue is a special case, each planetary orbit is close to a perfect circle.

3. *The orbits of the planets all lie in nearly the same plane.* The planes swept out by the planets' orbits are accurately aligned to within a few degrees. Again, Mercury is a slight exception.

4. *The direction in which the planets orbit the Sun (counterclockwise as viewed from above Earth's North Pole) is the same as the direction in which the Sun rotates on its axis.* Virtually all the large-scale motions in the solar system (other than comets' orbits) are in the same plane and in the same sense. The plane is that of the Sun's equator, and the sense is that of the Sun's rotation.

5. *The direction in which most planets rotate on their axis is roughly the same as the direction in which the Sun rotates on its axis.* This property is less general than the one just described for revolution, as two planets—Venus and Uranus—do not share it.

6. *Most of the major moons revolve about their parent planets in the same direction that the planets rotate on their axes.* Only Neptune's large moon Triton orbits in a retrograde sense.

7. *Our planetary system is highly differentiated.* The inner, terrestrial planets are characterized by high densities, moderate atmospheres, slow rotation rates, and few or no moons. By contrast, the jovian planets, farther from the Sun, have low densities, thick atmospheres, rapid rotation rates, and many moons.

8. *The asteroids are very old and exhibit a range of properties not characteristic of either the inner or the outer planets or their moons.* The asteroid belt shares, in rough terms, the bulk orbital properties of the planets. However, it appears to be made of primitive, unevolved material, and the meteorites that strike Earth are the oldest rocks known.

9. *The Kuiper belt is a collection of asteroid-sized icy bodies orbiting beyond Neptune.* Pluto is the largest known member of this class.

10. *The Oort cloud comets are primitive, icy fragments that do not orbit in the plane of the ecliptic and reside primarily at large distances from the Sun.* While similar to the Kuiper belt in composition, the Oort cloud is a completely distinct part of the outer solar system.

All these observed facts, taken together, strongly suggest a high degree of order within our solar system, at least on large scales. The whole system is not a random assortment of objects spinning or orbiting this way or that. Rather, the overall organization points toward a single origin—an ancient, but one-time, event that occurred 4.6 billion years ago. ∞ (Sec. 14.4) A convincing theory that explains all the features just listed has been a goal of astronomers for centuries.

In Chapter 6, we outlined how the modern condensation theory can account for much of the basic architecture of our planetary system. ∞ (Sec. 6.7) In this chapter, we consider some of the details of that theory, outlined in Figure 11.1, and try to connect them with the growing list of planetary systems beyond our own.

## Planetary Irregularities

Our theory of the solar system must explain the facts just listed, but it is equally important to recognize what it does *not* have to explain. There is plenty of scope for planets to evolve after their formation, so things that may have happened after the initial state of the solar system was established need not be included in our list. Examples are Mercury's 3:2 spin–orbit coupling, Venus's runaway greenhouse effect, the Moon's synchronous rotation, the emergence of life on Earth and its apparent absence on Mars, the Kirkwood gaps in the asteroid belt, and the rings and atmospheric appearance of the jovian planets. There are many more. Since we have already provided an *evolutionary* explanation for these and other properties, our theory need not attempt to account for them at the outset.

In addition to its many regularities, our solar system has many notable *irregularities*, some of which we have already mentioned. Far from threatening our theory, however, these irregularities are important facts for us to consider in shaping our explanations. For example, it is necessary that the explanation for the solar system not insist that *all* planets rotate in the same sense or have *only* prograde moons, because that is not what we observe. Instead, the theory of the solar system should provide strong reasons for the observed planetary characteristics, yet be flexible enough to allow for and explain the deviations, too. And, of course, the existence of the asteroids, comets, and Kuiper belt objects that tell us so much about our past must be an integral part of the picture. That's quite a tall order, yet many researchers now believe that we are close to that level of understanding.

---

CONCEPT CHECK

✔ Why is it important that a theory of solar system formation make clear statements about how planets arose, yet not be too rigid in its predictions?

---

▶ FIGURE 15.1 **Solar System Formation** The condensation theory of planet formation. (a) An infalling interstellar cloud, actually very much larger than the resulting planetary system. (b) The solar nebula after it has contracted and flattened to form a spinning disk. The temperature is greatest in the center, near the red proto-Sun, and coolest at the edges. (c) Dust grains act as condensation nuclei, forming clumps of matter that collide, stick together, and grow into moon-sized (and larger) planetesimals. The composition of the grains and thus of any planetesimal depends on location within the nebula. (d) After a few million years, strong winds from the still-forming Sun begin expelling the nebular gas, and some massive planetesimals in the outer solar system have already accreted gas from the nebula. (e) With the gas ejected, planetesimals continue to collide and grow; the gas giant planets are already formed and the Sun has become a genuine star. (f) Over the course of a hundred million years or so, planetesimals are accreted or ejected, leaving a few large planets that travel in roughly circular orbits.

## 15.2 Formation of the Solar System

Armed with our new knowledge of the planets and their moons, let's take a closer look at how the solar system formed. Modern models trace the formative stages of our solar system along the following broad lines.

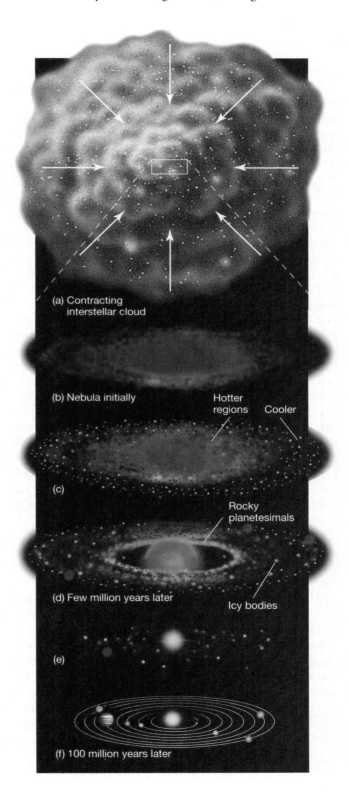

(a) Contracting interstellar cloud

(b) Nebula initially

Hotter regions    Cooler

(c)

Rocky planetesimals

(d) Few million years later

Icy bodies

(e)

(f) 100 million years later

ANIMATION/VIDEO    The Formation of the Solar System, Evolution of Protoplanetary Disk, Protoplanetary Disk, Protoplanetary Disks in the Orion Nebula, Protoplanetary Disk Destruction

The story starts with a dusty interstellar cloud fragment perhaps 100,000 AU—roughly a light-year—across (Figure 15.1a). Some external influence, such as the passage of another interstellar cloud or perhaps the explosion of a nearby star, starts the fragment contracting. As the cloud collapses, it rotates faster (because of the law of conservation of angular momentum) and begins to flatten. ∞ (More Precisely 6-2) By the time it has shrunk to a diameter of 100 AU, the *solar nebula*, as it is known at this stage, has formed an extended, rotating disk (Figure 15.1b).

As discussed in Chapter 6, if we now simply imagine that planets form within that disk, then we have a natural explanation for many of the observed bulk properties of our planetary system. This was the essence of Laplace's nebular theory. ∞ (Sec. 6.7)

## The Condensation Theory

Scientific theories must continually be tested and refined as new data become available. ∞ (Sec. 1.2) Unfortunately for the original nebular theory, although its description of the collapse and flattening of the solar nebula was basically correct, we now know that a disk of warm gas would *not* form clumps of matter that would subsequently evolve into planets. In fact, modern computer calculations predict just the opposite: Clumps in the gas would tend to disperse, not contract further. However, the **condensation theory**—the model favored by most astronomers—rests squarely on the old nebular theory, combining its basic physical reasoning with new information about interstellar chemistry to avoid most of the original theory's problems.

The key new ingredient is the presence of *interstellar dust* in the solar nebula. Astronomers now recognize that the space between the stars is strewn with microscopic dust grains, an accumulation of the ejected matter of many long-dead stars (see Chapter 22). These dust particles probably formed in the cool atmospheres of old stars and then grew by accumulating more atoms and molecules from the interstellar gas within the Milky Way Galaxy. The end result is that our entire Galaxy is littered with miniature chunks of icy and rocky matter, having typical sizes of about $10^{-5}$ m. Figure 15.2 shows one of many such dusty regions found in the vicinity of the Sun.

Dust grains play two important roles in the evolution of a gas cloud. First, dust helps to cool warm matter by efficiently radiating its heat away in the form of infrared radiation. ∞ (Sec. 3.4) As the cloud cools, its molecules move around more slowly, reducing the internal pressure and allowing the nebula to collapse more easily under the influence of gravity. ∞ (More Precisely 3-1) Second, by acting as **condensation nuclei**—microscopic platforms to which other atoms can attach, forming larger and larger balls of matter—the dust grains greatly speed up the process of collecting enough atoms to form a planet. This mechanism is similar to the way raindrops form in Earth's

atmosphere: Dust and soot in the air act as condensation nuclei around which water molecules cluster.

Thus, according to the modern condensation theory, once the solar nebula had formed and begun to cool (Figure 15.1b), dust grains formed condensation nuclei around which matter began to accumulate (Figure 15.1c). This vital step greatly hastened the critical process of forming the first small clumps of matter. Once these clumps formed, they grew rapidly by sticking to other clumps. (Imagine a snowball thrown through a fierce snowstorm, growing bigger as it encounters more snowflakes.) As the clumps grew larger, their surface areas increased and consequently, the rate at which they swept up new material accelerated. Gradually, this process of **accretion**—the gradual growth of small objects by collision and sticking—produced objects of pebble size, baseball size, basketball size, and larger.

Simulations indicate that, in perhaps as little as 100,000 years, accretion resulted in objects a few hundred kilometers across (Figure 15.1d)—the size of small moons. By that time, their gravitational pulls had become just strong enough to affect their neighbors. Astronomers call these objects **planetesimals**—the building blocks of the solar system (Figure 15.1e–f). Figure 15.3 shows two infrared views of relatively nearby stars thought to be surrounded by disks in which planetesimals are growing.

The planetesimals' gravity was now strong enough to sweep up material that would otherwise not have collided with them, and their rate of growth accelerated, allowing them to form still larger objects. Because larger bodies have stronger gravity, eventually (Figure 15.1e) almost all the original material was swept up into a few large **protoplanets**—the accumulations of matter that would in time evolve into the planets we know today (Figure 15.1f).

## Differentiation of the Solar System

The condensation theory just described can account—in broad terms, at least—for the formation of the planets and the large-scale architecture of the solar system. What does it say about the differences between the terrestrial and the jovian planets? Specifically, why are smaller, rocky planets found close to the Sun, while the larger gas giants orbit at much greater distances? To understand why a planet's composition depends on its location in the solar system, we must consider the *temperature* profile of the solar nebula. Indeed, it is in this context that the term *condensation* derives its true meaning.

As the primitive solar system contracted under the influence of gravity, it heated up as it flattened into a disk. The density and temperature were greatest near the center and much lower in the outlying regions. Detailed calculations indicate that the gas temperature near the core of the contracting system was several thousand kelvins. At a distance of 10 AU, out where Saturn now resides, the temperature was only about 100 K.

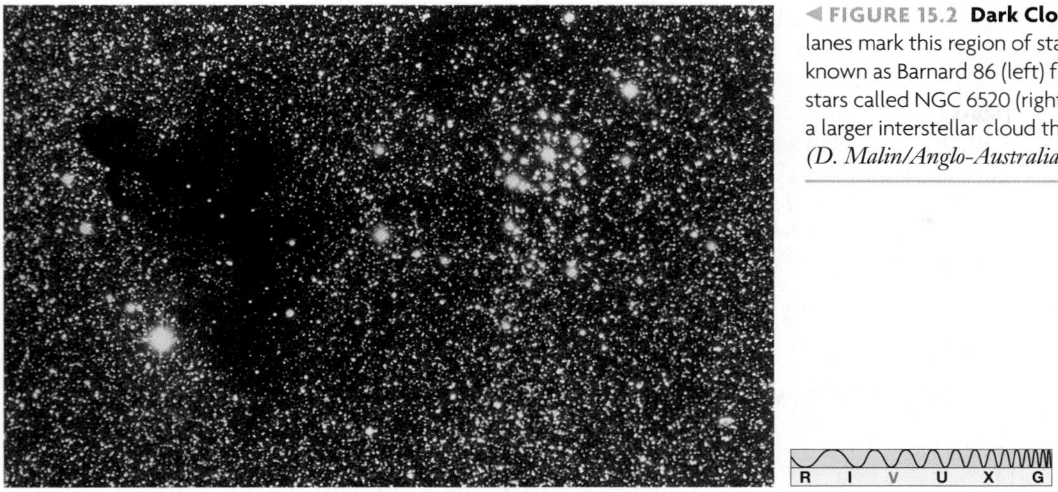


◀ FIGURE 15.2 **Dark Cloud** Interstellar gas and dark dust lanes mark this region of star formation. The dark cloud known as Barnard 86 (left) flanks a cluster of young blue stars called NGC 6520 (right). Barnard 86 may be part of a larger interstellar cloud that gave rise to these stars. *(D. Malin/Anglo-Australian Telescope)*

The high temperatures in the warmer regions of the cloud caused dust grains to break apart into molecules, which in turn split into excited atoms. Because the extent to which the dust was destroyed depended on temperature, it therefore also depended on location in the solar nebula. Most of the original dust in the inner solar system disappeared at this stage, whereas the grains in the outermost parts probably remained largely intact.

As the dusty nebula radiated away its heat, its temperature decreased everywhere except in the very core, where the Sun was forming. As the gas cooled, new dust grains began to condense (or crystallize) out, much as raindrops, snowflakes, and hailstones condense from moist, cooling air here on Earth. It may seem strange that although there was plenty of interstellar dust early on, it was partly destroyed, only to form again later. However, a critical change had occurred. Initially, the nebular gas was uniformly peppered with dust grains of all compositions; when the dust re-formed later, the distribution of grains was very different.

Figure 15.4 plots the temperature in various parts of the primitive solar system just before accretion began. At any given location, the only materials to condense out were those able to survive the temperature there. As marked

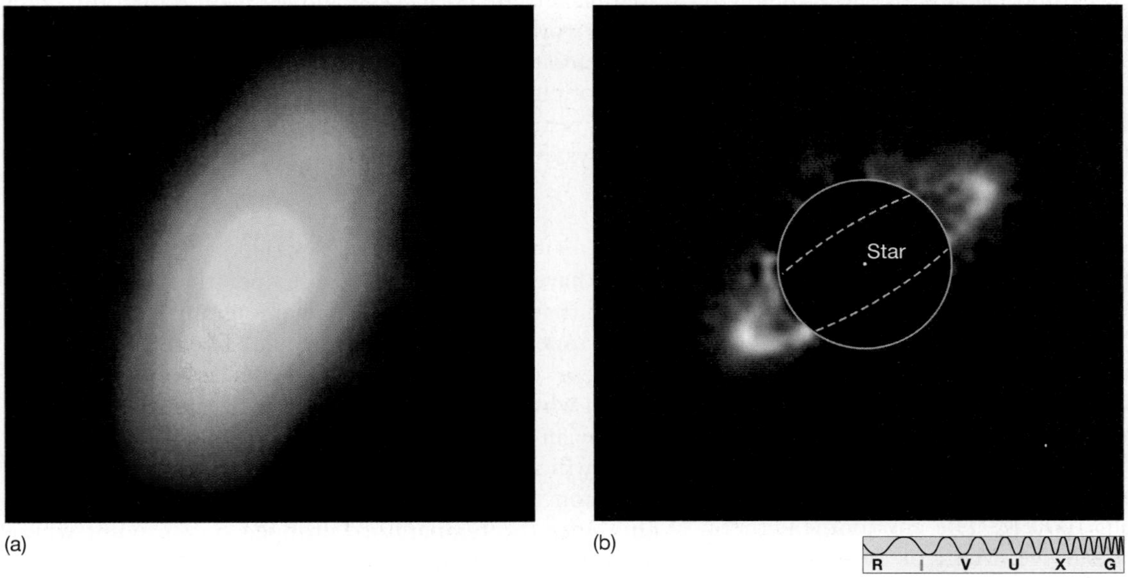


▲ FIGURE 15.3 **Newborn Solar Systems?** (a) This infrared image, taken by the *Spitzer Space Telescope*, of the bright star Fomalhaut, some 25 light-years from Earth, shows a circumstellar disk in which the process of accretion is underway. The star itself is well inside the yellowish blob at center. The outer disk, which is falsely colored orange to match the cooler dust emission, is about three times the diameter of our solar system. (b) This higher-resolution *Hubble Space Telescope* image blocks out the central (circled) parts of another such disk around a more distant star HR4796A, but shows the edges more clearly. *(NASA)*

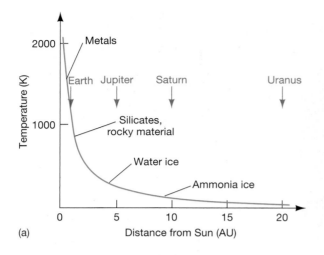

(a)

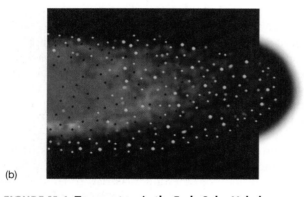

(b)

▲ FIGURE 15.4 **Temperature in the Early Solar Nebula**
(a) Theoretically computed variation of temperature across the primitive solar nebula illustrated in (b), which shows half of the disk in Figure 15.1(c). In the hot central regions, only metals could condense out of the gaseous state to form grains. At greater distances from the central proto-Sun, the temperature was lower, so rocky and icy grains could also form. The labels indicate the minimum radii at which grains of various types could condense out of the nebula.

on the figure, in the innermost regions, around Mercury's present orbit, only metallic grains could form; it was simply too hot for anything else to exist. A little farther out, at about 1 AU, it was possible for rocky, silicate grains to form, too. Beyond about 3 or 4 AU, water ice could exist, and so on, with the condensation of more and more material possible at greater and greater distances from the Sun.

In the inner regions of the primitive solar system, condensation from gas to solid began when the average temperature was about 1000 K. The environment there was too hot for ice to survive. Many of the more abundant heavier elements, such as silicon, iron, magnesium, and aluminum, combined with oxygen to produce a variety of rocky materials. The dust grains in the inner solar system were therefore predominantly rocky or metallic in nature, as were the protoplanets and planets they eventually became.

In the middle and outer regions of the primitive planetary system, beyond about 5 AU from the center, the

temperature was low enough for the condensation of several abundant gases—water ($H_2O$), ammonia ($NH_3$), and methane ($CH_4$)—into solid form. After hydrogen (H) and helium (He), the elements carbon (C), nitrogen (N), and oxygen (O) are the most common materials in the universe. As a result, wherever icy grains could form, they greatly outnumbered the rocky and metallic particles that condensed out of the solar nebula at the same location. Consequently, the objects that formed at these distances were formed under cold conditions out of predominantly low-density, icy material. These ancestral fragments were destined to form the cores of the jovian planets.

Note that, in this scenario, the composition of the outer solar system is much more typical of the universe as a whole than are the predominantly rocky and metallic inner planets. The outer solar system is not deficient in heavy elements. Rather, because of the conditions under which it formed, the inner solar system is *underrepresented in light material.*

CONCEPT CHECK
✔ Why was interstellar dust so important to the formation of our solar system?

## 15.3    Terrestrial and Jovian Planets

The condensation theory explains why the inner terrestrial planets are both *denser* and *less massive* than the outer jovian worlds. Armed with our new understanding of the differentiation of the early solar system, let's complete the story of planet formation by considering the sequences of events resulting in planet formation in the inner and outer solar system.

### Making the Inner Planets

As we have just seen, the high temperatures near the Sun meant that only rocky and metallic grains could survive in the inner part of the solar nebula. The relative scarcity of these dense materials elements is one important reason why the inner planets never became as massive as the jovian worlds—there was simply less raw material from which terrestrial planets could form. In addition, accretion could begin in the inner parts of the nebula only *after* the temperature dropped to the point where the grains appeared. In contrast, beyond the "snow line" at 3–4 AU, the temperature was low enough that condensation and accretion began much sooner—almost with the formation of the nebula itself—and with more resources to draw on. The larger masses of the outer planets reflect the significant "head start" they thus had in the formation process.

Figure 15.5 shows a computer simulation of accretion in the inner solar system over the course of about 100 million

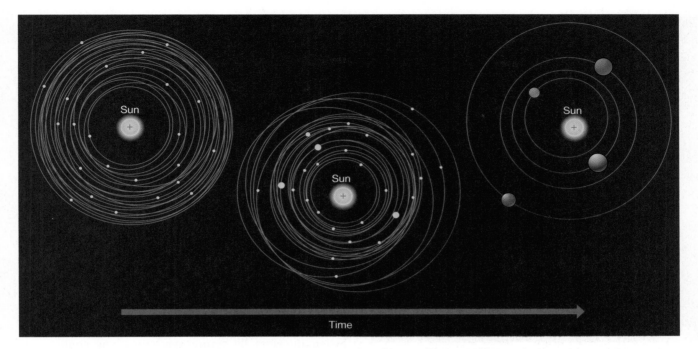

▲ **FIGURE 15.5** **Making the Inner Planets** Accretion in the inner solar system: Initially, many moon-sized planetesimals orbited the Sun. Over the course of about 100 million years, they gradually collided and coalesced, forming a few large planets in roughly circular orbits.

years, spanning parts (d–f) of Figure 15.1. Notice how, as the number of bodies decreases, the orbits of the remainder become more widely spaced and more nearly circular. The fact that this particular simulation produced exactly four terrestrial planets is strictly a matter of chance. The nature of the accretion process means that the eventual outcome is controlled by random events—which objects happen to collide and when. However, regardless of the precise number of planets formed, the computer models do generally reproduce both the planets' approximately circular orbits and their increasing orbital spacing as one moves outward from the Sun.

As the protoplanets grew, their strengthening gravitational fields produced many high-speed collisions between planetesimals and protoplanets. These collisions led to **fragmentation,** as small objects broke into still smaller chunks, which were then swept up by the larger protoplanets. Not only did the rich get richer in the early solar system, but the poor were mostly driven to destruction! Some of the fragments produced the intense meteoritic bombardment we know occurred during the early evolution of the planets and moons, as we have seen repeatedly in the last few chapters. ∞ (Secs. 8.9, 11.5, 13.5)

## Making the Jovian Worlds

The accretion picture just described has become the accepted model for the formation of the terrestrial planets. However, although similar processes may also have occurred in the outer solar system, the origin of the giant jovian worlds is decidedly less clear. Two somewhat different views, with potentially important consequences for our understanding of extrasolar planets, have emerged.

The first, more conventional, scenario is the chain of events illustrated in Figure 15.1. With raw material readily available in the form of abundant icy grains, protoplanets in the outer solar system grew quickly and soon became massive enough for their strong gravitational fields to capture large amounts of gas directly from the solar nebula. In this view, called the **core-accretion theory,** four large protoplanets became the cores of the jovian worlds; the captured gas became their atmospheres. ∞ (Secs. 11.1, 12.1, 13.2) The smaller, inner protoplanets never reached this stage— yet another reason why their masses remained relatively low.

Recently, some astronomers have highlighted a potentially serious snag in this picture: There may not have been enough *time* for these events to have taken place. Most young stars apparently go through a highly active evolutionary stage known as the *T Tauri* phase (see Chapter 19), in which their radiation and stellar winds become very intense. During this period, much of the nebular gas between the planets was blown away into interstellar space (Figure 15.6).

The problem is that the nebular disk was probably at most a few million years old when all this occurred, leaving very little time for the large jovian cores to grow and capture gas from the nebula before it was destroyed. Furthermore, some researchers argue that, in the relatively dense stellar environments in which most stars are born (see Chapter 19), close encounters between

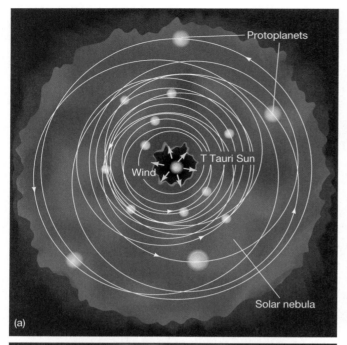

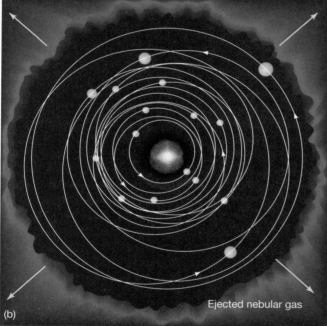

▲ FIGURE 15.6 **T Tauri Star** (a) Strong stellar winds from the newborn Sun sweep away the gas disk of the solar nebula, (b) leaving only giant planets and planetesimals behind. This stage of stellar evolution occurs only a few million years after the formation of the nebula.

still-forming stars may destroy many disks even sooner than that, giving giant planets perhaps as little as a few hundred thousand years in which to form.

The second formation scenario suggests that the giant planets formed through *instabilities* in the cool outer regions of the solar nebula, where portions of the cloud began to

collapse under their own gravity—a picture not so far removed from Laplace's basic idea—mimicking, on small scales, the collapse of the initial interstellar cloud. ∞ (Sec. 6.7) In this alternative **gravitational instability theory,** illustrated in Figure 15.7, the jovian protoplanets formed directly from the nebular gas, skipping the initial condensation-and-accretion stage and perhaps taking no more than a thousand years to acquire much of their mass. Right from the start, these first protoplanets had gravitational fields strong enough to scoop up more gas and dust from the solar nebula, allowing them to grow into the giants we see today before the gas supply dispersed.

If both these theories eventually lead to gas-rich jovian planets, how can we distinguish between them? One possible way involves the composition of the rocky jovian cores. Because the planets formed so quickly in the instability theory, computer models suggest that their cores should contain no more than about six Earth masses of rocky material. The core-accretion theory, by contrast, predicts much larger core masses—up to 20 times that of Earth. Detailed measurements of the jovian interiors by future space missions could settle the argument. Another decisive observation would be the detection of a Jupiter-sized extrasolar planet orbiting far from its parent star—out where Neptune is in our solar system. The accretion theory suggests that such a planet

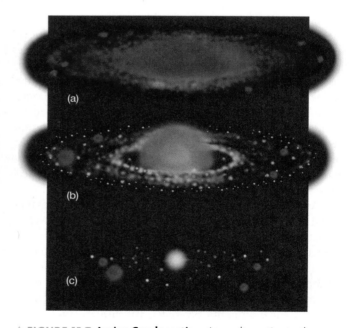

▲ FIGURE 15.7 **Jovian Condensation** As an alternative to the growth of massive protoplanetary cores followed by the accretion of nebular gas, it is possible that some or all of the giant planets formed directly through instabilities in the cool gas of the outer solar nebula. Part (a) shows the same instant as Figure 15.1(b). (b) Only a few thousand years later, four gas giants have already formed, preceding and circumventing the accretion process sketched in Figure 15.1. With the nebula gone (c), the giant planets have taken their place in the outer solar system. (See Figure 15.1e.)

would take too long (100 million years or more) to form, so finding one would argue strongly in favor of the instability model.

Many of the moons of the jovian planets presumably also formed through accretion, but on a smaller scale, in the gravitational fields of their parent planets. Once the nebular gas began to accrete onto the large jovian protoplanets, conditions probably resembled a miniature solar nebula, with condensation and accretion continuing to occur. The large moons of the outer planets (with the possible exception of Triton) almost certainly formed in this way. ∞ (Secs. 11.5, 13.5) The smaller moons are more likely captured planetesimals.

## Giant-Planet Migration

Many aspects of the formation of the giant planets remain unresolved. Interactions among the growing planets, and between the planets and their environment, probably played a critical role in determining just how and where the planets formed. One particularly intriguing scenario, accepted by some—but not all—planetary scientists, is the possibility that Jupiter—and maybe all four giant planets—formed considerably farther from the Sun than its present orbit and subsequently "migrated" inward. This supposed migration is indicated schematically in Figures 15.1 and 15.5 by the changing locations of the jovian protoplanets.

The idea of planetary migration has been around since the mid-1980s, when theorists realized that friction between massive planets and the nebula in which they moved would have caused just such an inward drift. Observational support came in 1999, when *Galileo* scientists announced much higher than expected concentrations of the gases nitrogen, argon, krypton, and xenon in Jupiter's atmosphere. ∞ (Sec. 11.2) These gases, which are thought to have been carried to the planet by captured planetesimals, could not have been retained in the planetesimal ice at temperatures typical of Jupiter's current orbit. Instead, they imply that the planetesimals—and, presumably, Jupiter too—formed at much lower temperatures. Either the nebula was cooler than previously thought, or Jupiter formed out in what is now the Kuiper belt!

## Formation Stops

The events just described did not take long, astronomically speaking. The giant planets formed within a few million years of the appearance of the flattened solar nebula—the blink of an eye compared with the 4.6-*billion*-year age of the solar system. At that point, intense radiation and strong winds from violent activity on the surface of the newborn Sun ejected the nebular gas, effectively halting further growth of the outer worlds. Accretion in the inner solar system proceeded more slowly, taking perhaps 100 million years to form the planets we know today (Figure 15.1f). The rocky asteroids and icy comets are all that remain of the matter that

originally condensed out of the solar nebula—the last surviving witnesses to the birth of our planetary system.

### CONCEPT CHECK

✔ Why is the rate at which the Sun formed important in a theory of the formation of the jovian planets?

## 15.4 Interplanetary Debris

Regardless of the precise chain of events that led to the formation of the gas-rich jovian planets, once the solar nebula was ejected into interstellar space, all that remained in orbit around the Sun were protoplanets and planetesimal fragments, ready to continue their long evolution into the solar system we know today. To place all these formative processes in perspective, Figure 15.8 presents a simplified time line of the first billion years after the formation of the solar nebula.

### The Asteroid Belt

In the inner solar system, planetesimal fragments that escaped capture by one of the terrestrial planets received repeated "gravity assists" from those bodies and were eventually boosted beyond the orbit of Mars. ∞ (*Discovery 6-1*) Roughly a billion years were required to sweep the inner solar system clear of interplanetary "trash." This was the period that saw the heaviest meteoritic bombardment, most evident on Earth's Moon and tapering off as the number of planetesimals decreased. ∞ (Sec. 8.5)

The myriad rocks of the asteroid belt between Mars and Jupiter failed to accumulate into a planet. Probably, nearby Jupiter's huge gravitational field caused them to collide too destructively to coalesce. Strong tidal forces from Jupiter on the planetesimals in the belt would also have hindered the development of a protoplanet. The result is a band of rocky planetesimals, still colliding and occasionally fragmenting, but never coalescing into a larger body.

### Comets and the Kuiper Belt

In the outer solar system, with the formation of the four giant jovian planets, the remaining planetesimals were subject to those planets' strong gravitational fields. Over a period of hundreds of millions of years, interactions with the giant planets, especially Uranus and Neptune, flung many of the outer region's interplanetary fragments into orbits taking them far from the Sun (Figure 15.9). Astronomers think that those icy bodies now make up the Oort cloud, whose members occasionally visit the inner solar system as comets. ∞ (Sec. 14.2)

A key prediction of this model is that some of the original planetesimals remained behind, forming the broad band known as the Kuiper belt, lying beyond

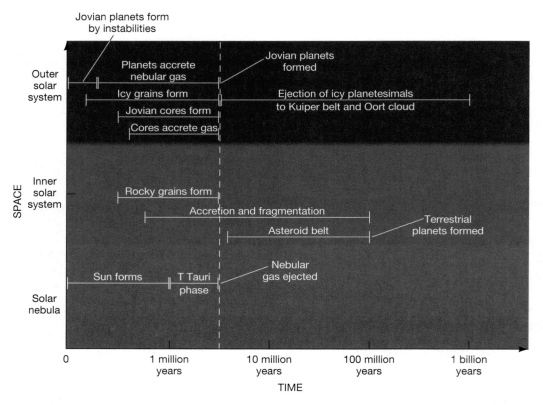

▲ **FIGURE 15.8 Solar System Formation** Schematic time line of some key events occurring during the first billion years of our solar system. The various tracks show the evolution of the Sun and the solar nebula, as well as that of the inner and outer solar system. Note that the tracks are intended to illustrate approximate relationships between events, not the precise times at which they occurred.

the orbit of Neptune, some 30 to 40 AU from the Sun. ∞ (Sec. 14.3) More than 1000 Kuiper belt objects, having diameters ranging from 50 km to more than 2000 km, are now known. Their existence lends strong support to the condensation theory of planetary formation.

Computer simulations reveal that the ejection of the planetesimals involved a remarkably complex interplay among the jovian planets, whereby Uranus and Neptune kicked some bodies out into the Kuiper belt, but deflected others *inward* toward Jupiter and Saturn, whose strong gravitational fields then propelled the planetesimals out into the distant Oort cloud. As shown in Figure 15.9, the orbits of all four giant planets were significantly modified by these interactions. By the time the outer solar system had been cleared of comets, Jupiter had moved slightly closer to the Sun, its orbital semimajor axis decreasing by a few tenths of an AU. The other giant planets moved outward—Saturn by about 1 AU, Uranus by 3 or 4 AU, and Neptune by some 7–10 AU. Note that these orbital changes occurred long *after* the supposed inward migrations mentioned earlier. Life as a jovian planet is far from simple!

Strong support for the preceding ideas comes from one curious feature of the Kuiper belt: A large fraction of its members—perhaps 15 percent of the entire belt—orbit in a 3:2 *resonance* with Neptune. That is, they orbit the Sun

twice for every three orbits of Neptune. Pluto shares this resonance, and the Kuiper belt objects orbiting in this manner have accordingly been dubbed **plutinos.** ∞ (Sec. 14.3) The presence of one object (Pluto) in such an exceptional orbit might conceivably be attributed to chance: It is possible for a planet such as Neptune to "capture" a Pluto-sized object into a resonant orbit, if the object's orbit started out close to resonance. However, to account for more than 100 similar orbits, a more comprehensive theory is needed.

The leading explanation of the plutinos is that, as Neptune's orbit moved slowly outward, the radius corresponding to the 3:2 resonance also swept outward through the surviving planetesimals. Apparently, this process was slow enough that many, if not most, of the planetesimals on near-resonant orbits (Pluto included) were captured and subsequently carried outward, locked forever in synchrony with Neptune as that planet's orbit drifted outward to its present location. These are the plutinos we see today.

During this period, many icy planetesimals were also deflected into the inner solar system, where they played an important role in the evolution of the inner planets. A long-standing puzzle in the condensation theory's account of the formation of the inner planets has been where the water and other volatile gases on Earth and elsewhere originated. At the inner planets' formation, their surface temperatures

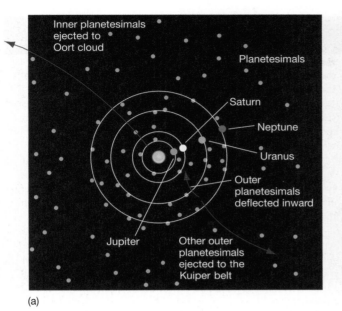
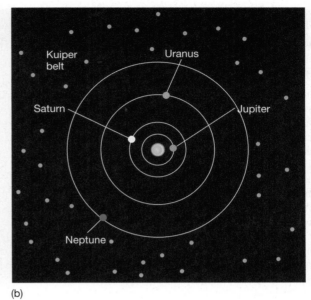

(a)                                                    (b)

▲ FIGURE 15.9  **Planetesimal Ejection**  The ejection of icy planetesimals to form the Oort cloud and Kuiper belt. (a) Initially, once the giant planets had formed, leftover planetesimals were found throughout the solar system. Interactions with Jupiter and Saturn apparently "kicked" planetesimals out to very large radii (the Oort cloud). Interactions with Uranus and especially Neptune tended to keep the Kuiper belt populated, but also deflected many planetesimals inward to interact with Jupiter and Saturn. (b) After hundreds of millions of years and as a result of the inward and outward "traffic," the orbits of all four giant planets were significantly modified by the time the planetesimals interior to Neptune's orbit had been ejected. As depicted here, Neptune was affected most and may have moved outward by as much as 10 AU.

were far too high, and their gravity too low, to capture or retain those gases. The most likely explanation seems to be that the water and other light gases found on Earth and elsewhere in the inner solar system arrived there in the form of comets from the outer solar system. Kicked into eccentric orbits as the gravitational fields of the jovian planets cleared the outer solar system of leftover planetesimals, these icy fragments bombarded the newborn terrestrial worlds, supplying them with water *after* their formation.

CONCEPT CHECK
✔ Might you expect to find comets and asteroids orbiting other stars?

## 15.5  Solar System Regularities and Irregularities

The modern theory of solar system formation—the condensation theory—accounts for the 10 "characteristic" points listed at the start of this chapter. Specifically, the planets' orbits are nearly circular (2), in the same plane (3), and in the same direction as the Sun's rotation on its axis (4) as a direct consequence of the nebula's shape and rotation. The rotation of the planets (5) and the orbits of the moon systems (6) are due to the tendency of the smaller

scale condensations to inherit the nebula's overall sense of rotation. The growth of planetesimals throughout the nebula, with each protoplanet ultimately sweeping up the material near it, accounts for (1) the fact that the planets are widely spaced (even if the theory does not quite explain the regularity of the spacing). The heating of the nebula and the Sun's eventual ignition resulted in the observed differentiation (7), and the debris from the accretion—fragmentation stage naturally accounts for the asteroids (8), the Kuiper belt (9), and the Oort cloud comets (10).

The condensation theory is an example of an *evolutionary* theory—one that describes the development of the solar system as a series of gradual and natural steps, understandable in terms of well-established physical principles. Evolutionary theories may be contrasted with *catastrophic* theories, which invoke accidental or unlikely celestial events to interpret observations. A good example of such a theory is the *collision hypothesis*, which imagines that the planets were torn from the Sun by a close encounter with a passing star. This hypothesis enjoyed some measure of popularity during the 19th century, due in part to the inability of the nebular theory to account for the observed properties of the solar system, but no scientist takes it seriously today (see *Discovery 15-1*). Aside from its extreme improbability, the collision hypothesis is completely unable to explain the orbits, the rotations, or the composition of the planets and their moons.

# DISCOVERY 15-1

## The Angular Momentum Problem

According to Laplace's nebular hypothesis, the Sun and solar system formed from a contracting, spinning cloud of interstellar gas. ∞ (Sec. 6.7) Conservation of angular momentum caused the cloud to spin faster as it collapsed, and this in turn caused it to flatten into a disk, from which the planets eventually formed. ∞ (*More Precisely 6-2*) This simple, yet elegant, idea underlies the modern condensation theory of planetary formation. However, not long after Laplace proposed his explanation for the basic architecture of the solar system, astronomers uncovered what appeared to be a serious flaw in the theory. It has to do with the distribution of angular momentum in the solar system.

Although our Sun contains about a thousand times more mass than all the planets combined, it possesses a mere 0.3 percent of the total angular momentum of the solar system. Jupiter, for example, has a lot more angular momentum than does our Sun. In fact, because of its large mass and great distance from the Sun, Jupiter holds about 60 percent of the solar system's total angular momentum. All told, the four jovian planets account for well over 99 percent of the angular momentum of the entire present-day solar system. (The lighter and closer terrestrial planets contribute negligibly to the total.)

The problem here is that the law of conservation of angular momentum predicts that the Sun should have been spinning very rapidly during the earliest epochs of the solar system. The Sun should command most of the solar system's angular momentum basically because it contains most of the mass. However, as we have just seen, the reverse is true. Indeed, if all the planets, with their large amounts of orbital angular momentum, were placed inside the Sun, it would spin on its axis about a hundred times faster than it does at present. Somehow, the Sun must have lost (or perhaps never gained) most of this angular momentum. This discrepancy between observations and theoretical expectations is known as the *angular momentum problem.*

Nineteenth-century scientists had no ready explanation of how the Sun could have shed its angular momentum, leading some to abandon the nebular theory even though it provides a natural explanation of many of the planets' orbital properties. For a time, the *collision hypothesis,* in which the planets (or preplanetary blobs) were ripped from the Sun by the gravitational pull of a passing star, gained some degree of prominence. The fact that this latter theory persisted despite the extreme improbability of such a stellar encounter, as well as the theory's own difficulties in explaining planetary orbits, is a measure of how seriously the angular momentum problem was taken by scientists of the time.

Although the details remain uncertain, astronomers today surmise that the Sun transferred much of its spin angular momentum to the orbital angular momentum of the planets via the solar nebular disk. Friction within the disk would have caused the rapidly spinning inner regions (i.e., the embryonic Sun) to slow, while causing the slowly spinning outer regions (where the planets would someday form) to speed up. The net effect was to move angular momentum outward from the Sun to the planets.

In addition, many researchers speculate that the solar wind, moving away from the Sun into interplanetary space, may have carried away much of the Sun's remaining angular momentum. ∞ (Sec. 6.5) The early Sun probably produced more of a dense solar gale than the relatively gentle "breezes" now measured by our spacecraft. High-velocity particles leaving the Sun followed the solar magnetic field lines. As the rotating magnetic field of the Sun tended to drag those particles around with it, they acted as a brake on the Sun's spin. The accompanying figure illustrates the process. This interaction between the solar wind and the Sun's magnetic field was completely unknown to 19th-century astronomers.

Although each particle that is boiled off the Sun carries only a tiny amount of the Sun's angular momentum with it, over the course of nearly 5 billion years the vast numbers of escaping particles have probably robbed the Sun of most of its initial spin momentum. Even today, our Sun's spin continues to slow. A similar mechanism operating while the Sun was expelling the relatively dense nebular disk could also have transported a lot of solar angular momentum into interstellar space.

Today, the angular momentum problem represents a minor source of uncertainty, rather than a fundamental challenge, to the nebular–condensation theory. With new physical insights, the problem has been greatly reduced in severity—if not completely explained away—and the nebular hypothesis is once again central to theories of solar system formation.

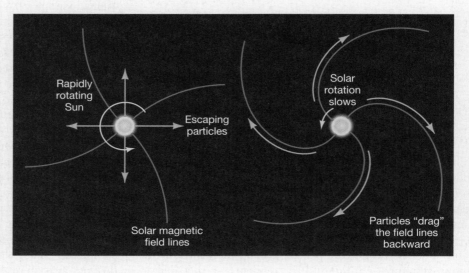

Rapidly rotating Sun

Escaping particles

Solar magnetic field lines

Solar rotation slows

Particles "drag" the field lines backward

Scientists generally try not to invoke catastrophes to explain the universe, but there are instances where pure chance really has played a critical role in determining the present state of the solar system. We stressed earlier the fact that an important aspect of any solar system theory is its capability to allow for the possibility of imperfections—deviations from the otherwise well-ordered scheme of things. In the condensation theory, that capability is provided by the randomness inherent in the encounters that ultimately combined the planetesimals into protoplanets. As the numbers of large bodies decreased and their masses increased, individual collisions acquired greater and greater importance. The effects of these collisions can still be seen today in many parts of the solar system—for example, the large craters on many of the moons we have studied thus far.

Having started with 10 regular points to explain, we end our discussion of solar system formation with the following eight irregular features that still fall within the theory's scope:

1. Mercury's exceptionally large nickel–iron core may be the result of a collision between two partially differentiated protoplanets. The cores may have merged, and much of the mantle material may have been lost. ∞ (Sec. 8.7)

2. Two large bodies could have merged to form Venus, giving it its abnormally low rotation rate. ∞ (Sec. 9.2)

3. The Earth–Moon system may have formed from a collision between the proto-Earth and a Mars-sized object. ∞ (Sec. 8.8)

4. A late collision with a large planetesimal may have caused Mars's curious north–south asymmetry and ejected much of the planet's atmosphere. ∞ (Sec. 10.4)

5. The tilted rotation axis of Uranus may have been caused by a grazing collision with a sufficiently large planetesimal or by a merger of two smaller planets. ∞ (Sec. 13.2)

6. Uranus's moon Miranda may have been almost destroyed by a planetesimal collision, accounting for its bizarre surface terrain. ∞ (Sec. 13.5)

7. Interactions between the jovian protoplanets and one or more planetesimals may account for the irregular moons of those planets and, in particular, Triton's retrograde motion. ∞ (Sec. 13.5)

8. The Pluto–Charon system and the other known binary Kuiper belt objects may have formed by collisions or near misses between two icy planetesimals before most were ejected by interactions with the jovian planets. ∞ (Sec. 14.3)

Note that it is impossible to test any of these assertions directly, but it is reasonable to suppose that some (or even all) of the preceding "odd" aspects of the solar system can be explained in terms of collisions late in the formative stages of the protoplanetary system. Not all astronomers agree with all of the explanations. However, most would accept at least some.

CONCEPT CHECK
✔ What is the key "random" element in the condensation theory?

## 15.6 Planets beyond the Solar System

The test of any scientific theory is how well it holds up in situations different from those in which it was originally conceived. ∞ (Sec. 1.2) With the discovery in recent years of numerous planets orbiting other stars, astronomers now have the opportunity—indeed, the scientific obligation—to test their theories of solar system formation.

### The Discovery of Extrasolar Planets

The detection of planets orbiting other stars has been a goal of astronomers for decades if not centuries. Many claims of extrasolar planets have been made since the middle of the 20th century, but before 1995 none had been confirmed, and most have been discredited. Only since the mid-1990s have we seen genuine advances in this fascinating area of astronomy. The advances have come, not because of dramatic scientific or technical breakthroughs, but rather through steady improvements in telescope and detector technology and in computerized data analysis.

Extrasolar planets are very faint and generally lie close to their parent stars, making them very hard to resolve with current equipment. Figure 15.10 shows an image, taken with the European *Very Large Telescope* in Chile, of a nearby system containing a planet a few times more massive than Jupiter. However, this is (to date) the only extrasolar planet detected by direct imaging. In all other cases, the techniques used to find the planets have been *indirect*, based on analyses of light from the parent stars, not from the planets themselves.

Figure 15.11 illustrates the basic approach. As a planet orbits a star, gravitationally pulling first one way and then the other, the star "wobbles" slightly as the planet and star orbit their common center of mass. ∞ (Sec. 2.8) The more massive the planet, or the less massive the star, the greater is the star's movement. If the wobble happens to occur along our line of sight to the star, then we see small fluctuations in the star's radial velocity, which can be measured using the Doppler effect. ∞ (Sec. 3.5) Those fluctuations allow us to estimate the planet's mass.

Figure 15.12 shows two sets of radial-velocity data that betray the presence of planets orbiting other stars.

▲ FIGURE 15.10 **Extrasolar Planet** Most known extrasolar planets are too faint to be detectable against the glare of their parent stars. However, in this system, called 2M1207, the parent itself (centered) is very faint—a so-called *brown dwarf* (see Chapter 19)—allowing the planet (lower left) to be detected in the infrared. This planet has a mass about 5 times that of Jupiter and orbits 55 AU from the star, which is 230 light-years away. *(ESO)*

Part (a) shows the line-of-sight velocity of the star 51 Pegasi, a near twin to our Sun lying some 40 light-years away. The data, acquired in 1994 by Swiss astronomers using the 1.9-m telescope at Haute-Provence Observatory in France, were the first substantiated evidence for an extrasolar planet orbiting a Sun-like star.*

The regular 50-m/s fluctuations in the star's velocity have since been confirmed by several groups of astronomers and imply that a planet of at least half the mass of Jupiter orbits 51 Pegasi in a circular orbit with a period of just 4.2 days. (For comparison, the corresponding fluctuation in the Sun's velocity due to Jupiter is roughly 12 m/s.) Note that we say "at least half" here because Doppler observations suffer from a fundamental limitation: They cannot distinguish between low-speed orbits seen edge-on and high-speed orbits seen almost face-on (so only a small component of the orbital motion contributes to the line-of-sight Doppler effect). As a result, only lower limits to planetary masses can be obtained.

Figure 15.12(b) shows another set of Doppler data, this time revealing one of the most complex system of planets discovered to date: a triple-planet system orbiting another nearby Sun-like star named Upsilon Andromedae. The three planets have minimum masses of 0.7, 2.1, and 4.3 times the mass of Jupiter and orbital semimajor axes of 0.06, 0.83, and 2.6 AU, respectively. Figure 15.12(c) sketches their orbits, with the orbits of the solar terrestrial planets shown for scale. Almost

*As we will see in Chapter 22, two other planets having masses comparable to Earth, and one planet with a mass comparable to that of Earth's Moon, had previously been detected orbiting a particular kind of collapsed star called a pulsar. However, their formation was the result of a chain of events very different from those that formed Earth and the solar system.*

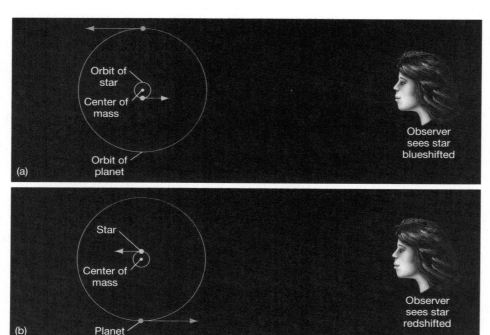

◀ FIGURE 15.11 **Detecting Extrasolar Planets** As a planet orbits its parent star, it causes the star to "wobble" back and forth. The greater the mass of the planet, the larger is the wobble. The center of mass of the planet–star system stays fixed. If the wobble happens to occur along our line of sight to the star, as shown by the yellow arrow, we can detect it by the Doppler effect. (In principle, side-to-side motion perpendicular to the line of sight is also measurable, although there are as yet no confirmed cases of planets being detected this way.)

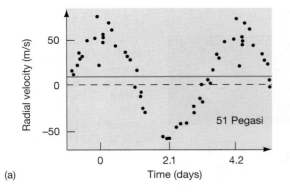

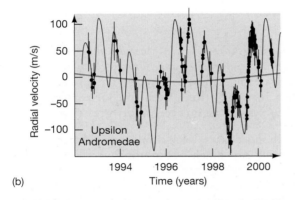

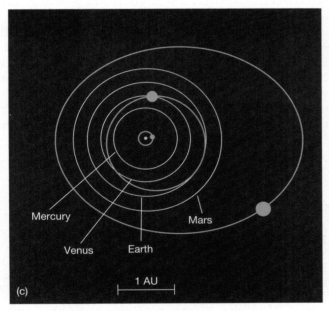

▲ FIGURE 15.12 **Planets Revealed** (a) Measurements of the Doppler shift of the star 51 Pegasi reveal a clear periodic signal indicating the presence of a planetary companion of mass at least half the mass of Jupiter. (b) Radial-velocity data for Upsilon Andromedae are much more complex, but are well fit (solid line) by a three-planet system orbiting the star. For reference in parts (a) and (b), the maximum possible signal produced by Jupiter orbiting the Sun (i.e., the wobble our Sun would display, as seen by a distant observer looking edge-on at our solar system) is shown in blue. (c) A sketch of the inferred orbits of three planets from the Upsilon Andromedae system (in orange), with the orbits of the terrestrial planets superimposed for comparison (in white).

200 planets have now been detected by means of radial-velocity searches.

If the wobble produced in a star's motion is predominantly *perpendicular* to our line of sight, then little or no Doppler effect will be observed, so the radial-velocity technique cannot be used to detect a planet. However, in this case, the star's *position* in the sky changes slightly from night to night, and, in principle, measuring this transverse motion provides an alternative means of detecting extrasolar planets. Unfortunately, these side-to-side wobbles have proven difficult to measure accurately, as the angles involved are very small and the star in question has to be quite close to the Sun for useful observation to be possible. On the basis of observations of this type, several candidate planetary systems have been proposed, but none has yet been placed on the "official" list of confirmed observations.

As just noted, the Doppler technique suffers from the limitation that the angle between the line of sight and the planet's orbital plane cannot be determined. However, in a few systems originally discovered through Doppler measurements, that is not the case. Figure 15.13 shows how observations of a distant solar-type star (known only by its catalog name of HD 209458 and lying some 150 light-years from Earth) reveal a clear drop in brightness each time its 0.6-Jupiter-mass companion, orbiting at a distance of just 7 million km, passes between the star and Earth (Figure 15.13). The drop in brightness is just 1.7 percent, but it occurs precisely on schedule every 3.5 days, the orbital period inferred from radial-velocity measurements.

Such **planetary transits,** similar to the transit of Mercury shown in Figure 2.18, are rare, as they require us to see the orbit almost exactly edge-on. When they do occur, however, they allow an unambiguous determination of the planet's radius, and hence its mean density. ∞ (Sec. 6.2) In this case, the planet's radius is found to be 100,000 km (1.4 times the radius of Jupiter), implying a density of just 200 kg/m$^3$, consistent with a high-temperature gas-giant planet orbiting very close to its parent star. Roughly half a dozen extrasolar planets are currently known to produce measurable fluctuations in brightness in their partner stars.

## Planetary Properties

As of mid-2007, more than 200 extrasolar planets have been detected orbiting some 180 stars within a few hundred light-years of the Sun. Overall, only a relatively small fraction (about 5 percent) of the nearby stars surveyed to date have shown evidence for planetary companions. Most observed planetary systems consist of a single massive planet orbiting its parent star. However, 14 two-planet systems, 4 three-planet systems (one of them shown in Figure 15.12c), and 2 four-planet systems have been confirmed.

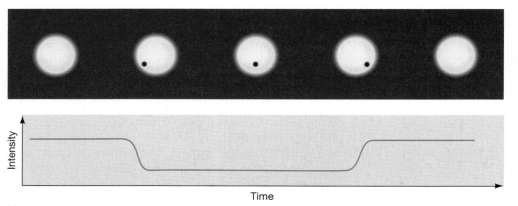

(a)

◀ FIGURE 15.13 **An Extrasolar Transit** (a) If an extrasolar planet happens to pass between us and its parent star, the light from the star dims in a characteristic way. (b) Artist's conception of the planet orbiting a Sun-like star known as HD 209458. The planet is 200,000 km across and transits every 3.5 days, blocking about 2 percent of the star's light each time it does so.

3. *The observed orbits are generally much more eccentric than those of Jupiter and Saturn.* Only about 30 percent of the planets detected so far have eccentricities less than 0.1. (Recall that no jovian planet in our solar system has an eccentricity greater than 0.06.)

Since, in most cases, we see only a single massive planet, we cannot say much about the overall properties of these planetary "systems" in comparison with our own. So far, no terrestrial planets nor any evidence of interplanetary matter has been found orbiting any of these stars.

Figure 15.14 shows the orbital semimajor axes and eccentricities of some of the known extrasolar planets. Each dot represents a planet, and we have added points corresponding to Earth and Jupiter in our own solar system. The tight orbits and broad range of eccentricities of the extrasolar planets contrast markedly with those in our own planetary system. The numerous

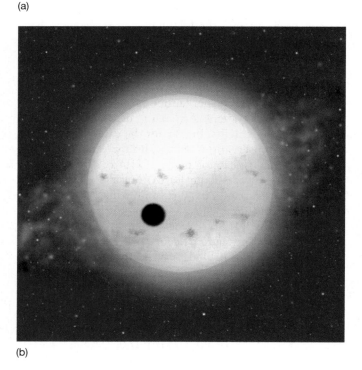

(b)

The basic properties of the "official" list of extrasolar planets may be summarized as follows:

1. *Most of the planets observed so far have masses comparable to that of Jupiter.* The majority of measured extrasolar planet masses are between one-third and 10 times the mass of Jupiter. However, planets at the low end of this range are far more common—most of the measured masses are less than twice that of Jupiter and, as observations improve, a few have masses comparable to Saturn. As of mid-2007, just two planets with masses comparable to that of Earth (5 and 8 Earth masses, respectively) have been detected—both orbiting the same low-mass star.

2. *The observed orbits are generally much smaller than those of Jupiter or Saturn—less than a few AU across.* A substantial fraction of these planets (about 15 percent) orbit very close to their parent star, with semimajor axes of 0.1 AU or less.

▲ FIGURE 15.14 **Extrasolar Orbital Parameters** Orbital semimajor axes and eccentricities of nearly half of the approximately 200 known extrasolar planets. Each point represents one planetary orbit, and to plot them all would make a mess. The corresponding points for Earth and Jupiter in our solar system are also shown. The known extrasolar planets generally move on smaller, much more eccentric orbits than do the planets circling the Sun.

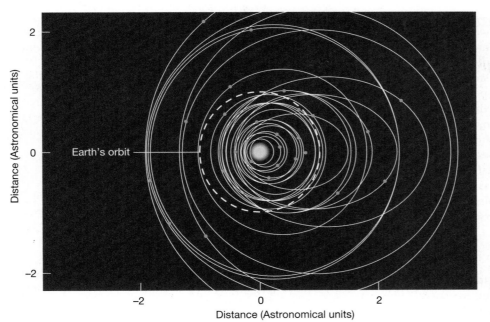

◀ FIGURE 15.15 **Extrasolar Orbits**
The orbits of many extrasolar planets residing more than 0.15 AU from their parent star, superimposed on a single plot, with Earth's orbit shown for comparison. All these extrasolar planets are comparable in mass to Jupiter. A plot of all known extrasolar planets would be very cluttered, but the message would be much the same: These planetary systems don't look much like ours!

jovian-mass planets orbiting close (within 0.1 AU) to their parent stars (at the left of the diagram) have come to be known as **hot Jupiters**. Figure 15.15 plots the actual orbits of some of the known extrasolar planets (excluding the hot Jupiters), with Earth's orbit superimposed for comparison.

Notice in Figure 15.14 that the hot Jupiters tend to have more circular (low-eccentricity) orbits. This is not a matter of chance. These bodies reside so close to their stars that their orbits are thought to have been circularized by tidal effects, similar to those controlling the orbits of Earth's Moon or the Galilean moons of Jupiter. ∞ (Secs. 8.4, 11.5) Thus, the shape of a hot-Jupiter orbit is a direct evolutionary consequence of the planet's proximity to the star. As we will see in the next section, the manner in which the planet got into such a tight orbit in the first place may provide a much-needed connection between extrasolar planetary systems and our own. Most of the Saturn-mass planets (as well as the super-Earth) also move in "hot" orbits, perhaps suggesting that we may be seeing different stages in the removal of a jovian envelope by the intense heat of the parent star, ultimately exposing the "terrestrial" core. ∞ (Sec. 11.3)

Finally, spectroscopic observations of the parent stars reveal what may be a crucial piece in the puzzle of extrasolar-planet formation. ∞ (Sec. 4.5) Stars having compositions similar to that of the Sun are statistically *much more likely* to have planets orbiting them than are stars containing smaller fractions of the key elements carbon, nitrogen, silicon, and iron. Because the elements found in a star reflect the composition of the nebula from which it formed, and the elements just listed are the main ingredients of interstellar dust, this finding provides strong support for the condensation theory. Dusty disks really are more likely to form planets.

## Are They Really Planets?

Given that these systems seem so alien from our own, for a time some astronomers questioned whether the mass measurements, and hence the identification of these objects as "planets," could be trusted. Eccentric orbits are known to be common among double-star systems (pairs of stars in orbit around one another—see Section 17.7), and some researchers suggested that many of the newly found planets were actually **brown dwarfs**—"failed stars" having insufficient mass to become true stars (Section 19.3).

The dividing line between genuine Jupiter-like planets and starlike brown dwarfs is unclear, but it is thought to be around 15 Jupiter masses. This number is comparable to the largest extrasolar planet mass yet measured. Planet proponents would say that this is not a coincidence, but rather indicates that planets up to the maximum possible mass have in fact been observed. Detractors would argue that we just happen to see the orbits almost face-on, greatly reducing the parent star's radial velocity and fooling us into thinking that we are observing low-mass planets instead of higher-mass brown dwarfs.

However, the latter view has a serious problem, which worsens with every new low-mass extrasolar planet reported: Since the orientations of the actual orbits are presumably random, it is extremely unlikely that we would just happen to see *all* of them face-on—and even if we did, there is no observational evidence for the many edge-on (and much easier to detect) systems we would then also expect to see on statistical grounds. Consequently, most astronomers agree that, though there may well be a few brown dwarfs lurking among the list of extrasolar planets, they probably do not constitute a significant fraction of the total.

## CONCEPT CHECK

✔ Describe three ways in which observed extrasolar planetary systems differ from the solar system.

# 15.7    Is Our Solar System Unusual?

Not so long ago, many astronomers argued that the condensation scenario described earlier in this chapter was in no way unique to our own system. The same basic processes could have occurred, and perhaps *did* occur, during the formative stages of many of the stars in our Galaxy, so planetary systems like our own should be common. Today we know that planetary systems *are* apparently quite common, but, by and large, the ones we see don't look at all like ours! We can thus legitimately ask whether our solar system really is as unusual as recent observations seem to imply and whether those observations undermine our current theory of solar system formation.

## Observational Limitations

Let's start by asking whether the planets we observe really are representative of extrasolar planets in general. The fact that we don't see many low-mass planets, or more massive planets on wide orbits, is not surprising. It is what astronomers call a **selection effect**: Lightweight or distant planets simply don't produce large enough velocity fluctuations for them to be easily detectable. The methods employed so far are heavily biased toward finding massive objects orbiting close to their parent stars. Those systems would be expected to give the strongest signal, and they are precisely what have been observed.

Almost all of the planets detected so far produce stellar radial velocities substantially greater than the 12 m/s that would be produced (under the best circumstances) by Jupiter's orbit around the Sun. Furthermore, although the Sun's wobble could be detected with current technology, it is close enough to the instrumental limits that several orbits—that is, several decades' worth of observations—would be needed before a definitive detection could be claimed.

In fact, as search techniques improve, astronomers are finding more and more Jupiter-mass (and lower) planets on wider and less eccentric orbits. Figure 15.16 shows evidence for one of the most "Jupiter-like" planets yet detected: a 2-Jupiter-mass planet moving on a roughly circular orbit around a near twin of our own Sun. The planet's period is 6 years. Planet hunters are quietly confident that advances during the next decade will either bring numerous detections of extrasolar planets in orbits comparable to those in the solar system or allow astronomers to conclude that systems like our own are indeed in the minority. Either way, the consequences are profound.

## Making Eccentric Jupiters

Are the extrasolar orbits we do see *inconsistent* with the condensation theory? Probably not. Current theory in fact provides many ways in which massive planets can

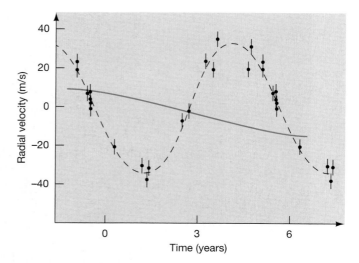

▲ FIGURE 15.16 **Jupiter-like Planet?** Velocity "wobbles" in the star HD 70642 reveal the presence of the extrasolar planet with the most "Jupiter-like" orbit yet discovered. The parent star is almost identical to the Sun, and the 2-Jupiter-mass planet orbits at a distance of 3.3 AU with an orbital eccentricity of 0.1. Again, the blue line marks the corresponding plot for Jupiter itself.

end up in short-period or eccentric orbits. Indeed, an important aspect of solar system formation not mentioned in our earlier discussion is the fact that many theorists worry about how Jupiter could have remained in a stable orbit after it formed in the protosolar disk! Jupiter-sized planets may be knocked into eccentric orbits by interactions with other Jupiter-sized planets or by the tidal effects of nearby stars. If these planets formed by gravitational instability, they could have eccentric orbits right from the start (and we then would have to explain how those orbits circularized in the case of the solar system).

Regardless of how a massive planet forms, gravitational interactions between it and the gas disk in which it moves tend to make the planet spiral inward, as mentioned earlier, and can easily deposit it in an orbit very close to the parent star (Figure 15.17). Interestingly, it now appears that the presence of Saturn may have helped stabilize Jupiter's orbit against this last effect. Isolated or particularly massive "Jupiters" are precisely the planets one would expect to find on "hot" orbits.

The orbits and masses of the observed extrasolar planets spell disaster for any low-mass "terrestrial" planets in these systems. In our solar system, the presence of a massive Jupiter on a nearly circular orbit is known to have a stabilizing influence on the other planetary orbits, tending to preserve the relative tranquility of our planetary environment. In the known extrasolar systems, not only is this stabilization absent, but having a

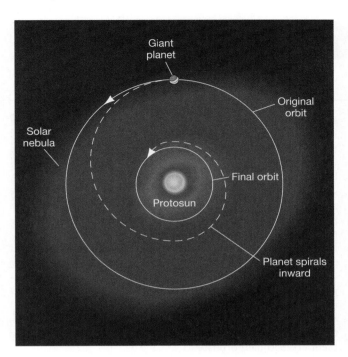

**▲ FIGURE 15.17 Sinking Planet** Friction between a giant planet and the nebular disk in which it formed tends to make the planet spiral inward. The process continues until the disk is dispersed by the wind from the central star, possibly leaving the planet in a "hot-Jupiter" orbit.

Jupiter-sized planet repeatedly plow through the inner parts of the system means that any terrestrial planets or planetesimals have almost certainly been ejected from the system.

## Searching for Earth-like Planets

To summarize the current state of extrasolar planet searches, only a small percentage (perhaps 5 percent) of the stars surveyed to date show evidence for extrasolar planets. The planets that are observed are precisely those that could have been detected, given today's technology, and their orbits are allowed, if not actually predicted, by current theories of planet formation. The properties of the systems we can see reveal little about those we can't, so for now, the planetary properties of the vast majority of stars remain a mystery. Still, there seems to be no pressing reason to conclude just yet that our solar system is unique, or even unusual, in its structure.

Assuming, then, that planetary systems like our own do exist, with Jupiter-like planets in circular, or at least nondisruptive, orbits and some terrestrial planets on stable orbits close to their parent star, how might we go about searching for planets like our own? Current estimates

suggest that looking for "wobbles" is not likely to be successful. Variations in radial velocities will in most cases be so small as to be lost within the natural internal motions of the parent star itself, and the side-to-side movement of a Sun-like star due to Earth-sized objects will be very hard to measure, even from space. Direct imaging of terrestrial planets remains a long-term goal of NASA, but tangible results are probably decades away.

Perhaps surprisingly, the approach given the best chance of success entails searching for planets *transiting* their parent stars. The effect of a transit on a star's brightness is tiny—less than one part in $10^4$ for an Earth-like planet crossing the face of the Sun—and also very unlikely to be detected, as only nearly edge-on systems will show it. Nevertheless, it is measurable with existing technology. From the change in brightness as the planet passes in front of the star, the planet's size can be calculated. The size of the orbit can be calculated from the orbital period and application of Newton's laws. ∞ (*More Precisely 2-3*) Once the distance from the planet to the star is known, the planet's surface temperature may be estimated (as in our own solar system) by determining the temperature at which the planet radiates as much energy back into space as it receives in the form of starlight. ∞ (Secs. 3.4, 7.2)

The European *CoRoT* mission (short for **Co**nvection, **Ro**tation, and planetary **T**ransits), launched in December 2006, will study some 120,000 Sun-like stars over a 2.5-year period, monitoring them for (among other things) brightness fluctuations caused by Earth-like planets. Under optimistic assumptions, a few tens of "super-Earth" transits are expected, along with several hundred hot Jupiters and perhaps a few dozen "cool" Jupiters at distances of a few AU from their parent star. NASA's planned *Kepler* mission, currently scheduled for launch in late 2008, will monitor a similar number of stars and may detect transits of up to 100 terrestrial-size planets.

The detection by *CoRoT* or *Kepler* of even one planet with properties similar to our own could revolutionize our view of the universe, as it would suggest that Earth-like planets are common in our Galaxy. Even more ambitious is NASA's *Terrestrial Planet Finder*, an interferometric system designed to image nearby terrestrial worlds. ∞ (Sec. 5.6) Nominally scheduled for launch in 2014, this mission is currently on hold as NASA reevaluates its budget priorities for the next decade.

## CONCEPT CHECK

✔ Why is it not too surprising that the extrasolar planetary systems detected thus far have properties quite different from those of planets in our solar system?

# CHAPTER REVIEW

## Summary

**1** Our solar system is an orderly place, making it unlikely that the planets were simply and "accidentally" captured by the Sun. The overall organization points toward the solar system's formation as the product of an ancient, one-time event 4.6 billion years ago. An ideal theory of the solar system should provide strong reasons for the observed characteristics of the planets, yet be flexible enough to allow for deviations. The conden-

sation theory provides just that, as the bulk properties of planetary orbits follow naturally from the assertion that they formed in a flattened, rotating disk surrounding a newly formed star. Many "odd" aspects of the solar system may conceivably be explained in terms of collisions late in the formation stages of the protoplanetary system.

**2** According to the condensation theory, as the solar nebula collapsed under its own gravity, it began to spin faster, eventually forming a disk. Small clumps of matter

appeared around condensation nuclei and grew by accretion, sticking together and expanding into moon-sized **planetesimals (p. 392)**, whose gravitational fields were strong enough to accelerate the accretion process. In the inner solar system, planetesimals collided and merged. Competing with accretion was **fragmentation (p. 395)**—the breaking up of small bodies following collisions with larger ones. Eventually, only a few planet-sized objects remained.

**3** At any given location, the temperature would determine which materials could condense out of the nebula; thus, temperature controlled the future composition of the planets. The condensation period ended when the Sun's strong winds expelled the nebular gas.

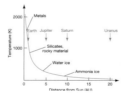

In the hot inner solar system, rocky and metallic planetesimals eventually formed the terrestrial planets.

**4** In the outer solar system, the nebula was cooler, and ices of water and ammonia could also form. According to the **gravitational instability theory (p. 396)**, the jovian planets formed directly and very rapidly through instabilities in the nebular disk. In the more standard **core-accretion theory**

**(p. 395)**, icy protoplanet cores became so large that they could capture hydrogen and helium gas from the nebula. Before the nebula was ejected, interactions between the giant planets and the gas probably caused the former to migrate inward from their initial orbits.

**5** The asteroid belt is a collection of rocky planetesimals that never managed to form a terrestrial planet because of Jupiter's gravitational influence. Many leftover planetesimals in the outer solar system were ejected into the Oort cloud and the Kuiper belt by the gravitational

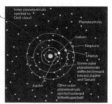

fields of the outer planets. Some now occasionally revisit our part of the solar system as comets. In the inner solar system, light elements such as hydrogen and helium would have escaped into space. Much, if not all, of Earth's water was carried to our world by comets deflected from the outer solar system. The expulsion of the icy planetesimals may have significantly changed the giant planets' orbits. Uranus and Neptune probably moved outward during this period. As Neptune's orbit expanded, it captured Pluto and the Kuiper belt **plutinos (p. 398)**, sending them into resonant orbits.

**6** Some 200 **extrasolar planets (p. 390)** are now known. All have been discovered by observing their parent star wobble back and forth as the planet orbits, although one has since been observed passing in front of the star, reducing the star's brightness slightly. Most systems found so far contain a single, massive planet comparable in

mass to Jupiter in an orbit taking the planet close to the central star. None look much like our solar system. Many contain **hot Jupiters (p. 405)**, jovian-sized planets orbiting within a fraction of an astronomical of the star. Several theories have been advanced to explain how hot Jupiters might form, including migration in the nebular disk.

**7** It is currently not known whether the observed extrasolar systems or our own solar system represent the norm in planetary systems. The fact that we see the extrasolar planets we do may well be a **selection effect (p. 406)**—these are the only planets we *can*

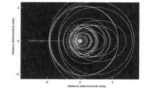

see with current techniques. The orbits of the observed Jupiter-mass planets would probably have ejected any terrestrial planets long ago. The observations indicate that stars containing larger fractions of "dusty" elements, such as carbon and silicon, are more likely to have planets, lending support to the condensation theory. Searches for terrestrial planetary systems are in the planning stage.

# Review and Discussion

1. List at least six properties of the solar system that any model of its formation must be able to explain.

2. Give three examples of present-day properties that our solar system model does not have to explain, and say why no explanation is necessary.

3. Explain the difference between evolutionary theories and catastrophic theories of the solar system's origin.

4. Describe the basic features of the condensation theory of solar system formation.

5. Why are the jovian planets so much larger than the terrestrial planets?

6. Describe two possible ways in which the jovian planets may have formed.

7. What role did the Sun play in governing the formation of the giant planets?

8. Why do giant planets "migrate?"

9. What solar system objects, still observable today, resulted from the process of fragmentation?

10. What influence did Earth's location in the solar nebula have on our planet's final composition?

11. Why could Earth not have formed out of material containing water? How might Earth's water have gotten here?

12. What happened to the outer planets as the solar system was cleared of icy planetesimals?

13. What are plutinos, and how did they come to be in their present orbits?

14. How did the Kuiper belt and the Oort cloud form?

15. Describe some ways in which random processes played a role in the determination of planetary properties.

16. Describe a possible history of a single comet now visible from Earth, starting with its birth in the solar nebula somewhere near the planet Jupiter.

17. How do astronomers set about looking for extrasolar planets?

18. In what ways do extrasolar planetary systems differ from our own solar system?

19. What is a "hot Jupiter?"

20. Do the observed extrasolar planets imply that Earth-like planets are rare?

# Conceptual Self-Test: True or False/Multiple Choice

1. Theory predicts that planets must always rotate in the same sense as the Sun's rotation.

2. Moons usually revolve in the same direction as their parent planet.

3. The asteroids were recently formed from the collision and breakup of an object orbiting within the asteroid belt.

4. The large number of leftover planetesimals beyond about 5 AU were destined to become asteroids.

5. The condensation theory offers an evolutionary explanation of the slow retrograde rotation of Venus.

6. Random collisions can explain many of the odd properties found in the solar system.

7. The ejection of the comets from the outer solar system caused all four giant planets to move inward toward the Sun.

8. Astronomers can detect extrasolar planets by observing the spectra of their parent stars.

9. Many of the extrasolar planets observed so far have masses and orbits similar to those of Jupiter.

10. Astronomers have no theoretical explanation for the "hot Jupiters" observed orbiting some other stars.

11. A successful scientific model of the origin of planetary systems must account for all of the following solar system features, except for (a) intelligent life; (b) the roughly circular planetary orbits; (c) the roughly coplanar planetary orbits; (d) the extremely distant orbits of the comets.

12. The initial gas cloud that formed the Sun and solar system must have been about as large as (a) the present-day Sun; (b) Jupiter's orbit; (c) Pluto's orbit; (d) one light-year across.

13. If interstellar dust did not exist, then (a) the terrestrial planets would not have formed; (b) the solar nebula would not have collapsed; (c) there would be many more jovian planets; (d) new stars would serve as condensation nuclei.

14. After the central star has formed, the gas making up the original nebular cloud is (a) blown away into interstellar space; (b) absorbed into the star; (c) ignited by the star's heat; (d) incorporated into jovian planet atmospheres.

15. The inner planets formed (a) when the Sun's heat destroyed all the smaller bodies in the inner solar system; (b) in the outer solar system and then were deflected inward by interactions with Jupiter and Saturn; (c) by collisions and mergers of planetesimals; (d) when a larger planet broke into pieces.

16. Water on Earth (a) was transported there by comets; (b) was accreted from the solar nebula; (c) was outgassed from volcanoes in the form of steam; (d) was created by chemical reactions involving hydrogen and oxygen shortly after Earth formed.

17. Using the standard model of planetary system formation, scientists invoke catastrophic events to explain why (a) Mercury has no moon; (b) Pluto is not a gas giant; (c) Uranus has an extremely tilted rotation axis; (d) there is no planet between Mars and Jupiter.

**18.** Astronomers have confirmed the existence of at least (a) one; (b) ten; (c) one hundred; (d) one thousand planets beyond our own solar system.

**19.** So far, most of the planets discovered orbiting other stars have orbits that (a) are larger than that of Pluto; (b) are larger than that of Saturn; (c) are very similar to that of Jupiter; (d) approach within 1 AU of the parent star.

**20.** Astronomers have not yet detected any Earth-like planets orbiting other stars because (a) there are none; (b) they are not detectable with current technology; (c) no nearby stars are of the type expected to have Earth-like planets; (d) the government is preventing them from reporting their discovery.

## Problems

 *Algorithmic versions of these Problems are available in the Practice Problems module of the Companion Website.* *The number of dots preceding each Problem indicates its approximate level of difficulty.*

**1.** •• The orbital angular momentum of a planet in a circular orbit is simply the product of the planet's mass, its orbital speed, and its distance from the Sun. ∞ *(More Precisely 6-2)* (a) Compare the orbital angular momentum of Jupiter, Saturn, and Earth. (b) Calculate the orbital angular momentum of an Oort cloud comet with mass $10^{13}$ kg, moving in a circular orbit 50,000 AU from the Sun.

**2.** ••• We can make a rough model of accretion in the inner solar nebula by imagining a 1-km-diameter body moving at a relative speed of 500 m/s through a collection of similar bodies having a roughly uniform spatial density of $10^{-10}$ body per cubic kilometer. Neglecting any gravitational forces (and hence assuming that the body moves in a straight line until it collides with something), estimate how much time, on average, it will take before the body collides.

**3.** • If Neptune formed by gravitational instability in just a thousand years at a distance of 25 AU from the Sun, how many orbits did Neptune complete while it was forming?

**4.** • How many 100-km-diameter rocky (3000 kg/m³) planetesimals would have been needed to form Earth?

**5.** •• Two asteroids, each of mass $10^{18}$ kg, orbit near the center of the asteroid belt in the plane of the ecliptic on circular paths of radii 2.80 and 2.81 AU, respectively. Calculate the gravitational force between them at closest approach, and compare it with the tidal force exerted by Jupiter if that planet were also at closest approach at the time.

6. •• The temperature in the early solar nebula at a distance of 1 AU from the Sun was about 1100 K. On the basis of the discussion of surface temperature in Chapter 7, estimate the factor by which the Sun's current energy output would have to increase in order for Earth's present temperature to have that value. ∞ (Sec. 7.2)

7. • A typical comet contains some $10^{13}$ kg of water ice. How many comets would have to strike Earth in order to account for the roughly $2 \times 10^{21}$ kg of water presently found on our planet? If this amount of water accumulated over a period of 0.5 billion years, how frequently must Earth have been hit by comets during that time?

8. • Use the data given in the text to calculate Neptune's orbital period before interactions with planetesimals expanded the orbit to its present size.

9. •• What was the period of a plutino's orbit when Neptune's orbital semimajor axis was 25 AU?

10. • How many comet-sized planetesimals (10 km in diameter, density of 100 kg/m$^3$) would have been needed to form Pluto?

11. •• The two planets orbiting the nearby star Gliese 876 are observed to be in a 2:1 resonance (i.e., the period of one is twice that of the other). The inner planet has an orbital period of 30 days. If the star's mass is the mass of the Sun, calculate the semimajor axis of the outer planet's orbit.

12. •• The planet orbiting star HD187123 has a semimajor axis of 0.042 AU. If the star's mass is 1.06 times the mass of the Sun, calculate how many times the planet has orbited its star since the paper announcing its discovery was published on December 1, 1998.

13. ••• Given that Mercury's noontime surface temperature at perihelion is 700 K, estimate the temperature of a "hot Jupiter" moving on a 3-day circular orbit around a Sun-like star.

14. •• How close to the Sun would an identical star have to come in order to exert a 0.1 percent tidal perturbation on Jupiter (i.e., a tidal force equal to 0.1 percent of the Sun's gravitational attraction on the planet)?

15. ••• Not surprisingly, as the mass of a planet decreases, the planet becomes harder to detect. Current Doppler techniques can reliably detect a planet having 0.25 times the mass of Jupiter orbiting a Sun-like star only if the planet's *maximum* orbital speed exceeds 40 km/s. Under these conditions, what is the minimum detectable eccentricity for a planet with a 5-year orbital period?

*The Companion Website at www.aw-bc.com/chaisson provides algorithmically generated versions of each chapter's Problems, along with additional quizzes, an Animations & Videos gallery, an Images gallery, an interactive Glossary, and a full eBook.*

# PART 3

# STARS and STELLAR EVOLUTION

Portrait of Cecilia Payne-Gaposchkin (*Harvard*)

Life can be hard for graduate students in astronomy. They take some tough courses, many in physics, and they assist in teaching undergraduate courses, but mostly they strive to do original research. Ideally, on the (typically 5- or 6-year) road to their Ph.D. degree, they make a discovery or gain some unique insight that they then write up as part of their doctoral dissertation. The process is exhausting, and some leave the field after the grad school grind, never again to publish in a scientific journal. Others, though, also find it exhilarating, and go on to highly productive careers in astronomy.

Arguably one of the most brilliant doctoral theses in astronomy was written in 1925 by a student at Harvard—and she did it in 2 years. Cecilia Payne (1900–1979) was an English student who crossed the ocean to pursue graduate studies at Radcliffe College, and she quickly gravitated to the nearby Harvard Observatory, then perhaps the leading center for research on stars. It was also a place where women, though they often did not get the credit at the time, were making some of the most fundamental advances in stellar astronomy. It was Nirvana for her, and she never left.

Cecilia knew far more physics than most astronomers of the time. She was one of the first to apply the then revolutionary quantum theory of atoms to the spectra of stars, thereby ascertaining stellar temperatures and chemical abundances. Of fundamental importance, her work proved that hydrogen and helium—not the heavy elements, as was then supposed—are the most abundant elements in stars and, therefore, in the universe. Her findings were so revolutionary that the leading theorist of the time, Henry Norris Russell of Princeton, declared her work to be "clearly impossible." It took years to convince the astronomical community that hydrogen is about a million times more abundant in stars than are most of the common elements found on Earth. Yet we take these findings for granted today.

In collaboration with her husband, the exiled Russian astronomer Sergei Gaposchkin, Cecilia spent decades making literally millions of observations of thousands of star clusters, variable stars, and galactic novae. Her analysis provided a firm theoretical basis for many properties of stars and their use as distance indicators in the universe. Much of her work has stood the test of time. Despite a flood of new data and new theoretical ideas, it remains the bedrock of modern astronomy.

Women "computers" at work: Cecilia at inclined desk (*Harvard*)

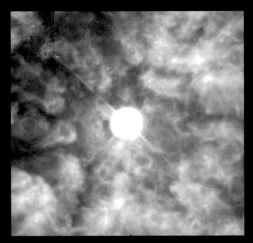

Variable Star Wr 124 (*STScI*)

Pistol Star (*STScI*)

Today, the landscape of stellar research is even richer than Cecilia Payne-Gaposchkin knew. We now see stars much more clearly, their spectra in much finer detail, and all of it to much greater distances. Not only do we know what stars are made of, in most cases we also know *why* their composition is as it is and how many of those stars are born, live, and die. Yet the picture is still very much unfinished, as 21st-century astronomy continues to uncover new and exciting features of stars and stellar systems.

In 1976, well past her retirement, Cecilia was chosen (perhaps ironically, given Russell's initial reaction to her work half a century earlier) as the Henry Norris Russell Lecturer, the highest accolade of the American Astronomical Society. Her talk was an enthusiastic summary of a lifetime of astronomical research—an encyclopedic talk with no notes and no prompts, given in perfectly punctuated English. It was very clear that she knew some individual giant stars as well as she knew her best friends. Her talk ended with the following advice to astronomers young and old:

"The reward of the young scientist is the emotional thrill of being the first person in the history of the world to see something or to understand something. Nothing can compare with that experience. The reward of the old scientist is the sense of having seen a vague sketch grow into a masterly landscape. Not a finished picture, of course; a picture that is still growing in scope and detail with the application of new techniques and new skills. The old scientist cannot claim that the masterpiece is his own work. He may have roughed out part of the design, laid on a few strokes, but he has learned to accept the discoveries of others with the same delight that he experienced on his own when he was young."

Illustrated on this page are some recent findings in stellar research—work that undoubtedly would have caused Cecilia to express more of her trademark enthusiasm. Today's research also would have made her justly proud, for so much of it relies on the insights gained by her and her colleagues during the first half of the 20th century.

Rosebud Nebula (*JPL*)

Henize Nebula (*JPL*)

413

*The Sun is our star—the main source of energy that powers weather, climate, and life on Earth. Humans simply would not exist without the Sun. Although we take it for granted each and every day, the Sun is of great importance to us in the cosmic scheme of things. This spectacular image shows a small piece of the Sun "up close"—a high resolution photograph taken from Earth, revealing its granulated surface and dark spots, all of which are gas. The region shown is 50,000 kilometers on a side, equal to about four times the size of Earth, but only a few percent of the Sun's complete surface.*
(ROYAL SWEDISH ACADEMY OF SCIENCES) ▶

# THE SUN
## Our Parent Star

Living in the solar system, we have the chance to study at close range perhaps the most common type of cosmic object—a star. Our Sun is a star, and a fairly average one at that, but with a unique feature: It is very close to us—some 300,000 times closer than our next nearest neighbor, Alpha Centauri. Whereas Alpha Centauri is 4.3 light-years distant, the Sun is only 8 light-minutes away from us. Consequently, astronomers know far more about the properties of the Sun than about any of the other distant points of light in the universe.

A good fraction of all our astronomical knowledge is based on modern studies of the Sun—from the production of seemingly boundless energy in its core to the surprisingly complex activity in its atmosphere. Just as we studied our parent planet, Earth, to set the stage for our exploration of the solar system, we now examine our parent star, the Sun, as the next step in our exploration of the universe.

## LEARNING GOALS

*Studying this chapter will enable you to*

1   Summarize the overall properties and internal structure of the Sun.
2   Describe the concept of luminosity, and explain how it is measured.
3   Explain how studies of the solar surface tell us about the Sun's interior.
4   List and describe the outer layers of the Sun.
5   Discuss the nature and variability of the Sun's magnetic field.
6   Describe the various types of solar activity and their relation to solar magnetism.
7   Outline the process by which energy is produced in the Sun's interior.
8   Explain how observations of the Sun's core changed our understanding of fundamental physics.

 Visit www.aw-bc.com/chaisson for additional images, animations, quizzes, and eBook for this chapter.

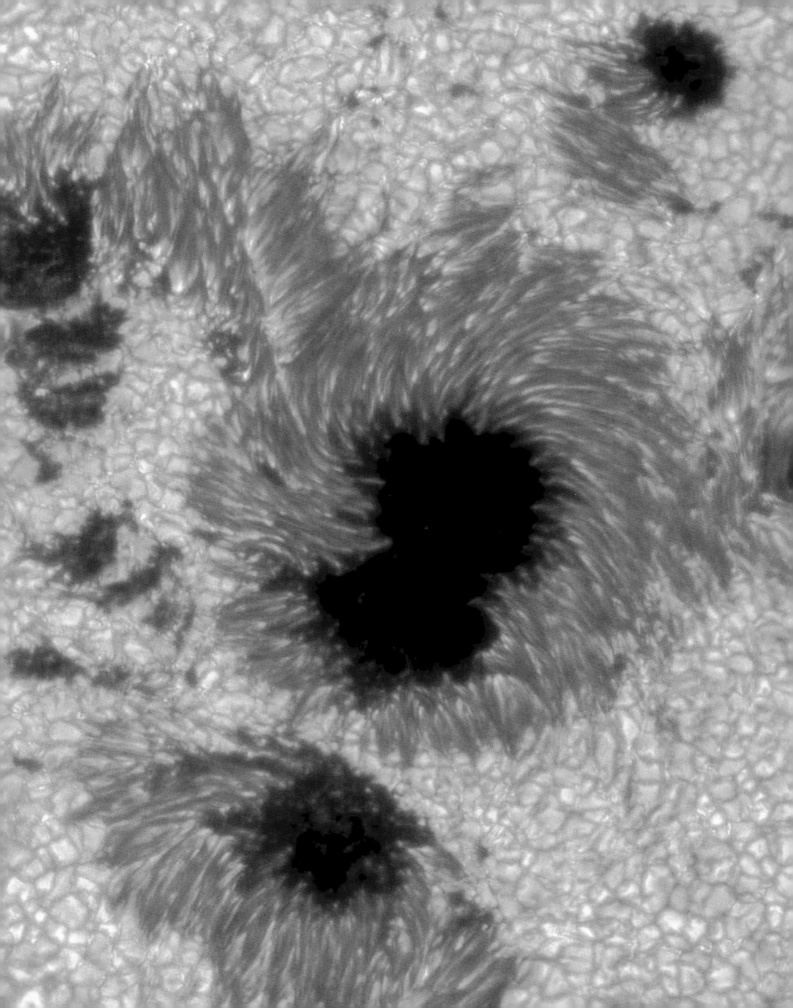

## 16.1 Physical Properties of the Sun

The Sun is the sole source of light and heat for the maintenance of life on Earth. The Sun is a **star**—a glowing ball of gas held together by its own gravity and powered by nuclear fusion at its center. In its physical and chemical properties, the Sun is similar to most other stars, regardless of when and where they formed. Indeed, our Sun appears to be a rather typical star, lying right in the middle of the observed ranges of stellar mass, radius, brightness, and composition. Far from detracting from our interest in the Sun, this very mediocrity is one of the main reasons that astronomers study it—they can apply knowledge of solar phenomena to many other stars in the universe.

### Overall Properties

The Sun Data box at the right lists some basic orbital and physical solar data. The Sun's radius, roughly 700,000 km, is determined most directly by measuring the angular size (0.5°) of the Sun and then employing elementary geometry. ∞ (Sec. 1.7) The Sun's mass, $2.0 \times 10^{30}$ kg, follows from Newton's laws of motion and gravity, applied to the observed orbits of the planets. ∞ (*More Precisely 2-3*) The average solar density derived from its mass and volume, approximately 1400 kg/m³, is quite similar to that of the jovian planets and about one-quarter the average density of Earth.

Solar rotation can be measured by timing sunspots and other surface features as they traverse the solar disk. ∞ (Sec. 2.4) These observations indicate that the Sun rotates in about a month, but it does not do so as a solid body. Instead, it spins *differentially*, like Jupiter and Saturn—faster at the equator and slower at the poles. ∞ (Sec. 11.1) The equatorial rotation period at the equator is about 25 days. Sunspots are never seen above latitude 60° (north or south), but at that latitude they indicate a 31-day period. Other measurement techniques, such as those discussed in Section 16.2, reveal that the Sun's rotation period continues to increase as we approach the poles. The polar rotation period is not known with certainty, but it may be as long as 36 days.

The Sun's surface temperature is measured by applying the radiation laws to the observed solar spectrum. ∞ (Sec. 3.4) The distribution of solar radiation has the approximate shape of a blackbody curve for an object at about 5800 K. The average solar temperature obtained in this way is known as the Sun's *effective temperature.*

Having a radius of more than 100 Earth radii, a mass of more than 300,000 Earth masses, and a surface temperature well above the melting point of any known material, the Sun is clearly a body that is very different from any other we have encountered so far.

### Solar Structure

The Sun has a surface of sorts—not a solid surface (the Sun contains no solid material), but rather that part of the brilliant gas ball we perceive with our eyes or view

## SUN DATA

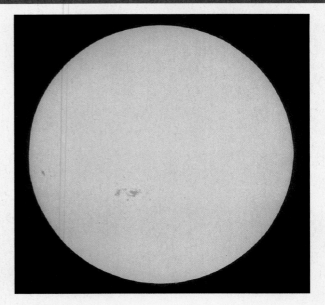

| Greatest angular diameter, as seen from Earth | 32.5' |
|---|---|
| Mass | $1.99 \times 10^{30}$ kg<br>332,000 (Earth = 1) |
| Equatorial radius | 696,000 km<br>109 (Earth = 1) |
| Mean density | 1410 kg/m³<br>0.255 (Earth = 1) |
| Surface gravity | 274 m/s²<br>28.0 (Earth = 1) |
| Escape speed | 618 km/s |
| Sidereal rotation period | 25.1 solar days (equator)<br>30.8 solar days (60° latitude)<br>36 solar days (poles)<br>26.9 solar days (interior) |
| Axial tilt | 7.25° (relative to ecliptic) |
| Surface temperature | 5780 K<br>(effective temperature) |
| Luminosity | $3.84 \times 10^{26}$ W |

through a heavily filtered telescope. This "surface"—the part of the Sun that emits the radiation we see—is called the **photosphere.** Its radius (listed in the Data box as the equatorial radius of the Sun) is about 700,000 km. However, the thickness of the photosphere is probably no more than 500 km, less than 0.1 percent of the radius, which is why we perceive the Sun as having a well-defined, sharp edge (Figure 16.1).

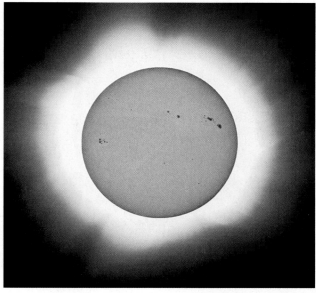

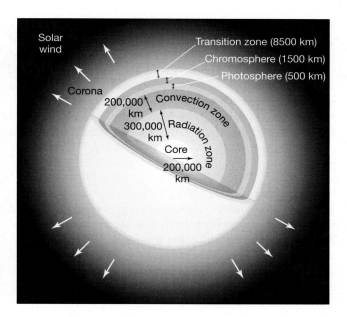

▲ **FIGURE 16.2 Solar Structure** The main regions of the Sun, not drawn to scale, with some physical dimensions labeled. The photosphere is the visible "surface" of the Sun. Below it lie the convection zone, the radiation zone, and the core. Above the photosphere, the solar atmosphere consists of the chromosphere, the transition zone, and the corona.

▲ **FIGURE 16.1 The Sun** The inner part of this composite, filtered image of the Sun shows a sharp solar edge, although our star, like all stars, is made of a gradually thinning gas. The edge appears sharp because the solar photosphere is so thin. The outer portion of the image is the solar corona, normally too faint to be seen, but visible during an eclipse, when the light from the solar disk is blotted out. Note the blemishes; they are sunspots. ∞ (Sec. 2.4) *(NOAO)*

The main regions of the Sun are illustrated in Figure 16.2 and summarized in Table 16.1. We will discuss them all in more detail later in the chapter. Just above the photosphere is the Sun's lower atmosphere, called the **chromosphere,** about 1500 km thick. From 1500 km to 10,000 km above the top of the photosphere

lies a region called the **transition zone,** in which the temperature rises dramatically. Above 10,000 km, and stretching far beyond, is a tenuous (thin), hot upper atmosphere: the solar **corona.** At still greater distances, the corona turns into the **solar wind,** which flows away from the Sun and permeates the entire solar system. ∞ (Sec. 6.5) Extending down some 200,000 km below the photosphere is the **convection zone,** a region

**TABLE 16.1 The Standard Solar Model**

| Region | Inner Radius (km) | Temperature (K) | Density (kg/m³) | Defining Properties |
|---|---|---|---|---|
| Core | 0 | 15,000,000 | 150,000 | Energy generated by nuclear fusion |
| Radiation zone | 200,000 | 7,000,000 | 15,000 | Energy transported by electromagnetic radiation |
| Convection zone | 496,000* | 2,000,000 | 150 | Energy carried by convection |
| Photosphere | 696,000* | 5800 | $2 \times 10^{-4}$ | Electromagnetic radiation can escape—the part of the Sun we see |
| Chromosphere | 696,500* | 4500 | $5 \times 10^{-6}$ | Cool lower atmosphere |
| Transition zone | 698,000* | 8000 | $2 \times 10^{-10}$ | Rapid increase in temperature |
| Corona | 706,000* | 3,000,000 | $10^{-12}$ | Hot, low-density upper atmosphere |
| Solar wind | 10,000,000 | >1,000,000 | $10^{-23}$ | Solar material escapes into space and flows outward through the solar system |

*These radii are based on the accurately determined radius of the photosphere. The other radii quoted are approximate, round numbers.*

where the material of the Sun is in constant convective motion. Below the convection zone lies the **radiation zone,** in which solar energy is transported toward the surface by radiation rather than by convection. The term *solar interior* is often used to mean both the radiation and convection zones. The central **core,** roughly 200,000 km in radius, is the site of powerful nuclear reactions that generate the Sun's enormous energy output.

## Luminosity

The properties of size, mass, density, rotation rate, and temperature are familiar from our study of the planets. But the Sun has an additional property, perhaps the most important of all from the point of view of life on Earth: The Sun *radiates* a great deal of energy into space, uniformly (we assume) in all directions. By holding a light-sensitive device—a photoelectric cell, perhaps—perpendicular to the Sun's rays, we can measure how much solar energy is received per square meter of surface area every second. Imagine our detector as having a surface area of 1 square meter (1 m$^2$) and as being placed at the top of Earth's atmosphere. The amount of solar energy reaching this surface each second is a quantity known as the **solar constant,** whose value is approximately 1400 watts per square meter (W/m$^2$).

About 50 to 70 percent of the incoming energy from the Sun reaches Earth's surface; the rest is intercepted by the atmosphere (30 percent) or reflected away by clouds (0 to 20 percent). Thus, on a clear day, a sunbather's body having a total surface area of about 0.5 m$^2$ receives solar energy at a rate of roughly 1400 W/m$^2$ × 0.70 (70 percent) × 0.5 m$^2$ ≈ 500 W, equivalent to the output of a small electric room heater or five 100-watt lightbulbs.

Let us now ask about the *total* amount of energy radiated in all directions from the Sun, not just the small fraction intercepted by our detector or by Earth. Imagine a three-dimensional sphere is centered on the Sun and just large enough that its surface intersects Earth's center (Figure 16.3). The sphere's radius is 1 AU, and its surface area is therefore $4\pi \times (1 \text{ AU})^2$, or approximately $2.8 \times 10^{23}$ m$^2$. Multiplying the rate at which solar energy falls on each square meter of the sphere (i.e., the solar constant) by the total surface area of our imaginary sphere, we can determine the total rate at which energy leaves the Sun's surface. This quantity is known as the **luminosity** of the Sun. It turns out to be just under $4 \times 10^{26}$ W.

The Sun is an enormously powerful source of energy. *Every second*, it produces an amount of energy equivalent to the detonation of about 10 billion 1-megaton nuclear bombs. Six seconds worth of solar energy output, suitably focused, would evaporate all of Earth's oceans. Three minutes would melt our planet's crust. The scale on which

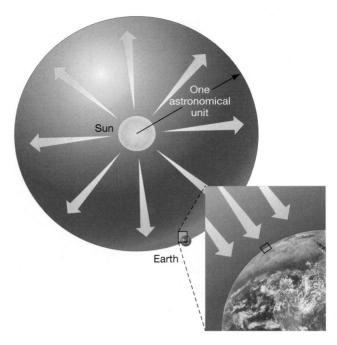

▲ FIGURE 16.3 **Solar Luminosity** We can draw an imaginary sphere around the Sun so that the sphere's surface passes through Earth's center. The radius of this imaginary sphere equals 1 AU. The "solar constant" is the amount of power striking a 1-m$^2$ detector at Earth's distance, as suggested by the inset. By multiplying the sphere's surface area by the solar constant, we can measure the Sun's luminosity—the amount of energy it emits each second.

the Sun operates simply defies earthly comparison. Let's begin our more detailed study with a look at where all this energy comes from.

CONCEPT CHECK

✔ Why must we assume that the Sun radiates equally in all directions when we compute the solar luminosity from the solar constant?

## 16.2   The Solar Interior

How do astronomers know about conditions in the interior of the Sun? As we have just seen, the fact that the Sun shines tells us that its center must be very hot, but our direct knowledge of the solar interior is actually quite limited. (See Section 16.7 for a discussion of one important "window" we do have into the solar core.) Lacking direct measurements, researchers must use other means to probe the inner workings of our parent star. To this end, they construct *mathematical models* of the Sun, combining all available data with theoretical insight into solar physics to find the model that agrees most closely with observations. ∞ (Sec. 1.2) Recall from Chapter 11 how similar techniques are used to infer the structures of the jovian planets. ∞ (Sec. 11.3)

The result in the case of the Sun is the **standard solar model,** which has gained widespread acceptance among astronomers.

## Modeling the Structure of the Sun

The Sun's bulk properties—its mass, radius, temperature, and luminosity—do not vary much from day to day or from year to year. Although we will see in Chapter 20 that stars like the Sun do change significantly over periods of *billions* of years, for our purposes here this slow evolution may be ignored. On "human" time scales, the Sun may reasonably be thought of as unchanging.

Based on this simple observation, as illustrated in Figure 16.4, theoretical models generally begin by assuming that the Sun is in a state of **hydrostatic equilibrium,** in which pressure's outward push exactly counteracts gravity's inward pull. This stable balance between opposing forces is the basic reason that the Sun neither collapses under its own weight nor explodes into interstellar space. ∞ (*More Precisely 8-1*) The assumption of hydrostatic equilibrium, coupled with our knowledge of some basic physics, then lets us predict the density and temperature in the solar interior. This information, in turn, allows the model to make predictions about other observable solar properties—luminosity, radius, spectrum, and so on—and the internal details of the model are fine-tuned until the predictions agree with observations. This is the scientific method at work; the standard solar model is the result. ∞ (Sec. 1.2)

Hydrostatic equilibrium has an important consequence for the solar interior. Because the Sun is very massive, its gravitational pull is very strong, so very high internal pressure is needed to maintain the balance. This high pressure in turn requires a very high central temperature, a fact crucial to our understanding of solar energy generation (Section 16.6). Indeed, calculations of this sort carried out by British astrophysicist Sir Arthur Eddington around 1920 provided astronomers with the first inkling that fusion might be the process that powers the Sun.

To test and refine the standard solar model, astronomers are eager to obtain information about the solar interior. However, with so little direct information about conditions below the photosphere, we must rely on more indirect techniques. In the

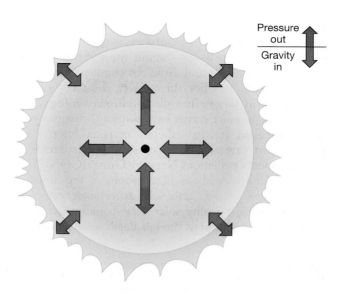

▲ **FIGURE 16.4 Hydrostatic Equilibrium** In the interior of a star such as the Sun, the outward pressure of hot gas exactly balances the inward pull of gravity. This is true at every point within the star, guaranteeing its stability.

1960s, measurements of the Doppler shifts of solar spectral lines revealed that the surface of the Sun oscillates, or vibrates, like a complex set of bells. ∞ (Secs. 3.5, 4.5) These vibrations, illustrated in Figure 16.5(a), are the result of internal pressure waves (somewhat like sound waves in air) that reflect off the photosphere and repeatedly cross the solar interior (Figure 16.5b). Because the waves can penetrate deep inside the Sun, analysis of their

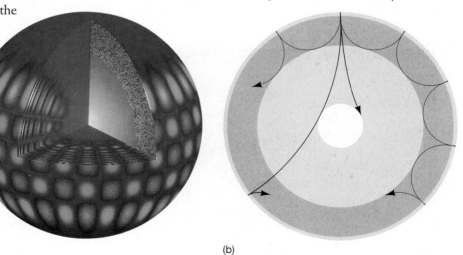

(a)                                    (b)

▲ **FIGURE 16.5 Solar Oscillations** (a) The Sun has been found to vibrate in a very complex way. By observing the motion of the solar surface, scientists can determine the wavelengths and the frequencies of the individual waves and deduce information about the Sun not obtainable by other means. The alternating patches represent gas moving down (red) and up (blue). (See also *Discovery 16-1*.) (b) Depending on their initial directions, the waves contributing to the observed oscillations may travel deep inside the Sun, providing vital information about the solar interior. The wave shown closest to the surface here corresponds approximately to the vibration pattern depicted in part (a). (*National Solar Observatory*)

surface patterns allows scientists to study conditions far below the Sun's surface. The process is similar to the way in which seismologists learn about the interior of Earth by observing the P- and S-waves produced by earthquakes. ∞ (Sec. 7.3) For this reason, the study of solar surface patterns is usually called **helioseismology**, even though solar pressure waves have nothing whatever to do with solar seismic activity—there is no such thing.

The most extensive study of solar oscillations is the ongoing Global Oscillations Network Group (GONG) project. By making continuous observations of the Sun from many clear sites around Earth, solar astronomers can obtain uninterrupted high-quality solar data spanning many days and even weeks—almost as though Earth were not rotating

and the Sun never set. The *Solar and Heliospheric Observatory (SOHO)*, launched by the European Space Agency in 1995 and now permanently stationed between Earth and the Sun some 1.5 million km from our planet (see *Discovery 16-1*), also provides continuous monitoring of the Sun's surface and atmosphere. Analysis of ground- and space-based data provides important additional information about the temperature, density, rotation, and convective state of the solar interior, allowing detailed comparisons between theory and reality to be made. Direct comparison is possible throughout a large portion of the Sun, and the agreement between model and observations is spectacular: The frequencies and wavelengths of observed solar oscillations are within 0.1 percent of the predictions of the standard solar model.

# DISCOVERY 16-1

## *SOHO:* Eavesdropping on the Sun

Throughout the few decades of the Space Age, various nations, led by the United States, have sent spacecraft to all but one of the major bodies in the solar system. That unexplored body is the Sun. Currently, the next best thing to a dedicated reconnoitering spacecraft is the *Solar and Heliospheric Observatory (SOHO)*, which has radioed back to Earth volumes of new data—and more than a few new puzzles—about our parent star since the spacecraft's launch in 1995.

*SOHO* is a billion-dollar mission operated primarily by the European Space Agency. The 2-ton robot is now on station about 1.5 million km sunward of Earth—about 1 percent of the distance from Earth to the Sun. This is the so-called

$L_1$ Lagrangian point, where the gravitational pull of the Sun and Earth are precisely equal—a good place to park a monitoring platform. ∞ (Sec. 14.1) There, *SOHO* looks unblinkingly at our star 24 hours a day, eavesdropping on the Sun's surface, atmosphere, and interior. The automated vehicle carries a dozen instruments capable of measuring almost everything from the Sun's corona and magnetic field to its solar wind and internal vibrations. The accompanying figure shows a false-color image of the Sun's lower corona, obtained by combining *SOHO* data captured at three different ultraviolet wavelengths.

*SOHO* is positioned just beyond Earth's magnetosphere, so its instruments can cleanly study the charged particles of the solar wind—high-speed matter escaping from the Sun, flowing outward from the corona. Coordinating these on-site measurements with *SOHO* images of the Sun itself, astronomers now think they can follow solar magnetic field loops expanding and breaking as the Sun prepares itself for mass ejections several days before they actually occur (see Section 16.5). Given that such coronal storms can wreak havoc on communications, power grids, satellite electronics, and other human activities, the prospect of having accurate forecasts of disruptive solar events is a welcome development.

Section 16.2 discusses how astronomers can "take the pulse" of the Sun by measuring its complex rhythmic motions. *SOHO* also has the ability to study the weak sound waves that echo and resonate inside the Sun and can map these vibrations with much higher resolution than was previously possible. It does so not by sensing sound itself, but by watching the Sun's surface move up and down ever so slightly. The Sun's "loudest" vibrations are extremely low pitched (0.003 Hz, or one oscillation every 5 minutes), more like rolling rumbles, just as we might expect from such a huge and massive object.

As of late 2006, *SOHO* is still operating, 8 years beyond its planned lifetime, having survived assaults from both the Sun and Earth. Twice so far, skillful engineers have brought the spacecraft back from apparent death after mission controllers mistakenly sent incorrect commands from the ground. It is a good thing that they did, for this remarkable spacecraft has radioed back to Earth a wealth of new scientific insight into our parent star.

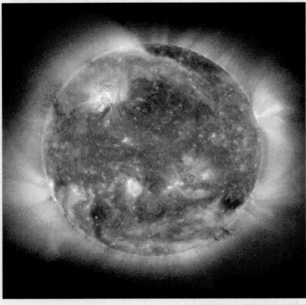

(NASA/ESA)

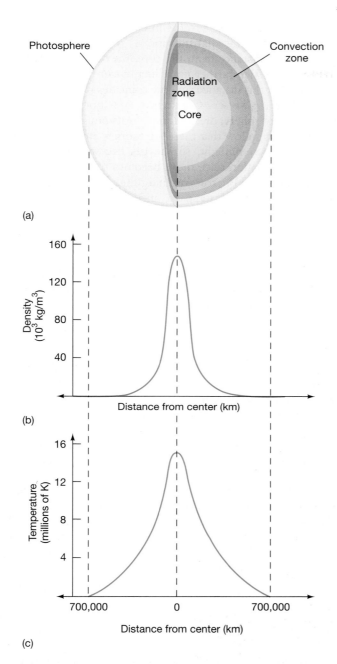

(a)

(b)

(c)

▲ FIGURE 16.6 **Solar Interior** (a) A cross-sectional cut through the middle of the Sun, with corresponding graphs of (b) density and (c) temperature across the cut, according to the standard solar theoretical model.

of about 1000 kg/m$^3$, the density of water, to an extremely small photospheric value of $2 \times 10^{-4}$ kg/m$^3$, 10,000 times less dense than air at the surface of Earth. Because the density is so high in the core, roughly 90 percent of the Sun's mass is contained within the inner half of its radius. The solar density continues to decrease out beyond the photosphere, reaching values as low as $10^{-23}$ kg/m$^3$ in the far corona—about as thin as the best vacuum physicists can create in laboratories on Earth.

The solar temperature also decreases with increasing radius in the solar interior, but not as rapidly as the density. Computer models indicate a temperature of about 15 million K at the core, consistent with the minimum 10 million K needed to initiate the nuclear reactions known to power most stars, decreasing to the observed value of about 5800 K at the photosphere.

As the data improve and old mysteries are resolved, new ones often emerge. For example, helioseismology indicates that the Sun's rotation speed varies with depth—perhaps not too surprising, given the surface differential rotation mentioned earlier and the fact that similar behavior has been noted in the outer planets. What is puzzling, though, is the *complexity* of the differential motion. The surface layers show a "zonal flow" of sorts, with alternating bands of higher- and lower-than-average rotation rates. Just below the surface are wide "rivers" of lower speed (at the equator) and higher speed (polar) rotation. The material at the base of the convection zone appears to oscillate in rotation speed, sometimes moving faster (by about 10 percent) than the surface layers, sometimes slower, with a period of about 1.3 years. Deeper still, the radiative interior rotates more or less as a solid body, once every 26.9 days. A full explanation of the Sun's rotation currently eludes theorists.

## Energy Transport

The very hot solar interior ensures violent and frequent collisions among gas particles. Particles move in all directions at high speeds, bumping into one another unceasingly. In and near the core, the extremely high temperatures guarantee that the gas is completely ionized. Recall from Chapter 4 that, under less extreme conditions, atoms absorb photons that can boost their electrons to more excited states. ∞ (Sec. 4.2) With no electrons left on atoms to capture the photons, however, the deep solar interior is relatively transparent to radiation. Only occasionally does a photon encounter and scatter off of a free electron or proton. The energy produced by nuclear reactions in the core travels outward toward the surface in the form of radiation with relative ease.

As we move outward from the core, the temperature falls, atoms collide less frequently and less violently, and more and more electrons manage to remain bound to their parent nuclei. With more and more atoms retaining electrons that can absorb the outgoing radiation, the gas in the

Figure 16.6 shows the solar density and temperature, plotted as functions of distance from the Sun's center, according to the standard solar model. Notice how the density drops rather sharply at first and then decreases more slowly near the solar photosphere, some 700,000 km from the center. The variation in density is large, ranging from a core value of about 150,000 kg/m$^3$, 20 times the density of iron, to an intermediate value (at 350,000 km)

interior changes from being relatively transparent to being almost totally opaque. By the outer edge of the radiation zone, roughly 500,000 km from the center (actually, 496,000 km, according to the best available *SOHO* data), *all* the photons produced in the Sun's core have been absorbed. Not one of them reaches the surface. But what happens to the energy they carry?

The photons' energy must travel beyond the Sun's interior: That we see sunlight—visible energy—proves that energy escapes. The escaping energy reaches the surface by *convection*—the same basic physical process we saw in our study of Earth's atmosphere, although it operates in a very different environment in the Sun. ∞ (Sec. 7.2) Hot solar gas moves outward while cooler gas above it sinks, creating a characteristic pattern of convection cells. All through the convection zone, energy is transported to the surface by physical motion of the solar gas. (Note that this actually represents a departure from hydrostatic equilibrium, as defined above, but it can still be handled within the standard solar model.) Remember that there is no physical movement of material when radiation is the energy-transport mechanism; convection and radiation are *fundamentally different* ways in which energy can be transported from one place to another.

Figure 16.7 is a schematic diagram of the solar convection zone. There is a hierarchy of convection cells, organized in tiers of many different sizes at different depths. The deepest tier, lying approximately 200,000 km below the photosphere, is thought to contain large cells some tens of thousands of kilometers in diameter. Heat is then successively carried upward through a series of progressively smaller cells, stacked one on another, until, at a depth of about 1000 km, the individual cells are about 1000 km across. The top of this uppermost tier of convection is the visible surface of the Sun, where astronomers can directly observe the cell sizes. Information about

convection below that level is inferred mostly from computer models of the solar interior.

At some distance from the core, the solar gas becomes too thin to sustain further upwelling by convection. Theory suggests that this distance roughly coincides with the photospheric surface we see. Convection does not proceed into the solar atmosphere; there is simply not enough gas there—the density is so low that there are too few atoms or ions to intercept much sunlight, so the gas becomes transparent again and radiation once more becomes the mechanism of energy transport. Photons reaching the photosphere escape more or less freely into space, and the photosphere emits thermal radiation, like any other hot object. The photosphere is narrow, and the "edge" of the Sun sharp, because this transition from opacity to complete transparency is very rapid. Just below the bottom of the photosphere the gas is still convective, and radiation does not reach us directly. A few hundred kilometers higher, the gas is too thin to emit or absorb any significant amount of radiation.

## Granulation

Figure 16.8 is a high-resolution photograph of the solar surface. The visible surface is highly mottled, or **granulated,** with regions of bright and dark gas known as *granules.* Each bright granule measures about 1000 km across—comparable in size to a continent on Earth—and has a lifetime of between 5 and 10 minutes. Together, several million granules constitute the top layer of the convection zone, immediately below the photosphere.

Each granule forms the topmost part of a solar convection cell. Spectroscopic observation within and around the bright regions shows direct evidence for the upward motion of gas as it "boils" up from within—evidence that convection really does occur just below the photosphere. Spectral lines detected from the bright granules appear slightly bluer than normal, indicating Doppler-shifted matter approaching us at about 1 km/s. ∞ (Sec. 3.5) Spectroscopes focused on the darker portions of the granulated photosphere show the same spectral lines to be redshifted, indicating matter moving away from us.

The variations in brightness of the granules result strictly from differences in temperature. The upwelling gas is hotter and therefore emits more radiation than the cooler, downward-moving gas. The adjacent bright and dark gases appear to contrast considerably, but in reality their

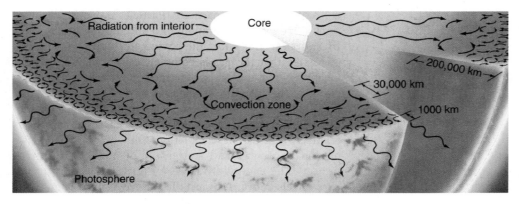

▲ FIGURE 16.7 **Solar Convection** Physical transport of energy in the Sun's convection zone. The upper-interior region is visualized as a boiling, seething sea of gas. Near the surface, each convective cell is about 1000 km across. The sizes of the convective cells become progressively larger at greater depths, reaching some 30,000 km in diameter at the base of the convection zone, 200,000 km below the photosphere. (This is a highly simplified diagram; there are many different cell sizes, and they are not so neatly arranged.)

ANIMATION/VIDEO   Solar Granulation

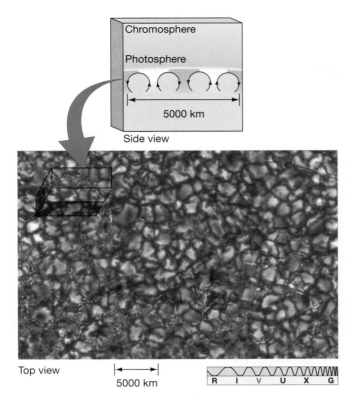

**▲ FIGURE 16.8 Solar Granulation** A photograph of the granulated solar photosphere, taken with the 1-m Swedish Solar Telescope looking directly down on the Sun's surface. Typical solar granules are comparable in size to Earth's continents. The bright portions of the image are regions where hot material is upwelling from below, as illustrated in Figure 16.7. The darker (redder) regions correspond to cooler gas that is sinking back down into the interior. The inset drawing shows a perpendicular cut through the solar surface. (SST)

temperature difference is less than about 500 K. Careful measurements also reveal a much larger-scale flow on the solar surface. **Supergranulation** is a flow pattern quite similar to granulation, except that supergranulation cells measure some 30,000 km across. As with granulation, material upwells at the center of the cells, flows across the surface, then sinks down again at the edges. Scientists suspect that supergranules are the imprint on the photosphere of a deeper tier of large convective cells, like those depicted in Figure 16.7.

CONCEPT CHECK

✔ What are the two distinct ways in which energy moves outward from the solar core to the photosphere?

## 16.3 The Sun's Atmosphere

Astronomers can glean an enormous amount of information about the Sun from an analysis of the absorption lines that arise in the photosphere and lower atmosphere. ∞ (Sec. 4.5) Figure 16.9 (see also Figure 4.4) is a detailed spectrum of the Sun spanning a range of wavelengths from 360 to 690 nm. Notice the intricate dark Fraunhofer absorption lines superposed on the background continuous spectrum.

Tens of thousands of spectral lines have been observed and cataloged in the solar spectrum. In all, some 67 elements have been identified in the Sun in various states of ionization and excitation. ∞ (Sec. 4.2) More elements

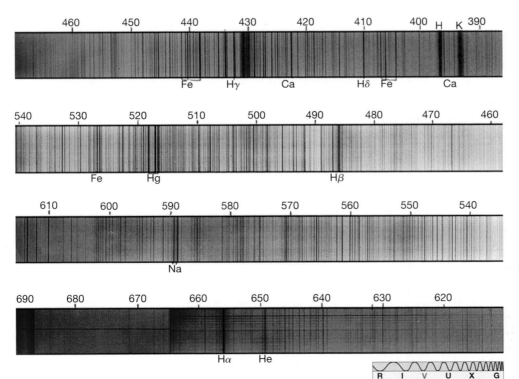

**◄ FIGURE 16.9 Solar Spectrum** A detailed visible spectrum of our Sun shows thousands of dark Fraunhofer (absorption) spectral lines indicating the presence of 67 different elements in various stages of excitation and ionization in the lower solar atmosphere. The numbers give wavelengths, in nanometers. (Palomar Observatory/Caltech)

probably exist there, but they are present in such small quantities that our instruments are simply not sensitive enough to detect them. Table 16.2 lists the 10 most common elements in the Sun. Notice that hydrogen is by far the most abundant element, followed by helium. This distribution is just what we saw on the jovian planets, and it is what we will find for the universe as a whole.

## Solar Spectral Lines

As discussed in Chapter 4, spectral lines arise when electrons in atoms or ions make transitions between states of well-defined energies, emitting or absorbing photons of specific energies (i.e., wavelengths or colors) in the process. ∞ (Sec. 4.2) However, to explain the spectrum of the Sun (and, indeed, the spectra of all stars), we must slightly modify our earlier description of the formation of absorption lines. We explained these lines in terms of cool foreground gas intercepting light from a hot background source. In actuality, both the bright background and the dark absorption lines in Figure 16.9 form at roughly the same locations in the Sun—the solar photosphere and lower chromosphere. To understand how these lines are formed, consider again the solar energy emission process in a little more detail.

Below the photosphere, the solar gas is sufficiently dense, and interactions among photons, electrons, and ions sufficiently common, that radiation cannot escape directly into space. In the solar atmosphere, however, the probability that a photon will escape without further interaction with matter depends on the photon's energy. Recall from Chapter 4 that an atom or ion can absorb a photon only if that photon's energy has just the right value to cause an electron to jump from one energy level to another. ∞ (Sec. 4.3) Hence, if the photon energy

happens to correspond to some electronic transition in an atom or ion in the gas, then the photon may be absorbed again before it can travel very far—the more elements present of the type suitable for absorption, the lower is the escape probability. Conversely, if the photon's energy does not coincide with any such transition, then the photon cannot interact further with the gas, and it leaves the Sun headed for interstellar space or perhaps the detector of an astronomer on Earth.

Thus, as illustrated in Figure 16.10, when we look at the Sun, we are actually peering down into the solar atmosphere to a depth that depends on the wavelength of the light under consideration. Photons with wavelengths far from any absorption feature (i.e., having energies far from any atomic transition) are less likely to interact with matter as they travel through the solar gas and so tend to come from deep in the photosphere. However, photons with wavelengths near the centers of absorption lines are much more likely to be captured by an atom or ion and therefore escape mainly from higher (and cooler) levels. The lines are darker than their surroundings because the temperature at the level of the atmosphere where they form is lower than the 5800-K temperature at the base of the photosphere, where most of the continuous emission originates. (Recall that, by Stefan's law, the brightness of a radiating object depends on its temperature—the cooler the gas, the less energy it radiates.) ∞ (Sec. 3.4) Thus, the existence of Fraunhofer lines is direct evidence that the temperature in the Sun's atmosphere decreases with height above the photosphere.

Strictly speaking, spectral analysis allows us to draw conclusions only about the part of the Sun where the lines form—the photosphere and chromosphere. However, most astronomers think that, with the exception of the solar core (where nuclear reactions are steadily changing the composition—see Section 16.6), the data in Table 16.2 are representative of the entire Sun. That assumption is strongly supported by the excellent agreement between the standard solar model, which makes the same assumption, and helioseismological observations of the solar interior.

## The Chromosphere

Above the photosphere lies the cooler chromosphere, the inner part of the solar atmosphere. This region emits very little light of its own and cannot be observed visually under normal conditions. The photosphere is just too bright, dominating the chromosphere's radiation. The relative dimness of the chromosphere results from its low density—large numbers of photons simply cannot be emitted by a tenuous gas containing very few atoms per unit volume. Still, although it is not normally seen, astronomers have long been aware of the chromosphere's existence. Figure 16.11 shows the Sun during an eclipse

| TABLE 16.2 | The Composition of the Sun | |
| --- | --- | --- |
| Element | Percentage of Total Number of Atoms | Percentage of Total Mass |
| Hydrogen | 91.2 | 71.0 |
| Helium | 8.7 | 27.1 |
| Oxygen | 0.078 | 0.97 |
| Carbon | 0.043 | 0.40 |
| Nitrogen | 0.0088 | 0.096 |
| Silicon | 0.0045 | 0.099 |
| Magnesium | 0.0038 | 0.076 |
| Neon | 0.0035 | 0.058 |
| Iron | 0.0030 | 0.14 |
| Sulfur | 0.0015 | 0.040 |

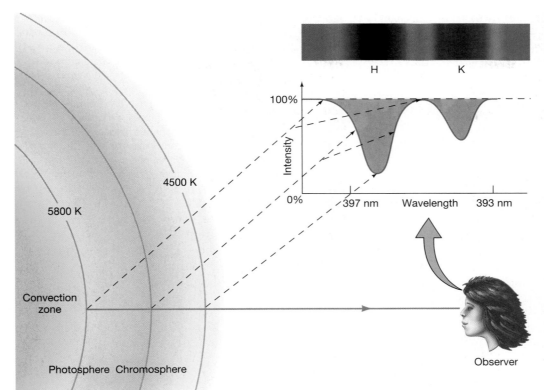

in which the photosphere—but not the chromosphere—is obscured by the Moon. The chromosphere's characteristic reddish hue is plainly visible. This coloration is due to the red Hα (hydrogen alpha) emission line of hydrogen, which dominates the chromospheric spectrum. ∞ (*More Precisely 4-1*)

The chromosphere is far from tranquil. Every few minutes, small solar storms erupt, expelling jets of hot matter known as *spicules* into the Sun's upper atmosphere (Figure 16.12). These long, thin spikes of matter leave the Sun's surface at typical speeds of about 100 km/s and reach several thousand kilometers above the photosphere. Spicules are not spread evenly across the solar surface. Instead, they cover only about 1 percent of the total area, tending to accumulate around the edges of supergranules. The Sun's magnetic field is also known to be somewhat stronger than average in those regions. Scientists speculate that the downward-moving material there tends to strengthen the solar magnetic field, and spicules are the result of magnetic disturbances in the Sun's churning outer layers.

## The Transition Zone and the Corona

During the brief moments of an eclipse, if the Moon's angular size is large enough that both the photosphere and the chromosphere are blocked, the ghostly solar corona can be seen (Figure 16.13). With the photospheric light

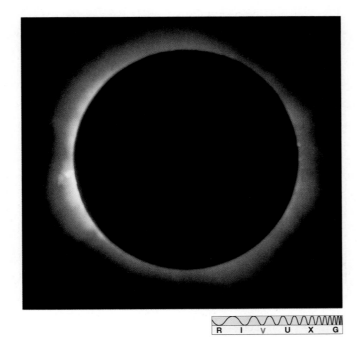

▲ **FIGURE 16.11** **Solar Chromosphere** This photograph of a total solar eclipse shows the solar chromosphere a few thousand kilometers above the Sun's surface. (*G. Schneider*)

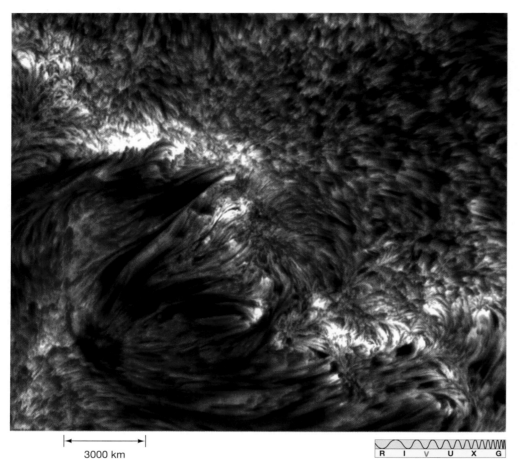

```
R  I  V  U  X  G
```

◀ FIGURE 16.12 **Solar Spicules** Short-lived, narrow jets of gas that typically last mere minutes can be seen sprouting up from the solar chromosphere in this Hα image of the Sun. These so-called spicules are the thin, dark, spikelike regions. They appear dark against the face of the Sun because they are cooler than the underlying photosphere. *(SST)*

3000 km

removed, the pattern of spectral lines changes dramatically. The intensities of the usual lines alter (suggesting changes in composition or temperature, or both), the spectrum shifts from absorption to emission, and an entirely new set of spectral lines suddenly appears. The shift from absorption to emission is entirely in accordance with Kirchhoff's laws, because we see the corona against the blackness of space, not against the bright continuous spectrum from the photosphere below. ∞ (Sec. 4.1)

These new coronal (and in some cases chromospheric) lines were first observed during eclipses in the 1920s. For years afterward, some researchers (for want of any better explanation) attributed them to a new nonterrestrial element, which they dubbed "coronium." We now recognize that these new spectral lines do not indicate any new kind of atom. Coronium does not exist. Rather, the new lines arise because atoms in the corona have lost several more electrons than atoms in the photosphere—that is, the coronal atoms are much more highly ionized. Therefore, their internal electronic structures, and hence their spectra, are quite different from the structure and spectra of atoms and ions in the photosphere. For example, astronomers have identified coronal lines corresponding to iron ions with as many as 13 of their normal 26 electrons missing. In the

```
R  I  V  U  X  G
```

▲ FIGURE 16.13 **Solar Corona** When both the photosphere and the chromosphere are obscured by the Moon during a solar eclipse, the faint corona becomes visible. This photograph clearly shows the emission of radiation from a relatively inactive solar corona. *(Bencho Angelov)*

photosphere, most iron atoms have lost only 1 or 2 of their electrons.

The cause of this extensive electron stripping is the high coronal *temperature*. The degree of ionization inferred from spectra observed during solar eclipses tells us that the temperature of the upper chromosphere exceeds that of the photosphere. Furthermore, the temperature of the solar corona, where even more ionization is seen, is higher still. Figure 16.14 shows how the temperature of the Sun's atmosphere varies with altitude. The temperature decreases to a minimum of about 4500 K some 500 km above the photosphere, after which it rises steadily. About 1500 km above the photosphere, in the transition zone, the temperature begins to rise rapidly, reaching more than 1 million K at an altitude of 10,000 km. Thereafter, in the corona, the temperature remains roughly constant at around 3 million K, although *SOHO* and other orbiting instruments have detected coronal "hot spots" having temperatures many times higher than this average value.

The cause of the rapid temperature rise is not fully understood. The temperature profile runs contrary to intuition: Moving away from a heat source, we would normally expect the heat to diminish, but this is not the case in the lower atmosphere of the Sun. The corona must have another energy source. Astronomers now think that magnetic disturbances in the solar photosphere are ultimately responsible for heating the corona (Section 16.5).

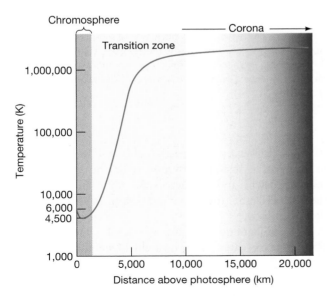

▲ FIGURE 16.14 **Solar Atmospheric Temperature** The change of gas temperature in the lower solar atmosphere is dramatic. The temperature, indicated by the blue line, reaches a minimum of 4500 K in the chromosphere and then rises sharply in the transition zone, finally leveling off at around 3 million K in the corona.

### The Solar Wind

Electromagnetic radiation and fast-moving particles—mostly protons and electrons—escape from the Sun all the time. The radiation moves away from the photosphere at the speed of light, taking 8 minutes to reach Earth. The particles travel more slowly, although at the still considerable speed of about 500 km/s, reaching Earth in a few days. This constant stream of escaping solar particles is the solar wind.

The solar wind results from the high temperature of the corona. About 10 million km above the photosphere, the coronal gas is hot enough to escape the Sun's gravity, and it begins to flow outward into space. At the same time, the solar atmosphere is continuously replenished from below. If that were not the case, the corona would disappear in about a day. The Sun is, in effect, "evaporating"—constantly shedding mass through the solar wind. The wind is an extremely thin medium, however. Even though it carries away roughly 2 million tons of solar matter each second, less than 0.1 percent of the Sun's mass has been lost this way since the solar system formed 4.6 billion years ago.

CONCEPT CHECK
✔ Describe two ways in which the spectrum of the solar corona differs from that of the photosphere.

## 16.4 Solar Magnetism

The Sun has a powerful and complex magnetic field. Discovered in 1908 by American astronomer George Ellery Hale, the solar field still presents puzzles to scientists today. The structure of the Sun's magnetic field lines is crucial to understanding many aspects of the Sun's appearance and surface activity, yet the details of the field-line geometry, and even the mechanism responsible for generating and sustaining the entire solar field, remain subjects of intense research. Curiously, the keys to understanding many aspects of solar magnetism lie in a phenomenon first observed nearly three centuries before Hale's groundbreaking discovery.

### Sunspots

Figure 16.15 is an optical photograph of the entire Sun, showing numerous dark blemishes on its surface. First studied in detail by Galileo around 1613, these "spots" provided one of the first clues that the Sun was not a perfect, unvarying creation, but rather a place of constant change. ∞ (Sec. 2.4) The dark areas are called **sunspots** and typically measure about 10,000 km across, approximately the size of Earth. As shown in the figure, they often occur in groups. At any given time, the Sun may have hundreds of sunspots, or it may have none at all.

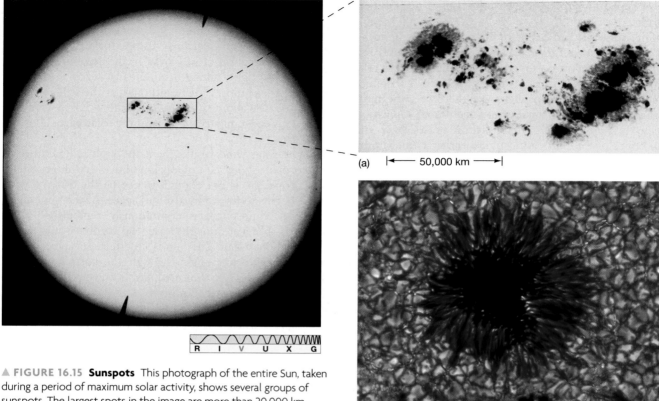

R I V U X G

▲ FIGURE 16.15 **Sunspots** This photograph of the entire Sun, taken during a period of maximum solar activity, shows several groups of sunspots. The largest spots in the image are more than 20,000 km across, nearly twice the diameter of Earth. Typical sunspots are only about half that size. (*Palomar Observatory/Caltech*)

(a) |←— 50,000 km —→|

(b) |←— 10,000 km —→|

R I V U X G

▲ FIGURE 16.16 **Sunspots, Up Close** (a) An enlarged photograph of the largest pair of sunspots in Figure 16.15 shows how each spot consists of a cool, dark inner region called the umbra surrounded by a warmer, brighter region called the penumbra. The spots appear dark because they are slightly cooler than the surrounding photosphere. (b) A high-resolution image of a single typical sunspot—about the size of Earth—shows details of its structure as well as the surface granules surrounding it. See also the full-page opening photo at the start of this chapter. (*Palomar Observatory/Caltech; SST/Royal Swedish Academy of Science*)

Studies of sunspots show an **umbra,** or dark center, surrounded by a grayish **penumbra.** The close-up views in Figure 16.16 show each of these dark areas and the brighter undisturbed photosphere nearby. This gradation in darkness is really a gradual change in photospheric temperature—sunspots are simply *cooler* regions of the photospheric gas. The temperature of the umbra is about 4500 K, compared with the penumbra's 5500 K. The spots, then, are certainly composed of hot gases. They seem dark only because they appear against an even brighter background (the 5800 K photosphere). If we could magically remove a sunspot from the Sun (or just block out the rest of the Sun's emission), the spot would glow brightly, just like any other hot object having a temperature of roughly 5000 K.

## The Sun's Magnetic Field

What causes a sunspot? Why is it cooler than the surrounding photosphere? The answers to these questions are closely tied to the structure of the Sun's magnetic field. We saw in Chapter 4 that analysis of spectral lines can yield detailed information about the magnetic field at the location where the lines originate. ∞ (Sec. 4.5) Indeed, Hale's discovery of solar magnetism was made through observations of the Zeeman effect (broadening or splitting

of spectral lines by a magnetic field) in Hα lines observed in sunspots. Most importantly, both the *strength* of the magnetic field and the *orientation* of a field line along the line of sight (toward or away from the observer) can be determined.

The magnetic field in a typical sunspot is about 1000 times greater than the field in neighboring, undisturbed photospheric regions (which is itself several times stronger than Earth's magnetic field). Furthermore, the field lines are not randomly oriented, but instead are directed roughly perpendicular to (out of or into) the Sun's surface. Scientists think that sunspots are cooler than their surroundings because these abnormally strong fields tend to block (or redirect) the convective flow of hot gas, which is normally toward the surface of the Sun.

The **polarity** of a sunspot simply indicates which way its magnetic field is directed relative to the solar surface. We conventionally label spots where field lines emerge from the interior as "S" and those where the lines dive below the photosphere as "N" (so field lines above the surface always run from S to N, as on Earth). Sunspots almost always come in pairs whose members lie at roughly the same latitude and have opposite magnetic polarities.

Figure 16.17(a) illustrates how magnetic field lines emerge from the solar interior through one member (S) of a sunspot pair, loop through the solar atmosphere, and then reenter the photosphere through the other member (N). As in Earth's magnetosphere, charged particles tend to follow the solar magnetic field lines. ∞ (Sec. 7.5) Figure 16.17(b) shows an actual image of solar magnetic loops, revealing high-temperature gas flowing along a

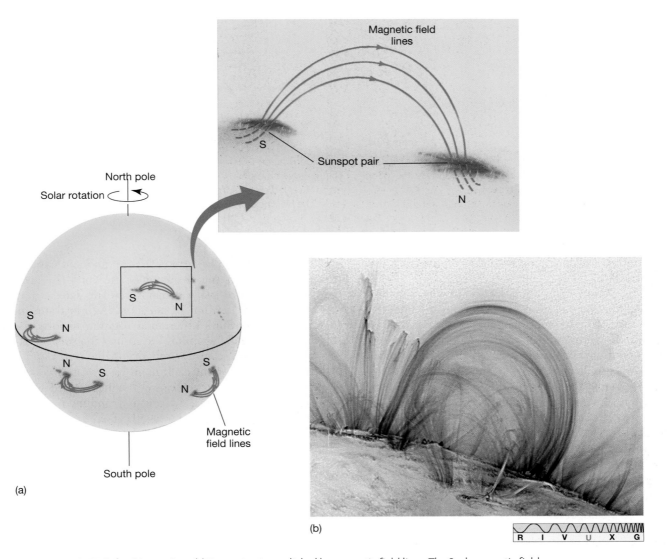

(a)

(b)

R I V U X G

▲ FIGURE 16.17 **Solar Magnetism** (a) Sunspot pairs are linked by magnetic field lines. The Sun's magnetic field lines emerge from the surface through one member of a pair and reenter the Sun through the other member. The leading members of all sunspot pairs in the solar northern hemisphere have the same polarity (labeled N or S, as described in the text). If the magnetic field lines are directed into the Sun in one leading spot, they are inwardly directed in all other leading spots in that hemisphere. The same is true in the southern hemisphere, except that the polarities are always opposite those in the north. The entire magnetic field pattern reverses itself roughly every 11 years. (b) A far-ultraviolet image taken by the *Transition Region and Coronal Explorer (TRACE)* satellite in 1999, showing magnetic field lines arching between two sunspot groups. Note the complex structure of the field lines, which are seen here via the radiation emitted by superheated gas flowing along them. Resolution here is about 700 km. In this negative image (which shows the lines more clearly), the darkest regions have temperatures of about 2 million K. *(NASA)*

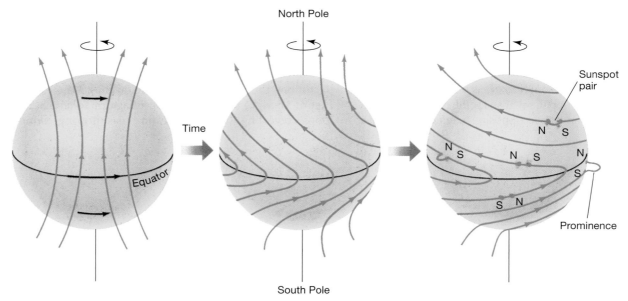

North Pole

Time

Equator

South Pole

Sunspot pair

Prominence

▲ FIGURE 16.18 **Solar Rotation** (a, b) The Sun's differential rotation wraps and distorts the solar magnetic field. (c) Occasionally, the field lines burst out of the surface and loop through the lower atmosphere, thereby creating a sunspot pair. The underlying pattern of the solar field lines explains the observed pattern of sunspot polarities. If the loop happens to occur on the edge of the Sun and is seen against the blackness of space, we see a phenomenon called a prominence, described in Section 16.5. (See Figure 16.22.)

complex network of magnetic field lines connecting two sunspot groups.

Despite the irregular appearance of the sunspots themselves, there is a great deal of order in the underlying solar field. *All* the sunspot pairs in the same solar hemisphere (north or south) at any instant have the *same* magnetic configuration. That is, if the leading spot (measured in the direction of the Sun's rotation) of one pair has N polarity, as shown in the figure, then all leading spots in that hemisphere have the same polarity. What's more, in the other hemisphere at the same time, all sunspot pairs have the *opposite* magnetic configuration (S polarity leading). To understand these regularities in sunspot polarities, we must look at the Sun's magnetic field in a little more detail.

The combination of differential rotation and convection radically affects the character of the Sun's magnetic field, which in turn plays a major role in determining the numbers and location of sunspots. As illustrated in Figure 16.18, the Sun's differential rotation distorts the solar magnetic field, "wrapping" it around the solar equator and eventually causing any originally north–south magnetic field to reorient itself in an east–west direction. At the same time, convection causes the magnetized gas to well up toward the surface, twisting and tangling the magnetic field pattern. In some places, the field lines become kinked like a twisted garden hose, causing the field strength to increase. Occasionally, the field becomes so strong that it overwhelms the Sun's gravity, and a "tube" of field lines bursts out of the surface and loops through the

▼ FIGURE 16.19 **Sunspot Rotation** The evolution of some sunspots and lower chromospheric activity over a period of 12 days. The sequence runs from left to right. An Hα filter was used to make these photographs, taken from the *Skylab* space station. An arrow follows one set of sunspots over the course of a week as they are carried around the Sun by its rotation. *(NASA)*

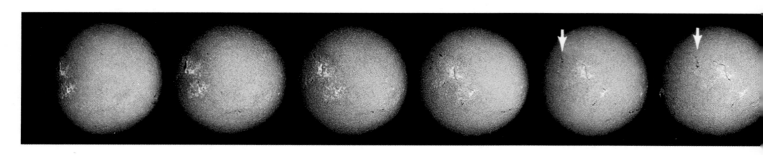

lower atmosphere, forming a sunspot pair. The general east–west orientation of the underlying solar field accounts for the observed polarities of the resulting sunspot pairs in each hemisphere.

## The Solar Cycle

Sunspots are not steady. Most change their size and shape, and all come and go. Figure 16.19 shows a time sequence in which a number of spots varied—sometimes growing, sometimes dissipating—over a period of several days. Individual spots may last anywhere from 1 to 100 days; a large group typically lasts 50 days. Not only do sunspots come and go with time, but their numbers and distribution across the face of the Sun also change fairly regularly. Centuries of observations have established a clear **sunspot cycle.** Figure 16.20(a) shows the number of sunspots observed each year during the 20th century. The average number of spots reaches a maximum every 11 or so years and then falls off almost to zero before the cycle begins afresh.

The latitudes at which sunspots appear vary as the sunspot cycle progresses. Individual sunspots do not move up or down in latitude, but new spots appear closer to the equator as older ones at higher latitudes fade away. Figure 16.20(b) is a plot of observed sunspot latitude as a function of time. At the start of each cycle, at *solar minimum*, only a few spots are seen, and these are generally confined to two narrow zones about 25° to 30° north and south of the solar equator. Approximately four years into the cycle, around *solar maximum*, the number of spots has increased markedly, and they are found within about 15° to 20° of the equator. Finally, by the end of the cycle, at solar

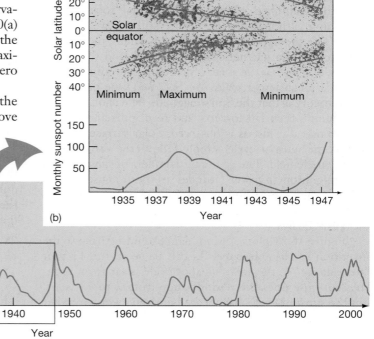

(b)

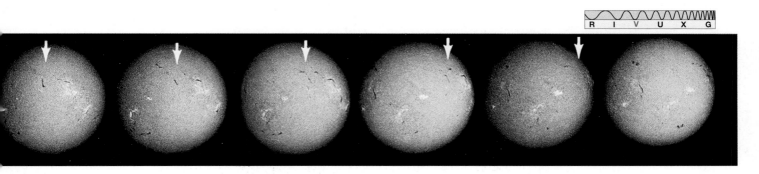

(a)

▲ **FIGURE 16.20 Sunspot Cycle** (a) Monthly number of sunspots throughout the 20th century, showing yearly averages of the data to make long-term trends more evident. The (roughly) 11-year solar cycle is clearly visible. At the time of minimum solar activity, hardly any sunspots are seen. About 4 years later, at maximum solar activity, about 100 to 200 spots are observed per month. The most recent solar maximum occurred in 2001. (b) Sunspots cluster at high latitudes when solar activity is at a minimum. They appear at lower and lower latitudes as the number of sunspots peaks. They are again prominent near the Sun's equator as solar minimum is approached once more. The blue lines in the upper plot indicate how the "average" sunspot latitude varies over the course of the cycle.

minimum, the number has fallen again, and most sunspots lie within about 10° of the solar equator. The beginning of each new cycle appears to overlap the end of the last.

Complicating this picture further, the 11-year sunspot cycle is actually only half of a longer 22-year **solar cycle.** During the first 11 years of the cycle, the leading spots of all the pairs in the northern hemisphere have the same polarity, while spots in the southern hemisphere have the opposite polarity (Figure 16.17). These polarities then reverse their signs for the next 11 years, so the full solar cycle takes 22 years.

Astronomers think that the Sun's magnetic field is both generated and amplified by the constant stretching, twisting, and folding of magnetic field lines that results from the combined effects of differential rotation and convection, although the details are still not well understood. The theory is similar to the "dynamo" theory that accounts for the magnetic fields of Earth and the jovian planets, except that the solar dynamo operates much faster and on a much larger scale. ∞ (Sec. 7.5) One prediction of this theory is that the Sun's magnetic field should rise to a maximum, then fall to zero, and reverse itself, more or less periodically, just as is observed. Solar surface activity, such as the sunspot cycle, simply follows the variations in the magnetic field. The changing numbers of sunspots and their migration to lower latitudes are both consequences of the strengthening and eventual decay of the field lines as they become more and more tightly wrapped around the solar equator.

Figure 16.21 plots sunspot data extending back to the invention of the telescope. As can be seen, the 11-year "periodicity" of the solar sunspot cycle is far from regular. Not only does the period range from 7 to 15 years, but the sunspot cycle disappeared entirely over a number of years in the relatively recent past. The lengthy period of solar inactivity that extended from 1645 to 1715 is called the **Maunder minimum,** after the British astronomer who drew attention to these historical records. The corona was apparently also less prominent during

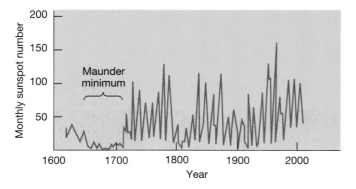

▲ **FIGURE 16.21 Maunder Minimum** Average number of sunspots occurring each month over the past four centuries. Note the absence of spots during the late 17th century.

total solar eclipses around that time, and Earth aurorae were sparse throughout the late 17th century. Lacking a complete understanding of the solar cycle, we cannot easily explain how it could shut down entirely. Most astronomers suspect changes in the Sun's convection zone or rotation pattern, but the specific causes of the Sun's century-long variations, as well as the details of the connection between solar activity and Earth's climate, remain a mystery (see *Discovery 16-2*).

**CONCEPT CHECK**

✔ What do observations of sunspot polarities tell us about the solar magnetic field?

## 16.5 The Active Sun

Most of the Sun's luminosity results from continuous emission from the photosphere. However, superimposed on this steady, predictable aspect of our star's energy output is a much more irregular component, characterized by explosive and unpredictable surface activity. Solar activity contributes little to the Sun's total luminosity and probably has no significant bearing on the evolution of the Sun, but it does affect us here on Earth. The size and duration of coronal holes are strongly influenced by the level of solar activity. Hence, so is the strength of the solar wind, and that in turn directly affects Earth's magnetosphere.

### Active Regions

The photosphere surrounding a pair or group of sunspots can be a violent place, sometimes erupting explosively, spewing forth large quantities of energetic particles into the corona. The sites of these energetic events are known as **active regions.** Most groups of sunspots have active regions associated with them. Like all other aspects of solar activity, these phenomena tend to follow the solar cycle and are most frequent and violent around the time of solar maximum.

Figure 16.22 shows two large solar **prominences**—loops or sheets of glowing gas ejected from active regions on the solar surface, moving through the inner parts of the corona under the influence of the Sun's magnetic field. Magnetic instabilities in the strong fields found in and near sunspot groups may cause the prominences, although the details are not fully understood. The arching magnetic field lines in and around the active region are also easily seen (see also Figure 16.17b). The rapidly changing structure of the field lines and the fact that they can quickly transport mass and energy from one part of the solar surface to another, possibly tens of thousands of kilometers away, make the theoretical study of active regions an extraordinarily difficult task.

*Quiescent prominences* persist for days or even weeks, hovering high above the photosphere, suspended by the Sun's magnetic field. *Active prominences* come and go much more erratically, changing their appearance in a matter of hours or surging up from the solar photosphere and then immediately falling back on themselves. A typical solar prominence measures some 100,000 km in extent, nearly 10 times the diameter of planet Earth. Prominences as large as the one shown in Figure 16.22(a) (which traversed almost half a million kilometers of the solar surface) are less common and usually appear only at times of greatest solar activity. The largest prominences can release up to $10^{25}$ joules of energy, counting both particles and radiation—not much compared with the total solar luminosity of $4 \times 10^{26}$ W, but still enormous by terrestrial standards. (All the power plants on Earth would take a billion years to produce that much energy.)

**Flares** are another type of solar activity observed low in the Sun's atmosphere near active regions. Also the result of magnetic instabilities, flares, like that shown in Figure 16.23, are even more violent (and even less well understood) than prominences. They often flash across a region of the Sun in minutes, releasing enormous amounts of energy as they go. Space-based observations indicate that X-ray and ultraviolet emissions are especially intense in the extremely compact hearts of flares, where temperatures can reach 100 million K.

So energetic are these cataclysmic explosions that some researchers have likened flares to bombs exploding in the lower regions of the Sun's atmosphere. A major flare can release as much energy as the largest prominences, but in a matter of minutes or hours rather than days or weeks. Unlike the gas that makes up the characteristic loop of a prominence, the particles produced by a flare are so energetic that the Sun's magnetic field is unable to hold them and shepherd them back to the

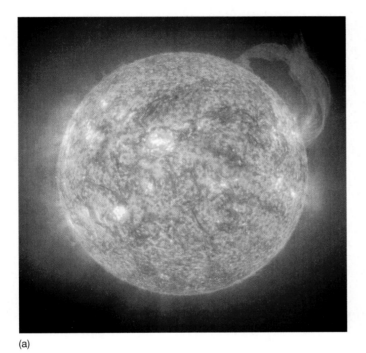

(a)

(b)

◀ **FIGURE 16.22 Solar Prominences** (a) This particularly large solar prominence was observed by ultraviolet detectors aboard the *SOHO* spacecraft in June 2002. (b) Like a phoenix rising from the solar surface, this filament of hot gas measures more than 100,000 km in length. Earth could easily fit between its outstretched "arms." Dark regions in this *TRACE* image have temperatures less than 20,000 K; the brightest regions are about 1 million K. The ionized gas follows the solar magnetic field lines away from the Sun. Most of the gas will subsequently cool and fall back into the photosphere. *(NASA)*

R I V U X G

# DISCOVERY 16-2

## Solar–Terrestrial Relations

Our Sun has often been worshipped as a god with power over human destinies. Obviously, the steady stream of solar energy arriving at our planet every day is essential to our lives, but over the past century there have also been repeated claims of a correlation between the Sun's activity and Earth's weather. Only recently, however, has the subject become scientifically respectable—that is, more natural than supernatural.

In fact, there do seem to be some correlations between the 22-year solar cycle (two sunspot cycles with oppositely directed magnetic fields) and periods of climatic dryness here on Earth. For example, near the start of the past eight cycles, there have been droughts in North America—at least within the middle and western plains from South Dakota to New Mexico. Another possible Sun–Earth connection is a link between solar activity and increased atmospheric circulation on our planet. As circulation increases, terrestrial storm systems deepen, extend over wider ranges of latitude, and carry more moisture. The relationship is complex and the subject controversial, because no one has yet shown any physical mechanism (other than the Sun's heat, which does not vary much during the solar cycle) that would allow solar activity to stir our terrestrial atmosphere. Without a better understanding of the physical mechanism involved, none of these effects can be incorporated into our weather-forecasting models.

Solar activity may also influence long-term climate on Earth. For example, the Maunder minimum (see Section 16.6) seems to correspond fairly well with the coldest years of the so-called Little Ice Age that chilled northern Europe and North America during the late 1600s. The accompanying "winter" scene actually captured one summer season in 17th-century Holland. How the active Sun and its abundance of sunspots may affect Earth's climate is a frontier problem in terrestrial climatology.

Measurements of the solar constant made over the past two decades indicate that the Sun's energy output varies with the solar cycle. Paradoxically, the Sun's luminosity is greatest when many dark sunspots cover its surface! Thus, the Maunder minimum does correspond to an extended period of lower-than-average solar emission. However, recent observed changes in the Sun's luminosity have been small—no more than 0.2 or 0.3 percent. It is not known by how much, if at all, the Sun's output declined during the Maunder minimum, nor how large a change would be needed to account for the alterations in climate that occurred.

One correlation that is definitely established, and also better understood, is that between solar activity and geomagnetic disturbances at Earth. The extra radiation and particles thrown off by flares or coronal mass ejections impinge on Earth's environment, overloading the Van Allen belts, thereby causing brilliant auroras in our atmosphere and degrading our communication networks. We are only beginning to understand how the radiation and particles emitted by solar phenomena also interfere with terrestrial radars, power networks, and other technological equipment. Some power outages on Earth are actually caused, not by increased customer demand or malfunctioning equipment, but by weather on the Sun!

We cannot yet predict just when and where solar flares or coronal mass ejections will occur. However, it would certainly be to our advantage to be able to do so, as that aspect of the active Sun affects our lives. This is a highly fertile area of astronomical research and one with clear terrestrial applications.

*(Rijksmuseum, Amsterdam, Holland/The Bridgman Art Library)*

surface. Instead, the particles are simply blasted into space by the violence of the explosion.

Figure 16.24 shows a **coronal mass ejection** from the Sun. Sometimes (but not always) associated with flares and prominences, these phenomena are giant magnetic "bubbles" of ionized gas that separate from the rest of the solar atmosphere and escape into interplanetary space. Such ejections occur about once per week at times of sunspot minimum, but up to two or three times per day at solar maximum. Carrying an enormous amount of energy, they can—if their fields are properly oriented—merge with Earth's magnetic field via a process known as *reconnection*, dumping some of their energy into the magnetosphere and potentially causing widespread communications and power disruptions on our planet (Figure 16.24b; see also *Discovery 16-2*).

## The Sun in X Rays

Unlike the 5800K photosphere, which emits most strongly in the visible part of the electromagnetic spectrum, the hot coronal gas radiates at much higher frequencies—primarily in the X-ray range. ∞ (Sec. 3.4) For this reason, X-ray telescopes have become important tools in the study of the solar corona. Figure 16.25(a) shows several X-ray images of the Sun. The full corona extends well beyond the regions shown, but the density of coronal particles emitting the radiation diminishes rapidly with distance from the Sun. The intensity of X-ray radiation farther out is too dim to be seen here.

In the mid-1970s, instruments aboard NASA's *Skylab* space station revealed that the solar wind escapes mostly through solar "windows" called **coronal holes.** The dark area moving from left to right in Figure 16.25(a), which

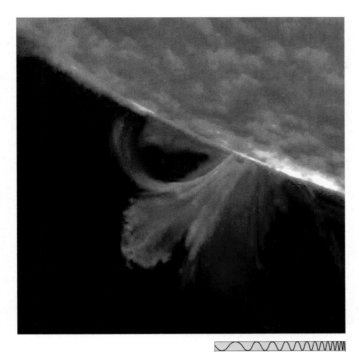

▲ FIGURE 16.23 **Solar Flare** Much more violent than a prominence, a solar flare is an explosion on the Sun's surface that sweeps across an active region in a matter of minutes, accelerating solar material to high speeds and blasting it into space. *(USAF)*

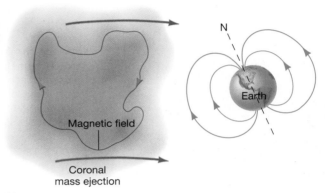

(b)

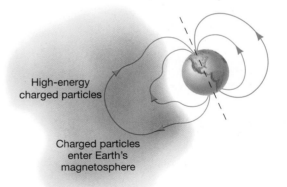

(c)

(a)

◄ FIGURE 16.24 **Coronal Mass Ejection** (a) A few times per week, on average, a giant magnetized "bubble" of solar material detaches itself from the Sun and rapidly escapes into space, as shown in this *SOHO* image taken in 2002. The circles are artifacts of an imaging system designed to block out the light from the Sun itself and exaggerate faint features at larger radii. (b) Should a coronal mass ejection encounter Earth with its magnetic field oriented opposite to our own, as illustrated, the field lines can join together as in part (c), allowing high-energy particles to enter and possibly severely disrupt our planet's magnetosphere. By contrast, if the fields are oriented differently, the coronal mass ejection can slide by Earth with little effect. *(NASA/ESA)*

ANIMATION/VIDEO  Coronal Mass Ejections

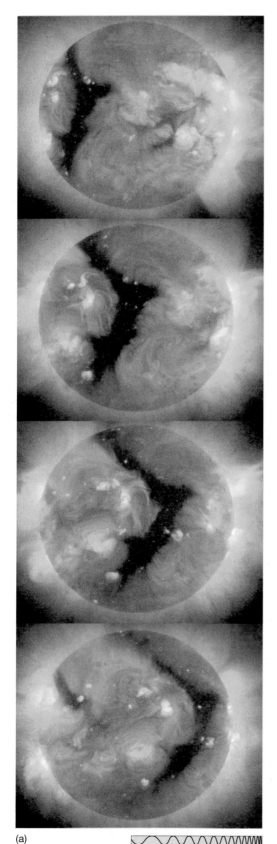

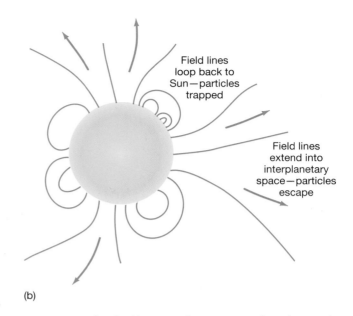

(b)

◀ **FIGURE 16.25 Coronal Hole** (a) Images of X-ray emission from the Sun observed by the *Yohkoh* satellite. These frames were taken at roughly 2-day intervals, starting at the top. Note the dark, V-shaped coronal hole traveling from left to right, where the X-ray observations outline in dramatic detail the abnormally thin regions through which the high-speed solar wind streams forth. (b) Charged particles follow magnetic field lines that compete with gravity. When the field is trapped and loops back toward the photosphere, the particles are also trapped; otherwise, they can escape as part of the solar wind. (*ISAS/Lockheed Martin*)

shows more recent data from the Japanese *Yohkoh* X-ray solar observatory, represents a coronal hole. Not really holes, such structures are simply deficient in matter—vast regions of the Sun's atmosphere where the density is about 10 times lower than the already tenuous, normal corona. Note that the underlying solar photosphere looks black in these images because it is far too cool to emit X rays in any significant quantity.

Coronal holes are lacking in matter because the gas there is able to stream freely into space at high speeds, driven by disturbances in the Sun's atmosphere and magnetic field. Figure 16.25(b) illustrates how, in coronal holes, the solar magnetic field lines extend from the surface far out into interplanetary space. Because charged particles tend to follow the field lines, they can escape, particularly from the Sun's polar regions, according to findings from *SOHO* and NASA's *Ulysses* spacecraft, which flew high above the ecliptic plane to explore the Sun's polar regions. In other regions of the corona, the solar magnetic field lines stay close to the Sun, keeping charged particles near the surface and inhibiting the outward flow of the solar wind (just as Earth's magnetic field tends to prevent the incoming solar wind from striking Earth), so the density remains relatively high. Because of the "open" field structure in coronal holes, flares and other magnetic activity (which, as we have seen, are related to magnetic loops near the solar photosphere) tend to be suppressed there.

The largest coronal holes, like that shown in Figure 16.25(a), can be hundreds of thousands of kilometers across and may survive for many months. Structures of this size are seen only a few times per decade. Smaller holes—perhaps only a few tens of thousand kilometers in size—are much more common, appearing every few hours.

Coronal holes appear to be an integral part of the process by which the Sun's large-scale field reverses and replenishes itself over the course of the solar cycle. Long-lived holes persist at the Sun's polar regions over much of the magnetic cycle, and the numbers and locations of other holes appear to change in step with solar activity. However, like many aspects of the solar magnetic field, the structure and evolution of coronal holes are not fully understood; they are currently the subject of intense research.

## The Changing Solar Corona

The solar corona varies with the sunspot cycle. The photograph of the corona in Figure 16.13 shows the quiet Sun, at sunspot minimum. At such times, the corona is fairly regular in appearance and seems to surround the Sun more or less uniformly. Compare that image with Figure 16.26, which was taken in 1994 near a peak in the sunspot cycle. The active corona is much more irregular in appearance and extends farther from the solar surface. The "streamers" of coronal material pointing away from the Sun are characteristic of this phase.

Astronomers think that the corona is heated primarily by solar surface activity, which can inject large amounts of energy into the upper solar atmosphere. *SOHO* observations have implicated a "magnetic carpet" as the source of much of the heating. Using the craft's Doppler imager, astronomers have observed small magnetic loops perpetually sprouting up and then disappearing all over the solar photosphere—resembling a kind of celestial shag carpet. The whole carpet replenishes itself rapidly, roughly every 40 hours or so, but the loops don't just sink back beneath the surface. Rather, they tend to break open, dumping vast amounts of energy into the lower solar atmosphere. Along with the myriad spicules, these small-scale magnetic loops probably provide most of the energy needed to heat the corona. In addition, more extensive disturbances often move through the corona above an active site in the photosphere, distributing the energy throughout the coronal gas. Given this connection, it is hardly surprising that both the appearance of the corona and the strength of the solar wind are closely correlated with the solar cycle.

## CONCEPT CHECK

✔ Why is solar activity important to life on Earth?

◀ FIGURE 16.26 **Active Corona** Photograph of the solar corona during the July 1994 eclipse, near the peak of the sunspot cycle. At these times, the corona is much less regular and much more extended than at sunspot minimum (compare to Figure 16.13). Astronomers think that coronal heating is caused by surface activity on the Sun. The changing shape and size of the corona is the direct result of variations in prominence and flare activity over the course of the solar cycle. (*NCAR High Altitude Observatory*)

R  I  V  U  X  G

## 16.6 The Heart of the Sun

What powers the Sun? What forces are at work in the Sun's core to produce its enormous luminosity? By what process does the Sun shine, day after day, year after year, eon after eon? Answers to these questions are central to all astronomy. Without them, we can understand neither the physical existence of stars and galaxies in the universe nor the biological existence of life on Earth.

### Solar Energy Production

In round numbers, the Sun's luminosity is $4 \times 10^{26}$ W and its mass is $2 \times 10^{30}$ kg. We can quantify how efficiently the Sun generates energy by dividing the solar luminosity by the solar mass:

$$\frac{\text{solar luminosity}}{\text{solar mass}} = 2 \times 10^{-4}\,\text{W/kg}.$$

This simply means that, on average, every kilogram of solar material yields about 0.2 milliwatt of energy—0.0002 joule (J) of energy every second.

This is not much energy—a piece of burning wood generates about a million times more energy per unit mass per unit time than does our Sun, so the equivalent solar luminosity could (in principle) be created by a pile of burning logs comparable in mass to planet Earth. But there is one very important difference: The logs cannot continue to burn at this rate for billions of years.

To appreciate the magnitude of the energy generated by our Sun, we must consider not the ratio of the solar luminosity to the solar mass, but instead the total amount of energy generated by each gram of solar matter *over the entire lifetime of the Sun as a star.* This is easy to do. We simply multiply the rate at which the Sun generates energy by the age of the Sun, about 5 billion years. We obtain a value of $3 \times 10^{13}$ J/kg. This is the average amount of energy radiated by every kilogram of solar material since the Sun formed. It represents a *minimum* value for the total energy radiated by the Sun, for more energy will be needed for every additional day the Sun shines. Should the Sun endure for another 5 billion years (as is predicted by theory), we would have to double this value.

Either way, this energy-to-mass ratio is very large. At least 60 trillion joules (on average) of energy must arise from every kilogram of solar matter to power the Sun throughout its lifetime. But the Sun's generation of energy is not explosive, releasing large amounts of energy in a short period. Instead, it is slow and steady, providing a uniform and long-lived rate of energy production. Only one known energy-generation mechanism can conceivably power the Sun in this way: **nuclear fusion**—the combining of light nuclei into heavier ones.

### Nuclear Fusion

We can represent a typical fusion reaction symbolically as

$$\text{nucleus 1} + \text{nucleus 2} \rightarrow \text{nucleus 3} + \text{energy}.$$

For powering the Sun and other stars, the most important piece of this equation is the energy produced.

The essential point here is that, during a fusion reaction, the total mass *decreases*—the mass of nucleus 3 is less than the combined masses of nuclei 1 and 2. Where does this mass go? It is converted into energy in accordance with Einstein's famous equation of **mass-energy equivalence**

$$E = mc^2,$$

or

$$\text{energy} = \text{mass} \times (\text{speed of light})^2.$$

This equation expresses the discovery, made by Albert Einstein at the beginning of the 20th century, that matter and energy are interchangeable—one can be converted into the other. To determine the amount of energy corresponding to a given mass, simply multiply the mass by the square of the speed of light ($c$ in the equation). For example, the energy equivalent of 1 kg of matter is $1 \times (3 \times 10^8)^2$, or $9 \times 10^{16}$ J. The speed of light is so large that even small amounts of mass translate into enormous amounts of energy.

The production of energy by a nuclear fusion reaction is an example of the **law of conservation of mass and energy,** which states that the *sum* of mass and energy (suitably converted into the same units, using Einstein's equation) must always remain *constant* during any physical process. *There are no known exceptions.* According to this law, an object can literally disappear, provided that some energy appears in its place. If magicians really made rabbits disappear, the result would be a flash of energy equaling the product of the rabbit's mass and the square of the speed of light—enough to destroy the magician, everyone in the audience, and probably all of the surrounding state as well! In the case of fusion reactions in the solar core, the energy is produced primarily in the form of electromagnetic radiation. The light we see coming from the Sun means that the Sun's mass must be slowly, but steadily, decreasing with time.

### Charged Particle Interactions

All atomic nuclei are positively charged, so they repel one another. Furthermore, by the inverse-square law, the closer two nuclei come to one another, the greater is the repulsive force between them (Figure 16.27a). ∞ (Sec. 3.2) How, then, do nuclei—two protons, say—ever manage to fuse into anything heavier? The answer is that if they collide at high enough speeds, one proton can momentarily plow deep into the other, eventually

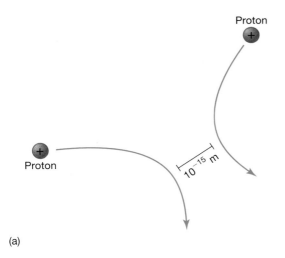

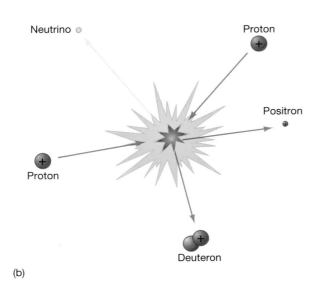

▲ **FIGURE 16.27 Proton Interactions** (a) Since like charges repel, two low-speed protons veer away from one another, never coming close enough for fusion to occur. (b) Sufficiently high-speed protons may succeed in overcoming their mutual repulsion, approaching close enough for the strong force to bind them together—in which case they collide violently, triggering nuclear fusion that ultimately powers the Sun.

coming within the exceedingly short range of the **strong nuclear force,** which binds nuclei together (see *More Precisely 16-1*). At distances less than about $10^{-15}$ m, the attraction of the nuclear force overwhelms the electromagnetic repulsion, and fusion occurs. Speeds in excess of a few hundred kilometers per second, corresponding to a gas temperature of $10^7$ K or more, are needed to slam protons together fast enough to initiate fusion. Such conditions are found in the core of the Sun and at the centers of all stars.

The fusion of two protons is illustrated schematically in Figure 16.27(b). In effect, one of the protons turns into a neutron, creating new particles in the process, and combines with the other proton to form a **deuteron,** the nucleus of a special form of hydrogen called *deuterium.* Deuterium (also referred to as "heavy hydrogen") differs from ordinary hydrogen by virtue of an extra neutron in its nucleus. We can represent the reaction as follows:

proton + proton → deuteron + positron + neutrino.

The **positron** in this reaction is a positively charged electron. Its properties are identical to those of a normal, negatively charged electron, except for the positive charge. Scientists call the electron and the positron a "matter–antimatter pair"—the positron is said to be the *antiparticle* of the electron. The newly created positrons find themselves in the midst of a sea of electrons with which they interact immediately and violently. The particles and antiparticles annihilate (destroy) one another, producing pure energy in the form of gamma-ray photons.

The final product of the reaction is a particle known as a **neutrino,** a word derived from the Italian for "little neutral one." Neutrinos carry no electrical charge and are of very low mass—at most 1/100,000 the mass of an electron, which itself has only 1/2000 the mass of a proton. (The exact mass of the neutrino remains uncertain, although experimental evidence strongly suggests that it is not zero.) Neutrinos move at almost the speed of light and interact with hardly anything. They can penetrate, without stopping, several light-years of lead (a very dense material, widely used in terrestrial laboratories as an effective shield against radiation). Their interactions with matter are governed by the **weak nuclear force,** described in *More Precisely 16-1.* Despite their elusiveness, neutrinos can be detected with carefully constructed instruments. In the final section of this chapter we discuss some rudimentary neutrino "telescopes" and the important contribution they have made to solar astronomy.

Nuclei such as normal hydrogen and deuterium, containing the same number of protons, but different numbers of neutrons, represent different forms of the same element—they are known as **isotopes** of that element. Usually, there are about as many neutrons in a nucleus as protons, but the exact number of neutrons can vary, and most elements exist in a number of isotopic forms. To avoid confusion when talking about isotopes of the same element, nuclear physicists attach a number to the symbol representing the element. This number indicates the total number of particles (protons plus neutrons) in the nucleus of an atom of the element. Thus, ordinary hydrogen is denoted by $^1$H, deuterium by $^2$H. Normal helium (two protons plus two neutrons) is $^4$He (also referred to as helium-4), and so on. We will adopt this convention for the rest of the book.

$$4 \, (^1\text{H}) \rightarrow \, ^4\text{He} + 2 \text{ neutrinos} + \text{energy}.$$

As discussed in more detail in *More Precisely 16-2*, to fuel the Sun's present energy output, hydrogen must be fused into helium in the core at a rate of 600 million tons per second—a lot of mass, but only a tiny fraction of the total amount available. The Sun will be able to sustain this rate of core burning for about another 5 billion years (see Chapter 20).

luminosity. Note how, at each stage of the chain, more massive, complex nuclei are created from simpler, lighter ones. We will see in Chapters 20 and 21 that not all stars are powered by hydrogen fusion. Nevertheless, *nuclear* fusion—the slow, but steady, transformation of light elements into heavier ones, creating energy in the process—is responsible for virtually *all* of the starlight we see.

## MORE PRECISELY 16-1

### Fundamental Forces

Our studies of nuclear reactions have uncovered new ways in

weak nuclear force governs the emission of radiation from some radioactive atoms; the emission of a neutrino during the first stage of the proton-proton reaction (Figure 16.27) is the

## MORE PRECISELY 16-2

### Energy Generation in the Proton–Proton Chain

Let's look in a little more detail at the energy produced by fusion in the solar core and compare it with the energy needed to account for the Sun's luminosity. Using the notation presented in the text, the proton–proton chain may be compactly described by the following reactions:

proton fusion:    $^1H + {}^1H \rightarrow {}^2H +$ positron + neutrino.    (I)

deuterium fusion: $^2H + {}^1H \rightarrow {}^3He +$ energy.    (II)

helium-3 fusion:  $^3He + {}^3He \rightarrow {}^4He + {}^1H + {}^1H +$ energy. (III)

As discussed in the text and illustrated below, the net effect of the fusion process is that four protons combine to produce a nucleus of helium-4, in the process creating two neutrinos and two positrons (which are quickly converted into energy by annihilation with electrons).

We can calculate the total amount of energy released by accounting carefully for the total masses of the nuclei involved and applying Einstein's famous formula $E = mc^2$. This

in turn allows us to relate the Sun's total luminosity to the consumption of hydrogen fuel in the core. Careful laboratory experiments have determined the masses of all the particles involved in the above reaction: The total mass of the protons is $6.6943 \times 10^{-27}$ kg, the mass of the helium-4 nucleus is $6.6466 \times 10^{-27}$ kg, and the neutrinos are virtually massless. We omit the positrons here—their masses will end up being counted as part of the total energy released. The difference between the total mass of the four protons and that of the final helium-4 nucleus, $0.0477 \times 10^{-27}$ kg, is not great, but it is easily measurable.

Multiplying the vanished mass by the square of the speed of light yields $0.0477 \times 10^{-27}$ kg $\times (3.00 \times 10^8$ m/s$)^2 = 4.28 \times 10^{-12}$ J. This is the energy produced in the form of radiation when $6.69 \times 10^{-27}$ kg (the rounded-off mass of the four protons) of hydrogen fuses to helium. It follows that fusion of 1 kg of hydrogen generates $4.28 \times 10^{-12}/6.69 \times 10^{-27} = 6.40 \times 10^{14}$ J. Put another way, the process converts about 0.71 percent of the original mass into energy. A negligible fraction (actually, about 2 percent) of this energy is carried away by the neutrinos. The rest appears in the form of gamma rays and ultimately is radiated away from the solar photosphere—that is, it becomes the solar luminosity.

Thus we have established a direct connection between the Sun's energy output and the consumption of hydrogen in the core. The Sun's luminosity of $3.84 \times 10^{26}$ W (see Table 16.1), or $3.84 \times 10^{26}$ J/s (joules per second), implies a mass consumption rate of $3.84 \times 10^{26}$ J/s$/6.40 \times 10^{14}$ J/kg $= 6.00 \times 10^{11}$ kg/s—600 million tons of hydrogen every second (1 ton = 1000 kg). A mass of 600 million tons sounds like a lot—the mass of a small mountain—but it represents only a few million million millionths of the total mass of the Sun. Put another way, of this 600 million tons, roughly 600 million tons/s $\times 0.0071 = 4.3$ million tons per second of solar matter is converted into radiation—comparable to the mass carried away by the solar wind. Our parent star will be able to sustain this loss rate for a very long time.

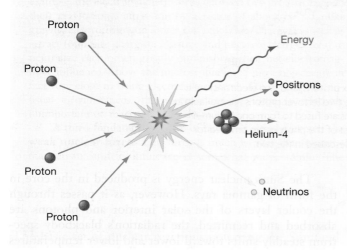

### CONCEPT CHECK

✔ Why does the fact that we see sunlight imply that the Sun's mass is slowly decreasing?

## 16.7  Observations of Solar Neutrinos

Theorists are quite sure that the proton–proton chain operates in the Sun. However, because the gamma-ray energy created in the solar core chain has been transformed into visible and infrared radiation by the time it emerges from the photosphere, astronomers have no direct electromagnetic evidence of the nuclear reactions in the solar core. Fortunately, there are other ways of probing the Sun's deep interior.

### Neutrino Detectors

Electromagnetic radiation is the source of virtually all the information presented in this text. However, the *neutrinos* created in the proton–proton chain are our best bet for learning about conditions in the heart of the Sun. Core neutrinos travel cleanly out of the Sun, interacting with virtually nothing, and escape into space a few seconds after being created. They arrive at Earth's orbit about eight minutes later. Of course, the fact that they can pass through the entire Sun without interacting also makes neutrinos difficult to detect here on Earth! Nevertheless, with some knowledge of neutrino physics, it is possible to construct neutrino detectors.

Over the past four decades, several major experiments have been designed to detect solar neutrinos reaching Earth's surface. They fall into two basic categories:

- In a *neutrino capture detector*, the goal is to measure the change that occurs when a neutrino is absorbed by the nucleus of an atom, converting a neutron into a proton and thereby changing the atom into a new element. The elements chlorine (which is converted into argon) and gallium (converted into germanium) happen to be somewhat more likely than most to interact with neutrinos and so have been favored in these instruments. The newly created nucleus is radioactive, and the detection of the radiation produced by its decay signals the neutrino capture.

- *Neutrino scattering detectors* use a somewhat different approach. They also look for telltale light resulting from the passage of a neutrino through the detector, but in this case the light is produced when a high-energy neutrino occasionally collides with ("scatters" from) an electron in a water molecule, accelerating the electron to almost the speed of light. Large photomultiplier tubes (light-amplification devices) detect the resultant faint glow that betrays the neutrino's passage.

In either case, the probability of a given neutrino actually interacting with matter in the detector is extraordinarily small. For example, the chlorine experiment described below "saw" just 1 in $10^{16}$ of the neutrinos passing through the apparatus. Large amounts (tons) of target material and long-duration experiments (months or even years) are needed to obtain accurate measurements.

The pioneering neutrino capture experiment (which launched the entire field of observational neutrino astronomy) was built in the late 1960s. Sited 1.5 km below ground near the bottom of the Homestake gold mine in South Dakota, to shield it from interference due to cosmic rays and other sources, most of which cannot penetrate Earth's crust to such depths, it used some 300 tons of a chlorine-containing chemical—dry-cleaning fluid—as a detector. It ran until 1993, detecting, on average, about 2 neutrinos per week, roughly one-third the predicted value. During the 1990s, two gallium-based experiments—the Soviet–American Gallium Experiment (SAGE) and the U.S.–European GALLEX collaboration—came online. Gallium is a lot more expensive than chlorine, but it is much more likely to interact with neutrinos of the sort expected from the Sun. Both experiments found a roughly 50 percent shortfall in the numbers of neutrinos detected, relative to theoretical expectations.

Figure 16.29(a) shows a large neutrino-scattering detector located in Kamioka, Japan, constructed in the 1980s and upgraded in the late 1990s to greatly increase its sensitivity. During several years of operation before a freak accident in 2001 caused one of its detectors to implode, creating a shock wave and chain reaction that destroyed several thousand more detectors, the "Super Kamiokande" device, like the gallium experiments, detected just half the number of neutrinos predicted by theory. Repairs were completed in 2006 and it is now up and running again. Figure 16.29(b) shows the SNO (Sudbury Neutrino Observatory) detector in Ontario, Canada, which is similar in design to the Super Kamiokande instrument. Its important contribution to the solar neutrino puzzle is described below.

Interestingly, by careful measurement of the detected light pattern, these scattering detectors can determine the *direction* of motion of the electron producing the radiation and hence estimate the direction from which the original neutrino came. In other words, they can function as *neutrino telescopes*, albeit of very low resolution. The inset to Figure 16.29(a) shows an "image" of the Sun produced by the Kamiokande instrument. As technology improves, such directional detectors will likely come to play a central role in neutrino astronomy.

## The Solar Neutrino Problem

The designs of the detectors just discussed differ widely. They are sensitive to neutrinos of very different energies, and they disagree somewhat on the details of their findings, but they all agree on one very important point: The number of solar neutrinos detected on Earth is substantially less—by some 50 to 70 percent—than the prediction of the standard solar model. This discrepancy is known as the **solar neutrino problem.**

For a time, it seemed that theorists might have some "wiggle room" in their attempts to explain this disagreement. The Homestake and Kamioka experiments were actually sensitive not to neutrinos produced in reaction (I) in Section 16.6 (and in Figure 16.28)—the initial step in the proton–proton chain—but to those created by a much less probable sequence of events, occurring only about 0.25 percent of the time. Perhaps the problem lay in our understanding of that less likely branch, not in the main proton–proton chain itself. However, that possibility was effectively cut off by SAGE and GALLEX, which did detect neutrinos produced by reaction (I), providing a much more direct probe of energy generation in the solar core.

Thus, by the end of the 1990s, the inescapable conclusion was that, although solar neutrinos were observed (and in fact their measured energies did lie in the range predicted by the standard solar model), there was a real discrepancy between the Sun's theoretical neutrino output and the number of neutrinos actually detected on Earth.

## Neutrino Physics and the Standard Model

How can we explain the discrepancy posed by the solar neutrino problem? The broad agreement among several independent, well-designed, and thoroughly tested experiments implies to most scientists that experimental

(a)

(b)

◄ **FIGURE 16.29 Neutrino Telescopes** (a) This swimming pool-sized "neutrino telescope" is buried beneath a mountain near Tokyo, Japan. Called Super Kamiokande, it is filled (in operation) with 50,000 tons of purified water, and contains 13,000 individual light detectors (some shown here being inspected by technicians) to sense the telltale signature—a brief burst of light—of a neutrino passing through the apparatus. The inset shows an "image" of the Sun made by the Kamiokande instrument. Current neutrino detectors have very low resolution, and the field of view here is about 90° across. The small black dot indicates the actual size of the Sun. (b) The Sudbury Neutrino Observatory (SNO), situated 2 km underground in Ontario, Canada. The SNO detector is similar in design to the Kamiokande device, but by using "heavy" water (with hydrogen replaced by deuterium) instead of ordinary water, and adding two tons of salt, it also becomes sensitive to other neutrino types. The device contains 10,000 light-sensitive detectors arranged on the inside of the large sphere shown here. *(ICRR, SNO)*

reducing the theoretical number of neutrinos by postulating a lower temperature in the solar core. However, the nuclear reactions described earlier are just too well known, and the agreement between the standard solar model and helioseismological observations is far too close to permit conditions in the core to deviate much from the predictions of the model.

Instead, the answer involves the properties of the neutrinos themselves and has caused scientists to rethink some very fundamental concepts in particle physics. If neutrinos have even a minute amount of mass, theory indicates that it should be possible for them to change their properties—even to transform into other particles—during their 8-minute flight from the solar core to Earth through a process known as **neutrino oscillations.** In this picture, neutrinos are produced in the Sun at the rate required by the standard solar model, but some turn into something else—actually, other types of neutrinos—on their way to Earth and hence go undetected in the experiments just described. (In the jargon of the field, the neutrinos are said to "oscillate" into other particles.)

In 1998 the Japanese group operating the Super Kamiokande detector (Figure 16.29a) reported the first experimental evidence of neutrino oscillations (and hence of nonzero neutrino masses), although the observed oscillations did not involve neutrinos of the type produced in the Sun. In 2001, measurements made by SNO (Figure 16.29b) revealed strong evidence for the "other" neutrinos into which the Sun's neutrinos have been transformed. As mentioned earlier, the SNO detector is similar in design to the Kamioka device, but, by using "heavy" water (with hydrogen replaced by

error is not the cause. In that case, there are really only two possibilities: Either *solar neutrinos are not produced as frequently as we think*, or *not all of them make it to Earth*.

Theorists have long considered it unlikely that the resolution of the solar neutrino problem lies in the physics of the Sun's interior. For instance, we might imagine

deuterium) instead of ordinary water (H₂O), it also becomes sensitive to other neutrino species.

Further SNO observations with a modified detector, reported in 2002, confirmed these result. The *total* numbers of neutrinos observed are completely consistent with the standard solar model. The solar neutrino problem has at last been solved—and neutrino astronomy can claim its first major triumph! The lead scientists of the

Homestake and Kamioka experiments received the 2002 Nobel Prize for their work.

## CONCEPT CHECK

✔ Why do neutrinos give us direct information about conditions in the solar core, whereas electromagnetic radiation does not?

# CHAPTER REVIEW

## Summary

**1** Our Sun is a **star (p. 416)**, a glowing ball of gas held together by its own gravity and powered by nuclear fusion at its center. The **photosphere (p. 416)** is the region at the Sun's surface from which virtually all the visible light is emitted. The main interior regions of the Sun are the **core (p. 418)**, where nuclear reactions generate energy; the **radiation zone (p. 418)**, where the energy travels outward in the form of electromagnetic radiation; and the **convection zone (p. 417)**, where the Sun's matter is in constant convective motion.

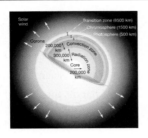

**2** The amount of solar energy reaching a 1 m² at the top of Earth's atmosphere each second is a quantity known as the **solar constant (p. 418)**. The Sun's **luminosity (p. 418)** is the total amount of energy radiated from the solar surface per second. It is determined by multiplying the solar constant by the area of an imaginary sphere of radius 1 AU.

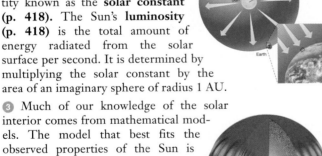

**3** Much of our knowledge of the solar interior comes from mathematical models. The model that best fits the observed properties of the Sun is the **standard solar model (p. 419)**. **Helioseismology (p. 420)**—the study of vibrations of the solar surface caused by pressure waves in the interior—provides further insight into the Sun's structure. The effect of the solar convection zone can be seen on the surface in the form of **granulation (p. 422)** of the photosphere. Lower levels in the convection zone also leave their mark on the photosphere in the form of larger transient patterns called **supergranulation (p. 423)**.

**4** Above the photosphere lies the **chromosphere (p. 417)**, the Sun's lower atmosphere. Most of the absorption lines seen in the solar spectrum are produced in the upper photosphere and the chromosphere. In the **transition zone (p. 417)** above the

chromosphere, the temperature increases from a few thousand to around a million kelvins. Above the transition zone is the Sun's thin, hot upper atmosphere, the solar **corona (p. 417)**. At a distance of about 15 solar radii, the gas in the corona is hot enough to escape the Sun's gravity, and the corona begins to flow outward as the **solar wind (p. 417)**.

**5** **Sunspots (p. 427)** are Earth-sized regions on the solar surface that are a little cooler than the surrounding photosphere. They are regions of intense magnetism. Both the numbers and locations of sunspots vary in a roughly 11-year **sunspot cycle (p. 431)** as the Sun's magnetic field rises and falls. The overall direction of the field reverses from one sunspot cycle to the next. The 22-year cycle that results when the direction of the field is taken into account is called the **solar cycle (p. 432)**.

**6** Solar activity tends to be concentrated in **active regions (p. 432)** associated with groups of sunspots. **Prominences (p. 432)** are looplike or sheetlike structures produced when hot gas ejected by activity on the solar surface interacts with the Sun's magnetic field. The more intense **flares (p. 433)** are violent surface explosions that blast particles and radiation into interplanetary space. **Coronal mass ejections (p. 434)** are huge blobs of magnetized gas escaping into interplanetary space. Most of the solar wind flows outward from low-density regions of the corona called **coronal holes (p. 435)**.

**7** The Sun generates energy by converting hydrogen to helium in its core by the process of **nuclear fusion (p. 438)**. Nuclei are held together by the **strong nuclear force (p. 439)**. When four protons are converted to a helium nucleus in the **proton–proton chain (p. 440)**, some mass is lost. The **law of conservation of mass and energy (p. 438)** requires that this mass appear as energy, eventually resulting in the light we see. Very high temperatures are needed for fusion to occur.

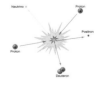

**8** **Neutrinos (p. 439)** are nearly massless particles that are produced in the proton–proton chain and escape from the Sun. They interact via the **weak nuclear force (p. 439)**. Despite their elusiveness, it is possible to detect a small fraction of the neutrinos streaming  from the Sun. Observations over several decades led to the **solar neutrino problem (p. 443)**—substantially fewer neutrinos are observed than are predicted by theory. The accepted explanation, supported by recent observational evidence, is that **neutrino oscillations (p. 444)** convert some neutrinos to other (undetected) particles en route from the Sun to Earth.

## Review and Discussion

1. Name and briefly describe the main regions of the Sun.
2. How massive is the Sun, compared with Earth?
3. How hot is the solar surface? The solar core?
4. What is luminosity, and how is it measured in the case of the Sun?
5. What fuels the Sun's enormous energy output?
6. What are the ingredients and the end result of the proton–proton chain in the Sun?
7. Why is energy released in the proton–proton chain?
8. How do scientists construct models of the Sun?
9. What is helioseismology, and what does it tell us about the Sun?
10. How do observations of the Sun's surface tell us about conditions in the solar interior?
11. Describe how energy generated in the solar core eventually reaches Earth.
12. Why does the Sun appear to have a sharp edge?
13. Give the history of "coronium," and tell how it increased our understanding of the Sun.
14. What is the solar wind?
15. Why do we say that the solar cycle is 22 years long?
16. What is the cause of sunspots, flares, and prominences?
17. Describe how coronal mass ejections may influence life on Earth.
18. Why are scientists trying so hard to detect solar neutrinos?
19. What is the most likely solution to the solar neutrino problem?
20. What would we observe on Earth if the Sun's internal energy source suddenly shut off? How long do you think it might take—minutes, days, years, or millions of years—for the Sun's light to begin to fade? Repeat the question for solar neutrinos.

## Conceptual Self-Test: True or False/Multiple Choice

1. The Sun is a rather normal star.
2. Nuclei are held together by the strong nuclear force.
3. The proton–proton chain releases energy because mass is created in the process.
4. The most abundant element in the Sun is hydrogen.
5. Convection involves cool gas rising toward the solar surface and hot gas sinking into the interior.
6. There are as many absorption lines in the solar spectrum as there are elements present in the Sun.
7. The density and temperature in the solar corona are much higher than in the photosphere.
8. Sunspots appear dark because they are hotter than the surrounding gas of the photosphere.
9. Flares are caused by magnetic disturbances in the lower atmosphere of the Sun.
10. Neutrinos have never been detected experimentally.
11. Compared with Earth's diameter, the Sun's diameter is about (a) the same; (b) ten times larger; (c) one hundred times larger; (d) one million times larger.
12. Overall, the Sun's average density is roughly the same as that of (a) rain clouds; (b) water; (c) silicate rocks; (d) iron–nickel meteorites.
13. The Sun spins on its axis roughly once each (a) hour; (b) day; (c) month; (d) year.
14. If astronomers lived on Venus instead of on Earth, the solar constant they measure would be (a) larger; (b) smaller; (c) the same.
15. The primary source of the Sun's energy is (a) fusion of light nuclei to make heavier ones; (b) fission of heavy nuclei into lighter ones; (c) the slow release of heat left over from the Sun's formation; (d) the solar magnetic field.
16. According to the standard model of the Sun (Figure 16.6), as the distance from the center increases, the density decreases (a) at about the same rate as the temperature decreases; (b) faster than the temperature decreases; (c) more slowly than the temperature decreases; (d) but the temperature increases.
17. A typical solar granule is about the size of (a) a U.S. city; (b) a large U.S. state; (c) the Moon; (d) Earth.

18. As we move to greater and greater distances above the solar photosphere, the temperature in the Sun's atmosphere (a) steadily increases; (b) steadily decreases; (c) first decreases and then increases; (d) stays the same.

19. The time between successive sunspot maxima is about (a) a month; (b) a year; (c) a decade; (d) a century.

20. The solar neutrino problem is that (a) we detect more solar neutrinos than we expect; (b) we detect fewer solar neutrinos than we expect; (c) we detect the wrong type of neutrinos; (d) we can't detect solar neutrinos.

## Problems

 *Algorithmic versions of these Problems are available in the Practice Problems module of the Companion Website. The number of dots preceding each Problem indicates its approximate level of difficulty.*

1. • Use the reasoning presented in Section 16.1 to calculate the value of the "solar constant" (a) on Mercury at perihelion, (b) on Jupiter.

2. • Use Wien's law to determine the wavelength corresponding to the peak of the blackbody curve (a) in the core of the Sun, where the temperature is $10^7$ K, (b) in the solar convection zone ($10^5$ K), and (c) just below the solar photosphere ($10^4$ K). ∞ (Sec. 3.4). What form (visible, infrared, X ray, etc.) does the radiation take in each case?

3. •• The largest-amplitude solar pressure waves have periods of about 5 minutes and move at the speed of sound in the outer layers of the Sun, roughly 10 km/s. (a) How far does such a wave move during one wave period? (b) Approximately how many wavelengths are needed to completely encircle the Sun's equator? (c) Compare the wave period with the orbital period of an object moving just above the solar photosphere. ∞ (*More Precisely 2-3*)

4. • If convected solar material moves at 1 km/s, how long does it take to flow across the 1000-km expanse of a typical granule? Compare your answer with the roughly 10-minute lifetimes observed for most solar granules.

5. •• Use Stefan's law (flux ∝ $T^4$, where $T$ is the temperature in kelvins; to calculate how much less energy (as a fraction) is emitted per unit area of a 4500-K sunspot than from the surrounding 5800-K photosphere. ∞ (Sec. 3.4)

6. •• The Sun's differential rotation is responsible for wrapping the solar magnetic field around its orb (Figure 16.18). Using the data presented in the Sun Data box on p. 416, calculate how long it takes for material at the solar equator to "lap" the material near the poles—that is, to complete one extra trip around the Sun's rotation axis.

7. • Use the data presented in this chapter and in *More Precisely 8-1* to estimate the radius at which the speed of protons in the corona first exceeds the solar escape speed.

8. • The solar wind carries mass away from the Sun at a rate of about 2 million tons/s (1 ton = 1000 kg). At this rate, how long would it take for all of the Sun's mass to escape?

9. • How long does it take for one Earth mass of material to escape the Sun as solar wind? (See Problem 8.)

10. • A large coronal mass ejection was observed by *SOHO* as the material left the Sun. The ejected material reached Earth roughly 36 hours later. What was its average speed? If such ejections occur, on average, three times per week and contain roughly 1 billion tons of matter, calculate the average rate at which they carry off solar mass. How does your answer compare with the rate of mass lost due to the solar wind?

11. •• How long does it take for the Sun to convert one Earth mass of hydrogen into helium?

12. •• (a) Assuming constant luminosity, calculate how much equivalent mass (relative to the current mass of the Sun) the Sun has radiated into space in the 4.6 billion years since it formed. How much hydrogen has been consumed? (b) How long would it take the Sun to radiate its entire mass into space?

13. ••• The entire reaction sequence shown in Figure 16.28 generates $4.3 \times 10^{-12}$ J of electromagnetic energy and releases two neutrinos. Assuming that neutrino oscillations transform half of these neutrinos into other particles by the time they travel 1 AU from the Sun, estimate the total number of solar neutrinos passing through Earth each second.

14. •• Assuming that (1) the solar luminosity has been constant since the Sun formed and (2) the Sun was initially of uniform composition throughout, as described in Table 16.2, estimate how long it would take the Sun to convert all of its original hydrogen into helium.

15. ••• The nearby star Sirius has a luminosity of $8.4 \times 10^{27}$ W and a mass 2.1 times the mass of the Sun. (a) How much hydrogen does Sirius convert into helium each second? (b) At this rate, how long would it take for Sirius to convert all of its hydrogen into helium? (Assume the same composition as the Sun, as in Problem 14.)

*The Companion Website at www.aw-bc.com/chaisson provides algorithmically generated versions of each chapter's Problems, along with additional quizzes, an Animations & Videos gallery, an interactive Glossary, and a full eBook.*

CHAPTER

# 28

*Is there intelligent life elsewhere in the universe? This fanciful painting by noted space artist Dana Berry suggests a plurality of civilizations—some perhaps extinct—on alien worlds well beyond our own solar system. Despite blockbuster movies, science-fiction novels, and a host of claims for extraterrestrial contact, astronomers have found no unambiguous evidence for life, intelligent or otherwise, anywhere else in the universe.*
*(Galaxyrise Over Alien Planet, courtesy D. Berry)* ▶

# LIFE IN THE UNIVERSE
## Are We Alone?

Are we unique? Is life on our planet the only example of life in the universe? These are difficult questions: Earth is the only place in the universe where we know for certain that life exists, and we have no direct evidence for extraterrestrial life of any kind. Yet these are important questions, with profound implications for our species.

In this chapter we take a look at how humans evolved on Earth and then consider whether those evolutionary steps might have happened elsewhere. We then assess the likelihood of our having galactic neighbors and consider how we might learn about them if they do exist.

## LEARNING GOALS

*Studying this chapter will enable you to*

1  Summarize the process of cosmic evolution as it is currently understood.
2  Evaluate the chances of finding life elsewhere in the solar system.
3  Summarize the various probabilities used to estimate the number of advanced civilizations that might exist in the Galaxy.
4  Discuss some of the techniques we might use to search for extraterrestrials and to communicate with them.

 Visit www.aw-bc.com/chaisson for additional images, animations, quizzes, and eBook for this chapter.

## 28.1 Cosmic Evolution

In our study of the universe we have been very careful to avoid any inference or conclusion that places Earth in a special place in the cosmos. This Copernican principle, or principle of mediocrity, has been our invaluable guide in helping us define our place in the "big picture." However, when discussing life in the universe, we face a problem: ours is the only planet we know of on which life and intelligence have evolved, making it hard for any discussion of intelligent life *not* to treat humans as special cases.

Accordingly, in this final chapter we adopt a decidedly different approach. We first describe the chain of events leading to the only technologically proficient, intelligent civilization we know—us. Then we try to assess the likelihood of finding and communicating with intelligent life elsewhere in the cosmos.

### Life In the Universe

With this human-centered view clearly evident, Figure 28.1 identifies seven major evolutionary phases that have contributed to development of life on our planet: *particulate, galactic, stellar, planetary, chemical, biological,* and *cultural* evolution. Matter formed from energy in the early universe, then cooled and clumped to form galaxies and stars. Within galaxies, generation after generation of stars formed and died, seeding the interstellar medium with heavy elements so that, when our Sun formed billions of years after the first star blazed, the rocky planet Earth formed along with it. Eventually, on Earth, life appeared and slowly evolved into the diverse environment we see today.

Together, these evolutionary phases represent the grand sweep of **cosmic evolution**—the continuous transformation of matter and energy that has led to the appearance of life and civilization on our planet. The first four represent, in reverse order, the contents of this book. We now expand our field of view beyond astronomy to include the other three.

From the Big Bang, to the formation of galaxies, to the birth of the solar system, to the emergence of life, to the evolution of intelligence and culture, the universe has evolved from simplicity to complexity. We are the result of an incredibly complex chain of events that spanned billions of years. Were those events random, making us unique, or are they in some sense *natural*, so that *technological civilization*—which, as a practical matter, we will take to mean "civilization capable of off-planet communication by electromagnetic or other means"—is inevitable? Put another way, are we alone in the universe, or are we just one among countless other intelligent life-forms in our Galaxy?

Before embarking on our study, we need a working definition of *life*. Defining life, however, is not an easy task: The distinction between the living and the nonliving is not as obvious as we might at first think. Although most physicists would agree on the definitions of matter and energy, biologists have not arrived at a clear-cut definition of life. Generally speaking, scientists regard the following as characteristics of living organisms: (1) They can *react* to their environment and can often heal themselves when damaged; (2) they can *grow* by taking in nourishment from their surroundings and processing it into energy; (3) they can *reproduce*, passing along some of their own characteristics to their offspring; and (4) they have the capacity for genetic change and can therefore *evolve* from generation to generation and adapt to a changing environment.

These rules are not strict, and there is great leeway in interpreting them. Stars, for example, react to the gravity of their neighbors, grow by accretion, generate energy, and "reproduce" by triggering the formation of new stars, but no one would suggest that they are alive. By contrast,

▼ **FIGURE 28.1 Arrow of Time** Some highlights of cosmic history, as it relates to the emergence of life on Earth, are noted along this arrow of time, from the beginning of the universe to the present. Along the bottom of the arrow are seven "windows" outlining the major phases of cosmic evolution: evolution of primal energy into elementary particles; of atoms into galaxies and stars; of stars into heavy elements; of elements into solid, rocky planets; of those same elements into the molecular building blocks of life; of those molecules into life itself; and of advanced life forms into intelligence, culture, and technological civilization. *(D. Berry)*

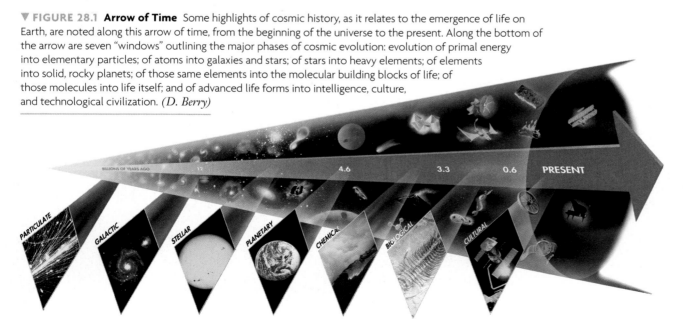

# DISCOVERY 28-1

## The Virus

The central idea of chemical evolution is that life evolved from nonliving molecules. But aside from insight based on biochemical knowledge and laboratory simulations of some key events on primordial Earth, do we have any direct evidence that life could have developed from nonliving molecules? The answer is yes.

The smallest and simplest entity that sometimes appears to be alive is a virus. We say "sometimes" because viruses seem to have the attributes of both nonliving molecules and living cells. *Virus* is the Latin word for "poison," an appropriate name, since viruses often cause disease. Although they come in many sizes and shapes—a representative example is the polio virus, shown here magnified 300,000 times—all viruses are smaller than the size of a typical modern cell. Some are made of only a few thousand atoms. In terms of size, then, viruses seem to bridge the gap between cells that are living and molecules that are not.

Viruses contain some proteins and genetic information (in the form of DNA or the closely related molecule RNA, the two molecules responsible for transmitting genetic characteristics from one generation to the next), but not much else—none of the material by which living organisms normally grow and reproduce. How, then, can a virus be considered alive? Indeed, alone, it cannot; a virus is absolutely lifeless when it is isolated from living organisms. But when it is inside a living system, a virus has all the properties of life.

Viruses come alive by transferring their genetic material into living cells. The genes of a virus seize control of a cell and establish themselves as the new master of chemical activity. Viruses grow and reproduce copies of themselves by using the genetic machinery of the invaded cell, often robbing the cell of its usual function. Some viruses multiply rapidly and wildly, spreading disease and—if unchecked—eventually killing the invaded organism. In a sense, then, viruses exist within the gray area between the living and the nonliving.

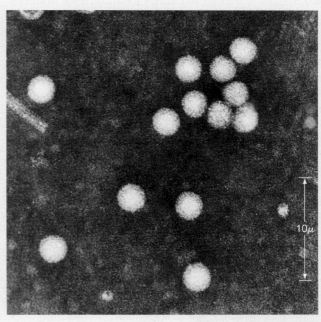

10μ

(R. Williams)

R  I  V  U  X  G

a virus (see *Discovery 28-1*) is inert when isolated from living organisms, but once inside a living system, it exhibits all the properties of life, seizing control of a living cell and using the cell's own genetic machinery to grow and reproduce. Most researchers now think that the distinction between living and nonliving matter is more one of structure and complexity than a simple checklist of rules.

The general case in favor of extraterrestrial life is summed up in what are sometimes called the *assumptions of mediocrity*: (1) Because life on Earth depends on just a few basic molecules, and (2) because the elements that make up these molecules are (to a greater or lesser extent) common to all stars, and (3) if the laws of science we know apply to the entire universe, as we have supposed throughout this book, then—given sufficient time—life must have originated elsewhere in the cosmos. The opposing view maintains that intelligent life on Earth is the product of a series of extremely fortunate accidents—astronomical, geological, chemical, and biological events unlikely to have occurred anywhere else in the universe. The purpose of this chapter is to examine some of the arguments for each of these viewpoints.

## Chemical Evolution

What information do we have about the earliest stages of planet Earth? Unfortunately, not very much. Geological hints about the first billion years or so were largely erased by violent surface activity, as volcanoes erupted and meteorites bombarded our planet; subsequent erosion by wind and water has seen to it that little evidence of that era has survived to the present. Scientists think that the early Earth was barren, with shallow, lifeless seas washing upon grassless, treeless continents. Gases emanating from our planet's interior through volcanoes, fissures, and geysers produced an atmosphere rich in hydrogen, nitrogen, and carbon compounds and poor in free oxygen. As Earth cooled, ammonia, methane, carbon dioxide, and water formed. The stage was set for the appearance of life.

The surface of the young Earth was a very violent place. Natural radioactivity, lightning, volcanism, solar ultraviolet radiation, and meteoritic impacts all provided large amounts of energy that eventually shaped the ammonia, methane, carbon dioxide, and water on our planet into more complex molecules known as **amino acids** and **nucleotide bases**—organic (carbon-based) molecules that are the building blocks of life as we know it. Amino acids build *proteins*, and proteins control metabolism, the daily utilization of food and energy by means of which organisms stay alive and carry out their vital activities. Sequences of nucleotide bases form *genes*—parts of the DNA molecule—which direct the synthesis of proteins and thus determine the characteristics of the organism (Figure 28.2). These same genes, via the DNA contained within every cell in the organism, transfer hereditary characteristics from one generation to the next through reproduction. In all living creatures on Earth—from bacteria to amoebas to humans—genes mastermind life and proteins maintain it.

The idea that complex molecules could have evolved naturally from simpler ingredients found on the primitive Earth has been around since the 1920s. The first experimental verification was provided in 1953, when scientists Harold Urey and Stanley Miller, using laboratory equipment somewhat similar to that shown in Figure 28.3. The **Urey-Miller** experiment took a mixture of the materials thought to be present on Earth long ago—a "primordial soup" of water, methane, carbon dioxide, and ammonia—and energized it by passing an electrical discharge ("lightning") through the gas. After a few days, they analyzed their mixture and found that it contained many of the same amino acids found in all living things on Earth.

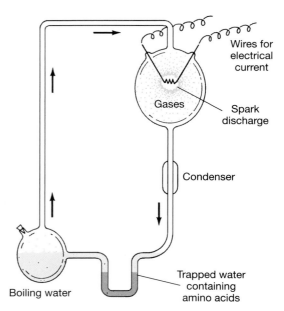

▲ **FIGURE 28.3 Urey-Miller Experiment** This chemical apparatus is designed to synthesize complex biochemical molecules by energizing a mixture of simple chemicals. A mixture of gases (ammonia, methane, carbon dioxide, and water vapor) is placed in the upper bulb to simulate Earth's primordial atmosphere and then energized by spark-discharge electrodes to simulate lightning. After about a week, amino acids and other complex molecules are found in the trap at the bottom, which simulates the primordial oceans into which heavy molecules produced in the overlying atmosphere would have fallen.

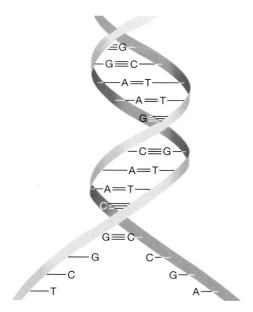

▲ **FIGURE 28.2 DNA Molecule** DNA (deoxyribonucleic acid) is the molecule containing all the genetic information needed for a living organism to reproduce and survive. Often consisting of literally tens of billions of individual atoms, its double-helix structure allows it to "unzip," exposing its internal structure to control the creation of proteins needed for a cell to function. The ordering of its constituent parts is unique to each individual organism. The letters represent the four types of nucleotide bases that makeup DNA: adenine, cytosine, guanine, and thymine.

About a decade later, scientists succeeded in constructing nucleotide bases in a similar manner. These experiments have been repeated in many different forms, with more realistic mixtures of gases and a variety of energy sources, but always with the same basic outcomes.

Although none of these experiments has ever produced a living organism, or even a single strand of DNA, they do demonstrate conclusively that "biological" molecules—the molecules involved in the functioning of living organisms—can be synthesized by strictly *non*biological means, using raw materials available on the early Earth. More advanced experiments, in which amino acids are united under the influence of heat, have fashioned proteinlike blobs that behave to some extent like true biological cells. Such near-protein material resists dissolution in water (so it would remain intact when it fell from the primitive atmosphere into the ocean) and tends to cluster into small droplets called microspheres—a little like oil globules floating on the surface of water. Figure 28.4 shows some of these laboratory-made proteinlike microspheres, whose walls permit the inward passage of small molecules, which then combine within the droplet to construct more complex molecules that are too large to pass back out through the walls. As the droplets "grow," they tend to "reproduce," forming smaller droplets.

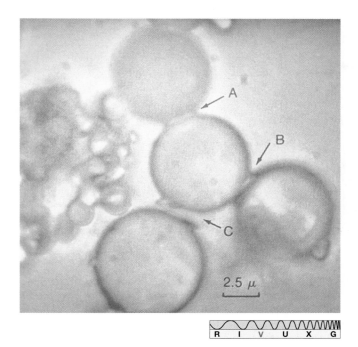

▲ FIGURE 28.4 **Chemical Evolution** These carbon-rich, proteinlike droplets display the clustering of as many as a billion amino acid molecules in a liquid. The droplets can "grow," and parts of them can separate from their "parent" droplet to become new individual droplets (as at A, B, and C). The scale of 2.5 microns noted here is 1/4000 of a centimeter. *(S. Fox)*

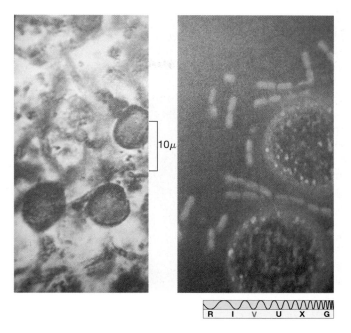

▲ FIGURE 28.5 **Primitive Cells** The photograph on the left, taken through a microscope, shows a fossilized organism found in sediments radioactively dated as 2 billion years old. This primitive system possesses concentric spheres or walls connected by smaller spheroids. The roundish fossils here measure about a thousandth of a centimeter in diameter. The photograph on the right, on approximately the same scale, displays modern blue-green algae. *(E. Barghoorn)*

Can we consider these proteinlike microspheres to be alive? Almost certainly not. Most biochemists would say that the microspheres are not life itself, but they contain many of the basic ingredients needed to form life. The microspheres lack the hereditary DNA molecule. However, as illustrated in Figure 28.5, they do have similarities to ancient cells found in the fossil record, which in turn have many similarities to modern organisms (such as blue-green algae). Thus, while no actual living cells have yet been created "from scratch" in any laboratory, many biochemists feel that the chain of events leading from simple nonbiological molecules almost to the point of life itself has been amply demonstrated.

## An Interstellar Origin?

Recently, a dissenting view has emerged. Some scientists have argued that Earth's primitive atmosphere might *not* in fact have been a particularly suitable environment for the production of complex molecules. These scientists say that there may not have been sufficient energy available to power the necessary chemical reactions and the early atmosphere may not have contained enough raw material for the reactions to have become important in any case. They suggest instead that much, if not all, of the organic (carbon-based) material that combined to form the first living cells was produced in *interstellar space* and subsequently

arrived on Earth in the form of comets, interplanetary dust, and meteors that did not burn up during their descent through the atmosphere.

Several pieces of evidence support this idea. Interstellar molecular clouds are known to contain complex molecules—indeed, there have even been reports (still unconfirmed) of at least one amino acid (glycine) in interstellar space. ∞ (Sec. 18.5)

To test the interstellar space hypothesis, NASA researchers have carried out their own version of the Urey–Miller experiment in which they exposed an icy mixture of water, methanol, ammonia, and carbon monoxide—representative of many interstellar grains—to ultraviolet radiation to simulate the energy from a nearby newborn star. As shown in Figure 28.6, when they later placed the irradiated ice in water and examined the results, they found that the ice had formed droplets surrounded by membranes and containing complex organic molecules. As with the droplets found in earlier experiments, no amino acids, proteins, or DNA were observed in the mix, but the results, repeated numerous times, clearly show that even the harsh, cold vacuum of interstellar space can be a suitable medium in which complex molecules and primitive cellular structures can form.

As described in Chapter 15, these icy interstellar grains are thought to have formed the comets in our own solar system. ∞ (Sec. 15.3) Large amounts of organic material

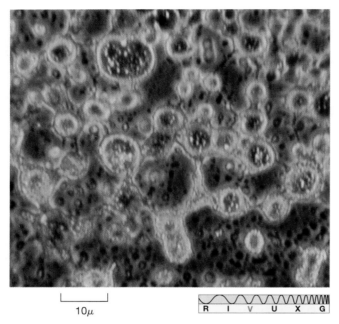

10μ

R I V U X G

▲ FIGURE 28.6 **Interstellar Globules** These oily, hollow droplets rich in organic molecules were made by exposing a freezing mixture of primordial matter to harsh ultraviolet radiation. The larger ones span about 10 microns across and, when immersed in water, show cell-like membrane structure. Although they are not alive, they bolster the idea that life on Earth could have come from space. *(NASA)*

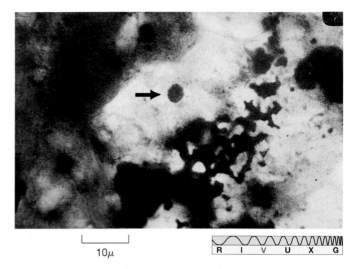

10μ

R I V U X G

▲ FIGURE 28.7 **Murchison Meteorite** The Murchison meteorite contains relatively large amounts of amino acids and other organic material, indicating that chemical evolution of some sort has occurred beyond our own planet. In this magnified view of a fragment from the meteorite, the arrow points to a microscopic sphere of organic matter. *(CfA)*

were detected on comet Halley by space probes when Halley last visited the inner solar system, and similarly complex molecules have been observed on many other well-studied comets, such as Hale–Bopp. ∞ (Sec. 14.2) Also as discussed in Chapter 15, there is reason to suspect that cometary impacts were responsible for most of Earth's water, and it is perhaps a small step to imagining that this water already contained the building blocks for life. ∞ (Sec. 15.3)

In addition, a small fraction of the meteorites that survive the plunge to Earth's surface—including perhaps the controversial "Martian meteorite" discussed in Chapter 10—contain organic compounds. ∞ *(Discovery 10-1)* The Murchison meteorite (Figure 28.7), which fell near Murchison, Australia, in 1969, is a particularly well-studied example. Located soon after crashing to the ground, this meteorite has been shown to contain 12 of the amino acids

normally found in living cells, although the detailed structures of these molecules indicate potentially important differences between those found in space and those found on Earth. At the very least, though, these discoveries argue that complex molecules can form in an interplanetary or interstellar environment and that they could have reached Earth's surface unscathed after their fiery descent.

Thus, the hypothesis that organic matter is constantly raining down on Earth from space in the form of interplanetary debris is quite plausible. However, whether this was the *primary* means by which complex molecules first appeared in Earth's oceans remains unclear.

## Diversity and Culture

However the basic materials appeared on Earth, we know that life *did* appear. The fossil record chronicles how life on Earth became widespread and diversified over the course of time. The study of fossil remains shows the

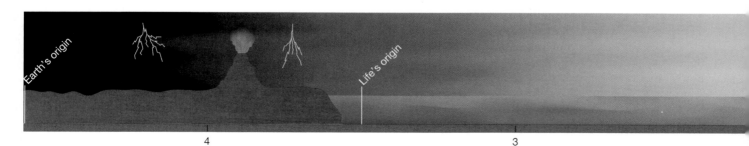

Earth's origin

Life's origin

initial appearance of simple one-celled organisms such as blue-green algae more than 3.5 billion years ago. These were followed by more complex one-celled creatures, such as the amoeba, about 2 billion years ago. Multicellular organisms such as sponges did not appear until about 1 billion years ago, after which there flourished a wide variety of increasingly complex organisms—insects, reptiles, and mammals. Figure 28.8 illustrates some of the key developments in the evolution of life on our planet.

The fossil record leaves no doubt that biological organisms have changed over time—all scientists accept the reality of *biological evolution*. As conditions on Earth shifted and Earth's surface evolved, those organisms that could best take advantage of their new surroundings succeeded and thrived—often at the expense of organisms that could not make the necessary adjustments and consequently became extinct.

What led to these changes? Chance. An organism that happened to have a certain useful genetically determined trait—for example, the ability to run faster, climb higher, or even hide more easily—would find itself with the upper hand in a particular environment. That organism was therefore more likely to reproduce successfully, and its advantageous characteristic would then be more likely to be passed on to the next generation. The evolution of the rich variety of life on our planet, including human beings, occurred as chance mutations—changes in genetic structure—led to changes in organisms over millions of years.

What about the development of intelligence? Many anthropologists think that, like any other highly advantageous trait, intelligence *is* strongly favored by natural selection. As humans learned about fire, tools, and agriculture, the brain became more and more elaborate. The social cooperation that went with coordinated hunting efforts was another important competitive advantage that developed as brain size increased.

Perhaps most important of all was the development of language. Indeed, some anthropologists have gone so far as to suggest that human intelligence *is* human language. Through language, individuals could signal one another while hunting for food or seeking protection. Even more importantly, now our ancestors could share ideas as well as food and shelter. Experience, stored in the brain as memory, could be passed down from generation to generation. A new kind of evolution had begun, namely, *cultural evolution*, the changes in the ideas and behavior of society. Within only the past 10,000 years or so, our more recent ancestors have created the entirety of human civilization.

To put all this into historical perspective, let's imagine the entire lifetime of Earth to be 46 years rather than 4.6 billion years. On this scale, we have no reliable record of the first decade of our planet's existence. Life originated at least 35 years ago, when Earth was about 10 years old. Our planet's middle age is largely a mystery, although we can be sure that life continued to evolve and that generations of mountain chains and oceanic trenches came and went. Not until about 6 years ago did abundant life flourish throughout Earth's oceans. Life came ashore about 4 years ago, and plants and animals mastered the land only about 2 years ago. Dinosaurs reached their peak about 1 year ago, only to die suddenly about 4 months later. ∞ (*Discovery 11-1*) Humanlike apes changed into apelike humans only last week, and the latest ice ages occurred only a few days ago. *Homo sapiens*—our species—did not emerge until about 4 hours ago. Agriculture was invented within the last hour, and the Renaissance—along with all of modern science—is just 3 minutes old!

## CONCEPT CHECK

✔ Has chemical evolution been verified in the laboratory?

▼ FIGURE 28.8 **Life on Earth** A simplified timeline of the origin and evolution of life on our planet. The scale begins at left with the origin of Earth about 4.6 billion years ago and extends linearly to the present day at right. Notice how life forms most familiar to us emerged relatively recently in the history of our planet.

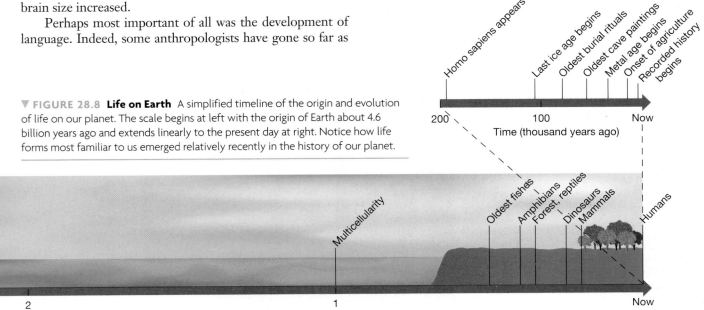

## 28.2 Life in the Solar System

Simple one-celled life-forms reigned supreme on Earth for most of our planet's history. It took time—a great deal of time—for life to emerge from the oceans, to evolve into simple plants, to continue to evolve into complex animals, and to develop intelligence, culture, and technology. Have those (or similar) events occurred elsewhere in the universe? Let's try to assess what little evidence we have on the subject.

### Life as We Know It

"Life as we know it" is generally taken to mean carbon-based life that originated in a liquid water environment—in other words, life on Earth. Might such life exist elsewhere in our solar system?

The Moon and Mercury lack liquid water, protective atmospheres, and magnetic fields, so these two bodies are subjected to fierce bombardment by solar ultraviolet radiation, the solar wind, meteoroids, and cosmic rays. Simple molecules could not survive in such hostile environments. Venus, by contrast, has *far too much* protective atmosphere! Its dense, dry, scorchingly hot atmospheric blanket effectively rules it out as an abode for life, at least like us.

The jovian planets have no solid surfaces (although some researchers have suggested that life might have evolved in their atmospheres), and Pluto and most of the moons of the outer planets are too cold. However, the possibility of liquid water below Europa's icy surface has refueled speculation about the development of life there, making this moon of Jupiter a prime candidate for future exploration. ∞ (Sec. 11.5) For now at least, Europa is high on the priority list for missions by both NASA and the European Space Agency. Saturn's moon Titan, with its atmosphere of methane, ammonia, and nitrogen, and possibly with some liquid on its surface, is conceivably a site where life might have arisen. The most recent results from the *Cassini-Huygens* mission suggest that Titan's frigid surface conditions are inhospitable for anything familiar to us, but (as we discuss below) scientists are finding more and more examples of terrestrial organisms able to survive and thrive in environments once regarded as uninhabitable. ∞ (Sec. 12.5)

The planet most likely to harbor life (or to have harbored it in the past) seems to be Mars. The Red Planet is harsh by Earth standards: Liquid water is scarce, the atmosphere is thin, and the absence

of magnetism and an ozone layer allows solar high-energy particles and ultraviolet radiation to reach the surface unabated. But the Martian atmosphere was thicker, and the surface probably warmer and much wetter, in the past. ∞ (Sec. 10.5) Indeed, there is strong photographic evidence from orbiters such as *Viking* and *Mars Global Surveyor* for flowing and standing water on Mars in the distant (and perhaps even relatively recent) past. In 2004, the European *Mars Express* orbiter confirmed the long-hypothesized presence of water ice at the Martian poles, and NASA's *Opportunity* rover reported strong geological evidence that the region around its landing site was once "drenched" with water for an extended period.

All of these lines of reasoning strongly suggest that Mars—at least at some time in its past—harbored large amounts of liquid water. However, none of the Mars landers has detected anything that might be interpreted as the remains (fossilized or otherwise) of large plants or animals, and only the *Viking* landers carried equipment capable of performing the detailed biological analysis needed to detect bacterial life (or its fossil remnants). The *Viking* robots scooped up Martian soil (Figure 28.9) and tested it for the presence of life by conducting chemical experiments designed to detect the waste gases and other products of metabolic activity, but no unambiguous evidence of Martian life has emerged. ∞ (*Discovery 10-1*) Of course, we might argue that the landers touched down on the safest Martian terrain, not in the most interesting regions, such as near the moist polar caps.

Some scientists have suggested that a different type of biology may be operating on the Martian surface. They propose that Martian microbes capable of eating and digesting oxygen-rich compounds in the Martian soil could also explain the *Viking* results. This speculation would be greatly strengthened if recent announcements of fossilized bacteria in meteorites originating on Mars were confirmed (although it seems that the weight of scientific opinion is currently running against that interpretation of the data). The consensus among biologists and chemists today is that

▶ FIGURE 28.9 **Search for Martian Life**
Several trenches were dug by the *Viking* robots, such as *Viking 2* seen here on Mars's Utopia Planitia. Soil samples were scooped up and taken inside the robot, where instruments tested them for chemical composition and any signs of life. (*NASA*)

Mars does not house any life similar to that on Earth, but a solid verdict regarding past life on Mars will not be reached until we have thoroughly explored our intriguing neighbor.

In considering the emergence of life under adversity, we should perhaps not be too quick to rule out an environment based solely on its extreme properties. Figure 28.10 shows a very hostile environment on a deep-ocean floor, where hydrothermal vents spew forth boiling hot water from vertical tubes a few meters tall. The conditions are quite unlike anything on our planet's surface, yet life thrives in an environment rich in sulfur, poor in oxygen, and completely dark. Such underground hot springs might conceivably exist on alien worlds, raising the possibility of life forms with much greater diversity over a much wider range of conditions than those known to us on Earth.

In recent years scientists have discovered many instances of so-called **extremophiles**—life-forms that have adapted to live in extreme environments. The superheated hydrothermal vents in Figure 28.10 are one example, but extremophiles have also been found in frigid lakes buried deep under the Antarctic glaciers, in the dark, oxygen-poor and salt-rich floor of the Mediterranean Sea, in the mineral-rich superalkaline environment of California's Mono lake, and even in the hydrogen-rich volcanic darkness far below Earth's crust. In many cases these organisms have evolved to create the energy they need by purely chemical means, using *chemosynthesis* instead of *photosynthesis*, the process whereby plants turn sunlight into energy. These environments may present conditions not so different from those found on Mars, Europa, or Titan, suggesting that even "life as we know it" might well be able to thrive in these hostile, alien worlds.

## Alternative Biochemistries

Conceivably, some types of biology might be so different from life on Earth that we would not recognize them and would not know how to test for them. What might these other biologies be?

Some scientists have pointed out that the abundant element silicon has chemical properties somewhat similar to those of carbon and have suggested silicon as a possible alternative to carbon as the basis for living organisms. Ammonia (made of the common elements hydrogen and nitrogen) is sometimes put forward as a possible liquid medium in which life might develop, at least on a planet cold enough for ammonia to exist in the liquid state. Together or separately, these alternatives would surely give rise to organisms with biochemistries (the basic biological and chemical processes responsible for life) radically different from those we know on Earth. Conceivably, we might have difficulty even identifying these organisms as alive.

Although the possibility of such alien life-forms is a fascinating scientific problem, most biologists would argue that chemistry based on carbon and water is the one most likely to give rise to life. Carbon's flexible chemistry and water's wide liquid temperature range are just what are needed for life to develop and thrive. Silicon and ammonia seem unlikely to fare as well as bases for advanced life-forms. Silicon's chemical bonds are weaker than those of carbon and may not be able to form complex molecules—an apparently essential aspect of carbon-based life. Also, the colder the environment, the less energy there is to drive biological processes. The low temperatures necessary for ammonia to remain liquid might inhibit or even completely prevent the chemical reactions leading to the equivalent of amino acids and nucleotide bases.

Still, we must admit that we know next to nothing about noncarbon, nonwater biochemistries, for the very good reason that there are no examples of them to study experimentally. We can speculate about alien life-forms and try to make general statements about their characteristics, but we can say little of substance about them.

### CONCEPT CHECK

✔ Which solar system bodies (other than Earth) are the leading candidates in the search for extraterrestrial life?

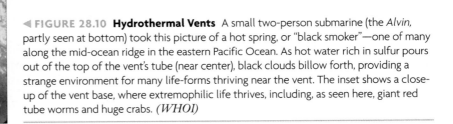

◀ FIGURE 28.10 **Hydrothermal Vents** A small two-person submarine (the *Alvin*, partly seen at bottom) took this picture of a hot spring, or "black smoker"—one of many along the mid-ocean ridge in the eastern Pacific Ocean. As hot water rich in sulfur pours out of the top of the vent's tube (near center), black clouds billow forth, providing a strange environment for many life-forms thriving near the vent. The inset shows a close-up of the vent base, where extremophilic life thrives, including, as seen here, giant red tube worms and huge crabs. *(WHOI)*

## 28.3 Intelligent Life in the Galaxy

With humans apparently the only intelligent life in the solar system, we must broaden our search for extraterrestrial intelligence to other stars and perhaps even other galaxies. At such distances, though, we have little hope of actually detecting life with current equipment. Instead, we must ask, "How likely is it that life in any form—carbon based, silicon based, water based, ammonia based, or something we cannot even dream of—exists?" Let's look at some numbers to develop estimates of the probability of life elsewhere in the universe.

### The Drake Equation

An early approach to this problem is known as the **Drake equation** below, after the U.S. astronomer who pioneered the analysis (below). It attempts to express the probability of life in our Galaxy in terms of specific factors with roots in astronomy, biology, and anthropology.

Of course, several of the factors in this formula are largely a matter of opinion. We do not have nearly enough information to determine—even approximately—every factor in the equation, so the Drake equation cannot give us a hard-and-fast answer. Its real value is that it subdivides a large and difficult question into smaller pieces that we can attempt to answer separately. The equation provides the framework within which the problem can be addressed and parcels out the responsibility for the final solution among many different scientific disciplines. Figure 28.11 illustrates how, as our requirements become more and more stringent, only a small fraction of star systems in the Milky Way is likely to generate the advanced qualities specified by the combination of factors on the right-hand side of the equation.

Let's examine the factors in the equation one by one and make some educated guesses about their values. Bear in mind, though, that if you ask two scientists for their best estimates of any given factor, you will likely get two very different answers!

### Rate of Star Formation

We can estimate the average number of stars forming each year in the Galaxy simply by noting that at least 100 billion stars now shine in the Milky Way. Dividing this number by the 10-billion-year lifetime of the Galaxy, we obtain a formation rate of 10 stars per year. This rate may be an

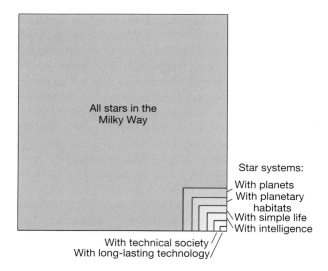

▲ FIGURE 28.11 **Drake Equation** Of all the star systems in our Milky Way Galaxy (represented by the largest box), progressively fewer and fewer have each of the qualities typical of a long-lasting technological society (represented by the smallest box at the lower right corner).

overestimate, because we think that fewer stars are forming now than formed at earlier epochs of the Galaxy, when more interstellar gas was available. However, we do know that stars are forming today, and our estimate does not include stars that formed in the past and have since died, so our value of 10 stars per year is probably reasonable when averaged over the lifetime of the Milky Way.

### Fraction of Stars Having Planetary Systems

Many astronomers regard planet formation as a natural result of the star-formation process. If the condensation theory (Chapter 15) or some variant of it is correct, and if there is nothing special about our Sun, as we have argued throughout this book, then we would expect many stars to have at least one planet. ∞ (Sec. 15.2) Indeed, as we have seen, increasingly sophisticated observations indicate the presence of disks around young stars. Could these disks be protosolar systems? The condensation theory suggests that they are, and the short (theoretical) lifetimes of disks imply the existence of many planet-forming systems in the neighborhood of the Sun.

No planets like our own have yet been *seen* orbiting any other star. The light reflected by an Earth-like planet circling

| number of technological, intelligent civilizations now present in the Galaxy | | rate of star formation, averaged over the lifetime of the Galaxy | fraction of stars having planetary systems | average number of habitable planets within those planetary systems | fraction of those habitable planets on which life arises | fraction of those life-bearing planets on which intelligence evolves | fraction of those intelligent-life planets that develop technological society | average lifetime of a technologically competent civilization. |
|---|---|---|---|---|---|---|---|---|
| | = | | × | × | × | × | × | × |

even the closest star would be too faint to detect even with the best equipment. The light would be lost in the glare of the parent star. Large orbiting telescopes may soon be able to detect Jupiter-sized planets orbiting the nearest stars, but even those huge planets will be barely visible. However, as described in Chapter 15, there is now overwhelming evidence for planets orbiting other stars. ∞ (Sec. 15.5) The planets found so far are Jupiter sized rather than Earth sized and generally have rather eccentric orbits, but astronomers are confident that an Earth-like planet (in an Earth-like orbit) will one day be detected, and plans to build the necessary equipment are well underway.

Accepting the condensation theory and its consequences, and without being either too conservative or naïvely optimistic, we assign a value near unity to this factor—that is, we think that nearly all stars form with planetary systems of some sort.

## Number of Habitable Planets per Planetary System

What determines the feasibility of life on a given planet? *Temperature* is perhaps the single most important factor, although the possibility of catastrophic external events, such as cometary impacts or even distant supernovae, must also be considered. ∞ (*Discovery 11-1*, Sec. 21.3)

The surface temperature of a planet depends on two things: the planet's distance from its parent star and the thickness of the planet's atmosphere. Planets with a nearby parent star (but not too close) and some atmosphere (though not too thick) should be reasonably warm, like Earth or Mars. Planets far from the star and with no atmosphere, like Pluto, will surely be cold by our standards. And planets too close to the star and with a thick atmosphere, like Venus, will be very hot indeed.

Figure 28.12 illustrates how a three-dimensional **habitable zone** of "comfortable" temperatures surrounds every star. This *stellar habitable zone* represents the range of distances within which a planet of mass and composition similar to Earth's would have a surface temperature between the freezing and boiling points of water. (Our Earth-based bias is again clear here!) The hotter the star, the larger is this zone. For example, A- and F-type stars have a rather large habitable zone, but the size of the zone diminishes rapidly as we proceed through G-, K-, and M-type stars (although we note that one of the "Earth-mass" planets mentioned in Chapter 15 in fact lies within the habitable zone of its M-type parent star). ∞ (Sec. 15.6) O- and B-type stars are not considered in this regard because they are not expected to last long enough for life to develop, even if they do have planets. M-type stars, despite their large numbers, have such small habitable zones and, in any case, are thought to be prone to such violent surface activity, that they are not generally considered likely hosts to life-bearing planets. ∞ (Sec. 17.8)

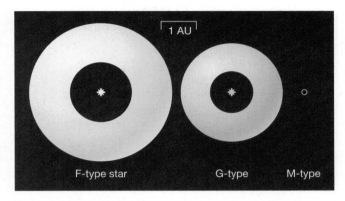

▲ FIGURE 28.12 **Stellar Habitable Zones** The extent of the habitable zone is much larger around a hot star than around a cool one. For a star like the Sun (a G-type star), the zone extends from about 0.85 to 2.0 AU. For a hotter F-type star, the range is 1.2 to 2.8 AU. For a cool M-type star, only Earth-like planets orbiting between about 0.02 and 0.06 AU would be habitable.

Three planets—Venus, Earth, and Mars—reside in or near the habitable zone surrounding our Sun. Venus is too hot because of its thick atmosphere and proximity to the Sun. Mars is a little too cold because its atmosphere is too thin and it is too far from the Sun. But if the orbits of Venus and Mars were swapped—not inconceivable, since chance played such a large role in the formation of the terrestrial planets—then both of these nearby planets might conceivably have evolved surface conditions resembling those on Earth. ∞ (Secs. 10.5, 15.4) In that case our solar system would have had three habitable planets instead of one. Proximity to a giant planet may also render a moon (such as Europa) habitable, the planet's tidal heating making up for the lack of sunlight. ∞ (Sec. 11.5)

A planet moving on a "habitable" orbit may still be rendered uninhabitable by external events. Many scientists think that the outer planets in our own solar system are critical to the habitability of the inner worlds, both by stabilizing their orbits and by protecting them from cometary impacts, deflecting would-be impactors away from the inner part of the solar system. The theory presented in Chapter 15 suggests that a star with inner terrestrial planets on stable orbits would probably also have the jovian worlds needed to safeguard their survival. ∞ (Sec. 15.3) However, observations of extrasolar planets are not yet sufficiently refined to determine the fraction of stars having "outer planet" systems like our own. ∞ (Sec. 15.6)

Other external forces may also influence a planet's survival. Some researchers have suggested that there is a *galactic habitable zone* for stars in general, outside of which conditions are unfavorable for life (see Figure 28.13). Far from the Galactic center, the star formation rate is low and few cycles of star formation have occurred, so there are insufficient heavy elements to form terrestrial planets or populate them with technological civilizations if any should form. ∞ (Sec. 21.5) Too close, and the radiation

High density and intense radiation in here

Sun's orbit

Too few heavy elements out here

Galactic habitable zone

◀ FIGURE 28.13 **Galactic Habitable Zone** Some regions of the Galaxy may be more conducive to life than others. Too far from the Galactic center, there may not be enough heavy elements for terrestrial planets to form or technological society to evolve. Too close, the radiative or gravitational effects of nearby stars may render life impossible. The result is a ring-shaped habitable zone, colored here in green. This zone presumably encloses the orbit of the Sun, but its radial extent in both directions is uncertain.

think that, on average, there is one potentially habitable planet for every 10 planetary systems that might exist in our Galaxy. Single F-, G-, and K-type stars are the best candidates.

## Fraction of Habitable Planets on Which Life Actually Arises

The number of possible combinations of atoms is incredibly large. If the chemical reactions that led to the complex molecules that make up living organisms occurred completely at random, then it

from bright stars and supernovae in the crowded inner part of the Galaxy might be detrimental to life. More importantly, the gravitational effects of nearby stars may send frequent showers of comets from the counterpart of the Oort cloud into the inner regions of a planetary system, striking the terrestrial planets and terminating any chain of evolution that might lead to intelligent life.

Thus, to estimate the number of habitable planets per planetary system, we must first take inventory of how many stars of each type shine in the Galactic habitable zone, then calculate the sizes of their stellar habitable zones, and, finally, estimate the number of planets likely to be found there. We must eliminate almost all of the 10 percent of surveyed stars around which planets have so far been observed, and, presumably, a similar fraction of stars in general, because the large jovian planets seen in most cases have eccentric orbits that would destabilize the motion of any inner terrestrial world, either ejecting it completely from the system or making conditions so extreme that the chances for the development of life are severely reduced. ∞ (Sec. 15.6)

Finally, we exclude the majority of binary-star systems, because a planet's orbit within the habitable zone of a binary would be unstable in many cases, as illustrated in Figure 28.14. Given the observed properties of binaries in our Galaxy, we expect that most habitable planetary orbits would be unstable, so there would not be time for life to develop.

The inner and outer radii of the Galactic habitable zone are not known with any certainty, and we still have insufficient data about most stars to make any definitive statement about their planetary systems. Taking these uncertainties into account as best we can, we assign a value of 1/10 to this factor in our equation. In other words, we

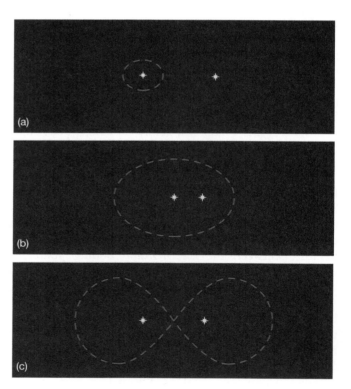

(a)

(b)

(c)

▲ FIGURE 28.14 **Binary-Star Planets** In binary-star systems, planets are restricted to only a few kinds of orbits that are gravitationally stable. (a) The orbit is stable only if the planet lies very close to its parent star, so that the gravity of the other star is negligible. (b) A planet circulating at a great distance about both stars in an elliptical orbit, in which case the orbit is stable only if it lies far from both stars. (c) Another possible but unstable path interweaves between the two stars in a figure-eight pattern.

is extremely unlikely that those molecules could have formed at all. In that case, life is extraordinarily rare, this factor is close to zero, and we are probably alone in the Galaxy, perhaps even in the entire universe.

However, laboratory experiments (such as the Urey–Miller experiment described earlier) seem to suggest that certain chemical combinations are strongly favored over others—that is, the reactions are *not* random. Of the billions upon billions of basic organic groupings that could occur on Earth from the random combination of all sorts of simple atoms and molecules, only about 1500 actually do occur. Furthermore, these 1500 organic groups of terrestrial biology are made from only about 50 simple "building blocks" (including the amino acids and nucleotide bases mentioned earlier). This suggests that molecules critical to life are not assembled by chance alone; apparently, additional factors are at work on the microscopic level. If a relatively small number of chemical "evolutionary tracks" are likely to exist, then the formation of complex molecules—and hence, we assume, life—becomes much more likely, given sufficient time.

To assign a very low value to this factor in the equation is to think that life arises randomly and rarely. To assign a value close to unity is to think that life is inevitable, given the proper ingredients, a suitable environment, and a long enough time. No simple experiment can distinguish between these extreme alternatives, and there is little or no middle ground. To many researchers, the discovery of life (past or present) on Mars, Europa, Titan, or some other object in our solar system would convert the appearance of life from an unlikely miracle to a virtual certainty throughout the Galaxy. We will take the optimistic view and adopt a value of unity.

## Fraction of Life-Bearing Planets on Which Intelligence Arises

As with the evolution of life, the appearance of a well-developed brain is a highly unlikely event if only chance is involved. However, biological evolution through natural selection is a mechanism that generates apparently highly improbable results by singling out and refining useful characteristics. Organisms that profitably use adaptations can develop more complex behavior, and complex behavior provides organisms with the *variety* of choices needed for more advanced development.

One school of thought maintains that, given enough time, intelligence is inevitable. On this view, assuming that natural selection is a universal phenomenon, at least one organism on a planet will always rise to the level of "intelligent life." If this is correct, then the fifth factor in the Drake equation equals or nearly equals unity.

Others argue that there is only one known case of intelligence: human beings on Earth. For 2.5 billion years—from the start of life about 3.5 billion years ago to the first appearance of multicellular organisms about 1 billion years ago—life did not advance beyond the one-celled stage. Life remained simple and dumb, but it survived. If this latter view is correct, then the fifth factor in our equation is very

small, and we are faced with the depressing prospect that humans may be the smartest form of life anywhere in the Galaxy. As with the previous factor, we will be optimistic and simply adopt a value of unity here.

## Fraction of Planets on Which Intelligent Life Develops and Uses Technology

To evaluate the sixth factor of our equation, we need to estimate the probability that intelligent life eventually develops technological competence. Should the rise of technology be inevitable, this factor is close to unity, given a long enough time. If it is not inevitable—if intelligent life can somehow "avoid" developing technology—then this factor could be much less than unity. The latter scenario envisions a universe possibly teeming with intelligent civilizations, but very few among them ever becoming technologically competent. Perhaps only one managed it—ours.

Again, it is difficult to decide conclusively between these two views. We don't know how many prehistoric Earth cultures failed to develop simple technology or rejected its use. We do know that the roots of our present civilization arose independently at several different places on Earth, including Mesopotamia, India, China, Egypt, Mexico, and Peru. Because so many of these ancient cultures originated at about the same time, it is tempting to conclude that the chances are good that some sort of technological society will inevitably develop, given some basic intelligence and enough time.

If technology is inevitable, then why haven't other life-forms on Earth also found it useful? Possibly the competitive edge given by intellectual and technological skills to humans, the first species to develop them, allowed us to dominate so rapidly that other species—gorillas and chimpanzees, for example—simply haven't had time to catch up. The fact that only one technological society exists on Earth does not imply that the sixth factor in our Drake equation must be very much less than unity. On the contrary, it is precisely because *some* species will probably always fill the niche of technological intelligence that we will take this factor to be close to unity.

## Average Lifetime of a Technological Civilization

The reliability of the estimate of each factor in the Drake equation declines markedly from left to right. For example, our knowledge of astronomy enables us to make a reasonably good stab at the first factor, namely, the rate of star formation in our Galaxy, but it is much harder to evaluate some of the later factors, such as the fraction of life-bearing planets that eventually develop intelligence. The last factor on the right-hand side of the equation, the longevity of technological civilizations, is totally unknown. There is only one known example of such a civilization: humans on planet Earth. Our own civilization has survived in its "technological" state for only about 100 years, and how long we will be around before a natural or human-made catastrophe ends it all is impossible to tell. ∞ (*Discovery 11-1*)

## Number of Technological Civilizations in the Galaxy

One thing is certain: If the correct value for *any one factor* in the equation is very small, then few technological civilizations now exist in the Galaxy. In other words, the pessimistic view of the development of life or of intelligence is correct, then we are unique, and that is the end of our story. However, if both life and intelligence are inevitable consequences of chemical and biological evolution, as many scientists think, and if intelligent life always becomes technological, then we can plug the higher, more optimistic values into the Drake equation. In that case, combining our estimates for the other six factors (and noting that $10 \times 1 \times 1/10 \times 1 \times 1 \times 1 = 1$), we have

| number of technological, intelligent civilizations now present in the Milky Way Galaxy | = | average lifetime of a technologically competent civilization in years. |
|---|---|---|

Thus, if civilizations typically survive for 1000 years, there should be 1000 of them currently in existence scattered throughout the Galaxy. If they live for a million years, on average, we would expect there to be a million advanced civilizations in the Milky Way, and so on.

Note that, even setting aside language and cultural issues, the sheer size of the Galaxy presents a significant hurdle to communication between technological civilizations. The minimum requirement for a two-way conversation is that we can send a signal and receive a reply in a time shorter than our own lifetime. If the lifetime is short, then civilizations are literally few and far between—small in number, according to the Drake equation, and scattered over the vastness of the Milky Way—and the distances between them (in light years) are much greater than their lifetimes (in years). In that case, two-way communication, even at the speed of light, will be impossible. However, as the lifetime increases, the distances get smaller as the Galaxy becomes more crowded, and the prospects improve.

Taking into account the size, shape, and distribution of stars in the Galactic disk (why do we exclude the halo?), and under the optimistic assumptions just made, we find that, unless the life expectancy of a civilization is *at least* a few thousand years, it is unlikely to have time to communicate with even its nearest neighbor.

### CONCEPT CHECK

✔ How does the Drake equation assist astronomers in refining their search for extraterrestrial life?

## 28.4 The Search for Extraterrestrial Intelligence

Let us continue our optimistic assessment of the prospects for life and assume that civilizations enjoy a long stay on their parent planet once their initial technological "teething problems" are past. In that case, intelligent, technological, and perhaps also communicative cultures are likely to be plentiful in the Galaxy. How might we become aware of their existence? The ongoing search for extraterrestrial intelligence (known to many by its acronym, **SETI**) is the topic of this final section.

### Meeting Our Neighbors

For definiteness, let's assume that the average lifetime of a technological civilization is 1 million years—only 1 percent of the reign of the dinosaurs, but 100 times longer than human civilization has survived thus far. Given the size and shape of our Galaxy and the known distribution of stars in the Galactic disk, we can then estimate the average *distance* between these civilizations to be some 30 pc, or about 100 light-years. Thus, any two-way communication with our neighbors—using signals traveling at or below the speed of light—will take at least 200 years (100 years for the message to reach the planet and another 100 years for the reply to travel back to us).

One obvious way to search for extraterrestrial life would be to develop the capability to travel far outside our solar system. However, that may never be a practical possibility. At a speed of 50 km/s, the speed of the fastest space probes operating today, the round-trip to even the nearest Sun-like star, Alpha Centauri, would take about 50,000 years. The journey to the nearest technological neighbor (assuming a distance of 30 pc) and back would take 600,000 years—almost the entire lifetime of our species! Interstellar travel at these speeds is clearly not feasible. Speeding up our ships to near the speed of light would reduce the travel time, but doing that is far beyond our present technology.

Actually, our civilization has already launched some interstellar probes, although they have no specific stellar destination. Figure 28.15 is a reproduction of a plaque mounted on board the *Pioneer 10* spacecraft launched in the mid-1970s and now well beyond the orbit of Pluto, on its way out of the solar system. Similar information was included aboard the *Voyager* probes launched in 1978. Although these spacecraft would be incapable of reporting back to Earth the news that they had encountered an alien culture, scientists hope that the civilization on the other end would be able to unravel most of its contents using the universal language of mathematics. The caption to Figure 28.15 notes how the aliens might discover from where and when the *Pioneer* and *Voyager* probes were launched.

Setting aside the many practical problems that arise in trying to establish direct contact with extraterrestrials, some scientists have argued that it might not even be a

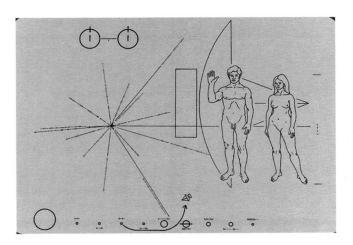

▲ **FIGURE 28.15** *Pioneer 10* **Plaque** A replica of a plaque mounted on board the *Pioneer 10* spacecraft. Included are scale drawings of the spacecraft, a man, and a woman; a diagram of the hydrogen atom undergoing a change in energy (top left); a starburst pattern representing various pulsars and the frequencies of their radio waves that can be used to estimate when the craft was launched (middle left); and a depiction of the solar system, showing that the spacecraft departed the third planet from the Sun and passed the fifth planet on its way to outer space (bottom). All the drawings have computer-coded (binary) markings from which actual sizes, distances, and times can be derived. *(NASA)*

particularly good idea. Our recent emergence as a technological civilization implies that we must be one of the least advanced technological intelligences in the entire Galaxy. Any other civilization that discovers us will almost surely be more advanced than us. Consequently, a healthy degree of caution may be warranted. If extraterrestrials behave even remotely like human civilizations on Earth, then the most advanced aliens may naturally try to dominate all others. The behavior of the "advanced" European cultures toward the "primitive" races they encountered on their voyages of discovery in the seventeenth, eighteenth, and nineteenth centuries should serve as a clear warning of the possible undesirable consequences of contact. Of course, the aggressiveness of Earthlings may not apply to extraterrestrials, but given the history of the one intelligent species we know, the cautious approach may be in order.

## Radio Communication

A cheaper and much more practical alternative to direct contact is to try to communicate with extraterrestrials by using only electromagnetic radiation, the fastest known means of transferring information from one place to another. Because light and other high-frequency radiation is heavily scattered while moving through dusty interstellar space, long-wavelength radio radiation seems to be the natural choice. We would not attempt to broadcast to *all* nearby candidate stars, however—that would be far too expensive and inefficient. Instead, radio telescopes on Earth

would listen *passively* for radio signals emitted by other civilizations. Indeed, some preliminary searches of selected nearby stars are now underway, thus far without success.

In what direction should we aim our radio telescopes? The answer to this question, at least, is fairly easy: On the basis of our earlier reasoning, we should target all F-, G-, and K-type stars in our vicinity. But are extraterrestrials broadcasting radio signals? If they are not, this search technique will obviously fail. And even if they are, how do we distinguish their artificially generated radio signals from signals naturally emitted by interstellar gas clouds? To what frequency should we tune our receivers? The answer to this question depends on whether the signals are produced deliberately or are simply "waste radiation" escaping from a planet.

Consider how Earth would look at radio wavelengths to extraterrestrials. Figure 28.16 shows the pattern of radio signals we emit into space. From the viewpoint of a distant observer, the spinning Earth emits a bright flash of radio radiation every few hours. In fact, Earth is now a more intense radio emitter than the Sun. The flashes result from the periodic rising and setting of hundreds of FM radio stations and television transmitters. Each station broadcasts mostly parallel to Earth's surface, sending a great "sheet" of electromagnetic radiation into interstellar space, as illustrated in Figure 28.16(a). (The more common AM broadcasts are trapped below our ionosphere, so those signals never leave Earth.)

Because the great majority of these transmitters are clustered in the eastern United States and western Europe, a distant observer would detect periodic blasts of radiation from Earth as our planet rotates each day (Figure 28.16b). This radiation races out into space and has been doing so since the invention of these technologies more than seven decades ago. Another civilization at least as advanced as ours might have constructed devices capable of detecting these blasts of radiation. If any sufficiently advanced (and sufficiently interested) civilization resides on a planet orbiting any of the thousand or so stars within about 70 light-years (20 pc) of Earth, then we have already broadcast our presence to them.

Of course, it may very well be that, having discovered cable and fiber-optics technology, most civilizations' indiscriminate transmissions cease after a few decades. In that case, radio silence becomes the hallmark of intelligence, and we must find an alternative means of locating our neighbors.

## The Water Hole

Now let us suppose that a civilization has decided to assist searchers by actively broadcasting its presence to the rest of the Galaxy. At what frequency should we listen for such an extraterrestrial beacon? The electromagnetic spectrum is enormous; the radio domain alone is vast. To hope to detect a signal at some unknown radio frequency is like searching for a needle in a haystack. Are some frequencies more likely than others to carry alien transmissions?

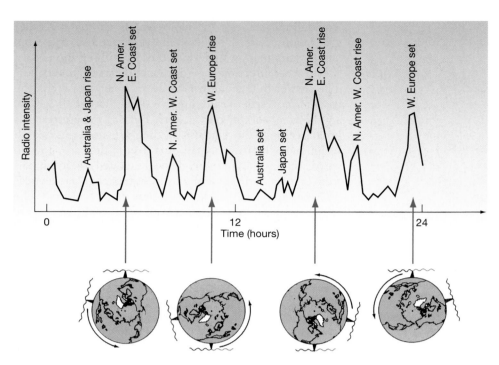

Distant
observer

◀ FIGURE 28.16 **Leakage** Radio radiation leaks from Earth into space because of the daily activities of our technological civilization. FM radio and television transmitters broadcast their energy parallel to Earth's surface, producing the strongest signal in any given direction when they lie on Earth's horizon as seen from afar. (The more common AM signals are trapped below our ionosphere and never leave Earth.) Because most transmitters are clustered in the eastern United States and western Europe, a distant observer would detect blasts of radio radiation from Earth as our planet rotates.

Some basic arguments suggest that civilizations might communicate at a wavelength near 20 cm. As we saw in Chapter 18, the basic building blocks of the universe, namely, hydrogen atoms, radiate naturally at a wavelength of 21 cm. ∞ (Sec. 18.4) Also, one of the simplest molecules, hydroxyl (OH), radiates near 18 cm. Together, these two substances form water ($H_2O$). Arguing that water is likely to be the interaction medium for life anywhere and that radio radiation travels through the disk of our Galaxy with the least absorption by interstellar gas and dust, some researchers have proposed that the interval between 18 and 21 cm is the best range of wavelengths for civilizations to transmit or monitor. Called the **water hole,** this radio interval might serve as an "oasis" where all advanced galactic civilizations would gather to conduct their electromagnetic business.

The water-hole frequency interval is only a guess, of course, but it is supported by other arguments as well. Figure 28.17 shows the water hole's location in the electromagnetic spectrum and plots the amount of natural emission from our Galaxy and from Earth's atmosphere. The 18- to 21-cm range lies within the quietest part of the spectrum, where the galactic "static" from stars and interstellar clouds happens to be minimized. Furthermore, the atmospheres of typical planets are also expected to interfere least at these wavelengths. Thus, the water hole seems like a good choice for the frequency of an interstellar beacon, although we cannot be sure of this reasoning until contact is actually achieved.

A few radio searches are now in progress at frequencies in and around the water hole. One of the most sensitive and comprehensive SETI projects was Project Phoenix, carried out during the late 1990s. Large radio antennas, such as that

in Figure 28.18, were used to search millions of channels simultaneously in the 1- to 3-GHz spectrum. Actually, in these searches, computers do most of the "listening;" humans get involved only if the signals look intriguing. The inset in Figure 28.18 shows what a typical narrowband, 1-Hz signal—a potential "signature" of an intelligent transmission—would look like on a computer monitor. However, this observation

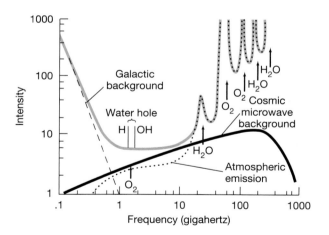

▲ FIGURE 28.17 **Water Hole** The "water hole" is bounded by the natural emission frequencies of the hydrogen (H) atom (21-cm wavelength) and the hydroxyl (OH) molecule (18-cm wavelength). ∞ (Secs. 18.4, 18.5) The topmost solid (blue) curve sums the natural emissions of our Galaxy (dashed line on left side of diagram, labeled "Galactic background") and Earth's atmosphere (dotted line on right side of diagram, denoted by various chemical symbols). ∞ (Sec. 26.7) This sum is minimized near the water-hole frequencies. Perhaps all intelligent civilizations conduct their interstellar communications within this quiet "electromagnetic oasis."

◀ **FIGURE 28.18 Project Phoenix** This large radio telescope at the National Radio Astronomy Observatory in Green Bank, West Virginia, was used by Project Phoenix during the 1990s to search for extraterrestrial intelligent signals. The inset shows a typical recording of an alien signal—here, as a test, the Doppler-shifted broadcast from the *Pioneer 10* spacecraft, now well beyond the orbit of Neptune. The diagonal line across the computer monitor, in contrast to the random noise in the background, betrays the presence of an intelligent signal in the incoming data stream. *(NRAO; SETI Institute)*

was merely a test to detect the weak, redshifted radio signal emitted by the *Pioneer 10* robot, now receding into the outer realm of our solar system—a sign of intelligence, but one that we put there. Nothing resembling an extraterrestrial signal has yet been detected.

Right now, the space surrounding all of us could be flooded with radio signals from extraterrestrial civilizations. If only we knew the proper direction and frequency, we might be able to make one of the most startling discoveries of all time. The result would likely provide whole new opportunities to study the cosmic evolution of energy, matter, and life throughout the universe.

CONCEPT CHECK

✔ Why do many researchers regard the "water hole" as a likely place to search for extraterrestrial signals?

# CHAPTER REVIEW

## Summary

1 **Cosmic evolution (p. 760)** is the continuous process that has led to the appearance of galaxies, stars, planets, and life on Earth. Living organisms may be characterized by their ability to react to their environment, to grow by taking in nutrition from their surroundings, and to reproduce, passing along some of their own characteristics to their offspring. Organisms that can best take advantage of their new surroundings succeed at the expense of those organisms that cannot make the necessary adjustments. Intelligence is strongly favored by natural selection.

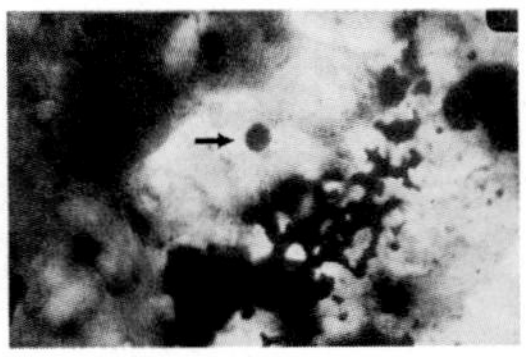

2 Powered by natural energy sources, reactions between simple molecules in the oceans of the primitive Earth are thought to have led to the formation of **amino acids (p. 761)** and **nucleotide bases (p. 762),** the basic molecules of life. Alternatively, some complex molecules may have been formed in interstellar space and then delivered to Earth by meteors or comets. The best hope for life beyond Earth in the solar system is the planet Mars, although no evidence for living organisms has been found there. Jupiter's moon Europa and Saturn's Titan may also be possibilities, but conditions on those bodies are harsh by terrestrial standards.

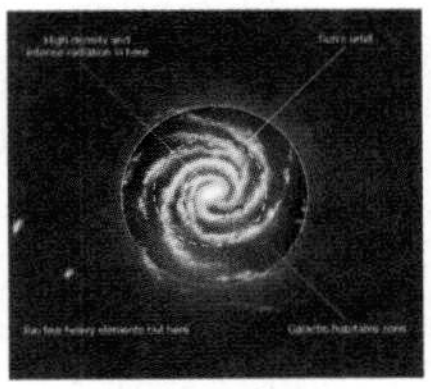

3 The **Drake equation (p. 768)** provides a means of estimating the probability of intelligent life in the Galaxy. The astronomical factors in the equation are the galactic star-formation rate, the likelihood of planets, and the number of habitable planets. Chemical and biological factors are the probability that life appears and the probability that it subsequently develops intelligence. Cultural and political factors are the probability that intelligence leads to technology and the lifetime of a civilization in the technological state. Taking an optimistic view of the development of life and intelligence leads to the conclusion that the total number of technologically competent civilizations in the Galaxy is approximately equal to the lifetime of a typical civilization, expressed in years. Even with optimistic assumptions, the distance to our nearest intelligent neighbor is likely to be many hundreds of parsecs.

④ Currently, space travel is not a feasible means of searching for intelligent life. Existing programs to discover extraterrestrial intelligence involve scanning the electromagnetic spectrum for signals. So far, no intelligible broadcasts have been received. A technological civilization would probably "announce" itself to the universe by the radio and television signals it emits

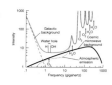

into space. Observed from afar, our planet would appear as a radio source with a 24-hour period, as different regions of the planet rise and set. The **"water hole" (p. 774)** is a region in the radio range of the electromagnetic spectrum, near the 21-cm line of hydrogen and the 18-cm line of hydroxyl, where natural emissions from the Galaxy happen to be minimized. Many researchers regard this region as the best part of the spectrum for communication purposes.

## Review and Discussion

1. Why is life difficult to define?

2. What is chemical evolution?

3. What is the Urey–Miller experiment? What important organic molecules were produced in that experiment?

4. What other experiments have attempted to produce organic molecules by inorganic means?

5. What are the basic ingredients from which biological molecules formed on Earth?

6. Why do some scientists think life might have originated in space?

7. How do we know anything at all about the early episodes of life on Earth?

8. What is the role of language in cultural evolution?

9. Where else, besides Earth, have organic molecules been found?

10. Where—besides Earth and the planet Mars—might we hope to find signs of life in our solar system?

11. Do we know whether Mars ever had life at any time during its past? What argues in favor of the position that it may once have harbored life?

12. What is generally meant by "life as we know it"? What other forms of life might be possible?

13. How many of the factors in the Drake equation are known with any degree of certainty? Which factor is least well known?

14. What factors determine the suitability of a star as the parent of a planet on which life might arise?

15. What is the relationship between the average lifetime of galactic civilizations and the possibility of our someday communicating with them?

16. How would Earth appear at radio wavelengths to extraterrestrial astronomers?

17. Do you think that advanced civilizations would continue to emit large amounts of radio energy as they evolved?

18. What are the advantages in using radio waves for communication over interstellar distances?

19. What is the "water hole"? What advantages does it offer for interstellar communication?

20. If you were designing a SETI experiment, what parts of the sky would you monitor?

## Conceptual Self-Test: True or False/Multiple Choice

1. The definition of life requires only that, to be considered "alive," you must be able to reproduce.

2. Organic molecules exist only on Earth.

3. Laboratory experiments have created living cells from non-biological molecules.

4. Dinosaurs lived on Earth for a thousand times longer than human civilization has existed to date.

5. The *Viking* landers on Mars discovered microscopic evidence of life, but found no large fossil evidence.

6. We have no direct evidence for Earth-like planets orbiting other stars.

7. Our civilization has already launched probes into interstellar space and broadcast our presence to our neighbors.

8. The Drake equation estimates the number of Earth-like planets in the Milky Way Galaxy.

9. Planets in binary-star systems are not considered habitable, because the planetary orbits are usually unstable.

10. The development of life and intelligence on Earth are extremely unlikely if chance is the only evolutionary factor involved.

11. The "assumptions of mediocrity" suggest that (a) life should be common throughout the cosmos; (b) lower forms of life

must evolve to higher forms; (c) lower forms of life have lower intelligence; (d) viruses are actually life-forms.

12. The chemical elements that form the basic molecules needed for life are found (a) in the cores of Sun-like stars; (b) commonly throughout the cosmos; (c) only on planets that have liquid water; (d) only on Earth.

13. Fossil records of early life-forms on Earth suggest that life began about (a) 6000 years ago; (b) 65 million years ago; (c) 3.5 billion years ago; (d) 14 billion years ago.

14. The discovery of bacteria on another planet would be an important discovery because bacteria (a) can easily survive in high temperatures; (b) are the only life-form to exist on Earth for most of the planet's history; (c) are the lowest form of life known to exist; (d) eventually evolve into intelligent beings.

15. The least-well-known factor in the Drake equation is (a) the rate of star formation; (b) the average number of habitable planets within planetary systems; (c) the average lifetime of a technologically competent civilization; (d) the diameter of the Milky Way Galaxy.

16. Although the habitable zone around a large B-class star is large, we don't often look for life on planets there because the star (a) has too much gravity; (b) is too short lived for

life to evolve; (c) is at too low a temperature to sustain life; (d) would have only gas giant planets.

17. NASA's Space Shuttle orbits Earth at about 17,500 mph. If it traveled to the next Sun-like star at that speed, the trip would take at least (a) one week; (b) one decade; (c) one century; (d) 100 millennia.

18. If Figure 28.16 ("Earth's Radio Leakage") were to be redrawn for a planet spinning twice as fast, the new jagged line would be (a) unchanged; (b) taller; (c) stretched out horizontally; (d) compressed horizontally.

19. Radio telescopes cannot simply scan the skies looking for signals, because (a) astronomers don't know what frequencies alien civilizations might use; (b) many nonliving objects emit radio signals naturally; (c) Earth's radio communications drown out extraterrestrial signals; (d) inclement weather in the winter prevents the use of radio telescopes.

20. The strongest radio-wavelength emitter in the solar system is (a) human-made signals from Earth; (b) the Sun; (c) the Moon; (d) Jupiter.

## Problems

*Algorithmic versions of these Problems are available in the Practice Problems module of the Companion Website.*
*The number of dots preceding each Problem indicates its approximate level of difficulty.*

1. • If Earth's 4.6-billion-year age were compressed to 46 years, as described in the text, what would be your age, in seconds? On that scale, how long ago was the end of World War II? The Declaration of Independence? Columbus's discovery of the New World? The extinction of the dinosaurs?

2. ••• According to the inverse-square law, a planet receives energy from its parent star at a rate proportional to the star's luminosity and inversely proportional to the square of the planet's distance from the star. ∞ (Sec. 17.2) According to Stefan's law, the rate at which the planet radiates energy into space is proportional to the fourth power of its surface temperature. ∞ (Sec. 3.4) In equilibrium, the two rates are equal. Based on this information, and given the fact that (taking into account the greenhouse effect) the Sun's habitable zone extends from 0.6 AU to 1.5 AU, estimate the extent of the habitable zone surrounding a K-type main-sequence star of the luminosity of the Sun.

3. •• Using the data in the previous problem, how would the inner and outer radii of the Sun's habitable zone change if the solar luminosity increased by a factor of four?

4. •• The outer edge of the habitable zone corresponds roughly to the freezing point of water (273 K). Using the data in problem 2, how would this radius change if life could survive at temperatures as low as 150 K?

5. ••• Using the data in problem 2, what would be the orbital period of a planet orbiting at the outer edge of the habitable zone of an F-type main-sequence star of 1.5 solar masses and luminosity five times that of the Sun?

6. • Based on the numbers presented in the text, and assuming an average lifetime of 5 billion years for suitable stars, estimate the total number of habitable planets in the Galaxy.

7. •• A planet orbits one component of a binary-star system at a distance of 1 AU. (See Figure 28.14a.) If both stars have the same mass and their orbit is circular, estimate the minimum distance between the stars for the tidal force due to the companion not to exceed a "safe" 0.01 percent of the gravitational force between the planet and its parent star.

8. • Suppose that each of the fractional factors in the Drake equation turns out to have a value of 1/10, that stars form at an average rate of 20 per year, and that each star has exactly one habitable planet orbiting it. Estimate the present number of technological civilizations in the Milky Way Galaxy if the average lifetime of a civilization is (a) 100 years; (b) 10,000 years; (c) 1 million years.

9. ••• If we adopt the estimate from the text that the number of technological civilizations in the Milky Way Galaxy is equal to the average lifetime of a civilization, it follows that the distance to our nearest neighbor decreases as the average lifetime increases. Assuming that civilizations are uniformly spread over a two-dimensional Galactic disk of radius 15 kpc and that all have the same lifetime, calculate the *minimum* lifetime for which two-way radio communication with our nearest neighbor would be possible before our civilization ends. Is the assumption that civilizations are uniformly spread across the disk a reasonable one?

10. ••• Repeat the calculation in the previous question for a round-trip personal visit, using current-technology spacecraft that travel at 50 km/s.

11. • How fast would a spacecraft have to travel in order to complete the trip from Earth to Alpha Centauri (a distance of 1.3 pc) and back in less than an average human lifetime (80 years, say)?

12. • Assuming that there are 10,000 FM radio stations on Earth, each transmitting at a power level of 50 kW, calculate the total radio luminosity of Earth in the FM band. Compare this value with the roughly $10^6$ W radiated by the Sun in the same frequency range.

13. • Convert the water hole's wavelengths to frequencies. For practical reasons, any search of the water hole must be broken up into channels, much like those you find on a television, except that the water-hole's channels are very narrow in radio frequency, about 100 Hz wide. How many channels must astronomers search in the water hole?

14. • At what wavelength does the microwave background radiation peak, and what is the corresponding frequency? How does this frequency compare with the water-hole's frequency range?

15. • There are 20,000 stars within 100 light-years that are to be searched for radio communications. How long will the search take if 1 hour is spent looking at each star? What if 1 day is spent per star?

*The Companion Website at www.aw-bc.com/chaisson provides algorithmically generated versions of each chapter's Problems, along with additional quizzes, an Animations & Videos gallery, an Images gallery, an interactive Glossary, and a full eBook.*

# APPENDIX 1
## Scientific Notation

The objects studied by astronomers range in size from the smallest particles to the largest expanse of matter we know—the entire universe. Subatomic particles have sizes of about 0.000000000000001 meter, while galaxies (like that shown in Figure 1.3) typically measure some 1,000,000,000,000,000,000,000 meters across. The most distant known objects in the universe lie on the order of 100,000,000,000,000,000,000,000,000 meters from Earth.

Obviously, writing all those zeros is both cumbersome and inconvenient. More important, it is also very easy to make an error—write down one zero too many or too few and your calculations become hopelessly wrong! To avoid this, scientists always write large numbers using a short-hand notation in which the number of zeros following or preceding the decimal point is denoted by a superscript power, or *exponent*, of 10. The exponent is simply the number of places between the first significant (nonzero) digit in the number (reading from left to right) and the decimal point. Thus, 1 is $10^0$, 10 is $10^1$, 100 is $10^2$, 1000 is $10^3$, and so on. For numbers less than 1, with zeros between the decimal point and the first significant digit, the exponent is negative: 0.1 is $10^{-1}$, 0.01 is $10^{-2}$, 0.001 is $10^{-3}$, and so on. Using this notation we can shorten the number describing subatomic particles to $10^{-15}$ meter, and write the number describing the size of a galaxy as $10^{21}$ meters.

More complicated numbers are expressed as a combination of a power of 10 and a multiplying factor. This factor is conventionally chosen to be a number between 1 and 10, starting with the first significant digit in the original number. For example, 150,000,000,000 meters (the distance from Earth to the Sun, in round numbers) can be more concisely written at $1.5 \times 10^{11}$ meters, 0.000000025 meters as $2.5 \times 10^{-8}$ meter, and so on. The exponent is simply the number of places the decimal point must be moved *to the left* to obtain the multiplying factor.

Some other examples of scientific notation are:

- the approximate distance to the Andromeda Galaxy
  = 2,500,000 light-years = $2.5 \times 10^6$ light-years
- the size of a hydrogen atom
  = 0.00000000005 meter = $5 \times 10^{-11}$ meter
- the diameter of the Sun
  = 1,392,000 kilometers = $1.392 \times 10^6$ kilometers

- the U.S. national debt (as of July 1, 2007)
  = \$9,049,632,000,000.00 = \$9.049632 trillion
  = $9.049632 \times 10^{12}$ dollars.

In addition to providing a simpler way of expressing very large or very small numbers, this notation also makes it easier to do basic arithmetic. The rule for multiplication of numbers expressed in this way is simple: Just multiply the factors and add the exponents. Similarly for division: Divide the factors and subtract the exponents. Thus, $3.5 \times 10^{-2}$ multiplied by $2.0 \times 10^3$ is simply $(3.5 \times 2.0) \times 10^{-2+3} = 7.0 \times 10^1$—that is, 70. Again, $5 \times 10^6$ divided by $2 \times 10^4$ is just $(5/2) \times 10^{6-4}$, or $2.5 \times 10^2$ (= 250). Applying these rules to unit conversions, we find, for example, that 200,000 nanometers is $200,000 \times 10^{2-9}$ meter (since 1 nanometer = $10^{-9}$ meter; see Appendix 2), or $2 \times 10^5 \times 10^{-9}$ meter, or $2 \times 10^{5-9} = 2 \times 10^{-4}$ meter = 0.2 mm. Verify these rules for yourself with a few examples of your own. The advantages of this notation when considering astronomical objects will soon become obvious.

Scientists often use "rounded-off" versions of numbers, both for simplicity and for ease of calculation. For example, we will usually write the diameter of the Sun as $1.4 \times 10^6$ kilometers, instead of the more precise number given earlier. Similarly, Earth's diameter is 12,756 kilometers, or $1.2756 \times 10^4$ kilometers, but for "ballpark" estimates we really don't need so many digits and the more approximate number $1.3 \times 10^4$ kilometers will suffice. Very often, we perform rough calculations using only the first one or two significant digits in a number, and that may be all that is necessary to make a particular point. For example, to support the statement, "The Sun is much larger than Earth," we need only say that the ratio of the two diameters is roughly $1.4 \times 10^6$ divided by $1.3 \times 10^4$. Since 1.4/1.3 is close to 1, the ratio is approximately $10^6/10^4 = 10^2$, or 100. The essential fact here is that the ratio is much larger than 1; calculating it to greater accuracy (to get 109.13) would give us no additional *useful* information. This technique of stripping away the arithmetic details to get to the essence of a calculation is very common in astronomy, and we use it frequently throughout this text.

# APPENDIX 2
Astronomical Measurement

Astronomers use many different kinds of units in their work, simply because no single system of units will do. Rather than the *Système Internationale* (SI), or meter-kilogram-second (MKS), metric system used in most high school and college science classes, many professional astronomers still prefer the older centimeter-gram-second (CGS) system. However, astronomers also commonly introduce new units when convenient. For example, when discussing stars, the mass and radius of the Sun are often used as reference points. The solar mass, written as $M_\odot$, is equal to $2.0 \times 10^{33}$ g, or $2.0 \times 10^{30}$ kg (since 1 kg = 1000 g). The solar radius, $R_\odot$, is equal to 700,000 km, or $7.0 \times 10^{8}$ m (1 km = 1000 m). The subscript $\odot$ always stands for Sun. Similarly, the subscript $\oplus$ always stands for Earth. In this book, we try to use the units that astronomers commonly use in any given context, but we also give the "standard" SI equivalents where appropriate.

Of particular importance are the units of length astronomers use. On small scales, the *angstrom* (1 Å = $10^{-10}$ m = $10^{-8}$ cm), the *nanometer* (1 nm = $10^{-9}$ m = $10^{-7}$ cm), and the *micron* (1 $\mu$m = $10^{-6}$ m = $10^{-4}$ cm) are used. Distances within the solar system are usually expressed in terms of the *astronomical unit* (AU), the mean distance between Earth and the Sun. One AU is approximately equal to 150,000,000 km, or $1.5 \times 10^{11}$ m. On larger scales, the *light-year* (1 ly = $9.5 \times 10^{15}$ m = $9.5 \times 10^{12}$ km) and the *parsec* (1 pc = $3.1 \times 10^{16}$ m = $3.1 \times 10^{13}$ km = 3.3 ly) are commonly used. Still larger distances use the regular prefixes of the metric system: *kilo* for one thousand and *mega* for one million. Thus 1 kiloparsec (kpc) = $10^3$ pc = $3.1 \times 10^{19}$ m, 10 megaparsecs (Mpc) = $10^7$ pc = $3.1 \times 10^{23}$ m, and so on.

Astronomers use units that make sense within a context, and as contexts change, so do the units. For example, we might measure densities in grams per cubic centimeter (g/cm$^3$), in atoms per cubic meter (atoms/m$^3$), or even in solar masses per cubic megaparsec ($M_\odot$/Mpc$^3$), depending on the circumstances. The important thing to know is that once you understand the units, you can convert freely from one set to another. For example, the radius of the Sun could equally well be written as $R_\odot = 6.96 \times 10^3$ m, or $6.96 \times 10^{10}$ cm, or 109 $R_\oplus$, or $4.65 \times 10^{-3}$ AU, or even $7.363 \times 10^{-5}$ ly—whichever happens to be most useful. Some of the more common units used in astronomy, and the contexts in which they are most likely to be encountered, are listed below.

| Length: | | |
|---|---|---|
| 1 angstrom (Å) | = $10^{-10}$ m | |
| 1 nanometer (nm) | = $10^{-9}$ m | atomic physics, spectroscopy |
| 1 micron ($\mu$m) | = $10^{-6}$ m | interstellar dust and gas |
| 1 centimeter (cm) | = 0.01 m | |
| 1 meter (m) | = 100 cm | in widespread use throughout all astronomy |
| 1 kilometer (km) | = 1000 m = $10^5$ cm | |
| Earth radius ($R_\oplus$) | = 6378 km | planetary astronomy |
| Solar radius ($R_\odot$) | = $696 \times 10^8$ m | |
| 1 astronomical unit (AU) | = $1.496 \times 10^{11}$ m | solar system, stellar evolution |
| 1 light-year (ly) | = $9.46 \times 10^{15}$ m = 63,200 AU | |
| 1 parsec (pc) | = $3.09 \times 10^{16}$ m = 206,000 AU | galactic astronomy, stars and star clusters |
| | = 3.26 ly | |
| 1 kiloparsec (kpc) | = 1000 pc | |
| 1 megaparsec (Mpc) | = 1000 kpc | galaxies, galaxy clusters, cosmology |
| **Mass:** | | |
| 1 gram (g) | | |
| 1 kilogram (kg) | = 1000 g | in widespread use in many different areas |
| Earth mass ($M_\oplus$) | = $5.98 \times 10^{24}$ kg | planetary astronomy |
| Solar mass ($M_\odot$) | = $1.99 \times 10^{30}$ kg | "standard" unit for all mass scales larger than Earth |
| **Time:** | | |
| 1 second (s) | | in widespread use throughout astronomy |
| 1 hour (h) | = 3600 s | |
| 1 day (d) | = 86,400 s | planetary and stellar scales |
| 1 year (yr) | = $3.16 \times 10^7$ s | virtually all processes occurring on scales larger than a star |

# APPENDIX 3
## Tables

| TABLE 1 Some Useful Constants and Physical Measurements* | |
|---|---|
| astronomical unit | $1 \text{ AU} = 1.496 \times 10^8 \text{ km} (1.5 \times 10^8 \text{ km})$ |
| light-year | $1 \text{ ly} = 9.46 \times 10^{12} \text{ km} (10^{13} \text{ km 6 trillion miles})$ |
| parsec | $1 \text{ pc} = 3.09 \times 10^{13} \text{ km} = 206{,}000 \text{ AU} = 3.3 \text{ ly}$ |
| speed of light | $c = 299{,}792.458 \text{ km/s} (3 \times 10^5 \text{ km/s})$ |
| Stefan-Boltzmann constant | $a = 5.67 \times 10^{-8} \text{ W/m}^2 \cdot \text{K}^4$ |
| Planck's constant | $h = 6.63 \times 10^{-34} \text{ J s}$ |
| gravitational constant | $G = 6.67 \times 10^{-11} \text{ Nm}^2/\text{kg}^2$ |
| mass of Earth | $M_\oplus = 5.98 \times 10^{24} \text{ kg} (6 \times 10^{24} \text{ kg}, \text{ about 6000 billion billion tons})$ |
| radius of Earth | $R_\oplus = 6378 \text{ km} (6500 \text{ km})$ |
| mass of the Sun | $M_\odot = 1.99 \times 10^{30} \text{ kg} (2 \times 10^{30} \text{ kg})$ |
| radius of the Sun | $R_\odot = 6.96 \times 10^5 \text{ km} (7 \times 10^5 \text{ kw})$ |
| luminosity of the Sun | $L_\odot = 3.90 \times 10^{26} \text{ W} (4 \times 10^{26} \text{ W})$ |
| effective temperature of the Sun | $T_\odot = 5778 \text{ K} (5800 \text{ K})$ |
| Hubble's constant | $H_0 = 70 \text{ km/s/Mpc}$ |
| mass of an electron | $m_e = 9.11 \times 10^{-31} \text{ kg}$ |
| mass of a proton | $m_p = 1.67 \times 10^{-27} \text{ kg}$ |

*The rounded-off values used in the text are shown above in parentheses.*

### Conversions Between Common English and Metric Units

| English | Metric |
|---|---|
| 1 inch | = 2.54 centimeters (cm) |
| 1 foot (ft) | = 0.3048 meters (m) |
| 1 mile | = 1.609 kilometers (km) |
| 1 pound (lb) | = 453.6 grams (g) or 0.4536 kilograms (kg) [on Earth] |

## TABLE 2    Periodic Table of Elements

Key:

| 2 | Atomic number |
|---|---|
| He | Symbol of element |
| 4.003 | Atomic weight |
| Helium | Name of element |

| Group | 1 | 2 | 3 | 4 | 5 | 6 | 7 | 8 | 9 | 10 | 11 | 12 | 13 | 14 | 15 | 16 | 17 | 18 |
|---|---|---|---|---|---|---|---|---|---|---|---|---|---|---|---|---|---|---|
| **Period 1** | 1 H 1.0080 Hydrogen | | | | | | | | | | | | | | | | | 2 He 4.003 Helium |
| **Period 2** | 3 Li 6.939 Lithium | 4 Be 9.012 Beryllium | | | | | | | | | | | 5 B 10.81 Boron | 6 C 12.011 Carbon | 7 N 14.007 Nitrogen | 8 O 15.9994 Oxygen | 9 F 18.998 Fluorine | 10 Ne 20.183 Neon |
| **Period 3** | 11 Na 22.990 Sodium | 12 Mg 24.31 Magnesium | | | | | | | | | | | 13 Al 26.98 Aluminum | 14 Si 28.09 Silicon | 15 P 30.974 Phosphorus | 16 S 32.064 Sulfur | 17 Cl 35.453 Chlorine | 18 Ar 39.948 Argon |
| **Period 4** | 19 K 39.102 Potassium | 20 Ca 40.08 Calcium | 21 Sc 44.96 Scandium | 22 Ti 47.90 Titanium | 23 V 50.94 Vanadium | 24 Cr 52.00 Chromium | 25 Mn 53.94 Manganese | 26 Fe 55.85 Iron | 27 Co 58.93 Cobalt | 28 Ni 58.71 Nickel | 29 Cu 63.54 Copper | 30 Zn 65.37 Zinc | 31 Ga 69.72 Gallium | 32 Ge 72.59 Germanium | 33 As 74.92 Arsenic | 34 Se 78.96 Selenium | 35 Br 79.909 Bromine | 36 Kr 83.80 Krypton |
| **Period 5** | 37 Rb 85.47 Rubidium | 38 Sr 87.62 Strontium | 39 Y 88.91 Yttrium | 40 Zr 91.22 Zirconium | 41 Nb 92.91 Niobium | 42 Mo 95.94 Molybdenum | 43 Tc (99) Technetium | 44 Ru 101.1 Ruthenium | 45 Rh 102.90 Rhodium | 46 Pd 106.4 Palladium | 47 Ag 107.87 Silver | 48 Cd 112.40 Cadmium | 49 In 114.82 Indium | 50 Sn 118.69 Tin | 51 Sb 121.75 Antimony | 52 Te 127.60 Tellurium | 53 I 126.9 Iodine | 54 Xe 131.30 Xenon |
| **Period 6** | 55 Cs 132.91 Cesium | 56 Ba 137.34 Barium | 71 Lu 174.97 Lutetium * | 72 Hf 178.49 Hafnium | 73 Ta 180.95 Tantalum | 74 W 183.85 Tungsten | 75 Re 186.2 Rhenium | 76 Os 190.2 Osmium | 77 Ir 192.2 Iridium | 78 Pt 195.09 Platinum | 79 Au 197.0 Gold | 80 Hg 200.59 Mercury | 81 Tl 204.37 Thallium | 82 Pb 207.19 Lead | 83 Bi 208.98 Bismuth | 84 Po (210) Polonium | 85 At (210) Astantine | 86 Rn (222) Radon |
| **Period 7** | 87 Fr (223) Francium | 88 Ra 226.05 Radium | 103 Lw (257) Lawrencium ** | 104 Rf (261) Rutherfordium | 105 Db (262) Dubnium | 106 Sg (263) Seaborgium | 107 Bh (262) Bohrium | 108 Hs (265) Hassium | 109 Mt (266) Meitnerium | 110 Uun (269) Ununnilium | 111 Uuu (272) Unununium | 112 Uub (277) Ununbium | 113 Uut (undiscovered) Unununtrium | 114 Uuq (285) Ununquadium | 115 Uup (288) Ununpentium | 116 Uuh (292) Ununhexium | | 118 Uuo (294) Ununoctium |

| * | 57 La 138.91 Lanthanum | 58 Ce 140.12 Cerium | 59 Pr 140.91 Praseodymium | 60 Nd 144.24 Neodymium | 61 Pm (147) Promethium | 62 Sm 150.35 Samarium | 63 Eu 151.96 Europium | 64 Gd 157.25 Gadolinium | 65 Tb 158.92 Terbium | 66 Dy 162.50 Dysprosium | 67 Ho 164.93 Holmium | 68 Er 167.26 Erbium | 69 Tm 168.93 Thulium | 70 Yb 173.04 Ytterbium |
|---|---|---|---|---|---|---|---|---|---|---|---|---|---|---|
| ** | 89 Ac (227) Actinium | 90 Th 232.04 Thorium | 91 Pa (231) Protactinium | 92 U 238.03 Uranium | 93 Np (237) Neptunium | 94 Pu (242) Plutonium | 95 Am (243) Americium | 96 Cm (247) Curium | 97 Bk (249) Berkelium | 98 Cf (251) Californium | 99 Es (254) Einsteinium | 100 Fm (253) Fermium | 101 Md (256) Mendelevium | 102 No (254) Nobelium |

*Preliminary evidence has been reported for elements 114, 115, and 116. Element 118 was "discovered" in 1999, retracted in 2002, and reported again in 2006.*

## TABLE 3A  Planetary Orbital Data

| Planet | Semi-Major Axis (AU) | Semi-Major Axis (10⁶ km) | Eccentricity (e) | Perihelion (AU) | Perihelion (10⁶ km) | Aphelion (AU) | Aphelion (10⁶ km) |
|---|---|---|---|---|---|---|---|
| Mercury | 0.39 | 57.9 | 0.206 | 0.31 | 46.0 | 0.47 | 69.8 |
| Venus | 0.72 | 108.2 | 0.007 | 0.72 | 107.5 | 0.73 | 108.9 |
| Earth | 1.00 | 149.6 | 0.017 | 0.98 | 147.1 | 1.02 | 152.1 |
| Mars | 1.52 | 227.9 | 0.093 | 1.38 | 206.6 | 1.67 | 249.2 |
| Jupiter | 5.20 | 778.4 | 0.048 | 4.95 | 740.7 | 5.46 | 816 |
| Saturn | 9.54 | 1427 | 0.054 | 9.02 | 1349 | 10.1 | 1504 |
| Uranus | 19.19 | 2871 | 0.047 | 18.3 | 2736 | 20.1 | 3006 |
| Neptune | 30.07 | 4498 | 0.009 | 29.8 | 4460 | 30.3 | 4537 |

| Planet | Mean Orbital Speed (km/s) | Sidereal Period (tropical years) | Synodic Period (days) | Inclination to the Ecliptic (degrees) | Greatest Angular Diameter as Seen from Earth (arc seconds) |
|---|---|---|---|---|---|
| Mercury | 47.87 | 0.24 | 115.88 | 7.00 | 13 |
| Venus | 35.02 | 0.62 | 583.92 | 3.39 | 64 |
| Earth | 29.79 | 1.00 | — | 0.01 | — |
| Mars | 24.13 | 1.88 | 779.94 | 1.85 | 25 |
| Jupiter | 13.06 | 11.86 | 398.88 | 1.31 | 50 |
| Saturn | 9.65 | 29.42 | 378.09 | 2.49 | 21 |
| Uranus | 6.80 | 83.75 | 369.66 | 0.77 | 4.1 |
| Neptune | 5.43 | 163.7 | 367.49 | 1.77 | 2.4 |

## TABLE 3B    Planetary Physical Data

| Planet | Equatorial Radius | | Mass | | Mean Density | Surface Gravity | Escape Speed |
|--------|------|-----------|------------------------|------------|-----------|-----------|----------|
|        | (km) | (Earth = 1) | (kg) | (Earth = 1) | (kg/m$^*$) | (Earth = 1) | (km/s) |
| Mercury | 2440 | 0.38 | $3.30 \times 10^{23}$ | 0.055 | 5430 | 0.38 | 4.2 |
| Venus | 6052 | 0.95 | $4.87 \times 10^{24}$ | 0.82 | 5240 | 0.91 | 10.4 |
| Earth | 6378 | 1.00 | $5.97 \times 10^{24}$ | 1.00 | 5520 | 1.00 | 11.2 |
| Mars | 3394 | 0.53 | $6.42 \times 10^{23}$ | 0.11 | 3930 | 0.38 | 5.0 |
| Jupiter | 71,492 | 11.21 | $1.90 \times 10^{27}$ | 317.8 | 1330 | 2.53 | 60 |
| Saturn | 60,268 | 9.45 | $5.68 \times 10^{26}$ | 95.16 | 690 | 1.07 | 36 |
| Uranus | 25,559 | 4.01 | $8.68 \times 10^{25}$ | 14.54 | 1270 | 0.91 | 21 |
| Neptune | 24,766 | 3.88 | $1.02 \times 10^{26}$ | 17.15 | 1640 | 1.14 | 24 |

| Planet | Sidereal Rotation Period (solar days)* | Axial Tilt (degrees) | Surface Magnetic Field (Earth = 1) | Magnetic Axis Tilt (degrees relative to rotation axis) | Albedo [†] | Surface Temperature[‡] (k) | Number of Moons** |
|--------|------|------|--------|-------|------|---------|----|
| Mercury | 58.6 | 0.0 | 0.011 | <10 | 0.11 | 100–700 | 0 |
| Venus | −243.0 | 177.4 | <0.001 | | 0.65 | 730 | 0 |
| Earth | 0.9973 | 23.45 | 1.0 | 11.5 | 0.37 | 290 | 1 |
| Mars | 1.026 | 23.98 | 0.001 | | 0.15 | 180–270 | 2 |
| Jupiter | 0.41 | 3.08 | 13.89 | 9.6 | 0.52 | 124 | 16 |
| Saturn | 0.44 | 26.73 | 0.67 | 0.8 | 0.47 | 97 | 18 |
| Uranus | −0.72 | 97.92 | 0.74 | 58.6 | 0.50 | 58 | 27 |
| Neptune | 0.67 | 29.6 | 0.43 | 46.0 | 0.5 | 59 | 13 |

*A negative sign indicates retrograde rotation.
[†]Fraction of sunlight reflected from surface.
[‡]Temperature is effective temperature for jovian planets.
**Moons more than 10 km in diameter.

**TABLE 4 The Twenty Brightest Stars in Earth's Night Sky**

| Name | Star | Spectral Type* A | B | Parallax (arc seconds) | Distance (pc) | Apparent Visual Magnitude* A | B |
|---|---|---|---|---|---|---|---|
| Sirius | α CMa | A1V | wd[†] | 0.379 | 2.6 | −1.44 | +8.4 |
| Canopus | α Car | F0Ib–II | | 0.010 | 96 | −0.62 | |
| Arcturus | α Boo | K2III | | 0.089 | 11 | −0.05 | |
| Rigel Kentaurus (Alpha Centauri) | α Gen | G2V | K0V | 0.742 | 1.3 | −0.01 | +1.4 |
| Vega | α Lyr | A0V | | 0.129 | 7.8 | +0.03 | |
| Capella | α Aur | GIII | M1V | 0.077 | 13 | +0.08 | +10.2 |
| Rigel | β Ori | B8Ia | B9 | 0.0042 | 240 | +0.18 | +6.6 |
| Procyon | α CMi | F5IV–V | wd[†] | 0.286 | 3.5 | +0.40 | +10.7 |
| Betelgeuse | α Ori | M2Iab | | 0.0076 | 130 | +0.45 | |
| Achernar | α Eri | B5V | | 0.023 | 44 | +0.45 | |
| Hadar | β Cen | B1III | ? | 0.0062 | 160 | +0.61 | +4 |
| Altair | α Aq1 | A7IV–V | | 0.194 | 5.1 | +0.76 | |
| Acrux | α Cru | B1IV | B3 | 0.010 | 98 | +0.77 | +1.9 |
| Aldebaran | α Tau | K5III | M2V | 0.050 | 20 | +0.87 | +13 |
| Spica | α Vir | B1V | B2V | 0.012 | 80 | +0.98 | 2.1 |
| Antares | α Sco | M1Ib | B4V | 0.005 | 190 | +1.06 | +5.1 |
| Pollux | β Gem | K0III | | 0.097 | 10 | +1.16 | |
| Formalhaut | α PsA | A3V | ? | 0.130 | 7.7 | +1.17 | +6.5 |
| Deneb | α Cyg | A2Ia | | 0.0010 | 990 | +1.25 | |
| Mimosa | β Cru | B1IV | | 0.0093 | 110 | +1.25 | |

| Name | Visual Luminosity* (Sun = 1) A | B | Absolute Visual Magnitude A | B | Proper Motion (arc seconds/yr) | Transverse Velocity (km/s) | Radial Velocity (km/s) |
|---|---|---|---|---|---|---|---|
| Sirius | 22 | 0.0025 | +1.5 | +11.3 | 1.33 | 16.7 | −7.6[‡] |
| Canopus | $1.4 \times 10^4$ | | −5.5 | | 0.02 | 9.1 | 20.5 |
| Arcturus | 110 | | −0.3 | | 2.28 | 119 | −5.2 |
| Rigel Kentaurus | 1.6 | 0.45 | +4.3 | +5.7 | 3.68 | 22.7 | −24.6 |
| Vega | 50 | | +0.6 | | 0.34 | 12.6 | −13.9 |
| Capella | 130 | 0.01 | −0.5 | +9.6 | 0.44 | 27.1 | 30.2[‡] |
| Rigel | $4.1 \times 10^4$ | 110 | −6.7 | −0.3 | 0.00 | 1.2 | 20.7[‡] |
| Procyon | 7.2 | 0.0006 | +2.7 | +13.0 | 1.25 | 20.7 | −3.2[‡] |
| Betelgeuse | 9700 | | −5.1 | | 0.03 | 18.5 | 21.0[‡] |
| Achernar | 1100 | | −2.8 | | 0.10 | 20.9 | 19 |
| Hadar | $1.3 \times 10^4$ | 560 | −5.4 | −2.0 | 0.04 | 30.3 | −12[‡] |
| Altair | 11 | | +2.2 | | 0.66 | 16.3 | −26.3 |
| Acrux | 4100 | 2200 | −4.2 | −3.5 | 0.04 | 22.8 | −11.2 |
| Aldebaran | 150 | 0.002 | −0.6 | +11.5 | 0.20 | 19.0 | 54.1 |
| Spica | 2200 | 780 | −3.5 | −2.4 | 0.05 | 19.0 | 1.0[‡] |
| Antares | $1.1 \times 10^4$ | 290 | −5.3 | −1.3 | 0.03 | 27.0 | −3.2 |
| Pollux | 31 | | +1.1 | | 0.62 | 29.4 | 3.3 |
| Formalhaut | 17 | 0.13 | +1.7 | +7.1 | 0.37 | 13.5 | 6.5 |
| Deneb | $2.6 \times 10^5$ | | −8.7 | | 0.003 | 14.1 | −4.6[‡] |
| Mimosa | 3200 | | −3.9 | | 0.05 | 26.1 | — |

*Energy output in the visible part of the spectrum; A and B columns identify individual components of binary-star systems.
[†]"wd" stands for "white dwarf."
[‡]Average value of variable velocity.

## TABLE 5  The Twenty Nearest Stars

| Name | Spectral Type A | Spectral Type B | Parallax (arc seconds) | Distance (pc) | Apparent Visual Magnitude* A | Apparent Visual Magnitude* B |
|---|---|---|---|---|---|---|
| Sun | G2V | | | | −26.74 | |
| Proxima Centauri | M5 | | 0.772 | 1.30 | +11.01 | |
| Alpha Centauri | G2V | K1V | 0.742 | 1.35 | −0.01 | +1.35 |
| Barnard's Star | M5V | | 0.549 | 1.82 | +9.54 | |
| Wolf 359 | M8V | | 0.421 | 2.38 | +13.53 | |
| Lalande 21185 | M2V | | 0.397 | 2.52 | +7.50 | |
| UV Ceti | M6V | M6V | 0.387 | 2.58 | +12.52 | +13.02 |
| Sirius | A1V | wd[†] | 0.379 | 2.64 | −1.44 | +8.4 |
| Ross 154 | M5V | | 0.345 | 2.90 | +10.45 | |
| Ross 248 | M6V | | 0.314 | 3.18 | +12.29 | |
| $\epsilon$ Eridani | K2V | | 0.311 | 3.22 | +3.72 | |
| Ross 128 | M5V | | 0.298 | 3.36 | +11.10 | |
| 61 Cygni | K5V | K7V | 0.294 | 3.40 | +5.22 | +6.03 |
| $\epsilon$ Indi | K5V | | 0.291 | 3.44 | +4.68 | |
| Grm 34 | M1V | M6V | 0.290 | 3.45 | +8.08 | +11.06 |
| Luyten 789-6 | M6V | | 0.290 | 3.45 | +12.18 | |
| Procyon | F5IV–V | wd[†] | 0.286 | 3.50 | +0.40 | +10.7 |
| $\Sigma$ 2398 | M4V | M5V | 0.285 | 3.55 | +8.90 | +9.69 |
| Lacaille 9352 | M2V | | 0.279 | 3.58 | +7.35 | |
| G51-15 | MV | | 0.278 | 3.60 | +14.81 | |

| Name | Visual Luminosity* (Sun = 1) A | Visual Luminosity* (Sun = 1) B | Absolute Visual Magnitude* A | Absolute Visual Magnitude* B | Proper Motion (arc seconds/yr) | Transverse Velocity (km/s) | Radial Velocity (km/s) |
|---|---|---|---|---|---|---|---|
| Sun | 1.0 | | +4.83 | | | | |
| Proxima Centauri | $5.6 \times 10^{-5}$ | | +15.4 | | 3.86 | 23.8 | −16 |
| Alpha Centauri | 1.6 | 0.45 | +4.3 | +5.7 | 3.68 | 23.2 | −22 |
| Barnard's Star | $4.3 \times 10^{-4}$ | | +13.2 | | 10.34 | 89.7 | −108 |
| Wolf 359 | $1.8 \times 10^{-5}$ | | +16.7 | | 4.70 | 53.0 | +13 |
| Lalande 21185 | 0.0055 | | +10.5 | | 4.78 | 57.1 | −84 |
| UV Ceti | $5.4 \times 10^{-5}$ | 0.00004 | +15.5 | +16.0 | 3.36 | 41.1 | +30 |
| Sirius | 22 | 0.0025 | +1.5 | +11.3 | 1.33 | 16.7 | −8 |
| Ross 154 | $4.8 \times 10^{-4}$ | | +13.3 | | 0.72 | 9.9 | −4 |
| Ross 248 | $1.1 \times 10^{-4}$ | | +14.8 | | 1.58 | 23.8 | −81 |
| $\epsilon$ Eridani | 0.29 | | +6.2 | | 0.98 | 15.3 | +16 |
| Ross 128 | $3.6 \times 10^{-4}$ | | +13.5 | | 1.37 | 21.8 | −13 |
| 61 Cygni | 0.082 | 0.039 | +7.6 | +8.4 | 5.22 | 84.1 | −64 |
| $\epsilon$ Indi | 0.14 | | +7.0 | | 4.69 | 76.5 | −40 |
| Grm 34 | 0.0061 | 0.00039 | +10.4 | +13.4 | 2.89 | 47.3 | +17 |
| Luyten 789-6 | $1.4 \times 10^{-4}$ | | +14.6 | | 3.26 | 53.3 | −60 |
| Procyon | 7.2 | 0.00055 | +2.7 | +13.0 | 1.25 | 2.8 | −3 |
| $\Sigma$ 2398 | 0.0030 | 0.0015 | +11.2 | +11.9 | 2.28 | 38.4 | +5 |
| Lacaille 9352 | 0.013 | | +9.6 | | 6.90 | 117 | +10 |
| G51-15 | $1.1 \times 10^{-5}$ | | +17.0 | | 1.26 | 21.5 | — |

*A and B columns identify individual components of binary-star systems.
[†]"wd" stands for "white dwarf."

# GLOSSARY

*Key terms that are boldface in the text are followed by a page reference in the Glossary.*

## A

**A ring** One of three Saturnian rings visible from Earth. The A ring is farthest from the planet and is separated from the B ring by the Cassini Division. (p. 315)

**aberration of starlight** Small shift in the observed direction to a star, caused by Earth's motion perpendicular to the line of sight. (p. 44)

**absolute brightness** The apparent brightness a star would have if it were placed at a standard distance of 10 parsecs from Earth. (p. 441)

**absolute magnitude** The apparent magnitude a star would have if it were placed at a standard distance of 10 parsecs from Earth. (p. 455)

**Absolute Zero** The lowest possible temperature that can be obtained; all thermal motion ceases at this temperature.

**absorption line** Dark line in an otherwise continuous bright spectrum, where light within one narrow frequency range has been removed. (p. 88)

**abundance** Relative amount of different elements in a gas.

**acceleration** The rate of change of velocity of a moving object. (p. 53)

**accretion** Gradual growth of bodies, such as planets, by the accumulation of other, smaller bodies. (p. 392)

**accretion disk** Flat disk of matter spiraling down onto the surface of a neutron star or black hole. Often, the matter originated on the surface of a companion star in a binary-star system. (p. 556)

**active galactic nucleus** Region of intense emission at the center of an active galaxy, responsible for virtually all of the galaxy's nonstellar luminosity. (p. 664)

**active galaxies** The most energetic galaxies, which can emit hundreds or thousands of times more energy per second than the Milky Way, mostly in the form of long-wavelength nonthermal radiation. (p. 664)

**active optics** Collection of techniques used to increase the resolution of ground-based telescopes. Minute modifications are made to the overall configuration of an instrument as its temperature and orientation change; used to maintain the best possible focus at all times. (p. 123)

**active region** Region of the photosphere of the Sun surrounding a sunspot group, which can erupt violently and unpredictably. During sunspot maximum, the number of active regions is also a maximum. (p. 432)

**active Sun** The unpredictable aspects of the Sun's behavior, such as sudden explosive outbursts of radiation in the form of prominences and flares.

**adaptive optics** Technique used to increase the resolution of a telescope by deforming the shape of the mirror's surface under computer control while a measurement is being taken; used to undo the effects of atmospheric turbulence. (p. 124)

**aerosol** Suspension of liquid or solid particles in air.

**alpha particle** A helium-4 nucleus.

**alpha process** Process occurring at high temperatures, in which high-energy photons split heavy nuclei to form helium nuclei.

**ALSEP** Acronym for Apollo Lunar Surface Experiments Package.

**altimeter** Instrument used to determine altitude.

**amino acids** Organic molecules that form the basis for building the proteins that direct metabolism in living creatures. (p. 761)

**Amor asteroid** Asteroid that crosses only the orbit of Mars.

**amplitude** The maximum deviation of a wave above or below zero point. (p. 65)

**angstrom** Distance unit equal to 0.1 nanometers, or one ten-billionth of a meter.

**angular diameter** Angle made between the top (or one edge) of an object, the observer and the bottom (or opposite edge) of the object.

**angular distance** Angular separation between two objects as seen by some observer.

**angular momentum** Tendency of an object to keep rotating; proportional to the mass, radius, and rotation speed of the body. (p. 163)

**angular resolution** The ability of a telescope to distinguish between adjacent objects in the sky. (p. 117)

**annular eclipse** Solar eclipse occurring at a time when the Moon is far enough away from Earth that it fails to cover the disk of the Sun completely, leaving a ring of sunlight visible around its edge. (p. 23)

**antiparallel** Configuration of the electron and proton in a hydrogen (or other) atom when their spin axes are parallel but the two rotate in opposite directions.

**antiparticle** A particle of the same mass but opposite in all other respects (e.g., charge) to a given particle; when a particle and its antiparticle come into contact, they annihilate and release energy in the form of gamma rays.

**aphelion** The point on the elliptical path of an object in orbit about the Sun that is most distant from the Sun.

**Apollo asteroid** *See* Earth-crossing asteroid.

**apparent brightness** The brightness that a star appears to have, as measured by an observer on Earth. (p. 454)

**apparent magnitude** The apparent brightness of a star, expressed using the magnitude scale. (p. 455)

**association** Small grouping of (typically 100 or less) bright stars, spanning up to a few tens of parsecs across, usually rich in very young stars. (p. 518)

**Assumption of Mediocrity** Statements suggesting that the development of life on Earth did not require any unusual circumstances, suggesting that extraterrestrial life may be common.

**asteroid** One of thousands of very small members of the solar system orbiting the Sun between the orbits of Mars and Jupiter. Often referred to as "minor planets." (p. 360)

**asteroid belt** Region of the solar system, between the orbits of Mars and Jupiter, in which most asteroids are found. (p. 360)

**asthenosphere** Layer of Earth's interior, just below the lithosphere, over which the surface plates slide. (p. 181)

**astrology** Pseudoscience that purports to use the positions of the planets, sun, and moon to predict daily events and human destiny.

**astronomical unit (AU)** The average distance of Earth from the Sun. Precise radar measurements yield a value for the AU of 149,603,500 km. (p. 49)

**astronomy** Branch of science dedicated to the study of everything in the universe that lies above Earth's atmosphere. (p. 4)

**asymptotic giant branch** Path on the Hertzsprung–Russell diagram corresponding to the changes that a star undergoes after helium burning ceases in the core. At this stage, the carbon core shrinks and drives the expansion of the envelope, and the star becomes a swollen red giant for a second time. (p. 535)

**Aten asteroid** Earth-crossing asteroid with semimajor axis less than 1 AU.

**atmosphere** Layer of gas confined close to a planet's surface by the force of gravity. (p. 170)

**atom** Building block of matter, composed of positively charged protons and neutral neutrons in the nucleus surrounded by negatively charged electrons. (p. 90)

**atomic epoch** Period after decoupling when the first simple atoms and molecules formed.

**aurora** Event that occurs when atmospheric molecules are excited by incoming charged particles from the solar wind, then emit energy as they fall back to their ground states. Aurorae generally occur at high latitudes, near the north and south magnetic poles. (p. 191)

**autumnal equinox** Date on which the Sun crosses the celestial equator moving southward, occurring on or near September 22. (p. 16)

**B**

**B ring** One of three Saturnian rings visible from Earth. The B ring is the brightest of the three, and lies just past the Cassini Division, closer to the planet than the A ring. (p. 315)

**background noise** Unwanted light in an image, from unresolved sources in the telescope's field of view, scattered light from the atmosphere, or instrumental "hiss" in the detector itself.

**barred-spiral galaxy** Spiral galaxy in which a bar of material passes through the center of the galaxy, with the spiral arms beginning near the ends of the bar. (p. 652)

**basalt** Solidified lava; an iron-magnesium-silicate mixture.

**baseline** The distance between two observing locations used for the purposes of triangulation measurements. The larger the baseline, the better the resolution attainable. (p. 26)

**belt** Dark, low-pressure region in the atmosphere of a jovian planet, where gas flows downward. (p. 283)

**Big Bang** Event that cosmologists consider the beginning of the universe, in which all matter and radiation in the entire universe came into being. (p. 714)

**Big Crunch** Point of final collapse of a bound universe.

**binary asteroid** Asteroid with a partner in orbit around it.

**binary pulsar** Binary system in which both components are pulsars.

**binary-star system** A system that consists of two stars in orbit about their common center of mass, held together by their mutual gravitational attraction. Most stars are found in binary-star systems. (p. 469)

**biological evolution** Change in a population of biological organisms over time.

**bipolar flow** Jets of material expelled from a protostar perpendicular to the surrounding protostellar disk. (p. 515)

**blackbody curve** The characteristic way in which the intensity of radiation emitted by a hot object depends on frequency. The frequency at which the emitted intensity is highest is an indication of the temperature of the radiating object. Also referred to as the Planck curve. (p. 74)

**black dwarf** The endpoint of the evolution of an isolated, low-mass star. After the white-dwarf stage, the star cools to the point where it is a dark "clinker" in interstellar space. (p. 541)

**black hole** A region of space where the pull of gravity is so great that nothing—not even light—can escape. A possible outcome of the evolution of a very massive star. (p. 593)

**blazar** Particularly intense active galactic nucleus in which the observer's line of sight happens to lie directly along the axis of a high-speed jet of particles emitted from the active region. (p. 670)

**blue giant** Large, hot, bright star at the upper-left end of the main sequence on the Hertzsprung–Russell diagram. Its name comes from its color and size. (p. 465)

**blueshift** Motion-induced changes in the observed wavelength from a source that is moving toward us. Relative approaching motion between the object and the observer causes the wavelength to appear shorter (and hence bluer) than if there were no motion at all.

**blue straggler** Star found on the main sequence of the Hertzsprung–Russell diagram, but which should already have evolved off the main sequence, given its location on the diagram; thought to have formed from mergers of lower mass stars.

**blue supergiant** The very largest of the large, hot, bright stars at the uppermost-left end of the main sequence on the Hertzsprung–Russell diagram. (p. 465)

**Bohr model** First theory of the hydrogen atom to explain the observed spectral lines. This model rests on three ideas: that there is a state of lowest energy for the electron, that there is a maximum energy beyond which the electron is no longer bound to the nucleus, and that within these two energies the electron can only exist in certain energy levels.

**Bok globule** Dense, compact cloud of interstellar dust and gas on its way to forming one or more stars.

**boson** Particle that exerts or mediates forces between elementary particles in quantum physics. (p. 738)

**bound trajectory** Path of an object with launch speed too low to escape the gravitational pull of a planet.

**brown dwarf** Fragments of collapsing gas and dust that did not contain enough mass to initiate core nuclear fusion. Such an object is then frozen somewhere along its pre-main-sequence contraction phase, continually cooling into a compact dark object. Because of their small size and low temperature brown dwarfs are extremely difficult to detect observationally. (pp. 405, 510)

**brown oval** Feature of Jupiter's atmosphere that appears only at latitudes near 20 degrees N, this structure is a long-lived hole in the clouds that allows us to look down into Jupiter's lower atmosphere. (p. 288)

## C

**C ring** One of three Saturnian rings visible from Earth. The C ring lies closest to the planet and is relatively thin compared to the A and B rings. (p. 315)

**caldera** Crater that forms at the summit of a volcano.

**capture theory (Moon)** Theory suggesting that the Moon formed far from Earth but was later captured by it.

**carbonaceous asteroid** The darkest, or least reflective, type of asteroid, containing large amounts of carbon.

**carbon-based molecule** Molecule containing atoms of carbon.

**carbon-detonation supernova** *See* Type I supernova. (p. 561)

**cascade** Process of deexcitation in which an excited electron moves down through energy states one at a time.

**Cassegrain telescope** A type of reflecting telescope in which incoming light hits the primary mirror and is then reflected upward toward the prime focus, where a secondary mirror reflects the light back down through a small hole in the main mirror into a detector or eyepiece. (p. 112)

**Cassini Division** A relatively empty gap in Saturn's ring system, discovered in 1675 by Giovanni Cassini. It is now known to contain a number of thin ringlets. (p. 315)

**cataclysmic variable** Collective name for novae and supernovae.

**catalyst** Something that causes or helps a reaction to occur, but is not itself consumed as part of the reaction.

**catastrophic theory** A theory that invokes statistically unlikely accidental events to account for observations.

**celestial coordinates** Pair of quantities—right ascension and declination—similar to longitude and latitude on Earth, used to pinpoint locations of objects on the celestial sphere. (p. 12)

**celestial equator** The projection of Earth's equator onto the celestial sphere. (p. 11)

**celestial mechanics** Study of the motions of bodies, such as planets and stars, that interact via gravity.

**celestial pole** Projection of Earth's North or South pole onto the celestial sphere. (p. 10)

**celestial sphere** Imaginary sphere surrounding Earth to which all objects in the sky were once considered to be attached. (p. 10)

**Celsius** Temperature scale in which the freezing point of water is 0 degrees and the boiling point of water is 100 degrees.

**center of mass** The "average" position in space of a collection of massive bodies, weighted by their masses. For an isolated system this point moves with constant velocity, according to Newtonian mechanics. (p. 56)

**centigrade** *See* Celsius.

**centripetal force** (literally "center seeking") Force directed toward the center of a body's orbit.

**centroid** Average position of the material in an object; in spectroscopy, the center of a spectral line.

**Cepheid variable** Star whose luminosity varies in a characteristic way, with a rapid rise in brightness followed by a slower decline. The period of a Cepheid variable star is related to its luminosity, so a determination of this period can be used to obtain an estimate of the star's distance. (p. 621)

**Chandrasekhar mass** Maximum possible mass of a white dwarf.

**chaotic rotation** Unpredictable tumbling motion that nonspherical bodies in eccentric orbits, such as Saturn's satellite Hyperion, can exhibit. No amount of observation of an object rotating chaotically will ever show a well-defined period.

**charge-coupled device (CCD)** An electronic device used for data acquisition; composed of many tiny pixels, each of which records a buildup of charge to measure the amount of light striking it. (p. 118)

**chemical bond** Force holding atoms together to form a molecule.

**chemosynthesis** Analog of photosynthesis that operates in total darkness.

**chromatic aberration** The tendency for a lens to focus red and blue light differently, causing images to become blurred.

**chromosphere** The Sun's lower atmosphere, lying just above the visible atmosphere. (p. 417)

**circumnavigation** Traveling all the way around an object.

**cirrus** High-level clouds composed of ice or methane crystals.

**closed universe** Geometry that the universe as a whole would have if the density of matter is above the critical value. A closed universe is finite in extent and has no edge, like the surface of a sphere. It has enough mass to stop the present expansion and will eventually collapse. (p. 721)

**CNO cycle** Chain of reactions that converts hydrogen into helium using carbon, nitrogen, and oxygen as catalysts.

**cocoon nebula** Bright infrared source in which a surrounding cloud of gas and dust absorb ultraviolet radiation from a hot star and reemits it in the infrared.

**cold dark matter** Class of dark-matter candidates made up of very heavy particles, possibly formed in the very early universe. (p. 752)

**collecting area** The total area of a telescope capable of capturing incoming radiation. The larger the telescope, the greater its collecting area, and the fainter the objects it can detect. (p. 113)

**collisional broadening** Broadening of spectral lines due to collisions between atoms, most often seen in dense gases.

**color index** A convenient method of quantifying a star's color by comparing its apparent brightness as measured through different filters. If the star's radiation is well described by a blackbody spectrum, the ratio of its blue intensity (B) to its visual intensity (V) is a measure of the object's surface temperature.

**color-magnitude diagram** A way of plotting stellar properties, in which absolute magnitude is plotted against color index. (p. 464)

**coma** An effect occurring during the formation of an off-axis image in a telescope. Stars whose light enters the telescope at a large angle acquire comet-like tails on their images. The brightest part of a comet, often referred to as the "head." (p. 367)

**comet** A small body, composed mainly of ice and dust, in an elliptical orbit about the Sun. As it comes close to the Sun, some of its material is vaporized to form a gaseous head and extended tail. (p. 367)

**common envelope** Outer layer of gas in a contact binary.

**comparative planetology** The systematic study of the similarities and differences among the planets, with the goal of obtaining deeper insight into how the solar system formed and has evolved in time. (p. 147)

**composition** The mixture of atoms and molecules that make up an object.

**condensation nuclei** Dust grains in the interstellar medium which act as seeds around which other material can cluster. The presence of dust was very important in causing matter to clump during the formation of the solar system. (p. 392)

**condensation theory** Currently favored model of solar system formation which combines features of the old nebular theory with new information about interstellar dust grains, which acted as condensation nuclei. (pp. 162, 392)

**conjunction** Orbital configuration in which a planet lies in the same direction as the Sun, as seen from Earth.

**conservation of angular momentum** *See* law of conservation of angular momentum. (p. 163)

**conservation of mass and energy** *See* law of conservation of mass and energy.

**constellation** A human grouping of stars in the night sky into a recognizable pattern. (p. 8)

**constituents** *See* composition.

**contact binary** A binary-star system in which both stars have expanded to fill their Roche lobes and the surfaces of the two stars merge. The binary system now consists of two nuclear burning stellar cores surrounded by a continuous common envelope. (p. 549)

**continental drift** The movement of the continents around Earth's surface.

**continuous spectrum** Spectrum in which the radiation is distributed over all frequencies, not just a few specific frequency ranges. A prime example is the blackbody radiation emitted by a hot, dense body. (p. 86)

**convection** Churning motion resulting from the constant upwelling of warm fluid and the concurrent downward flow of cooler material to take its place. (p. 171)

**convection zone** Region of the Sun's interior, lying just below the surface, where the material of the Sun is in constant convection motion. This region extends into the solar interior to a depth of about 20,000 km. (p. 417)

**co-orbital satellites** Satellites sharing the same orbit around a planet.

**Copernican Principle** The removal of Earth from any position of cosmic significance.

**Copernican revolution** The realization, toward the end of the sixteenth century, that Earth is not at the center of the universe. (p. 42)

**core** The central region of Earth, surrounded by the mantle. (p. 170); The central region of any planet or star. (p. 170)

**core-accretion theory** Theory that the jovian planets formed when icy protoplanetary cores became massive enough to capture gas directly from the solar nebula. *See* gravitational instability theory. (p. 395)

**core-collapse supernova** *See* Type II supernova. (p. 560)

**core hydrogen burning** The energy burning stage for main-sequence stars, in which the helium is produced by hydrogen fusion in the central region of the star. A typical star spends up to 90 percent of its lifetime in hydrostatic equilibrium brought about by the balance between gravity and the energy generated by core hydrogen burning. (p. 528)

**cornea (eye)** The curved transparent layer covering the front part of the eye.

**corona** One of numerous large, roughly circular regions on the surface of Venus, thought to have been caused by upwelling mantle material causing the planet's crust to bulge outward (plural, *coronae*). (p. 240); The tenuous outer atmosphere of the Sun, which lies just above the chromosphere and, at great distances, turns into the solar wind. (p. 417)

**coronal hole** Vast region of the Sun's atmosphere where the density of matter is about 10 times lower than average. The gas there streams freely into space at high speeds, escaping the Sun completely. (p. 435)

**coronal mass ejection** Giant magnetic "bubble" of ionized gas that separates from the rest of the solar atmosphere and escapes into interplanetary space. (p. 434)

**corpuscular theory** Early particle theory of light.

**cosmic density parameter** Ratio of the universe's actual density to the critical value corresponding to zero curvature.

**cosmic distance scale** Collection of indirect distance-measurement techniques that astronomers use to measure distances in the universe. (p. 26)

**cosmic evolution** The collection of the seven major phases of the history of the universe; namely particulate, galactic, stellar, planetary, chemical, biological, and cultural evolution. (p. 760)

**cosmic microwave background** The almost perfectly isotropic radio signal that is the electromagnetic remnant of the Big Bang. (p. 727)

**cosmic ray** Very energetic subatomic particle arriving at Earth from elsewhere in the Galaxy.

**cosmological constant** Quantity originally introduced by Einstein into general relativity to make his equations describe a static universe. Now one of several candidates for the repulsive "dark energy" force responsible for the observed cosmic acceleration. (p. 723)

**cosmological distance** Distance comparable to the scale of the universe.

**cosmological principle** Two assumptions that make up the basis of cosmology, namely that the universe is homogeneous and isotropic on sufficiently large scales. (p. 711)

**cosmological redshift** The component of the redshift of an object that is due only to the Hubble flow of the universe. (p. 662)

**cosmology** The study of the structure and evolution of the entire universe. (p. 710)

**cosmos** The universe.

**coudé focus** Focus produced far from the telescope using a series of mirrors. Allows the use of heavy and/or finely tuned equipment to analyze the image.

**crater** Bowl-shaped depression on the surface of a planet or moon, resulting from a collision with interplanetary debris. (p. 203)

**crescent** Appearance of the Moon (or a planet) when less than half of the body's hemisphere is visible from Earth.

**crest** Maximum departure of a wave above its undisturbed state.

**critical density** The cosmic density corresponding to the dividing line between a universe that recollapses and one that expands forever. (p. 718)

**critical universe** Universe in which the density of matter is exactly equal to the critical density. The universe is infinite in extent and has zero curvature. The expansion will continue forever, but will approach an expansion speed of zero. (p. 721)

**crust** Layer of Earth which contains the solid continents and the seafloor. (p. 170)

**C-type asteroid** *See* carbonaceous asteroid.

**cultural evolution** Change in the ideas and behavior of a society over time.

**current sheet** Flat sheet on Jupiter's magnetic equator where most of the charged particles in the magnetosphere lie due to the planet's rapid rotation.

**D**

**D ring** Collection of very faint, thin rings, extending from the inner edge of the C ring down nearly to the cloud tops of Saturn. This region contains so few particles that it is completely invisible from Earth. (p. 319)

**dark dust cloud** A large cloud, often many parsecs across, which contains gas and dust in a ratio of about 1012 gas atoms for every dust particle. Typical densities are a few tens or hundreds of millions of particles per cubic meter. (p. 490)

**dark energy** Generic name given to the unknown cosmic force field thought to be responsible for the observed acceleration of the Hubble expansion, (p. 723)

**dark halo** Region of a galaxy beyond the visible halo where dark matter is believed to reside. (p. 637)

**dark matter** Term used to describe the mass in galaxies and clusters whose existence we infer from rotation curves and other techniques, but that has not been confirmed by observations at any electromagnetic wavelength. (p. 637)

**dark matter particle** Particle undetectable at any electromagnetic wavelength, but can be inferred from its gravitational influence.

**Daughter/Fission theory** Theory suggesting that the Moon originated out of Earth.

**declination** Celestial coordinate used to measure latitude above or below the celestial equator on the celestial sphere.

**decoupling** Event in the early universe when atoms first formed, after which photons could propagate freely through space. (p. 745)

**deferent** A construct of the geocentric model of the solar system which was needed to explain observed planetary motions. A deferent is a large circle encircling Earth, on which an epicycle moves. (p. 40)

**degree** Unit of angular measure. There are 360 degrees in one complete circle.

**density** A measure of the compactness of the matter within an object, computed by dividing the mass of the object by its volume. Units are kilograms per cubic meter ($kg/m^3$), or grams per cubic centimeter ($g/cm^3$). (p. 149)

**detached binary** Binary system where each star lies within its respective Roche lobe.

**detector noise** Readings produced by an instrument even when it is not observing anything; produced by the electronic components within the detector itself.

**deuterium** A form of hydrogen with an extra neutron in its nucleus.

**deuterium bottleneck** Period in the early universe between the start of deuterium production and the time when the universe was cool enough for deuterium to survive.

**deuteron** An isotope of hydrogen in which a neutron is bound to the proton in the nucleus. Often called "heavy hydrogen" because of the extra mass of the neutron. (p. 439)

**differential rotation** The tendency for a gaseous sphere, such as a jovian planet or the Sun, to rotate at a different rate at the equator than at the poles. More generally, a condition where the angular speed varies with location within an object. (p. 281)

**differentiation** Variation in the density and composition of a body, such as Earth, with low density material on the surface and higher density material in the core. (p. 180)

**diffraction** The ability of waves to bend around corners. The diffraction of light establishes its wave nature. (p. 72)

**diffraction grating** Sheet of transparent material with many closely spaced parallel lines ruled on it, designed to separate white light into a spectrum.

**diffraction-limited resolution** Theoretical resolution that a telescope can have due to diffraction of light at the telescope's aperture. Depends on the wavelength of radiation and the diameter of the telescope's mirror. (p. 119)

**direct motion** *See* prograde motion.

**distance modulus** Difference between the apparent and absolute magnitude of an object; equivalent to distance, by the inverse-square law.

**diurnal motion** Apparent daily motion of the stars, caused by Earth's rotation.

**DNA** Deoxyribonucleic acid, the molecule that carries genetic information and determine the characteristics of a living organism.

**Doppler effect** Any motion-induced change in the observed wavelength (or frequency) of a wave. (p. 78)

**double-line spectroscopic binary** Binary system in which spectral lines of both stars can be distinguished and seen to shift back and forth as the stars orbit one another.

**double-star system** System containing two stars in orbit around one another.

**Drake equation** Expression that gives an estimate of the probability that intelligence exists elsewhere in the galaxy, based on a number of supposedly necessary conditions for intelligent life to develop. (p. 768)

**dust grain** An interstellar dust particle, roughly $10^{-7}$ m in size, comparable to the wavelength of visible light. (p. 480)

**dust lane** A lane of dark, obscuring interstellar dust in an emission nebula or galaxy. (p. 484)

**dust tail** The component of a comet's tail that is composed of dust particles. (p. 369)

**dwarf** Any star with radius comparable to, or smaller than, that of the Sun (including the Sun itself). (p. 462)

**dwarf elliptical** Elliptical galaxy as small as 1 kiloparsec across, containing only a few million stars.

**dwarf galaxy** Small galaxy containing a few million stars.

**dwarf irregular** Small irregular galaxy containing only a few million stars.

**dynamo theory** Theory that explains planetary and stellar magnetic fields in terms of rotating, conducting material flowing in an object's interior. (p. 192)

**E**

**E ring** A faint ring, well outside the main ring system of Saturn, which was discovered by *Voyager* and is believed to be associated with volcanism on the moon Enceladus. (p. 319)

**Earth-crossing asteroid** An asteroid whose orbit crosses that of Earth. Earth-crossing asteroids are also called Apollo asteroids, after the first asteroid of this type discovered. (p. 363)

**earthquake** A sudden dislocation of rocky material near Earth's surface. (p. 143) (p. 171)

**eccentricity** A measure of the flatness of an ellipse, equal to the distance between the two foci divided by the length of the major axis. (p. 47)

**eclipse** Event during which one body passes in front of another, so that the light from the occulted body is blocked. (p. 21)

**eclipse season** Time of the year when the Moon lies in the same plane as Earth and Sun, so that eclipses are possible. (p. 25)

**eclipse year** Time interval between successive orbital configurations in which the line of nodes of the Moon's orbit points toward the Sun.

**eclipsing binary** Rare binary-star system that is aligned in such a way that from Earth we observe one star pass in front of the other, eclipsing the other star. (p. 470)

**ecliptic** The apparent path of the Sun, relative to the stars on the celestial sphere, over the course of a year. (p. 15)

**effective temperature** Temperature of a blackbody of the same radius and luminosity as a given star or planet.

**ejecta (planetary)** Material thrown outward by a meteoroid impact.

**ejecta (stellar)** Material thrown into space by a nova or supernova.

**electric field** A field extending outward in all directions from a charged particle, such as a proton or an electron. The electric field determines the electric force exerted by the particle on all other charged particles in the universe; the strength of the electric field decreases with increasing distance from the charge according to an inverse-square law. (p. 67)

**electromagnetic energy** Energy carried in the form of rapidly fluctuating electric and magnetic fields.

**electromagnetic force** Force (electric or magnetic) exerted between any two charged particles. (p. 440)

**electromagnetic radiation** Another term for light, electromagnetic radiation transfers energy and information from one place to another. (p. 64)

**electromagnetic spectrum** The complete range of electromagnetic radiation, from radio waves to gamma rays, including the visible spectrum. All types of electromagnetic radiation are basically the same phenomenon, differing only by wavelength, and all move at the speed of light. (p. 70)

**electromagnetism** The union of electricity and magnetism, which do not exist as independent quantities but are in reality two aspects of a single physical phenomenon. (p. 68)

**electron** An elementary particle with a negative electric charge; one of the components of the atom. (p. 67)

**electron degeneracy pressure** The pressure produced by the resistance of electrons to further compression once they are squeezed to the point of contact. (p. 534)

**electrostatic force** Force between electrically charged objects.

**electroweak force** Unification of the weak electromagnetic forces. (p. 440)

**element** Matter made up of one particular atom. The number of protons in the nucleus of the atom determines which element it represents. (p. 97)

**elementary particle** Technically, a particle that cannot be subdivided into component parts; however, the term is also often used to refer to particles such as protons and neutrons, which are themselves made up of quarks.

**ellipse** Geometric figure resembling an elongated circle. An ellipse is characterized by its degree of flatness, or eccentricity, and the length of its long axis. In general, bound orbits of objects moving under gravity are elliptical. (p. 47)

**elliptical galaxy** Category of galaxy in which the stars are distributed in an elliptical shape on the sky, ranging from highly elongated to nearly circular in appearance. (p. 653)

**elongation** Angular distance between a planet and the Sun.

**emission line** Bright line in a specific location of the spectrum of radiating material, corresponding to emission of light at a certain frequency. A heated gas in a glass container produces emission lines in its spectrum. (p. 86)

**emission nebula** A glowing cloud of hot interstellar gas. The gas glows as a result of one or more nearby young stars which ionize the gas. Since the gas is mostly hydrogen, the emitted radiation falls predominantly in the red region of the spectrum, because of the hydrogen-alpha emission line. (p. 483)

**emission spectrum** The pattern of spectral emission lines produced by an element. Each element has its own unique emission spectrum.

**empirical** Discovery based on observational evidence (rather than from theory).

**Encke gap** A small gap in Saturn's A ring. (p. 315)

**energy flux** Energy per unit area per unit time radiated by a star (or recorded by a detector).

**epicycle** A construct of the geocentric model of the solar system which was necessary to explain observed planetary motions. Each planet rides on a small epicycle whose center in turn rides on a larger circle (the deferent). (p. 40)

**epoch of inflation** Short period of unchecked cosmic expansion early in the history of the universe. During inflation, the universe swelled in size by a factor of about $10^{50}$. (p. 747)

**equinox** *See* vernal equinox, autumnal equinox.

**equivalence principle** There is no experimental way to distinguish between a gravitational field and an accelerated frame of reference. (p. 598)

**escape speed** The speed necessary for one object to escape the gravitational pull of another. Anything that moves away from a gravitating body with more than the escape speed will never return. (p. 58)

**euclidean geometry** Geometry of flat space.

**event horizon** Imaginary spherical surface surrounding a collapsing star, with radius equal to the Schwarzschild radius, within which no event can be seen, heard, or known about by an outside observer. (p. 594)

**evolutionary theory** A theory which explains observations in a series of gradual steps, explainable in terms of well-established physical principles.

**evolutionary track** A graphical representation of a star's life as a path on the Hertzsprung–Russell diagram. (p. 506)

**excited state** State of an atom when one of its electrons is in a higher energy orbital than the ground state. Atoms can become excited by absorbing a photon of a specific energy, or by colliding with a nearby atom. (p. 91)

**extinction** The dimming of starlight as it passes through the interstellar medium. (p. 480)

**extrasolar planet** Planet orbiting a star other than the Sun. (p. 390)

**extremophilic** Adjective describing organisms that can survive in very harsh environments. (p. 767)

**eyepiece** Secondary lens through which an observer views an image. This lens is often chosen to magnify the image.

**F**

**F ring** Faint narrow outer ring of Saturn, discovered by *Pioneer 11* in 1979. The F ring lies just inside the Roche limit of Saturn and was found by *Voyager 1* to be made up of several ring strands apparently braided together. (p. 320)

**Fahrenheit** Temperature scale in which the freezing point of water is 32 degrees and the boiling point of water is 212 degrees.

**false vacuum** Region of the universe that remained in the "unified" state after the strong and electroweak forces separated; one possible cause of cosmic inflation at very early times.

**fault line** Dislocation on a planet's surface, often indicating the boundary between two plates.

**field line** Imaginary line indicating the direction of an electric or magnetic field.

**fireball** Large meteor that burns up brightly and sometimes explosively in Earth's atmosphere.

**firmament** Old-fashioned term for the heavens (i.e., the sky).

**fission** *See* nuclear fission.

**flare** Explosive event occurring in or near an active region on the Sun. (p. 433)

**flatness problem** One of two conceptual problems with the standard Big Bang model, which is that there is no natural way to explain why the density of the universe is so close to the critical density. (p. 747)

**fluidized ejecta** The ejecta blankets around some Martian craters, which apparently indicate that the ejected material was liquid at the time the crater formed.

**fluorescence** Phenomenon where an atom absorbs energy, then radiates photons of lower energy as it cascades back to the ground state; in astronomy, often produced as ultraviolet photons from a hot young star are absorbed by a neutral gas, causing some of the gas atoms to become excited and give off an optical (red) glow.

**flyby** Unbound trajectory of a spacecraft around a planet or other body.

**focal length** Distance from a mirror or the center of a lens to the focus.

**focus** One of two special points within an ellipse, whose separation from each other indicates the eccentricity. In a bound orbit, planets orbit in ellipses with the Sun at one focus. (p. 47)

**forbidden line** A spectral line seen in emission nebulae, but not seen in laboratory experiments because, under laboratory conditions, collisions kick the electron in question into some other state before emission can occur.

**force** Action on an object that causes its momentum to change. The rate at which the momentum changes is numerically equal to the force. (p. 52)

**fragmentation** The breaking up of a large object into many smaller pieces (for example, as the result of high-speed collisions between planetesimals and protoplanets in the early solar system). (p. 395)

**Fraunhofer lines** The collection of over 600 absorption lines in the spectrum of the Sun, first categorized by Joseph Fraunhofer in 1812.

**frequency** The number of wave crests passing any given point in a unit time. (p. 65)

**full** When the full hemisphere of the Moon or a planet can be seen from Earth.

**full Moon** Phase of the Moon in which it appears as a complete circular disk in the sky.

**fusion** *See* nuclear fusion.

## G

**G ring** Faint, narrow ring of Saturn, discovered by *Pioneer 11* and lying just outside the F ring. (p. 319)

**galactic bulge** Thick distribution of warm gas and stars around the galactic center. (p. 618)

**galactic cannibalism** A galaxy merger in which a larger galaxy consumes a smaller one.

**galactic center** The center of the Milky Way, or any other galaxy. The point about which the disk of a spiral galaxy rotates. (p. 624)

**galactic disk** Flattened region of gas and dust that bisects the galactic halo in a spiral galaxy. This is the region of active star formation. (p. 618)

**galactic epoch** Period from 100 million to 3 billion years after the Big Bang when large agglomerations of matter (galaxies and galaxy clusters) formed and grew.

**galactic habitable zone** Region of a galaxy in which conditions are conducive to the development of life.

**galactic halo** Region of a galaxy extending far above and below the galactic disk, where globular clusters and other old stars reside. (p. 618)

**galactic nucleus** Small, central, high-density region of a galaxy. Almost all the radiation from active galaxies is generated within the nucleus. (p. 642)

**galactic rotation curve** Plot of rotation speed versus distance from the center of a galaxy.

**galactic year** Time taken for objects at the distance of the Sun (about 8 kpc) to orbit the center of the Galaxy, roughly 225 million years.

**galaxy** Gravitationally bound collection of a large number of stars. The Sun is a star in the Milky Way Galaxy. (p. 618)

**galaxy cluster** A collection of galaxies held together by their mutual gravitational attraction. (p. 659)

**Galilean moons** The four brightest and largest moons of Jupiter (Io, Europa, Ganymede, Callisto), named after Galileo Galilei, the seventeenth-century astronomer who first observed them. (p. 280)

**Galilean satellites** *See* Galilean moons.

**gamma ray** Region of the electromagnetic spectrum, far beyond the visible spectrum, corresponding to radiation of very high frequency and very short wavelength. (p. 64)

**gamma-ray burst** Object that radiates tremendous amounts of energy in the form of gamma rays, possibly due to the collision and merger of two neutron stars initially in orbit around one another. (p. 588)

**gamma-ray spectrograph** Spectrograph designed to work at gamma-ray wavelengths. Used to map the abundances of certain elements on the Moon and Mars.

**gaseous** Composed of gas.

**gas-exchange experiment** Experiment to look for life on Mars. A nutrient broth was offered to Martian soil specimens. If there were life in the soil, gases would be created as the broth was digested.

**gene** Sequence of nucleotide bases in the DNA molecule that determines the characteristics of a living organism.

**general relativity** Theory proposed by Einstein to incorporate gravity into the framework of special relativity. (p. 598)

**general theory of relativity** Theory proposed by Einstein to incorporate gravity into the framework of special relativity (p. 598).

**geocentric model** A model of the solar system that holds that Earth is at the center of the universe and all other bodies are in orbit around it. The earliest theories of the solar system were geocentric. (p. 40)

**giant** A star with a radius between 10 and 100 times that of the Sun. (p. 462)

**giant elliptical** Elliptical galaxy up to a few megaparsecs across, containing trillions of stars.

**gibbous** Appearance of the Moon (or a planet) when more than half (but not all) of the body's hemisphere is visible from Earth.

**globular cluster** Tightly bound, roughly spherical collection of hundreds of thousands, and sometimes millions, of stars spanning about 50 parsecs. Globular clusters are distributed in the halos around the Milky Way and other galaxies. (p. 519)

**gluon** Particle that exerts or mediates the strong force in quantum physics.

**gradient** Rate of change of some quantity (e.g., temperature or composition) with respect to location in space.

**Grand Unified Theories** Class of theories describing the behavior of the single force that results from unification of the strong, weak, and electromagnetic forces in the early universe. (p. 474) (p. 738)

**granite** Igneous rock, containing silicon and aluminum, that makes up most of Earth's crust.

**granulation** Mottled appearance of the solar surface, caused by rising (hot) and falling (cool) material in convective cells just below the photosphere. (p. 422)

**gravitational force** Force exerted on one body by another due to the effect of gravity. The force is directly proportional to the masses of both bodies involved and inversely proportional to the square of the distance between them. (pp. 53, 440)

**gravitational instability theory** Theory that the jovian planets formed directly from the solar nebula via instabilities in the gas leading to gravitational contraction. *See* core-accretion theory. (p. 396)

**gravitational lensing** The effect induced on the image of a distant object by a massive foreground object. Light from the distant object is bent into two or more separate images. (p. 639)

**gravitational radiation** Radiation resulting from rapid changes in a body's gravitational field.

**gravitational redshift** A prediction of Einstein's general theory of relativity. Photons lose energy as they escape the gravitational field of a massive object. Because a photon's energy is proportional to its frequency, a photon that loses energy suffers a decrease in frequency, which corresponds to an increase, or redshift, in wavelength. (p. 600)

**graviton** Particle carrying the gravitational field in theories attempting to unify gravity and quantum mechanics.

**gravity** The attractive effect that any massive object has on all other massive objects. The greater the mass of the object, the stronger its gravitational pull. (p. 53)

**gravity assist** Using gravity to change the flight path of a satellite or spacecraft.

**gravity wave** Gravitational counterpart of an electromagnetic wave.

**great attractor** A huge accumulation of mass in the relatively nearby universe (within about 200 Mpc of the Milky Way).

**Great Dark Spot** Prominent storm system in the atmosphere of Neptune observed by *Voyager 2*, near the equator of the planet. The system was comparable in size to Earth. (p. 343)

**Great Red Spot** A large, high-pressure, long-lived storm system visible in the atmosphere of Jupiter. The Red Spot is roughly twice the size of Earth. (p. 283)

**great wall** Extended sheet of galaxies measuring at least 200 megaparsecs across; one of the largest known structures in the universe.

**greenhouse effect** The partial trapping of solar radiation by a planetary atmosphere, similar to the trapping of heat in a greenhouse. (p. 173)

**greenhouse gas** Gas (such as carbon dioxide or water vapor) that efficiently absorbs infrared radiation. (p. 175)

**ground state** The lowest energy state that an electron can have within an atom. (p. 90)

**GUT epoch** Period when gravity separated from the other three forces of nature.

**gyroscope** System of rotating wheels that allows a spacecraft to maintain a fixed orientation in space.

**H**

**habitable zone** Three-dimensional zone of comfortable temperature (corresponding to liquid water) that surrounds every star. (p. 769)

**half-life** The amount of time it takes for half of the initial amount of a radioactive substance to decay into something else. (p. 184)

**hayashi track** Evolutionary track followed by a protostar during the final pre-main-sequence phase before nuclear fusion begins.

**heat** Thermal energy, the energy of an object due to the random motion of its component atoms or molecules.

**heat death** End point of a bound universe, in which all matter and life are destined to be incinerated.

**heavy element** In astronomical terms, any element heavier than hydrogen and helium.

**heliocentric model** A model of the solar system that is centered on the Sun, with Earth in motion about the Sun. (p. 41)

**helioseismology** The study of conditions far below the Sun's surface through the analysis of internal "sound" waves that repeatedly cross the solar interior. (p. 420)

**helium-burning shell** Shell of burning helium gas surrounding a non-burning stellar core of carbon ash.

**helium capture** The formation of heavy elements by the capture of a helium nucleus. For example, carbon can form heavier elements by fusion with other carbon nuclei, but it is much more likely to occur by helium capture, which requires less energy. (p. 570)

**helium flash** An explosive event in the post-main-sequence evolution of a low-mass star. When helium fusion begins in a dense stellar core, the burning is explosive in nature. It continues until the energy released is enough to expand the core, at which point the star achieves stable equilibrium again. (p. 534)

**helium precipitation** Mechanism responsible for the low abundance of helium of Saturn's atmosphere. Helium condenses in the upper layers to form a mist, which rains down toward Saturn's interior, just as water vapor forms into rain in the atmosphere of Earth. (p. 314)

**helium shell flash** Condition in which the helium-burning shell in the core of a star cannot respond to rapidly changing conditions within it, leading to a sudden temperature rise and a dramatic increase in nuclear reaction rates.

**Hertzsprung–Russell (H–R) diagram** A plot of luminosity against temperature (or spectral class) for a group of stars. (p. 464)

**HI region** Region of space containing primarily neutral hydrogen. (p. 486)

**high-energy astronomy** Astronomy using X- or gamma-ray radiation rather than optical radiation.

**high-energy telescope** Telescope designed to detect X- and gamma radiation. (p. 134)

**highlands** Relatively light-colored regions on the surface of the Moon that are elevated several kilometers above the maria. Also called terrae. (p. 203)

**high-mass star** Star with a mass more than 8 times that of the Sun; progenitor of a neutron star or black hole.

**HII region** Region of space containing primarily ionized hydrogen. (p. 486)

**homogeneity** Assumed property of the universe such that the number of galaxies in an imaginary large cube of the universe is the same no matter where in the universe the cube is placed. More generally, "the same everywhere." (p. 711)

**horizon problem** One of two conceptual problems with the standard Big Bang model, which is that some regions of the universe that have very similar properties are too far apart to have exchanged information within the age of the universe. (p. 746)

**horizontal branch** Region of the Hertzsprung–Russell diagram where post-main-sequence stars again reach hydrostatic equilibrium. At this point, the star is burning helium in its core and fusing hydrogen in a shell surrounding the core. (p. 534)

**hot dark matter** A class of candidates for the dark matter in the universe, composed of lightweight particles such as neutrinos, much less massive than the electron. (p. 752)

**hot Jupiter** A massive, gaseous planet orbiting very close to its parent star. (p. 405)

**hot longitudes (Mercury)** Two opposite points on Mercury's equator where the Sun is directly overhead at perihelion.

**Hubble classification scheme** Method of classifying galaxies according to their appearance, developed by Edwin Hubble. (p. 650)

**Hubble diagram** Plot of galactic recession velocity versus distance; evidence for an expanding universe.

**Hubble flow** Universal recession described by the Hubble diagram and quantified by the Hubble Law.

**Hubble's constant** The constant of proportionality that gives the relation between recessional velocity and distance in Hubble's law. (p. 662)

**Hubble's law** Law that relates the observed velocity of recession of a galaxy to its distance from us. The velocity of recession of a galaxy is directly proportional to its distance away. (p. 661)

**hydrocarbon** Molecule consisting solely of hydrogen and carbon.

**hydrogen envelope** An invisible sheath of gas engulfing the coma of a comet, usually distorted by the solar wind and extending across millions of kilometers of space. (p. 368)

**hydrogen shell burning** Fusion of hydrogen in a shell that is driven by contraction and heating of the helium core. Once hydrogen is depleted in the core of a star, hydrogen burning stops and the core contracts due to gravity, causing the temperature to rise, heating the surrounding layers of hydrogen in the star, and increasing the burning rate there. (p. 531)

**hydrosphere** Layer of Earth that contains the liquid oceans and accounts for roughly 70 percent of Earth's total surface area. (p. 170)

**hydrostatic equilibrium** Condition in a star or other fluid body in which gravity's inward pull is exactly balanced by internal forces due to pressure. (p. 419)

**hyperbola** Curve formed when a plane intersects a cone at a small angle to the axis of the cone.

**hypernova** Explosion where a massive star undergoes core collapse and forms a black hole and a gamma-ray burst. *See* supernova.

## I

**igneous** Type of rock formed from molten material.

**image** The optical representation of an object produced when from the object is reflected or refracted by a mirror or lens. (p. 108)

**Impact Theory (Moon)** Combination of the Capture and Daughter theories, suggesting that the Moon formed after an impact which dislodged some of Earth's mantle and placed it in orbit. (p. 222)

**inertia** The tendency of an object to continue moving at the same speed and in the same direction, unless acted upon by a force. (p. 52)

**inferior conjunction** Orbital configuration in which an inferior planet (Mercury or Venus) lies closest to Earth.

**inflation** *See* epoch of inflation. (p. 747)

**infrared** Region of the electromagnetic spectrum just outside the visible range, corresponding to light of a slightly longer wavelength than red light. (p. 64)

**infrared telescope** Telescope designed to detect infrared radiation. Many such telescopes are designed to be lightweight so that they can be carried above (most of) Earth's atmosphere by balloons, airplanes, or satellites. (p. 132)

**infrared waves** Electromagnetic radiation with wavelength in the infrared part of the spectrum. (p. 64)

**inhomogeneity** Deviation from perfectly uniform density; in cosmology, inhomogeneities in the universe are ultimately due to quantum fluctuations before inflation.

**inner core** The central part of Earth's core, believed to be solid, and composed mainly of nickel and iron. (p. 177)

**instability strip** Region of the Hertzsprung–Russell diagram where pulsating post-main-sequence stars are found.

**intensity** A basic property of electromagnetic radiation that specifies the amount or strength of the radiation.

**intercloud medium** Superheated bubbles of hot gas extending far into interstellar space.

**intercrater plains** Regions on the surface of Mercury that do not show extensive cratering but are relatively smooth. (p. 217)

**interference** The ability of two or more waves to interact in such a way that they either reinforce or cancel each other. (p. 72)

**interferometer** Collection of two or more telescopes working together as a team, observing the same object at the same time and at the same wavelength. The effective diameter of an interferometer is equal to the distance between its outermost telescopes. (p. 129)

**interferometry** Technique in widespread use to dramatically improve the resolution of radio and infrared maps. Several telescopes observe the object simultaneously, and a computer analyzes how the signals interfere with each other. (p. 129)

**intermediate-mass black hole** Black hole having a mass 100–1000 times greater than the mass of the Sun.

**interplanetary matter** Matter in the solar system that is not part of a planet or moon—cosmic "debris." (p. 153)

**interstellar dust** Microscopic dust grains that populate space between the stars, having their origins in the ejected matter of long-dead stars.

**interstellar gas cloud** A large cloud of gas found in the space among the stars.

**interstellar medium** The matter between stars, composed of two components, gas and dust, intermixed throughout all of space. (p. 480)

**intrinsic variable** Star that varies in appearance due to internal processes (rather than, say, interaction with another star).

**inverse-square law** The law that a field follows if its strength decreases with the square of the distance. Fields that follow the inverse-square law decrease rapidly in strength as the distance increases, but never quite reach zero. (p. 54)

**Io plasma torus** Doughnut-shaped region of energetic ionized particles, emitted by the volcanoes on Jupiter's moon Io and swept up by Jupiter's magnetic field.

**ion** An atom that has lost one or more of its electrons.

**ionization state** Term describing the number of electrons missing from an atom: I refers to a neutral atom, II refers to an atom missing one electron, and so on.

**ionized** State of an atom or molecule that has lost one or more of its electrons. (p. 90)

**ionosphere** Layer in Earth's atmosphere above about 100 km where the atmosphere is significantly ionized and conducts electricity. (p. 171)

**ion tail** Thin stream of ionized gas that is pushed away from the head of a comet by the solar wind. It extends directly away from the Sun. Often referred to as a plasma tail. (p. 368)

**irregular galaxy** A galaxy that does not fit into any of the other major categories in the Hubble classification scheme. (p. 654)

**isotopes** Nuclei containing the same number of protons but different numbers of neutrons. Most elements can exist in several isotopic forms. A common example of an isotope is deuterium, which differs from normal hydrogen by the presence of an extra neutron in the nucleus. (p. 439)

**isotropy** Looking the same in every direction. Often applied to the universe as part of the cosmological principle. (p. 711)

**J**

**jet stream** Relatively strong winds in the upper atmosphere channeled into a narrow stream by a planet's rotation. Normally refers to a horizontal, high-altitude wind.

**joule** The SI unit of energy.

**jovian planet** One of the four giant outer planets of the solar system, resembling Jupiter in physical and chemical composition. (p. 152)

**K**

**Kelvin Scale** Temperature scale in which absolute zero is at 0 K; a change of 1 kelvin is the same as a change of 1 degree Celsius.

**Kelvin-Helmholtz contraction phase** Evolutionary track followed by a star during the protostar phase.

**Kepler's laws of planetary motion** Three laws, based on precise observations of the motions of the planets by Tycho Brahe, that summarize the motions of the planets about the Sun.

**kinetic energy** Energy of an object due to its motion.

**Kirchhoff's laws** Three rules governing the formation of different types of spectra. (p. 89)

**Kirkwood gaps** Gaps in the spacings of orbital semimajor axes of asteroids in the asteroid belt, produced by dynamical resonances with nearby planets, especially Jupiter. (p. 365)

**Kuiper belt** A region in the plane of the solar system outside the orbit of Neptune where most short-period comets are thought to originate. (p. 374)

**Kuiper-belt object** Small icy body orbiting in the Kuiper belt. (p. 375)

**L**

**labeled-release experiment** Experiment to look for life on Mars. Radioactive carbon compounds were added to Martian soil specimens. Scientists looked for indications that the carbon had been eaten or inhaled.

**Lagrangian point** One of five special points in the plane of two massive bodies orbiting one another, where a third body of negligible mass can remain in equilibrium. (p. 331)

**lander** Spacecraft that lands on the object it is studying.

**laser ranging** Method of determining the distance to an object by firing a laser beam at it and measuring the time taken for the light to return.

**lava dome** Volcanic formation formed when lava oozes out of fissures in a planet's surface, creating the dome, and then withdraws, causing the crust to crack and subside. (p. 240)

**law of conservation of angular momentum** A fundamental law of physics which states that the total angular momentum of a system always remains constant in any physical process. This law ensures that a spinning gas cloud must spin faster as it contracts.

**law of conservation of mass and energy** A fundamental law of modern physics which states that the sum of mass and energy must always remain constant in any physical process. In fusion reactions, the lost mass is converted into energy, primarily in the form of electromagnetic radiation. (p. 438)

**laws of planetary motion** Three laws derived by Kepler describing the motion of the planets around the Sun. (p. 47)

**leap year** Year in which an additional day is inserted into the calendar in order to keep the calendar year synchronized with Earth's orbit around the Sun. (p. 19)

**lens** (eye) The part of the eye that refracts light onto the retina.

**lens** Optical instrument made of glass or some other transparent material, shaped so that, as a parallel beam of light passes through it, the rays are refracted so as to pass through a single focal point.

**lepton** From the Greek word for light, referring in particle physics to low-mass particles such as electrons, muons, and neutrinos that interact via the weak force.

**lepton epoch** Period when the light elementary particles (leptons) were in thermal equilibrium with the cosmic radiation field.

**lidar** Light Detection and Ranging—a device that uses laser-ranging to measure distance.

**light** *See* electromagnetic radiation.

**light curve** The variation in brightness of a star with time. (p. 470)

**light element** In astronomical terms, hydrogen and helium.

**light-gathering power** Amount of light a telescope can view and focus, proportional to the area of the primary mirror.

**lighthouse model** The leading explanation for pulsars. A small region of the neutron star, near one of the magnetic poles, emits a steady stream of radiation that sweeps past Earth each time the star rotates. The period of the pulses is the star's rotation period. (p. 581)

**light pollution** Unwanted, upward-directed light from streets, homes and businesses, that scatters back to Earth from dust in the air, obscuring the faint objects that astronomers want to observe. (p. 122)

**light-year** The distance that light, moving at a constant speed of 300,000 km/s, travels in one year. One light-year is about 10 trillion kilometers. (p. 4)

**limb** Edge of a lunar, planetary, or solar disk.

**line of nodes** Intersection of the plane of the Moon's orbit with Earth's orbital plane.

**line of sight technique** Method of probing interstellar clouds by observing their effects on the spectra of background stars.

**linear momentum** Tendency of an object to keep moving in a straight line with constant velocity; the product of the object's mass and velocity.

**lithosphere** Earth's crust and a small portion of the upper mantle that make up Earth's plates. This layer of Earth undergoes tectonic activity. (p. 181)

**Local Bubble** The particular low-density intercloud region surrounding the Sun.

**Local Group** The small galaxy cluster that includes the Milky Way Galaxy. (p. 659)

**local supercluster** Collection of galaxies and clusters centered on the Virgo cluster. *See also* supercluster.

**logarithm** The power to which 10 must be raised to produce a given number.

**logarithmic scale** Scale using the logarithm of a number rather than the number itself; commonly used to compress a large range of data into more manageable form.

**look-back time** Time in the past when an object emitted the radiation we see today.

**Lorentz contraction** Apparent contraction of an object in the direction in which it is moving. (p. 596)

**low-mass star** Star with a mass less than 8 times that of the Sun; progenitor of a white dwarf.

**luminosity class** A classification scheme that groups stars according to the width of their spectral lines. For a group of stars with the same temperature, luminosity class differentiates between supergiants, giants, main-sequence stars, and subdwarfs. (p. 468)

**lunar dust** *See* regolith.

**luminosity** One of the basic properties used to characterize stars, luminosity is defined as the total energy radiated by star each second, at all wavelengths. (p. 418)

**lunar eclipse** Celestial event during which the moon passes through the shadow of Earth, temporarily darkening its surface. (p. 21)

**lunar phase** The appearance of the Moon at different points along its orbit. (p. 19)

**Lyman-alpha forest** Collection of lines in an object's spectra starting at the redshifted wavelength of the object's own Lyman-alpha emission line and extending to shorter wavelengths; produced by gas in galaxies along the line of sight.

**M**

**macroscopic** Large enough to be visible by the unaided eye.

**Magellanic Clouds** Two small irregular galaxies that are gravitationally bound to the Milky Way Galaxy. (p. 654)

**magnetic field** Field that accompanies any changing electric field and governs the influence of magnetized objects on one another. (p. 68)

**magnetic poles** Points on a planet where the planetary magnetic field lines intersect the planet's surface vertically.

**magnetism** The presence of a magnetic field.

**magnetometer** Instrument that measures magnetic field strength.

**magnetopause** Boundary between a planet's magnetosphere and the solar wind.

**magnetosphere** A zone of charged particles trapped by a planet's magnetic field, lying above the atmosphere. (p. 170)

**magnitude scale** A system of ranking stars by apparent brightness, developed by the Greek astronomer Hipparchus. Originally, the brightest stars in the sky were categorized as being of first magnitude, while the faintest stars visible to the naked eye were classified as sixth magnitude. The scheme has since been extended to cover stars and galaxies too faint to be seen by the unaided eye. Increasing magnitude means fainter stars, and a difference of five magnitudes corresponds to a factor of 100 in apparent brightness. (p. 270) (p. 455)

**main sequence** Well-defined band on the Hertzsprung–Russell diagram on which most stars are found, running from the top left of the diagram to the bottom right. (p. 464)

**main-sequence turnoff** Special point on the Hertzsprung–Russell diagram for a cluster, indicative of the cluster's age. If all the stars in the cluster are plotted, the lower mass stars will trace out the main sequence up to the point where stars begin to evolve off the main sequence toward the red giant branch. The point where stars are just beginning to evolve off is the main-sequence turnoff. (p. 546)

**major axis** The long axis of an ellipse.

**mantle** Layer of Earth just interior to the crust. (p. 170)

**mare** Relatively dark-colored and smooth region on the surface of the Moon (plural: *maria*). (p. 203)

**marginally bound universe** Universe that will expand forever but at an increasingly slow rate.

**mass** A measure of the total amount of matter contained within an object. (p. 52)

**mass-energy equivalence** Principle that mass and energy are not independent, but can be converted from one to the other according to Einstein's formula $E = mc^2$. (p. 438)

**mass function** Relation between the component masses of a single-line spectroscopic binary.

**mass transfer** Process by which one star in a binary system transfers matter onto the other.

**massive compact halo object (MACHO)** Collective name for "stellar" candidates for dark matter, including brown dwarfs, white dwarfs, and low-mass red dwarfs.

**mass-luminosity relation** The dependence of the luminosity of a main-sequence star on its mass. The luminosity increases roughly as the mass raised to the third power.

**mass–radius relation** The dependence of the radius of a main-sequence star on its mass. The radius rises roughly in proportion to the mass. (p. 459)

**mass-transfer binary** *See* semi-detached binary. (p. 549)

**matter** Anything having mass.

**matter era** (Current) era following the Radiation era when the universe is larger and cooler, and matter is the dominant constituent of the universe.

**matter-antimatter annihilation** Reaction in which matter and antimatter annihilate to produce high energy gamma rays. *See also* antiparticle.

**matter-dominated universe** A universe in which the density of matter exceeds the density of radiation. The present-day universe is matter dominated. (p. 735)

**Maunder minimum** Lengthy period of solar inactivity that extended from 1645 to 1715. (p. 432)

**mean solar day** Average length of time from one noon to the next, taken over the course of a year—24 hours. (p. 18)

**medium** Material through which a wave, such as sound, travels.

**meridian** An imaginary line on the celestial sphere through the north and south celestial poles, passing directly overhead at a given location. (p. 17)

**mesosphere** Region of Earth's atmosphere lying between the stratosphere and the ionosphere, 50–80 km above Earth's surface. (p. 171)

**Messier object** Member of a list of "fuzzy" objects compiled by astronomer Charles Messier in the eighteenth century.

**metabolism** The daily utilization of food and energy by which organisms stay alive.

**metallic** Composed of metal or metal compounds.

**metamorphic** Rocks created from existing rocks exposed to extremes of temperature or pressure.

**meteor** Bright streak in the sky, often referred to as a "shooting star," resulting from a small piece of interplanetary debris entering Earth's atmosphere and heating air molecules, which emit light as they return to their ground states. (p. 380)

**meteor shower** Event during which many meteors can be seen each hour, caused by the yearly passage of Earth through the debris spread along the orbit of a comet. (p. 381)

**meteorite** Any part of a meteoroid that survives passage through the atmosphere and lands on the surface of Earth. (p. 380)

**meteoroid** Chunk of interplanetary debris prior to encountering Earth's atmosphere. (p. 380)

**meteoroid swarm** Pebble-sized cometary fragments dislodged from the main body, moving in nearly the same orbit as the parent comet. (p. 381)

**microlensing** Gravitational lensing by individual stars in a galaxy. (p. 702)

**micrometeoroid** Relatively small chunk of interplanetary debris ranging in size from a dust particle to a pebble. (p. 381)

**microquasar** Stellar sized source of energetic X and gamma radiation, powered by accretion onto a neutron star or black hole, somewhat like a quasar but on a much smaller scale. (p. 586)

**microsphere** Small droplet of protein-like material that resists dissolution in water.

**midocean ridge** Place where two plates are moving apart, allowing fresh magma to well up. (p. 186)

**Milky Way Galaxy** The spiral galaxy in which the Sun resides. The disk of our Galaxy is visible in the night sky as the faint band of light known as the Milky Way. (p. 618)

**millisecond pulsar** A pulsar whose period indicates that the neutron star is rotating nearly 1000 times each second. The most likely explanation for these rapid rotators is that the neutron star has been spun up by drawing in matter from a companion star. (p. 586)

**molecular cloud** A cold, dense interstellar cloud which contains a high fraction of molecules. It is widely believed that the relatively high density of dust particles in these clouds plays an important role in the formation and preservation of the molecules. (p. 495)

**molecular cloud complex** Collection of molecular clouds that spans as much as 50 parsecs and may contain enough material to make millions of Sun-size stars. (p. 497)

**molecule** A tightly bound collection of atoms held together by the atoms' electromagnetic fields. Molecules, like atoms, emit and absorb photons at specific wavelengths. (p. 98)

**molten** In liquid form due to high temperatures.

**moon** A small body in orbit around a planet.

**mosaic (photograph)** Composite photograph made up of many smaller images.

**M-type asteroid** Asteroid containing large fractions of nickel and iron.

**multiple star system** Group of two or more stars in orbit around one another.

**muon** A type of lepton (along with the electron and tau).

**N**

**naked singularity** A singularity that is not hidden behind an event horizon.

**nanobacterium** Very small bacterium with diameter in the nanometer range.

**nanometer** One billionth of a meter.

**neap tide** The smallest tide, occurring when the Earth-Moon line is perpendicular to the Earth-Sun line at the first and third quarters.

**nebula** General term used for any "fuzzy" patch on the sky, either light or dark. (p. 484)

**nebular theory** One of the earliest models of solar system formation, dating back to Descartes, in which a large cloud of gas began to collapse under its own gravity to form the Sun and planets. (p. 162)

**nebulosity** "Fuzziness," usually in the context of an extended or gaseous astronomical object.

**neon-oxygen white dwarf** White dwarf formed from a low-mass star with a mass close to the "high-mass" limit, in which neon and oxygen form in the core.

**neutrino** Virtually massless and chargeless particle that is one of the products of fusion reactions in the Sun. Neutrinos move at close to the speed of light, and interact with matter hardly at all. (p. 439)

**neutrino oscillations** Likely solution to the solar neutrino problem, in which the neutrino has a very tiny mass. In this case, the correct number of neutrinos can be produced in the solar core, but on their way to Earth some can "oscillate," or become transformed into other particles, and thus go undetected. (p. 444)

**neutron** An elementary particle with roughly the same mass as a proton, but which is electrically neutral. Along with protons, neutrons form the nuclei of atoms. (p. 97)

**neutron capture** The primary mechanism by which very massive nuclei are formed in the violent aftermath of a supernova. Instead of fusion of like nuclei, heavy elements are created by the addition of more and more neutrons to existing nuclei. (p. 571)

**neutron degeneracy pressure** Pressure due to the Pauli exclusion principle, arising when neutrons are forced to come into close contact. (p. 580)

**neutron spectrometer** Instrument designed to search for water ice by looking for hydrogen.

**neutron star** A dense ball of neutrons that remains at the core of a star after a supernova explosion has destroyed the rest of the star. Typical neutron stars are about 20 km across, and contain more mass than the Sun. (p. 580)

**neutronization** Process occurring at high densities, in which protons and electrons are crushed together to form neutrons and neutrinos.

**new Moon** Phase of the moon during which none of the lunar disk is visible.

**Newtonian mechanics** The basic laws of motion, postulated by Newton, which are sufficient to explain and quantify virtually all of the complex dynamical behavior found on Earth and elsewhere in the universe. (p. 52)

**Newtonian telescope** A reflecting telescope in which incoming light is intercepted before it reaches the prime focus and is deflected into an eyepiece at the side of the instrument. (p. 111)

**nodes** Two points on the Moon's orbit when it crosses the ecliptic.

**nonrelativistic** Speed that is much less than the speed of light.

**nonthermal spectrum** Continuous spectrum not well described by a blackbody.

**normal galaxy** Galaxy whose overall energy emission is consistent with the summed light of many stars. (p. 664)

**north celestial pole** Point on the celestial sphere directly above Earth's North Pole.

**Northern and Southern Lights** Colorful displays produced when atmospheric molecules, excited by charged particles from the Van Allen belts, fall back to their ground state.

**nova** A star that suddenly increases in brightness, often by a factor of as much as 10,000, then slowly fades back to its original luminosity. A nova is the result of an explosion on the surface of a white-dwarf star, caused by matter falling onto its surface from the atmosphere of a binary companion, (p. 556)

**nuclear binding energy** Energy that must be supplied to split an atomic nucleus into neutrons and protons.

**nuclear epoch** Period when protons and neutrons fused to form heavier nuclei.

**nuclear fission** Mechanism of energy generation used in nuclear reactors on Earth, in which heavy nuclei are split into lighter ones, releasing energy in the process. (p. 559)

**nuclear fusion** Mechanism of energy generation in the core of the Sun, in which light nuclei are combined, or fused, into heavier ones, releasing energy in the process. (p. 438)

**nuclear reaction** Reaction in which two nuclei combine to form other nuclei, often releasing energy in the process. *See also* fusion.

**nucleotide base** An organic molecule, the building block of genes that pass on hereditary characteristics from one generation of living creatures to the next. (p. 762)

**nucleus** Dense, central region of an atom, containing both protons and neutrons, and orbited by one or more electrons. (p. 90); The solid region of ice and dust that composes the central region of the head of a comet. (p. 367); The dense central core of a galaxy. (p. 642)

**O**

**obscuration** Blockage of light by pockets of interstellar dust and gas.

**Olbers's paradox** A thought experiment suggesting that if the universe were homogeneous, infinite, and unchanging, the entire night sky would be as bright as the surface of the Sun. (p. 714)

**Oort cloud** Spherical halo of material surrounding the solar system out to a distance of about 50,000 AU; where most comets reside. (p. 374)

**opacity** A quantity that measures a material's ability to block electromagnetic radiation. Opacity is the opposite of transparency.

**open cluster** Loosely bound collection of tens to hundreds of stars, a few parsecs across, generally found in the plane of the Milky Way. (p. 518)

**open universe** Geometry that the universe would have if the density of matter were less than the critical value. In an open universe there is not enough matter to halt the expansion of the universe. An open universe is infinite in extent. (p. 462) (p. 721)

**opposition** Orbital configuration in which a planet lies in the opposite direction from the Sun, as seen from Earth.

**optical double** Chance superposition in which two stars appear to lie close together but are actually widely separated.

**optical telescope** Telescope designed to observe electromagnetic radiation at optical wavelengths.

**orbital** One of several energy states in which an electron can exist in an atom. (p. 91)

**orbital period** Time taken for a body to complete one full orbit around another.

**orbiter** Spacecraft that orbits an object to make observations.

**organic compound** Chemical compound (molecule) containing a significant fraction of carbon atoms; the basis of living organisms.

**outer core** The outermost part of Earth's core, believed to be liquid and composed mainly of nickel and iron. (p. 177)

**outflow channel** Surface feature on Mars, evidence that liquid water once existed there in great quantity; believed to be the relics of catastrophic flooding about 3 billion years ago. Found only in the equatorial regions of the planet. (p. 260)

**outgassing** Production of atmospheric gases (carbon dioxide, water vapor, methane, and sulphur dioxide) by volcanic activity.

**ozone layer** Layer of Earth's atmosphere at an altitude of 20–50 km where incoming ultraviolet solar radiation is absorbed by oxygen, ozone, and nitrogen in the atmosphere. (p. 172)

**P**

**pair production** Process in which two photons of electromagnetic radiation give rise to a particle-antiparticle pair. (p. 735)

**parallax** The apparent motion of a relatively close object with respect to a more distant background as the location of the observer changes. (p. 28)

**parsec** The distance at which a star must lie in order for its measured parallax to be exactly 1 arc second; 1 parsec equals 206,000 AU. (p. 451)

**partial eclipse** Celestial event during which only a part of the occulted body is blocked from view. (p. 21)

**particle** A body with mass but of negligible dimension.

**particle accelerator** Device used to accelerate subatomic particles to relativistic speeds.

**particle–antiparticle pair** Pair of particles (e.g., an electron and a positron) produced by two photons of sufficiently high energy.

**particle detector** Experimental equipment that allows particles and antiparticles to be detected and identified.

**Pauli exclusion principle** A rule of quantum mechanics that prohibits electrons in dense gas from being squeezed too close together.

**penumbra** Portion of the shadow cast by an eclipsing object in which the eclipse is seen as partial. (p. 22); The outer region of a sunspot, surrounding the umbra, which is not as dark and not as cool as the central region. (p. 428)

**perihelion** The closest approach to the Sun of any object in orbit about it.

**period** The time needed for an orbiting body to complete one revolution about another body. (pp. 48, 65)

**period–luminosity relation** A relation between the pulsation period of a Cepheid variable and its absolute brightness. Measurement of the pulsation period allows the distance of the star to be determined. (p. 623)

**permafrost** Layer of permanently frozen water ice believed to lie just under the surface of Mars. (p. 262)

**phase** Appearance of the sunlit face of the Moon at different points along its orbit, as seen from Earth. (p. 19)

**photodisintegration** Process occurring at high temperatures, in which individual photons have enough energy to split a heavy nucleus (e.g., iron) into lighter nuclei.

**photoelectric effect** Emission of an electron from a surface when a photon of electromagnetic radiation is absorbed. (p. 94)

**photoevaporation** Process in which a cloud in the vicinity of a newborn hot star is dispersed by the star's radiation.

**photometer** A device that measures the total amount of light received in all or part of the image. (p. 121)

**photometry** Branch of observational astronomy in which the brightness of a source is measured through each of a set of standard filters. (p. 121) (p. 459)

**photomicrograph** Photograph taken through a microscope.

**photon** Individual packet of electromagnetic energy that makes up electromagnetic radiation. (p. 91)

**photosphere** The visible surface of the Sun, lying just above the uppermost layer of the Sun's interior, and just below the chromosphere. (p. 416)

**photosynthesis** Process by which plants manufacture carbohydrates and oxygen from carbon dioxide and water using chlorophyll and sunlight as the energy source.

**pixel** One of many tiny picture elements, organized into a two-dimensional array, making up a digital image. (p. 118)

**Planck curve** *See* blackbody curve.

**Planck epoch** Period from the beginning of the Universe to roughly $10^{-43}$ second, when the laws of physics are not understood.

**Planck's constant** Fundamental physical constant relating the energy of a photon to its radiation frequency (color).

**planet** One of eight major bodies that orbit the Sun, visible to us by reflected sunlight.

**planetary nebula** The ejected envelope of a red-giant star, spread over a volume roughly the size of our solar system. (p. 537)

**planetary ring system** Material organized into thin, flat rings encircling a giant planet, such as Saturn.

**planetary transit** Orbital configuration in which an inferior planet (e.g. Mercury or Venus, as seen from Earth) is observed to pass directly in front of the Sun.

**planetesimal** Term given to objects in the early solar system that had reached the size of small moons, at which point their gravitational fields were strong enough to begin influencing their neighbors. (p. 392)

**plasma** A gas in which the constituent atoms are completely ionized.

**plate tectonics** The motions of regions of Earth's lithosphere, which drift with respect to one another. Also known as continental drift. (p. 181)

**plutino** Kuiper-belt object whose orbital period (like that of Pluto) is in a 3:2 resonance with the orbit of Neptune. (p. 398)

**polarity** A measure of the direction of the solar magnetic field in a sunspot. Conventionally, lines coming out of the surface are labeled "S" while those going into the surface are labeled "N." (p. 429)

**polarization** The alignment of the electric fields of emitted photons, which are generally emitted with random orientations. (p. 72)

**polar vortex** Long-lived pattern of circulating winds around the pole of a planet. (p. 243)

**Population I and II stars** Classification scheme for stars based on the abundance of heavy elements. Within the Milky Way, Population I refers to young disk stars and Population II refers to old halo stars.

**positron** Atomic particle with properties identical to those a negatively charged electron, except for its positive charge. This positron is the antiparticle of the electron. Positrons and electrons annihilate one another when they meet, producing pure energy in the form of gamma rays. (p. 439)

**prebiotic compound** Molecule that can combine with others to form the building blocks of life.

**precession** The slow change in the direction of the rotation axis of a spinning object, caused by some external gravitational influence. (p. 16)

**primary atmosphere** The chemical components that would have formed Earth's atmosphere.

**primary mirror** Mirror placed at the prime focus of a telescope (*see* prime focus).

**prime focus** The point in a reflecting telescope where the mirror focuses incoming light to a point. (p. 108)

**prime-focus image** Image formed at the prime focus of a telescope.

**primordial matter** Matter created during the early, hot epochs of the universe.

**primordial nucleosynthesis** The production of elements heavier than hydrogen by nuclear fusion in the high temperatures and densities that existed in the early universe. (p. 743)

**principle of cosmic censorship** A proposition to separate the unexplained physics near a singularity from the rest of the well-behaved universe. The principle states that nature always hides any singularity, such as a black hole, inside an event horizon, which insulates the rest of the universe from seeing it.

**progenitor** "Ancestor" star of a given object; e.g., the star that existed before a supernova explosion is the supernova's progenitor.

**prograde motion** Motion across the sky in the eastward direction.

**prominence** Loop or sheet of glowing gas ejected from an active region on the solar surface, which then moves through the inner parts of the corona under the influence of the Sun's magnetic field. (p. 432)

**proper motion** The angular movement of a star across the sky, as seen from Earth, measured in seconds of arc per year. This movement is a result of the star's actual motion through space. (p. 268) (p. 439)

**protein** Molecule made up of amino acids that controls metabolism.

**proton** An elementary particle carrying a positive electric charge, a component of all atomic nuclei. The number of protons in the nucleus of an atom dictates what type of atom it is. (p. 67)

**proton–proton chain** The chain of fusion reactions, leading from hydrogen to helium, that powers main-sequence stars. (p. 440)

**protoplanet** Clump of material, formed in the early stages of solar system formation, that was the forerunner of the planets we see today.

**protostar** Stage in star formation when the interior of a collapsing fragment of gas is sufficiently hot and dense that it becomes opaque to its own radiation. The protostar is the dense region at the center of the fragment. (p. 506)

**protostellar disk** Swirling disk of gas and dust within which a star (and possibly a planetary system) forms. The "solar nebula," in the case of our Sun. (p. 507)

**protosun** The central accumulation of material in the early stages of solar system formations, the forerunner of the present-day Sun. (p. 386)

**Ptolemaic model** Geocentric solar system model, developed by the second century astronomer Claudius Ptolemy. It predicted with great accuracy the positions of the then known planets. (p. 40)

**pulsar** Object that emits radiation in the form of rapid pulses with a characteristic pulse period and duration. Charged particles, accelerated by the magnetic field of a rapidly rotating neutron star, flow along the magnetic field lines, producing radiation that beams outward as the star spins on its axis. (p. 581)

**pulsating variable star** A star whose luminosity varies in a predictable, periodic way. (p. 621)

**P-waves** Pressure waves from an earthquake that travel rapidly through liquids and solids.

**pyrolific-release experiment** Experiment to look for life on Mars. Radioactively tagged carbon dioxide was added to Martian soil specimens. Scientists looked for indications that the radioactive material had been absorbed.

**Q**

**quantization** The fact that light and matter on small scales behave in a discontinuous manner, and manifest themselves in the form of tiny "packets" of energy, called quanta. (p. 91)

**quantum fluctuation** Temporary random change in the amount of energy at a point in space.

**quantum gravity** Theory combining general relativity with quantum mechanics.

**quantum mechanics** The laws of physics as they apply on atomic scales. (p. 91)

**quark** A fundamental matter particle that interacts via the strong force; basic constituent of protons and neutrons. (p. 740)

**quark epoch** Period when all heavy elementary particles (composed of quarks) were in thermal equilibrium with the cosmic radiation field.

**quarter Moon** Lunar phase in which the Moon appears as a half disk.

**quasar** Starlike radio source with an observed redshift that indicates an extremely large distance from Earth. The brightest nucleus of a distant active galaxy. (p. 671)

**quasi-stellar object (QSO)** See quasar. (p. 671)

**quiescent prominence** Prominence that persists for days or weeks, hovering high above the solar photosphere.

**quiet Sun** The underlying predictable elements of the Sun's behavior, such as its average photospheric temperature, which do not significantly change in time. (p. 425)

**R**

**radar** Acronym for Radio Detection and Ranging. Radio waves are bounced off an object, and the time taken for the echo to return indicates its distance. (p. 51)

**radial motion** Motion along a particular line of sight, which induces apparent changes in the wavelength (or frequency) of radiation received.

**radial velocity** Component of a star's velocity along the line of sight.

**radian** Angular measure equivalent to $180/\pi = 57.3$ degrees.

**radiant** Constellation from which a meteor shower appears to come.

**radiation** A way in which energy is transferred from place to place in the form of a wave. Light is a form of electromagnetic radiation. (p. 62)

**radiation darkening** The effect of chemical reactions that result when high-energy particles strike the icy surfaces of objects in the outer solar system. The reactions lead to a buildup of a dark layer of material. (p. 347)

**radiation era** The first few thousand years after the Big Bang when the universe was small, dense, and dominated by radiation.

**radiation zone** Region of the Sun's interior where extremely high temperatures guarantee that the gas is completely ionized. Photons only occasionally interact with electrons, and travel through this region with relative ease. (p. 418)

**radiation-dominated universe** Early epoch in the universe, when the equivalent density of radiation in the cosmos exceeded the density of matter. (p. 735)

**radio** Region of the electromagnetic spectrum corresponding to the longest-wavelength radiation.

**radioactivity** The release of energy by rare, heavy elements when their nuclei decay into lighter nuclei. (pp. 181, 184)

**radio galaxy** Type of active galaxy that emits most of its energy in the form of long-wavelength radiation. (p. 667)

**radiograph** Image made from observations at radio wavelengths.

**radio lobe** Roundish extended region of radio-emitting gas, lying well beyond the center of a radio galaxy. (p. 667)

**radio telescope** Large instrument designed to detect radiation from space at radio wavelengths. (p. 125)

**radio wave** Electromagnetic radiation with wavelength in the radio part of the spectrum (p. 64)

**radius–luminosity–temperature relationship** A mathematical proportionality, arising from Stefan's Law, which allows astronomers to indirectly determine the radius of a star once its luminosity and temperature are known. (p. 462)

**rapid mass transfer** Mass transfer in a binary system that proceeds at a rapid and unstable rate, transferring most of the mass of one star onto the other.

**ray** The path taken by a beam of radiation.

**Rayleigh scattering** Scattering of light by particles in the atmosphere.

**recession velocity** Rate at which two objects are separating from one another.

**recurrent nova** Star that "goes nova" several times over the course of a few decades.

**reddening** Dimming of starlight by interstellar matter, which tends to scatter higher-frequency (blue) components of the radiation more efficiently than the lower-frequency (red) components. (p. 481)

**red dwarf** Small, cool faint star at the lower-right end of the main sequence on the Hertzsprung–Russell diagram. (p. 465)

**red giant** A giant star whose surface temperature is relatively low so that it glows red. (p. 462)

**red-giant branch** The section of the evolutionary track of a star corresponding to intense hydrogen shell burning, which drives a steady expansion and cooling of the outer envelope of the star. As the star gets larger in radius and its surface temperature cools, it becomes a red giant. (p. 531)

**red-giant region** The upper-right corner of the Hertzsprung–Russell diagram, where red-giant stars are found. (p. 466)

**redshift** Motion-induced change in the wavelength of light emitted from a source moving away from us. The relative recessional motion causes the wave to have an observed wavelength longer (and hence redder) than it would if it were not moving.

**redshift survey** Three-dimensional survey of galaxies, using redshift to determine distance.

**red supergiant** An extremely luminous red star. Often found on the asymptotic-giant branch of the Hertzsprung–Russell diagram. (p. 462)

**reflecting telescope** A telescope which uses a mirror to gather and focus light from a distant object. (p. 108)

**reflection nebula** Bluish nebula caused by starlight scattering from dust particles in an interstellar cloud located just off the line of sight between Earth and a bright star.

**refracting telescope** A telescope that uses a lens to gather and focus light from a distant object. (p. 108)

**refraction** The tendency of a wave to bend as it passes from one transparent medium to another. (p. 108)

**regolith** Surface dust on the moon, several tens of meters thick in places, caused by billions of years of meteoritic bombardment.

**relativistic** Speeds comparable to the speed of light.

**relativistic fireball** Leading explanation of a gamma-ray burst in which an expanding region of superhot gas radiates in the gamma-ray part of the spectrum.

**remnant** The object left behind after a supernova explosion. Can refer to (1) the expanding and cooling shell of glowing gas resulting from the event, or (2) the neutron star or black hole that remains at the center of the explosion. (p. 580)

**residual cap** Portion of Martian polar ice caps that remains permanently frozen, undergoing no seasonal variations. (p. 263)

**resonance** Circumstance in which two characteristic times are related in some simple way, e.g., an asteroid with an orbital period exactly half that of Jupiter. (p. 207)

**retina (eye)** The back part of the eye onto which light is focused by the lens.

**retrograde motion** Backward, westward loop traced out by a planet with respect to the fixed stars. (p. 39)

**revolution** Orbital motion of one body about another, such as Earth about the Sun. (p. 13)

**revolving** See revolution.

**Riemannian geometry** Geometry of positively curved space (such as the surface of a sphere).

**right ascension** Celestial coordinate used to measure longitude on the celestial sphere. The zero point is the position of the Sun at the vernal equinox.

**rille** A ditch on the surface of the Moon where molten lava flowed in the past. (p. 216)

**ring** See planetary ring system.

**ringlet** Narrow region in Saturn's planetary ring system where the density of ring particles is high. *Voyager* discovered that the rings visible from Earth are actually composed of tens of thousands of ringlets. (p. 317)

**Roche limit** Often called the tidal stability limit, the Roche limit gives the distance from a planet at which the tidal force (due to the planet) between adjacent objects exceeds their mutual attraction. Objects within this limit are unlikely to accumulate into larger objects. The rings of Saturn occupy the region within Saturn's Roche limit. (p. 317)

**Roche lobe** An imaginary surface around a star. Each star in a binary system can be pictured as being surrounded by a teardrop-shaped zone of gravitational influence, the Roche lobe. Any material within the Roche lobe of a star can be considered to be part of that star. During evolution, one member of the binary system can expand so that it overflows its own Roche lobe and begins to transfer matter onto the other star. (p. 549)

**rock cycle** Process by which surface rock on Earth is continuously redistributed and transformed from one type into another. (p. 189)

**rock** Material made predominantly from compounds of silicon and oxygen.

**rotation** Spinning motion of a body about an axis. (p. 10)

**rotation curve** Plot of the orbital speed of disk material in a galaxy against its distance from the galactic center. Analysis of rotation curves of spiral galaxies indicates the existence of dark matter. (p. 637)

**R-process** "Rapid" process in which many neutrons are captured by a nucleus during a supernova explosion.

**RR Lyrae star** Variable star whose luminosity changes in a characteristic way. All RR Lyrae stars have more or less the same average luminosity. (p. 607)

**RR Lyrae variable** Variable star whose luminosity changes in a characteristic way. All RR Lyrae stars have more or less the same average luminosity. (p. 621)

**runaway greenhouse effect** A process in which the heating of a planet leads to an increase in its atmosphere's ability to retain heat and thus to further heating, causing extreme changes in the temperature of the surface and the composition of the atmosphere. (p. 246)

**runoff channel** River-like surface feature on Mars, evidence that liquid water once existed there in great quantities. They are found in the southern highlands and are thought to have been formed by water that flowed nearly 4 billion years ago. (p. 260)

**S**

**S0 galaxy** Galaxy that shows evidence of a thin disk and a bulge, but which has no spiral arms and contains little or no gas. (p. 654)

**Sagittarius A/Sgr A** Strong radio source corresponding to the supermassive black hole at the center of the Milky Way. (p. 642)

**Saros cycle** Time interval between successive occurrences of the "same" solar eclipse, equal to 18 years, 11.3 days.

**satellite** A small body orbiting another larger body.

**SB0 galaxy** An S0-type galaxy whose disk shows evidence of a bar. (p. 654)

**scarp** Surface feature on Mercury believed to be the result of cooling and shrinking of the crust forming a wrinkle on the face of the planet. (p. 219)

**Schmidt telescope** Type of telescope having a very wide field of view, allowing large areas of the sky to be observed at once.

**Schwarzschild radius** The distance from the center of an object such that, if all the mass were compressed within that region, the escape speed would equal the speed of light. Once a stellar remnant collapses within this radius, light cannot escape and the object is no longer visible. (p. 593)

**scientific method** The set of rules used to guide science, based on the idea that scientific "laws" be continually tested, and modified or replaced if found inadequate. (p. 7)

**scientific notation** Expressing large and small numbers using power-of-10 notation.

**seasonal cap** Portion of Martian polar ice caps that is subject to seasonal variations, growing and shrinking once each Martian year. (p. 263)

**seasons** Changes in average temperature and length of day that result from the tilt of Earth's (or any planet's) axis with respect to the plane of its orbit. (p. 16)

**secondary atmosphere** The chemicals that composed Earth's atmosphere after the planet's formation, once volcanic activity outgassed chemicals from the interior.

**sedimentary** Type of rock formed from the buildup of sediment.

**seeing** Term used to describe the ease with which good telescopic observations can be made from Earth's surface, given the blurring effects of atmospheric turbulence. (p. 122)

**seeing disk** Roughly circular region on a detector over which a star's point-like images is spread, due to atmospheric turbulence. (p. 122)

**seismic wave** A wave that travels outward from the site of an earthquake through Earth. (p. 177)

**seismology** The study of earthquakes and the waves they produce in Earth's interior.

**seismometer** Equipment designed to detect and measure the strength of earthquakes (or quakes on any other planet).

**selection effect** Observational bias in which a measured property of a collection of objects is due to the way in which the measurement was made, rather than being intrinsic to the objects themselves. (p. 406)

**self-propagating star formation** Mode of star formation in which shock waves produced by the formation and evolution of one generation of stars triggers the formation of the next. (p. 634)

**semi-detached** Binary system where one star lies within its Roche lobe but the other fills its Roche lobe and is transferring matter onto the first star.

**semimajor axis** One-half of the major axis of an ellipse. The semimajor axis is the way in which the size of an ellipse is usually quantified. (p. 47)

**SETI** Acronym for Search for Extraterrestrial Intelligence. (p. 772)

**Seyfert galaxy** Type of active galaxy whose emission comes from a very small region within the nucleus of an otherwise normal-looking spiral system. (p. 666)

**shepherd satellite** Satellite whose gravitational effect on a ring helps preserve the ring's shape. Examples are two satellites of Saturn, Prometheus and Pandora, whose orbits lie on either side of the F ring. (p. 320)

**shield volcano** A volcano produced by repeated nonexplosive eruptions of lava, creating a gradually sloping, shield-shaped low dome. Often contains a caldera at its summit. (p. 239)

**shock wave** Wave of matter, which may be generated by a newborn star or supernova, that pushes material outward into the surrounding molecular cloud. The material tends to pile up, forming a rapidly moving shell of dense gas. (p. 516)

**short-period comet** Comet with an orbital period of less than 200 years.

**SI** Système International, the international system of metric units used to define mass, length, time, etc.

**sidereal day** The time needed between successive risings of a given star. (p. 13)

**sidereal month** Time required for the Moon to complete one trip around the celestial sphere. (p. 21)

**sidereal year** The time required for the constellations to complete one cycle around the sky and return to their starting points, as seen from a given point on Earth. Earth's orbital period around the Sun is one sidereal year. (p. 16)

**single-line spectroscopic binary** Binary system in which one star is too faint for its spectrum to be distinguished, so only the spectrum of the brighter star can be seen to shift back and forth as the stars orbit one another.

**singularity** A point in the universe where the density of matter and the gravitational field are infinite, such as at the center of a black hole. (p. 602)

**sister/coformation theory (Moon)** Theory suggesting that the Moon formed as a separate object close to Earth.

**soft landing** Use of rockets, parachutes, or packaging to break the fall of a space probe as it lands on a planet.

**solar constant** The amount of solar energy reaching Earth's atmosphere per unit area per unit time, approximately 1400 $W/m^2$. (p. 418)

**solar core** The region at the center of the Sun, with a radius of nearly 200,000 km, where powerful nuclear reactions generate the Sun's energy output.

**solar cycle** The 22-year period that is needed for both the average number of spots and the Sun's magnetic polarity to repeat themselves. The Sun's polarity reverses on each new 11-year sunspot cycle. (p. 432)

**solar day** The period of time between the instant when the Sun is directly overhead (i.e., noon) to the next time it is directly overhead. (p. 13)

**solar eclipse** Celestial event during which the new Moon passes directly between Earth and the Sun, temporarily blocking the Sun's light. (p. 21)

**solar interior** The region of the Sun between the solar core and the photosphere.

**solar maximum** Point of the sunspot cycle during which many spots are seen. They are generally confined to regions in each hemisphere, between about 15 and 20 degrees latitude.

**solar minimum** Point of the sunspot cycle during which only a few spots are seen. They are generally confined to narrow regions in each hemisphere at about 25–30 degrees latitude.

**solar nebula** The swirling gas surrounding the early Sun during the epoch of solar system formation, also referred to as the primitive solar system. (p. 162)

**solar neutrino problem** The discrepancy between the theoretically predicted flux of neutrinos streaming from the Sun as a result of fusion reactions in the core and the flux that is actually observed. The observed number of neutrinos is only about half the predicted number. (p. 443)

**solar system** The Sun and all the bodies that orbit it—Mercury, Venus, Earth, Mars, Jupiter, Saturn, Uranus, Neptune, their moons, the asteroids, the comets, and trans-Neptunian objects. (p. 146)

**solar wind** An outward flow of fast-moving charged particles from the Sun. (p. 417)

**solstice** *See* summer solstice, winter solstice.

**sotropic** Assumed property of the universe such that the universe looks the same in every direction.

**south celestial pole** Point on the celestial sphere directly above Earth's South Pole. (p. 6) (p. 8)

**spacetime** Single entity combining space and time in special and general relativity. (p. 594)

**spatial resolution** The dimension of the smallest detail that can be seen in an image.

**special theory of relativity** Theory proposed by Einstein to deal with the preferred status of the speed of light. (p. 594)

**speckle interferometry** Technique whereby many short-exposure images of a star are combined to make a high-resolution map of the star's surface.

**spectral class** Classification scheme, based on the strength of stellar spectral lines, which is an indication of the temperature of a star. (p. 461)

**spectral window** Wavelength range in which Earth's atmosphere is transparent.

**spectrograph** Instrument used to produce detailed spectra of stars. Usually, a spectrograph records a spectrum on a CCD detector, for computer analysis.

**spectrometer** Instrument used to produce detailed spectra of stars. Usually, a spectrograph records a spectrum on a photographic plate, or more recently, in electronic form on a computer. (p. 121)

**spectroscope** Instrument used to view a light source so that it is split into its component colors. (p. 86)

**spectroscopic binary** A binary-star system which appears as a single star from Earth, but whose spectral lines show back-and-forth Doppler shifts as two stars orbit one another. (p. 469)

**spectroscopic parallax** Method of determining the distance to a star by measuring its temperature and then determining its absolute brightness by comparing with a standard Hertzsprung–Russell diagram. The absolute and apparent brightness of the star give the star's distance from Earth. (p. 467)

**spectroscopy** The study of the way in which atoms absorb and emit electromagnetic radiation. Spectroscopy allows astronomers to determine the chemical composition of stars. (p. 89)

**spectrum** The separation of light into its component colors.

**speed** Distance moved per unit time, independent of direction. *See also* velocity.

**speed of light** The fastest possible speed, according to the currently known laws of physics. Electromagnetic radiation exists in the form of waves or photons moving at the speed of light. (p. 69)

**spicule** Small solar storm that expels jets of hot matter into the Sun's lower atmosphere.

**spin–orbit resonance** State that a body is said to be in if its rotation period and its orbital period are related in some simple way.

**spiral arm** Distribution of material in a galaxy forming a pinwheel-shaped design, beginning near the galactic center. (p. 633)

**spiral density wave** (1) A wave of matter formed in the plane of planetary rings, similar to ripples on the surface of a pond, which wrap around the rings forming spiral patterns similar to grooves in a record disk. Spiral density waves can lead to the appearance of ringlets. (p. 633) (2) Proposed explanation for the existence of galactic spiral arms, in which coiled waves of gas compression move through the galactic disk, triggering star formation.

**spiral galaxy** Galaxy composed of a flattened, star-forming disk component which may have spiral arms and a large central galactic bulge. (pp. 620, 650)

**spiral nebula** Historical name for spiral galaxies, describing their appearance.

**spring tide** The largest tides, occurring when the Sun, Moon, and Earth are aligned at new and full Moon.

**S-process** "Slow" process in which neutrons are captured by a nucleus; the rate is typically one neutron capture per year.

**standard candle** Any object with an easily recognizable appearance and known luminosity, which can be used in estimating distances. Supernovae, which all have the same peak luminosity (depending on type) are good examples of standard candles and are used to determine distances to other galaxies. (p. 657)

**standard solar model** A self-consistent picture of the Sun, developed by incorporating the important physical processes that are believed to be important in determining the Sun's internal structure into a computer program. The results of the program are then compared with observations of the Sun and modifications are made to the model. The standard solar model, which enjoys widespread acceptance, is the result of this process. (p. 419)

**standard time** System of dividing Earth's surface into 24 time zones, with all clocks in each zone keeping the same time. (p. 18)

**star** A glowing ball of gas held together by its own gravity and powered by nuclear fusion in its core. (p. 416)

**starburst galaxy** Galaxy in which a violent event, such as near-collision, has caused an intense episode of star formation in the recent past. (p. 689)

**star cluster** A grouping of anywhere from a dozen to a million stars which formed at the same time from the same cloud of interstellar gas. Stars in clusters are useful to aid our understanding of stellar evolution because, within a given cluster, stars are all roughly the same age and chemical composition and lie at roughly the same distance from Earth. (p. 518)

**Stefan's law** Relation that gives the total energy emitted per square centimeter of its surface per second by an object of a given temperature. Stefan's law shows that the energy emitted increases rapidly with an increase in temperature, proportional to the temperature raised to the fourth power. (p. 75)

**stellar epoch** Most recent period, when stars, planets, and life have appeared in the universe.

**stellar nucleosynthesis** The formation of heavy elements by the fusion of lighter nuclei in the hearts of stars. Except for hydrogen and helium, all other elements in our universe resulted from stellar nucleosynthesis. (p. 568)

**stellar occultation** The dimming of starlight produced when a solar system object such as a planet, moon, or ring passes directly in front of a star. (p. 352)

**stratosphere** The portion of Earth's atmosphere lying above the troposphere, extending up to an altitude of 40–50 km. (p. 171)

**string theory** Theory that interprets all particles and forces in terms of particular modes of vibration of submicroscopic strings. (p. 741)

**strong nuclear force** Short-range force responsible for binding atomic nuclei together. The strongest of the four fundamental forces of nature. (p. 439)

**S-type asteroid** Asteroid made up mainly of silicate or rocky material.

**subatomic particle** Particle smaller than the size of an atomic nucleus.

**subduction zone** Place where two plates meet and one slides under the other. (p. 186)

**subgiant** Star on the subgiant branch of the Hertzsprung-Russell diagram. (p. 531)

**subgiant branch** The section of the evolutionary track of a star corresponding to changes that occur just after hydrogen is depleted in the core, and core hydrogen burning ceases. Shell hydrogen burning heats the outer layers of the star, which causes a general expansion of the stellar envelope. (p. 531)

**sublimation** Process by which element changes from the solid to the gaseous state, without becoming liquid.

**summer solstice** Point on the ecliptic where the Sun is at its northernmost point above the celestial equator, occurring on or near June 21. (p. 15)

**sunspot** An Earth-sized dark blemish found on the surface of the Sun. The dark color of the sunspot indicates that it is a region of lower temperature than its surroundings. (p. 427)

**sunspot cycle** The fairly regular pattern that the number and distribution of sunspots follows, in which the average number of spots reaches a maximum every 11 or so years, then fall off to almost zero. (p. 431)

**supercluster** Grouping of several clusters of galaxies into a larger, but not necessarily gravitationally bound, unit. (p. 697)

**superforce** An attempt to combine the strong and electroweak forces into one single force.

**supergiant** A star with a radius between 100 and 1000 times that of the Sun. (p. 462)

**supergranulation** Large-scale flow pattern on the surface of the Sun, consisting of cells measuring up to 30,000 km across, believed to be the imprint of large convective cells deep in the solar interior. (p. 423)

**superior conjunction** Orbital configuration in which an inferior planet (Mercury or Venus) lies farthest from Earth (on the opposite side of the Sun).

**supermassive black hole** Black hole having a mass a million to a billion times greater than the mass of the Sun; usually found in the central nucleus of a galaxy.

**supernova** Explosive death of a star, caused by the sudden onset of nuclear burning (Type I), or an enormously energetic shock wave (Type II). One of the most energetic events of the universe, a supernova may temporarily outshine the rest of the galaxy in which it resides. (p. 560)

**supernova remnant** The scattered glowing remains from a supernova that occurred in the past. The Crab Nebula is one of the best-studied supernova remnants. (p. 562)

**supersymmetric relic** Massive particle that should have been created in the Big Bang if supersymmetry is correct.

**surface gravity** The acceleration due to gravity at the surface of a star or planet.

**S-waves** Shear waves from an earthquake, which can travel only through solid material and move more slowly than p-waves.

**synchronous orbit** State of an object when its period of rotations is exactly equal to its average orbital period. The Moon is in a synchronous orbit, and so presents the same face toward Earth at all times. (p. 206)

**synchrotron radiation** Type of nonthermal radiation produced by high-speed charged particles, such as electrons, as they are accelerated in a strong magnetic field. (p. 676)

**synodic month** Time required for the Moon to complete a full cycle of phases. (p. 21)

**synodic period** Time required for a body to return to the same apparent position relative to the Sun, taking Earth's own motion into account; (for a planet) the time between one closest approach to Earth and the next. (p. 233)

## T

**T Tauri star** Protostar in the late stages of formation, often exhibiting violent surface activity. T Tauri stars have been observed to brighten noticeably in a short period of time, consistent with the idea of rapid evolution during this final phase of stellar formation. (p. 508)

**tail** Component of a comet that consists of material streaming away from the main body, sometimes spanning hundreds of millions of kilometers. May be composed of dust or ionized gases. (p. 367)

**tau** A type of lepton (along with the electron and muon).

**tectonic fracture** Crack on a planet's surface, in particular on the surface of Mars, caused by internal geological activity.

**telescope** Instrument used to capture as many photons as possible from a given region of the sky and concentrate them into a focused beam for analysis. (p. 108)

**temperature** A measure of the amount of heat in an object, and an indication of the speed of the particles that comprise it. (p. 73)

**tenuous** Thin, having low density.

**terminator** The line separating night from day on the surface of the Moon or a planet.

**terrae** *See* highlands.

**terrestrial planet** One of the four innermost planets of the solar system, resembling Earth in general physical and chemical properties. (p. 152)

**theoretical model** An attempt to construct a mathematical explanation of a physical process or phenomenon, within the assumptions and confines of a given theory. In addition to providing an explanation of the observed facts, the model generally also makes new predictions that can be tested by further observation or experimentation. (p. 6)

**theories of relativity** Einstein's theories, on which much of modern physics rests. Two essential facts of the theory are that nothing can travel faster than the speed of light and that everything, including light, is affected by gravity. (p. 584)

**theory** A framework of ideas and assumptions used to explain some set of observations and make predictions about the real world. (p. 6)

**thermal equilibrium** Condition in which new particle-antiparticle pairs are created from photons at the same rate as pairs annihilate one another to produce new photons.

**thick disk** Region of a spiral galaxy where an intermediate population of stars resides, younger than the halo stars but older than stars in the disk.

**threshold temperature** Critical temperature above which pair production is possible and below which pair production cannot occur.

**tidal bulge** Elongation of Earth caused by the difference between the gravitational force on the side nearest the Moon and the force on the side farthest from the Moon. The long axis of the tidal bulge points toward the Moon. More generally, the deformation of any body produced by the tidal effect of a nearby gravitating object. (p. 193)

**tidal force** The variation in one body's gravitational force from place to place across another body—for example, the variation of the Moon's gravity across Earth. (p. 193)

**tidal locking** Circumstance in which tidal forces have caused a moon to rotate at exactly the same rate at which it revolves around its parent planet, so that the moon always keeps the same face turned toward the planet.

**tidal stability limit** The minimum distance within which a moon can approach a planet before being torn apart by the planet's tidal force.

**tides** Rising and falling motion of terrestrial bodies of water, exhibiting daily, monthly, and yearly cycles. Ocean tides on Earth are caused by the competing gravitational pull of the Moon and Sun on different parts of Earth. (p. 192)

**time dilation** A prediction of the theory of relativity, closely related to the gravitational reshift. To an outside observer, a clock lowered into a strong gravitational field will appear to run slow. (p. 601)

**time zone** Region on Earth in which all clocks keep the same time, regardless of the precise position of the Sun in the sky, for consistency in travel and communications. (p. 18)

**total eclipse** Celestial event during which one body is completely blocked from view by another. (p. 21)

**transit** *See* planetary transit.

**transition zone** The region of rapid temperature increases that separates the Sun's chromosphere from the corona. (p. 417)

**trans-Neptunian object** A small, icy body orbiting beyond the orbit of Neptune. Pluto and Eris are the largest currently known examples.

**transverse motion** Motion perpendicular to a particular line of sight, which does not result in Doppler shift in radiation received.

**transverse velocity** Component of star's velocity perpendicular to the line of sight.

**triangulation** Method of determining distance based on the principles of geometry. A distant object is sighted from two well-separated locations. The distance between the two locations and the angle between the line joining them and the line to the distant object are all that are necessary to ascertain the object's distance. (p. 26)

**triple-alpha process** The creation of carbon-12 by the fusion of three helium-4 nuclei (alpha particles). Helium-burning stars occupy a region of the Hertzsprung–Russell diagram known as the horizontal branch. (p. 533)

**triple star system** Three stars that orbit one another, bound together by gravity.

**Trojan asteroid** One of two groups of asteroids which orbit at the same distance from the Sun as Jupiter, 60 degrees ahead of and behind the planet. (p. 364)

**tropical year** The time interval between one vernal equinox and the next. (p. 16)

**troposphere** The portion of Earth's atmosphere from the surface to about 15 km. (p. 171)

**trough** Maximum departure of a wave below its undisturbed state.

**true space motion** True motion of a star, taking into account both its transverse and radial motion according to the Pythagorean theorem.

**Tully-Fisher relation** A relation used to determine the absolute luminosity of a spiral galaxy. The rotational velocity, measured from the broadening of spectral lines, is related to the total mass, and hence the total luminosity. (p. 658)

**turnoff mass** The mass of a star that is just now evolving off the main sequence in a star cluster.

**21-centimeter line** Spectral line in the radio region of the electromagnetic spectrum, associated with a change of spin of the electron in a hydrogen atom.

**21-centimeter radiation** Radio radiation emitted when an electron in the ground state of a hydrogen atom flips its spin to become parallel to the spin of the proton in the nucleus. (p. 494)

**twin quasar** Quasar that is seen twice at different locations in the sky due to gravitational lensing.

**Type I supernova** One possible explosive death of a star. A white dwarf in a binary-star system can accrete enough mass that it cannot support its own weight. The star collapses and temperatures become high enough for carbon fusion to occur. Fusion begins throughout the white dwarf almost simultaneously and an explosion results. (p. 561)

**Type II supernova** One possible explosive death of a star, in which the highly evolved stellar core rapidly implodes and then explodes, destroying the surrounding star. (p. 561)

**U**

**ultraviolet** Region of the electromagnetic spectrum, just beyond the visible range, corresponding to wavelengths slightly shorter than blue light. (p. 64)

**ultraviolet telescope** A telescope that is designed to collect radiation in the ultraviolet part of the spectrum. Earth's atmosphere is partially opaque to these wavelengths, so ultraviolet telescopes are put on rockets, balloons, and satellites to get high above most or all of the atmosphere. (p. 133)

**umbra** Central region of the shadow cast by an eclipsing body. (p. 22); The central region of a sunspot, which is its darkest and coolest part. (p. 428)

**unbound** An orbit which does not stay in a specific region of space, but where an object escapes the gravitational field of another. Typical unbound orbits are hyperbolic in shape. (p. 58)

**unbound trajectory** Path of an object with launch speed high enough that it can escape the gravitational pull of a planet.

**uncompressed density** The density a body would have in the absence of any compression due to its own gravity.

**universal time** Mean solar time at the Greenwich meridian. (p. 18)

**universe** The totality of all space, time, matter, and energy.

**unstable nucleus** Nucleus that cannot exist indefinitely, but rather must eventually decay into other particles or nuclei.

**upwelling** Upward motion of material having temperature higher than the surrounding medium.

**V**

**Van Allen belt** One of at least two doughnut-shaped regions of magnetically trapped, charged particles high above Earth's atmosphere. (p. 190)

**variable star** A star whose luminosity changes with time. (p. 621)

**velocity** Displacement (distance plus direction) per unit time. *See* also speed. (p. 52)

**vernal equinox** Date on which the Sun crosses the celestial equator moving northward, occurring on or near March 21. (p. 16)

**visible light** The small range of the electromagnetic spectrum that human eyes perceive as light. The visible spectrum ranges from about 400–700 nm, corresponding to blue through red light. (p. 64)

**visible spectrum** The small range of the electromagnetic spectrum that human eyes perceive as light. The visible spectrum ranges from about 4000–7000 angstroms, corresponding to blue through red light.

**visual binary** A binary-star system in which both members are resolvable from Earth. (p. 469)

**void** Large, relatively empty region of the universe around which superclusters and "walls" of galaxies are organized. (p. 699)

**volcano** Upwelling of hot lava from below Earth's crust to the planet's surface.

## W

**wane** (referring to the Moon or a planet) To shrink. The Moon appears to wane, or shrink in size, for two weeks after full Moon.

**warm longitudes (Mercury)** Two opposite points on Mercury's equator where the Sun is directly overhead at aphelion. Cooler than the hot longitudes by 150 degrees.

**water hole** The radio interval between 18 cm and 21 cm, the respective wavelengths at which hydroxyl (OH) and hydrogen (H) radiate, in which intelligent civilizations might conceivably send their communication signals. (p. 774)

**water volcano** Volcano that ejects water (molten ice) rather than lava (molten rock) under cold conditions.

**watt/kilowatt** Unit of power: 1 watt (W) is the emission of 1 joule (J) per second; 1 kilowatt (kW) is 1000 watts.

**wave** A pattern that repeats itself cyclically in both time and space. Waves are characterized by the speed at which they move, their frequency, and their wavelength. (p. 64)

**wave period** The amount of time required for a wave to repeat itself at a specific point in space.

**wave theory of radiation** Description of light as a continuous wave phenomenon, rather than as a stream of individual particles. (p. 69)

**wavelength** The distance from one wave crest to the next, at a given instant in time. (p. 65)

**wax** (referring to the Moon or a planet) To grow. The Moon appears to wax, or grow in size, for two weeks after new Moon.

**weak nuclear force** Short-range force, weaker than both electromagnetism and the strong force, but much stronger than gravity; responsible for certain nuclear reactions and radioactive decays. (p. 439)

**weakly interacting massive particle (WIMP)** Class of subatomic particles that might have been produced early in the history of the universe; dark-matter candidates.

**weight** The gravitational force exerted on you by Earth (or the planet on which you happen to be standing). (p. 52)

**weird terrain** A region on the surface of Mercury with oddly rippled features. This feature is thought to be the result of a strong impact which occurred on the other side of the planet, and sent seismic waves traveling around the planet, converging in the weird region.

**white dwarf** A dwarf star with sufficiently high surface temperature that it glows white. (p. 462) (p. 539)

**white-dwarf region** The bottom-left corner of the Hertzsprung–Russell diagram, where white-dwarf stars are found. (p. 466)

**white oval** Light-colored region near the Great Red Spot in Jupiter's atmosphere. Like the red spot, such regions are apparently rotating storm systems. (p. 200) (p. 288)

**Wien's law** Relation between the wavelength at which a blackbody curve peaks and the temperature of the emitter. The peak wavelength is inversely proportional to the temperature, so the hotter the object, the bluer its radiation. (p. 75)

**winter solstice** Point on the ecliptic where the Sun is at its southernmost point below the celestial equator, occurring on or near December 21. (p. 16)

**wispy terrain** Prominent light-colored streaks on Rhea, one of Saturn's moons.

## X

**X ray** Region of the electromagnetic spectrum corresponding to radiation of high frequency and short wavelength, far beyond the visible spectrum. (p. 64)

**X-ray burster** X-ray source that radiates thousands of times more energy than our Sun in short bursts lasting only a few seconds. A neutron star in a binary system accretes matter onto its surface until temperatures reach the level needed for hydrogen fusion to occur. The result is a sudden period of rapid nuclear burning and release of energy. (p. 585)

**X-ray nova** Nova explosion detected at X-ray wavelengths.

## Z

**Zeeman effect** Broadening or splitting of spectral lines due to the presence of a magnetic field.

**zero-age main sequence** The region on the Hertzsprung–Russell diagram, as predicted by theoretical models, where stars are located at the onset of nuclear burning in their cores. (p. 509)

**zodiac** The twelve constellations on the celestial sphere through which the Sun appears to pass during the course of a year. (p. 15)

**zonal flow** Alternating regions of westward and eastward flow, roughly symmetrical about the equator of Jupiter, associated with the belts and zones in the planet's atmosphere. (p. 284)

**zone** Bright, high-pressure region in the atmosphere of a jovian planet, where gas flows upward. (p. 283)

# ANSWERS TO CONCEPT CHECK QUESTIONS

## Chapter 1

**1.1** (*p. 8*) A theory can never become proven "fact," because it can always be invalidated, or forced to change, by a single contradictory observation. However, once a theory's predictions have been repeatedly verified by experiments over many years, it is often widely regarded as "true." **1.2** (*p. 13*) (1) Because the celestial sphere provides a natural means of specifying the locations of stars on the sky. Celestial coordinates are directly related to Earth's orientation in space, but are independent of Earth's rotation. (2) Distance information is lost. **1.3** (*p. 17*) (1) In the Northern Hemisphere, summer occurs when the Sun is near its highest (northernmost) point on the celestial sphere, or, equivalently, when Earth's North Pole is "tipped" toward the Sun, the days are longest and the Sun is highest in the sky. Winter occurs when the Sun is lowest in the sky (near its southernmost point on the celestial sphere) and the days are shortest. (2) We see different constellations because Earth has moved halfway around its orbit between these seasons and the darkened hemisphere faces an entirely different group of stars. **1.4** (*p. 19*) No, because the time shown by your watch (if it is accurate) is the standard time in your time zone, and noon is not directly related to the time when the Sun passes overhead at your location. **1.5** (*p. 26*) (1) The angular size of the Moon would remain the same, but that of the Sun would be halved, making it easier for the Moon to eclipse the Sun. We would expect to see total or partial eclipses, but no annular ones. (2) If the distance is halved, the angular size of the Sun would double and total eclipses would never be seen, only partial or annular ones. **1.6** (*p. 30*) Because astronomical objects are too distant for direct ("measuring tape") measurements, so we must rely on indirect means and mathematical reasoning.

## Chapter 2

**2.1** (*p. 42*) In the geocentric view, retrograde motion is the real backward motion of a planet as it moves on its epicycle. In the heliocentric view, the backward motion is only apparent, caused by Earth "overtaking" the planet in its orbit. **2.2** (*p. 45*) The discovery of the phases of Venus could not be reconciled with the geocentric model, and observations of Jupiter's moons proved that some objects in the universe did not orbit Earth. **2.3** (*p. 50*) Because the laws were derived using only the orbits of the planets Mercury through Saturn, before the outermost planets were known. **2.4** (*p. 51*) Because Kepler determined the overall geometry of the solar system by triangulation using Earth's orbit as a baseline, so all distances were known only relative to the scale of Earth's orbit—the astronomical unit. **2.5** (*p. 59*) In the absence of any force, a planet would move in a straight line with constant velocity (Newton's first law) and therefore tends to move along the tangent to its orbital path. The Sun's gravity causes the planet to accelerate toward the Sun (Newton's second law), bending its trajectory into the orbit we observe.

## Chapter 3

**3.1** (*p. 70*) Light is an electromagnetic wave produced by accelerating charged particles. All waves on the list are characterized by their wave periods, frequencies, and wavelengths, and all carry energy and information from one location to another. Unlike waves in water or air, however, light waves require no physical medium in which to propagate. **3.2** (*p. 73*) All are electromagnetic radiation and travel at the speed of light. In physical terms, they differ only in frequency (or wavelength), although their effects on our bodies (or our detectors) differ greatly. **3.3** (*p. 77*) As the switch is turned and the temperature of the filament rises, the bulb's brightness increases rapidly, by Stefan's law, and its color shifts, by Wien's law, from invisible infrared to red to yellow to white. **3.4** (*p. 80*) According to *More Precisely 2-3*, measuring masses in astronomy usually entails measuring the orbital speed of one object—a companion star, or a planet, perhaps—around another. In most cases, the Doppler effect is an astronomer's only way of making such measurements.

## Chapter 4

**4.1** (*p. 89*) They are characteristic frequencies (wavelengths) at which matter absorbs or emits photons of electromagnetic radiation. They are unique to each atom or molecule, and thus provide a means of identifying the gas producing them. **4.2** (*p. 95*) Electron orbits can occur only at certain specific energies and there is a ground state which has the lowest possible energy. Planetary orbits can have any energy. Planets can stay in any orbit indefinitely; in an atom, the electron must eventually fall to the ground state, emitting electromagnetic radiation in the process. Planets may reasonably be thought of as having specific trajectories around the Sun—there is no ambiguity as to "where" a planet is. Electrons in an atom are smeared out into an electron cloud, and we can only talk of the electron's location in probabilistic terms. **4.3** (*p. 98*) Spectral lines correspond to transitions between specific orbitals within an atom. The structure of an atom determines the energies of these orbitals, hence the possible transitions, and hence the energies (colors) of the photons involved. **4.4** (*p. 99*) In addition to changes involving electron energies, changes involving a molecule's vibration or rotation can also result in emission or absorption of radiation. **4.5** (*p. 103*) Because with few exceptions, spectral analysis is the only way we have of determining the physical conditions—composition, temperature, density, velocity, etc.—in a distant object. Without spectral analysis, astronomers would know next to nothing about the properties of stars and galaxies.

## Chapter 5

**5.1** (*p. 113*) Because reflecting telescopes are easier to design, build, and maintain than refracting instruments. **5.2** (*p. 117*) The need to gather as much light as possible, and the need to achieve the highest possible angular resolution. **5.3** (*p. 122*) First, photographic plates are inefficient; CCDs are almost always used. Second, because they want to make detailed measurements of brightness, variability, and spectra, which require the use of other, non-imaging detectors. **5.4** (*p. 125*) To reduce or overcome the effects of atmospheric absorption, instruments are placed on high mountains or in space. To compensate for atmospheric turbulence, adaptive optics techniques probe the air above the observing site and adjust the mirror surface accordingly to try to recover the undistorted image. **5.5** (*p. 129*) Radio observations allow us to see objects whose visible light is obscured by intervening matter, or which simply do not emit most of their energy in the visible portion of the spectrum. **5.6** (*p. 131*) The long wavelength of radio radiation. Astronomers use the largest radio telescopes possible and interferometry, which combines the signals from two or more separate telescopes to create the effect of a single instrument of much larger diameter. **5.7** (*p. 137*) Benefits: they are above the atmosphere, so they are unaffected by seeing or absorption; they can also make round-the-clock observations. Drawbacks: cost, smaller size, inaccessibility, vulnerability to damage by radiation and cosmic rays.

## Chapter 6

**6.1** (*p. 147*) They provide new tests of our theories of solar system formation, and new examples of the sorts of planetary systems that are possible. **6.2** (*p. 149*) By applying the laws of geometry and Newtonian mechanics to observations made from Earth and (more recently) by visiting spacecraft. **6.3** (*p. 151*) The planets orbit in very nearly the same plane—the ecliptic. Viewed from outside the system, only the orbit of Mercury would deviate noticeably from this plane. **6.4** (*p. 153*) Because the two classes of planet differ in almost every physical property—orbit, mass, radius, composition, existence of rings, and number of moons. **6.5** (*p. 154*) Because it is thought to be much less evolved than material now found in planets, and hence a better indicator of conditions in the early solar system. **6.6** (*p. 164*) The shapes and orientations of the orbits were set by the fact that the planets formed in a spinning disk; the composition differences result from the differing materials available to form planets at various distances from the heat of the Sun.

## Chapter 7

**7.1** (*p. 175*) It raises Earth's average surface temperature above the freezing point of water, which was critical for the development of life on our planet. Should the greenhouse effect continue to strengthen, however, it may conceivably cause catastrophic climate changes on Earth.   **7.2** (*p. 181*) We would have far less detailed direct (from volcanoes) or indirect (from seismic studies following earthquakes) information on our planet's interior.   **7.3** (*p. 190*) Convection currents in the upper mantle cause portions of Earth's crust—plates—to slide around on the surface. As the plates move and interact, they are responsible for volcanism, earthquakes, the formation of mountain ranges and ocean trenches, and the creation and destruction of oceans and continents.   **7.4** (*p. 192*) It tells us that the planet has a conducting, liquid core in which the magnetic field is continuously generated.   **7.5** (*p. 194*) A tidal force is the *variation* in one body's gravitational force from place to place across another. Tidal forces tend to deform a body, rather than causing an overall acceleration, and they decrease proportional to the inverse cube, rather than the inverse square, of the distance.

## Chapter 8

**8.1** (*p. 203*) Both bodies have substantially lower escape speeds than Earth, and any atmosphere they may have once had has escaped into space long ago due to their high temperatures.   **8.2** (*p. 204*) The maria are younger, denser, and much less heavily cratered than the highlands.   **8.3** (*p. 211*) In the case of the Moon, Earth's tidal force has slowed the spin to the point where the rotation rate is now exactly synchronized with the Moon's orbital period around Earth. For Mercury, the Sun's tidal force has caused the rotation period to become exactly $\frac{2}{3}$ of the orbital period, and the rotation axis to be exactly perpendicular to the planet's orbit plane. A synchronous orbit is not possible because of Mercury's eccentric orbit around the Sun.   **8.4** (*p. 217*) Heavy bombardment long ago created the basins which later filled with lava to form the maria. Subsequent impacts created irtually all the lunar craters, large and small, we see today. Meteoritic bombardment is the main agent of lunar erosion, although the rate is much smaller than the erosion rate on Earth.   **8.5** (*p. 220*) They are thought to have formed when the planet's core cooled and contracted, causing the crust to crumple. Faults on Earth are the result of tectonic activity.   **8.6** (*p. 222*) Generation of a magnetic field is thought to require a rapidly rotating body with a conducting liquid core. Neither the Moon nor Mercury rotates rapidly and, while Mercury's core may still be partly liquid, the Moon's probably is not.   **8.7** (*p. 223*) The impact theory holds that a collision created the Moon essentially from Earth's mantle, accounting for the composition similarities. If the collision occurred after Earth had already differentiated and the dense material had formed a core, relatively little of this material would have found its way into the newborn Moon.

## Chapter 9

**9.1** (*p. 232*) It is very slow and retrograde. The reason is unknown, but may simply be a matter of chance.   **9.2** (*p. 235*) Because

they were observing in the optical and could not see the surface. Their measurements pertained to the upper atmosphere, above the planet's reflective cloud layers.   **9.3** (*p. 242*) No—they are mostly shield volcanoes, where lava upwells through a "hot spot" in the crust. There appears to be no plate tectonic activity on Venus.   **9.4** (*p. 246*) Given Venus's other bulk similarities to Earth, the planet might well have had an Earth-like climate. **9.5** (*p. 247*) The dynamo model of planetary magnetism implies that both a conducting liquid core and rapid rotation are needed to generate a magnetic field. Venus's rotation is the slowest of any planet in the solar system.

## Chapter 10

**10.1** (*p. 253*) Because they occur when Martian opposition happens to coincide approximately with Martian perihelion. Such an alignment happens every 7 Martian synodic years.   **10.2** (*p. 255*) Mars does have seasons, since its rotation axis is inclined to its orbit plane in much the same way as Earth's is. However, the Martian seasons are also affected by the planet's eccentric orbit. The appearance of the planet changes seasonally, although the changes have nothing to do with growing cycles, as was once thought.   **10.3** (*p. 259*) The lowlands are much less heavily cratered, implying that they have been resurfaced by volcanism (or smoothed by erosion) since the highlands formed.   **10.4** (*p. 267*) Some of it may have escaped into space. Of the rest, some exists in the form of permafrost (or perhaps liquid water) below the surface. The rest is contained in the Martian polar caps.   **10.5** (*p. 272*) Some of it was lost due to the planet's weak gravity, aided by violent meteoritic impacts. The rest became part of the planet's surface rocks, the polar caps, or the subsurface permafrost.   **10.6** (*p. 273*) The planet's small size, which allowed the planet's internal heat to escape, effectively shutting down the processes that drive mantle convection, volcanism, and plate tectonics.   **10.7** (*p. 274*) They are much smaller, and orbit much closer to the parent planet. In addition, they seem to have a different formation history—their composition differences from Mars suggest that they were captured by the planet long after they (and Mars) formed; for Earth's Moon, this scenario does not seem to work.

## Chapter 11

**11.1** (*p. 283*) The magnetic field is generated by the motion of electrically conducting liquid in the deep interior, and therefore presumably shares the rotation of that region of the planet.   **11.2** (*p. 289*) Like weather systems on Earth, the belts and zones are regions of high and low pressure and are associated with convective motion. However, unlike storms on Earth, they wrap all the way around the planet because of Jupiter's rapid rotation. In addition, the clouds are arranged in three distinct layers and the bright colors are the result of cloud chemistry unlike anything operating in Earth's atmosphere. Jupiter's spots are somewhat similar to hurricanes on Earth, but they are far larger and longer-lived.   **11.3** (*p. 291*) By constructing theoretical models contrained by detailed spacecraft observations of the planet's

bulk properties.   **11.4** (*p. 292*) Because Jupiter rotates more rapidly than Earth and because the volume of conducting fluid responsible for the field is much greater. Both these factors result in a much stronger planetary magnetic field.   **11.5** (*p. 303*) (1) Jupiter's gravitational field, via its tidal effect on the moons. (2) Because liquid water is thought to have played a critical role in the appearance of life on Earth, and is presumed to be similarly important elsewhere in the solar system.

## Chapter 12

**12.1** (*p. 308*) Because Earth crossed Saturn's ring plane in 1995. The images from before then see the rings from above, those from later see the rings from below.   **12.2** (*p. 313*) Saturn's cloud and haze layers are thicker than those on Jupiter because of Saturn's lower gravity, so we usually see only the upper level of the atmosphere even though the same basic features and cloud layers are there.   **12.3** (*p. 315*) Roughly half of it has precipitated into the planet's interior, reducing the helium fraction in the outer layers (and releasing gravitational energy as it fell).   **12.4** (*p. 322*) The Roche limit is the radius inside of which a moon will be torn apart by tidal forces. Planetary rings are found inside the Roche limit, (most) moons outside. Orbital resonances between ring particles and moons are responsible for much of the fine structure in the rings. They may also maintain the orbits of shepherd satellites that keep some ring features sharp.   **12.5** (*p. 328*) It has a thick atmosphere (denser than Earth's), unlike any other moon in the solar system, and liquid methane on its surface.   **12.6** (*p. 332*) Because they are tidally locked by Saturn's gravity into synchronous orbits, and therefore all have permanently leading and trailing faces that interact differently with the environment around the planet.

## Chapter 13

**13.1** (*p. 339*) Uranus's orbit was observed to deviate from a perfect ellipse, leading astronomers to try to compute the mass and location of the body responsible for those discrepancies. That body was Neptune.   **13.2** (*p. 341*) For unknown reasons, Uranus rotates "on its side"—with its rotation axis almost in the plane of the ecliptic.   **13.3** (*p. 344*) Neptune lies far from the Sun and is very cold, so the source of the energy for these atmospheric phenomena is not known. Uranus, closer to the Sun but with a similar atmospheric temperature, shows much less activity.   **13.4** (*p. 346*) In each case, the field is significantly offset from the planet's center and inclined to the rotation axis.   **13.5** (*p. 351*) Triton shows evidence for surface activity—nitrogen geysers and water volcanoes—that seems to have erased most of its impact craters.   **13.6** (*p. 354*) Both are thin rings that require shepherd satellites to prevent them from spreading out and dispersing.

## Chapter 14

**14.1** (*p. 363*) Similarities: all orbit in the inner solar system, and are solid bodies of generally "terrestrial" composition.

Differences: asteroids are much smaller than the terrestrial planets, and their orbits are much less regular.   **14.2** (*p. 366*) Because most Earth-crossing asteroids will eventually come very close to or even collide with our planet, with potentially catastrophic results.   **14.3** (*p. 374*) (1) Asteroids are generally rocky. Comets are predominantly made of ice, with some dust and other debris mixed in. (2) Most comets never come close enough to the Sun for us to see them.   **14.4** (*p. 380*) Because of its similarities to the known Kuiper belt objects, as well as to the icy moons of the outer planets, it is now regarded as the largest member of the Kuiper belt, rather than as the smallest planet.   **14.5** (*p. 384*) They are fragments of comets or asteroids orbiting the Sun in interplanetary space. They are important to planetary scientists because they generally consist of ancient material, and contain vital information on conditions in the early solar system.

## Chapter 15

**15.1** (*p. 391*) Because it must explain certain general features of solar system architecture, while accommodating the fact that there are exceptions to many of them.   **15.2** (*p. 394*) Because the dust grains formed condensation nuclei that began the process leading to the growth of planetesimals and eventually planets.   **15.3** (*p. 397*) Because the planets could start to form only after the solar nebula had formed, but their formation was terminated when the Sun reached the T Tauri phase and dispersed the disk.   **15.4** (*p. 399*) Yes—if a star formed with a disk of matter around it then the basic processes of condensation and accretion would probably have occurred there too, even if it doesn't have planets like Earth.   **15.5** (*p. 401*) The fact that larger objects were built up through collisions between smaller objects. The general trends are predictable, but the details, at the level of individual collisions, are not.   **15.6** (*p. 405*) (1) Only massive planets have so far been found; (2) The planets' orbits are often quite eccentric; (3) Many of the observed planets orbit very close to their parent star.   **15.7** (*p. 407*) Because the search techniques are most sensitive to massive planets orbiting close to their parent stars, which are exactly what have been observed.

## Chapter 16

**16.1** (*p. 418*) When we simply multiply the solar constant by the total area to obtain the solar luminosity, we are implicitly assuming that the same amount of energy reaches every square meter of the large sphere in Figure 16.3.   **16.2** (*p. 423*) The energy may be carried in the form of (1) radiation, where energy travels in the form of light, and (2) convection, where energy is carried by physical motion of upwelling solar gas.   **16.3** (*p. 427*) The spectrum shows (1) emission lines of (2) highly ionized elements, implying high temperature.   **16.4** (*p. 432*) There is a strong field, with a well-defined east–west organization, just below the surface. The field direction in the southern hemisphere is opposite that in the north. However, the details of the fields are very complex.

**16.5** (*p. 437*) Because high-energy particles from flares and coronal mass ejections can affect Earth's magnetosphere and atmosphere, disrupting communications and power transmission on our planet. On longer time scales, it appears that long-term changes in the solar cycle may also be correlated with climate variations on Earth.   **16.6** (*p. 442*) Because the Sun shines by nuclear fusion, which converts mass into energy as hydrogen turns into helium.   **16.7** (*p. 445*) Almost all of the neutrinos produced in the solar core reach Earth, with very little chance of interaction with intervening matter. Electromagnetic energy, produced in the form of gamma radiation, interacts frequently with solar matter, taking possibly thousands of years to reach the surface.

## Chapter 17

**17.1** (*p. 454*) (1) Because stars are so far away that their parallaxes relative to any baseline on Earth are too small to measure accurately. (2) Because the transverse component must be determined by measuring the star's proper motion, which decreases as the star's distance increases, and is too small to measure for most distant stars.   **17.2** (*p. 456*) Nothing—we need to know their distances before the luminosities (or absolute magnitudes) can be determined.   **17.3** (*p. 461*) Because temperature controls which excited states the star's atoms and ions are in, and hence which atomic transitions are possible.   **17.4** (*p. 462*) Yes, using the radius–luminosity–temperature relationship, but only if we can find a method of determining the luminosity that doesn't depend on the inverse-square law (Sec. 17.3).   **17.5** (*p. 466*) Because giants are intrinsically very luminous, and can be seen to much greater distances than the more common main-sequence stars or white dwarfs.   **17.6** (*p. 469*) All stars would be further away, but their measured spectral types and apparent brightnesses would be unchanged, so their luminosities would be greater than previously thought. The main sequence would therefore move vertically upward in the H–R diagram. (We would then use larger luminosities in the method of spectroscopic parallax, so distances inferred by that method would also increase.)   **17.7** (*p. 474*) We don't—we assume that their masses are the same as similar stars found in binaries.

## Chapter 18

**18.1** (*p. 483*) Because the scale of interstellar space is so large that even very low densities can add up to a large amount of obscuring matter along the line of sight to a distant star.   **18.2** (*p. 489*) Because the UV light is absorbed by hydrogen gas in the surrounding cloud, ionizing it to form the emission nebulae. The red light is Hα radiation; part of the visible hydrogen spectrum emitted as electrons and protons recombine to form hydrogen atoms.   **18.3** (*p. 493*) By studying absorption lines caused by atoms and molecules in the clouds, and the general extinction due to dust, it is possible to map out a cloud's properties—so long as enough stars are conveniently located behind it.   **18.4** (*p. 495*) Because most of the interstellar matter in the galactic disk is made

up of atomic hydrogen, and 21-centimeter radiation provides a probe of that gas largely unaffected by interstellar absorption.   **18.5** (*p. 497*) Because the main constituent, hydrogen, is very hard to observe, so astronomers must use other molecules as tracers of the cloud's properties.

## Chapter 19

**19.1** (*p. 504*) The competing effects of gravity, which tends to make an interstellar cloud collapse, and heat (pressure), which opposes that collapse.   **19.2** (*p. 509*) (1) The existence of a photosphere, meaning that the inner part of the cloud becomes opaque to its own radiation, signaling the slowing of the collapse phase. (2) Nuclear fusion in the core and equilibrium between pressure and gravity.   **19.3** (*p. 510*) No—different parts of the main sequence correspond to stars of different masses. A typical star stays at roughly the same location on the main sequence of most of its lifetime.   **19.4** (*p. 515*) We assume that we observe objects at many different evolutionary stages, and that the snapshot therefore provides a representative sample of the evolutionary stages that stars go through.   **19.5** (*p. 517*) Star formation may be triggered by some external event, which might cause several interstellar clouds to start contracting at once. Alternatively, the shock wave produced when an emission nebula forms may be sufficient to send another nearby part of the same cloud into collapse.   **19.6** (*p. 523*) Because stars form at different rates—high-mass stars reach the main sequence and start disrupting the parent cloud long before lower-mass stars have finished forming.

## Chapter 20

**20.1** (*p. 529*) Hydrostatic equilibrium means that, if some property of the Sun changes a little, the star's structure adjusts to compensate. Small changes in the Sun's internal temperature or pressure will not lead to large changes in its radius or luminosity.   **20.2** (*p. 536*) Because the nonburning inner core, unsupported by fusion, begins to shrink, releasing gravitational energy, heating the overlying layers and causing them to burn more vigorously, thus increasing the luminosity.   **20.3** (*p. 542*) Because the core's contraction is halted by the pressure of degenerate (tightly packed) electrons before it reaches a temperature high enough for the next stage of fusion to begin. In fact, this statement is true whether the "next round" is hydrogen fusion (brown dwarf), helium fusion (helium white dwarf), or carbon fusion (carbon-oxygen white dwarf).   **20.4** (*p. 545*) Fusion in high-mass stars is not halted by electron degeneracy pressure. Temperatures are always high enough that each new burning stage can start before degeneracy becomes important. Such stars continue to fuse more and more massive nuclei, faster and faster, eventually exploding in a supernova.   **20.5** (*p. 548*) Because a star cluster gives us a "snapshot" of stars of many different masses, but of the same age and initial composition, allowing us to directly test the predictions of the theory.   **20.6** (*p. 551*) Because many, if not most, stars are found in binaries, and stars in binaries can follow evolutionary paths quite different from those they would follow if single.

## Chapter 21

**21.1** (*p. 557*) No, because it is of low mass and not a member of a binary-star system. **21.2** (*p. 560*) Because iron cannot fuse to produce energy. As a result, no further nuclear reactions are possible, and the core's equilibrium cannot be restored. **21.3** (*p. 566*) Because the two types of supernova differ in their spectra and their light curves, making it impossible to explain them in terms of a single phenomenon. **21.4** (*p. 572*) Because they are readily formed by helium capture, a process common in evolved stars. Other elements (with masses not multiples of four) had to form via less common reactions involving proton and neutron capture. **21.5** (*p. 574*) Because it is responsible for creating and dispersing all the heavy elements out of which we are made. In addition, it may also have played an important role in triggering the collapse of the interstellar cloud from which our solar system formed.

## Chapter 22

**22.1** (*p. 581*) No—only Type II supernovae. According to theory, the rebounding central core of the original star in a Type II supernova becomes a neutron star. **22.2** (*p. 584*) Because (1) not all supernovae form neutron stars, (2) the pulses are beamed, so not all pulsing neutron stars are visible from Earth, and (3) pulsars spin down and become too faint to observe after a few tens of millions of years. **22.3** (*p. 588*) Some X-ray sources are binaries containing accreting neutron stars, which may be in the process of being spun up to form millisecond pulsars. **22.4** (*p. 592*) They are energetic bursts of gamma rays, roughly isotropically distributed in the sky, occurring about once per day. They pose a challenge because they are very distant, and hence extremely luminous, but their energy originates in a region less than a few hundred kilometers across. **22.5** (*p. 599*) Newton's theory describes gravity as a force produced by a massive object that influences all other massive objects. Einstein's relativity describes gravity as a curvature of space-time produced by a massive object; that curvature then determines the trajectories of all particles—matter or radiation—in the universe. **22.6** (*p. 602*) Because the object would appear to take infinitely long to reach the event horizon, and its light would be infinitely redshifted by the time it got there. **22.7** (*p. 609*) By observing their gravitational effects on other objects, and from the X rays emitted as matter plunges toward the event horizon.

## Chapter 23

**23.1** (*p. 618*) The Milky Way is the thin plane of the Galactic disk, seen from within. When our line of sight lies in the plane of the Galaxy, we see many stars blurring into a continuous band. In other directions, we see darkness. **23.2** (*p. 626*) No, because even the brightest Cepheids are unobservable at distances of more than a kiloparsec or so through the obscuration of interstellar dust. **23.3** (*p. 629*) Because the compositions, ages, and orbits of

the two classes of stars are quite different from one another. **23.4** (*p. 631*) The halo formed early in our Galaxy's history, before gas and dust had formed a spinning, flattened disk. Disk stars are still forming today. Consequently, halo stars are old and move in more or less random three-dimensional orbits, while the disk contains stars of all ages, all moving in roughly circular orbits around the Galactic center. **23.5** (*p. 635*) Because differential rotation would destroy the spiral structure within a few hundred million years. **23.6** (*p. 640*) Its emission has not been observed at any electromagnetic wavelength. Its presence is inferred from its gravitational effect on stars and gas orbiting the Galactic center. **23.7** (*p. 644*) Observations of rapidly moving stars and gas, and the variability of the radiation emitted suggest the presence of a roughly 4 million-solar-mass black hole.

## Chapter 24

**24.1** (*p. 656*) Most galaxies are not large spirals—the most common galaxy types are dwarf ellipticals and dwarf irregulars. **24.2** (*p. 661*) Because distance-measurement techniques ultimately rely upon the existence of very bright objects whose luminosities can be inferred by other means. Such objects become increasingly hard to find and calibrate the farther we look out into intergalactic space. **24.3** (*p. 664*) It doesn't use the inverse-square law. The other methods all provide a way of determining the luminosity of a distant object, which then is converted to a distance using the inverse-square law. Hubble's law gives a direct connection between redshift and distance. **24.4** (*p. 670*) It means that the energy source cannot simply be the summed energy of a huge number of stars—some other mechanism must be at work. **24.5** (*p. 672*) They often appear starlike, but it was at first not realized that their unusual spectra were actually highly redshifted. **24.6** (*p. 676*) The energy is generated in an accretion disk in the central nucleus of the visible galaxy, then transported by jets out of the galaxy and into the lobes, where it is eventually emitted by the synchrotron process in the form of radio waves.

## Chapter 25

**25.1** (*p. 685*) First, that the galaxies are gravitationally bound to the cluster. Second, and more fundamentally, that the laws of physics as we know them in the solar system—gravity, atomic structure, the Doppler effect—all apply on very large scales, and to systems possibly containing a lot of dark matter. **25.2** (*p. 693*) Stars form by collapse and fragmentation of a large interstellar cloud and subsequently evolve largely in isolation. Galaxies form by mergers of smaller objects, and interactions with other galaxies play a major role in their evolution. **25.3** (*p. 696*) Probably not. There may be galaxies (especially the smaller ones) which do not harbor supermassive central black holes, and in any case, only galaxies in clusters are likely to experience the encounters that trigger activity. **25.4** (*p. 704*) Light traveling from these objects to Earth is influenced by cosmic structure all along the line of sight. Light rays are deflected by the gravitational

field of intervening concentrations of mass, and gas along the line of sight produces absorption lines whose redshifts tell us the distance at which each feature formed. The light received on Earth thus gives astronomers a "core sample" through the universe, from which detailed information can be extracted.

## Chapter 26

**26.1** (*p. 712*) On very large scales—more than 300 Mpc—the distribution of galaxies seems to be roughly the same everywhere and in all directions. **26.2** (*p. 717*) Because, tracing the motion backwards in time, it implies that all galaxies, and in fact everything in the entire universe, were located at a single point at the same instant in the past. **26.3** (*p. 719*) The universe can expand forever, in which case we die a cold death in which all activity gradually fades away, or the expansion can stop and the universe will recollapse to a fiery Big Crunch. **26.4** (*p. 721*) A low-density universe has negative curvature, a critical density universe is spatially flat (Euclidean), and a high-density universe has positive curvature (and is finite in extent). **26.5** (*p. 725*) There doesn't seem to be enough matter to halt the collapse and, in addition, the observed cosmic acceleration suggests the existence of a large-scale repulsive force in the cosmos that also opposes recollapse. **26.6** (*p. 727*) Dark energy tends to accelerate the expansion of the universe, competing with the effect of gravity, which tends to slow it. The age of the universe that we infer based on present-day measurements depends on the outcome of that competition. **26.7** (*p. 729*) At the time of the Big Bang. It is the electromagnetic remnant of the primeval fireball.

## Chapter 27

**27.1** (*p. 737*) It means that the total mass-energy density of the universe, which today is made up almost entirely of matter and dark energy, was once comprised almost entirely of radiation. We know this because, going back in time toward the Big Bang, the dark energy density stays constant as the universe contracts, the matter density increases because the volume shrinks, but the radiation density increases even faster because of the cosmological redshift. Thus, at sufficiently early times, radiation was the dominant component of the cosmos. **27.2** (*p. 742*) Once the temperature of the expanding, cooling universe dropped below the point where particle–antiparticle pairs could no longer be created from the radiation background, the particles separated out of the radiation field. Particles and antiparticles annihilated one another, and any leftover "frozen out" matter has survived to the present day. **27.3** (*p. 746*) Because the amount of deuterium observed in the universe today implies that the present density of normal matter is at most a few percent of the critical value—much less than the density of dark matter inferred from dynamical studies. **27.4** (*p. 750*) Inflation implies that the universe is flat, and hence that the total cosmic density equals the critical value. However, the matter density seems to be only about one third of the critical value, and the density of electromagnetic radiation (the microwave background) is a tiny fraction of the critical density. The remaining density may be in the form of the "dark energy" thought to be powering the accelerating cosmic expansion (Section 27.3).   **27.5** (*p. 752*) Because it could clump to form the large density fluctuations that would ultimately become galaxies and clusters, without leaving a correspondingly large imprint in the microwave background. After it decoupled from the radiation background, normal matter flowed into the dark matter structures, forming the galaxies we see.   **27.6** (*p. 755*) They allow us to measure the value of $\Omega_0$—and in fact imply that it is very close to 1.

### Chapter 28

**28.1** (*p. 765*) The formation of complex molecules from simple ingredients by nonbiological processes has been repeatedly demonstrated, but no living cell or self-replicating molecule has ever been created. **28.2** (*p. 767*) Mars remains the most likely site, although Europa and Titan also have properties that might have been conducive to the emergence of living organisms.

**28.3** (*p. 772*) It breaks a complex problem up into simpler "astronomical," "biochemical," "anthropological," and "cultural" pieces, which can be analyzed separately. It also identifies the types of stars where a search might be most fruitful.   **28.4** (*p. 775*) It is in the radio part of the spectrum, where Galactic absorption is least, at a region where natural Galactic background "static" is minimized, and in a portion of the spectrum characterized by lines of hydrogen and hydroxyl, both of which would likely have significance to a technological civilization.

# ANSWERS TO SELF-TEST QUESTIONS

**Chapter 1**

*True-False* 1.1 T, 1.2 F, 1.3 T, 1.4 F, 1.5 T, 1.6 F, 1.7 T, 1.8 F, 1.9 T, 1.10 F
*Multiple Choice* 1.11 b, 1.12 b, 1.13 d, 1.14 a, 1.15 c, 1.16 a, 1.17 c, 1.18 c, 1.19 a, 1.20 d
*Odd-Numbered Problems* 1.1 (c) the Moon (384,000 km) 1.3 it would decrease by roughly eight minutes 1.5 (a) 7.14 solar days; (b) infinite 1.7 1.02 km/s 1.9 (a) 57,300 km; (b) $3.44 \times 10^6$ km; (c) $2.06 \times 10^8$ km 1.11 $6.5 \times 10^{-5}$ arc sec 1.13 391 1.15 0°.

**Chapter 2**

*True-False* 2.1 F, 2.2 T, 2.3 F, 2.4 F, 2.5 F, 2.6 F, 2.7 F, 2.8 T, 2.9 F, 2.10 T
*Multiple Choice* 2.11 a, 2.12 d, 2.13 b, 2.14 c, 2.15 c, 2.16 a, 2.17 c, 2.18 b, 2.19 c, 2.20 a
*Odd-Numbered Problems* 2.1 (a) 110 km; (b) 44,000 km; (c) 370,000 km 2.3 54,800,000 km 2.5 146 days 2.7 8.1″, if Mercury is at aphelion and Earth is at perihelion at the point of closest approach 2.9 $9.57 \times 10^{-4}$ times the mass of the Sun = $1.9 \times 10^{27}$ kg 2.11 10,000 AU, 16 million years 2.13 velocities = 7.8 km/s, 7.3 km/s, 4.9 km/s; accelerations = 9.42 m/s², 7.27 m/s², 1.48 m/s²; equal in all cases 2.15 1.7 km/s; 2.4 km/s.

**Chapter 3**

*True-False* 3.1 T, 3.2 F, 3.3 F, 3.4 F, 3.5 F, 3.6 T, 3.7 T, 3.8 T, 3.9 T, 3.10 T
*Multiple Choice* 3.11 a, 3.12 c, 3.13 b, 3.14 b, 3.15 d, 3.16 a, 3.17 b, 3.18 d, 3.19 a, 3.20 b
*Odd-Numbered Problems* 3.1 1480 m/s 3.3 23.5 Hz; radio 3.5 9.4 cm; radio 3.7 310 K; 9.4 microns; infrared 3.9 2.9 microns 3.11 $6.32 \times 10^7$ W/m²; $3.9 \times 10^{26}$ W 3.13 300 km/s away 3.15 $1.94 \times 10^{27}$ kg.

**Chapter 4**

*True-False* 4.1 F, 4.2 T, 4.3 F, 4.4 T, 4.5 F, 4.6 F, 4.7 F, 4.8 T, 4.9 T, 4.10 T
*Multiple Choice* 4.11 c, 4.12 c, 4.13 d, 4.14 c, 4.15 b, 4.16 b, 4.17 b, 4.18 b, 4.19 b, 4.20 d
*Odd-Numbered Problems* 4.1 2.8 eV, 6.2 eV 4.3 620 nm, 12,400 nm, 0.25 nm 4.5 $2.4 \times 10^8$ 4.7 51.8 microns $5.7 \times 10^{12}$ Hz, infrared; 4.5 cm, 6.5 GHz, radio; 46 m, 6.6 MHz, radio 4.9 the first six Balmer lines, ranging in wavelength from 656 nm (Hα) to 410 nm (Hζ) 4.11 137 km/s, approaching 4.13 0.052 nm, to two significant figures 4.15 0.24 rev/day.

**Chapter 5**

*True-False* 5.1 F, 5.2 F, 5.3 F, 5.4 F, 5.5 T, 5.6 T, 5.7 T, 5.8 T, 5.9 F, 5.10 F
*Multiple Choice* 5.11 c, 5.12 d, 5.13 d, 5.14 b, 5.15 c, 5.16 c, 5.17 a, 5.18 d, 5.19 b, 5.20 c

*Odd-Numbered Problems* 5.1 0.29 arc seconds; 6.8 pixels 5.3 6.7 minutes; 1.7 minutes 5.5 (a) 0.022″; (b) 0.062″ 5.7 155 light-years 5.9 5.5 km, 92 m, 1.8 m. 5.11 roughly 8 times 5.13 (a) 0.003 arc seconds; (b) 0.005 arc seconds 5.15 No; it is not. The wavelength of a 1-keV photon is 1.2 nm, which would give an angular resolution of about 0.00025″ for the quoted diameter. The effective resolution is complicated by other design factors.

**Chapter 6**

*True-False* 6.1 T, 6.2 F, 6.3 T, 6.4 T, 6.5 F, 6.6 T, 6.7 T, 6.8 F, 6.9 F, 6.10 T
*Multiple Choice* 6.11 a, 6.12 d, 6.13 a, 6.14 b, 6.15 c, 6.16 a, 6.17 a, 6.18 b, 6.19 b, 6.20 d
*Odd-Numbered Problems* 6.1 18 seconds of arc; 17 days 6.3 Mercury: 0.31 AU 0.47 AU, Mars: 1.38 AU, 1.67 AU 6.5 $7 \times 10^{20}$ kg, 0.012 percent of Earth's mass 6.7 68 AU, beyond Pluto 6.9 0.21, 1.26 AU, 0.71 years 6.11 — 6.13 (a) 62.5 years; (b) 9 days 6.15 increases proportional to the radius; surface gravity doubles, escape speed doubles.

**Chapter 7**

*True-False* 7.1 T, 7.2 T, 7.3 F, 7.4 F, 7.5 T, 7.6 F, 7.7 F, 7.8 T, 7.9 F, 7.10 F
*Multiple Choice* 7.11 b, 7.12 d, 7.13 a, 7.14 c, 7.15 b, 7.16 b, 7.17 c, 7.18 b, 7.19 d, 7.20 b
*Odd-Numbered Problems* 7.1 Answers are in data table 7.3 $5.0 \times 10^{18}$ kg = $8.4 \times 10^{-7}$ times Earth's mass 7.5 21 m 7.7 (a) $8.4 \times 10^{-3}$; (b) 0.16; (c) 0.82; (d) 0.014 7.9 1.9 billion years 7.11 $3.07 \times 10^{-6}$ m/s², $3.07 \times 10^{-7}$ of surface gravity 7.13 $1.6 \times 10^{-7}$ 7.15 smaller by a factor of $6.0 \times 10^{-6}$. No!

**Chapter 8**

*True-False* 8.1 T, 8.2 F, 8.3 T, 8.4 F, 8.5 F, 8.6 F, 8.7 T, 8.8 T, 8.9 T, 8.10 T
*Multiple Choice* 8.11 a, 8.12 b, 8.13 b, 8.14 c, 8.15 b, 8.16 a, 8.17 b, 8.18 d, 8.19 a, 8.20 a
*Odd-Numbered Problems* 8.1 8.5 minutes 8.3 125 lbs (57 kg equivalent) 8.5 1.71°, 1.13° 8.7 1.7 hours 8.9 (a) 60.4 km/s, or 6.5°/day, (b) 39.7 km/s, or 4.3°/day 8.11 $4.2 \times 10^{23}$ kg, or roughly 26 percent greater than it is now 8.13 need at least $4.8 \times 10^5$ such craters, requiring at least 4.8 trillion years; rate would have to increase by a factor of about 1000 8.15 (a) 4 million years for a 2 cm deep bootprint; (b) 4 million years, accepting literally the factor of 10,000 given in the text; (c) 1.6 trillion years, assuming Reinhold is 8 km deep.

**Chapter 9**

*True-False* 9.1 F, 9.2 F, 9.3 F, 9.4 F, 9.5 T, 9.6 F, 9.7 T, 9.8 T, 9.9 F, 9.10 T
*Multiple Choice* 9.11 a, 9.12 c, 9.13 c, 9.14 b, 9.15 b, 9.16 c, 9.17 b, 9.18 c, 9.19 a, 9.20 a

*Odd-Numbered Problems* 9.1 (a) 35.1 arc seconds; (b) 22.9 arc second; (c) 9.7 arc seconds 9.3 470 seconds, 280 seconds 9.5 4.1°/day; 1.6°/day; 145 (Earth) days 9.7 tidal acceleration due to Earth is $3.66 \times 10^{-12}$ times Venus's surface gravity; tidal acceleration due to the Sun is $7.18 \times 10^{-8}$ times Venus's surface gravity; unlikely that Earth's tidal influence would cause the resonance, as it is much smaller than the tidal force due to the Sun 9.9 yes—the smallest detectable feature would be about 20 km across, much smaller than the largest impact features 9.11 400 km/h, 250 mph 9.13 35 times 9.15 197 minutes, 94 minutes.

**Chapter 10**

*True-False* 10.1 T, 10.2 F, 10.3 F, 10.4 F, 10.5 F, 10.6 F, 10.7 F, 10.8 T, 10.9 F, 10.10 T
*Multiple Choice* 10.11 b, 10.12 c, 10.13 b, 10.14 c, 10.15 b, 10.16 d, 10.17 a, 10.18 c, 10.19 b, 10.20 a
*Odd-Numbered Problems* 10.1 780 Earth days 10.3 41° 10.5 2.2 minutes longer 10.7 59 lbs, assuming you weigh 150 lbs on Earth 10.9 $3.2 \times 10^{16}$ kg, taking Earth's atmospheric mass to be $5.0 \times 10^{18}$ kg; 2.8 times the seasonal polar cap mass of $1.1 \times 10^{16}$ kg 10.11 $1.3 \times 10^{20}$ kg; 3900 times the current atmospheric mass 10.13 $2.9 \times 10^{17}$ kg; $4.5 \times 10^{-7}$ times the mass of Mars 10.15 Phobos: 10.3 arc minutes; Deimos: 2.34 arc minutes; no.

**Chapter 11**

*True-False* 11.1 F, 11.2 T, 11.3 F, 11.4 T, 11.5 T, 11.6 T, 11.7 F, 11.8 F, 11.9 T, 11.10 T
*Multiple Choice* 11.11 c, 11.12 b, 11.13 a, 11.14 c, 11.15 a, 11.16 d, 11.17 b, 11.18 a, 11.19 a, 11.20 c
*Odd-Numbered Problems* 11.1 2.5 times greater 11.3 62 days 11.5 1.2% of the actual mass 11.7 4800 times greater; ratio for Earth–Moon is 81 11.9 — 11.11 0.001, or 0.1% 11.13 Europa's pull is 13,800 times weaker 11.15 Io: 36′, Europa: 18′, Ganymede: 18′, Callisto: 9′, Sun: 6′. Yes.

**Chapter 12**

*True-False* 12.1 F, 12.2 F, 12.3 T, 12.4 F, 12.5 T, 12.6 T, 12.7 T, 12.8 T, 12.9 F, 12.10 F
*Multiple Choice* 12.11 b, 12.12 c, 12.13 d, 12.14 c, 12.15 d, 12.16 d, 12.17 b, 12.18 c, 12.19 a, 12.20 c
*Odd-Numbered Problems* 12.1 47″ 12.3 $7.3 \times 10^{22}$ kg; $1.3 \times 10^{-4}$ times the actual mass, 1.2 percent of Earth's mass 12.5 74 K 12.7 $1.1 \times 10^{18}$ particles 12.9 1.33 m/s² about $\frac{1}{7}$ of Earth's; 2.6 km/s 12.11 142,000 km; 117,000 km; 142,000 km; 237,000 km 12.13 Saturn's tidal acceleration is $7.9 \times 10^{-5}$ times Titan's surface gravity; no—the tidal acceleration is much smaller than what Jupiter does to the Galilean moons 12.15 approximately 54 km.

## Chapter 13

*True-False* 13.1 T, 13.2 T, 13.3 T, 13.4 F, 13.5 T, 13.6 T, 13.7 T, 13.8 T, 13.9 T, 13.10 F
*Multiple Choice* 13.11 b, 13.12 b, 13.13 c, 13.14 d, 13.15 c, 13.16 d, 13.17 d, 13.18 a, 13.19 a, 13.20 d
*Odd-Numbered Problems* 13.1 (a) 45.3 years (a Neptune year is 3.6 times longer); (b) 171 years (about 1 Neptune year) 13.3 1.7′, 13.2′ yes 13.5 10,000 km/h—it moves halfway around the planet in just over eight hours; this is half the rotation period, it's mostly rotation. 13.7 0.42 Earth Moon masses 13.9 yes, at 37 K, an escape speed of 1.08 km/s would be needed; Triton's escape speed is 1.46 km/s 13.11 0.24 degrees in one week 13.13 6 times farther out 13.15 8.3 hours, 15,000 km.

## Chapter 14

*True-False* 14.1 T, 14.2 F, 14.3 F, 14.4 T, 14.5 F, 14.6 F, 14.7 F, 14.8 T, 14.9 T, 14.10 T
*Multiple Choice* 14.11 c, 14.12 d, 14.13 a, 14.14 c, 14.15 a, 14.16 d, 14.17 b, 14.18 d, 14.19 d, 14.20 d
*Odd-Numbered Problems* 14.1 (a) a 150-lb person would weigh 4.8 lbs; (b) 0.41 km/s 14.3 230 km diameter 14.5 1.1 AU, 2.0 AU; not necessarily—the orbital inclination could differ from Earth's 14.7 21 hours, taking the middle of the stated mass range and assuming a circular orbit 14.9 (a) 4.0 million years; (b) 49 AU 14.11 $3 \times 10^{12}$ kg; 0.06 percent 14.13 5 trillion 14.15 1.9 billion times (2 crossings per orbit).

## Chapter 15

*True-False* 15.1 F, 15.2 T, 15.3 F, 15.4 F, 15.5 F, 15.6 T, 15.7 F, 15.8 T, 15.9 F, 15.10 F
*Multiple Choice* 15.11 a, 15.12 d, 15.13 a, 15.14 a, 15.15 c, 15.16 a, 15.17 c, 15.18 c, 15.19 d, 15.20 b
*Odd-Numbered Problems* 15.1 (a) Jupiter: $1.9 \times 10^{43}$; Saturn: $7.8 \times 10^{42}$, and Earth: $2.7 \times 10^{40}$ kg m$^2$/s; (b) $1.0 \times 10^{31}$ kg m$^2$/s 15.3 8 revolutions 15.5 $3.0 \times 10^7$ newtons; tidal force = $8.1 \times 10^9$ newtons, 270 times larger 15.7 $2 \times 10^8$; once every 2.5 years 15.9 188 years 15.11 0.3 AU 15.13 2210 K 15.15 0.81.

## Chapter 16

*True-False* 16.1 T, 16.2 T, 16.3 F, 16.4 T, 16.5 F, 16.6 F, 16.7 F, 16.8 F, 16.9 T, 16.10 F
*Multiple Choice* 16.11 c, 16.12 b, 16.13 c, 16.14 a, 16.15 a, 16.16 b, 16.17 b, 16.18 c, 16.19 c, 16.20 b
*Odd-Numbered Problems* 16.1 14,600 W/m$^2$; 52 W/m$^2$ 16.3 (a) 3000 km, (b) 1460; (c) 167 minutes 16.5 36% as much light as the photosphere 16.7 10,800,000 km, for a coronal temperature of 1,000,000 K 16.9 95 million years 16.11 310,000 years 16.13 $4 \times 10^{28}$ neutrinos 16.15 (a) 13 billion tons; (b) 7.2 billion years.

## Chapter 17

*True-False* 17.1 F, 17.2 F, 17.3 T, 17.4 F, 17.5 T, 17.6 F, 17.7 T, 17.8 F, 17.9 T, 17.10 T

*Multiple Choice* 17.11 a, 17.12 d, 17.13 d, 17.14 b, 17.15 c, 17.16 b, 17.17 c, 17.18 b, 17.19 c, 17.20 d
*Odd-Numbered Problems* 17.1 83 pc; 0.36 arc seconds 17.3 80 solar luminosities 17.5 B is three times farther away 17.7 $3.3 \times 10^{-10}$ W/m$^2$; $2.3 \times 10^{-13}$ times the solar constant 17.9 M = $-1.0$ 17.11 (a) 100 pc; (b) 4200 pc; (c) 170,000 pc; (d) 1,000,000 pc 17.13 3.5, 2.3 solar masses 17.15 (a) 3.4 solar masses; (b) 0.22 solar masses.

## Chapter 18

*True-False* 18.1 F, 18.2 F, 18.3 T, 18.4 T, 18.5 F, 18.6 T, 18.7 F, 18.8 F, 18.9 F, 18.10 T
*Multiple Choice* 18.11 a, 18.12 d, 18.13 c, 18.14 a, 18.15 d, 18.16 c, 18.17 b, 18.18 d, 18.19 a, 18.20 a
*Odd-Numbered Problems* 18.1 1.9 grams 18.3 $7.1 \times 10^{20}$ m$^3$, or a cube of side 8900 km 18.5 9.03 magnitudes 18.7 100,000 pc, 1.3 magnitudes/kpc 18.9 1.8 kpc 18.11 escape speeds (km/s): 1.8, 1.1, 1.1, 0.80; average molecular speeds: 13.6, 14.0, 14.6, 14.2; no 18.13 frequency: 1419.65–1421.24 MHz, wavelength: 21.1053–21.0965 cm 18.15 photon energy = $9.4 \times 10^{-25}$ J; $2.6 \times 10^{17}$ W.

## Chapter 19

*True-False* 19.1 F, 19.2 F, 19.3 T, 19.4 F, 19.5 T, 19.6 F, 19.7 F, 19.8 T, 19.9 T, 19.10 T
*Multiple Choice* 19.11 c, 19.12 a, 19.13 d, 19.14 b, 19.15 b, 19.16 a, 19.17 a, 19.18 b, 19.19 a, 19.20 b
*Odd-Numbered Problems* 19.1 yes, barely: the escape speed is 0.93 km/s, molecular speed is 0.35 km/s 19.3 luminosity decreases by a factor of 7900; absolute magnitude increases by 9.7 19.5 M = 2.3 19.7 8.3 magnitudes 19.9 10$^{-6}$ solar luminosities 19.11 3930 years 19.13 m = 20.25 19.15 37 pc.

## Chapter 20

*True-False* 20.1 F, 20.2 T, 20.3 T, 20.4 T, 20.5 F, 20.6 F, 20.7 F, 20.8 F, 20.9 F, 20.10 T
*Multiple Choice* 20.11 c, 20.12 b, 20.13 a, 20.14 b, 20.15 a, 20.16 c, 20.17 a, 20.18 b, 20.19 b, 20.20 d
*Odd-Numbered Problems* 20.1 $9.9 \times 10^{26}$ kg destroyed, $8.9 \times 10^{43}$ J; emitted 20.3 (a) 3370; (b) 84,300 solar luminosities 20.5 roughly a factor of 90–100; factor of 10 smaller 20.7 $5 \times 10^{-5}$; not noticeable to the naked eye, but (just) measurable with a telescope and photometer 20.9 2.9 years, 26,000 years 20.11 increases by a factor of 25 20.13 (a) 2.9 solar masses; (b) 1.7 solar masses 20.15 (a) 2.29 AU; (b) 1.97 AU, 1.6 yr.

## Chapter 21

*True-False* 21.1 F, 21.2 T, 21.3 T, 21.4 T, 21.5 F, 21.6 T, 21.7 T, 21.8 T, 21.9 T, 21.10 T
*Multiple Choice* 21.11 a, 21.12 d, 21.13 d, 21.14 b, 21.15 d, 21.16 c, 21.17 c, 21.18 b, 21.19 b, 21.20 c
*Odd-Numbered Problems* 21.1 15 solar radii 21.3 m = 20, 3.2 Mpc 21.5 0.44 pc; no—there

are no O or B stars (in fact no stars at all) within that distance of us. For the Moon: 320 pc; yes, but rarely—see problem 14 21.7 $1.2 \times 10^{44}$ J; about 10 times greater than a supernova's electromagnetic output, and about $\frac{1}{10}$ the output in the form of neutrinos 21.9 roughly 1000 km/s; not a very good assumption—the nebula is moving too fast to be affected much by gravity, although it is probably slowing down as it runs into the interstellar medium 21.11 5.6 pc 21.13 one per 270 years 21.15 0.48 percent, $9.5 \times 10^{27}$ kg or 1600 Earth masses.

## Chapter 22

*True-False* 22.1 T, 22.2 T, 22.3 T, 22.4 T, 22.5 F, 22.6 T, 22.7 F, 22.8 F, 22.9 T, 22.10 F
*Multiple Choice* 22.11 b, 22.12 c, 22.13 a, 22.14 b, 22.15 c, 22.16 d, 22.17 b, 22.18 b, 22.19 c, 22.20 b
*Odd-Numbered Problems* 22.1 1 million revolutions per day, or 11.6 revolutions per second 22.3 $1.9 \times 10^{12}$ m/s$^2$, or 200 billion Earth gravities; 190,000 km/s, or 64 percent of the speed of light; 306,000 km/s 22.5 $2.4 \times 10^{44}$ J, roughly twice the Sun's total energy output; $2.4 \times 10^{34}$ J; $1.4 \times 10^{25}$ J 22.7 63,000 km/s or 21% of the speed of light; 136,000 km/s or 45% of the speed of light 22.9 (a) 26 microarc sec; (b) 0.016 arc sec; (c) 4.1 arc min; (d) 8100 AU = 0.039 pc 22.11 $2 \times 10^{10}$ m/s$^2$ = $2 \times 10^9$ g; $2 \times 10^{-3}$ g; $2 \times 10^{-9}$ g 22.13 14,000 solar masses 22.15 30 million km = 0.2 AU, tidal acceleration is $1.97 \times 10^6$ m/s$^2$ and star's surface gravity acceleration is 829 m/s$^2$, so the tidal acceleration is much larger.

## Chapter 23

*True-False* 23.1 T, 23.2 T, 23.3 F, 23.4 T, 23.5 T, 23.6 T, 23.7 F, 23.8 F, 23.9 F, 23.10 T
*Multiple Choice* 23.11 d, 23.12 d, 23.13 d, 23.14 b, 23.15 b, 23.16 c, 23.17 a, 23.18 b, 23.19 d, 23.20 c
*Odd-Numbered Problems* 23.1 2.0″, much less than the angular diameter of Andromeda 23.3 100 kpc 23.5 M = $-6.4$; 17 Mpc 23.7 0.014″/yr; proper motion has in fact been measured for several globular clusters 23.9 (a) 3.9 kpc; (b) 19.7 kpc 23.11 530 million years (assuming a rotation speed of 240 km/s at 15 kpc); 500 million years (assuming a rotation speed of 200 km/s at 5 kpc) 23.13 1 kpc 23.15 124 AU; no.

## Chapter 24

*True-False* 24.1 F, 24.2 F, 24.3 T, 24.4 F, 24.5 T, 24.6 T, 24.7 F, 24.8 T, 24.9 T, 24.10 T
*Multiple Choice* 24.11 c, 24.12 c, 24.13 a, 24.14 c, 24.15 c, 24.16 b, 24.17 b, 24.18 b, 24.19 a, 24.20 b
*Odd-Numbered Problems* 24.1 320 Mpc 24.3 14,000 km/s, 57 Mpc; 12,000 km/s, 67 Mpc; 16,000 km/s, 50 Mpc 24.5 58 arc seconds 24.7 M = $-22.5$; $8.5 \times 10^{10}$ solar luminosities 24.9 $1.5 \times 10^8$ solar masses 24.11 14.3 degrees, 29 times the size of the Full Moon. 24.13 2.2 million years 24.15 $2.2 \times 10^{12}$ solar luminosities.

## Chapter 25

*True-False*   25.1 T,   25.2 F,   25.3 F,   25.4 F,
25.5 F,   25.6 T,   25.7 T,   25.8 T,   25.9 F,
25.10 T
*Multiple Choice*   25.11 d,   25.12 b,   25.13 c,
25.14 a,   25.15 a,   25.16 b,   25.17 a,
25.18 d,   25.19 b,   25.20 a
*Odd-Numbered Problems*   25.1 6.5 billion years
25.3 1.5 nm, taking the rotation speed to be
350 km/s   25.5 $2.6 \times 10^{14}$ solar masses;
reasonable to within a factor of perhaps 2–3
25.7 700 km/s; 650 km/s—comparable
25.9 $9.4 \times 10^{10}$ solar masses   25.11 $1.8 \times 10^{12}$
solar masses   25.13 21 Mpc   25.15 $3.6 \times 10^{12}$
solar masses.

## Chapter 26

*True-False*   26.1 F,   26.2 F,   26.3 F,
26.4 T,   26.5 F,   26.6 F,   26.7 F,   26.8 T,
26.9 T,   26.10 T
*Multiple Choice*   26.11 a,   26.12 d,   26.13 b,
26.14 b,   26.15 b,   26.16 d,   26.17 c,
26.18 c,   26.19 d,   26.20 d

*Odd-Numbered Problems*   26.1 1000 Mpc
26.3 $4 \times 10^{8}$   26.5 1200 km/s
26.7 (a) 30,000 tons; (b) 2.8 pc   26.9 (a) 82;
(b) 52 km/s/Mpc   26.11 7.0 km/s/Mpc; no
26.13 (a) 1.1 mm; (b) 0.01 mm; (c) 0.0001 mm;
(d) 0.000001 mm   26.15 8.0 kpc.

## Chapter 27

*True-False*   27.1 F,   27.2 T,   27.3 T,
27.4 F,   27.5 F,   27.6 T,   27.7 F,   27.8 F,
27.9 T,   27.10 T
*Multiple Choice*   27.11 a,   27.12 d,   27.13 a,
27.14 b,   27.15 a,   27.16 c,   27.17 b,
27.18 a,   27.19 b,   27.20 c
*Odd-Numbered Problems*   27.1 16 kpc
27.3 18.9 K; $2.7 \times 10^{-24}$ kg/m$^3$
27.5 $2.4 \times 10^{9}$ K, less than half the temperature
in the text   27.7 $3 \times 10^{-12}$ m; hard X ray/
gamma ray   27.9 12.7 Mpc   27.11 166 times
27.13 74 kpc   27.15 220 kpc, $1.0 \times 10^{18}$ solar
masses in a cube of that size.

## Chapter 28

*True-False*   28.1 F,   28.2 F,   28.3 F,   28.4 T,
28.5 F,   28.6 T,   28.7 T,   28.8 F,   28.9 T,
28.10 T
*Multiple Choice*   28.11 a,   28.12 b,   28.13 c,
28.14 c,   28.15 c,   28.16 b,   28.17 d,
28.18 d,   28.19 b,   28.20 a
*Odd-Numbered Problems*   28.1 6.3 seconds
(for a 20 year old reader); 18.9 seconds
(in 2005); 73 seconds; 2.7 minutes; 240 days
28.3 They would both increase by a factor of
two   28.5 five years   28.7 27 AU
28.9 3370 years; no, civilizations are more
likely to be found within the galactic habitable
zone   28.11 32,000 km/s
28.13 $1.43 \times 10^{9}$ to $1.67 \times 10^{9}$ Hz; $2.4 \times 10^{6}$
channels   28.15 2.3 years; 55 years.

# INDEX

Page numbers followed by "n" indicate a footnote reference.

# STAR CHARTS

Have you ever become lost in an unfamiliar city or state? Chances are you used two things to get around: a map and some signposts. In much the same way, these two items can help you find your way around the night sky in any season. Fortunately, in addition to the seasonal Star Charts on the following pages, the sky provides us with two major signposts. Each seasonal description will talk about the Big Dipper—a group of seven bright stars that dominates the constellation Ursa Major the Great Bear. Meanwhile, the constellation Orion the Hunter plays a key role in finding your way around the sky from late autumn until early spring.

Each chart portrays the sky as seen from near 35° north latitude at the times shown at the top of the page. Located just outside the chart are the four directions: north, south, east, and west. To find stars above your horizon, hold the map overhead and orient it so a direction label matches the direction you're facing. The stars above the map's horizon now match what's in the sky.

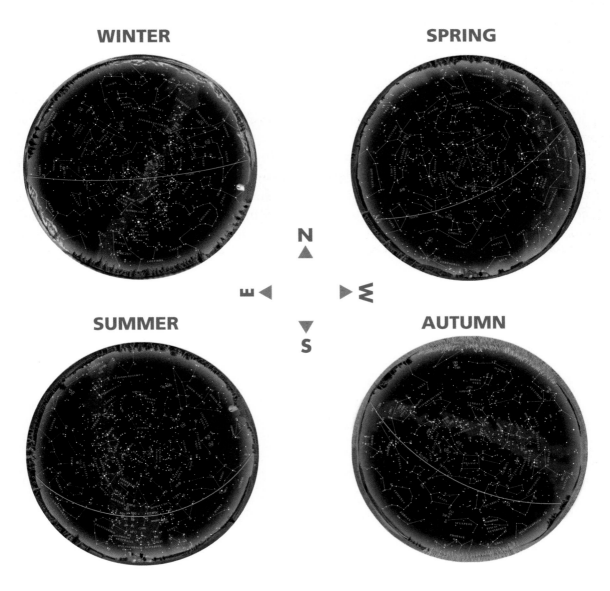

# E x p l o r i n g   t h e   W i n t e r   S k y

Winter finds the Big Dipper climbing the northeastern sky, with the three stars of its handle pointing toward the horizon and the four stars of its bowl standing highest. The entire sky rotates around a point near Polaris, a 2nd-magnitude star found by extending a line from the uppermost pair of stars in the bowl across the sky to the left of the Dipper. Polaris also performs two other valuable functions: The altitude of the star above the horizon equals your latitude north of the equator, and a straight line dropped from the star to the horizon.

Turn around with your back to the Dipper and you'll be facing the diamond-studded winter sky. The second great signpost in the sky, Orion the Hunter, is central to the brilliant scene. Three closely spaced, 2nd-magnitude stars form a straight line that represents the unmistakable belt of Orion. Extending the imaginary line joining these stars to the upper right leads to Taurus the Bull and its orangish 1st-magnitude star, Aldebaran. Reverse the direction of your gaze to the belt's lower left and you cannot miss Sirius the Dog Star—brightest in all the heavens at magnitude −1.5.

Now move perpendicular to the belt from its westernmost star, Mintaka, and find at the upper left of Orion the red supergiant star Betelgeuse. Nearly a thousand times the Sun's diameter, Betelgeuse marks one shoulder of Orion. Continuing this line brings you to a pair of bright stars, Castor and Pollux. Two lines of fainter stars extend from this pair back toward Orion—these represent Gemini the Twins. At the northeastern corner of this constellation lies the beautiful open star cluster M35. Head south of the belt instead and your gaze will fall on the blue supergiant star Rigel, Orion's other luminary.

Above Orion and nearly overhead on winter evenings is brilliant Capella, in Auriga the Charioteer. Extending a line through the shoulders of Orion to the east leads you to Procyon in Canis Minor the Little Dog. Once you have these principal stars mastered, using the chart to discover the fainter constellations will be a whole lot easier. Take your time, and enjoy the journey. Before leaving Orion, however, aim your binoculars at the line of stars below the belt. The fuzzy "star" in the middle is actually the glorious Orion Nebula (M42), a stellar nursery illuminated by bright, newly formed stars.

## WINTER
*2 a.m. on December 1; midnight on January 1; 10 p.m. on February 1*

# Exploring the Spring Sky

The Big Dipper, our signpost in the sky, swings high overhead during the spring and lies just north of the center of the chart. This season of rejuvenation encourages us to move outdoors with the milder temperatures, and with the new season a new set of stars beckons us.

Follow the arc of stars outlining the handle of the Dipper away from the bowl and you will land on brilliant Arcturus. This orangish star dominates the spring sky in the kite-shaped constellation Boötes the Herdsman. Well to the west of Boötes lies Leo the Lion. You can find its brightest star, Regulus, by using the pointers of the Dipper in reverse. Regulus lies at the base of a group of stars shaped like a sickle or backward question mark, which represents the head of the lion.

Midway between Regulus and Pollux in Gemini, which is now sinking in the west, is the diminutive group Cancer the Crab. Centered in this group is a hazy patch of light that binoculars reveal as the Beehive star cluster (M44).

To the southeast of Leo lies the realm of the galaxies and the constellation Virgo the Maiden. Virgo's brightest star, Spica, shines at magnitude 1.0.

During springtime, the Milky Way lies level with the horizon, and it's easy to visualize that we are looking out of the plane of our Galaxy. In the direction of Virgo, Leo, Coma Berenices, and Ursa Major lie thousands of galaxies whose light is unhindered by intervening dust in our own Galaxy. However, all these galaxies are elusive to the untrained eye and require binoculars or a telescope to be seen.

Boötes lies on the eastern border of this galaxy haven. Midway between Arcturus and Vega, the bright "summer" star rising in the northeast, is a region where no star shines brighter than 2nd magnitude. A semicircle of stars represents Corona Borealis the Northern Crown, and adjacent to it is a large region that houses Hercules the Strongman, the fifth-biggest constellation in the sky. It is here we can find the northern sky's brightest globular star cluster, M13. A naked-eye object from a dark site, it looks spectacular when viewed through a telescope.

Returning to Ursa Major, check the second-to-last star in the Dipper's handle. Most people will see it as double, while binoculars show this easily. The pair is called Mizar and Alcor, and they lie just 0.2° apart. A telescope reveals Mizar itself to be a double. Its companion star shines at magnitude 4.0 and lies 14 arcseconds away.

# SPRING

*1 a.m. on March 1; 11 p.m. on April 1; 9 p.m. on May 1. Add one hour for daylight-saving time*

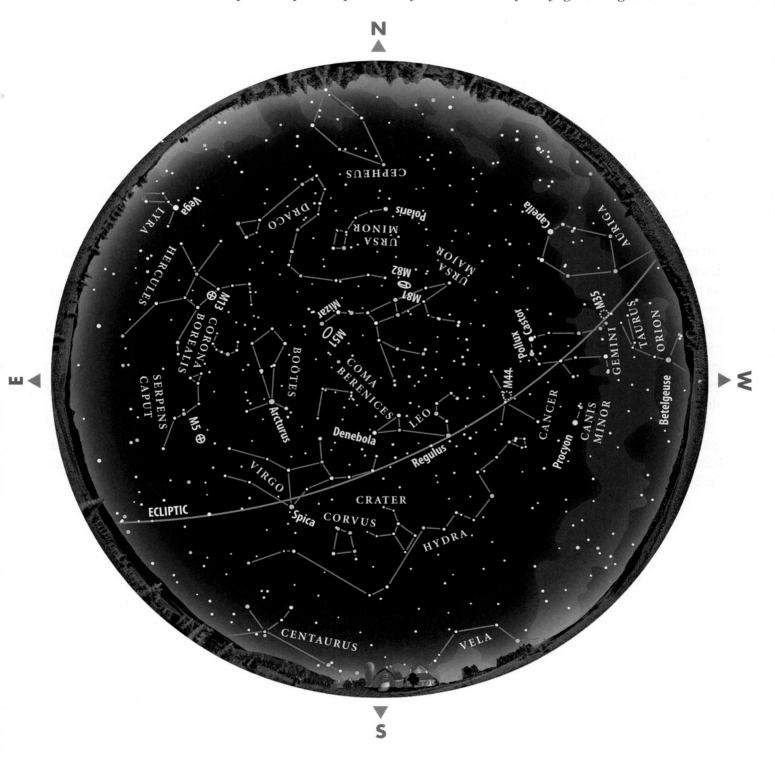

# Exploring the Summer Sky

The richness of the summer sky is exemplified by the splendor of the Milky Way. Stretching from the northern horizon in Perseus, through the cross-shaped constellation Cygnus overhead, and down to Sagittarius in the south, the Milky Way is packed with riches. These riches include star clusters, nebulae, double stars, and variable stars.

Let's start with the Big Dipper, our perennial signpost, which now lies in the northwest with its handle still pointing toward Arcturus. High overhead, and the first star to appear after sunset, is Vega in Lyra the Harp. Vega forms one corner of the summer triangle, a conspicuous asterism of three stars. Near Vega lies the famous double-double, Epsilon ($\epsilon$) Lyrae. Two 5th-magnitude stars lie just over 3 arcminutes apart and can be split when viewed through binoculars. Each of these two stars is also double, but you need a telescope to split them.

To the east of Vega lies the triangle's second star: Deneb in Cygnus the Swan (some see a cross in this pattern). Deneb marks the tail of this graceful bird, the cross represents its outstretched wings, and the base of the cross denotes its head, which is marked by the incomparable double star Albireo. Albireo matches a 3rd-magnitude yellow star and a 5th-magnitude blue star and offers the finest color contrast anywhere in the sky. Deneb is a supergiant star that pumps out enough light to equal 60,000 Suns. Also notice that the Milky Way splits into two parts in Cygnus, a giant rift caused by interstellar dust blocking starlight from beyond.

Altair, the third star of the summer triangle and the one farthest south, is the second brightest of the three. Lying 17 light-years away, it's the brightest star in the constellation Aquila the Eagle.

Frequently overlooked to the north of Deneb lies the constellation Cepheus the King. Shaped rather like a bishop's hat, the southern corner of Cepheus is marked by a compact triangle of stars that includes Delta ($\Delta$) Cephei. This famous star is the prototype of the Cepheid variable stars used to determine the distances to some of the nearer galaxies. It varies regularly from magnitude 3.6 to 4.3 and back again with a 5.37-day period.

Hugging the southern horizon, the constellations Sagittarius the Archer and Scorpius the Scorpion lie in the thickest part of the Milky Way. Scorpius's brightest star, Antares, is a red supergiant star whose name means "rival of Mars" and derives from its similarity to the planet in both color and brightness.

## SUMMER

*1 a.m. on June 1; 11 p.m. on July 1; 9 p.m. on August 1. Add one hour for daylight-saving time*

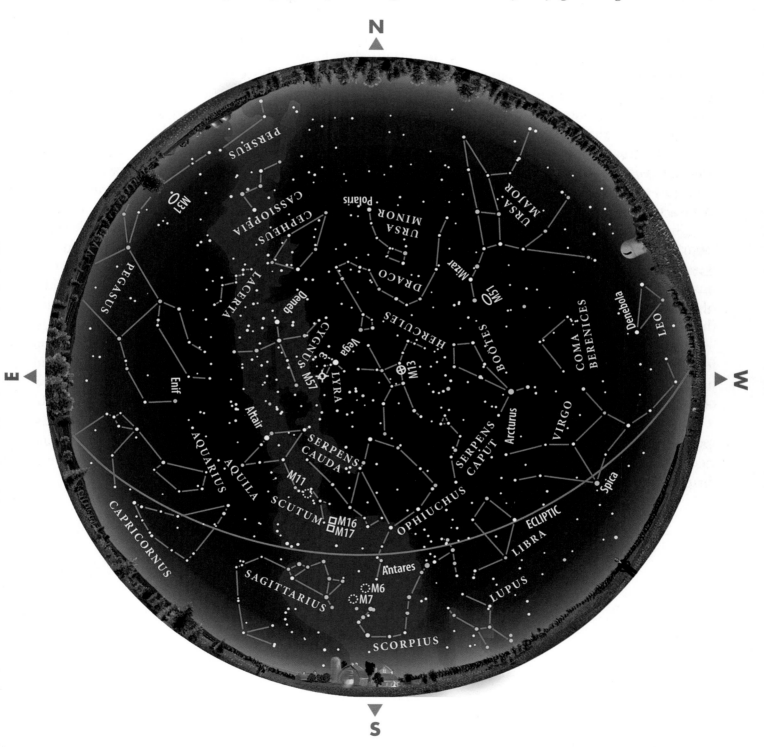

# Exploring the Autumn Sky

The cool nights of autumn are here to remind us the chill of winter is not far off. Along with the cool air, the brilliant stars of the summer triangle descend in the west to be replaced with a rather bland-looking region of sky. But don't let initial appearances deceive you. Hidden in the fall sky are gems equal to summertime.

The Big Dipper swings low this season, and for parts of the Southern United States it actually sets. Cassiopeia the Queen, a group of five bright stars in the shape of a "W" or "M," reaches its highest point overhead, the same spot the Big Dipper reached six months ago. To the east of Cassiopeia, Perseus the Hero rises high. Nestled between these two groups is the wondrous Double Cluster—NGC 869 and NGC 884—a fantastic sight in binoculars or a low-power telescope.

Our view to the south of the Milky Way is a window out of the plane of our Galaxy in the opposite direction to that visible in spring. This allows us to look at the Local Group of galaxies. Due south of Cassiopeia is the Andromeda Galaxy (M31), a 4th-magnitude smudge of light that passes directly overhead around 9 p.m. in mid-November. Farther south, between Andromeda and Triangulum, lies M33, a sprawling face-on spiral galaxy best seen in binoculars or a rich-field telescope.

The Great Square of Pegasus passes just south of the zenith. Four 2nd- and 3rd-magnitude stars form the square, but few stars can be seen inside of it. If you draw a line between the two stars on the west side of the square and extend it southward, you'll find 1st-magniude Fomalhaut in Piscis Austrinus the Southern Fish. Fomalhaut is the solitary bright star low in the south. Using the eastern side of the square as a pointer to the south brings you to Diphda in the large, faint constellation of Cetus the Whale.

To the east of the Square lies the Pleiades star cluster (M45) in Taurus, which reminds us of the forthcoming winter. By late evening in October and early evening in December, Taurus and Orion have both cleared the horizon and Gemini is rising in the northeast. In concert with the reappearance of winter constellations, the view to the northwest finds summertime's Cygnus and Lyra about to set. The autumn season is a great transition period, both on Earth and in the sky, and a fine time to experience the subtleties of these constellations.